PROTECTION TECHNOLOGIES OF ULTRA-HIGH-VOLTAGE AC TRANSMISSION SYSTEMS

PROTECTION TECHNOLOGIES OF ULTRA-HIGH-VOLTAGE AC TRANSMISSION SYSTEMS

BIN LI

Professor, School of Electrical and Information Engineering, Tianjin University, Tianjin, China

JIALI HE

Professor, School of Electrical and Information Engineering, Tianjin University, Tianjin, China

YONGLI LI

Professor, School of Electrical and Information Engineering, Tianjin University, Tianjin, China

YUPING ZHENG

Director, State Key Laboratory of Smart Grid Protection and Operation Control, China

ELSEVIER

ACADEMIC PRESS

An imprint of Elsevier

中国电力出版社

CHINA ELECTRIC POWER PRESS

Academic Press is an imprint of Elsevier
125 London Wall, London EC2Y 5AS, United Kingdom
525 B Street, Suite 1650, San Diego, CA 92101, United States
50 Hampshire Street, 5th Floor, Cambridge, MA 02139, United States
The Boulevard, Langford Lane, Kidlington, Oxford OX5 1GB, United Kingdom

Library of Congress Cataloging-in-Publication Data
A catalog record for this book is available from the Library of Congress

British Library Cataloguing-in-Publication Data
A catalogue record for this book is available from the British Library

ISBN: 978-0-12-816205-7

For information on all Academic Press publications visit our website at https://www.elsevier.com/books-and-journals

Publisher: Joe Hayton
Acquisitions Editor: Lisa Reading
Editorial Project Manager: John Leonard
Production Project Manager: Sruthi Satheesh
Cover Designer: Matthew Limbert

Typeset by TNQ Technologies

Contents

CHAPTER 1

Introduction

Contents

1.1 Development of the ultra-high-voltage AC transmission system

The power industry is a basic industry related to national economies and their people's livelihoods. Power grid construction plays an important role in optimizing the allocation of energy resources, ensuring national energy security, and promoting the development of the national economy. In the 1980s, the Soviet Union built and put into operation the world's first 1150 kV transmission line across Eurasia with a total length of more than 2000 km. Its purpose is to transport energy from the energy base to the load center over long distances. In order to save the route corridor, limit short-circuit current, and transport the eastern and northern nuclear power to Tokyo, Japan has built a 1000 kV ultra-high-voltage (UHV) transmission line of 380 km, but it is still at the level of decompression operation. In addition, the United States, Italy, Canada, Brazil, and other countries have conducted long-term research and experiments on UHV transmission technology. In many countries around the world, the geographical distribution of primary energy and the electrical loads used to generate electricity is not balanced. For example, while building large coal-fired power bases and large hydropower bases, China has also built a national energy transmission channel with UHV backbone grids as its core, enabling long-distance and large-capacity west—east power transmission, north—south mutual supply, and nationwide networking. In 2009, China's "Jindongnan—Nanyang—Jingmen" UHV AC transmission line demonstration project was put into operation. In 2011, the first UHV demonstration line achieved a full capacity of 5 million kW.

Protection Technologies of Ultra-High-Voltage AC Transmission Systems
ISBN 978-0-12-816205-7
https://doi.org/10.1016/B978-0-12-816205-7.00001-7

Compared with UHV AC transmission, UHV DC transmission has great advantages in point-to-point power transmission, time-division economic power exchange, and grid fault isolation. However, its disadvantages include difficult point selection, high reactive power consumption, and expensive converter stations. UHV AC transmission is conducive to networking and improving power transmission economics, but its stability and reliability problems are not easy to solve. With technology development, the advantages and disadvantages of AC and DC transmission may also be mutually transformed. The two transmission technologies will coexist for a long time and have fierce competition. This book focuses on the key technical issues of relay protection for UHV AC transmission systems. Therefore, "UHV" as written in the book refers to AC UHV.

In the 1960s, the need for electricity was rapidly increasing in some industrialized countries with large electricity demand—the Soviet Union, the United States, Canada, Brazil, etc.—due to industrial development and the improvement of people's lives. At the same time, due to the vast territory of these countries, energy production areas (mainly hydropower resources) and load centers were far away, and 400–500 kV transmission lines could not meet the needs of long-distance, large-capacity power transmission, so study and design began for UHV transmission lines, namely 735, 750, 765, and 800 kV transmission lines. In 1963, the Soviet Union put into operation the first 750 kV test transmission line, which is 90 km long, through Konakovo–Moscow [1]. The line was decompression-operated for 4 years and was officially operated at 750 kV in 1967 [2]. After that, a 1118 km long 750 kV line was built through Donbass–Dnepr–Vinnitsa–Western Ukraine, and a 525 km long 750 kV line was built through Leningrad–Konakovo [2,3]. In 1970, a 750 kV transmission line of 479 km through Western Ukraine–Albertirsa was built [4].

The number of 750 kV lines from Ukraine to Romania, Hungary, Czechoslovakia, and Poland totals 15, with a length of 120–500 km. Hydro Quebec of Canada was completed and put into operation as the first 735 kV transmission line in 1965 [5]. By 1986, about 21 735 kV lines had been built [6], most of which were double-circuit lines. In order to improve the line transmission capacity and system stability, series compensation capacitors were installed on these lines [7], and 37 sets of series-compensated capacitors were installed, for a total compensation capacity of 15,000 Mvar. In 1966, the United States approved the construction of a 765 kV transmission line in the American Electric Power Service Company (AEP) system. In February 1969, the first 765 kV transmission line was put into operation. By 1973, ten

765 kV lines with a total length of approximately 1760 km had been built. At present, AEP's 765 kV line is about 3000 km long.

The Itaipu hydropower station built by Brazil and Paraguay has eighteen 700 MW units. With a total capacity of 12,600 MW, it was put into operation in 1985. Among them, nine units are 60 Hz, and power is transmitted by three 750 kV transmission lines of about 900 km each to the load centers of the Brazilian capital and São Paulo. The other nine units are 50 Hz, and three 750 kV transmission lines are used to send power to the Paraguay load center. Each line is about 570 km long. Most of these 750 kV lines have series capacitor compensation, which was completed and put into operation in 1985—1990 [8].

The power industry in Venezuela is developing well. Six 765 kV lines were built from 1981 to 1984. The 765 kV line from the large power plant with ten 700 MVA generator sets in Guri to the load centers in San Geronimo, La Arenosa, and La Horqueta was 1249 km long in total. Five additional 765 kV lines were built after that, and power was sent from San Geronimo to four areas [9].

In addition to the aforementioned countries that first built 750 kV lines, China, India, South Korea, Japan, South Africa, and other countries have also built 750 kV transmission lines. Before the construction of 750 kV lines, they carried out long-term research on many technical problems, equipment problems, radio interference caused by the 750 kV line corona, and possible impacts on ground organisms. It can be said that the technology of the 750 kV transmission line has been basically mature and perfect with rich operational experience. These scientific research achievements and operational experience have laid a solid foundation for UHV transmission.

Since the 1970s, some industrialized countries such as the United States, the Soviet Union, Canada, Japan, and Italy have realized that due to the rapid growth of power loads, 750 kV transmission lines will soon be unable to meet the needs of long-distance and large-capacity power transmission. The need for the development of higher-level voltage transmission led to the study of 1000—1500 kV transmission technology [10]. From 1975 to 1978, the Soviet Union and United States jointly established the US—USSR Joint Committee on Cooperation in the Field of Energy and cooperated in research on UHV transmission technology to promote development of the technology.

In 1975, the US—USSR Joint Committee on Cooperation in the Field of Energy held a seminar on UHV AC transmission in Washington, DC, and in 1978, another seminar was held in Tashkent, the Soviet Union [11]. The seminar held in 1975 mainly discussed the results of research on 750−1150 kV lines including various structural types of 750−1150 kV transmission equipment, the main parameters of 750−1150 kV line design, measurement methods of electric field strength, protection principles and methods concerning operating and lightning overvoltages, insulation distances of 750−1150 kV lines, substation insulation and environmental protection, load conditions and standards of UHV transmission lines, calculation of foundation stress on the tower, wire, lightning protection lines, wire icing and wind load, wind load and vibration, vibration of split wires, and corona and its interference with radio waves. It involved nearly all problems of UHV transmission. The seminar held in 1978 focused on various issues regarding 1100−1200 kV transmission lines including measures to limit 1200 kV line overcurrent to 1.3, 1.5, and 1.8 times the rated voltage, protection measures for overvoltage caused by 1200 kV transmission line single-phase fault removal, 1100 kV (maximum voltage 1200 kV) line active and reactive power adjustment methods, long-term dielectric strength of 1200 kV tower, insulator and air gap, 1200 kV transmission line tower design considering environmental protection, minimum material consumption, the reduction of electrical distance and line corridor conditions, the impacts of power frequency electromagnetic fields on organisms, and other research results.

In the 1960s, the Soviet Union built the 450 MW Bratsk Grand Hydropower Station on the Ankara River in Siberia, and in the 1970s built a thermal power plant group in Krasnoyarsk. In order to transport the electricity to the Ural Heavy Industries Base and the European Load Center, construction was started in the 1980s from Itatsk, Siberia, 440 km to Barnaul, then 693 km to Ekibastuz in Kazakhstan, 494 km to Kokshetau, 396 km to Kostanay, and finally 321 km to the 1150 kV transmission line of Chelyabinsk in Ural. In total, the line was approximately 2500 km long and capable of delivering 6000 MW of power. This is the world's first UHV transmission line and has been tested for quite a long time.

Japan has a high load density. In order to save power transmission corridors, limit system short-circuit capacity, and improve system stability, Japanese researchers have studied transmission schemes of four voltage levels—800, 1100, 1200, and 1500 kV—in light of national conditions. Considering factors such as far−near integration, system structure, construction cost, operating cost, and environmental impact [12], it was determined that a UHV transmission system with a rated voltage of

1000 kV and a maximum voltage of 1100 kV should be constructed within the Tokyo Power Grid. The Tokyo Electric Power Company has built four 1000 kV transmission lines to send power from power plants in Fukushima and Kashiwabara to the 500 kV power grid outside the Tokyo area. In addition to the Soviet Union, Japan, and the United States, Canada, France, Italy, India, and other countries have carried out research on various problems of 1000—1200 kV transmission lines and established test circuits. Therefore, experience with 1000—1200 kV AC transmission technology is considerable and quite mature.

When installed system capacity reaches a certain level, the backbone grid's voltage level must be increased. Studies in Ref. [13] have shown that transmission energy consumption of the 345 kV main network is lower than transmission energy consumption of the main network of 230 kV, and transmission energy consumption of the 500 kV main network is lower than transmission energy consumption of the 345 kV main network. If the main network is 1100 kV, energy consumption can be further reduced.

The unit transmission capacity corridor width of a 1000 kV AC line is only one-third that of a 500 kV line, and the comprehensive cost of unit transmission capacity is less than three-fourths that of the 500 kV transmission scheme [14]. Due to increasingly tight land resources, the construction cost of transmission networks is becoming increasingly higher. UHV double-circuit/multicircuit/compact transmission lines will also become an inevitable trend in the development of UHV power grids. In the United States, the 220 and 345 kV voltage levels use common-tower double-transmission lines to reach 47% and 52% respectively. All UK 400 kV grades use common-tower double-transmission lines. More than 90% of the 500 kV lines in Japan are common-tower double-transmission. All three 1000 kV UHV systems that have been built us common-tower double-transmission lines. Following the idea of common-tower double-transmission lines, many countries have studied and built multiple transmission lines from the same tower to further improve transmission capacity under the unit corridor.

In general, the construction of UHV power grids is the fundamental guarantee for meeting power demand for continued future growth and is conducive to the realization of a wide range of resource optimization configurations. It is the basic way to improve the efficiency of energy development and utilization and is conducive to cracking the contradiction of power grid investment imbalance. It is an inevitable choice for improving the overall efficiency of the power industry and is conducive to achieving close coordination between power grids in various regions, possessing many advantages in terms of the safety, reliability, flexibility, and economy of large power grids.

1.2 Key technology of the ultra-high-voltage AC transmission system

(1) Voltage level
The highest operating voltage is the basis for the power system's design and equipment selection. There have been four schemes for the highest operating voltage of UHV transmission in the world, namely 1050, 1100, 1200 and 1600 kV. In terms of the highest operating voltages, China has carried out a comparative evaluation of network operational characteristics, line loss and corona loss, external insulation of the equipment at high altitude, transmission capacity and transient stability, manufacturing difficulty of UHV equipment, the impact of costs and pollution degree, etc. A large number of research calculations show that with the highest operating voltage of 1100 kV, system stability, operational flexibility, and resource allocation capabilities are all higher than they are for a 1050 kV system, and the total transmission loss and unit power investment costs are lower than they are for 1050 kV [15]. For these reasons, China has put forward the voltage standard of a nominal voltage of 1000 kV and a maximum voltage of 1100 kV, which has become the IEC voltage standard [16]. UHV transmission has a high voltage level and great influence on the system. Therefore, it is particularly important to strengthen voltage control, which has a great influence on the power system's security, reliability, and economy. The voltage control of a UHV transmission system mainly includes the suppression of power frequency overvoltage, switching overvoltage, and lightning overvoltage as well as control of system operating voltage.

(2) Overvoltage
Overvoltage for extra-high-voltage (EHV) transmission systems includes power frequency overvoltage, switching overvoltage, lightning overvoltage, and system operating voltage [16].

Power frequency overvoltage refers to the voltage increase caused by load shedding in the case of no fault or fault. The amplitude depends mainly on line length, high voltage shunt reactor configuration, system structure, power flow distribution, etc. Duration is generally considered on the order of seconds. The power frequency overvoltage level has a direct impact on equipment selection. In view of the long EHV transmission line, large capacity-increase effect, and strict requirements for power frequency overvoltage limitation, the amplitude of power frequency overvoltage is effectively limited by research and development as well as the installation of large-capacity EHV reactors. At the same time, in special cases, intertrip protection measures are taken to limit the power frequency overvoltage duration.

Switching overvoltage mainly refers to transient overvoltage caused by tripping and closing operations of circuit breakers or disconnectors. Its amplitude mainly depends on line parameters, system structure, operating mode, and restraint measures. Duration is generally considered in milliseconds. For EHV systems, the switching overvoltage level is the most critical indicator affecting the equipment internal insulation level, external insulation coordination, tower, and line design of the equipment. The switching overvoltage level of the EHV system is generally controlled by 2.0 pu. For EHV, if considered according to this standard—that is, the switching overvoltage reaches 1796 kV (peak)—the distance of air gap needs to be greatly increased, which creates great difficulty for insulation design and equipment manufacture. Therefore, when installing a large-capacity reactor, design must break through conventional practices that the line side arrester residual voltage is higher than the bus side arrester, and two equivalent high-performance lightning arresters with large discharge current capacity and low residual voltage level are installed on the bus side and line side of the UHV substation. At the same time, the closing resistor is installed on the EHV circuit breaker to further suppress switching overvoltage.

Lightning overvoltage refers to overvoltage generated by the lightning invasion wave on the transmission line, power transmission, and transformation equipments in the process of lightning back flashover and shielding failure, whose amplitude depends on the lightning current amplitude, waveform, ground lightning point, system structural parameters, arrester configuration, etc. Duration is generally considered as microsecond. The tower of the UHV line tower is relatively high, lightning shielding failure probability increases significantly, and the mean time between failure stipulated by regulations increases from 800 years of EHV to 1500 years of UHV [16], which puts forward higher requirements for UHV lightning protection design and lightning overvoltage suppression.

The system's operating voltage is a key indicator of its operating state. The operating voltage of China's UHV system is generally controlled between 1000 and 1100 kV. In actual operation, the voltage level of the power grid mainly depends on system power flow and reactive power balance. Considering factors such as line loss, it is advisable to optimize system reactive power allocation according to the principle of local balance. The UHV system has high voltage, long lines, large capacitive reactive power, and large transmission power variation. The capacitive reactive power of every 100 km line reaches 530 Mvar. The system's reactive power balance is more complex than that of the UHV power grid.

(3) Insulation

External insulation includes phase-to-ground and interphase air gaps as well as solid insulating medium surfaces such as insulators and bushings. The switching overvoltage and lightning overvoltage of the UHV system are high, and the external insulation coordination is more difficult [17]. First, the withstanding voltage of the air gap no longer increases linearly with an increase in gap distance, showing a significant saturation effect, which greatly increases the external insulation scale of the UHV system. Second, UHV transmission may cross high-altitude areas. The withstanding ability of a solid insulation surface decreases with an increase in pollution degree and altitude. At the same time, the air gap withstanding voltage level decreases significantly with altitude, further aggravating the difficulty of external insulation coordination.

(4) Electromagnetic environment

In 1994, China built its first megvolt UHV transmission research line. The tower is a true analog-type guyed V tower with eight bundle conductors whose bundle diameter is 1.04 m in a three-phase horizontal arrangement. In 1997, China carried out research on the use of power frequency test equipment to generate long wave front switching impulse wave. The modified power frequency test device can generate long-wave front with a voltage of 2250 kV and a wave front time of 2800–5000 μs. Research results regarding the UHV line's environmental impact show that using eight bundle conductors with a bundle diameter of 1 m, the ground electrostatic induction level of the UHV line is basically the same as that of the 500 kV transmission line, and the radio interference level is less than that of the 500 kV transmission line. The audible noise is in the accepted range for public noise levels.

(5) UHV transmission equipment

UHV AC equipment includes dozens of kinds of equipment such as transformers and circuit breakers. The nominal parameters are high, and the coordination of multiple physical fields such as electricity, magnetism, heat, and force is complicated.

The UHV transformer adopts the split structure of a UHV main transformer and a regulating and compensating transformer, which effectively reduces transportation weight and significantly improves the operational reliability; The proposed three-column parallel winding current distribution and magnetic circuit balancing method effectively control heat generation and vibration. The proposed ground potential high field intensity zone insulation control method, main longitudinal insulation, and outlet insulation structural scheme realize no partial discharge design. China has successfully developed an integral UHV transformer with a nominal capacity of 1000 MVA. The total capacity and single-column capacity are the largest, and the no-load loss and capacity per unit load loss are the smallest [18].

UHV reactors can effectively suppress the line capacitance rise effect. China's UHV lines use large-capacity high-voltage shunt reactors. Taking the Changzhi—Nanyang—Jingmen UHV demonstration project as an example, the shunt reactors on both sides of the Changzhi—Nanyang line have a capacity of 960 Mvar and 720 Mvar, and the shunt reactors on both sides of the Nanyang—Jingmen line have a capacity of 720 Mvar and 600 Mvar. At the same time, China has also developed UHV controllable reactors, which laid the foundation for the construction of longer-distance UHV projects.

A UHV gas insulation system adopts gas-insulated metal-enclosed switch technology to reduce covering area by approximately 50%. The development of double-break and four-break high-efficiency arc-extinguishing chambers has significantly improved current-breaking capability. The disconnector with a damping resistor has effectively suppressed the extremely fast transient overvoltage caused by switching.

In addition, China has developed a metal oxide surge arrester with a high discharge current capability and low residual voltage ratio, high strength, and strong hydrophobic composite insulators and bushings.

1.3 Relay protection of the ultra-high-voltage AC transmission system

UHV transmission lines are the framework of a combined system or large cross-region power grid. Their safe and reliable operation plays a decisive role in the safe and reliable operation of the whole system, so the requirements for the performance and reliability of their relay protection are extremely high [19,20]. Therefore, various possible measures should be taken to increase trip speed, sensitivity, selectivity, and reliability (including dependability and safety). However, UHV long-distance transmission lines have many characteristics such as large capacitance currents, line parameters with significant distribution characteristics, long secondary arc extinction time, and severe fault and switching overvoltage that bring new challenges to relay protection and reclosing. The traditional protection principle has a defect when applied to UHV transmission lines.

(1) Large transmission line capacitance current

The UHV transmission line has a large distributed capacitance and a long transmission distance, so full-line capacitance current is large. For typical EHV/UHV line parameters, the relationship between capacitance current and its natural power current is shown in Table 1.1. Where I_e is the load current per phase at natural power, and I_{cap} is the capacitance current per phase per one hundred kilometers.

It can be seen from Table 1.1 that the capacitance current of the EHV transmission line is significantly higher than that of the UHV line. Moreover, since the EHV transmission line will realize transregional and transbasin transmission, its transmission distance is very long. For example, the capacitance current of a 600 km long megavolt level EHV transmission line is as high as 63.88% that of the line for natural power current. It can be seen that the capacitor current under normal operating conditions of EHV long-distance transmission lines cannot be ignored. This is only the capacitive current under steady-state operation. When a fault or reclosing is performed, the transient charging and discharging current will increase several times. In

Table 1.1 Comparison of capacitance current and natural power current.

Voltage level	500 kV	765 kV	1150 kV	1500 kV
I_e	1070.9 A	1666 A	2901.1 A	2903.7 A
I_{cap}	113.36 A	178.45 A	308.89 A	309.16 A
$(I_e/I_{cap}) \times 100\%$	10.59%	10.71%	10.65%	10.65%

particular, many high-frequency components are in the voltage in transient state. Capacitance current is proportional to frequency, which generates a larger high-frequency capacitor current, which severely distorts the waveform, amplitude, and phase of the fault current.

The main protection of UHV and EHV transmission lines is pilot protection. So far, the principle of pilot protection of transmission lines mainly includes directional pilot protection that compares the power direction of the two ends of the line, differential pilot protection using phase comparators that compares the current phase at both ends of the line, and current differential pilot protection that compares the full current at both ends. These principles have their own advantages and disadvantages. The phase-to-phase current differential principle has obvious advantages. Phase-to-phase current differential protection based on Kirchhoff's current law is a simple, reliable, and applicable protection method for any form of transmission line. Compared with other types of transmission line protection, the main advantages are no influence of system oscillation; no influence of system non-full-phase operation and single-side power supply operation; no influence of parallel-line mutual inductance; phase-selection ability, etc. At present, the research hotspots of current differential protection mainly focus on three aspects: the application fault component, sample value differential, and long-distance line differential protection.

Current differential protection based on the fault component is beneficial for detecting high-resistance ground faults and greatly improving the sensitivity of current differential protection to reflect internal faults. Current differential protection based on the sampled value is beneficial for improving the protection action's speed and sensitivity. In long-distance transmission line current differential protection, when using conventional current differential protection, the distributed capacitance is approximately equivalent to a lumped capacitance element placed on both sides of the line, considering the influence of the distributed capacitance; the phase currents at both ends of the line are compensated by capacitive current. This method of approximate compensation of capacitive current cannot completely overcome the influence of capacitive current in principle. Especially in the case of EHV long line faults, capacitive current greatly affects the safety of current differential protection in the event of an external fault as well as the reliability of the internal fault. At the same time, in order to ensure that protection of the no-load operation line and a fault outside the area do not malfunction, the protection setting value must be raised, which greatly reduces the sensitivity of protection during internal faults.

(2) Significant distribution parameter characteristics of lines

EHV transmission lines often transmit power across regions and river basins, and the transmission distance is very long, so the line has significant distribution parameter characteristics. According to the electromagnetic wave propagation equation, the functional relationship between power frequency electrical quantities between any two points of the line can be analyzed and derived. For example, when a three-phase short circuit occurs at the end of the line, the measured impedance at the beginning and fault distance are in a hyperbolic tangent function rather than a linear relationship.

However, as one of the most widely used protection principles for power transmission lines, distance protection equalizes line impedance to concentrated impedance, and the measured impedance of distance protection is considered proportional to the fault distance under different fault types, thus achieving line protection with a certain protection-length range. Distance protection is used as backup protection for UHV and EHV transmission lines. Some lines also use pilot distance protection as the main protection for UHV and EHV lines. The distance protection principle currently applied successfully on transmission lines mainly includes conventional distance protection and distance protection based on superimposed components. The advantage of distance protection is that the protection range is basically fixed and not affected by changes in the system's operational mode. Various action characteristics can be adopted according to line conditions and protection targets. However, in principle, distance protection is greatly affected by system oscillation. The line may shorten or extend the range of protection in the case of fault through transition resistance. The distance protection based on superimposed components proposed by the literature [21] is not affected by system oscillation and has excellent characteristics such as strong adaptive ability, a simple criterion, and easy implementation via microcomputer. It is widely used in China's high-voltage power grid. Operational experience over the years shows that distance protection based on superimposed components has some advantages.

However, the distribution parameter characteristics of EHV long lines change the theoretical basis of distance protection. Strictly speaking, in principle it is not correct to apply conventional distance protection to EHV long lines. With the construction of EHV power grids, operating characteristics for conventional distance protection and distance protection based on superimposed components on EHV long lines are urgently needed. The research shows that when conventional distance protection is applied to an EHV long line, the measured impedance of the phase-to-phase distance component has a hyperbolic tangent function relationship with the fault distance, and the zero-sequence compensation coefficient of the grounding distance component is calculated by the concentrated impedance, so the measured impedance does not conform to the line distribution parameter model. Simulation verification shows that even if the distance protection setting is made according to the hyperbolic function, the traditional principle has a transcendence problem on the EHV long line. In addition, distance protection based on superimposed components is only effective at the beginning of the fault, so it is not suitable for the backup protection scheme of the EHV long line. Therefore, line distribution parameter characteristics create principle defects with the conventional distance protection and distance protection based on superimposed components applied to EHV long lines, and they cannot reliably meet the requirements of line pilot distance main protection or line backup protection.

(3) Long-term secondary arc-extinguishing time

With the increase of the insulation distance between the tower lines and the corresponding distance and size of the outdoor power distribution device of the substation, the electrical strength of the interphase insulation of overhead lines increases due to atmospheric and switching overvoltage. The operating experience of the EHV transmission line in the Soviet Union shows that single-phase ground faults account for more than 90% of all faults, and polyphase short-circuit faults rarely occur. In view of this, single-phase auto reclosing has broad application prospects for EHV lines.

Auto reclosing plays an important role in improving the stability of the parallel system operation and ensuring power supply continuity. However, hazards such as short-circuit current surges brought to the system in the reclosing of a permanent failure are enormous. Especially for EHV transmission lines, the submersible current is large, the recovery voltage is high, and the secondary arc is difficult to extinguish (the arc burning time of the EHV transmission line may exceed 0.5—0.9 s), which affects the success rate of single-phase reclosing. Studies have shown that the use of neutral reactors with small-reactance shunt reactors or high-speed grounding switches can accelerate extinction of the secondary arc.

In the case of a single-phase fault on UHV and EHV transmission lines, only the fault phase is cut off, and the sound phase continues to run. Therefore, it is of great significance to study adaptive reclosing that can correctly distinguish permanent and transient faults [22]. The essence of adaptive reclosing is to correctly identify transient or permanent faults before deciding whether to reclose. From the perspective of improving the success rate of single-phase reclosing, reducing overvoltage, and ensuring stable system operation, it is necessary to study single-phase the adaptive reclosing theory that avoids reclosing to permanent faults for the reliable operation of EHV transmission lines. Domestic and foreign scholars have done significant research on single-phase adaptive reclosing using a voltage criterion method [22], a beat frequency characteristic method [23], the current amplitude ratio of the shunt reactor, small neutral reactance as the discriminating basis for judging transient and permanent faults, etc. [24]. These methods make the single-phase adaptive reclosing theory mature. It is necessary to combine the research results of domestic and foreign scholars to study and perfect the single-phase adaptive reclosing theory from a practical point of view and to promote practical application of the technology.

Since the three-phase trips in the case of a phase-to-phase failure, there is no electrical condition for determining the nature of the fault after tripping. Therefore, three-phase adaptive reclosing has always been a difficult research field in studying adaptive coincidence. Some experts have proposed that the line with the shunt reactor compensation can be used to determine the fault nature using beat frequency characteristics after tripping [25]. Some scholars have studied the use of fault-transient high-frequency signal characteristics to achieve a three-phase adaptive coincidence criterion [26]. However, the sensitivity and reliability of these methods are currently limited. However, for EHV long-distance transmission lines, three-phase coincidence overvoltage is the decisive factor of system insulation, and serious overvoltage will occur if the coincidence is in permanent fault. Therefore, the study of three-phase adaptive reclosing theory is of great theoretical and practical significance for the safe and reliable operation of UHV lines.

(4) Serious overvoltage problem
The EHV transmission line has a high voltage level, and insulation cost accounts for a large proportion of total transmission line construction and operating costs. Therefore, EHV grids need to take various measures to limit overvoltage. Studies have shown that time for relay protection at both ends of EHV transmission lines should be as short as possible to avoid overvoltage from penetrating the insulation equipment. The reclosing overvoltage is the decisive factor of the insulation level of the EHV grid. When reclosing is coincident with a permanent fault, the most serious overvoltage is generated. Therefore, in addition to using primary equipment to suppress overvoltage, it is an advanced and desirable method to study three-phase adaptive reclosing to improve the success rate of reclosing, reduce the short-circuit impact after reclosing, improve system operating stability, and reduce the overvoltage level.

(5) Impact of fault transient process
The EHV transmission line has a large transmission capacity, and the short-circuit current is quite large when the line outlet fails. During the fault transient process, the decay time constant of the decaying DC component is large, and the high-frequency harmonic content is large. Long-term decaying of the DC component is likely to cause transient overreach of traditional distance protection and also causes the action time of distance protection to be too long when the line terminal fails. The large number of high-frequency harmonics greatly affects the calculation accuracy of line protection, thus affecting protection performance. In addition, the short-circuit current is too large and the decaying time constant is too long, which makes the current transformer saturation problem prominent.

A complete theoretical system and research results in the field of EHV line protection at domestic and foreign levels are very limited. Therefore, it is necessary to conduct in-depth research and exploration on the principle of EHV long-line protection and reclosing and provide more powerful technical support for the protection of EHV lines.

References

[1] Akapian AA, et al. The 750kV experimental-commercial transmission line Konokovo-Moscow. ClGRE 1964. paper 413.
[2] Lyskov YuI, Rovotyen SS. General direction of 750 ~ 1500kV AC power-transmission development and utilization of various types of equipment of these power transmissions. Proceeding of the Symposium on EHV A.C Power Transmission. Held by

The Joint American-Soviet Committee on Cooperation in The Field of Energy 1975. Washington D.C. Feb. 17~27.
[3] Kuzmitcheva KI, et al. Suppression of switching overvoltages in 750kV transmissions by synchronous closing of mass-produced air breakers. (Elektritchestvo, 1975, No.7 in Russian).
[4] Lyskov YI, Rovotyen SS. Problems of single-pole reclosing on long EHV transmission lines. Proceeding of the Symposium on EHV A.C Power Transmission. Held by The Joint American-Soviet Committee on Cooperation in The Field of Energy 1982. paper 33-10.
[5] International seminar on planning, design and operational experience on 800kV system. 1990. Held in Bombay (India) 22 Jan.
[6] Saint-Latour Jean G.Y, Landry M, Belanger J (IREQ), Huynh H, [Czech P (Hydro-Quebec)]. Solution to the problem of separation of the Hydro-Quebec 735kV system by switchable metal-oxide surge arresters.
[7] Que Bui-Van, Landry M, Khodabakhchian B. Performance of series-compensated line circuit breakers under delayed current-zero conditions. Hydro-Quebec, Canada IEEE Power Delivery 1997;12(1).
[8] Justs CP. The furnas 800kV transmission system. International seminar on planning, design and operational experience on 800kV system. January 22 , 1990 [Bombay India].
[9] Muratet JP. Experience on design, control system architecture, corona, swelling clays, compressed air, electromagnetic compatibility, earthquake performance for the 800kV network in Venezuela. International Seminar on Planning, Design and Operational Experience on 800kV System 22 Jan, 1990. Bombay, India.
[10] John G, Joseph R, Doyle EJ. Project UHV-Construction and Plan Anderson. O'Brien general electric Company Pittsfield, Massachusetts IEEE PAS; 1972. p. 204.
[11] Proceedings of the symposium on EHV AC power transmission. Held by the Joint American-Soviet Committee on cooperation in the field of energy. Washington D.C. Feb. 17–27, 1975.
[12] Workgroup No36.01 of CIGRE. Electric field and magnetic field generated by transmission systems – phenomenon sketch & practical calculation guideline.
[13] Zhenya Liu. Ultra-high voltage grid. Beijing: China Economic Publishing House; 2005.
[14] Shu Y. Research and application of 1000kV UHV AC transmission technology. Power System Technology 2005;(19):9–14.
[15] Xu Z, Dong H, Huang H. Debates on ultra-high-voltage synchronous power grid: the future super grid in China. IET Generation, Transmission & Distribution 2015;9(8):740–7.
[16] Liu Z. Innovation of UHVAC transmission technology in China. Power System Technology 2013;37(3):567–74.
[17] Huang D, Shu Y, Ruan J, Hu Y. Ultra high voltage transmission in China: developments, current status and future prospects. Proceedings of the IEEE 2009;97(3):555–83.
[18] O'Donnell R. Prolog to ultra-high-voltage transmission in China: developments, current status, and future prospects. Proceedings of the IEEE 2009;97(3):553–4.
[19] He J, Li Y, Guo Z, Li B, Dong X. Relay protection for UHV transmission lines: Part I-construction and operation characteristics. Automation of Electric Power Systems 2002;(23):1–6.
[20] He J, Li Y, Li B, Guo Z, Dong X. Relay protection for UHV transmission lines: part II- disposition of relay protection. Automation of Electric Power Systems 2002;(24):1–6.
[21] Shen G. A new directional relay based on the variation of power frequency components. Automation of Electric Power Systems 1983;1:28–38.

[22] Ge Y, Sui F, Yuan X. Prediction methods for preventing single-phase re-closing on permanent fault. IEEE Transactions on Power Delivery 1989;4(1):114–21.
[23] Li B, Li Y, Sheng K, Zeng Z. The study on single-pole adaptive reclosure of EHV transmission lines with the shunt reactor. Proceedings of the CSEE 2004;24(5):52–6.
[24] Suonan J, Sun D, Fu W, Wang X, Liu W, Jiao Z. Identification of permanent faults for single-phase Auto-reclosure on transmission lines with shunt reactors. Proceedings of the CSEE 2006;(11):75–81.
[25] Ge Y, Xiao Y. Adaptive three-phase automatic reclosure of EHV transmission lines. Electric Power Automation Equipment 1995;54(2):10–8.
[26] Bo ZQ, Aggarwal RK, Johns AT, et al. A new concept in transmission line reclosure using high frequency fault transients. IEE Proceedings - Generation, Transmission and Distribution 1997;144(5):1–6.

CHAPTER 2

Electromagnetic environment

Contents

2.1 Structural parameters of ultra-high-voltage transmission lines

2.1.1 Mechanical structural parameters

In some countries, the 750 kV voltage level is called ultra-high-voltage (UHV) because the overvoltage, equipment, operation mode of the shunt reactor, requirements for relay protection, and impact on system operation of 750 kV transmission lines differ much from those of 500 kV transmission lines. But most countries still consider 750 kV extra-high-voltage (EHV). Soviets call 1150 kV EHV as well. After 750 kV transmission technology became mature and was popularized in the 1970s, China built a 750 kV transmission network, so it is reasonable to call 750 kV transmission EHV.

Protection Technologies of Ultra-High-Voltage AC Transmission Systems
ISBN 978-0-12-816205-7
https://doi.org/10.1016/B978-0-12-816205-7.00002-9

The current, so-called UHV refers to voltage levels of 1000 kV and above. China has selected 1000 kV as the lowest UHV level.

As to 1000−1150 kV single-circuit lines, the structural parameters are as follows:

(1) Radius of the conductor, r_0;
(2) Number of bundled subconductors, n;
(3) Subconductor spacing;
(4) Arrangement form of the phase conductor (e.g., regular triangle, horizontal, quasihorizontal, etc.);
(5) Suspension height of the conductor;
(6) Structure and height of the tower.

These fundamental constraints determine the aforementioned parameters:

(1) Minimum electric field intensity, $E_{\mathrm{d.y.}min}$, when the corona is generated;
(2) Economical current density, J_{jj};
(3) Maximum allowable electric field intensity, $E_{\mathrm{d.m.}max}$, and magnetic field intensity on the ground.

2.1.2 Initial electric field intensity of the corona

Because of cosmic rays, a few air molecules are often ionized as ions with positive and negative charges. There is high electric field intensity on the surface of high-voltage transmission lines. Under the effect of electric field intensity, the charged ions obtain kinetic energy and move rapidly. During this movement, like ions are repelled and move away from the conductor while opposite ions are attracted and move close to the conductor. The collision of the charged ions at high velocity and atoms in the air dissociates the atoms into charged ions. When opposite charges are encountered, they are neutralized. The neutralization leads to energy release and blue light emission, which is actually the corona. Therefore, the air atoms around the conductor are being neutralized and ionized at the same time. The kinetic energy of ionization is supplied by transmission lines, and the energy released in neutralization generates light; these result in transmission line power loss. In addition, the electromagnetic waves generated during ionization and neutralization cause radio interference. Generation of the corona depends on the electric field intensity on the surface of the conductor. The minimum electric field intensity when the corona is generated is $E_{d.y.min}$, and according to experimental research on two parallel conductors by F. W. Peek, an American engineer, we know as follows [1]:

$$E_{\text{d.y.min}} = 30.3\delta m_1 m_2 \left(1 + \frac{0.298}{\sqrt{r_0 \delta}}\right) \text{kV/cm} \tag{2.1}$$

where

r_0 is the radius of the conductor (cm),

δ is the relative density of air,

m_1 is the roughness factor on the surface of the conductor for a single conductor with a smooth surface, $m_1 = 1$;

m_2 is meteorological factor, which is set from 0.8 to 1.0.

The Soviet *Electrical Equipment Installation Procedure* [2] gives the empirical equation as follows:

$$E_{\text{d.y.min}} = 14\delta \left(1 + \frac{0.62}{r_0^{0.38}\delta^{0.3}}\right) \text{kV/cm} \tag{2.2}$$

The meanings of the symbols in Eq. (2.2) are the same as those in Eq. (2.1). Meanwhile, it is recommended by the above procedure that $E_{d.y.min} = 20\text{–}21$ kV/cm can be used to approximate when there is no specific information. Because Eq. (2.2) is calculated for aluminum-conductor steel-reinforced cable under bad meteorological conditions, such as dense fog or light rain, there are no factors for meteorological condition and roughness of the conductor surface.

Example:

If we set $m_1 = 0.7$, $m_2 = 0.8$, $\delta = 0.8$, for a conductor with a sectional area of 300 mm^2:

$$r_0 = \sqrt{\frac{300}{\pi}} = 9.77(\text{mm}) = 0.977(\text{cm})$$

then

$$E_{\text{d.y.min}} = 30.3 \times 0.7 \times 0.8 \times 0.8 \times \left(1 + \frac{0.298}{\sqrt{0.977 \times 0.8}}\right)$$
$$= 18.13(\text{kV/cm})$$

From Eq. (2.1), we can get

$$E_{\text{d.y.min}} = 14 \times 0.8 \times \left(1 + \frac{0.62}{0.977^{0.38} \times 0.8^{0.3}}\right) = 18.7(\text{kV/cm})$$

It can be seen that these two empirical equations are basically consistent. We can either do the experimental research on our own or calculate it by Eqs. (2.1) and (2.2) according to local meteorological conditions and the relative density of air. In plains areas, its recommended value, 20—21 kV/cm, can be used as well.

2.1.3 Economical current density

The economical current density, J_{jj}, is related not only to the value of energy loss and construction investment of transmission lines but also to the operational condition of the transmission line (i.e., maximum transmitting power utilization hours) and other factors. Therefore, its value varies in a wide range, normally 0.6—1.5 A/mm^2 [3].

2.2 Electric field intensity under transmission lines

A power frequency electric field is generated between the high voltage of high-voltage transmission lines and the ground. And the large current on transmission lines generates a power frequency magnetic field in the space between the conductor and the ground. The power frequency electric field and power frequency magnetic field induces voltage and current on human bodies and objects. People may feel uncomfortable when the value of induced voltage and current in the body is small. But when the value of induced voltage and current is large, it can be life-threatening. Power frequency electric fields and magnetic fields also have negative effects on crops. Therefore, all countries stipulate the allowable maximum of electric field and magnetic field intensities within a 1—1.5 m height above ground under the high-voltage transmission line. In China, the regulation of electric field and magnetic field intensities of 500 kV transmission line is stipulated as follows [1]. For general areas, the electric field intensity limit is 7 kV/m where the line is crossing the highway, 10 kV/m crossing the cropland. For the lines close to the houses, the root-mean-square limit of the maximum undistorted electric field intensity is 4 kV/m at 1 m height off the ground at the house location. As for the power frequency magnetic field intensity, the regulation maximum is 100 μT, and 40 μT at most in practice.

The regulation of the maximum electric field and magnetic field intensities of a UHV transmission line is the same as it is with a 500 kV transmission line—i.e., to ensure the safety of people, animals, and plants.

2.2.1 Power frequency electric field intensity

As the subconductor spacing of the bundled subconductors is very small compared with the line height, the charges on each transmission line conductor can be assumed to be concentrated on the axis of the conductors in making calculations. That is to say, we can assume it as line charge and calculate the electric field intensity by the equivalent charge equation [1,4].

Assuming that the transmission line is infinite and parallel to the ground and the ground is a good conductor, we can calculate using the image method. Suppose the electric charge column vector constituted by the equivalent charge on the unit length conductor of the transmission line consisting of several conductors is $\dot{Q}$, the voltage column vector constituted by the voltage of each conductor to the ground is $\dot{U}$. The relationship between the two column vectors above is as follows:

$$\dot{U} = \lambda \dot{Q} \tag{2.3}$$

or

$$\dot{Q} = \lambda^{-1} \dot{U} \tag{2.4}$$

where

λ is the potential factor matrix.

The column vectors of voltage and charge are represented in complex form as follows:

$$\dot{U} = \dot{U}_R + j\dot{U}_I, \dot{Q} = \dot{Q}_R + j\dot{Q}_I$$

where

$\dot{U}_R$ and $\dot{Q}_R$ are the real part of the column vectors of voltage and charge, and

$\dot{U}_I$ and $\dot{Q}_I$ are the imaginary part of the column vectors of voltage and charge.

It can be obtained as follows:

$$\dot{Q}_R = \lambda^{-1} \dot{U}_R, \dot{Q}_I = \lambda^{-1} \dot{U}_I \tag{2.5}$$

On the basis of the Gauss theorem, the electric field lines that an isolated point charge Q emits is in a radial symmetry in all directions, and the surfaces of the sphere with different point-charge-centric radii are equipotential surfaces. The electric field intensity, E, at any point from the center is proportional to Q and inversely proportional to the square of distance L between the point and Q—i.e., $E = \frac{Q}{4\pi\varepsilon_0 L^2}$. As for the line charge of an infinitely long uniform conductor, suppose the charge in per-unit length (e.g., 1 m) is Q, and its equipotential surface is a cylindrical surface with different radii with the axis of the conductor. The electric field intensity of the point with a vertical distance L to the conductor is $E = \frac{Q}{2\pi\varepsilon_0 L}$, and Q is the linear density of the charge.

As for a three-phase transmission line consisting of bundled conductors, we assume that the charge of each conductor is concentrated on the axis (line charge), the charge in per-unit length of conductor i is $\dot{Q}_i$, and the electric field intensity $\dot{E}_i$ generated by all the charges on the conductor at the point $P(x,y)$ outside the transmission line, as well as its horizontal component $\dot{E}_{xi}$ and vertical component $\dot{E}_{yi}$, can be obtained by the Gauss theorem, method of images, and the superposition principle [4].

The electric field intensity $\dot{E}_i$ as well as its horizontal component $\dot{E}_{xi}$ and vertical component $\dot{E}_{yi}$ generated by charge $\dot{Q}_i$ of conductor i is shown in Fig. 2.1. Suppose the x-axis is the ground and is parallel to the conductor, and the Y axis is perpendicular to the ground. Assume that the ground is a good conductor in general, which is like a plane mirror for an electrostatic field. $+\dot{Q}_i$ is the charge of per-unit length of conductor i, and its location is $+Q_i(x_i,y_i)$. Its mirror charge, $-\dot{Q}_i$, is underground. Whether it is symmetrical to $+\dot{Q}_i$ on the axis of the ground depends on the conductivity of the ground. Its location is $-Q_i\left(x_i', -y_i'\right)$, $|x_i| = |x'|$, $\left|y_i'\right| \geq |y_i|$. The distances from $+\dot{Q}_i$ and $-\dot{Q}_i$ to the calculated point $P(x,y)$ are respectively L_i and L_i'. According to Gauss theorem, the electric field intensity $\dot{E}'$ and its

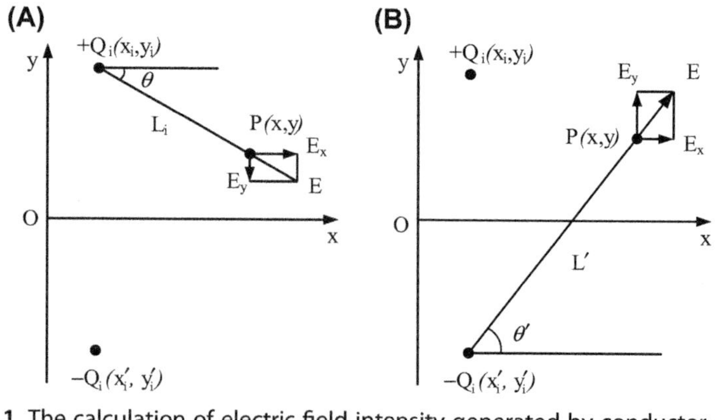

Figure 2.1 The calculation of electric field intensity generated by conductor *i* at point *P* with the image method. (A) The electric field intensity generated by conductor (I) (B) The electric field intensity generated by the mirror image of conductor *i*.

horizontal component $\dot{E}'_{xi}$ and vertical component $\dot{E}'_{yi}$ generated by $+\dot{Q}_i$ at point $P(x,y)$ are respectively as follows:

$$\dot{E}'_i = \frac{\dot{Q}_i}{2\pi\varepsilon_0 L_i}$$

$$\dot{E}'_{xi} = \dot{E}'_i \cos\theta = \frac{\dot{Q}_i}{2\pi\varepsilon_0 L_i} \bullet \frac{x - x_i}{L_i} = \frac{\dot{Q}_i}{2\pi\varepsilon_0} \bullet \frac{x - x_i}{L_i^2}$$

$$\dot{E}'_{yi} = \dot{E}'_i \sin\theta = \frac{\dot{Q}_i}{2\pi\varepsilon_0 L_i} \bullet \frac{y - y_i}{L_i} = \frac{\dot{Q}_i}{2\pi\varepsilon_0} \bullet \frac{y - y_i}{L_i^2}$$

The electric field intensity $\dot{E}''_i$ as well as its horizontal component $\dot{E}''_{xi}$ and vertical component $\dot{E}''_{yi}$ generated by the mirror charge, $-\dot{Q}_i$, are respectively as follows:

$$\dot{E}''_i = \frac{-\dot{Q}_i}{2\pi\varepsilon_0 L'_i}$$

$$\dot{E}''_{xi} = \dot{E}''_i \cos\theta' = \frac{-\dot{Q}_i}{2\pi\varepsilon_0 L'_i} \bullet \frac{x - x'_i}{L'_i} = \frac{-\dot{Q}_i}{2\pi\varepsilon_0} \bullet \frac{x - x'_i}{L'^2_i}$$

$$\dot{E}''_{yi} = \dot{E}''_i \sin\theta' = \frac{-\dot{Q}_i}{2\pi\varepsilon_0 L'_i} \bullet \frac{y + y'_i}{L'_i} = \frac{-\dot{Q}_i}{2\pi\varepsilon_0} \bullet \frac{y + y'_i}{L'^2_i}$$

The resultant horizontal component and vertical component are

$$\dot{E}_{xi} = \dot{E}'_{xi} + \dot{E}''_{xi} = \frac{\dot{Q}_i}{2\pi\varepsilon_0}\left(\frac{x - x_i}{L_i^2} - \frac{x - x'_i}{L'^2_i}\right)$$

$$\dot{E}_{yi} = \dot{E}'_{yi} + \dot{E}''_{yi} = \frac{\dot{Q}_i}{2\pi\varepsilon_0}\left(\frac{y - y_i}{L_i^2} - \frac{y + y'_i}{L'^2_i}\right)$$

Using the superposition principle, the resultant electric field intensity generated by all the conductors at point P can be calculated by Eq. (2.6) [1]:

$$\left.\begin{aligned} \dot{E}_x &= \frac{1}{2\pi\varepsilon_0}\sum_{i=1}^{m}\dot{Q}_i\left(\frac{x - x_i}{L_i^2} - \frac{x - x'_i}{L_i'^2}\right) \\ \dot{E}_y &= \frac{1}{2\pi\varepsilon_0}\sum_{i=1}^{m}\dot{Q}_i\left(\frac{y - y_i}{L_i^2} - \frac{y + y'_i}{L_i'^2}\right) \end{aligned}\right\} \tag{2.6}$$

where

x_i, y_i are the coordinates of conductor i, $i = 1,2,\ldots,m$

x_i', y_i' are the coordinates of the mirror image of conductor i,
m is the number of conductors,
L_i, L_i' are the distances from viewpoint (x,y) to conductor i and its mirror image conductor.

On the ground there is $y=0$, $x_i' = x_i$, $y_i' = -y_i$, $L_i = L_i'$, $\dot{E}_x = 0$, so

$$\dot{E}_y = -\frac{1}{\pi\varepsilon}\sum_{i=1}^{m}\dot{Q}_i\frac{y_i}{L_i^2}$$

It has been assumed that the ground is a good conductor, so the electric field intensity vector above the ground is perpendicular to the ground with no horizontal component. Because L_i differs from L_i' at any point above the ground, hence $\dot{E}_x \neq 0$.

$\dot{E}_x$ and $\dot{E}_y$ are both functions of time and complex numbers, which can be represented by real and imaginary parts as follows:

$$\left.\begin{aligned}\dot{E}_x &= \sum_{i=1}^{m} E_{ix\mathrm{R}} + \mathrm{j}\sum_{i=1}^{m} E_{ix\mathrm{I}} = E_{x\mathrm{R}} + \mathrm{j}E_{x\mathrm{I}} \\ \dot{E}_y &= \sum_{i=1}^{m} E_{iy\mathrm{R}} + \mathrm{j}\sum_{i=1}^{m} E_{iy\mathrm{I}} = E_{y\mathrm{R}} + \mathrm{j}E_{y\mathrm{I}}\end{aligned}\right\} \tag{2.7}$$

where

$E_{ix\mathrm{R}}$ and $E_{x\mathrm{R}}$ are the horizontal vectors of electric field intensity generated by the real parts of charges on conductor i and all the conductors at point (x,y).

$E_{ix\mathrm{I}}$ and $E_{x\mathrm{I}}$ are the horizontal vectors of electric field intensity generated by the imaginary parts of charges on conductor i and all the conductors at point (x,y).

$E_{iy\mathrm{R}}$ and $E_{y\mathrm{R}}$ are the vertical vectors of electric field intensity generated by the real parts of charges on conductor i and all the conductors at point (x,y).

$E_{iy\mathrm{I}}$ and $E_{y\mathrm{I}}$ are the vertical vectors of electric field intensity generated by the imaginary parts of charges on conductor i and all the conductors at point (x,y).

The resultant electric field intensity (space vector) at point (x,y) is as follows:

$$\vec{E} = (E_{x\mathrm{R}} + \mathrm{j}E_{x\mathrm{I}})\vec{x} + (E_{y\mathrm{R}} + \mathrm{j}E_{y\mathrm{I}})\vec{y} = \vec{\dot{E}}_x + \vec{\dot{E}}_y \tag{2.8}$$

The amplitude and initial phases of $\overrightarrow{\dot{E}}_x$ and $\overrightarrow{\dot{E}}_y$ are respectively as follows:

$$\left.\begin{aligned} E_x &= \sqrt{E_{xR}^2 + E_{xI}^2} \\ \varphi_x &= \arctan\frac{E_{xI}}{E_{xR}} \end{aligned}\right\} \qquad \left.\begin{aligned} E_y &= \sqrt{E_{yR}^2 + E_{yI}^2} \\ \varphi_y &= \arctan\frac{E_{yI}}{E_{yR}} \end{aligned}\right\} \tag{2.9}$$

The components of electric field intensity at point (x,y) along the x and y axes both vary with time, usually $E_x \neq E_y$. Therefore, the amplitude of the spatial resultant electric field intensity varies with time, and its locus is an ellipse. In addition, E_x and E_y are related to the relative position between viewpoint (x,y) and each conductor. As a result, for the electric field intensity vectors at different locations, the major and minor axes of ellipse trajectory over time are different, and the maximum spatial resultant electric field intensity cannot be calculated simply by $\sqrt{E_x^2 + E_y^2}$.

In time domain, resultant electric field intensity are as follows:

$$\overrightarrow{E} = E_x \sin(\omega t + \varphi_x)\overrightarrow{x} + E_y \sin(\omega t + \varphi_y)\overrightarrow{y} \tag{2.10}$$

The spatial electric field intensity at any point (x,y) and any time can be obtained by Eq. (2.10), and its maximum and minimum values and their direction can also be derived.

2.2.2 Distribution of electric field intensity

According to the calculation and analysis of a 500 kV transmission line by Eq. (2.10), we can conclude that the direction of the maximum electric field intensity from 1 to 3 m to the ground is similar to the vertical direction [1,4]. In engineering, the vertical component of electric field intensity is usually used to represent the intensity of electrostatic induction. However, the vertical component of electric field intensity is quite different from that of maximum electric field intensity when the ground height is higher.

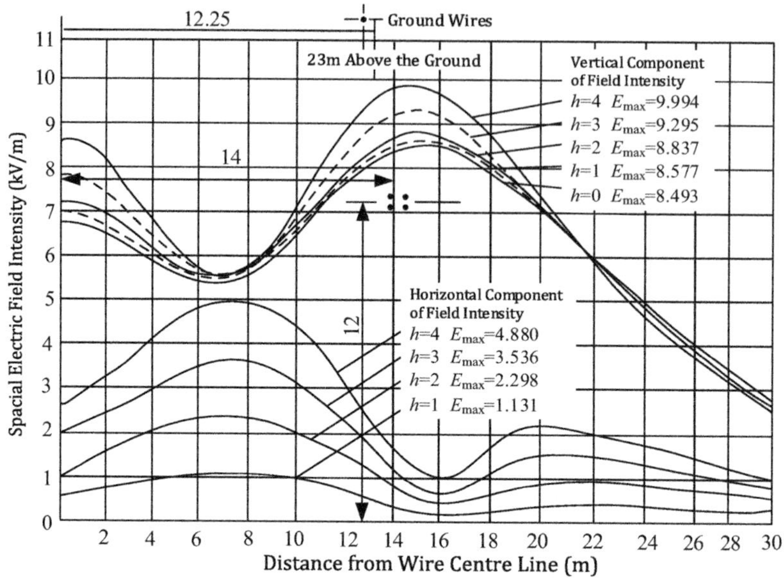

Figure 2.2 Transverse distribution of vertical and horizontal components of electric field intensity within 4 m of the ground under 500 kV transmission line.

Fig. 2.2 shows the transverse distribution of the vertical and horizontal components of electric field intensity within 4 m of the ground under the above-mentioned transmission line, and the five heights are given as $h = 0,1,2,3,4m$. From Fig. 2.2, we can conclude that the maximum vertical component at different heights appears near the minimum horizontal component, while the minimum vertical component appears near the maximum horizontal component. As the height increases, the maximum vertical component increases, but the variation appears obvious only under the centerline and outside the boundary line where maximum electric field intensity appears. Even in these two positions, the variation of electric field intensity is very little within 2 m of the ground. Fig. 2.3 shows the relationship between the maximum vertical component of electric field intensity under the transmission line and distance from the ground. As is shown, the maximum vertical component varies slightly within 4 m of the ground. Although the horizontal component varies greatly, its numerical value is low.

Fig. 2.4 shows the maximum electric field intensity of a horizontally arranged 1200 kV UHV AC transmission line within 3 m of the ground whose conductor-to-ground height is 28 m. Its horizontal component is

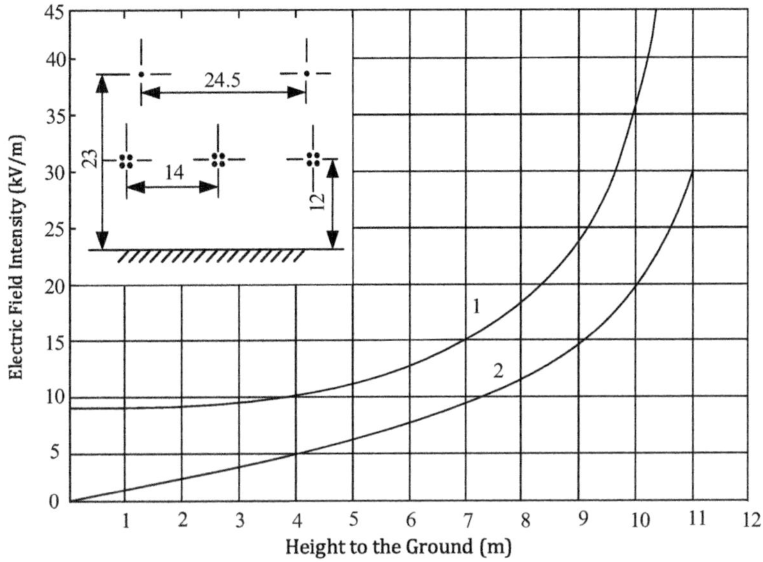

Figure 2.3 Relationship between the maximum vertical and horizontal components of electric field intensity under 500 kV transmission line and distance from the ground. (A) Maximum vertical field intensity. (B) Maximum horizontal field intensity.

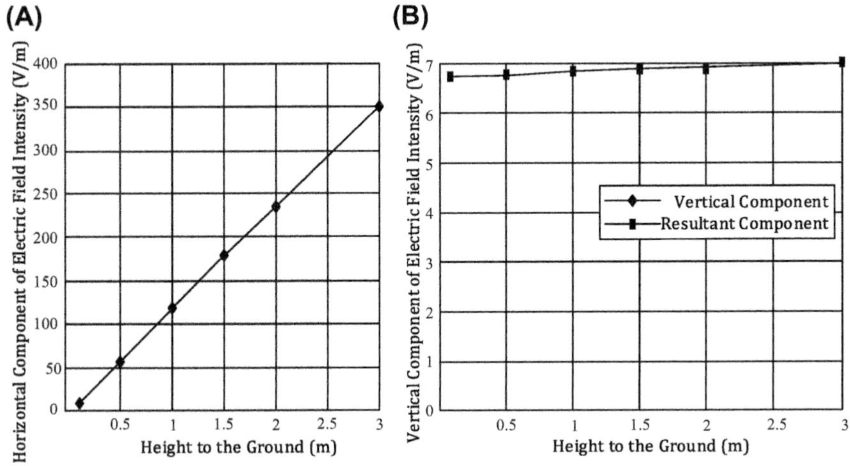

Figure 2.4 Maximum electric field intensity of a horizontally arranged 1200 kV UHV AC transmission line within 3 m of the ground. (A) Horizontal component. (B) Vertical component and resultant electric field intensity.

shown in Fig. 2.4A, and its vertical component and the resultant electric field intensity is shown in Fig. 2.4B. We can conclude that the vertical component and the resultant electric field intensity vary slightly. Although the horizontal component varies greatly, its value is low. This is similar to the case of the 500 kV transmission line.

It follows that for the electric field intensity within 2 m of the ground, the vertical component can be considered the only basis for environmental evaluation in actual projects if there is flat ground around the transmission line without other objects [1.4].

2.2.3 Influencing factors of power frequency electric field intensity

The magnitude and distribution of power frequency electric field intensity are affected by factors such as overhead ground wire, conductor height to ground, phase spacing, the number of bundled subconductors, and the arrangement of conductors. Among these factors, the influence of conductor height to ground is the most important. As for horizontally arranged 1200 kV AC single-circuit transmission lines (conductors:6 $\times$ 630 mm^2, phase spacing: 19 m), calculation indicates that to ensure an electric field intensity at 1 m above the ground of less than 10 kV/m, the minimum height of the phase conductor to the ground where the line sag is maximum is 22 m, while for 500 kV transmission lines, the minimum height is 11 m [1,4]. If the distance from conductor to ground is not long, increasing it (e.g., from 18 to 22 m, from 22 to 28 m) will significantly decrease the electric field intensity at 1 m above the ground (respectively from 14 kV/m to 10 kV/m, from 10 kV/m to 7 kV/m). However, electric field intensity decreases more slowly when the distance from conductor to ground is further increased, as shown in Fig. 2.5.

2.2.4 Electric field intensity under ultra-high-voltage transmission lines

In 1977—1980, measurements of a 1200 kV test line at the Bonneville Power Administration (BPA) showed a maximum electric field intensity of 7.5 kV/m at 1 m above the ground.

Chinese researchers have calculated and analyzed the power frequency electric field intensity of three kinds of typical Soviet, Japanese, and American UHV transmission lines and two kinds of compact transmission lines proposed by us. The first three kinds of towers used in the calculation are shown in Fig. 2.6, and the result of the calculation is shown in Fig. 2.7. In the calculation, the minimum distance from phase conductors to the

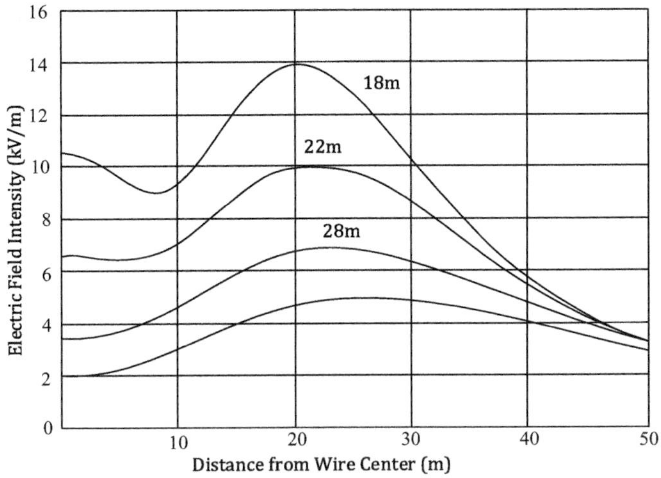

Figure 2.5 Transverse distribution of electric field intensity with different conductor-to-ground heights for horizontally arranged 1200 kV AC single-circuit transmission lines.

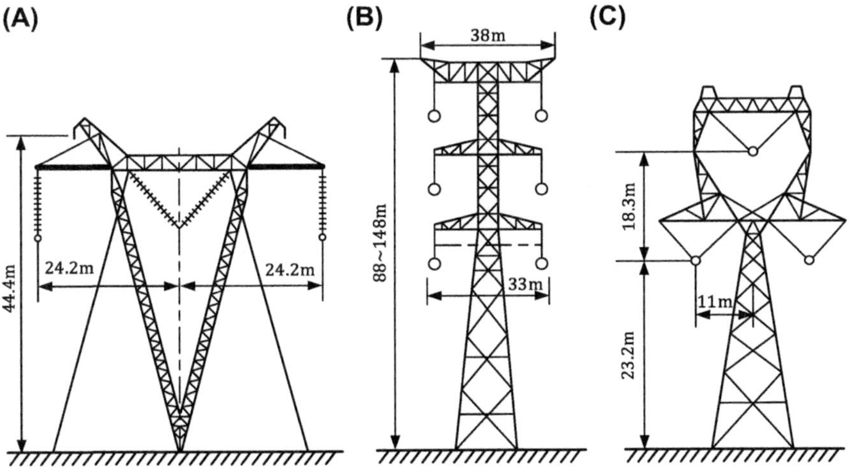

Figure 2.6 Towers of 1000 kV transmission lines of Soviet, Japanese, and American design. (A) Guyed V type Soviet tower. (B) Double-circuit tangent tower from Japan. (C) Cat form vertical tower from America.

ground is set for 22 m according to the horizontally arranged phase conductor from the Soviet Union, which is the minimum distance to the ground required to meet the 10 kV/m threshold value of the electric field intensity when transmission lines go across farmland. The voltage is set at 1050 kV when calculating. It can be seen from Fig. 2.7 that for power

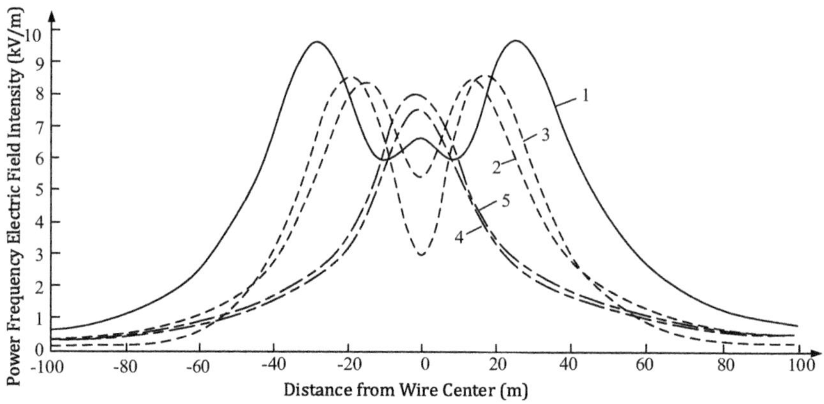

Figure 2.7 Power frequency electric field intensity distribution at 1 m above ground for five kinds of UHV transmission lines: (1) horizontal arrangement of phase conductors, (2) triangular arrangement of phase conductors, (3) double-circuit lines on the same tower, (4) compact type with 10-bundled subconductors, and (5) compact type with 12-bundled subconductors.

frequency electric field intensity 1 m off the ground, the Soviet horizontal arrangement scheme is the maximum, while all the other schemes can be assured to be below the allowable value of 10 kV/m.

2.3 Power frequency magnetic field of transmission lines

2.3.1 Power frequency magnetic field under transmission lines

The calculation method of the power frequency magnetic field is recommended by the 36-01 working group of the International Council on Large Electric Systems [4]. Because the power frequency magnetic field is generated only by the current passing through the conductor, the magnetic field generated by the current in each conductor can be calculated respectively by the ampere circuit law. Then the magnetic field intensity around the transmission line can be obtained by superimposing the calculated results above. The influence of ground is still calculated by method of images in the calculation. For the sake of accuracy, the depth of the image conductor underground is calculated through actual ground conductivity as follows:

$$d = 660\sqrt{\frac{\rho}{f}}(\mathrm{m}) \tag{2.11}$$

where

ρ is the ground conductivity,

f is frequency (Hz).

When the calculated d is very large, the effect of the image conductor current is very little and can be ignored. Then we can just calculate the power frequency magnetic field intensity generated by the current of the actual conductor, as shown in Fig. 2.8. At point A, a distance of r from conductor i, the magnetic field intensity H is

$$\dot{H} = \frac{\dot{I}}{2\pi\sqrt{h^2 + L^2}} \tag{2.12}$$

where

$\dot{I}$ is the current of conductor i,

L, h are respectively the lateral separation and longitudinal separation from conductor i.

The resultant power frequency magnetic field of the point can be obtained by superposition of the magnetic field vector generated by the current of each phase conductor. Similar to the power frequency electric field, the space trajectory of the resultant magnetic field vector generated by the time variable current of three-phase transmission lines is also an ellipse.

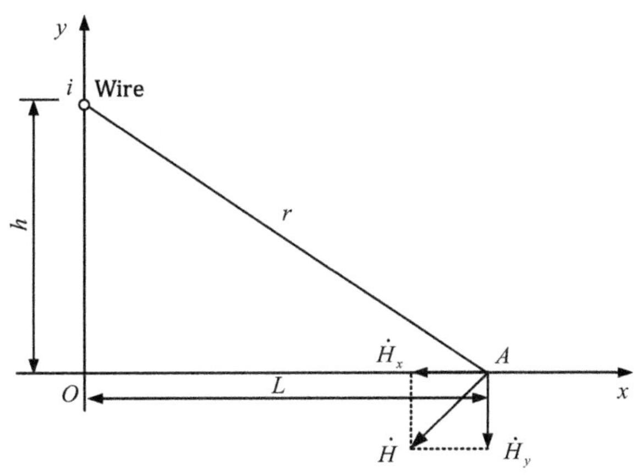

Figure 2.8 Calculation of magnetic field generated by current of single conductor.

2.3.2 Distribution of power frequency magnetic field

The power frequency magnetic field is generated by the current passing through the conductor. Except for the Soviet design, UHV test lines in other countries have not been operated with load, so they cannot be measured. At the same time, due to the small value of the power frequency magnetic field, it is rare to measure the power frequency magnetic field generated by actual UHV transmission lines.

Chinese researchers have calculated and analyzed the power frequency magnetic field of the above-mentioned domestic and overseas UHV transmission lines [1]. In the calculation, it is assumed that the arrangement and size of phase conductors as well as the number and subconductor spacing of bundled conductors remain unchanged for the three kinds of foreign UHV transmission lines, and the sections of the conductors are assumed to be the same. Setting the minimum distance from phase conductor to the ground of 22 m, the current of each phase conductor is 4000 A (current of surge impedance loading is 2880 A), and the voltages are all set for 1050 kV. Fig. 2.9 shows the result, where curves 1, 2, and 3 respectively represent the horizontal arrangement in the Soviet region, the erected triangular arrangement in America, and the double-circuit lines in Japan; and curves 4 and 5 respectively represent compact transmission lines in the operator arrangement with 10-bundled and 12-bundled subconductors. We can conclude that different arrangements of conductors lead to significant differences in power frequency magnetic field intensity under transmission lines. The maximum magnetic induction intensity

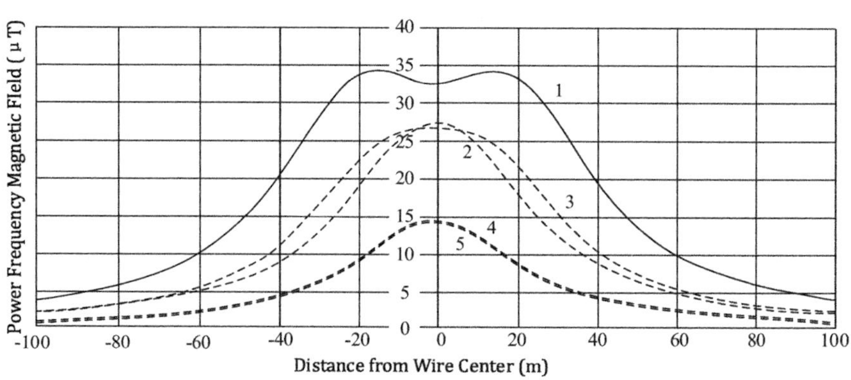

Figure 2.9 Comparison of power frequency magnetic field distribution at 1 m above the ground under five kinds of UHV transmission lines.

corresponding to curves 1−5 is respectively 34.45, 27.10, 26.63, 14.26, and 14.26 μT.

China Electric Power Research Institute and Wuhan High Voltage Research Institute (WHVRI) has calculated and analyzed the power frequency magnetic field of 1000 kV UHV AC transmission lines. With the lowest height of conductor to ground set in the range of 15−23 m in the calculation, the maximum magnetic induction intensity at 1 m above the ground with rated current is less than 35 μT for single-circuit and double-circuit lines on the same tower. As for 500 kV transmission lines, the general measurement results show that the typical value of the power frequency magnetic induction intensity is 20 μT, corresponding to current of 1000 A (equivalent to surge impedance loading) at 1 m above the ground.

Thus, the power frequency magnetic induction intensity under UHV transmission lines is slightly higher than that under EHV transmission lines but is still lower than the maximum limit of 100 μT stipulated by China.

2.4 Limits for electromagnetic fields and radio interference

2.4.1 Limits set by international organizations

The International Commission on Non-Ionizing Radiation Protection (ICNIRP) formally introduced the *ICNIRP Guidelines for Limiting Exposure to Time-Varying Electric, Magnetic, and Electromagnetic Fields (Up To 300 GHz)* in April 1998, in which the basic and derived limits are specified. The basic limit is the induced current of the human body that directly endangers the human body. When the induced current of the human body reaches 100 mA/m^2, the excitability of the human central nervous system changes dramatically and produces other acute reactions. Therefore, the basic limit for ordinary people (unprotected or uninformed) is 2 mA/m^2, 1/50 of the hazard threshold. For professionals (protected or informed), the basic limit is 10 mA/m^2. The derived electric field intensity limit for ordinary people is 4.2 kV/m, and that for professionals is 8.3 kV/m.

IEEE Std C95.6, *IEEE Standard for Safety Levels With Respect to Human Exposure to Electromagnetic Fields, 0~3 kHz*, specified a 50 Hz electric field intensity limit. The limit for controlled area (areas with protective measures) is 20 kV/m and that for ordinary people 5 kV/m. This limit for ordinary people is similar to the limit of ICNIRP.

2.4.2 Electric field intensity limits in various countries

Due to the different conditions in different countries, the limits of electric field intensity are different [1]. For example, with a vast territory and low population density, Soviet transmission lines cover long distances, so it would be a huge investment to raise the wire to reduce electric field intensity. Therefore, in order to reduce the investment in transmission lines, expansion of transmission corridors and strengthening of protective measures have been adopted. In accordance with this principle, the Soviet instruction for installation of electrical equipment stipulates setting up protective zones for transmission lines over 300 kV. The width of the protection zone is demarcated by an electric field intensity of 1 kV/m at the boundary. There are warning signs on the boundary of the protection zone that alert people not to stay in the area for too long. Necessary security measures must be taken when working in the area. For example, a temporary grounding wire of large vehicles or machines must be used when entering this area, and the induced current of objects near and under transmission lines must be less than 4 mA/m^2 [1,2,4].

After taking these measures, the maximum electric field intensity is specified as follows: 10 kV/m when crossing highways and similar locations; 15 kV/m for places where no people live but that people can get to; and 20 kV/m for hard-to-get-to places. UHV transmission lines are not allowed to pass through residential areas. And the absolutely safe threshold of the electric field intensity near a substation is 5 kV/m.

Conversely, Japan has a small geographical area and dense population. Therefore, high towers have been adopted in Japan to drastically hoist the height of conductors from the ground, and the average height of a 1000 kV UHV double-circuit tower is 110 m (the heights of Soviet single-circuit towers are 44.4 and 50 m). It is stipulated that the maximum ground electric field intensity in mountain areas, forests, and other places is 10 kV/m; for traffic, farms, and roads where human activities often occur, the maximum is set for 3 kV/m. It is also stipulated that building houses and planting trees are not allowed within 3 m outside the boundary projection, but growing crops and planting fruit trees are allowed.

The height of American 1200 kV test line towers is similar to the Soviet towers, which is about 50 m. Other countries, such as Italy, Canada, and France, have their own standards, which are similar and no longer listed here.

2.4.3 Radio interference

2.4.3.1 Causes of radio interference

For high-voltage transmission lines in service, the corona effect may cause radio interference, television interference, and audible noise [1,4]. These effects of UHV transmission lines are more intense. The intensity of these effects mainly depends on the circuit structure and climatic conditions. The main factors of circuit structure are the conductor structure, including the number of bundled subconductors per phase, the diameter of subconductors, phase spacing, and the height of conductors to the ground. As mentioned above, the main factor affecting the corona discharge of conductors is the surface electric field intensity of conductors. The influence of climate on corona discharge is very complex and is usually analyzed by experiment.

When the corona of transmission lines and some metal components discharges, it radiates electromagnetic waves into the space, which interferes with radio broadcasting, communications, and television reception. In addition, the external signal and radio noise act on the circuit and the tower and cause reflection, conduction, or reradiation to generate interference signals and may also interfere with radio reception. Radio and television interference generally refer to interference to amplitude-modulated broadcasts of 535–1600 kHz and television broadcasts at low-frequency stages of 48.5–92 MHz. Corona discharge is the main source of radio interference. The main characteristics of radio interference generated by transmission line coronas can be characterized by frequency spectrum, transversal attenuation, and statistical distribution characteristics.

2.4.3.2 Research on radio interference

America, Japan, Canada, the Soviet region, and some other countries have researched transmission line radio interference in basic research on UHV AC transmission lines. BPA chose bundled conductors of 8 × 41 mm and 7 × 41 mm to carry out test study under a test voltage of 1150 kV. The radio interference at 15 m outside the boundary projection of these two kinds of conductors is respectively 46 and 58 dB (0.5 MHz). The experimental results show that the radio interference level of the 1200 kV test line is similar to that of 550 kV transmission lines. Japan has tested steel-core aluminum-stranded bundled conductors of 8 × 34.2 mm and 8 × 34.8 mm, and the radio interference levels are respectively 59 and 53 dB. This corresponds to the radio interference levels of existing 500 kV transmission lines with 4 × 28.4 mm steel-core aluminum-stranded bundled conductors.

The Canada Quebec Laboratory tested four kinds of bundled conductors (6 × 46.3 mm, 6 × 50.7 mm, 8 × 41.4 mm, and 8 × 46.3 mm). After testing and considering various factors, it was agreed that the radio interference level of the 8 × 41.4 mm bundled conductors was acceptable. A Soviet test of the 1150 kV transmission line shows that the radio interference level of 8-bundled AC 330/43 conductors with subconductor spacing of 0.4 m is similar to that of 750 kV transmission lines. The results prove that appropriate selection of the number and diameter of the bundled subconductor could make the radio interference level of UHV transmission lines comparable to that of some EHV transmission lines that have already operated.

The conductor of UHV transmission lines is characterized by a large section and being multibundled (greater than 4-bundled). In the 1990s, China analyzed the environmental impact of UHV transmission lines and proposed a radio interference limit (0.5 MHz) for UHV transmission lines of 55—60 dB. It was put forward after calculation and analysis of wire parameters that may be adopted for some kinds of UHV transmission lines. The radio interference limit of 1000 kV UHV AC transmission lines in China is temporarily 58 dB with a reference frequency of 0.5 MHz and a reference point of 20 m outside the boundary projection. Compared with foreign limits, it is still relatively strict. According to the research results of WHVRI on the electromagnetic environment of 1000 kV AC transmission and its ecological impact, it is technically feasible to adopt this limit. This limit is also economically acceptable in areas where the impact of elevation does not need to be considered.

2.4.3.3 Characteristics of interference with television signals

The frequency of television signal is much higher than that of the broadcast signal. The frequency spectrum characteristic of the interference signal generated by the transmission line corona shows that the interference of transmission lines with television signals is much slighter than with the broadcast signal. Due to the particularity of the television signal, it is worth noting two aspects: (A) The electromagnetic noise generated by discharge causes interference to reception of the television signal, which leads to the deterioration of image quality. (B) The shielding and reflection of the television signal on transmission lines and towers lead to signal attenuation and ghosting. Television sound is a frequency modulation system with an intrinsic antiinterference characteristic, so it is not affected. There are two main sources of television signal interference:

(1) Intermittent discharge caused by fractured or damaged porcelain insulator. Iron rust causes insulation barriers not only in the splice position of the metal parts of each element but also between the steel leg bulb and gossan socket of the insulator. The electric field intensity here is very high, and it generates a continuous microarc discharge in alternating voltage.

(2) Strong discharge at hardware tips. Because the frequency range of spark discharge extends to the ultra-high-frequency section, it is the main source of interference with the television signal. But the impact of this interference is generally limited to a very small range, and it is easy to find the positions of this interference sources and eliminate them.

2.4.4 Audible noise

Audible transmission line noise can be directly heard and is caused by corona and spark discharge around wires, which belongs to audio interference [1,4]. In the past, due to the low voltage level, the audible noise caused by transmission lines was usually too low to draw attention. With the increase in transmission voltage level, the noise caused by corona discharge of UHV transmission lines affects the normal life and work of nearby residents. Therefore, it must be treated with caution in the development of UHV transmission.

On dry or sunny days, there is corona discharge caused by dust, insects, and the burr of the wire itself on the conductor, and the noise level is low. The most serious case is the rainy day. Because the probability of rain is higher than that of fog and snow, especially under drizzle, sprinkle, or moderate rain, the collision and aggregation of raindrops on wires lead to a large amount of corona discharge with random distribution along the conductor, and discharge at every turn bursts into noise. Therefore, the characteristic and limit of noise are usually estimated according to noise on rainy days. In addition, due to the complex process of the occurrence of corona noise in rainy days with a mass of random factors and great dispersion, it is difficult to derive an equation to predict corona noise theoretically and accurately. Therefore, the prediction of audible noise in various countries is derived from simulation in the corona cage or deduction on the basis of statistics and analysis of the long-term measured data of test lines.

According to international practice, the audible noise limit of transmission lines is 50% of the audible noise value on rainy days. Since the audible noise of AC transmission lines on sunny days is about 15–20 dB lower than on rainy days, it is actually the limitation of noise in the most severe cases, and the noise is much lower in other cases. If the audible noise generated by transmission lines during rainy days is limited to 55 dB, it is only 35–40 dB on sunny days. On rainy days, the audible noise caused by the transmission line corona increases, but the background noise during heavy rain is also loud, so the audible noise caused by the transmission line corona is submerged. When the rain is not heavy, the corona noise of transmission lines is more prominent. In consideration of this situation, noise should be reduced properly when the transmission line is adjacent to civil houses so as to avoid disturbing residents' rest.

References

[1] Ultra-high voltage grid. Zhenya liu. Beijing: China Economic Publishing House; 2005.
[2] Electrical equipment installation procedures. Beijing: China Nuclear Publishing House; 1993.
[3] AC power transmission. Alexander Rove, GH. Mark; 1998.
[4] Workgroup No. 36.01 of CIGRE. Electric field and magnetic field generated by transmission systems – phenomenon sketch & practical calculation guideline.

CHAPTER 3

Transmission line parameters

Contents

The parameters of ultra-high-voltage (UHV) power transmission lines include the positive sequence distributed capacitance C_0, distributed inductance L_0 and distributed resistance R_0 of the unit length of the line, conductor radius r_0, the number of bundled conductors n, circular radius of the bundled conductor r_p, three-phase arrangement form, the suspension height of the conductor, etc. As the UHV transmission line is well insulated, its leakage conductance G_0 is small enough to ignore.

3.1 Distributed parameters of transmission line

Construction of UHV power transmission lines requires a huge investment. In order to make the most of the investment and maximize a line's economic benefits, we should improve its transmission capacity as much as possible. The transmission line's rated transmission power shall be about the upper and lower limits of its natural power. When the transmission power is greater than the natural power, the voltage loss of the transmission line increases [1] and larger reactive power compensation is required, which leads to increased investment and reduces the line's economical efficiency. Therefore, increasing the natural power of transmission lines is the main way to increase transmission capacity.

Natural power means that when this power is transmitted, the reactive power loss L_0I^2 caused by the current on the line is compensated by the

Protection Technologies of Ultra-High-Voltage AC Transmission Systems
ISBN 978-0-12-816205-7
https://doi.org/10.1016/B978-0-12-816205-7.00003-0

capacitive reactive power $C_0U_{ph}^2$ generated by the line phase voltage U_{ph} through the line capacitance. That is, $L_0I^2 = C_0U_{ph}^2$

Now the line impedance becomes wave impedance Z_c—i.e.,

$$Z_c = \frac{U_{ph}}{I} = \sqrt{\frac{L_0}{C_0}} \tag{3.1}$$

Now the transmitted power is called natural power P_N—i.e.,

$$P_N = 3U_{ph}I = 3\frac{U_{ph}U_{ph}I}{U_{ph}} = \frac{U^2}{Z_c} \tag{3.2}$$

U is the rated line voltage in the above equation.

It follows that the natural power is inversely proportional to the wave impedance. To reduce the wave impedance, the per-unit-length inductance of the line should be reduced, and the per-unit-length capacitance should be increased. In the meantime, in order to reduce the power loss of the line, resistance should also be reduced. Multiple split wires can reduce resistance, increase wire surface area, and thereby increase capacitance. In cases with the same conductor radius and the same number of bundled conductors, line capacitance is related to maximum electric field intensity allowed on the conductor surface E_p and conductor surface area utilization factor k_{ly}, expressed as follows [1]:

$$C_0 = \frac{Q_0}{U_{ph}} = \frac{qS_0}{U_{ph}} = \frac{\varepsilon_0 E_{av} S_0}{U_{ph}} = \frac{\varepsilon_0 E_p k_{ly} S_0}{U_{ph}} \tag{3.3}$$

$$S_0 = 2n\pi r_0$$

$$q = \frac{Q_0}{S_0}$$

$$k_{ly} = \frac{E_{av}}{E_P} < 1$$

C_0—positive sequence capacitance of per-meter transmission line, F/m.

Q_0—charge of per-meter transmission line.

U_{ph}—phase voltage, kV.

S_0—total surface area of the phase conductor per meter, m^2.

q—per-unit-area charge on the transmission line's conductor surface—that is, the density of charge

n—number of bundled conductors.
r_0—radius of each conductor, m.
ε_0—vacuum (air) permittivity, equals 8.854×10^{-12}.
E_{av}—actual average electric field intensity on conductor surface.
k_{Ly}—conductor surface area utilization factor, typically less than 1.
E_p—allowed maximum electric field intensity on the conductor surface, with the condition of corona occurring.
E_p should be less than the minimum electric field intensity that can cause the corona—i.e., $E_P < E_{d.y.min}$—and its value is related to weather conditions and conductor surface conditions, as discussed in Chapter 2. In general, $E_{d.y.min} = 20 \sim 21 kV/cm$.

Let k_{bj} be the coefficient considering the uneven distribution of electric field intensity on the conductor surface. It is the ratio of actual maximum electric field intensity on the conductor surface E_{max} and average intensity of electric field E_{av} [1]—i.e.,

$$k_{bj} = \frac{E_{max}}{E_{av}} > 1 \tag{3.4}$$

Then,

$$k_{Ly} = \frac{E_{av}}{E_P} = \frac{E_{max}}{E_p} \times \frac{E_{av}}{E_{max}} = \frac{E_{max}}{E_p} \times \frac{1}{k_{bj}} \tag{3.5}$$

For general high-voltage lines, $E_{max} \ll E_p$, $E_{av} \ll E_{max} \ll E_p$, and the utilization coefficient of the conductor surface area is very low. For UHV transmission lines, E_{av} should be as close as possible to E_p so as to increase the area utilization coefficient of conductor surface areas as much as possible and maximize the economic benefits. Therefore, various measures must be taken to reduce the nonuniformity coefficient k_{bj} to improve the surface utilization coefficient k_{Ly}. As for UHV lines, k_{bj} should be minimized as far as possible, which can make k_{Ly} reach a range of 0.9—0.93.

The inhomogeneity of electric field intensity on conductor surface is caused by the shield and interactions among the phases and bundled conductors, and the difference of the distance to the ground. Only reasonably arranging each phase line and bundled conductor, the inhomogeneity of electric field intensity can be reduced.

Electromagnetic waves travel at approximately the same speed as light on overhead transmission line. That is:

$$v_B = 1/\sqrt{L_0 C_0} \approx v_c = 1/\sqrt{\varepsilon_0 \mu_0} \tag{3.6}$$

In above equation, $\mu_0 = 4\pi \times 10^{-7}$, which is the magnetic conductivity in a vacuum.

Therefore, the inductance and inductive reactance in per unit length of the line are as follows:

$$L_0 = \frac{1}{v_c^2 C_0} = \frac{U_{ph}}{v_c^2 \varepsilon_0 S_0 E_p k_{Ly}} = \frac{\varepsilon_0 \mu_0 U_{ph}}{\varepsilon_0 S_0 E_p k_{Ly}} = \frac{\mu_0 U_{ph}}{2\pi n r_0 E_p k_{Ly}} = \frac{2 \times 10^{-7} U_{ph}}{n r_0 E_p k_{Ly}} \tag{3.7}$$

$$X_0 = \omega L_0 = \frac{2 \times 10^{-7} \omega U_{ph}}{n r_0 E_p k_{Ly}} \approx \frac{6.28 \times 10^{-5} U_{ph}}{n r_0 E_p k_{Ly}} \tag{3.8}$$

The wave impedance is:

$$Z_c = \frac{U_{ph}}{I} = \sqrt{\frac{L_0}{C_0}} = v_B L_0 = \frac{1}{v_B C_0} = \frac{U_{ph}}{\varepsilon_0 v_B S_0 E_p k_{Ly}} \approx \frac{60 \times U_{ph}}{n r_0 E_p k_{Ly}} \tag{3.9}$$

It can be seen that the wave impedance is inversely proportional to the total surface area of phase conductors. In order to reduce the wave impedance, and increase the natural power, the total surface area of phase conductors needs to be enlarged—i.e., we should increase the number of bundled conductors and the radius of circumference made up of the bundled conductors.

The effective resistance in per unit length of phase conductor is:

$$R_0 = \frac{\rho}{n\pi r_0^2 \chi_e} \tag{3.10}$$

Here ρ is the resistance coefficient of aluminum cable steel reinforced, which is set 28.3 $\Omega \cdot \text{mm}^2/\text{km}$.

χ_e is the filling coefficient (proportion) of conducting material(aluminum) in the conductor, and it's less than 1. As for aluminum cable steel reinforced, we set $0.61 < \chi_e < 0.67$.

From Eqs. (3.8) and (3.10), we can get:

$$\frac{R_0}{X_0} = \frac{2\varepsilon_0 v_B^2 \rho E_p k_{Ly}}{r_0 \chi_e \omega U_{ph}} = \frac{10^7 \times \rho E_p k_{Ly}}{2\pi \omega r_0 \chi_e U_{ph}} \tag{3.11}$$

3.2 Conductors arrangement

3.2.1 Bundled conductors

The arrangement form of bundled conductors can be any shape [1]. Now we take circular arrangement for example to analyze, as shown in Fig. 3.1. Set n is the number of bundled conductors, r_0 is the radius of each bundled conductor, χ_e is the filling coefficient of conducting material (aluminum) in the cross-section of conductor, r_p is radius of circle by bundled conductors, so that the separation distance of conductors d is:

$$d = 2r_p \sin\frac{\pi}{n} \tag{3.12}$$

Considering the demand of mechanical stability [1,2], d cannot be less than 0.3m. However, it isn't the optimum value, about which is discussed in the optimization of the arrangement of bundled conductor in this section later.

The configuration of the phase conductor shall meet the following basic requirements:

(1) To limit the occurrence of corona and radio interference, we should set the actual maximum electric field intensity on the surface of the conductor $E_{max} \leq E_p$.

(2) When operating with economic current density under rated load, we set $J = J_{jj}$, so as to minimize the energy loss.

(3) Take full advantage of the surface area of conductors, making $k_{LY} \approx k_{ly.max}$.

According to Soviet experience [3] for UHV transmission lines the maximum permissible electric field intensity on conductor surface E_p can be set as follows:

$$E_p \leq 0.9E_{d.y.min} \tag{3.13}$$

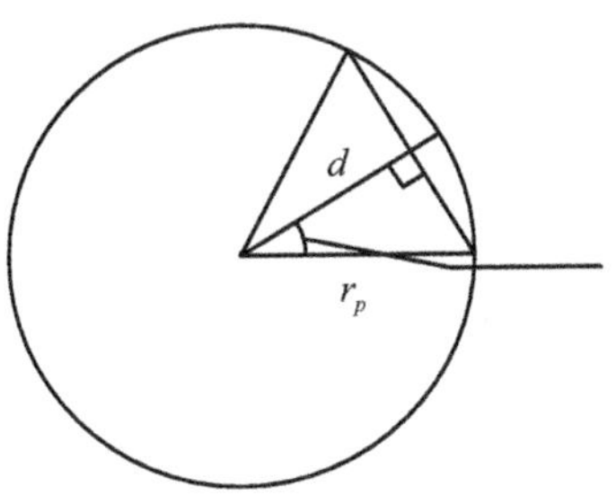

Figure 3.1 The calculation of separation distance of conductors$d$.

$E_{d.y.min}$ is the minimum electric field intensity on the surface when corona occurring, and its determining method has been discussed before.

Due to the differences of electric field intensity at different points on conductor surface, the ratio of maximum electric field intensity E_{max} to its average value E_{av} reflects the inhomogeneity of electric field intensity distribution. It's been defined that the nonuniform coefficient $K_{bj} = \frac{E_{max}}{E_{av}} > 1$, and we know

$$E_{av} = \frac{q}{\varepsilon_0} = \frac{Q_0}{\varepsilon_0 S_0} = \frac{Q_0}{\varepsilon_0 2n\pi r0} \tag{3.14}$$

Here q is the average charge density (charge on per unit surface area) on the conductor surface.

Therefore, we can arrive at the maximum electric field intensity on the conductor—i.e.:

$$E_{max} = k_{bj} E_{av} = \frac{Q_0 k_{bj}}{\varepsilon_0 2n\pi r_0} = \frac{C_0 U_{ph} k_{bj}}{\varepsilon_0 2n\pi r_0} = \frac{U_{ph} k_{bj}}{\varepsilon_0 2n\pi r_0 v_B Z_c} \tag{3.15}$$

$$E_{av} = \frac{U_{ph}}{\varepsilon_0 2n\pi r_0 v_B Z_c} \tag{3.16}$$

In which

$$C_0 = \sqrt{L_0 C_0}\sqrt{\frac{C_0}{L_0}} = \frac{1}{v_B Z_c}$$

The ratio of average electric field intensity E_{av} to the maximum permissible electric field intensity E_p on conductor surface, is called utilization factor of conductor surface K_{ly}. We can know from Eq. (3.16) as follows:

$$k_{Ly} = \frac{E_{av}}{E_p} = \frac{U_{ph}}{\varepsilon_0 2n\pi r_0 v_B Z_c E_P} \tag{3.17}$$

And we know from Eq. (3.5) as below:

$$k_{Ly} = \frac{E_{av}}{E_P} = \frac{E_{max}}{E_p} \times \frac{E_{av}}{E_{max}} = \frac{E_{max}}{E_p} \times \frac{1}{k_{bj}} \tag{3.18}$$

This shows that the more nonuniform the distribution of electric field intensity on the conductor surface is, the bigger k_{bj} may be, and the lower the utilization ratio of conductor surface area may be as well, making the utilization factor smaller. Consequently, the main purpose for studying the arrangement of the three phase conductors is to maximize the uniformity of electric field intensity on conductor surface, making E_{av} as close to E_{max} and E_p as possible. When $E_{av} = E_{max} = E_p$, $K_{ly} = 1$.

As to current density, we should make the actual current density J as close to economic current density J_{jj} as possible under rated load P.

$$J = \frac{P}{3U_{ph} n\pi r_0^2 \chi_e} = J_{jj} \tag{3.19}$$

Eq. (3.18) can also be written as follows:

$$\frac{U_{ph}}{\varepsilon_0 2n\pi r_0 v_B Z_c} = k_{Ly} E_p$$

Dividing above equation on both side by (3.19), we get

$$\frac{E_p k_{Ly}}{J_{jj}} = \frac{r_0 \chi_e}{2\varepsilon_0 v_B} \times \frac{3U_{ph}^2}{PZ_c}$$

Thus we get the equation for r_0 as

$$r_0 = 2\varepsilon_0 v_B \frac{P}{P_N} \frac{E_P k_{Ly}}{J_{jj} \chi_e} \approx \frac{PE_P k_{Ly}}{60\pi P_N J_{jj} \chi_e} \tag{3.20}$$

In this equation, the natural power is $P_N = \frac{3U_{ph}^2}{Z_c} = \frac{U^2}{Z_c}$. By putting the above equation of r_0 into Eq. (3.15), we get

$$n = \frac{U_{ph}}{\pi (2\varepsilon_0 v_B)^2 Z_c} \times \frac{P}{P_N} \times \frac{J_{jj} X_e}{(E_p k_{Ly})^2} \approx 3600\pi U_{ph} \frac{P_N}{P} \frac{J_{jj} X_e}{(E_p k_{Ly})^2} \tag{3.21}$$

From Eqs. (3.20) and (3.21), it is known that the radius r_0 of each conductor is irrelevant to rated voltage and only relevant to the ratio of transmission power to natural power and the economic current density. This is applicable to various voltage grades. However, the number of bundled conductors is relevant to rated voltage and transmission power. The selection of economic current density J_{jj} should minimize the sum of construction and operational costs (the cost of energy loss during transmission) of the power line, and it is complex to consider these factors—for example: the price of electricity, utilization hours under rated load, investment scale of the power line, cycle of construction, loan interest, and scheduled years of investment recovery. All these factors should be taken into account.

3.2.2 Three-phase conductors arrangement

For a single-circuit line, a three-phase horizontal arrangement is the simplest. We shall use this arrangement for an example to explain the optimization method. To improve the utilization ratio of the conductor surface area, the capacitances of the three phases in a symmetric state should be equal. In a three-phase horizontally arranged line, the geometric mean distance of the middle phase is less than those of the two side phases, and the capacitance of the middle phase is bigger than the capacitance of the side phases. This inhomogeneity reduces the utilization coefficient. For this kind of arrangement, two methods can make the capacitances of the three phases equal: ①making the height above ground of the middle phase a ΔH higher than that of the side phases; therefore, the height above ground of the middle phase is increased and its capacitance is reduced, as shown in Fig. 3.2; ②making the circumference of the two side phases a little more than that of the middle phase, so the surface areas of the side phase conductors are increased and their capacitances are increased. The two methods just mentioned can be used simultaneously.

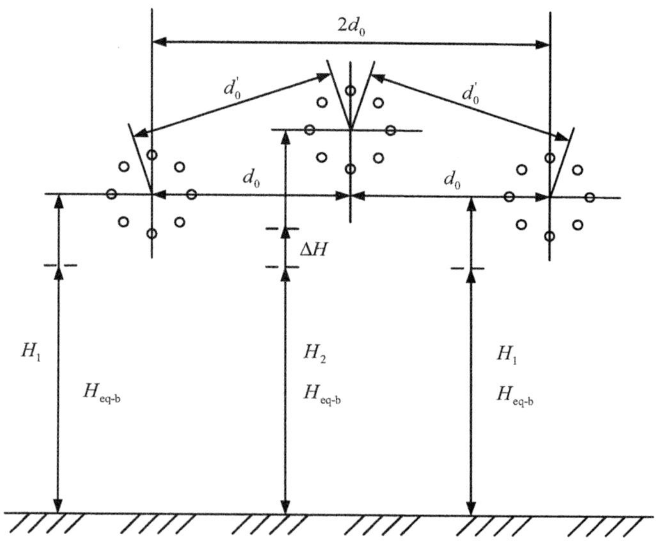

Figure 3.2 Optimization method of three-phase arrangement.

When the two methods are used simultaneously with the aim of making the three phase capacitances equal, we can get the relationship between the height ΔH by which the middle phase needs to be raised and the ratio of the side-phase circular equivalent radius to that of the middle phase by electromagnetic calculation. We can then select an appropriate solution [1]. Set r_{p1} and r_{p2}as the side-phase and middle-phase circular radii of the bundled conductors, $r_{ep.1}$ and $r_{ep.2}$, as the equivalent circular radius arranged on the side phase and middle phase bundled conductors with due regard to the interaction of the conductors. The following relation is valid:

$$r_{eq.1} = r_{p1}\sqrt[n]{\frac{nr_0}{r_{p1}}}$$

$$r_{eq.2} = r_{p2}\sqrt[n]{\frac{nr_0}{r_{p2}}}$$

$$H_{eq.1} = H_{1.\min} + \frac{1}{3}f_1$$

where $H_{1.\min}$ is the minimum height above ground of the side-phase conductor when arc sag is considered;

$H_{eq.1} = H_{eq.b}$ is the equivalent height of the side-phase conductor after considering the effect of arc sag;

H_1 is the actual suspension height of the side-phase conductor in the tower;

H_2 is the actual suspension height of the middle-phase conductor in the tower;

f_1 is the arc sag of the side-phase conductor.

Thus the ratio of $r_{ep.1}$ to $r_{ep.2}$ can be obtained as Eq. (3.22). Sequentially, we can get the side-phase and middle-phase circular radii of bundled conductors r_{p1} and r_{p2} that make the three phase capacitances equal. It should be noted that for the line segment close to the tower, the effect of the tower is equivalent to shortening the distance of the conductor to the ground, so capacitance is increased. Therefore, the circular radius of the bundled conductor should be reduced properly near the tower, where a shielding layer must be installed [1]:

$$\frac{r_{eq.2}}{r_{eq.1}} = \frac{2H_{eq.1}}{r_{eq.1}}\left(1+\frac{\Delta H}{H_{eq.1}}\right) \times exp\left[-\frac{\ln\left(\frac{2H_{eq.1}}{r_{eq.1}B}\right)\ln\left(\frac{2H_{eq.1}}{r_{eq.1}}B\sqrt{A}\right)+(\ln A)^2}{\ln\left(\frac{2H_{eq.1}}{r_{eq.1}B}\sqrt{B}\right)}\right] \tag{3.22}$$

In this equation, $A=\sqrt{1+\left(\frac{2H_{eq.1}}{D_0}\right)\left(1+\frac{\Delta H}{H_{eq.1}}\right)}$, $B=\sqrt{1+\frac{1}{\left(\frac{D_0}{H_{eq.1}}\right)^2-\left(\frac{\Delta H}{H_{eq.1}}\right)^2}}$.

3.2.3 Bundled conductors arrangement

Fig. 3.3 demonstrates the arrangement of the three phases following an equilateral triangle [1], which is the ideal arrangement form. This arrangement form can balance the three-phase capacitance and equalize the circular radii of three-phase bundled conductors. However, the difference of the conductor positions in the circumference of the bundled conductors leads to imbalance of the surface electric field intensity. In three phases, the facing conductors are nearer each other, so the surface electric field intensity between them is larger while that of the opposing ones is smaller. It is

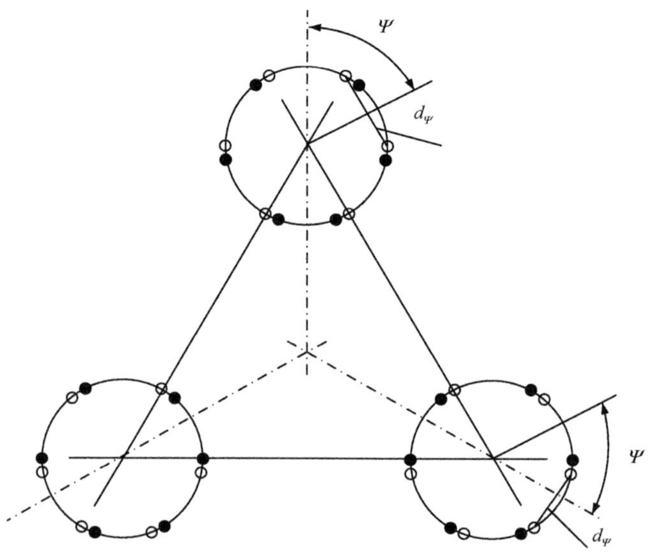

Figure 3.3 Optimization of bundled conductor arrangement.

known from Eq. (3.3) that with the same capacitance, electric field intensity is inversely proportional to surface area S_0. Therefore, if the surface area of the three-phase facing portions is reduced, the larger surface electric field strength can be compensated for, and the surface area of the back-to-back portion can be increased to compensate for the smaller surface electric field strength. On the symmetry axis of three angular bisectors of the equilateral triangle's three angles in Fig. 3.3, the conductors are arranged symmetrically. Meanwhile, inside the triangle the conductor distance near the symmetry axis is reduced—i.e., the surface area is reduced. Also, the conductor distance near the bisector of the exterior angles of the triangle is increased—i.e., the surface area is increased. As shown in the figure, hollow dots represent the conductor positions evenly distributed along the circumference, and the solid dots represent optimized conductor positions. How much the conductors should be moved must be specifically calculated.

The inhomogeneity of electric field intensity on the bundled conductor surface is caused by three factors:

(1) The relative distances of the three phases are asymmetrical, so the capacitances of the three phases are different. This inhomogeneity is denoted by $k_{bj.1}$.

$$k_{bj.1} = \frac{3C_{\max}}{C_A + C_B + C_C}$$

$C_{\max}$ is the capacitance of the largest phase of the capacitor.

C_A, C_B, and C_C are the capacitances of phases A, B and C.

It is thus clear that if the three phase capacitances are equal to the one that is maximum, $K_{bj.1} = 1$.

(2) The per-unit-area charge q differs in each bundled conductor. As shown in Fig. 3.2, the charges can basically be the same in each conductor, making $k_{bj.2} \approx 1$.

(3) When the bundled conductors are arranged in a circle, electric field intensity on the conductor surfaces is asymmetrical because of the interaction of the conductors. This inhomogeneity is denoted by $k_{bj.3}$:

$$k_{bj.3} = 1 + (n-1)\frac{r_0}{r_p}$$

In this equation, n stands for the number of bundled conductors. If $n = 1$—i.e., there is only one conductor for each phase—then $k_{bj.3} = 1$.

The total unevenness coefficient is $k_{bj} = k_{bj.1}k_{bj.2}k_{bj.3}$.

The above discussion is about the optimization method of the three-phase arrangement of bundled conductors. Each method requires careful analysis and calculations to be realized.

Due to the fully utilized conductor surface, the maximum electric field intensity on the conductor surface is close to the allowed maximum value, so its margin for withstanding overvoltage is much lower than that on a smaller-voltage line. In addition, the capacity of the shunt reactor is large. So inputting and cutting of shunt reactors must be coordinated closely with line protection. These characteristics should all be noted in the design of the relay protection configuration scheme.

3.3 Tower structure

The type and size of the transmission line tower have a great influence on line parameters and investment. The type and size of the three kinds of 1000 kV tower are introduced in Fig. 3.6, and some types of 1000 kV line towers are shown in Figs. 3.4–3.9 [4]. The structure and size of Soviet 750 and 1150 kV guyed and self-supporting compact towers are demonstrated in Figs. 3.8 and 3.9 [1–3,5]. In the 1150 kV line, a 12-bundled conductor is used. The conductor is 48-stranded aluminum-conductor steel-reinforced (ACSR), and the sectional area of each conductor is $300mm^2$.

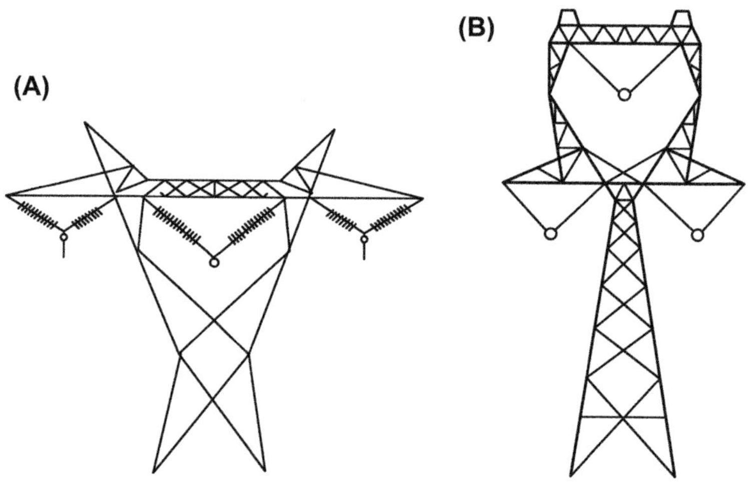

Figure 3.4 1000 kV pole towers of cat head type and wine glass type: (A) wine glass type; (B) cat head type.

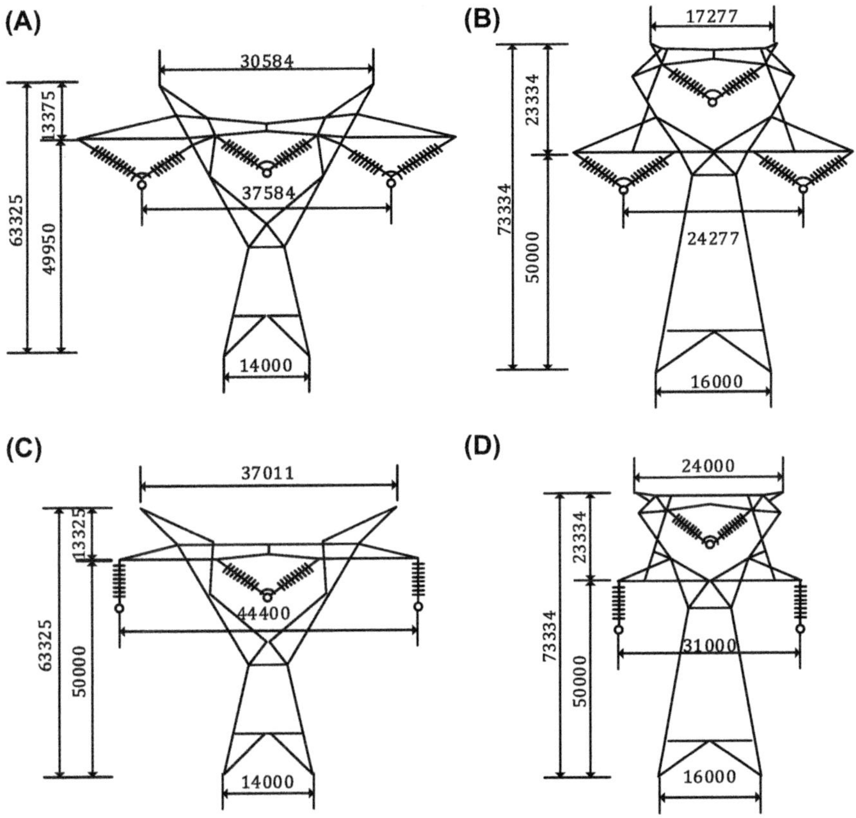

Figure 3.5 Straight pole tower for 1000 kV ultra-high-voltage transmission line: (A) 3V-type horizontal arrangement; (B) 3V-type triangle arrangement; (C) M-type horizontal arrangement; (D) M-type triangle arrangement.

For UHV transmission lines, compact tower structures should be adopted. The central idea of compact tower design is to try to reduce the phase distance without reducing the phase line to ground (steel frame) insulation gap. Thus, if the floor space of the line corridor, inductance, and wave impedance are reduced, the transmission capacity of the line is increased. The main way to do this is to design the special tower shape and insulator suspension method so that there is no ground component among the phases. The conductors should be suspended with a V-type insulator that can fix the conductors and may not be swayed by the wind. Thus the gaps, both between conductor and the ground component and among the phases, may be reduced. In consequence, the distance of the phases may be reduced. Practice has proven that based on the study on the structure of the

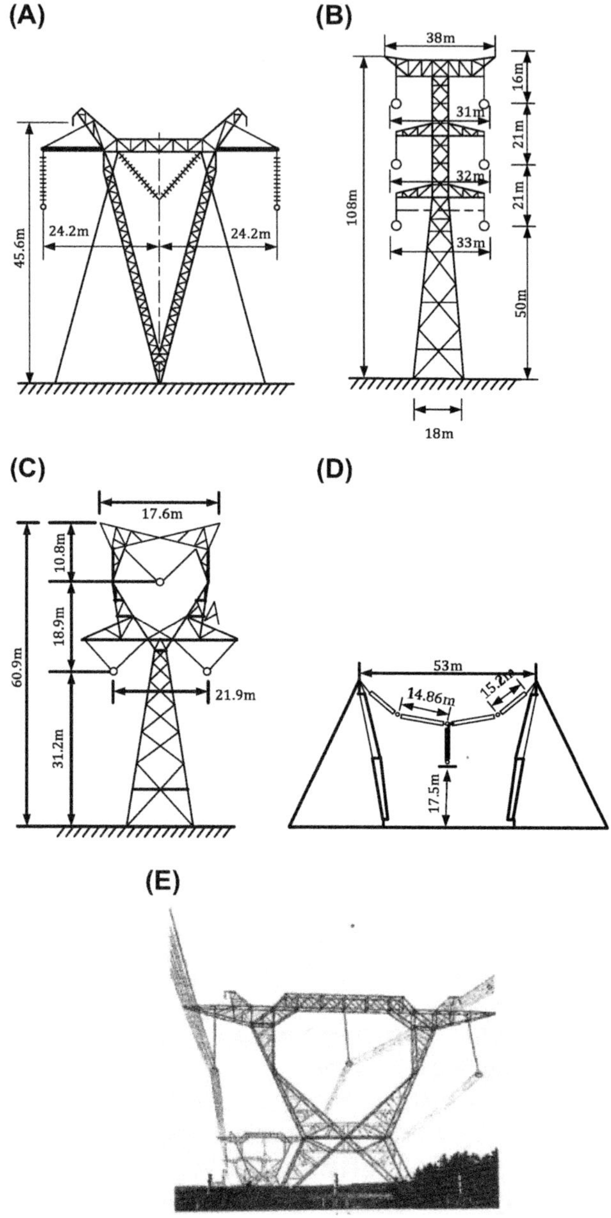

Figure 3.6 UHV tower types designed and adopted overseas: (A) and (E) Soviet; (B) Japanese; (C) American; (D) Italian.

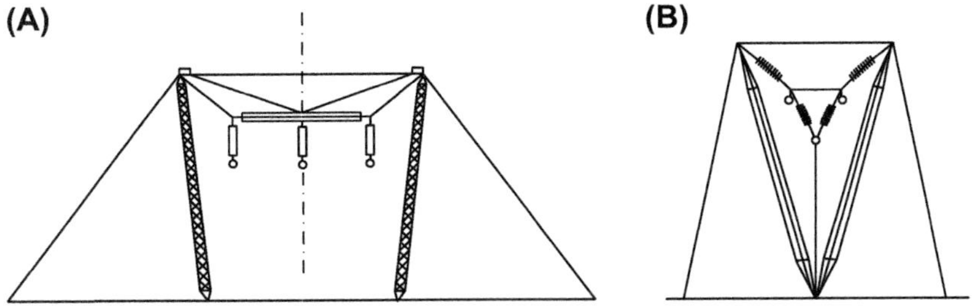

Figure 3.7 Compact tower: (A) 1150 kV suspension line tower; (B) Delta configuration pulling V-type tower.

tower and line conductor system, it is more economical to improve the transmission capacity of the line than to provide compensating devices.

By reference to a Brazil 500 kV line compact tower, the Northwest Power Grid Design Institute of the Soviet Union designed a new 1150 kV line self-supporting compact tower, as shown in Fig. 3.9. In this design, a 12-bundled conductor for each phase is adopted, and each conductor is 48-stranded ACSR with a sectional area of $300mm^2$(12AC 300/48). The circular radius of the bundled conductor is $r_p = 0.9$ m in the central span, the distance between the two conductors is $d = 0.47$ m, and $r_p' = 0.67$ m near the tower. Because of the increasing number of bundled conductors, according to the anticorona requirements, the minimum distance between phases can reach 11 m, and the span distance of the tower can reach 500 m. After the middle phase is elevated 8.05 m, the phase distance of the two sides is reduced to 15 m. However, the phase distance of the two sides in the established noncompact 1150 kV line tower is 46 m. Due to the reduction of the phase distance, the number of bundled conductors is

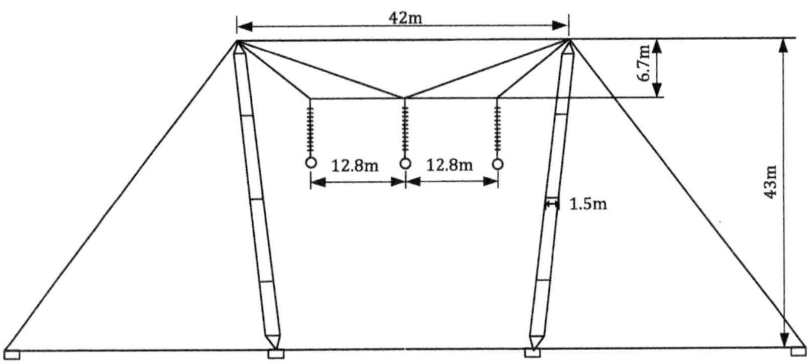

Figure 3.8 Soviet 750 kV line-pulling compact tower.

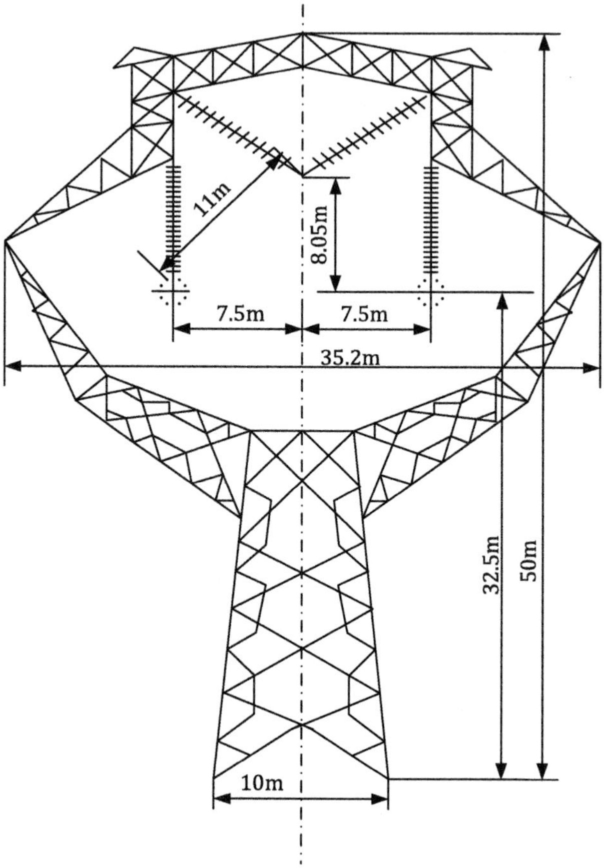

Figure 3.9 Soviet 1150 kV line self-supporting compact tower.

increased so that the wave impedance Z_C is reduced and natural power is increased by 0.5 times. The newly designed tower weighs 34 t with the same steel consumption as before, while its occupied area has been greatly reduced.

In addition, on this basis, it is an inevitable development trend of the UHV power grid to develop and construct UHV double-circuit/multi-circuit/compact transmission lines.

The research and construction of UHV double-circuit/multi-circuit/compact transmission lines are quite extensive abroad. In America, the proportion of transmission lines for which double-circuit lines have been adopted on the same tower in 220and 345 kV voltage grades is 47% and 52%, respectively. In Britain, double-circuit lines on the same tower are adopted on all 400 kV lines. In Japan, double-circuit lines on the same tower are adopted on 90% of 500 kV lines, and the structure of double-circuit lines on the same tower is adopted in all four established 1000 kV UHV lines. As for multiple circuits on the same tower, during the 1970s and 1980s, America studied and built transmission lines with multiple circuits on the same tower to improve transmission capacity in unit corridors. Furthermore, as pioneers of compact transmission technology, the Soviet Union and Japan successively built EHV (extra-high-voltage) multicircuit compact transmission lines on same tower.

Double-circuit lines are also widely used in China's EHV and UHV transmission systems. Fig. 3.10 shows the structure of a double-circuit compact transmission line on the same tower at the 750 kV voltage level, which is closest to UHV. However, there are still many problems to be solved for the tower structure for 1000 kV UHV double-circuit compact transmission on the same tower.

The operational experience of the Soviet Union's and Japan's UHV transmission lines shows that shielding failure and back flashover of lightning are the main causes of double-circuit fault and interline fault. The conventional double-circuit same-tower structure is beneficial to prevent sleeting jump, while the belt-umbrella tower has the advantage in lightning protection performance. The T-type compact double-circuit same-tower tower shown in Fig. 3.10B is widely used in Chinese 500 kV transmission systems. The distance between conductors of compact double-circuit

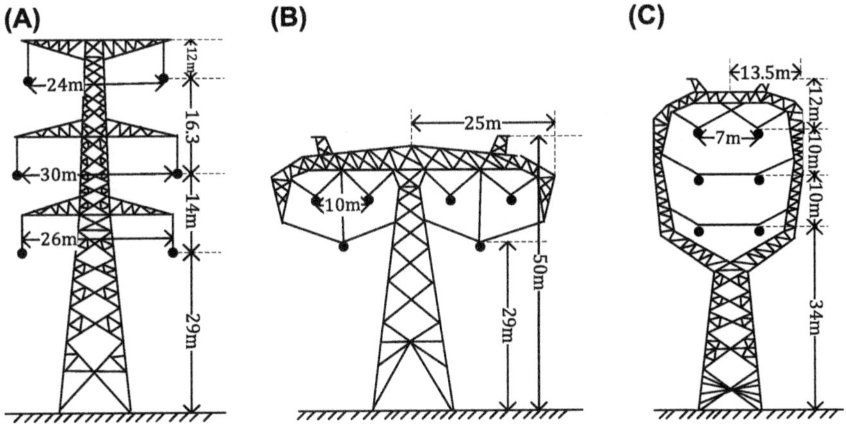

Figure 3.10 The 750 kV tower structure of a double-circuit same-tower line (A) 750 kV double-circuit same-tower routine type; (B) 750 kV compact double-circuit same-tower T-type; (C) 750 kV compact double-circuit same-tower with-same-window.

transmission lines is closer. Especially in the UHV electric field environment, the skin effect that occurs when alternating current circulates between conductor and earth makes the frequency characteristics of the line parameters much more obvious. The structure of a 750 kV compact double-circuit same-tower with same window proposed by China is shown in Fig. 3.10C. Furthermore, China has proposed a vertical, bitriangular arrangement, which is more compact with a better electromagnetic environment than the symmetrical inverted triangle arrangement. The structure of a Chinese 500 kV four-circuit on-same-tower transmission line is shown as Fig. 3.11.

It can be seen from Figs. 3.10 and 3.11, influenced by tower structure and conductor arrangement, the electromagnetic and static electricity couplings between the phases and conductors of double or multiple circuits in a line corridor are asymmetrical. Therefore, the imbalance of UHV double-circuit/multicircuit transmission lines should be researched and demonstrated.

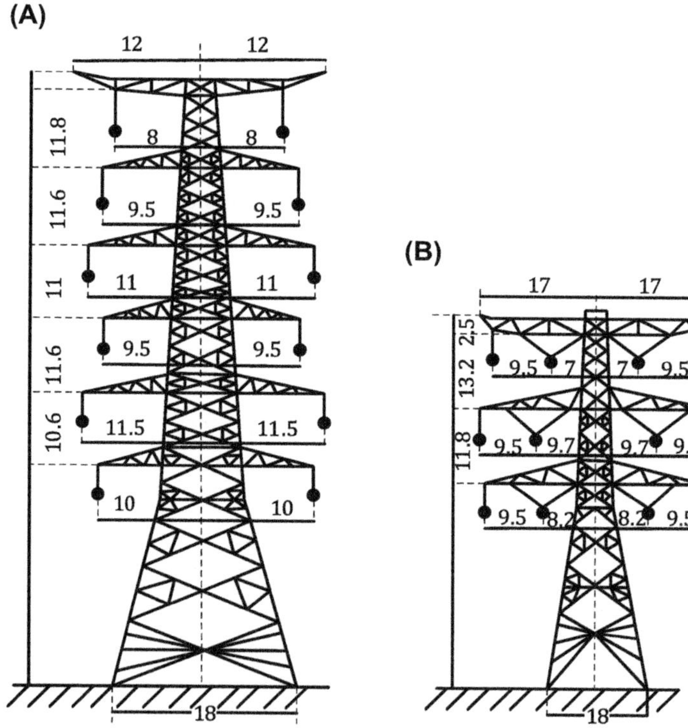

Figure 3.11 Tower structure of multicircuit lines: (A) 500 kV four-circuit on-same-tower d-type; (B) 500 kV four-circuit on-same-tower e-type.

References

[1] AC transmission. GH Mark; 1998.
[2] About the new voltage level of AC transmission. electricity. 1985. No.1. (in Russian).
[3] Electrical equipment installation procedures. Beijing: Electric and Nuclear Press; 1986.
[4] Ultra-high voltage grid. Zhenya Liu. Beijing: China Economic Publishing House; 2005.
[5] Design of ultra-high voltage transmission lines. GH Mark. Electric and Nuclear Press; 1986.

CHAPTER 4

Protection configuration

Contents

4.1 Requirements for protection

As the insulator, transformer, and switch equipment of an ultra-high-voltage (UHV) line can withstand little overvoltage, and overvoltage can break the line insulator and damage the equipment and overvoltage can break through the insulator, causing damage to the equipment, the basic requirement for the protection configuration of a UHV transmission line is to limit overvoltage. The economic loss caused by power failure when replacing insulators and equipment is much greater than the loss caused by loss of stability. Therefore, the following issues should be considered in the selection of a protection configuration scheme [1]:

(1) There should be two completely independent main protections as well as a backup protection that can quickly clear the fault all over the line. In any case, such as when a main protection is down for maintenance or refuses to operate, there will still be two sets of fast protection.

(2) The operation time of the main protection and the arc breaking time of the circuit breaker should meet the requirements of limiting overvoltage.

(3) Under normal operation or the fault condition, the cutoff time difference between the two terminals of the transmission line shall not exceed a certain value. A maximum value shall be specified according

Protection Technologies of Ultra-High-Voltage AC Transmission Systems
ISBN 978-0-12-816205-7

https://doi.org/10.1016/B978-0-12-816205-7.00004-2

to the overvoltage calculation and the insulator's overvoltage capacity, which imposes stringent requirements on the action time of the protection and disconnection time of the circuit breaker.

(4) In order to reduce overvoltage, the reclosing time difference between the fast automatic reclosures at both ends of the line should be defined as a reasonable value according to the overvoltage calculation. When the single-phase reclosing is unsuccessful, the ends of the other two phases should be cut off first to make the induced overvoltage smaller, which should be specified through the overvoltage calculation.

(5) Single-phase reclosure should be used in a single-phase fault and should be obtained by resonance overvoltage calculation in a two-phase operational state. If the resonance overvoltage exceeds the allowable value, the single-phase reclosure cannot be used.

(6) To ensure that the time difference between one end being input and the other end being removed is not greater than the specified value, the manual input should be semiautomated—i.e., manually closing one end while sending the closing signal through the channel to close the other end. The manual tripping is the same and should also be semiautomated.

(7) The operating mode of the shunt reactor should consider the overvoltage when the fault is cleared. The operating mode of the shunt reactor should consider the overvoltage during the fault clear. To reduce reactive power loss, the reactor would not add input to the transmission line when the line is carrying a large load. In this case, the reactor should be put into operation at the same time as the fault is cleared, and the neutral point arc suppression reactor should also be put into operation simultaneously. For shunt reactors, there should be one input/cut automatic device controlled by line protection. When a fault occurs inside the shunt reactor, the line circuit breaker is tripped by its protection, and the opposite side switch is switched on through the channel to prevent overvoltage.

4.2 Protection configuration

4.2.1 Complete dual with different principles

The configuration principle of UHV line protection is similar to that of 500 kV line protection but with a higher requirement [2]. It requires a complete, 100% duplicate configuration and local backup—that is to say, the two sets of main protections should have high independence, and the protection configuration should have certain redundancy. The two sets of main protection must be AC input from TA and TV, and the terminal box, cable, DC power supply, trip coil, protection screen, and communication channel should be completely separate and independent. There should be sufficient distance between the two main protective screens. The mode of signal transmission should also be different—for example, one could use a blocking mode and the other a permissive mode Backup protection must be at least completely separate from the set of primary protection. A configuration lacking a single one of the above-mentioned conditions cannot be regarded as having 100% duplication and local backup. In addition, it is also necessary to set remote backup and circuit breaker failure protection and to ensure the sensitivity of the remote backup as much as possible. Meanwhile, measures should be taken to minimize the extent of the failure to be removed—double busbars, separated operation, etc.

When the TA cannot be separated, the current loop should be connected to two different secondary windings of TA. As for CVT, if each phase has two coupling capacitors, it must be completely separated from the coupling capacitor. If each phase has only one coupling capacitor, it should be separated from the electromagnetic transformer, including the TV terminal box, the secondary cable, etc.

The DC power supply should have two sets of batteries or two independent DC buses, and the two sets of main protections are connected to different DC buses. When a DC bus fault occurs, all connected circuits to it should be switched automatically to the other bus. When a DC bus fault occurs, the blocking mode pilot protection connected to the bus must be able to send a blocking signal so that the protection does not misoperate on the other end, in case of an external fault, and must be able to send a permissive signal and make the protection refuse to move in case of an internal fault.

According to the above requirements and principles, UHV transmission lines should have two sets of main protection and a set of backup protection that can quickly remove the whole line fault and be used as backup protection for adjacent lines. In addition, when the single-phase reclosure is applied, there should be a protection set that may not misoperate in the case of a two-phase oscillation operating condition and can reliably respond to the other fault.

Because each protection principle has some shortcomings, the two main protections should apply different principles to complement each other. In microcomputer protection, a variety of protection principles can be realized at the same time. Therefore, two sets of devices with the same hardware and different software principles can be used when necessary for microprocessor-based protection. Transverse differential protection is a simple and effective protection principle for parallel double-circuit transmission lines, so it is possible to install transverse differential protection to increase the redundancy of protection for parallel double-circuit transmission lines. For microcomputer transverse differential protection, instead of cross-connecting the current loops of a double-circuit line in order to obtain differential current and check its direction, we can directly compare the power directional sign of protections of the double-circuit lines. When the double-circuit line is externally short-circuited, the power directions of the double-circuit line on the two ends are always the same. When there is an internal short circuit in the double-circuit line, the power direction of one end of the double-circuit line is inconsistent, so the internal fault can be judged and the fault line selected and cut off according to the direction of the fault. After one end is cut off, the power direction of the double-circuit line on the opposite end may inevitably become inconsistent, and the fault line can be cut off immediately. When one end has been judged to be an internal fault, the line breaker on the opposite end can be tripped by sending the tripping signal through the channels. In this way, the total removal time of the fault is less than the one following the other action.

4.2.2 Protection against misoperation

As there are many electronic components, nodes, and coils in integrated circuit protection and other nonmicrocomputer protection, and therefore the reliability problems of the above protection devices are mainly caused by bad contacts, disconnection of the relay coil, or damage to electronic components. Conversely, microcomputer protection has fewer electronic components with a smaller probability of damage, so its probability of misoperation due to interference is higher than refusal-operation. This is because the refusal action and unselected-action can only occur during the short period of the line fault, whereas misoperation may occur during the long period of normal operation. In addition, microcomputer protection continues to process and transmit a significant amount of digital code. In processing and transmission, when any bit of code, especially a high-order bit, is affected by external or internal disturbances and changed, the correct operation of protection may be affected accordingly. Therefore, for nonmicrocomputer protection, it is reasonable to output through the OR gate, but this is not the case for microcomputer protection. In some cases, it is more reasonable to output through an AND gate or voting mode, which requires sufficient redundancy of the protection device.

4.2.3 Coordination between automatic reclosure and protection

The coordination mode of automatic reclosure and protection has great influence on the overall performance of the protection configuration. To prevent operation overvoltage for UHV transmission lines, the cutoff time difference between the two ends of the circuit breakers should be limited to within 40—50 ms, and the maximum reclosing time difference should also be specified as a value that must be determined according to voltage calculation. Input to and cutoff of the transmission line should be semi-automatic. For example, it is necessary to manually input from one end (which should be subjected to an overvoltage calculation in advance, with the one that generates the lower overvoltage being input first), and the synchronous parallel device of the opposite end is automatically activated by the channel. At the same time, the synchronizing parallel device at the opposite end is input through the channel simultaneously and automatically. A manual cutoff cuts off manually from one end, and the other end is cut off by transmitting the trip signal through the channel. Automatic reclosing after fault tripping, and second tripping when the reclosing fails, should also be carried out in order.

In the case of a single-phase fault of the high-voltage transmission line, a single-phase reclosing should be preferred, because if the three-phase reclosing is used at this time, the end that is first cut off has not only the overvoltage produced by the capacitance effect but also the mutual inductance overvoltage produced by the fault phase current. Consequently, the overvoltage multiple may be higher. Single-phase reclosing is not allowed when the harmonic overvoltage multiple (considering the transient overvoltage in fault phase cutoff) generated in the non—phase state is greater than the allowable value. For lines with single-phase reclosing, the sequence and time interval of reclosing the two ends must be calculated in advance through the overvoltage calculation. In the case of unsuccessful single-phase reclosing, the sequence and time interval of the other two phases must be calculated in advance by overvoltage calculation. The capacity of the system connected by the UHV transmission lines is usually large with large inertia constants, and thus system oscillation develops slowly with a longer oscillation cycle. Therefore, it is possible to consider fast reclosing without a synchronization check—that is, the reclosing mode, besides single-phase, three-phase, and comprehensive, should include fast reclosing. Fast reclosing should be carried out at both ends simultaneously under channel coordination, avoiding the extra time for end input and end disconnection that could cause the allowable time to be exceeded and result in overvoltage. Only when the fault is cleared instantaneously from both ends is it allowed to be put into fast reclosing without a synchronization check.

Automatic reclosing should be configured according to the circuit breaker. The coordination relationship between automatic reclosing and each protection should be considered carefully to avoid the uncoordinated control of each protection for reclosing. For example, some protection starts fast reclosing, whereas other protection starts single-phase or three-phase reclosing. The closing time of single-phase reclosing should be adjusted adaptively according to the arc extinguishment situation. For example, according to the current passed by the arc-extinguishing reactor connected to the neutral point of the parallel reactor at the two ends of the transmission line, it is judged whether the arc-extinguishing reactor should be put into operation, and whether the arc in the short-circuit point is extinguished, so as to change the reclosing time of the single-phase reclosing. It is best to develop and adopt a new adaptive reclosing principle to improve the success rate of single-phase reclosing and reduce negative impacts on the power system.

4.2.4 Shunt reactor and its operation mode

In order to prevent overvoltage, the two ends of the UHV transmission line should be equipped with large-capacity shunt reactors. To prevent overvoltage in the middle of very long UHV transmission lines, a shunt reactor may be installed at the middle point in the line. Depending on operating conditions, the reactors at both ends may or may not be put into operation. However, the shunt reactor must be immediately put into operation when the circuit breaker trip to prevent overvoltage and ensure that the arc is extinguished. Therefore, special protection and automatic control devices should be installed for the shunt reactor and arc-extinguishing reactor. The circuit protection should continuously monitor the state of the circuit breaker of the shunt and arc-extinguishing reactors. (In order to reduce reactive power consumption of the transmission rated power, the shunt reactor of the UHV transmission line is preferably connected to the circuit through the circuit breaker. When the transmission line is carrying a large load, it is necessary to open the circuit breaker to cut off the shunt reactor from the line to reduce reactive power consumption and increase the maximum capacity of active power.) If both the shunt and arc-extinguishing reactors are not put into operation when line faults occur (if the arc-extinguishing reactor connected to the neutral point of the three-phase shunt reactor cannot bear the zero sequence current in the grounding short circuit, it may be bypassed by the circuit breaker in normal operation), line protection shall start protection and the automatic device of the shunt reactor while sending the trip order and immediately put the shunt and arc-extinguishing reactors into operation (Close the circuit side switch of the shunt reactor, and break off the bypass switch of the arc-extinguishing reactor.). At the same time, the single-phase automatic reclosing should monitor the current in the current transformer of the arc-extinguishing reactor. If there is current in the current transformer when the arc-extinguishing reactor is put into operation, it indicates that the arc-extinguishing reactor has been input, and the closing time for reclosing should be shortened immediately. On the contrary, no current in the transformer indicates that input was not successful, and the normal closing time should be adopted. At the same time, according to the magnitude of the current in the arc-extinguishing reactor, we can know the time of arc extinction so as to adjust the reclosing time adaptively and realize adaptive reclosure.

4.2.5 Circuit breaker failure protection

The circuit breaker failure protection principle of the UHV transmission line is the same as for a general high-voltage line. However, due to the long distance of the UHV transmission line, it is difficult to ensure the sensitivity of its failure protection starting element. Especially when line protection needs to be a remote backup protection for next-line protection, a requirement is that when a fault occurs at the end of the next line when next-line protection fails to respond and the local-line protection breaker also refuses to open, the starting element (element that detects that the fault has not disappeared) of failure protection should have enough sensitivity. Therefore, failure protection should adopt a variety of starting principles to ensure sensitive startup under various types of faults—for example, in addition to current and voltage of the fault phase, the starting parameters of negative sequence, zero sequence, impedance, and so on are needed. The failure protection startup element does not need to have a direction judgment function because in the case of reverse fault, line protection may not operate and failure protection may not start. The dead zone of the directional element can be relieved without a direction judgment function.

4.2.6 Fast fault clearance and overvoltage limitation

The UHV transmission line has strict requirements for the time difference between tripping and closing on both ends of the line, so we should make full use of communication channels to make the two end protections cooperate closely. The transceiver should be able to send out multiple different signals (locking signal, allowed signal, trip signal, signals of transmission state information and command, etc.) for various protection and reclosing types, and priority should be given according to the importance of their functions. For example, the signal of the straight-circuit or direct-coupled circuit breaker of the normal operation of a transmission line, or the signal of the local circuit breaker tripping, or the fault of the direct-breaking-end circuit breaker within range of the backup protection I segment shall be the highest priority. Next is the permissive tripping signal issued by the pilot protection action and the signals that start fast reclosing after quick protection clearing faults, lock fast reclosing at the other end when slow action protection tripping occurs, and exit protection in the two-phase operation state. It has been proved that the full use of communication channels can improve the performance of the protection, which may be easier to realize when the fiber channel is applied, so research in this area should be strengthened.

The foregoing are some requirements for the configuration scheme of relay protection for UHV transmission lines. For specific projects, we should also study the actual conditions of each line and formulate specific configuration programs.

4.3 Selection of protection and communication channels

As long as the characteristics of the UHV transmission line, such as large distribution capacitance, serious overvoltage, and high reliability requirements, have been fully considered and corresponding measures have been taken, the protection principles that have long been applied to extra-high-voltage transmission lines can in theory be applied to UHV transmission lines. Split-phase current differential protection, negative-sequence directional pilot protection, power frequency variation directional pilot protection, distance pilot protection, and phase voltage compensation directional pilot protection can all be applied to UHV transmission lines.

Each principle has advantages and disadvantages. Two kinds of main protection that are based on different principles and can complement each other's performance should be selected.

Local backup protection is the main backup protection. It should usable as the backup for main protection and be able to instantly clear any faults along the full length of the transmission line under the cooperation of protection communication channels. Multizone distance protection can meet this requirement. When the line is not too long, it can also be used as a remote backup for the full length of the next line. Any fault on the whole length of the line can be cleared instantaneously by transmitting the trip signal in distance I. In order to improve the reliable response of the ground fault, multistage zero-sequence current direction protections should be installed.

The performance and reliability of the communication channel play a decisive role in improving protection reliability. Two channels of different principles, such as fiber channel and carrier channel, should be used. Communication channels should be able to send out various kinds of signals, such as blocking, unblocking, allowing, and tripping, to meet all kinds of relay protection work requirements.

The development direction and ideal main protection of transmission lines should be traveling wave protection and no channel protection. But the two protection principles are still being tested and accumulated in operational experience and have not reached the practical level. The main protection of the UHV line should still rely on pilot protection with the aid of the communication channel.

The basic principle of pilot protection of transmission lines can in a broad sense be divided into two categories, unit protection scheme and nonunit protection scheme. So-called unit protection considers the transmission line as a protected unit, just like a transformer or generator. The position of the fault is judged by the current on each side of the unit. This protection scheme collects the sampled values of electrical quantities from each end of the transmission line and transmits them to other terminals through communication channels, directly comparing the sampling values at each end to decide whether the protection device should trip. According to this definition, both phase differential protection by comparison of current phase and current differential protection by comparison of current phase current waveform (amplitude and phase) belong to this category.

So-called nonunit protection also collects sampled values of one or several electrical quantities at each end of the transmission line; however, it does not transmit the sampled values directly to the other ends, but rather sends some obtained judgment results about the nature of the faults (direction of fault, position of fault, etc.) based on these measurements. This kind of protection includes directional comparison pilot protection, distance pilot protection, and so on. Directional comparison protection is based on different kinds of directional elements: negative-sequence directional protection, zero-sequence directional protection, directional protection based on power-frequency components, directional protection based on positive-sequence components, phase voltage compensation directional protection, energy directional protection, etc.

Distance pilot protection also belongs to the direction comparison protection category. But distance pilot protection not only judges the direction of the fault but also determines the fault location. In addition, it can be used not only as main and local backup protection but also as remote backup protection, so it should belong to one special class.

The advantage of unit protection is that there is no need for a voltage value, and it is simple and reliable with absolute selectivity. Its disadvantage is that it requires the sampling value of the current to be accurately transmitted through the communication channel, so it requires a high data transmission rate and communication channel reliability. The advantage of nonunit protection is that it only transmits the result of the judgment of fault direction or fault distance, for which the amount of data is small, so its communication rate demands are low. Its disadvantages are that it needs to collect voltage, and the calculation is more complicated. Two main protections based on different principles should be set up for UHV transmission lines: unit protection and nonunit main protection.

Distance protection is an ideal backup protection. It can be combined with the communication channel to become a distance pilot protection that can also quickly clear faults along the whole transmission line. Distance pilot protection has five schemes: overrange locking, overrange permissive, unlocking, underrange direct trip, and underrange permissive.

The overreaching blocking scheme is protection at each end of the line using two distance components, one pointing to the inside of the line, namely the positive directional distance element, and the other pointing to the outside of the line, which is the reverse directional distance element. The setting value of the positive directional distance element is greater than the length of the line (so-called overreaching), and the action range (setting value) of the reverse directional distance element should cover the range of the action range of the positive direction distance element of the other end.

When there is an external short circuit, the reverse directional element of the protection, which is close to the end of the short circuit, operates and sends out the locking signal to lock protections of all ends. When an external fault occurs within the range of a positive directional element that can operate, the reverse direction component of the local end can operate, and the latching signal is reliably issued. When an internal fault occurs, the positive directional elements at each end can operate, while the reverse directional elements do not operate, and both end protections can trip reliably. This is the same as the principle of the directional locking type, but the distance element is used as the directional element.

The overreaching permissive scheme is similar to the above, except that protection on each end of the line has only one positive directional distance element, which is set in overreach. When an internal fault occurs, the distance components of each end may all operate to send out the allowing signal to allow the protection to trip at the other end. When an external fault occurs, the distance components of the protection at each end may not operate, and the protection may not trip.

The structure of the unlocking scheme is similar to that of the overreaching permissive scheme, except that its positive directional overreaching elements at both ends of the protection send out continuous locking signals to the protection at the opposite end, which is also a monitoring signal for normal operation of the channel at the same time. When an internal fault occurs, the distance elements at each end operate, relieve the locking signal, and transmit the allowed signal instead so that protection is immediately tripped.

The underreaching direct trip scheme means that the action range of the positive directional distance elements of the protection at the two ends is set to be less than the total length of the line; for example, the distance of segment I is set to 85% of the total length of the line, and the overlap area of the ranges from each end to segment I is 70% of the total length of the transmission line. When an internal fault occurs, at least one end of distance I operates. When the action occurs, in addition to tripping the circuit breaker at the local end, the circuit breaker of the other end can be tripped directly through the channel.

The underreaching permissive type is similar to that previously discussed. But when distance I operates, it not only trips the circuit breaker at the local end but also sends out an enabling signal to the other end. And if segment II or III of the other end operates as well, it allows protection to trip. The underreaching permissive type is more reliable than the underreaching direct trip scheme, for it can avoid protection misoperation caused by interference.

All the foregoing types of distance pilot protection can be applied to UHV transmission lines. However, due to the high redundancy of UHV transmission line protection, its probability of rejection is low while its probability of misoperation is high. At the same time, the number of circuits in UHV transmission lines is less. Because of the great impact that protection misoperation can have on the system, the main protection should be mainly to prevent misoperation. For distance pilot protection, combination of the underreaching permissive and overreaching blocking schemes should be adopted. When an internal fault occurs, the reliability of the protection is high, and when an external fault occurs, owing to the lock signal, the protection may not misoperate as a result of interference.

References

[1] He J, Li Y, Guo Z, Li B, Dong X. Relay protection for UHV transmission lines: Part I-construction and operation characteristics. Automation of Electric Power Systems 2002;(23):1–6.

[2] He J, Li Y, Li B, Guo Z, Dong X. Relay protection for UHV transmission lines: Part II-disposition of relay protection. Automation of Electric Power Systems 2002;(24):1–6.

CHAPTER 5

Directional pilot protection

Contents

5.1 Configuration of main protection

The main protection principle applied by the Soviet Union on 1150 kV line was the same as that applied on the 750 kV line. It is negative- and zero-sequence directional pilot protection and has made great improvement in operating time, overvoltage prevention, filtering performance and reliability. This principle is universally applicable for UHV transmission lines. At the beginning of designing protection system, the Soviet Union adopted relay protection as an integral part of overvoltage control system and made following requirements for relay protection:

(1) The operating time of main protection should be 20 ms approximately. This is necessary for limiting the overvoltage caused by asynchronous trip of different ends and maintaining system stability.

Protection Technologies of Ultra-High-Voltage AC Transmission Systems
ISBN 978-0-12-816205-7
https://doi.org/10.1016/B978-0-12-816205-7.00005-4

(2) The principle of backup protection should be different from main protection and should be able to quickly isolate any faults of the whole line, so the operating time should be the same as that of main protection.
(3) In addition to the local backup protection, there should be remote backup protection for interphase faults and ground faults.
(4) Protections should operate correctly in transient conditions caused by fault occurrence, fault isolation, and various operating states.
(5) Protections can work correctly under various loads.

There should be two main protections based on different principles, and two sets of high-speed protections should work in any case. At the same time, the time difference between two ends tripping should not be more than 0.04 s. It is also stipulated that single-phase reclosing is not allowed if the steady state overvoltage is more than 1.4 times that of the open-phase condition. In the case of allowing single-phase reclosing, if reclosure is not successful, which end should be removed first and which next should be predetermined according to overvoltage calculation. It is stipulated that closing and opening of the line should be semiautomated, which ensures that the time of single-ended closing is less than 0.04 s.

With regard to protection principle and emergency control, the Soviet Union also made following requirements according to temporal technical conditions:

(1) The main protection is achieved by using carrier channel to transmit blocking signal.
(2) The backup protection should have three-zone interphase distance protection and four-zone zero-sequence directional overcurrent protection for ground faults. The underreach zone of backup protection can transmit the carrier permissible signal and accelerate tripping at the other end when internal short-circuiting occurs.
(3) The purpose of adding instantaneous overcurrent protection is
   a) to operate successfully when the fault occurs in dead zone of direction protection or when TV breakage causes blocking of distance protection;
   b) zero-sequence instantaneous overcurrent protection can quickly isolate ground faults of one circuit of double-circuit lines.
(4) Special circuit breaker failure protection is set up to isolate the fault in case of reject-operation of circuit breaker; it can also to isolate the fault of circuit breaker itself or the faults between circuit breaker and TA.
(5) To reduce protection delay, the breaker trip coil should be started using a thyratron with a fast-exit intermediate relay.

(6) Application of multifunction automatic reclosure is required.

It is proved by theoretical analysis that there are frequency components that are above 120 Hz and in the range of 30—45 Hz in electromagnetic transient caused by faults on 1000 km transmission line composed of several sections (without series compensation capacitor). These frequency components need to be filtered out by filters. In order not to increase the operating time delay, a special method is adopted to avoid the transient process of filters. Moreover, to eliminate the wave distortion caused by TA saturation under DC components, the TA with air gap is adopted. At the same time, research is also conducted on optical current transformer.

5.1.1 Principle and performance of the first main protection

According to the long-term operating experience of 750 and 1150 kV protection of Soviet Union, the main protection with combination of negative-sequence direction pilot protection and phase-difference pilot protection is reliable. Negative-sequence direction pilot protection reacts to all kinds of faults in the full-phase state. The phase-difference pilot protection comparing the phase of $(\dot{I}_1 + k\dot{I}_2)$ reacts to faults under open-phase operation after single-phase tripping. The phase-difference pilot protection is adopted only in the single-phase reclosing cycle (negative-sequence directional protection should exit at this time). The principle and configuration of this protection actually has been running for 6 years on the world's earliest 1150 kV line with rich operating experience.

Fig. 5.1 is the logical diagram of this compound main protection. In the figure, M_2 is the negative sequence directional element that can operate in double directions. When an backward fault occurs, M_2 moves and its "-" end outputs level "1," and the level "1" passes through memory elements "$BП_1$", "OR_1" and "NO_1" to start transmitter and send out a continuous blocking signal. In case of forward asymmetrical faults, the "+" end of M_2 outputs level "1," and the level "1" passes through the "OR_2", "NO_2", the 2 ms delay element "BC", and then the $BП_3$ time-delay relay trips. If an external forward fault occurs, the receiver receives the blocking signal from the other end and blocks the trip circuit through "NO_2." The reason why delay element "BC" opens the trip circuit with a time delay of 2 ms is to wait for blocking signal from the other end. If the frequency band of filter is more than 1 kHz, the delay of 2 ms can guarantee reliable blocking.

When the protection trips, the intermediate relay 1PП is started at the same time. After 1PП operates, 2PП is started, After 2PП operates, it keeps self-

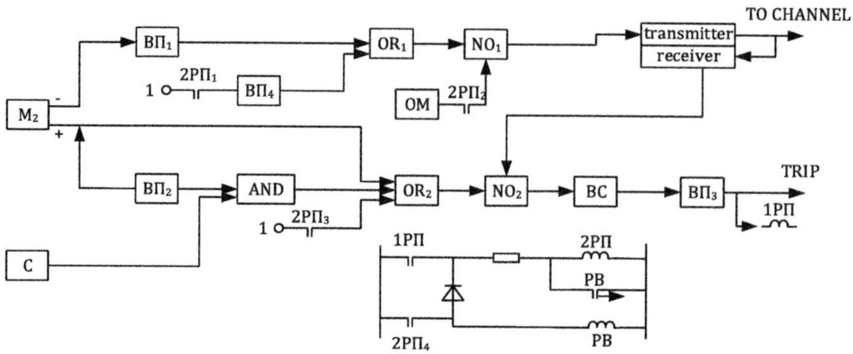

Figure 5.1 The logical diagram of the negative-sequence directional pilot protection and phase-difference pilot protection.

holding by its contact $2P\Pi_4$ until the PB's contact short-circuits its coils after the operation of time relay PB. After $2P\Pi$ operates, its contact $2P\Pi_1$ makes the transmitter continue transmitting signals through "BП1", "OR_1", "NO_1". When the protection first trips, its contact $2P\Pi_2$ closes and makes the transmitter controlled by the $(\dot{I}_1 + k\dot{I}_2)$ executive component. When $(\dot{I}_1 + k\dot{I}_2)$ is in negative half period, the transmitter is stopped by "NO_1". Therefore, after the first tripping of relay protection, the transmitter immediately changes to send out a half-period intermittent pilot signal, i.e. the protection is switched to phase-difference pilot protection. After contact $2P\Pi_2$ closes, the trip circuit is loaded with level "1", which passes through "OR_2", "NO_2", "BC", and "$B\Pi_3$" to trip. Whether the tripping circuit may trip depends on the width of pilot signal gap of receiver. If the width is larger than the sum of time delay of "BC" and "$B\Pi_3$", the trip circuit can start and trip, and "$B\Pi_3$" is used to adjust the blocking angle of phase-difference protection. This occurs if the sum of time delay of "BC" and "$B\Pi_3$" is equal to a certain length of time, such as 3 ms, so that block angle is equal to 55°. That is to say, if the width of pilot signal gap reaches or exceeds the block angle of 55°, it can trip. If the gap of pilot signal is less than 55°, it cannot trip in the case of external faults or open-phase operation. As a result, the switching of pilot protection with negative-sequence direction pilot protection and phase-difference pilot protection is simple and reliable.

In the case of a forward three-phase short circuit, short-time asymmetry may inevitably occur at the first moment of short circuit fault. Especially for UHV transmission lines, the interphase distance can reach 15–19 m, and the three-phase is unlikely to short-circuit completely at the same time. As long as there is more than 7 ms asymmetrical time at beginning of the short circuit, it is enough to make M_2 correctly judge fault direction. Because in this case the operation of M_2 is temporary, the memory element "BI_2" is used to memorize it. In addition, by parallel connection to interphase impedance element C, it passes through "AND", "OR_2" and "NO_2" to start trip circuit. The advantages of M_2 and C jointly starting the tripping are as follows: (1) The fault direction is checked by M_2, and the impedance element C can eliminate dead zone by its characteristic of shifting to the third quadrant. (2) M_2 may not be affected by system oscillation, and C does not need swing block. This kind of reaction to a three-phase short circuit, which has been tested for more than 30 years (including on 750 kV lines), has not been rejected in a three-phase short circuit. This method cannot react to a stable three-phase short circuit, for example, for the three-phase short circuit in case that ground line is not dismantled. For this kind of short circuit, C can only operate independently by manual switching or reclosing signal, or it can be isolated by backup distance protection (see the follow-up details). In China, there are more than 10,000 km of 500 kV lines, and not more than two three-phase short circuits have occurred during operation over 20 years [1]. As a result, the three-phase short circuit can be regarded as a fault type with minimal occurrence probability that can be handed over to high-speed backup protection (the distance protection zone I with cross range of transmitting trip signals), thus simplifying the main protection.

Then the principle and operation of blocking negative-sequence direction pilot protection are introduced.

Figs. 5.2–5.7 shows the operation of protection during external and internal short circuit and transceiver control logic.

5.1.1.1 Internal unsymmetrical fault

As shown in Fig. 5.5, there are two negative-sequence directional components, W_1 and W_2, which are used for the main and secondary trip channels respectively. For W_1 and W_2, there are forward and backward

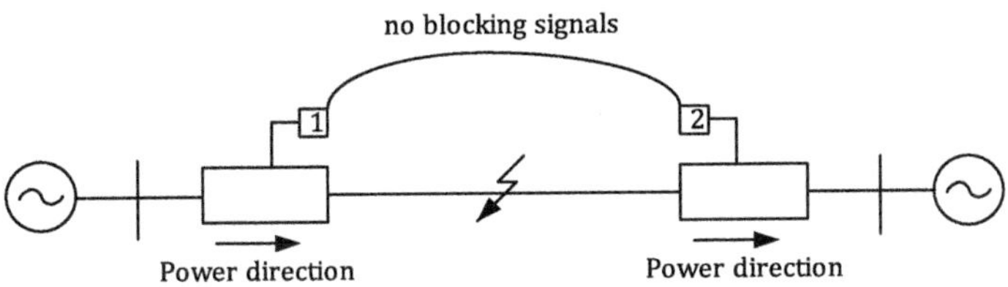

Figure 5.2 The principle diagram of high-frequency blocking negative-sequence pilot protection.

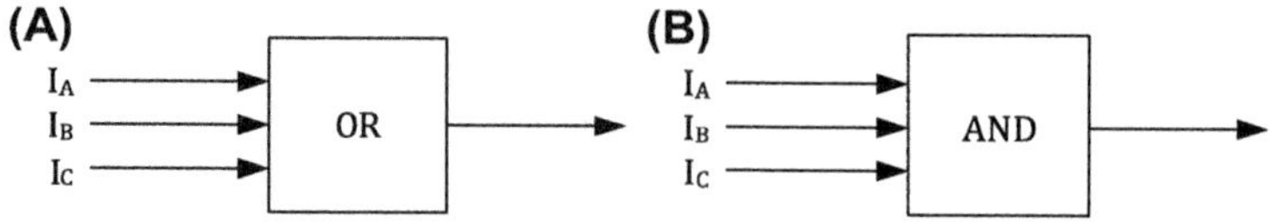

Figure 5.3 The connection of OR gate A_{min} and AND gate A_{max}.

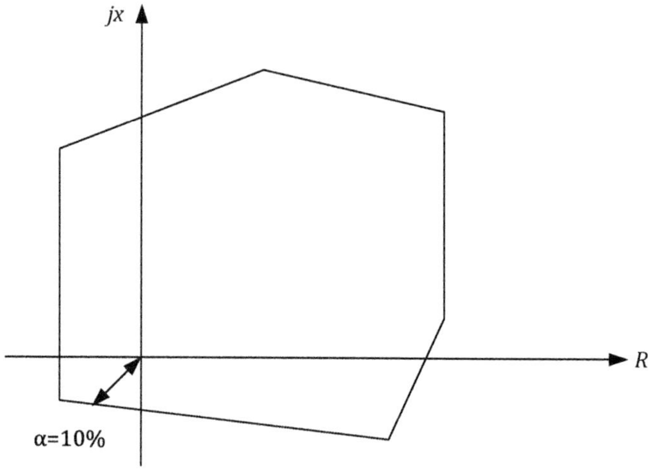

Figure 5.4 Operating characteristics of impedance element Z_{BC}.

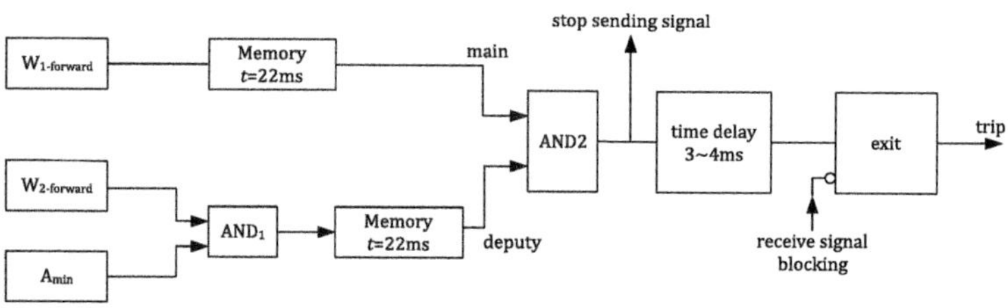

Figure 5.5 Operating logic under an internal unsymmetrical short circuit fault.

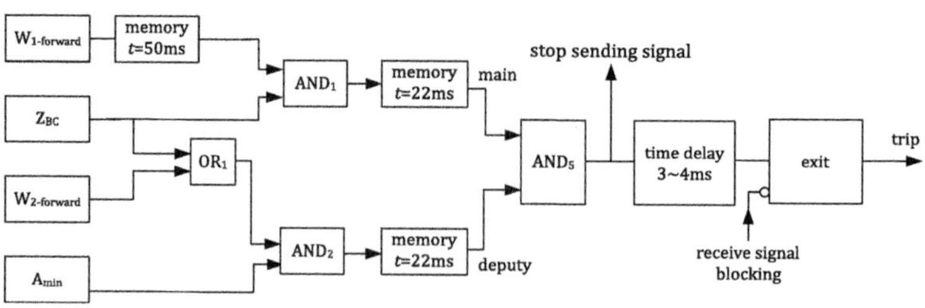

Figure 5.6 Operating logic under an internal symmetrical short circuit fault.

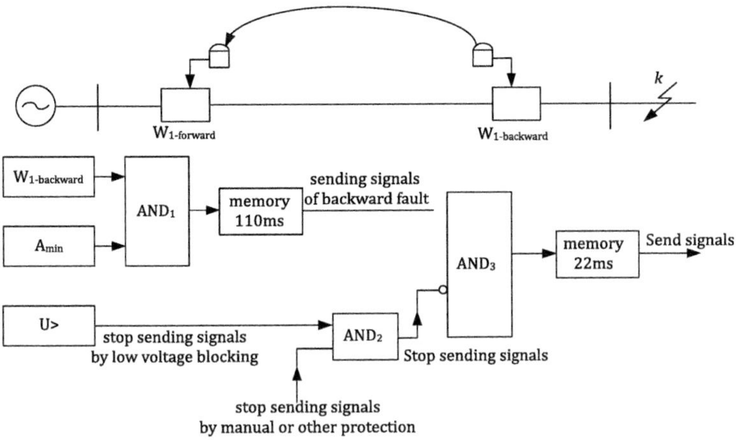

Figure 5.7 Control logic of transmitter under an external short circuit fault (in opposite direction).

directional components, respectively, $W_{1\text{-forward}}$ and $W_{1\text{-backward}}$, $W_{2\text{- forward}}$ and $W_{2\text{-backward}}$. The forward component operates on stop blocking signal and allow trip when forward fault occurs, while the backward element reflects backward fault and operates on pilot blocking signal, as shown in Fig. 5.7.

The operation of $W_{1\text{forward}}$ keeps memorizing for 22 ms, which is called the "main tripping channel". Fig. 5.3 represents an element that identifies open-phase state. A_{min} are three phase current relays that could output through an OR gate. When one or more current relays operate, A_{min} may have output, indicating that one or more of three-phase circuit breakers is in closing state. Another three-phase current relay outputs through an AND gate, which is represented as A_{max}, and it has output only when all three-phase breakers are closing. If A_{min} has output while A_{max} has not, a open-phase state is indicated.

$W_{2\text{-forward}}$ also reflects the forward faults. When both $W_{2\text{-forward}}$ and A_{min} operate, the AND_1 gate has output and keeps memorizing for 22 ms, which is called "auxiliary tripping channel." When both main and auxiliary tripping channel have output, the AND_2 gate has output and keeps memorizing for 3–4 ms, and if the receiver still has not received pilot blocking signal, the exit loop may output the tripping pulse. For UHV transmission lines, the internal protection device also needs to improve the reliability of anti-maloperation. So only when both of the two tripping channels have output and conditions are available, are trips allowed.

The voltage of direction element can be obtained directly from negative-sequence voltage U_2 or negative-sequence voltage $\dot{U}_2' = \dot{U}_2 - \dot{I}_2 Z_k$ compensated to some point of the line so as to improve the voltage sensitivity of direction element. Because the negative-sequence voltage contains $-\dot{I}_2 Z_k$ in case of backward three phase short circuit, the large current may lead to TA saturation, which makes the negative-sequence current filter produce an negative-sequence unbalanced current, causing negative-sequence voltage not tȯ be zero, so as to lead to mal-operation of the negative-sequence directional component. In order to eliminate this mal-operation, a current brake of the phase that has the minimum current in all three phases is added to the negative-sequence direction element. In case of backward three-phase short circuit, the three-phase current is basically the same without the minimum, so the brake is controlled by any phase current. In asymmetrical short circuits, no brake is needed. The phase with the minimum current is the non—fault phase, whose current is generally small. So the brake is basically not effective in asymmetrical short-circuits.

In case of normal operation, if at least one phase circuit breaker is closed, then A_{min} has output. In case of internal asymmetrical short circuit, both main and auxiliary tripping channels have output. The output lasts for 22 ms, then pass through an AND_2 gate after a delay of about 3—4 ms to wait blocking signal from the other end. Because there is no blocking signal in the internal short circuit, the exit circuit outputs the tripping pulse.

In case of backward asymmetrical short circuit, the negative sequence direction elements, $W_{1\text{-forward}}$ and $W_{2\text{-forward}}$, will not operate, so the protection at this end will not mal-operate. Meanwhile, the direction elements, $W1_{\text{-backward}}$ and $W_{2\text{-backward}}$, will send out blocking signal and block the protection at the other end, as Fig. 5.7 shows.

5.1.1.2 Internal symmetrical fault

As shown in Fig. 5.6, in case of forward symmetrical short circuit, the component $W_{1\text{-forward}}$, which reflects forward fault, will operate in a short time and memorize for 50 ms. At the same time, the impedance element Z_{BC} connected to BC phase operates, and the output passes through AND_1 and memorizes for 22 ms, which is the main tripping channel. When Z_{BC} operates (the stable three-phase short circuit) or $W_{2\text{-forward}}$ operates (the instantaneous asymmetry under three-phase short circuit), and if A_{min} has output at the same time, the output passes through AND_2 and delay for

22 ms, which forms the auxiliary tripping channel. When main and auxiliary tripping channels have both operated, the blocking signal is stopped. After passing through AND_3, the output is delayed for 3—4 ms, then go to exit circuit. In case of internal short circuit, the blocking signal is stopped at both ends. Therefore, without receiving blocking signal, it will trip.

As mentioned, there are two advantages of main tripping channel consists of W_1 and Z_{BC} passing through an AND_1 gate:

(1) Based on the characteristics of shifting to the third quadrant, Z_{BC} can eliminate dead zone of three-phase short circuit on outlet. Moreover, due to negative-sequence direction element checking, it may not mal-operate when a backward short circuit occurs.

(2) Because the negative-sequence element does not mal-operate under system oscillation, Z_{BC} does not need swing block.

Z_{BC} uses hexagonal characteristics and shifts to backward 10%, as shown in Fig. 5.4.

In case of backward three phase short circuit, the negative-sequence direction element $W_{1\text{-backward}}$, which relies on backward operation, operates to start the transmitter (see Fig. 5.7), and the blocking signal is sent out. At this time, the direction element only operates in the asymmetrical period of initial moment of three-phase short circuit, and its operating time is very short, so it is necessary to keep memorizing its operation for 110 ms. In case of backward three-phase short circuit, the large current may cause TA saturation and current waveform distortion, resulting in negative-sequence unbalanced current. $W_{1\text{-forward}}$ may mal-operate and stop the blocking signal, resulting in the mal-operation of protection. However, if the voltage is greatly reduced at this time, low voltage blocking can be adopted to continue to send the blocking signal for a short time to prevent the mal-operation, as shown in Fig. 5.8. When $U < 50\%U_n$, the "AND_2" gate in Fig. 5.7 is not accessible, and continue to send blocking signal for a short time. Memorizing for 110 ms is also in order to delay the vanishing of blocking signal after the isolation of external fault, so as to ensure that there is still a blocking signal before the reliable restoring of protection device to avoid mal-operation.

5.1.1.3 Closing from one terminal

When the no-load line is closing from one terminal, because the capacitance current is very large in transient process, so that it could cause protection to start and mal-operate. Because all the capacitance currents flow through protection of closing terminal when the line is closing from one terminal, it can only be closing with no-load when capacitance current compensation is

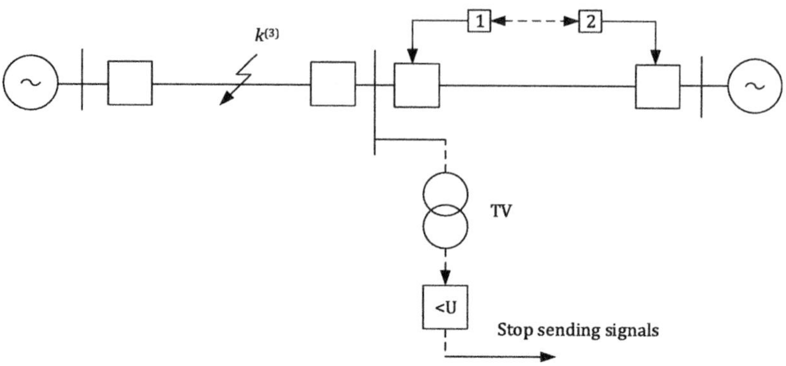

Figure 5.8 Low voltage blocking for stopping transmitting by mistake because of TA saturation.

1.25 ~ 1.4 of the whole line capacitance current. For this reason, the compensation factor is automatically switched to 1.25−1.4 in case of three-phase disconnection. So in condition of three-phase disconnection, it's possible that the compensation current in capacitance current compensation device leads to some undue negative-sequence voltage and current that make backward direction element mal-operate to start sending signal and keep memorizing, which causes the delay of operation at the other terminal when closing on faults. For this purpose, the starting transmitting circuit should be controlled by current relay. Only when at least one phase of circuit breaker is closing can A_{min} send blocking signals. The capacitance current of three-phase disconnection cannot cause transceiver to start.

When the three phases are disconnected, the capacitance current compensation is switched to 1.25 times the full capacitance current. In case of three-phase closing, all three phases have current, so there should be a 50 ms delay after the operation of A_{max} to let capacitance compensation return to 0.5 times the full capacitance current. The 50 ms delay is to elude the transition process when breaker is closing.

5.1.1.4 Closing from one terminal of no-load fault line

When the no-load line is closing on an asymmetrical fault, the forward directional element at closing end will operate. The backward directional element at the other end may mal-operate and attempt to send blocking signal (e.g., TV is on the line side). However, without the operation of A_{min}, no signal is sent, so the protection on closing end will not be blocked. Secondly, the three phases not being simultaneously closed may cause the operation of backward direction component on closing end to send blocking signal, and keep memorizing for 110 ms to delay fault isolation. Therefore, changing the memory function of sending blocking signal to 22 ms in no-load closing cases should be considered.

When the line is manually closing on a three-phase fault, the manual closing signal makes impedance relay uncontrolled by negative-sequence direction element and allows it to operate independently. The independent working time allows 150 ms, and then time relay started by A_{max} restores the impedance relay to original state and cancel its independent work. Or when there is a voltage after the closing, the voltage relay operates and restores the impedance relay to original state. The independent work of impedance relay in no-load closing (not controlled by direction element and pilot signal) may be helpful in two aspects:

1. to eliminate rejection of negative-sequence direction element due to the lack of asymmetrical time (such as in a no-load closing with three-phase ground line unremoved);
2. to eliminate rejection due to the transition process of closing that makes backward blocking operate to start sending signal.

5.1.1.5 Non-full-phase closing from one terminal (e.g., one- or two-phase refuse to close)

Three phases of the line should be isolated at this time. A_{min} operates while A_{max} does not, which is the sign of non-full-phase closing from one end. In this condition, the three phases should be tripped after 40 ms delay. This delay is greater than the maximum allowable time for three-phase circuit breakers not to be closed simultaneously.

5.1.1.6 Operation of differential protection

Negative-sequence directional pilot protection has no phase-selection function. When the single fault phase is isolated by phase selector, it is automatically switched to phase-difference pilot protection, which can reflect the fault of sound phase in open-phase operation state. There may be three-phase triping and three-phase reclosing under two- or three-phase short circuit, which has no effect on the switching of protection.

The UHV transmission line has a large phase-to-phase distance, and the three-phase fault often evolves from the initial asymmetrical fault to the three-phase fault. Therefore, it makes the negative-sequence direction pilot protection safely identify the fault in asymmetrical fault period. This experience is very important.

5.1.2 Second main protection and backup protection

As the second main protection (also as a backup protection for interphase short circuit), three-zone distance protection of permissible type or remote tripping type should be adopted.

5.1.2.1 The composition of the protection

The protection consists of three zones, all of which adopt quadrilateral features. Zone I passes through the origin. Zone II may be backward shifted or not shifted at all. Zone III can be backward shifted or to the first quadrant (for permissible or accelerated types). The operating characteristic are shown in Fig. 5.9. When the direct triping signal generated by underreach zone I of the other end is checked supplementary up, zone III can adopt backward shifted characteristic. For zone III, which sends an permissible signal to the other end, the forward shifted characteristic should be adopted in order to eliminate dead zone. For near faults, permissible signals can be sent by zone I or II. The swing block starting element ΠO is started by the incremental negative-sequence current, $\Delta \dot{I}_2$. Voltage loop break line blocking is achieved by star-shaped secondary winding voltage and the open-triangle-shaped secondary winding voltage of phase comparison TV.

5.1.2.2 The operations of each zone

Both zones I and II have two delays, "high-speed operating zone" and "time-delay operating zone." It is started by different starting elements ΠO

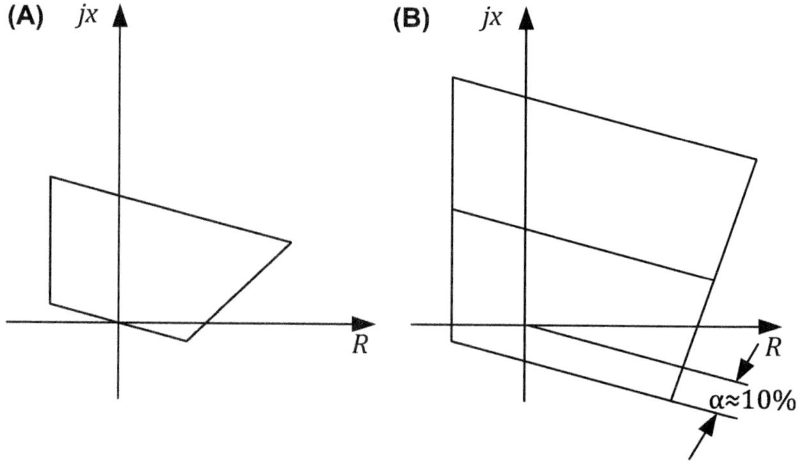

Figure 5.9 Operating characteristic of direction element: (A) zone I; (B) zones II and III.

of swing block ПОБ. In the time-delay operating zone, the mal-operation under oscillation can be avoided by time delay; the operation of swing block ПОБ is short, which should be memorized in all zones.

The operation of distance protection zone I:

(1) The operation is controlled by a three-phase current relay to prevent mal-operation due to unintended activation, interference, or component damage, and the setting of current relay should not affect the sensitivity of distance protection.

(2) When both zones I and II operate, zone II controls zone I to be self-holding (by instantaneous measurement) to prevent memory circuit from returning under dead zone short circuit or zone I from returning under high transition resistance. See Fig. 5.11.

(3) When zone I operates, the long-range trip signal can be sent directly to the opposite end to achieve triping.

For automatic reclosing or manual charging on faults, the reclosing or manual switching signal accelerates zones II and III.

An (permissible) acceleration signal is sent to the opposite end in the operation of no offset in zone II, which is checked by the opposite end. In

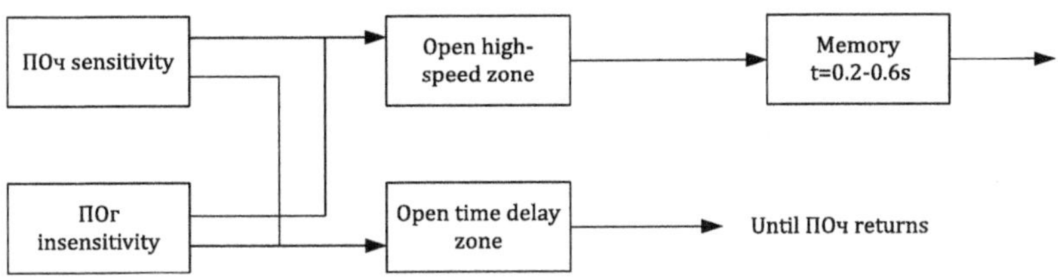

Figure 5.10 Operation logic of swing block.

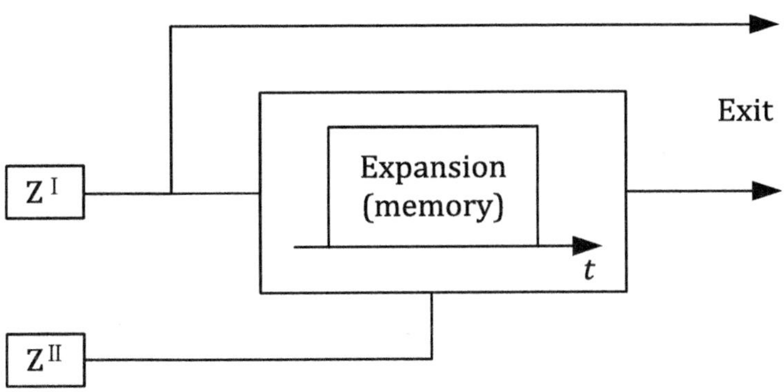

Figure 5.11 Instantaneous measurement of impedance zone I.

case of internal short circuit, it accelerates zone II (with no offset); it can also accelerate zone III, but at this time zone III must shift to the first quadrant to avoid mal-operation under backward short circuit.

5.1.2.3 Swing block (ПОБ)

For UHV transmission lines, the capacity of both ends is large as is the inertia constant. Therefore, the oscillating period in normal operation (destruction of static stability) is very long, about 8–10 s. The oscillating period under short circuit faults is short, about 2–2.5 s.

The protection of each zone is started by swing block starting element ПО, which is opened as follows due to the long oscillating period in normal operation. See Fig. 5.10.

(1) The high-speed zone can stay open for 0.2–0.6 s and then be blocked until swing block ПОБ returns.

(2) The delay zones (zones I, II, and III) can stay open until swing block ПОБ returns, for all these zones can elude the short oscillating period caused by short circuit. And all these zones should elude the oscillation caused by short circuit by time.

There are two starting elements in swing block starting element (ПОч and ПОГ):

(1) Sensitive starting element ПОч. ПОч should be able to operate under all kinds of short circuits, but it may start by mistake during system operation.

(2) Insensitive starting element ПОГ. ПОГ should be able to start under various short circuits, but it should be able to elude the transient process caused by system operation. If a second short circuit occurs, it should be able to start the blocked high-speed zone again and open it for 0.2–0.6 s in case ПОч has started but has not yet returned. The operation logic of swing block starting element ПО is shown in Fig. 5.10.

When the sensitive starting element ПОч operates, it starts the high-speed zone, opens it for 0.2–0.6 s, and then blocks it until ПОч returns. After 1.0–10 s (adjustable), when ПОч returns, the whole ПОБ returns. The delay zone should be opened at the same time as starting element ПОч operates, and it should be kept open as long as the sensitive starting element ПОч does not return.

The operating logic of insensitive starting element ПОГ is similar; it would not mal-operate under system operates, while it only operate under internal faults.

The conditions of first oscillating and then short circuit in normal operation are as follows:

(1) The fast zone is blocked, and the fault is isolated by delay zone.

(2) Criteria of system oscillation:

Criterion 1: Distance protection zone II operates but **ΠO** is not started. At this time, after a 0.05−0.15 s delay, the high-speed zone is blocked (not open), and the delay begins when the criterion is satisfied, as shown in Fig. 5.12. The delay of 0.05−0.15 s is to ensure that high-speed distance protection zone can not be blocked when the operation of **ΠO** is slow; The delay of 0.05−0.15 s exceeds the possible delay of **ΠO**.

Criterion 2 is the periodic operation and return of impedance protection zone II, and the high-speed zone is also blocked in this case. The block setting can be as small as the oscillating period, T_k=0.8 s, and if impedance zone II operates fairly frequently, the high-speed protection zone will be blocked at all times.

5.2 Zero-sequence current directional protection

Consisting of zero-sequence current direction protection, permissible directional pilot protection is the main protection for ground short circuit faults. Its zones III and IV work as backup protection of ground short circuit faults, which also includes interphase instantaneous overcurrent protection as supplementary protection to distance protection in export fault to avoid rejection because of circuit memory vanishing too quickly or a lack of memory when reclosing on export fault (e.g., when TV is installed on the side of line). The composition of the protection includes IV zone zero-sequence current direction protection reflecting ground short circuit fault, the instantaneous overcurrent protection reflecting interphase short

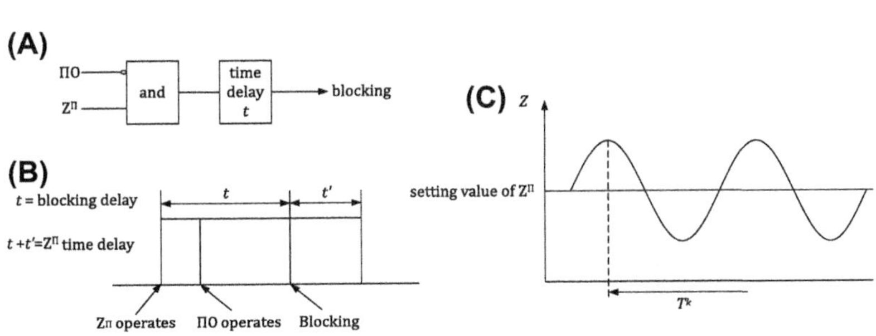

Figure 5.12 Detection and blocking of oscillation: (A) criterion 1; (B) blocking delay; (C) criterion 2.

circuit fault, and the open-phase protection. Its wiring and operating logic are detailed as follows:

(1) Four zones zero-sequence current direction protection is characterized by a zero-sequence direction element with two directions that can be connected as blocking-type export relay or permissible-type export relay. Which kind of connection should be adopted depends on the system.

The permissible connecting type is shown in Fig. 5.13B. Zero-sequence current direction element M_0 uses a normally open contact (the movable contact is upwards during operation). The contact is closed when forward faults occur, allowing the operation. Its disadvantage is that in a large capacity system, if there is a far forward ground short circuit, the voltage in direction element will be too small and the sensitivity not enough, which may lead to the rejection. In this case, it is better to use the blocking connecting type.

The blocking connecting type is shown in Fig. 5.13A. The zero-sequence current direction element M_0 uses a normally closed contact, which is closed instead of breaking when forward faults occur. When backward faults occur, the contact will be opened, and the protection is blocked. Its disadvantage is that when the zero-sequence impedance of opposite side system is too large, the zero-sequence current passing through the directional element in case of backward fault is too small, and the current sensitivity of the direction element is not enough so as to cause mal-operation. In this case, it is better to use the permissible connecting type.

(2) Zones I, II, and III of the zero-sequence current elements are all controlled by zero-sequence directional elements. Zones II and III exit in open-phase operation. Zones I and II can also be set to delay. If it does not need to elude the time in which three phase switches

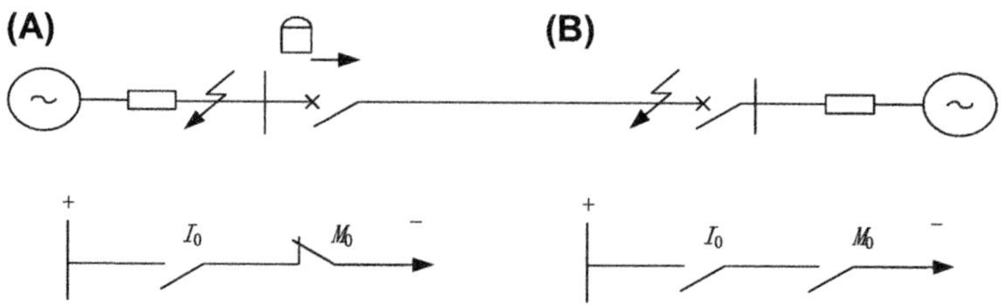

Figure 5.13 Types of zero-sequence current directional element: (A) blocking type, (B) permissible type. M_0 represents a zero-sequence power directional element, and the upward contact represents operation.

are not closed simultaneously when reclosing or manual closing, the delay of zone I can be canceled.

(3) Zone III of zero-sequence current elements constitutes permissible zero-sequence direction pilot protection. It is used to start transceiver AHKA, which can send and receive 14 kinds of signals. AHKA can send the permissible signal No. 4 or No. 14 (see below) and receives the permissible signal No. 4 or No. 14 from the other end. The permissible No. 4 signal allows the other end to trip single-phase, and the No. 14 signal allows the other end to trip three-phase. The zero-sequence current elements zone III is also used as an accelerating zone when manually closing or reclosing on fault. In order to prevent mal-operation after acceleration due to the inrush current of transformers and reactor, $3I_0$ of current component must have braking to the aperiodic component and secondary harmonic current.

Zero-sequence current elements zone III is also used as successive high-speed tripping protection for double-circuit lines, as shown in Fig. 5.14. When any external short circuit occurs on double-circuit lines, the operating directions of zero-sequence directional elements of two lines are always the same (ignoring the influence of zero-sequence circulating current of untransposed double circuit). If the operating directions of zero-sequence directional element of two lines at one end are different, it is sure that there is an internal ground short circuit. The fault line can be determined according to the operating direction and be isolated immediately. When one end achieves triping, the power direction will change, and the operating directions of directional elements of two lines on the other end must become different and will trip successively. This is called successive high-speed tripping,

Figure 5.14 Zero-sequence successive high-speed tripping protection of double-circuit line: (A) the operation of zero-sequence power direction under internal fault, (B) successive high-speed tripping of zero-sequence protection. S_0^{I}, S_0^{II}, S_0^{III}, and S_0^{IV} represent the zero-sequence power directional element of each protection, and the arrow represents the zero-sequence power direction. Q_1 and Q_2 represent circuit breakers.

which can isolate whole line faults by "successive operation" without the need of communication. Zero-sequence directional high-speed tripping can have a very small delay to elude the zero-sequence current caused by the asynchronous closing of three-phase circuit breaker, and for double-circuit lines it should also elude the time of power converse.

(4) Zero-sequence current elements zone IV can also work as backup open-phase protection to prevent the line from being open-phase state for a long time. It is characterized by A_{min} operating while A_{max} is not operating. In this case, the three phases are delayed to trip, and reclosure is prohibited.

Zero-sequence current elements zone IV can also detect whether there is zero-sequence current I_0 (in the case of arc suppression) after one phase triping. If I_0 does not vanish (greater than the maximum zero-sequence current setting value) within 0.1−0.6 s (adjustable), this means that the arc of fault point is not extinguished, and high-speed reclosing should be prohibited immediately.

(5) Zero-sequence power directional element. The current of zero-sequence power directional element should compensate for capacitance current. In order to improve the sensitivity, there must be zero-sequence current compensation for voltage, which can be calculated by:

$$\dot{U}_p = 3\dot{U}_0 - Z_k \cdot 3\dot{I}_0$$

Where Z_k is the compensation impedance, and its value is 1/2 of the line impedance.

After considering the above compensation, in case of backward three-phase short circuit or oscillation, there may be zero-sequence unbalanced current due to TA saturation, which may lead the maloperation of directional element. Therefore, the phase current which is the minimum should be used to brake. When this minimum current exceeds a certain value, it is necessary to consider braking.

(6) Operating logic

1) High-speed zone. The high-speed zone consists of interphase instantaneous overcurrent protection, zero-sequence protection zone I, successive high-speed tripping of double-circuit lines, and zone III, which has received the permissible signal No. 4 or No. 14. Its operating logic is shown in Fig. 5.15. Which operation it should perform is determined by the phase selector and signal sent from the other end.

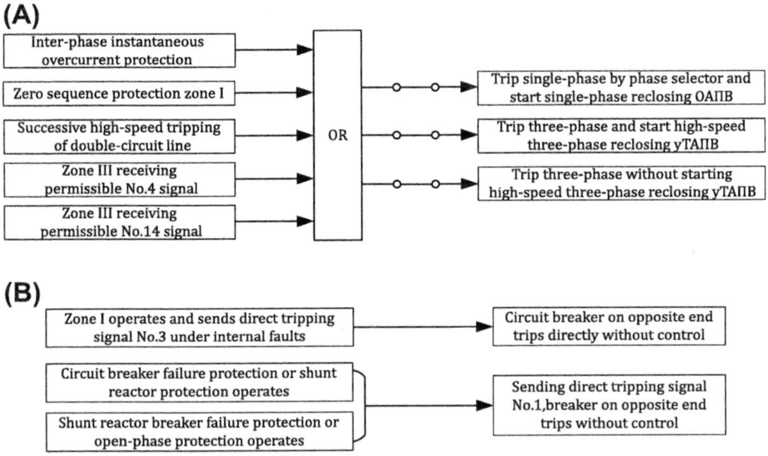

Figure 5.15 Operating logic of high-speed zone: (A) high-speed zone tripping by permissible signal, (B) high-speed zone tripping by tripping signal.

2) Time delay zone. The time delay zone consists of zones II and III. After the operation, three-phase triping is achieved and high-speed three-phase reclosing (YTAПB) is prohibited.
3) Long time delay zone. The long time delay zone is the zero-sequence protection zone IV. After the operation, three-phase triping is achieved and high-speed three-phase reclosing (YTAПB) is prohibited.
4) When open-phase protection, shunt reactor protection, circuit breaker failure protection, or reactor circuit breaker failure protection operates, tripping signal No. 1 is sent to directly trip three phases of the other end, and all kinds of three-phase reclosings are prohibited.
5) Zone III is accelerated when the line is a no-load closing or reclosing on fault, the three phases are tripped, and all kinds of three-phase reclosings are prohibited.

5.3 Configuration of automatic reclosure

5.3.1 Function of automatic reclosure

(1) In case of single-phase short circuit, the phase selector selects and trips fault phase and adopts single-phase reclosing (OAПB).
(2) In case of phase-to-phase short circuit, the phase selector should be bypassed to trip three phases and adopt three-phase reclosing (TAПB). In cases of a single-phase short circuit converting into a

multiphase short circuit or a short circuit fault occurs on sound phase during single-phase reclosing cycle, the phase selector should also be bypassed to trip three phases and adopt three-phase reclosing (TAПB).

(3) Trip three phases when single-phase reclosing fails and allow three-phase reclosing (TAПB).

(4) When the fault is isolated by high-speed protection, reclosing scheme that minimize power outage time (i.e., high-speed reclosing without synchronism check or three-phase reclosing with voltage and synchronism check) should be adopted. The bus can be recharged by reclosing automatically in order after failure protection operation and fault isolation and restore the operating status before the fault. The bus can also be charged automatically in order after overhaul.

5.3.2 Measure and control elements

(1) The ground distance relay with zero-sequence current compensation is used as the phase selector, whose operating characteristic is two intersecting circles, and its offset to the first and the third quadrants can be adjusted as shown in Fig. 5.16. In order to ensure the speed of the phase selector, the filter circuit should be provided with a zero initial condition to accelerate the reaction of filter.

After the phase selector trips fault phase, the current loop of impedance element is connected in series with contact of no-current checking current element (when the current is smaller than the setting value, the contact is disconnected, and the impedance relay returns). This is necessary because after the isolation of fault phase, due to the existence of zero-sequence current, zero-sequence current $3k\dot{I}_0$ of zero-sequence compensating

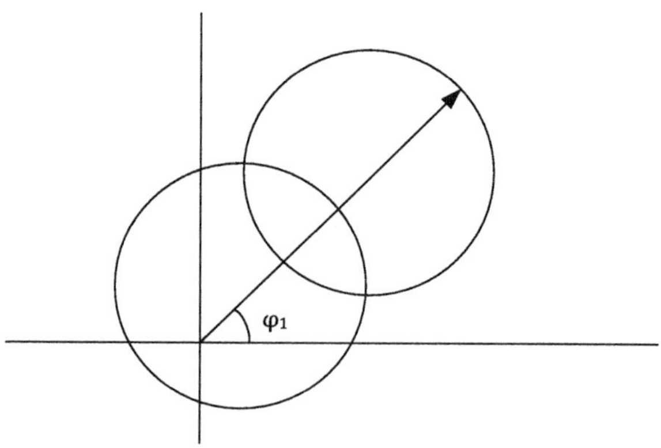

Figure 5.16 Operating characteristics of phase selector.

circuit of ground distance relay may prevent impedance relay from returning and may also lead to the mal-operation in open-phase oscillation or the failure of single-phase reclosing.

(2) Fast return zero-sequence current relay. After the isolation of a single-phase fault, the open-phase zero-sequence current makes fast-returning zero-sequence current relay PTH-БВ operate, and during the single-phase reclosing period it switches the impedance phase selector of isolated phase into independent ground distance protection, which works independently as auxiliary protection when reclosing on a permanent fault (there is also phase-difference pilot protection). Because there is contact of no-current checking current element connected in series in current loop, it will not mal-operate in oscillation of open-phase operation but will operate when reclosing on a permanent fault. However, after successful reclosing, the zero-sequence current will vanish, and PTH-БВ should return quickly to restore phase selector instead of working independently. As the independently working ground distance protection operates when reclosing on a permanent fault, it should have a time delay of 18 ms. Moreover, this delay should be greater than the return time of PTH-БВ to avoid the mal-operation under subsequent oscillation after the success of single-phase reclosing OAПB.

By switching the phase selector to independently working mode and then switching it quickly back to phase selector after the success of single-phase reclosing OAПB, the switching operation is achieved by zero-sequence current relay that returns quickly (this is easy to implement for microprocessor protection). In order to ensure that it can return quickly, special filtering measures should be taken to filter off aperiodic components when OAПB recloses successfully and the zero-sequence current vanishes. If ground distance protection does not mal-operate in open-phase oscillation, the phase selector of the sound phases can work independently during the reclosing of OAПB (Fig. 5.17).

5.3.3 Automatic reclosure mode

- Single-phase reclosing (OAПB)
- Three-phase reclosing (TAПB)
- High-speed three-phase reclosing without voltage check (yTAПB–БК)
- High-speed three-phase reclosing with no-voltage check (yTAПB–OH)

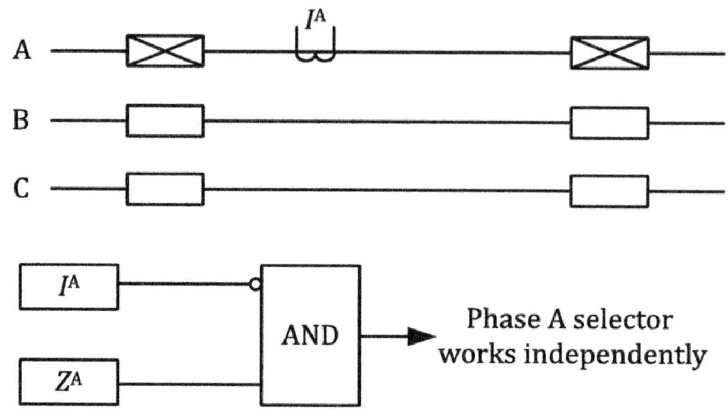

Figure 5.17 Schematic diagram of a phase selector controlled by current element.

- Three-phase reclosing with no-voltage check (ТАПВ–ОН)
- Three-phase reclosing with synchronism check (ТАПВ–КС)

5.4 Basic evaluation

The principle and configuration of this protection have many years of actual operating experience on the 1150 kV line and have proved reliable. As the negative- and zero-sequence components belong to fault components, they are not affected by load current system oscillation or transition resistance.

The advantages of this protection are as follows: (1) Because negative- and zero-sequence components are long-standing in asymmetrical short circuit and ground short circuit, the protection can respond to the whole process of faults. Another advantage of direction principle is that it is the least affected by the distributed capacitance of line, because the operating range of directional element is half plane, and the phase difference between the capacitance current and inductive short circuit current is almost 180°. In addition, it has a great influence on the amplitude of short circuit current while having less influence on the phase of short circuit current. This is the superior to distance protection, segregated current differential protection, and phase difference protection. This method of solving a three-phase short circuit in protection configuration is ingenious, and it has proved reliable in practice. The phase spacing of UHV lines is above 15 m, and the probability of three phases simultaneously being short-circuited is very small. Even if it were to happen, the second main protection (zone I has crossed distance protection that transmits a tripping signal) can also operate reliably.

(2) Because directional comparison protection belongs to "nonunit" protection, in which the data transmitted through the communication channel is the conclusion about the fault location judged by the two ends, rather than the electrical quantities, communication traffic is minor and less demanding on communication rate.

This protection has some disadvantages as well: (1) It doesn't have phase-selection function and must be equipped with an independent phase selector. (2) The protection was implemented in 1985 and achieved by the integrated circuit. But the principle has universal practicability and superiority. If the application is achieved by microcomputer, it is more beneficial to separate the second main protection and backup distance protection.

Reference

[1] Liu Z. Ultra-high voltage grid. Beijing: China Economic Publishing House; 2005.

CHAPTER 6

Transmission line model

Contents

6.1 Brief introduction

To improve the transmission capacity of UHV lines and reduce voltage and energy losses, it is necessary to reduce resistance, inductance, and leakage conductance as well as increase capacitance per unit length as much as possible. Therefore, the capacitance current of UHV long-distance transmission lines is large, and the line has significant distributed parameter characteristics. But the existing protection schematic fails to consider the distributed parameter characteristics of UHV long lines. In this way, it is difficult for the protection schematic to protect safe operation of the system, at least not reliably. Therefore, it is necessary to deeply analyze the line mode from wave propagation characteristics in the fault transient process to represent the essential characterization of UHV long lines. The protection schematic based on an accurate line model may certainly improve relay performance and other aspects.

The parameters of the actual circuit such as resistance, inductance, and capacitance are continuously distributed, but under certain conditions, distribution of circuit parameters can be neglected, and the concentrated parameter circuit can be approximately used as a model of the actual circuit. When the maximum linear size l of the part and the connecting line that compose the actual circuit can be compared with the wave length λ of the electromagnetic wave that spreads along the space around the circuit, the distribution of the circuit must be

Protection Technologies of Ultra-High-Voltage AC Transmission Systems
ISBN 978-0-12-816205-7
https://doi.org/10.1016/B978-0-12-816205-7.00006-6

considered. Generally, when $l \geq \lambda/100$, the distributed parameter circuit should be used as the actual circuit model.

The frequency of the AC transmission line is low (50 Hz), which means the electromagnetic wave wavelength is particularly long (6000 km). But as the voltage level increases and because the length of transmission lines is long, the influence of distributed inductance, distributed capacitance, leakage current, and other aspects must be considered, so the distributed parameter circuit must be used to study the essential characteristics of the UHV long line.

There is a time-varying electromagnetic field on and around the transmission line when it transmits electromagnetic energy. The general law of the time-varying electromagnetic field determines line voltage and current law as functions of time and space. Therefore, studying the electromagnetic process on the transmission line requires the theory of the distributed parameter circuit—the distributed capacitance is used to reflect the energy storage characteristic of the electric field in the space around the transmission line, and the distributed inductance is used to reflect the energy storage characteristic of the magnetic field in the space around the transmission line. What's more, the heat loss phenomenon caused by the current flowing through the metal conductor exists along the whole length of the transmission line, and the distributed resistance along the line is used to represent this process. The leakage current caused by imperfect insulation is also distributed along the line, and the distributed conductance is used to reflect the process. The line model is shown by Fig. 6.1, where r_0, L_0, g_0, C_0 respectively represent the resistance, inductance, conductance, and capacitance per unit length of the line.

The ground voltage and wire current of any point on the distributed parameter line are functions of distance x and time t. In this way, the

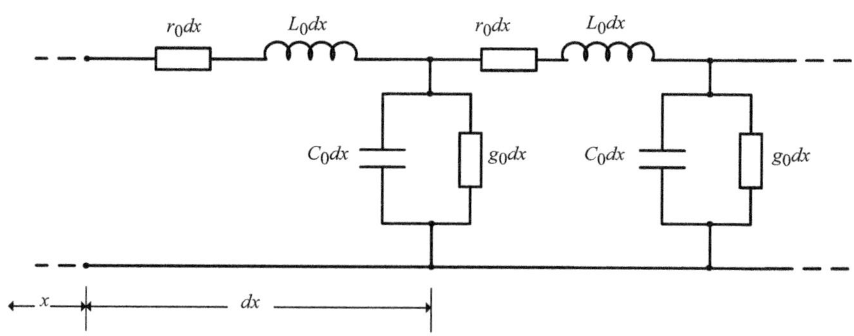

Figure 6.1 The per-unit-length model of the transmission line.

electromagnetic wave propagation process on the transmission line can be described by the following partial differential equation [1]:

$$\begin{cases} -\dfrac{\partial u}{\partial x} = L_0\dfrac{\partial i}{\partial t} + r_0 i \\ -\dfrac{\partial i}{\partial x} = C_0\dfrac{\partial u}{\partial t} + g_0 u \end{cases} \tag{6.1}$$

Eq. (6.1) shows that because of the potential decrease caused by the resistance and inductance continuously distributed along the line, voltage varies along the line, and because of the leakage current and displacement current caused by the leakage conductance and capacitance continuously distributed on the line, current varies along the line. When the initial conditions and boundary conditions are given, $u(x,t)$ and $i(x,t)$ can be determined by Eq. (6.1) uniquely. Obviously, Eq. (6.1) is suitable for any time-varying voltage and current, and it is the essential basis of study on the transmission line working state (steady state and transient state).

6.2 Long-line equation of transmission line

6.2.1 Sinusoidal steady state solution of transmission line

When the transmission line operates in steady state, the law of voltage and current that vary with time is the sinusoidal time function of the same frequency (50 Hz) as the excitation source. So voltage phasor $\dot{U}$ and current phasor $\dot{I}$ can be respectively used to represent the sinusoidal voltage $u(x,t)$ and sinusoidal current $i(x,t)$—that is, $u(x,t) = \text{Im}\left[\sqrt{2}\,\dot{U}e^{j\omega t}\right]$, $i(x,t) = \text{Im}\left[\sqrt{2}\,\dot{I}e^{j\omega t}\right]$.

Under the condition of sinusoidal steady state, the following phasor equation can be obtained according to Eq. (6.1):

$$\begin{cases} -\dfrac{d\dot{U}}{dx} = (r_0 + j\omega L_0)\dot{I} \\ -\dfrac{d\dot{I}}{dx} = (g_0 + j\omega C_0)\dot{U} \end{cases} \tag{6.2}$$

Eq. (6.2) is a constant coefficient differential equation without a time variable. Taking a derivative of Eq. (6.2), we can get the following result:

$$-\frac{d^2\dot{U}}{dx^2} = (r_0 + j\omega L_0)\frac{d\dot{I}}{dx} = -(r_0 + j\omega L_0)(g_0 + j\omega C_0)\dot{U} = -\gamma^2\dot{U} \quad (6.3)$$

where $\gamma = \alpha + j\beta = \sqrt{(r_0 + j\omega L_0)(g_0 + j\omega C_0)}$ is the propagation coefficient of the line, and α, β are the decay coefficient and the phase displacement coefficient, respectively. The general solution of the second order ordinary differential Eq. (6.3) is

$$\dot{U} = \dot{U}(x) = \dot{U}^{+}e^{-\gamma x} + \dot{U}^{-}e^{\gamma x} \quad (6.4)$$

$$\dot{I} = \dot{I}(x) = -\frac{1}{r_0 + j\omega L_0}\frac{d\dot{U}}{dx} = \frac{\dot{U}^{+}}{z_c}e^{-\gamma x} - \frac{\dot{U}^{-}}{z_c}e^{\gamma x} \quad (6.5)$$

where the complex constants $\dot{U}^{+}$, $\dot{U}^{-}$ are integral constants and $z_c = \sqrt{(r_0 + j\omega L_0)/(g_0 + j\omega C_0)}$ is the wave impedance of the transmission line. The integral constant $\dot{U}^{+}$, $\dot{U}^{-}$ is determined below by the boundary conditions. Suppose the voltage and current phasors at the head of the circuit are $\dot{U}_1$, $\dot{I}_1$ respectively, and the voltage phasor and current phasor at the end of the circuit are $\dot{U}_2$, $\dot{I}_2$ respectively.

(1) When $x = 0$, we can get

$$\begin{cases} \dot{U}^{+} = \frac{1}{2}(\dot{U}_1 + z_c\dot{I}_1) \\ \dot{U}^{-} = \frac{1}{2}(\dot{U}_1 - z_c\dot{I}_1) \end{cases} \quad (6.6)$$

Substituting Eq. (6.6) into Eqs. (6.4) and (6.5), we can get

$$\begin{bmatrix} \dot{U}_x \\ \dot{I}_x \end{bmatrix} = \begin{bmatrix} ch\gamma x & -z_c sh\gamma x \\ -\frac{sh\gamma x}{z_c} & ch\gamma x \end{bmatrix} \begin{bmatrix} \dot{U}_1 \\ \dot{I}_1 \end{bmatrix} \quad (6.7)$$

where x is the line length from the head of the circuit.

(2) When $x = l$, we can get

$$\begin{cases} \dot{U}^{+} = \frac{1}{2}(\dot{U}_2 + z_c\dot{I}_2)e^{\gamma l} \\ \dot{U}^{-} = \frac{1}{2}(\dot{U}_2 - z_c\dot{I}_2)e^{-\gamma l} \end{cases} \quad (6.8)$$

Substituting Eq. (6.8) into Eqs. (6.4) and (6.5), we can get

$$\begin{bmatrix} \dot{U}_y \\ \dot{I}_y \end{bmatrix} = \begin{bmatrix} ch\gamma y & z_c sh\gamma y \\ \dfrac{sh\gamma y}{z_c} & ch\gamma y \end{bmatrix} \begin{bmatrix} \dot{U}_2 \\ \dot{I}_2 \end{bmatrix} \tag{6.9}$$

where y is the line length from the end of the circuit, and $y = l - x$.

Eqs. (6.7) and (6.9) establish the functional relationship between the power frequency quantities along the line, which is called the long-line equation of the transmission line.

6.2.2 Long-line equation for three-phase transmission line

The equation for describing the wave process of a three-phase transmission line obtained from Eq. (6.1) is as follows (ignoring conductance):

$$\begin{cases} -\dfrac{\partial [u]}{\partial x} = [L]\dfrac{\partial [i]}{\partial t} + [R][i] \\ -\dfrac{\partial [i]}{\partial x} = [C]\dfrac{\partial [u]}{\partial t} \end{cases} \tag{6.10}$$

where $[u]$ and $[i]$ are the column vectors of voltage and current instantaneous values of the three-phase transmission line respectively, and $[R]$, $[L]$, and $[C]$ represent the resistance, inductance, and capacitance parameter matrices respectively.

All phases of the transmission line are coupled to one another, so the nondiagonal elements of the resistance, inductance, and capacitance parameter matrices are nonzero. Whether the line is completely transposed balanced or unbalanced, its parameter matrices can be completely diagonalized or approximately diagonalized by certain transformation matrices—that is, converted into module components, thereby forming the uncoupled module components. Each module component of the transmission line can satisfy the long line equation represented by Eqs. (6.7) and (6.9).

Suppose S and Q are the transformation matrices of the voltage and current column vectors. Then there is

$$\begin{cases} [u] = S[u_m] \\ [i] = Q[i_m] \end{cases} \tag{6.11}$$

where $[u_m]$ and $[i_m]$ are the voltage and current column vectors on the modulus respectively. For the completely transposed balanced line, $S = Q$, while for the incompletely transposed transmission line, $S = Q^{-T}$.

Substituting Eq. (6.11) into Eq. (6.10), we can get

$$\begin{cases} -\frac{\partial[u_m]}{\partial x} = [L_m]\frac{\partial[i_m]}{\partial t} + [R_m][i_m] \\ -\frac{\partial[i_m]}{\partial x} = [C_m]\frac{\partial[u_m]}{\partial t} \end{cases} \tag{6.12}$$

where $[R_m] = S^{-1}[R]Q$, $[L_m] = S^{-1}[L]Q$, $[C_m] = S^{-1}[C]Q$. $[R_m]$, $[L_m]$, and $[C_m]$ are diagonal matrices. After the above transformation, there is no coupling between each mode circuit. Therefore, the long-line equation can be used to calculate each mode circuit. But it should be pointed out that although the long-line equation based on an accurate line model establishes the functional relationship of the power frequency voltage and current along the line, in the fault transient process, voltage and current signals contain many harmonic components, and the filtering algorithm to extract the power frequency component may be influenced by them. Therefore, using the long-line equation to analyze and calculate the fault transient process has certain limitations.

The three commonly used divisions of module transformation matrices are as follows.

(1) Symmetrical component transformation matrix:

$$S = \begin{bmatrix} 1 & 1 & 1 \\ 1 & a^2 & a \\ 1 & a & a^2 \end{bmatrix}, S^{-1} = \frac{1}{3}\begin{bmatrix} 1 & 1 & 1 \\ 1 & a & a^2 \\ 1 & a^2 & a \end{bmatrix} \tag{6.13}$$

(2) Clarke transformation matrix:

$$S = \begin{bmatrix} 1 & 1 & 1 \\ 1 & -2 & 0 \\ 1 & 1 & -1 \end{bmatrix}, S^{-1} = \frac{1}{6}\begin{bmatrix} 2 & 2 & 2 \\ 1 & -2 & 1 \\ 3 & 0 & -3 \end{bmatrix} \tag{6.14}$$

(3) Karrenbauer transformation matrix:

$$S = \begin{bmatrix} 1 & 1 & 1 \\ 1 & -2 & 1 \\ 1 & 1 & -2 \end{bmatrix}, S^{-1} = \frac{1}{3}\begin{bmatrix} 1 & 1 & 1 \\ 1 & -1 & 0 \\ 1 & 0 & -1 \end{bmatrix} \tag{6.15}$$

For a nonuniformly transposed line, a fixed module transformation matrix cannot be used, and the module transformation matrix can be solved in the real or complex domain according to the line parameters.

6.3 Bergeron model of transmission line

The model proposed by Bergeron is a kind of relatively accurate transmission line model that represents the voltage and current relationship on both ends of the line when there is no fault through the transmission line (including steady operation and external faults). When there is an internal fault, it is equivalent to adding a node at the fault point on the transmission line, and the relationship is destroyed.

6.3.1 Bergeron model of single-phase lossless line

Ignoring the line loss, the partial differential equation of the lossless line can be obtained from partial differential Eq. (6.1) describing the single-wire wave process:

$$\begin{cases} \dfrac{\partial u}{\partial x} = -L_0 \dfrac{\partial i}{\partial t} \\ \dfrac{\partial i}{\partial x} = -C_0 \dfrac{\partial u}{\partial t} \end{cases} \tag{6.16}$$

Solving Eq. (6.16), a second order partial differential wave equation can be obtained as follows:

$$\begin{cases} \dfrac{\partial^2 u}{\partial x^2} = \dfrac{1}{v^2} \cdot \dfrac{\partial^2 i}{\partial t^2} \\ \dfrac{\partial^2 i}{\partial x^2} = \dfrac{1}{v^2} \cdot \dfrac{\partial^2 u}{\partial t^2} \end{cases} \tag{6.17}$$

where $v = \frac{1}{\sqrt{L_0 C_0}}$ is the propagation speed of the traveling wave along the line. For the lossless overhead line, v is approximately equal to the light speed—that is, the propagation speed of the electromagnetic wave in vacuum.

The solution of the voltage and current to the wave equation of the previously describe single lossless wire can be obtained as follows:

$$\begin{cases} u(x,t) = u_1\left(t - \dfrac{x}{v}\right) + u_2\left(t + \dfrac{x}{v}\right) \\ i(x,t) = i_1\left(t - \dfrac{x}{v}\right) + i_2\left(t + \dfrac{x}{v}\right) \end{cases} \tag{6.18}$$

where u_1, i_1 represent the forward voltage and current wave propagating along the $+x$ direction at speed v respectively, and u_2, i_2 represent the backward voltage and current wave propagating along the $-x$ direction at speed v respectively. The forward voltage wave and backward current wave, as well as the backward voltage wave and forward current wave, are connected by the wave impedance:

$$\begin{cases} i_1(x,t) = u_1(x,t)/z_c \\ i_2(x,t) = u_2(x,t)/z_c \end{cases} \tag{6.19}$$

where $z_c = \sqrt{\frac{L_0}{C_0}}$ is the wave impedance of the lossless line. Substituting Eq. (6.19) into Eq. (6.18) and eliminating u_1 and u_2, the following forward characteristic and backward characteristic equations can be obtained:

$$u(x,t) + z_c i(x,t) = 2u_1\left(t - \frac{x}{v}\right) \tag{6.20}$$

$$u(x,t) - z_c i(x,t) = 2u_2\left(t + \frac{x}{v}\right) \tag{6.21}$$

As is shown in Fig. 6.2, at the m end of the line, $x = 0$. Substituting it into Eq. (6.20), we can get

$$u_m(t) + z_c i_m(t) = 2u_1(t) \tag{6.22}$$

where u_m, i_m are the voltage and current at the m end. At the n end of the line, $x = l$. Substituting it into Eq. (6.20), we can get

$$u_n(t) - z_c i_n(t) = 2u_1\left(t - \frac{l}{v}\right) = 2u_1(t - \tau) \tag{6.23}$$

where u_n, i_n are the voltage and current at the n end and τ is the time it takes for the wave to propagate from the m end to the n end. Replacing t with $t - \tau$ in Eq. (6.22), we can get

$$u_m(t-\tau) + z_c i_m(t-\tau) = 2u_1(t-\tau) \tag{6.24}$$

Solving Eqs. (6.23) and (6.24) simultaneously, we can get

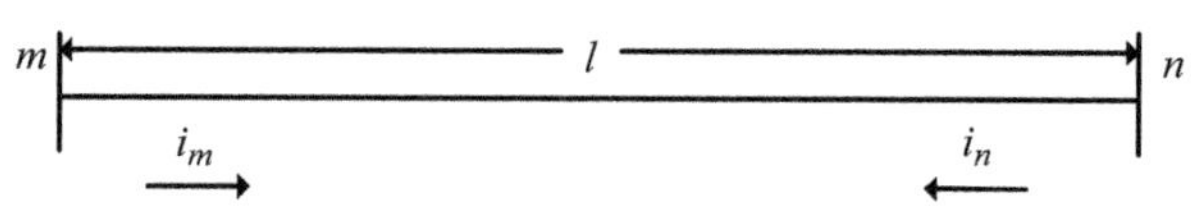

Figure 6.2 Single-phase uniform transmission line.

$$u_n(t) - z_c i_n(t) = u_m(t-\tau) + z_c i_m(t-\tau) \tag{6.25}$$

After transforming, we can get

$$i_n(t) = \frac{u_n(t)}{z_c} - \frac{u_m(t-\tau)}{z_c} - i_m(t-\tau) \tag{6.26}$$

Suppose
$I_{nm}(t-\tau) = -\frac{u_m(t-\tau)}{z_c} - i_m(t-\tau)$, $I_{mn}(t-\tau) = -\frac{u_n(t-\tau)}{z_c} - i_n(t-\tau)$,
there are

$$i_n(t) = \frac{u_n(t)}{z_c} + I_{nm}(t-\tau) \tag{6.27}$$

$$i_m(t) = \frac{u_m(t)}{z_c} + I_{mn}(t-\tau) \tag{6.28}$$

From Eqs. (6.27) and (6.28), the Bergeron equivalent calculation circuit of a single-phase lossless transmission line can be obtained, such as Fig. 6.3, where I_{mn} and I_{nm} are equivalent current sources that represent the reflected wave from the opposite end.

From Eq. (6.25) we can get

$$i_m(t-\tau) + \frac{u_m(t-\tau)}{z_c} = -i_n(t) + \frac{u_n(t)}{z_c} \tag{6.29}$$

Similar to the derivation of Eq. (6.29), we can obtain from Eq. (6.22)

$$-i_m(t) + \frac{u_m(t)}{z_c} = i_n(t-\tau) + \frac{u_n(t-\tau)}{z_c} \tag{6.30}$$

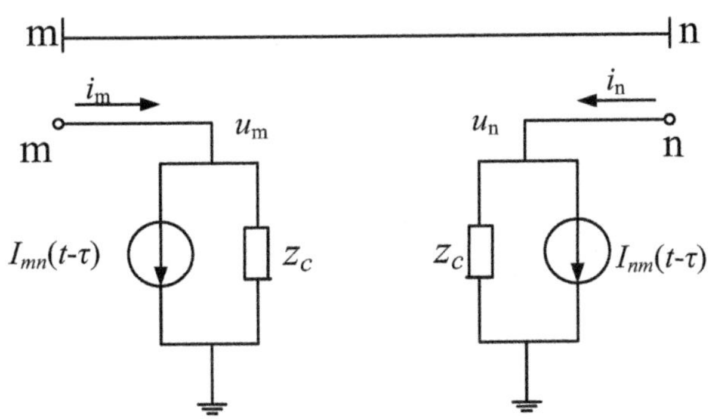

Figure 6.3 Bergeron equivalent calculation circuit of single-phase lossless transmission line.

6.3.2 Bergeron model considering power loss

In the derivation above, line resistance is neglected while resistance exists in the actual transmission lines and presents distributed characteristics. In engineering calculation, an approximate calculation method can be adopted—that is, the resistance of the line can be connected in series sectionally as a concentrated resistance, and each section of the line can still be treated as a lossless distributed parameter line.

When using the Bergeron model in engineering calculation, the length of line l is often divided into two parts, each part being $l/2$ long, as shown in Fig. 6.4.

Suppose the line is lossless, the concentrated resistance $r/4$ is connected in series at each end of the line, and resistance $r/2$ is connected in series in the middle, where r is the resistance of the whole line and $r=r_0l$. From Fig. 6.3, we can get the Bergeron model of a single-phase transmission line considering loss as shown in Fig. 6.5.

For the circuit shown in Fig. 6.5, nodes m', m'', n'', and n' can all be eliminated through theoretical derivation, after which a simplified Bergeron model for a lossy transmission line can be obtained as shown in Fig. 6.6.

In Fig. 6.6, Z is the equivalent wave impedance considering line loss, $Z=z_c+\frac{R}{4}$, $\tau=\frac{l}{v}=l\sqrt{L_0C_0}$. $I_{mn}(t-\tau)=-\frac{1-h}{2}\left[\frac{u_m(t-\tau)}{z_c}+hi_m(t-\tau)\right]-\frac{1+h}{2}\left[\frac{u_n(t-\tau)}{z_c}+hi_n(t-\tau)\right]$,

$I_{nm}(t-\tau)=-\frac{1-h}{2}\left[\frac{u_n(t-\tau)}{z_c}+hi_n(t-\tau)\right]-\frac{1+h}{2}\left[\frac{u_m(t-\tau)}{z_c}+hi_m(t-\tau)\right]$, where $h=\left(z_c-\frac{R}{4}\right)/\left(z_c+\frac{R}{4}\right)$.

From Fig. 6.6 we can get

$$i_m(t)=\frac{u_m(t)}{z_c}+I_{mn}(t-\tau) \tag{6.31}$$

$$i_n(t)=\frac{u_n(t)}{z_c}+I_{nm}(t-\tau) \tag{6.32}$$

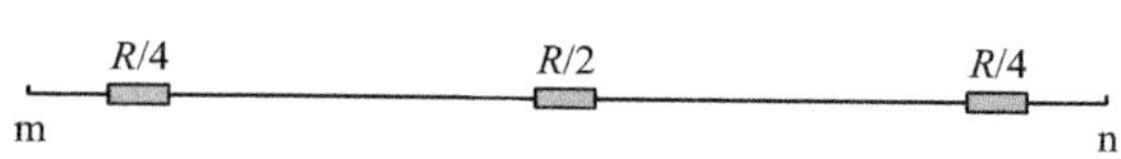

Figure 6.4 Transmission line considering resistance loss.

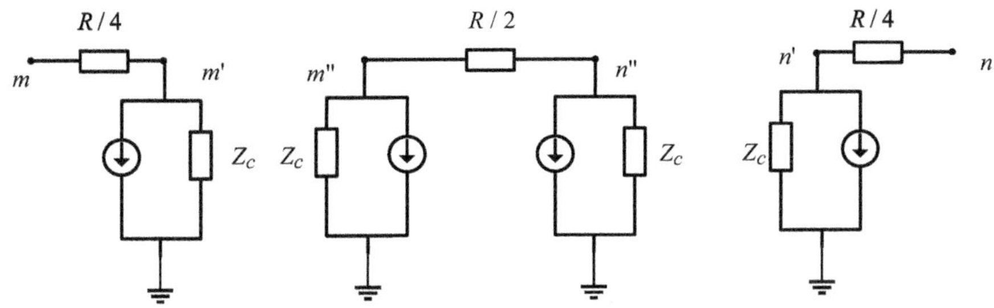

Figure 6.5 Bergeron model of a transmission line considering resistance loss.

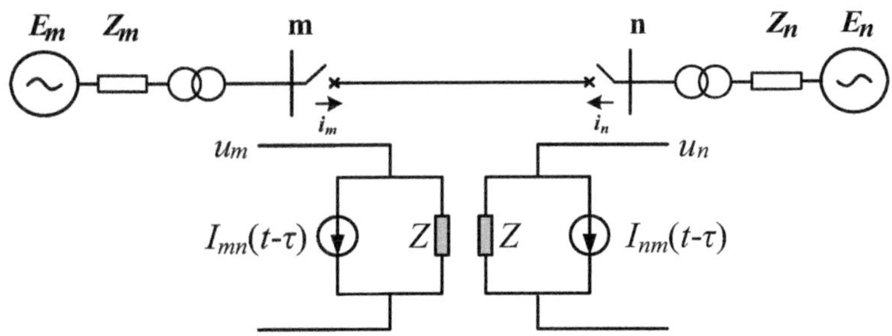

Figure 6.6 Simplified Bergeron model of transmission line considering loss.

Comparing Fig. 6.3 with Fig. 6.6, the basic structure of the circuit remains unchanged while the parameters change. According to the derivation above, the Bergeron model is a kind of relatively accurate line model that considers distributed parameter characteristics.

In the equation derivation ofBergeron model, line resistance with the same distributed characteristics is neglected, and the approximate calculation is hence adopted in the engineering calculation. Therefore, the equation based on the Bergeron model is not as accurate in describing the UHV long line as is the long-line equation. However, the long-line equation is based on frequency domain and only fits a certain frequency, which is 50 Hz for transmission lines. As the equation based on the Bergeron model is a relation-in-time domain, it is valid for all of frequency components. Consequently, the equation based on the Bergeron model is more suitable for a fault transient process, while the long-line equation is more suitable for phasor calculation in a steady state condition, which is an important difference in applications.

6.3.3 Bergeron model of three-phase transmission line

Similar to Eq. (6.1), the wave process of a three-phase transmission line can be described as follows (conductance can be ignored on UHV and UHV lines):

$$\left.\begin{aligned} -\frac{\partial \mathbf{u}}{\partial x} &= \mathbf{Ri} + \mathbf{L}\frac{\partial \mathbf{i}}{\partial t} \\ -\frac{\partial \mathbf{i}}{\partial x} &= \mathbf{C}\frac{\partial \mathbf{u}}{\partial t} \end{aligned}\right\} \tag{6.33}$$

In the above equation, **u** and **i** are respectively the instantaneous value column vector of the voltage and current of the three-phase transmission line, and **R**, **L**, and **C** respectively represent the resistance, inductance, and capacitance parameter matrix.

There are couplings between the phases of the transmission line. This is reflected in Eq. (6.33), where there are nonzero nondiagonal elements in the matrix of resistor **R**, capacitor **C**, and inductor **L**. Whether it is a fully transposed balanced line or an unbalanced line, the parameter matrix can be completely diagonalized or approximately diagonalized by a certain transformation matrix—that is, converted into a mode component, thereby forming module component that are not coupled with one another. Each component of the line satisfies the Bergeron model.

Let **S** and $\mathbf{S}^{-1}$ be the transformation matrix of the voltage and current column vectors **u** and **i**, respectively:

$$\left.\begin{aligned} \mathbf{u} &= \mathbf{S}\mathbf{u}_\mathrm{m} \\ \mathbf{i} &= \mathbf{Q}\mathbf{i}_\mathrm{m} \end{aligned}\right\} \tag{6.34}$$

This represents a fully transposed balanced line $\mathbf{S} = \mathbf{Q}$ to an incompletely transposed transmission line $\mathbf{S} = \mathbf{Q}^{-T}$, where $\mathbf{u}_\mathrm{m}$ and $\mathbf{i}_\mathrm{m}$ are respective voltage and current column vectors on the modulus.

Substituting Eq. (6.34) into Eq. (6.33),

$$\left.\begin{aligned} \frac{\partial \mathbf{u}_\mathrm{m}}{\partial x} &= -\mathbf{L}_\mathrm{m}\frac{\partial \mathbf{i}_\mathrm{m}}{\partial t} \\ \frac{\partial \mathbf{i}_\mathrm{m}}{\partial x} &= -\mathbf{C}_\mathrm{m}\frac{\partial \mathbf{u}_\mathrm{m}}{\partial t} \end{aligned}\right\} \tag{6.35}$$

In the equation: $\mathbf{R}_\mathrm{m} = \mathbf{S}^{-1}\mathbf{RQ}$, $\mathbf{L}_\mathrm{m} = \mathbf{S}^{-1}\mathbf{LQ}$, $\mathbf{C}_\mathrm{m} = \mathbf{Q}^{-1}\mathbf{C}_\mathrm{m}\mathbf{S}$.

Among the equations, $\mathbf{R_m}$, $\mathbf{L_m}$, and $\mathbf{C_m}$ are diagonal matrixes. Therefore, after the above transformation, there is no coupling between the respective modular circuits. For each modular circuit, the Bergeron model of Fig. 6.6 can be used for calculation.

The Karrenbauer transformation matrix (Eq. 6.15) is used to obtain the modulo component. For a nonuniform transposition line, a fixed modulo transformation matrix cannot be used, and the modulo transformation matrix can be solved in the real or complex domain according to the line parameters [2].

6.4 Dynamic phasor model of transmission line

If a time-varying dynamic phasor is imported into the power system, another kind of transmission line model can be formed—that is, a time-varying dynamic phasor model. The dynamic phasor model of the transmission line is shown in Fig. 6.7.

The calculation equation is

$$\begin{cases} I_m(t) = \dfrac{U_m(t)}{z_c} - \left[\dfrac{2U_n(t-\tau)\cdot e^{-j\theta}}{z_c} + I_n(t-2\tau)\bullet e^{-j\theta}\right] = \dfrac{U_m(t)}{z_c} + I_{mn}(t-\tau) \\ I_n(t) = \dfrac{U_n(t)}{z_c} - \left[\dfrac{2U_m(t-\tau)\cdot e^{-j\theta}}{z_c} + I_m(t-2\tau)\bullet e^{-j\theta}\right] = \dfrac{U_n(t)}{z_c} + I_{nm}(t-\tau) \end{cases} \tag{6.36}$$

where $\tau = l/v$ is the time it takes for the traveling wave to propagate from the m end to the n end, and $\theta = \omega\tau$. The time-varying dynamic phasor can qualitatively describe the electromagnetic transient process. For a distributed parameter transmission line, this model can analyze its wave process to some degree. The variable of a time-varying phasor model is phasor, so the calculation speed is fast. In this way the model can be used as an

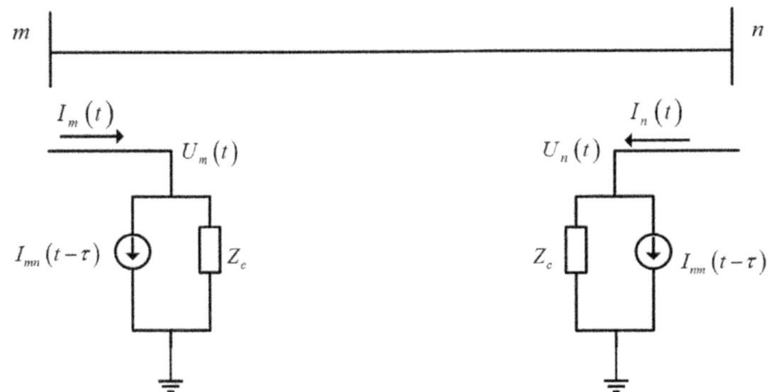

Figure 6.7 Time-varying dynamic phasor model.

intermediate model between the electromagnetic transient and steady state models. This model can satisfy the need for simulation, while its application in relay protection remains to be further studied.

The distributed parameter characteristics of the UHV transmission line needs to be considered, and its essential characteristics must be mastered and explored to research making the protection theory of the UHV long line more suitable. In this chapter, several kinds of line models with distributed parameter characteristics are recommended. For transmission lines, the long-line model, Bergeron model, and time-varying dynamic phasor model all have their own characteristics:

(1) The per-unit-length model of the transmission line accurately describes the distributed parameter characteristics of the lossy transmission line, but the voltage and current functional relationship at the two ends of the line established by the long-line equation can only be satisfied at a certain frequency. For power system, Eqs. (6.7) and (6.9) only represent the relationship of the power frequency electrical quantities at the two ends of the line. But in the fault transient process, there are plenty of harmonic components and decay DC components, while in practical calculation, a filtering algorithm cannot extract power frequency signals in a certain data window. Therefore, when this model is used in analysis and calculation of fault transient process, it is quite limited.

(2) The Bergeron line model is a kind of relatively accurate model proposed by Bergeron based on transmission line distributed parameter characteristics. It accurately describes the distributed parameter characteristics of the lossless transmission line. Eq. (6.25) represents the time domain functional relationship of voltage and current at the two ends of the line, so it can satisfy the whole frequency domain. But line resistance is neglected in derivation of the Bergeron model, so the model cannot describe the decay process of the fault transient signal accurately. The common solution is to segment the line into sections and solve the problem by equalizing the line resistance concentrated at both ends of each section of lossless line.

(3) The dynamic phasor model is similar to the Bergeron line model and also accurately describes the distributed parameter characteristics of the lossless transmission line. The variable of the model is phasor, so it is easy to calculate. The disadvantage is that it only satisfies the voltage and current functional relationship at the two ends of the line in power frequency. Because the phasor model imports time, it is a model between the electromagnetic transient model and the steady state model.

References

[1] He J, Ge Y. Ultra-high voltage transmission line fault analysis and relay protection. Beijing: Science Press; 1987.
[2] Li B. The theories and techniques of protection for UHV long distance transmission line- new theories of protection based on exact line model. Tianjin: Tianjin University; 2006 [in Chinese].

CHAPTER 7

Current differential protection

Contents

7.1 Basic principle

7.1.1 Differential current

Current differential protection is provided at both ends of the transmission line. Here one side (m side) is taken as an example to illustrate the working procedure of protection. On the m side, the three-phase voltage and current sampling values, $u_{m\varphi}$ and $i_{m\varphi}$ ($\varphi = a \cdot b \cdot c$), at various time of protection installation are obtained by sampling, as well as on the opposite side (n side). Sample values of voltage and current at the same time of protection installation, $u_{n\varphi}$ and $i_{n\varphi}$, can also be obtained after sampling synchronization through the optical fiber or microwave channel. Since the Bergeron model is established on the modulus value, before applying the Bergeron model calculation, the m side protection first converts the sample value of own side and the sample value transmitted by the opposite side into modulus value, u_{mi}, i_{mi}, u_{ni}, and i_{ni} (in the subscript, $i = 0$, α, β represents the code of the modulus value, the same below).

Using the three voltage modulus values of the current moment t on the m side and the six voltage and current modulus values of the three phases on the m side of the previous $(t - \tau)$ moment, as well as the six voltage and current modulus values of the opposite side (n side) of the three phases, the calculated values of the three current modulus value $i_{J\,mi}$ at the current moment at the m side are calculated by the Bergeron method Eq. (6.31) and converted into phase value $i_{J\,m\varphi}$. Then, the three measured values $i_{m\varphi}$ and three calculated values $i_{J\,m\varphi}$ are filtered by the half-wave differential Fourier

Protection Technologies of Ultra-High-Voltage AC Transmission Systems
ISBN 978-0-12-816205-7
https://doi.org/10.1016/B978-0-12-816205-7.00007-8

algorithm to obtain the respective fundamental wave vectors $\dot{I}_{m\varphi}$ and $\dot{I}_{Jm\varphi}$, and finally the measured and calculated values are compared according to traditional split-phase current differential protection. In this way, the action amount can be obtained [1].

Similar to this on the n side, first converting the voltage and current sample values on the n side and the voltage and current sample values transmitted from the opposite side into modulus values, then using the three voltage modulus values of the current moment t on the n side and the six voltage and current modulus values of the three phases on the m side of the previous $(t-\tau)$ moment as well as the six voltage and current modulus values of the opposite side (m side) of the three phases, the calculated values of the three current modulus values $i_{J\,ni}$ at the current moment at the m side are calculated by the Bergeron method, Eq. (6.32), and converted into phase value $i_{J\,n\varphi}$. Then, the three measured values $i_{n\varphi}$ and three calculated values $i_{J\,n\varphi}$ are filtered by the half-wave differential Fourier algorithm to obtain the respective fundamental wave vectors $\dot{I}_{n\varphi}$ and $\dot{I}_{Jn\varphi}$, and finally the measured values and calculated values are compared according to traditional split-phase current differential protection. In this way, the action amount can be obtained.

That is, on the m side there is:

$$dI_{ma}=\left|\dot{I}_{ma}-\dot{I}_{Jma}\right|,\quad dI_{mb}=\left|\dot{I}_{mb}-\dot{I}_{Jmb}\right|,\quad dI_{mc}=\left|\dot{I}_{mc}-\dot{I}_{Jmc}\right| \tag{7.1}$$

On the n side there is:

$$dI_{na}=\left|\dot{I}_{na}-\dot{I}_{Jna}\right|,\quad dI_{nb}=\left|\dot{I}_{nb}-\dot{I}_{Jnb}\right|,\quad dI_{nc}=\left|\dot{I}_{nc}-\dot{I}_{Jnc}\right| \tag{7.2}$$

Because the Bergeron model truly reflects the relationship between voltage and current on both sides of the line without internal faults (including steady-state operation and external faults), the theoretically calculated and measured values should be substantially equal when there is no internal fault. So, there are $\dot{I}_{m\varphi}=\dot{I}_{Jm\varphi}$ and $\dot{I}_{n\varphi}=\dot{I}_{Jn\varphi}$. Therefore, the action amount on both sides of the new split-phase current differential protection $dI_{m\varphi}$ and $dI_{n\varphi}$ should be equal to zero. If the line has an internal fault, the Bergeron model is destroyed, the fault phase action amount is not much different from the short circuit current, and the nonfault phase action amount is small relative to the fault phase, which can be seen from the specific equation of the action amount of each phase with the internal fault. The expression with details of the analysis can be seen in Section 7.1.2 below.

7.1.2 Relationship between differential current and fault current

Take single-phase lossless line as an example. When an internal fault occurs at point *f* as shown in Fig. 7.1, the voltage and current of the fault point can be presented as $u_f(t)$ and $i_f(t)$, so that

$$i_f(t) = i_{mf}(t) + i_{nf}(t) \tag{7.3}$$

Applying Eq. (6.30) to the line from m to f and taking the current direction into consideration, Eq. (7.4) can be obtained:

$$-i_m(t) + \frac{u_m(t)}{z_c} = -i_{mf}(t-\tau_m) + \frac{u_f(t-\tau_m)}{z_c} \tag{7.4}$$

Applying Eq. (6.30) to the line from f to n and taking the current direction into consideration, Eq. (7.5) can be obtained:

$$i_{nf}(t) + \frac{u_f(t)}{z_c} = i_n(t-\tau_n) + \frac{u_n(t-\tau_n)}{z_c} \tag{7.5}$$

Replacing t with $t-\tau_m$ in Eq. (6.24) and considering $\tau_m + \tau_n = \tau$, Eq. (7.6) can be obtained:

$$i_n(t-\tau) + \frac{u_n(t-\tau)}{z_c} = i_{nf}(t-\tau_m) + \frac{u_f(t-\tau_m)}{z_c} \tag{7.6}$$

Solving Eqs. (7.4) and (7.6) simultaneously and eliminating the fault point voltage u_f, Eq. (7.7) can be obtained:

$$\begin{aligned} i_m(t) - \frac{u_m(t)}{z_c} + i_n(t-\tau) + \frac{u_n(t-\tau)}{z_c} &= i_{mf}(t-\tau_m) + i_{nf}(t-\tau_m) \\ &= i_f(t-\tau_m) \end{aligned} \tag{7.7}$$

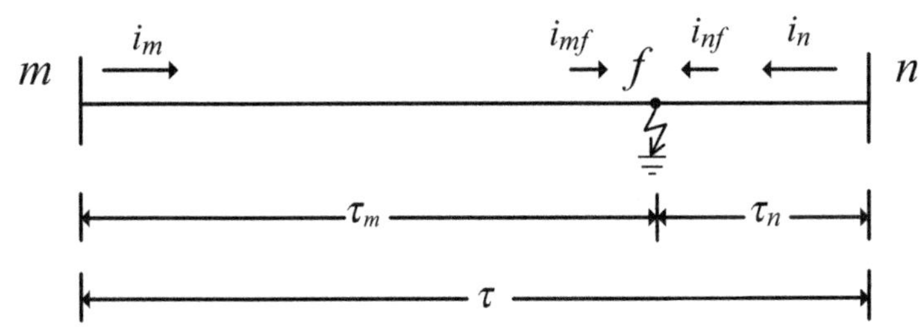

Figure 7.1 Single-phase lossless line when an internal fault occurs.

Similarly, from Eq. (6.29) Eq. (7.8) can be obtained:

$$i_n(t) - \frac{u_n(t)}{z_c} + i_m(t-\tau) + \frac{u_m(t-\tau)}{z_c} = i_f(t-\tau_n) \tag{7.8}$$

Eqs. (7.7) and (7.8) are the theoretical basis of split-phase current differential protection based on the Bergeron model and also are the theoretical basis of traveling wave differential protection [2].

The calculation of the m side protection action amount di_m can be expressed by the following equation:

$$di_m(t) = i_m(t) - i_{Jm}(t) \tag{7.9}$$

In the above equation, i_{Jm} represents the calculated value of i_m. Eq. (7.10) can be obtained from Eq. (6.28)

$$i_{Jm}(t) = \frac{u_m(t)}{z_c} + I_{mn}(t-\tau) = \frac{u_m(t)}{z_c} - \frac{u_n(t-\tau)}{z_c} - i_n(t-\tau) \tag{7.10}$$

Substituting Eq. (7.10) into Eq. (7.9) and using Eq. (7.7), Eq. (7.11) can be obtained:

$$di_m(t) = i_m(t) - i_{Jm}(t) = i_f(t-\tau_m) \tag{7.11}$$

Eq. (7.11) reflects the relationship between the action amount on a certain modulus component line and the short-circuit current at the fault point. In the three-phase line after phase-modulus transformation, the three modulus components satisfy the above equation and can be written in matrix form:

$$\begin{bmatrix} di_{m0}(t) \\ di_{m\alpha}(t) \\ di_{m\beta}(t) \end{bmatrix} = \begin{bmatrix} i_{f0}(t-\tau_{m0}) \\ i_{f\alpha}(t-\tau_{m\alpha}) \\ i_{f\beta}(t-\tau_{m\beta}) \end{bmatrix} \tag{7.12}$$

The subscripts 0, α, β in the above equation correspond to the three Karrenbauer mode components of the three-phase circuit after phase-mode transformation, the same as below. Using the phase-mode transformation relationship and Eq. (7.12), there is

$$\begin{bmatrix} di_{ma}(t) \\ di_{mb}(t) \\ di_{mc}(t) \end{bmatrix} = [\mathbf{S}] \begin{bmatrix} di_{m0}(t) \\ di_{m\alpha}(t) \\ di_{m\beta}(t) \end{bmatrix} = \begin{bmatrix} 1 & 1 & 1 \\ 1 & -2 & 1 \\ 1 & 1 & -2 \end{bmatrix} \begin{bmatrix} i_{f0}(t-\tau_{m0}) \\ i_{f\alpha}(t-\tau_{m\alpha}) \\ i_{f\beta}(t-\tau_{m\beta}) \end{bmatrix} \tag{7.13}$$

Eq. (7.13) reflects the relationship between the action amount of each phase and the modulus components of the short-circuit point current.

The phase-to-mode transformation relationship of the short-circuit point current when the A-phase is grounded is

$$\begin{bmatrix} i_{f0}(t) \\ i_{f\alpha}(t) \\ i_{f\beta}(t) \end{bmatrix} = [\mathbf{S}^{-1}] \begin{bmatrix} i_{fa}(t) \\ i_{fb}(t) \\ i_{fc}(t) \end{bmatrix} = \frac{1}{3}\begin{bmatrix} 1 & 1 & 1 \\ 1 & -1 & 0 \\ 1 & 0 & -1 \end{bmatrix}\begin{bmatrix} i_{fa}(t) \\ 0 \\ 0 \end{bmatrix} = \frac{1}{3}\begin{bmatrix} i_{fa}(t) \\ i_{fa}(t) \\ i_{fa}(t) \end{bmatrix} \tag{7.14}$$

The above equation is true for any time t. Use $(t-\tau_{m0})$, $(t-\tau_{m\alpha})$, and $(t-\tau_{m\beta})$ instead of t to get

$$\begin{bmatrix} i_{f0}(t-\tau_{m0}) \\ i_{f\alpha}(t-\tau_{m\alpha}) \\ i_{f\beta}(t-\tau_{m\beta}) \end{bmatrix} = \frac{1}{3}\begin{bmatrix} i_{fa}(t-\tau_{m0}) \\ i_{fa}(t-\tau_{m\alpha}) \\ i_{fa}(t-\tau_{m\beta}) \end{bmatrix} \tag{7.15}$$

Substituting the above equation into Eq. (7.13) and considering that $\tau_{m\alpha}=\tau_{m\beta}$ in the fully transposed three-phase line, Eq. (7.16) can be obtained:

$$\begin{bmatrix} di_{ma}(t) \\ di_{mb}(t) \\ di_{mc}(t) \end{bmatrix} = \frac{1}{3}\begin{bmatrix} i_{fa}(t-\tau_{m0}) + 2i_{fa}(t-\tau_{m\alpha}) \\ i_{fa}(t-\tau_{m0}) - i_{fa}(t-\tau_{m\alpha}) \\ i_{fa}(t-\tau_{m0}) - i_{fa}(t-\tau_{m\alpha}) \end{bmatrix} \tag{7.16}$$

Eq. (7.16) reflects the relationship between the action amount of each phase and the short-circuit point current when phase A is grounded. In addition, according to Eq. (7.1), there is the following equation:

$$\begin{bmatrix} d\dot{I}_{ma}(t) \\ d\dot{I}_{mb}(t) \\ d\dot{I}_{mc}(t) \end{bmatrix} = \begin{bmatrix} \dot{I}_{ma}(t) - \dot{I}_{Jma}(t) \\ \dot{I}_{mb}(t) - \dot{I}_{Jmb}(t) \\ \dot{I}_{mc}(t) - \dot{I}_{Jmc}(t) \end{bmatrix} = FT\begin{bmatrix} i_{ma}(t) - i_{Jma}(t) \\ i_{mb}(t) - i_{Jmb}(t) \\ i_{mc}(t) - i_{Jmc}(t) \end{bmatrix} = FT\begin{bmatrix} di_{ma}(t) \\ di_{mb}(t) \\ di_{mc}(t) \end{bmatrix} \tag{7.17}$$

In the above equation, FT represents the Fourier transform.

Substituting Eq. (7.16) into Eq. (7.17) and omitting the time symbol, according to the delay theorem of the Fourier transform Eq. (7.18) can be obtained:

$$\begin{bmatrix} d\dot{I}_{ma} \\ d\dot{I}_{mb} \\ d\dot{I}_{mc} \end{bmatrix} = \frac{1}{3}\begin{bmatrix} \dot{I}_{fa}e^{-j\omega\tau_{m0}} + 2\dot{I}_{fa}e^{-j\omega\tau_{m\alpha}} \\ \dot{I}_{fa}e^{-j\omega\tau_{m0}} - \dot{I}_{fa}e^{-j\omega\tau_{m\alpha}} \\ \dot{I}_{fa}e^{-j\omega\tau_{m0}} - \dot{I}_{fa}e^{-j\omega\tau_{m\alpha}} \end{bmatrix} \tag{7.18}$$

where $w = 2\pi f$.

Set $\dot{F}_1 = (e^{-j\omega\tau_{m0}} + 2e^{-j\omega\tau_{m\alpha}})/3$, $\dot{F}_2 = (e^{-j\omega\tau_{m0}} - e^{-j\omega\tau_{m\alpha}})/3$, so Eq. (7.18) can be written as

$$\begin{bmatrix} d\dot{I}_{ma} \\ d\dot{I}_{mb} \\ d\dot{I}_{mc} \end{bmatrix} = \begin{bmatrix} \dot{I}_{fa}\dot{F}_1 \\ \dot{I}_{fa}\dot{F}_2 \\ \dot{I}_{fa}\dot{F}_2 \end{bmatrix} = \begin{bmatrix} \dot{F}_1 & \dot{F}_2 & \dot{F}_2 \\ \dot{F}_2 & \dot{F}_1 & \dot{F}_2 \\ \dot{F}_2 & \dot{F}_2 & \dot{F}_1 \end{bmatrix} \begin{bmatrix} \dot{I}_{fa} \\ 0 \\ 0 \end{bmatrix} \tag{7.19}$$

The above equation is the expression of the action amount of each phase when the A phase is grounded and the relationship between the action amount of each phase and the short-circuit point current.

When both phases of A and B fail, after a similar derivation process, similar to Eq. (7.14), Eq. (7.20) can be obtained:

$$\begin{bmatrix} i_{f0}(t) \\ i_{f\alpha}(t) \\ i_{f\beta}(t) \end{bmatrix} = [\mathbf{S}^{-1}] \begin{bmatrix} i_{fa}(t) \\ i_{fb}(t) \\ i_{fc}(t) \end{bmatrix} = \frac{1}{3}\begin{bmatrix} 1 & 1 & 1 \\ 1 & -1 & 0 \\ 1 & 0 & -1 \end{bmatrix} \begin{bmatrix} i_{fa}(t) \\ i_{fb}(t) \\ 0 \end{bmatrix} = \frac{1}{3}\begin{bmatrix} i_{fa}(t) + i_{fb}(t) \\ i_{fa}(t) - i_{fb}(t) \\ i_{fa}(t) \end{bmatrix} \tag{7.20}$$

Similar to Eq. (7.16), Eq. (7.21) can be obtained:

$$\begin{bmatrix} di_{ma}(t) \\ di_{mb}(t) \\ di_{mc}(t) \end{bmatrix} = \frac{1}{3}\begin{bmatrix} i_{fa}(t-\tau_{m0}) + 2i_{fa}(t-\tau_{m\alpha}) + i_{fb}(t-\tau_{m0}) - i_{fb}(t-\tau_{m\alpha}) \\ i_{fb}(t-\tau_{m0}) + 2i_{fb}(t-\tau_{m\alpha}) + i_{fa}(t-\tau_{m0}) - i_{fa}(t-\tau_{m\alpha}) \\ i_{fa}(t-\tau_{m0}) - i_{fa}(t-\tau_{m\alpha}) + i_{fb}(t-\tau_{m0}) - i_{fb}(t-\tau_{m\alpha}) \end{bmatrix} \tag{7.21}$$

So there are

$$\begin{bmatrix} d\dot{I}_{ma} \\ d\dot{I}_{mb} \\ d\dot{I}_{mc} \end{bmatrix} = \frac{1}{3}\begin{bmatrix} \dot{I}_{fa}e^{-j\omega\tau_{m0}} + 2\dot{I}_{fa}e^{-j\omega\tau_{m\alpha}} + \dot{I}_{fb}e^{-j\omega\tau_{m0}} - \dot{I}_{fb}e^{-j\omega\tau_{m\alpha}} \\ \dot{I}_{fb}e^{-j\omega\tau_{m0}} + 2\dot{I}_{fb}e^{-j\omega\tau_{m\alpha}} + \dot{I}_{fa}e^{-j\omega\tau_{m0}} - \dot{I}_{fa}e^{-j\omega\tau_{m\alpha}} \\ \dot{I}_{fa}e^{-j\omega\tau_{m0}} - \dot{I}_{fa}e^{-j\omega\tau_{m\alpha}} + \dot{I}_{fb}e^{-j\omega\tau_{m0}} - \dot{I}_{fb}e^{-j\omega\tau_{m\alpha}} \end{bmatrix} \tag{7.22}$$

$$\begin{bmatrix} d\dot{I}_{ma} \\ d\dot{I}_{mb} \\ d\dot{I}_{mc} \end{bmatrix} = \begin{bmatrix} \dot{F}_1 & \dot{F}_2 & \dot{F}_2 \\ \dot{F}_2 & \dot{F}_1 & \dot{F}_2 \\ \dot{F}_2 & \dot{F}_2 & \dot{F}_1 \end{bmatrix} \begin{bmatrix} \dot{I}_{fa} \\ \dot{I}_{fb} \\ 0 \end{bmatrix} \tag{7.23}$$

When the two phases are short-circuited, $i_{fa} = -i_{fb}$, $d\dot{I}_{mc} = 0$.

Similarly, when there are faults in the three phases, the corresponding expressions are as follows:

$$\begin{bmatrix} di_{ma}(t) \\ di_{mb}(t) \\ di_{mc}(t) \end{bmatrix} = [\mathbf{S}] \begin{bmatrix} di_{m0}(t) \\ di_{m\alpha}(t) \\ di_{m\beta}(t) \end{bmatrix} = \begin{bmatrix} 1 & 1 & 1 \\ 1 & -2 & 1 \\ 1 & 1 & -2 \end{bmatrix} \begin{bmatrix} i_{f0}(t-\tau_{m0}) \\ i_{f\alpha}(t-\tau_{m\alpha}) \\ i_{f\beta}(t-\tau_{m\beta}) \end{bmatrix} \tag{7.24}$$

$$\begin{aligned} \begin{bmatrix} i_{f0}(t) \\ i_{f\alpha}(t) \\ i_{f\beta}(t) \end{bmatrix} = [\mathbf{S}^{-1}] \begin{bmatrix} i_{fa}(t) \\ i_{fb}(t) \\ i_{fc}(t) \end{bmatrix} &= \frac{1}{3} \begin{bmatrix} 1 & 1 & 1 \\ 1 & -1 & 0 \\ 1 & 0 & -1 \end{bmatrix} \begin{bmatrix} i_{fa}(t) \\ i_{fb}(t) \\ i_{fc}(t) \end{bmatrix} \\ &= \frac{1}{3} \begin{bmatrix} i_{fa}(t) + i_{fb}(t) + i_{fc}(t) \\ i_{fa}(t) - i_{fb}(t) \\ i_{fa}(t) - i_{fc}(t) \end{bmatrix} \end{aligned} \tag{7.25}$$

$$\begin{bmatrix} di_{ma}(t) \\ di_{mb}(t) \\ di_{mc}(t) \end{bmatrix} = \frac{1}{3} \begin{bmatrix} i_{fa}(t-\tau_{m0}) + 2i_{fa}(t-\tau_{m\alpha}) + i_{fb}(t-\tau_{m0}) - i_{fb}(t-\tau_{m\alpha}) + i_{fc}(t-\tau_{m0}) - i_{fc}(t-\tau_{m\alpha}) \\ i_{fb}(t-\tau_{m0}) + 2i_{fb}(t-\tau_{m\alpha}) + i_{fa}(t-\tau_{m0}) - i_{fa}(t-\tau_{m\alpha}) + i_{fc}(t-\tau_{m0}) - i_{fc}(t-\tau_{m\alpha}) \\ i_{fc}(t-\tau_{m0}) + 2i_{fc}(t-\tau_{m\alpha}) + i_{fa}(t-\tau_{m0}) - i_{fa}(t-\tau_{m\alpha}) + i_{fb}(t-\tau_{m0}) - i_{fb}(t-\tau_{m\alpha}) \end{bmatrix} \tag{7.26}$$

$$\begin{bmatrix} d\dot{I}_{ma} \\ d\dot{I}_{mb} \\ d\dot{I}_{mc} \end{bmatrix} = \frac{1}{3} \begin{bmatrix} \dot{I}_{fa}e^{-j\omega\tau_{m0}} + 2\dot{I}_{fa}e^{-j\omega\tau_{m\alpha}} + \dot{I}_{fb}e^{-j\omega\tau_{m0}} - \dot{I}_{fb}e^{-j\omega\tau_{m\alpha}} + \dot{I}_{fc}e^{-j\omega\tau_{m0}} - \dot{I}_{fc}e^{-j\omega\tau_{m\alpha}} \\ \dot{I}_{fb}e^{-j\omega\tau_{m0}} + 2\dot{I}_{fb}e^{-j\omega\tau_{m\alpha}} + \dot{I}_{fa}e^{-j\omega\tau_{m0}} - \dot{I}_{fa}e^{-j\omega\tau_{m\alpha}} + \dot{I}_{fc}e^{-j\omega\tau_{m0}} - \dot{I}_{fc}e^{-j\omega\tau_{m\alpha}} \\ \dot{I}_{fc}e^{-j\omega\tau_{m0}} + 2\dot{I}_{fc}e^{-j\omega\tau_{m\alpha}} + \dot{I}_{fa}e^{-j\omega\tau_{m0}} - \dot{I}_{fa}e^{-j\omega\tau_{m\alpha}} + \dot{I}_{fb}e^{-j\omega\tau_{m0}} - \dot{I}_{fb}e^{-j\omega\tau_{m\alpha}} \end{bmatrix} \tag{7.27}$$

$$\begin{bmatrix} d\dot{I}_{ma} \\ d\dot{I}_{mb} \\ d\dot{I}_{mc} \end{bmatrix} = \begin{bmatrix} \dot{F}_1 & \dot{F}_2 & \dot{F}_2 \\ \dot{F}_2 & \dot{F}_1 & \dot{F}_2 \\ \dot{F}_2 & \dot{F}_2 & \dot{F}_1 \end{bmatrix} \begin{bmatrix} \dot{I}_{fa} \\ \dot{I}_{fb} \\ \dot{I}_{fc} \end{bmatrix} \tag{7.28}$$

Other types of faults can be derived in this way and are not specifically given. To summarize Eqs. (7.19), (7.23), and (7.28), all fault types can be uniformly represented by Eq. (7.28) to indicate the relationship between the action amount and the short-circuit current at the fault point. When the internal phase φ ($\varphi = a, b, c$) is faulted, $\dot{I}_{f\varphi}$ is the short-circuit point current of the phase; otherwise, $\dot{I}_{f\varphi} = 0$. Eq. (7.28) has nothing to do with the

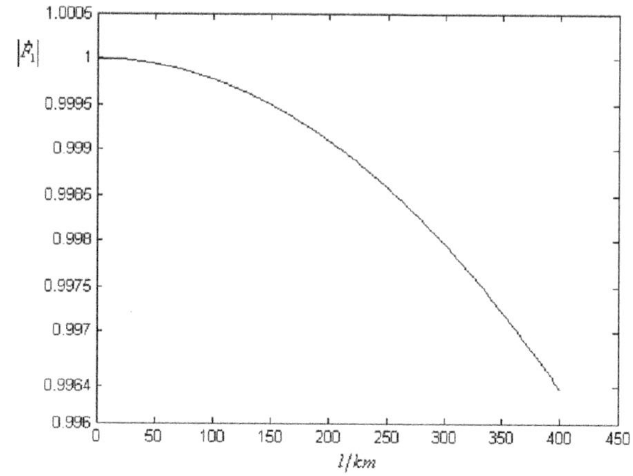

Figure 7.2 Comparing the measurement value and calculated value of the m-side current and the corresponding $|F_1|$ of different fault points.

transition resistance, which makes the protection criterion mentioned later less affected by the transition resistance.

The relative magnitudes of $|\dot{F}_1|$ and $|\dot{F}_2|$ are calculated by taking the 750 kV UHV line parameters used in the previous example, because their values are related to τ_{m0} and $\tau_{m\alpha}$—that is, related to the location of the fault point. Figs. 7.2 and 7.3 show the magnitudes of $|\dot{F}_1|$ and $|\dot{F}_2|$ for the m-side protection when the fault occurs at different locations. In the figure,

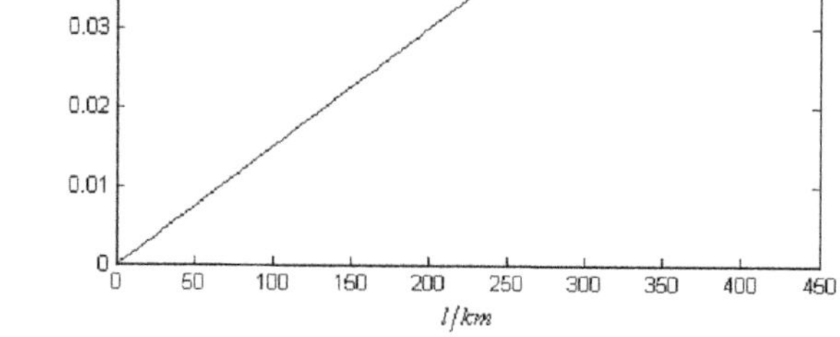

Figure 7.3 Comparing the measurement value and calculated value of the m-side current and the corresponding $|\dot{F}_2|$ of different fault points.

the abscissa l indicates the distance between the fault point and the m-side bus, the same as below.

It can be seen from Figs. 7.2 and 7.3 that the closer the fault point is to the m-side bus, the closer $|\dot{F}_1|$ is to 1, and the closer $|\dot{F}_2|$ is to 0; the farther the fault point is from the m-side bus, the smaller $|\dot{F}_1|$ is and the larger $|\dot{F}_2|$ is, regardless of the fault point, where $|\dot{F}_1|$ is not less than 0.9964, and $|\dot{F}_2|$ is not more than 0.06.

The above derivation is started from the lossless line. The actual lines all have resistance, and it is difficult to give a simple expression. However, the simulations in this paper show that line resistance has little effect on the calculation results of $|\dot{F}_1|$ and $|\dot{F}_2|$, and the fixed value of the protection can be used to avoid that.

In addition, the simulation shows that the nonfault phase current is small when the short-circuit current is small, and the calculation error in the transient process can be compared. Therefore, this problem should be considered in the protection criterion. For the fault phase with short-circuit current and the phase with large short-circuit current, the calculation error of the transient process accounts for a small proportion of the calculated value, and the influence is not significant.

7.1.3 Protection criteria

For phase A, define the braking current as follows:

$$I_{Ba} = K \cdot (I_{D\delta})_{\max} + I_S \tag{7.29}$$

where I_s is a fixed threshold, which can be taken as 300 A, δ represents the other two phases except phase A, $K \cdot (I_{D\delta})_{\max}$ is a floating threshold, and the floating threshold braking coefficient K satisfies

$$K = k_1 \cdot \frac{(|F_2|)_{\max}}{(|F_1|)_{\max}} \tag{7.30}$$

where k_1 is a reliability coefficient and can be taken as 2.

When the fault occurs in the area and the fault current is large, the floating threshold can ensure the correctness of the protection phase selection; the floating threshold can reduce the setting value of the fixed threshold and improve the ability to protect against transition resistance. When an external fault occurs, I_s can avoid the maximum unbalanced current that may occur on the line to avoid protection misoperation; when a fault occurs in the area and the fault current is small, I_s can avoid the unbalanced current of the nonfault phase, eliminating the phase selection error caused by the adverse effect of the error on the nonfault phase current and ensuring correct operation of the protection.

The braking current of the other two phases is available in the same way.

Under the aforementioned definitions of operating current and braking current, the typical operating logic of current differential protection is satisfied. For any phase, the protection acts when the operating current and braking current satisfy the following equation:

$$I_D > I_B \tag{7.31}$$

It can be seen from this analysis that phase-separated current differential protection based on the Bergeron model itself has the ability to select phases, and it is not necessary to install a phase selection component for phase selection.

In microcomputer protection, the procedure of calculating the currents of the two sides on the basis of the Bergeron model can be worked out according to Fig. 7.4.

A program to calculate the current from both ends to reference point k can also be worked out, which is omitted here.

7.2 Protection for transmission lines with shunt reactor

The two ends of a UHV long line must be equipped with a shunt reactor to limit overvoltage, and voltage distribution may be too high in the middle. In this case, a shunt reactor is also required in the middle of the line. This is equivalent to a lateral branch connected in the middle of the line [3].

Considering shunt reactor compensation, traditional segregated current differential protection can only use the approximate concentration equivalent method to compensate for capacitance current, which greatly affects current differential protection of UHV transmission lines on selectivity in the external fault and reliability in the internal fault. Tianjin University and Beijing Sifang Automation Co. Ltd. jointly put forward a current differential protection algorithm based on the Bergeron model with a shunt reactor in the middle of the line.

When there is a shunt reactor connected in the middle section of the line, the Bergeron equation is satisfied for the lines on both sides of the shunt reactor installation point. Therefore, for the line with the shunt reactor in the middle section of the line, it is assumed that the installation

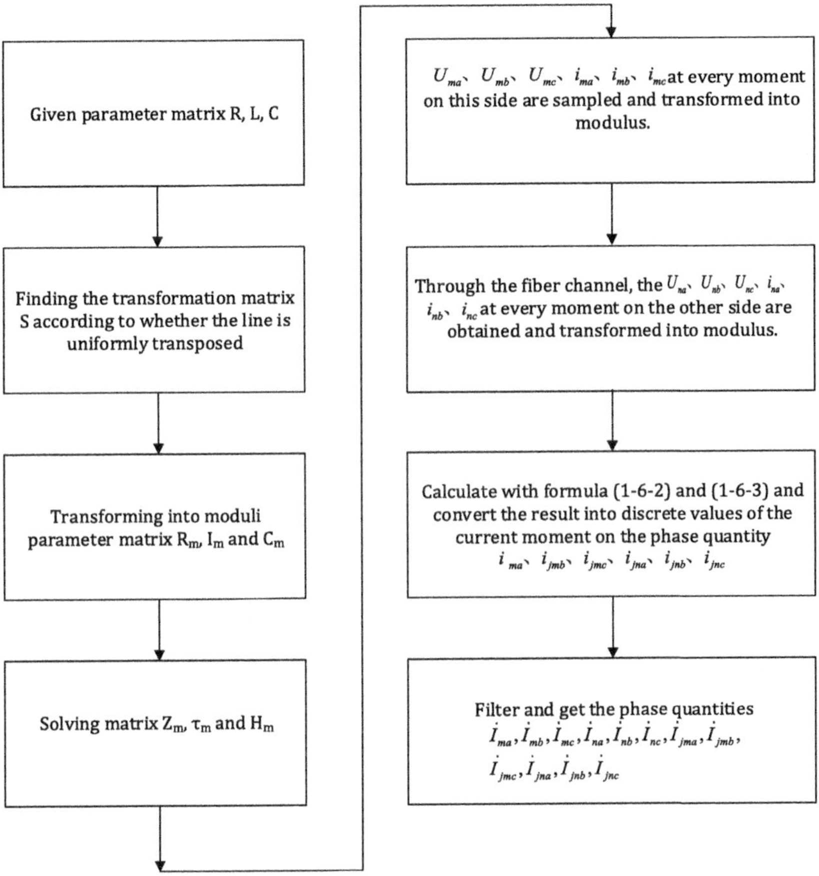

Figure 7.4 The procedure for calculating the currents of the two sides on the basis of the Bergeron model.

point of the shunt reactor is the calculating reference point k, as shown in Fig. 7.5. i_{mk}, i_{nk} are the module currents on both sides of the k point calculated by the voltage and current module components on both sides of transmission line; u_k is the voltage at the reactor mounting point; and shunt reactor current i_{sr} can be calculated by u_k/X_L. The positive direction of each current is defined as shown in Fig. 7.5.

The current differential protection of the line can be realized by judging whether the currents on the three sides of the reference point k meet Kirchhoff's current law. When there is no fault on the protected line, the Bergeron equation is established on the line between both the m and k points and the n and k points. Protection of the m side is derived by

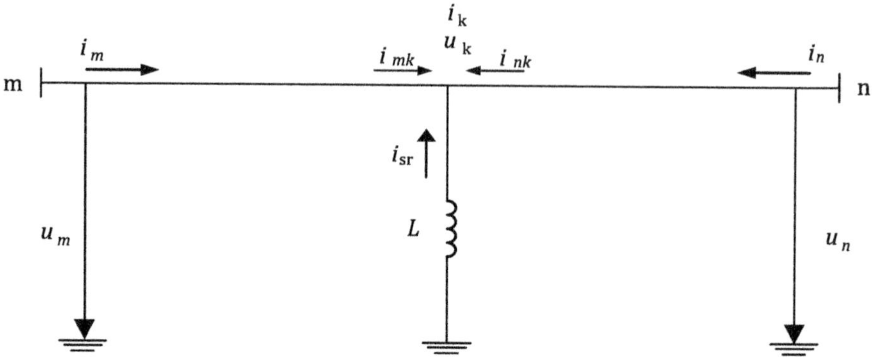

Figure 7.5 Bergeron equivalent circuit with a shunt reactor in the middle section of the line.

obtaining i_{mk}, u_k, and i_{sr} by converting the measured voltages and currents at both ends into modulus, and then getting i_{mkph} and i_{srph} by converting the above into phasors. In the same way, protection of the n side can be derived by obtaining i_{nkph} and transmitting it to the m-side protection through the communication channel. Using the half-wave Fourier algorithm, the current i_{mkph}, i_{srph}, and i_{nkph} can be filtered. And finally, the actuating quantity and operating criterion of segregated current differential protection on the m side can be obtained as follows [4]:

$$\left.\begin{aligned} dI_{\mathrm{ma}} &= \left|\dot{I}_{\mathrm{mka}} + \dot{I}_{\mathrm{nka}} + \dot{I}_{\mathrm{sra}}\right| \geq I_s \\ dI_{\mathrm{mb}} &= \left|\dot{I}_{\mathrm{mkb}} + \dot{I}_{\mathrm{nkb}} + \dot{I}_{\mathrm{srb}}\right| \geq I_s \\ dI_{\mathrm{mc}} &= \left|\dot{I}_{\mathrm{mkc}} + \dot{I}_{\mathrm{nkc}} + \dot{I}_{\mathrm{src}}\right| \geq I_s \end{aligned}\right\} \tag{7.32}$$

The criterion for segregated current differential protection on the n side is the same as that above. When there is no fault in the protected line (including when in steady-state operation and when with an external fault), the lines on both sides of the reference point may satisfy the Bergeron model. Theoretically, there exists $\dot{I}_{\mathrm{mkph}} + \dot{I}_{\mathrm{nkph}} + \dot{I}_{\mathrm{srph}} = 0$—namely, the actuating quantity of each phase is zero. Next, the actuating quantities of each phase when there is an internal fault on the line will be analyzed.

The calculation of the actuating quantity in Eq. (7.32) needs to calculate the current i_{sr} in the reactor—that is, the voltage u_k of the k point needs to be calculated. Taking a lossless line as an example, when the fault occurs at the f point on the line, the actuating quantity analysis at both ends of the line is shown in Fig. 7.6.

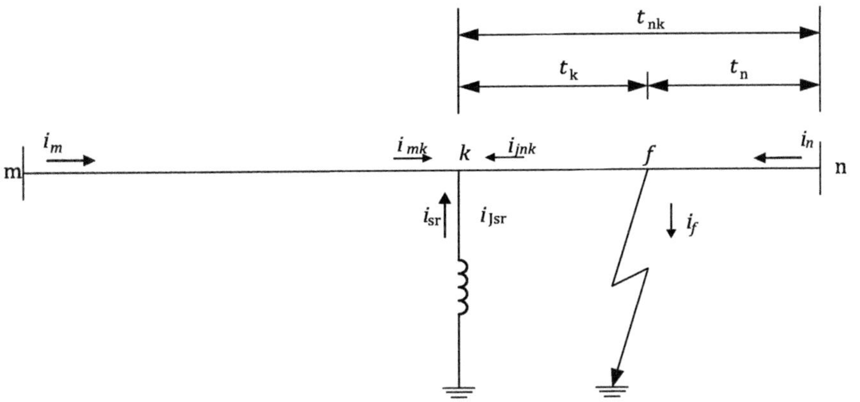

Figure 7.6 The time of the mode component waves transmitting from the fault point to each point during an internal fault.

For segregated current differential protection on the m side of the line, the Bergeron model for the line between the m point and k point has not been damaged, and therefore the real i_{mk} and $i_{sr}(u_k)$ can be calculated from the voltage and current of the m side. However, the Bergeron model for the line between the n and k points has been destroyed, so the real current i_{nk} cannot be obtained from the voltage and current of the n side. Using i_{Jnk} to represent the calculated current value (not real value) of the k point from the n side with the Bergeron method without considering fault, the instantaneous value of the actuating quantity in moduli for the differential protection on the m side of the line is

$$di_{\mathrm{m}}(t) = i_{\mathrm{mk}}(t) + i_{\mathrm{Jnk}}(t) + i_{\mathrm{sr}}(t) \tag{7.33}$$

Similarly, for segregated current differential protection on the n side of the line, because the real voltage i_{nk} and $i_{sr}(u_k)$ cannot be obtained from the voltage and current of the n side, i_{Jnk} and i_{Jsr} are used to represent the calculated current value from the n side to reference point k and the calculated current value in the reactor (not the real value). The calculated instantaneous value of the actuating quantity of n side differential protection on modulus is

$$di_{\mathrm{m}}(t) = i_{\mathrm{mk}}(t) + i_{\mathrm{Jnk}}(t) + i_{\mathrm{sr}}(t) \tag{7.34}$$

Thus, it can be seen that in the case of the line with a shunt compensating reactor, there is just the current i_{Jsr} in the reactor added to the actuating quantity, and i_{Jsr} can be calculated. Therefore, the analysis method of this protection is the same as that without a reactor. The simulation results prove the correctness of the principle. Specific analysis and simulation results are omitted here but can be found in Ref. [2].

References

[1] Guo Z. New principle of phase segregated current differential protection for UHV long transmission lines. Tianjin: Tianjin University; 2004 [in Chinese].

[2] He J, Guo Z. New principle of segregated phase current differential protection in long distance transmission line with series capacitor compensation. Relay 2005;33(1):1–9.

[3] Li B, He J, Guo Z, et al. Differential protection in transmission line with shunt reactor in stalled in the middle segment. Automation of Electric Power Systems 2006;30(7):31–6.

[4] Li B. The theories and techniques of protection for UHV long distance transmission line- new theories of protection based on exact line model. Tianjin: Tianjin University; 2006 [in Chinese].

CHAPTER 8

Distance protection

Contents

Distance protection is one of the most widely used protection principles in power system transmission lines. For extra-high-voltage (EHV)/ultra-high-voltage (UHV) transmission lines, distance protection is widely used as backup protection for the line. In some cases, pilot distance protection is also used as the main protection for EHV and UHV transmission lines.

At present, there are two main principles of distance protection successfully applied to transmission lines: (1) conventional distance protection, which reflects the fault distance by measured impedance and accordingly distinguishes internal and external faults; and (2) power frequency variation distance protection based on imposed fault components.

Conventional distance protection includes an interphase distance element and a grounding distance element. The line is equivalent to the concentrated impedance; under different fault types the measured impedance value of the impedance element is proportional to the fault distance, so line protection with a certain range of protection lengths can be realized. Distance protection has abundant operational experience resulting from the advantage that its protection range is basically fixed, and it is not affected by changes in the system operating mode, which can adopt a variety of different-acting characteristics according to line conditions and protection

Protection Technologies of Ultra-High-Voltage AC Transmission Systems
ISBN 978-0-12-816205-7
https://doi.org/10.1016/B978-0-12-816205-7.00008-X

objectives. However, the principle of distance protection is greatly affected by system oscillation, and the protection range may be shortened or extended when the line is in fault via transition resistance.

Distributed parameter characteristics of UHV long transmission lines make the measured impedance of conventional distance protection not directly proportional to fault distance. Therefore, the application of conventional distance protection in UHV transmission lines has great limitations. The operation characteristics of conventional distance protection and power frequency variation distance protection on UHV long transmission lines should be studied.

8.1 Influence of capacitance current

8.1.1 Conventional distance protection

When an interphase or a ground fault occurs on a transmission line, measured impedance Z_m of the interphase distance element, and the grounding distance element installed at the head of the line, can be respectively expressed as follows:

$$Z_m = \frac{\dot{U}_{ph-ph}}{\dot{I}_{ph-ph}} \tag{8.1}$$

$$Z_m = \frac{\dot{U}_{ph}}{\dot{I}_{ph} + K3\dot{I}_0} \tag{8.2}$$

In the above equation: $ph-ph$ means phase to phase, such as phase-A-to-phase-B, phase-B-to-phase-C, and phase-C-to-phase-A, where respectively, ph is phase A, B, or C; and $\dot{K} = \frac{Z_0 - Z_1}{3Z_1}$, wherein Z_0 and Z_1 are the zero-sequence and positive-sequence impedances of the unit length line, respectively. For general EHV and lower voltage level transmission lines, the line distribution parameter characteristics are not obvious, and the line is usually equivalent to a concentrated impedance consisting of a series of resistances and inductances. Therefore, the measured impedance calculated by Eqs. (8.1) and (8.2) is directly proportional to the fault distance, and this allows us to correctly distinguish between internal and external faults.

8.1.2 Performance of distance protection on ultra-high-voltage lines

When the fault passing through transition resistance R_g occurs at point f on the line, the distance protection's measured impedance changes due to the existence of the transition resistance, which may cause overrange action in the protection or shorten the protection range, resulting in misoperation or protection refusal-of-operation. This book is not detailed in this part, but the impact of UHV transmission line capacitive current on the distance protection may be specially analyzed. According to the superposition principle, the calculated current after faulting occurring on the line can be regarded as the superposition of the load current under normal operation circumstances and pure fault current components. Generally speaking, for unloaded line, consider its load current $I_{load} = 0$, that is, the influence of capacitive current can be neglected. But the capacitive current of UHV transmission line is very large. When there is a fault via transition resistance occurring on unloaded UHV transmission line, the short-circuit current is small, whose value can be closed to the capacitive current, which means that the capacitive current cannot be ignored. Therefore, it is of necessity to analyze the influence of capacitive current of UHV no-load long transmission line on the acting characteristics of distance protection.

8.1.2.1 Influence of capacitance current on grounding impedance elements

As shown in Fig. 8.1, the fault of phase A via transition resistance R_g occurs at point f on unloaded line, and the measured impedance is

$$Z_{ma} = \frac{\dot{U}}{\dot{I}_{ma} + K3\dot{I}_{mA0}} = Z_1 l_f + \frac{\dot{U}_{fA}}{\dot{I}_{mA} + K3\dot{I}_{mA0}} = Z_1 l_f + \Delta Z_A \tag{8.3}$$

$$\dot{I}_{mA} = \dot{I}_{load\cdot A} + \dot{I}_{mA1} + \dot{I}_{mA2} + \dot{I}_{mA0} = \dot{I}_{load\cdot A} + C_{1M}\dot{I}_{fA1} + C_{2M}\dot{I}_{fA2} + C_{0M}\dot{I}_{fA0}$$

In the above equation, $\dot{I}_{mA1}$, $\dot{I}_{mA2}$ and $\dot{I}_{mA0}\dot{I}_{fA1}$, $\dot{I}_{fA2}$, and $\dot{I}_{fA0}$ are respectively the positive-, negative-, and zero-sequence currents of phase A

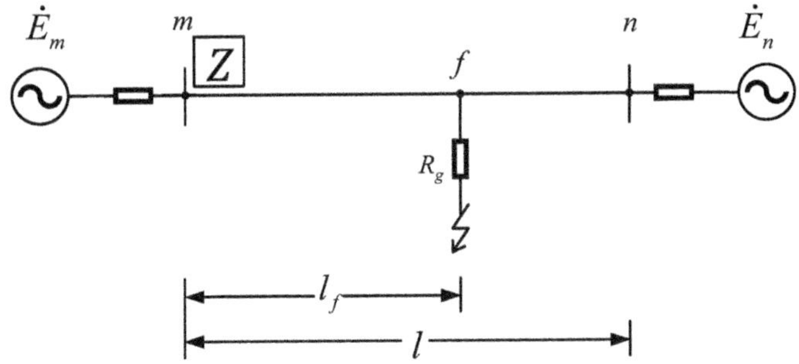

Figure 8.1 Fault diagram of two-terminal power transmission line.

at the protection installation point and the fault point on the m side. C_{1M}, C_{2M}, and C_{0M} are respectively the distribution coefficients of the positive-, negative-, and zero-sequence currents on the m side, and $C_{1M} = C_{2M}$.

To simplify the analysis, suppose KC_{1M}, C_{2M}, and C_{0M} are all real numbers. In a single-phase ground fault, $\dot{I}_{fA1} = \dot{I}_{fA2} = \dot{I}_{fA0}$, and the voltage drop on the transition resistance is $\dot{U}_f = 3\dot{I}_{fA0}R_g$. Thus the additional measured impedance can be obtained from Eq. (8.3) as follows:

$$\Delta Z_A = \frac{3R_g}{[2C_{1M} + (1+3K)C_{0M}] + \dfrac{\dot{I}_{load\cdot A}}{\dot{I}_{fA0}}} \tag{8.4}$$

For low-voltage lines or short EHV lines, I_{load} can generally be considered equal to 0—i.e., $I_{load} = 0$. Obviously, additional measured impedance ΔZ_A in this case appears to have a pure resistance characteristic. However, the capacitance current cannot be ignored on unloaded long UHV line, in which there is $\dot{I}_{load\cdot A} = \dot{I}_{cap\cdot A}$. Analyze the phase relation between the voltages and currents at the protection installation point and the fault point, as shown in Fig. 8.2.

In the figure, $\dot{U}_{fA[0]}$ is the voltage at point f under normal operation. Because the line is unloaded before the fault, $\dot{U}_{fA[0]}$ and $\dot{E}_{mA}$, $\dot{E}_{nA}$ are in phase. As the R_g value changes from 0 to infinity, $\dot{I}_{fA1}$ changes along the arc path. As shown in the figure, when a fault via R_g occurs, the short-circuit current $\dot{I}_{fA}$ is comparatively small and comparable to $\dot{I}_{cap\cdot A}$ in value. Obviously the capacitance current $\dot{I}_{cap\cdot A}$ may lead the phase of the resultant current phasor $\dot{I}_{mA} + K3\dot{I}_{mA0}$ to come close to or even exceed that of

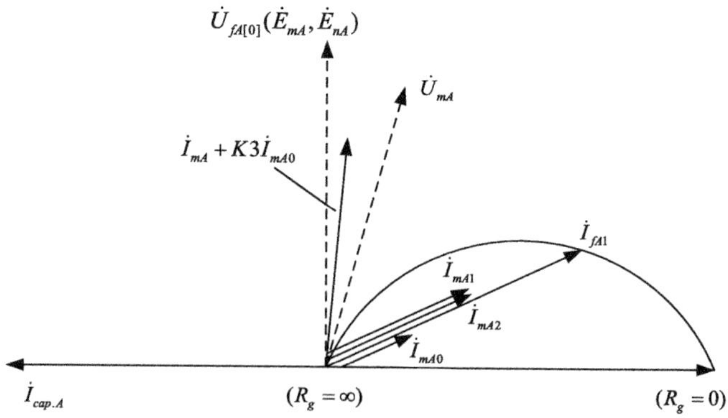

Figure 8.2 Voltage and current phase relation of the m side when an A phase grounding fault occurs.

voltage phasor $\dot{U}_{mA}$ so as to make the impedance angle of the measured impedance Z_m decrease or even fall into the fourth quadrant. From Eq. (8.4), it can also be obtained that since $\dot{I}_{cap\cdot A}$ is ahead of $\dot{I}_{fA0}$, the measured impedance ΔZ_A appears to be a capacitive characteristic. Therefore, when fault via transition resistance occurs outside the unloaded long-line area, the measured impedance may fall into the action zone due to the effect of the capacitor current, and the protection may be overreached; or the measured impedance may fall into the fourth quadrant when the fault occurs at the unloaded long-line exit. In serious cases, the measured impedance may fall outside the action area, and the protection may be rejected, as shown in Fig. 8.3.

8.1.2.2 Influence of capacitance current on interphase impedance element

When a fault via transition resistance R_g occurs between phase B and C at point f on unloaded line, the measured impedance of the interphase impedance element is as follows:

$$Z_{mBC} = \frac{\dot{U}_{mBC}}{\dot{I}_{mBC}} = Z_1 l_f + \frac{\dot{U}_{fBC}}{\dot{I}_{mBC}} = Z_1 l_f + \Delta Z_{BC} \tag{8.5}$$

Fault line voltage at fault point f is $\dot{U}_{fBC} = \dot{I}_{fB} R_g$. It can be determined from the fault-sequence network chart that $\dot{I}_{fB} = -j\sqrt{3}\dot{I}_{fA1}$, and the fault phase currents on the m side are respectively $\dot{I}_{mB} = \dot{I}_{load\cdot B} - j\sqrt{3}C_{1m}\dot{I}_{fA1}$,

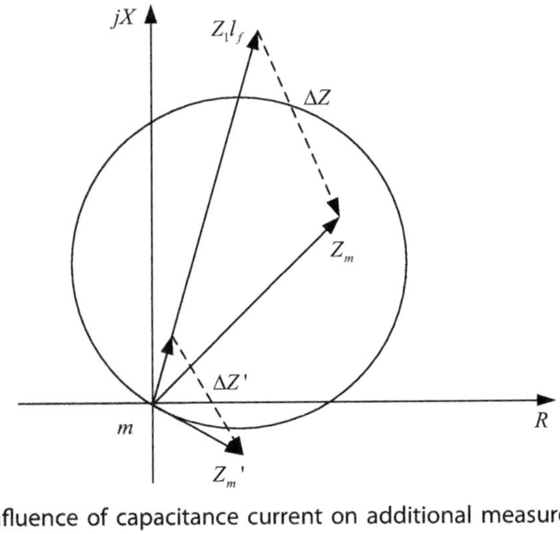

Figure 8.3 Influence of capacitance current on additional measured impedance.

$\dot{I}_{mC} = \dot{I}_{load\cdot C} - j\sqrt{3}C_{1m}\dot{I}_{fA1}$. Therefore, the measured impedance can be obtained as follows:

$$\Delta Z_B = \frac{\dot{U}_{fBC}}{\dot{I}_{mBC}} = \frac{R_g}{C_{1M} + j\dfrac{\dot{I}_{load\cdot BC}}{\sqrt{3}\dot{I}_{fA1}}} \tag{8.6}$$

For unloaded long transmission line, $\dot{I}_{load\cdot BC} = \dot{I}_{cap\cdot BC}$. When fault via transition resistance occurs on transmission line, the phase relation between the interphase capacitance current and the positive-sequence current at the fault point is analyzed as shown in Fig. 8.4.

It can be seen from Fig. 8.4 and Eq. (8.6) that $\dot{I}_{cap\cdot BC}$ is ahead of $\dot{I}_{fA1}$, and additional measured impedance ΔZ_{BC} appears to have capacitive characteristics—i.e., the influence of capacitance current on the interphase impedance element is the same as on the grounding impedance element. The analyses of two-phase short-circuit ground faults and three-phase short circuits are the same as above. Moreover, for the interphase impedance element, measured impedance is not affected by grounding transition resistance. At the same time, because transition resistance is generally small when an interphase fault occurs, and capacitive current $\dot{I}_{cap\cdot BC}$ is far less than fault short-circuit current $\dot{I}_{mBC}$, the influence of capacitance current on the interphase impedance element is relatively small.

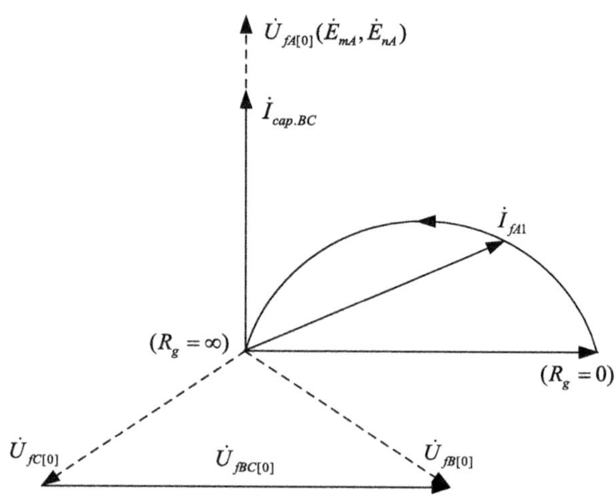

Figure 8.4 Phase relation of voltage and current during interphase fault between phases B and C.

8.2 Nonlinear measured impedance

8.2.1 Capacitance current

For UHV long-distance transmission lines, the distributed parameters characteristics of the transmission lines must be considered. From the differential equation describing the wave process of the transmission line, the relationship between voltage and current of the same sequence component at both ends of the line can be obtained as follows:

$$\begin{bmatrix} \dot{U}_1 \\ \dot{I}_1 \end{bmatrix} = \begin{bmatrix} ch\gamma l & z_c sh\gamma l \\ \dfrac{sh\gamma l}{z_c} & ch\gamma l \end{bmatrix} \begin{bmatrix} \dot{U}_2 \\ \dot{I}_2 \end{bmatrix} \tag{8.7}$$

In the equation, $\dot{U}_1$ and $\dot{I}_1$ are respectively voltage and current phasors at the head of the line; $\dot{U}_2$, $\dot{I}_2$ are respectively voltage and current phasors at the end of the line; z_c is the wave impedance of the line; and γ is the propagation constant of the line.

To increase the natural power of UHV transmission lines, wave impedance must be reduced; to reduce wave impedance, line inductance must be reduced and line capacitance must be increased. At the same time, the large number of bundled conductors on the UHV transmission line makes its capacitance large, which makes the capacitance current of the UHV transmission line very large. Taking the typical line parameters of EHV and UHV transmission lines as examples [1], the relationship between the capacitance current and its natural power current is calculated as shown in Table 1.1.

The capacitance current of EHV and UHV transmission lines is very large. For UHV transmission lines to achieve transregional or interbasin transmission, the transmission distance must be very long. For example, the capacitance current of a 600 km long 1150 kV UHV transmission line can reach 63.88% of the line's natural power current. It can be seen that the capacitance current of UHV long-distance transmission lines must not be ignored. Therefore, it is inaccurate to neglect line capacitance and equate the line with lumped impedance consisting of resistance and inductance in the theoretical analysis of UHV long-distance transmission lines, so is the protection principle based on this analysis.

8.2.2 Nonlinear measured impedance

The schematic diagram of a two-terminal power transmission line is shown as Fig. 8.5. In the figure, the distance protection is installed at the *m* side,

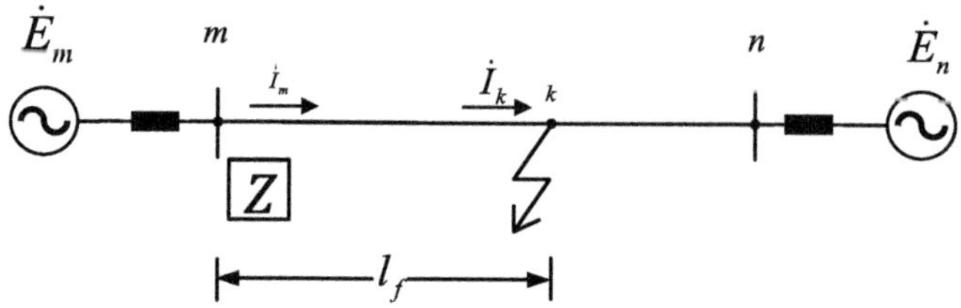

Figure 8.5 Schematic diagram of two-terminal power transmission line.

and the fault occurs at point k. $\dot{I}_m$ is the current of the head of the line, $\dot{I}_k$ is the m-side current flowing into the fault point, and l_f is the fault distance.

Considering the wave process of UHV long transmission lines, when there is a three-phase short circuit at the end of the long transmission line—i.e., $\dot{U}_2 = 0$—it can be known from Eq. (8.7) that the measured impedance measured at the protection installation point is as follows:

$$Z_m = \dot{U}_1/\dot{I}_1 = z_c \cdot th\gamma l \tag{8.8}$$

It can be seen that the measured impedance of the impedance element is a hyperbolic tangent rather than proportional function of the fault distance when a three-phase short circuit occurs on a long transmission line. Taking the parameters of a 1150 kV UHV transmission line as an example, the relationship between measured impedance at the protection installation and the line length is calculated as shown in Fig. 8.6.

Fig. 8.6 shows the resistance and inductance components of the measured impedance, where the solid line represents the relationship between the actual measured impedance and fault distance while the dotted line represents the relationship between the measured impedance

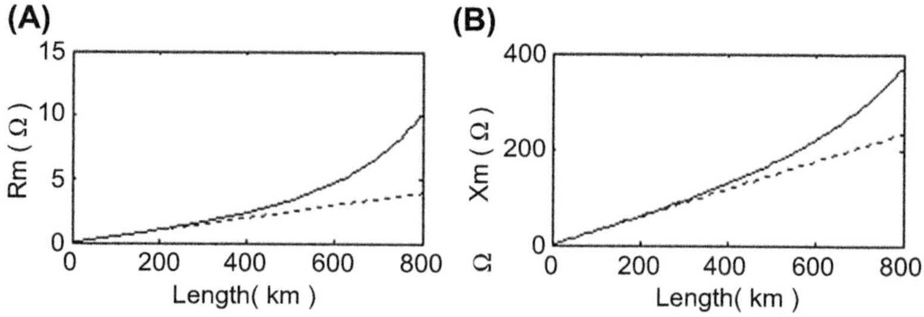

Figure 8.6 Relationship between measured impedance and line length during fault: (A) resistance; (B)reactance.

and distance when line impedance is linearly considered. It can be seen from the figure that measured impedance at the installation of distance protection is larger than line impedance considered linearly. Moreover, when the line is very long, a slight change of fault distance may greatly affect measured impedance. Considering transition resistance further, the error of the measured impedance of distance protection may be greater. For example, when a metallic fault occurs at the end of a 600 km long 1150 kV transmission line, the errors of the resistance and inductance components between the linearly set measured impedance and its actual measured impedance can reach 35.7% and 19.5% respectively. UHV lines are usually connected with shunt reactors, but in the transient process of faults, shunt reactors have very limited compensation ability to capacitance current, and this may be affected by fault location and many other factors. Therefore, the application of conventional distance protection to UHV long transmission lines is bound to have great errors.

8.3 Distance protection for long transmission lines

8.3.1 Measured impedance under phase-to-phase fault [2]

When an interphase fault or interphase ground fault occurs along the transmission line, after considering its distribution parameter characteristics, the A—B phase ground fault is taken as an example to analyze and deduce. (P.S. there is no zero-sequence component when no interphase ground fault occurs, in which case the conclusion is consistent with the following deduction.) As $\dot{U}_{kAB} = 0$, it is assumed that the parameters of the positive and negative sequences are the same, $\gamma_1 = \gamma_2$, $z_{c1} = z_{c2}$. The voltage and current at the protection installation can be obtained by Eq. (8.7):

$$\begin{aligned}\dot{U}_{mAB} &= \dot{U}_{kA1}\cdot ch\gamma_1 l_f + \dot{I}_{kA1}\cdot z_{c1}\cdot sh\gamma_1 l_f + \dot{U}_{kA2}\cdot ch\gamma_2 l_f + \dot{I}_{kA2}\cdot z_{c2}\cdot sh\gamma_2 l_f \\ &\quad + \dot{U}_{kA0}\cdot ch\gamma_0 l_f + \dot{I}_{kA0}\cdot z_{c0}\cdot sh\gamma_0 l_f - \dot{U}_{kB1}\cdot ch\gamma_1 l_f - \dot{I}_{kB1}\cdot z_{c1}\cdot sh\gamma_1 l_f \\ &\quad - \dot{U}_{kB2}\cdot ch\gamma_2 l_f - \dot{I}_{kB2}\cdot z_{c2}\cdot sh\gamma_2 l_f - \dot{U}_{kB0}\cdot ch\gamma_0 l_f - \dot{I}_{kB0}\cdot z_{c0}\cdot sh\gamma_0 l_f \\ &= \dot{U}_{kAB1}\cdot ch\gamma_1 l_f + \dot{I}_{kAB1}\cdot z_{c1}\cdot sh\gamma_1 l_f + \dot{U}_{kAB2}\cdot ch\gamma_2 l_f + \dot{I}_{kAB2}\cdot z_{c2}\cdot sh\gamma_2 l_f \\ &= \dot{I}_{kAB}\cdot z_{c1}\cdot sh\gamma_1 l_f\end{aligned} \tag{8.9}$$

$$
\begin{aligned}
\dot{I}_{mAB} &= \dot{U}_{kA1}\cdot\frac{sh\gamma_1 l_f}{z_{c1}} + \dot{I}_{kA1}\cdot ch\gamma_1 l_f + \dot{U}_{kA2}\cdot\frac{sh\gamma_2 l_f}{z_{c2}} + \dot{I}_{kA2}\cdot ch\gamma_2 l_f \\
&\quad + \dot{U}_{kA0}\cdot\frac{sh\gamma_0 l_f}{z_{c0}} + \dot{I}_{kA0}\cdot ch\gamma_0 l_f - \dot{U}_{kB1}\cdot\frac{sh\gamma_1 l_f}{z_{c1}} - \dot{I}_{kB1}\cdot ch\gamma_1 l_f \\
&\quad - \dot{U}_{kB2}\cdot\frac{sh\gamma_2 l_f}{z_{c2}} - \dot{I}_{kB2}\cdot ch\gamma_2 l_f - \dot{U}_{kB0}\cdot\frac{sh\gamma_0 l_f}{z_{c0}} - \dot{I}_{kB0}\cdot ch\gamma_0 l_f \\
&= \dot{U}_{kAB1}\cdot\frac{sh\gamma_1 l_f}{z_{c1}} + \dot{I}_{kAB1}\cdot ch\gamma_1 l_f + \dot{U}_{kAB2}\cdot\frac{sh\gamma_2 l_f}{z_{c2}} + \dot{I}_{kAB2}\cdot ch\gamma_2 l_f \\
&= \dot{I}_{kAB}\cdot ch\gamma_1 l_f
\end{aligned}
\tag{8.10}
$$

Therefore, the measured impedance of the interphase impedance element is

$$
Z_m = \frac{\dot{U}_{mAB}}{\dot{I}_{mAB}} = z_{c1}\cdot th\gamma_1 l_f \tag{8.11}
$$

It can be seen that the measured impedance of the interphase fault is the same as that of the three-phase short-circuit fault analyzed in Eq. (8.8), and the measured impedance is a hyperbolic tangent function of the fault distance rather than proportional to the fault distance. Therefore, the impedance setting calculation of the interphase distance elements must be set according to the hyperbolic tangent function of the distance.

8.3.2 Measured impedance under phase-to-ground fault [2]

When a single-phase ground fault occurs along the transmission line, considering the distributed parameter characteristics of the line and taking an A-phase ground fault as an example, since $\dot{U}_{kA} = \dot{U}_{kA1} + \dot{U}_{kA2} + \dot{U}_{kA0} = 0$, it is assumed that $\gamma_1 = \gamma_2$ and $z_{c1} = z_{c2}$. Therefore, from Eq. (8.7), voltage and current at the protection installation can be obtained as follows:

$$
\begin{aligned}
\dot{U}_{mA} &= \dot{U}_{kA1}\cdot ch\gamma_1 l_f + \dot{I}_{kA1}\cdot z_{c1}\cdot sh\gamma_1 l_f + \dot{U}_{kA2}\cdot ch\gamma_2 l_f \\
&\quad + \dot{I}_{kA2}\cdot z_{c2}\cdot sh\gamma_2 l_f + \dot{U}_{kA0}\cdot ch\gamma_0 l_f + \dot{I}_{kA0}\cdot z_{c0}\cdot sh\gamma_0 l_f \\
&= \dot{I}_{kA}\cdot z_{c1}\cdot sh\gamma_1 l_f + \dot{U}_{kA0}\cdot ch\gamma_0 l_f + \dot{I}_{kA0}\cdot z_{c0}\cdot sh\gamma_0 l_f \\
&\quad - \dot{U}_{kA0}\cdot ch\gamma_1 l_f - \dot{I}_{kA0}\cdot z_{c1}\cdot sh\gamma_1 l_f
\end{aligned}
\tag{8.12}
$$

It is known from Eq. (8.7) that $\dot{U}_{kA0}\cdot ch\gamma_0 l_f + \dot{I}_{kA0}\cdot z_{c0}\cdot sh\gamma_0 l_f = \dot{U}_{mA0}$. Let $\dot{U}_{kA0}\cdot ch\gamma_1 l_f + \dot{I}_{kA0}\cdot z_{c1}\cdot sh\gamma_0 l_f = \dot{U}'_{mA0}$, Eq. (8.12) can be changed as follows:

$$\dot{U}_{mA} = \dot{I}_{kA}\cdot z_{c1}\cdot sh\gamma_1 l_f + \dot{U}_{mA0} - \dot{U}'_{mA0} \tag{8.13}$$

Similarly, the fault phase current of the protective installation can be obtained as follows:

$$\begin{aligned}
\dot{I}_{mA} &= \dot{U}_{kA1}\cdot\frac{sh\gamma_1 l_f}{z_{c1}} + \dot{I}_{kA1}\cdot ch\gamma_1 l_f + \dot{U}_{kA2}\cdot\frac{sh\gamma_2 l_f}{z_{c2}} \\
&\quad + \dot{I}_{kA2}\cdot ch\gamma_2 l_f + \dot{U}_{kA0}\cdot\frac{sh\gamma_0 l_f}{z_{c0}} + \dot{I}_{kA0}\cdot ch\gamma_0 l_f \\
&= \left(\dot{U}_{kA1} + \dot{U}_{kA2} + \dot{U}_{kA0}\right)\cdot\frac{sh\gamma_1 l_f}{z_{c1}} + \left(\dot{I}_{kA1} + \dot{I}_{kA2} + \dot{I}_{kA0}\right)\cdot ch\gamma_1 l_f \\
&\quad + \dot{U}_{kA0}\cdot\frac{sh\gamma_0 l_f}{z_{c0}} + \dot{I}_{kA0}\cdot ch\gamma_0 l_f - \dot{U}_{kA0}\cdot\frac{sh\gamma_1 l_f}{z_{c1}} - \dot{I}_{kA0}\cdot ch\gamma_1 l_f \\
&= \dot{I}_{kA}\cdot ch\gamma_1 l_f + \dot{I}_{mA0} - \dot{I}'_{mA0}
\end{aligned} \tag{8.14}$$

In Eqs. (8.13) and (8.14), $\dot{U}_{mA0}$ and $\dot{I}_{mA0}$ are respectively the zero-sequence components obtained at the protection installation—i.e., at the head of the line. While $\dot{U}'_{mA0}$ and $\dot{I}'_{mA0}$ are respectively the zero-sequence components at the head of the line obtained by the line's positive-sequence two-port network calculation according to Eq. (8.7) from the fault point zero-sequence components $\dot{U}_{kA0}$ and $\dot{I}_{kA0}$. Fault point zero-sequence $\dot{U}_{kA0}$ can be represented as follows by the zero-sequence voltage and current $\dot{U}_{mA0}$ and $\dot{I}_{mA0}$, which can be measured at the head of the line:

$$\dot{U}_{kA0} = \dot{U}_{mA0}\cdot ch\gamma_0 l_f - \dot{I}_{mA0}\cdot z_{c0}\cdot sh\gamma_0 l_f = \dot{U}'_{mA0}\cdot ch\gamma_1 l_f - \dot{I}'_{mA0}\cdot z_{c1}\cdot sh\gamma_1 l_f \tag{8.15}$$

So the relationship between $\dot{U}_{mA0}$, $\dot{I}_{mA0}$ and $\dot{U}'_{mA0}$, $\dot{I}'_{mA0}$ can be represented as follows:

$$ch\gamma_1 l_f\left[\left(\dot{U}_{mA0} - \dot{U}'_{mA0}\right) + K_u\dot{U}_{mA0}\right] = z_{c1}\cdot sh\gamma_1 l_f\left[\left(\dot{I}_{mA0} - \dot{I}'_{mA0}\right) + K_i\dot{I}_{mA0}\right] \tag{8.16}$$

In the above equation, K_i and K_u are respectively the compensation coefficients of the zero-sequence current and voltage, whose values are $K_i = \frac{z_{c0} \cdot sh\gamma_0 l_f - z_{c1} \cdot sh\gamma_1 l_f}{z_{c1} \cdot sh\gamma_1 l_f}$ and $K_u = \frac{ch\gamma_0 l_f - ch\gamma_1 l_f}{ch\gamma_1 l_f}$.

Combining Eqs. (8.13), (8.14), and (8.16), the following equation can be obtained:

$$\begin{aligned} & ch\gamma_1 l_f\left(\dot{U}_{mA} + K_u \dot{U}_{mA0}\right) \\ & = ch\gamma_1 l_f\left[\dot{I}_{kA} \cdot z_{c1} \cdot sh\gamma_1 l_f + \left(\dot{U}_{mA0} - \dot{U}'_{mA0}\right) + K_u \dot{U}_{mA0}\right] \\ & = z_{c1} \cdot sh\gamma_1 l_f\left[\dot{I}_{kA} \cdot ch\gamma_1 l_f + \left(\dot{I}_{mA0} - \dot{I}'_{mA0}\right) + K_i \dot{I}_{mA0}\right] \\ & = z_{c1} \cdot sh\gamma_1 l_f\left[\dot{I}_{mA} + K_i \dot{I}_{mA0}\right] \end{aligned} \tag{8.17}$$

So, in a single-phase ground fault, the relationship between voltage and current is

$$\frac{\dot{U}_{mA} + K_u \dot{U}_{mA0}}{\dot{I}_{mA} + K_i \dot{I}_{mA0}} = z_{c1} \cdot th\gamma_1 l_f \tag{8.18}$$

Obviously, Eq. (8.18) shows that Eq. (8.2), which is the calculation for the conventional ground impedance element, cannot be used to calculate the measured impedance of the ground impedance relay. The theoretical basis of Eq. (8.2) considers the line equivalent to a concentrated impedance. Therefore, when a conventional ground impedance element is applied to a UHV long transmission line, there is no definite functional relationship between its measured impedance and the fault distance, and there is no strict theoretical basis at all.

The measured impedance of the ground impedance element represented by Eq. (8.18) is a hyperbolic tangent function of the fault distance that is the same as that of the interphase fault. But it should be noted that the compensation coefficient of zero-sequence current and voltage varies with a change in fault distance. Taking typical UHV transmission lines as an example, the variation curves of K_u and K_i with fault distance are shown in Fig. 8.7.

In the actual protection calculation, K_u and K_i must be substituted into Eq. (8.18) as the known quantities to solve for the measured impedance. Therefore, the approximate values of K_u and K_i can be substituted into Eq. (8.18) for the calculation, and the approximate value can be calculated as follows:

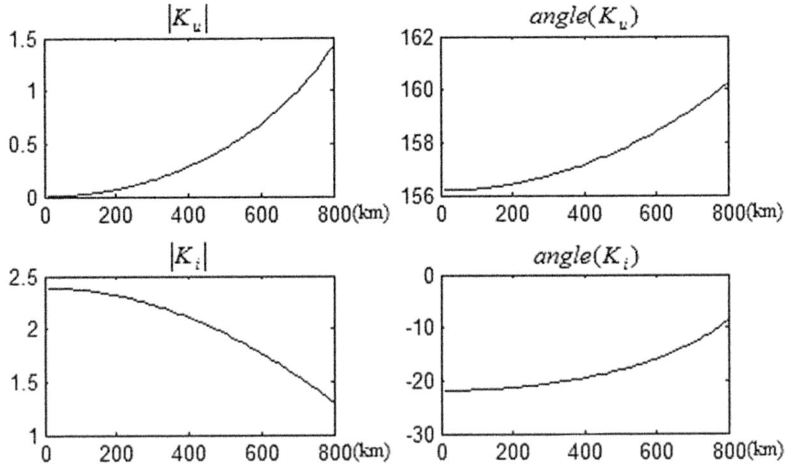

Figure 8.7 The variation curves of K_u and K_i with fault distance.

(1) K_u and K_i are set as the average values across the whole length of the protected line. This ensures that the absolute value of the maximum error of the measured impedance obtained by Eq. (8.18) is smallest when the fault occurs at different locations of the line.

(2) According to the line length of the protected range, values for K_u and K_i are taken when the fault occurs at the end of the protection range. This is because distance protection has a large margin when the near-end fault occurs, and the key problem is whether the calculation accuracy of the distance protection can be ensured when the fault occurs at the end of the protection range. The simulation results in the following are obtained by taking K_u and K_i as the corresponding values when the fault occurs at the end of the protection range.

(3) When the fault occurs, the fault distance should be estimated for approximating the values of K_u and K_i accordingly, which ensures that the calculation result of Eq. (8.18) is close to the actual line impedance value when the fault occurs at different locations.

Eq. (8.18) is an accurate equation for calculating the measured impedance of ground impedance, but there are many approximate methods for determining K_u and K_i. These methods can be selected according to the actual situation of the transmission lines as long as they can meet the calculation accuracy of distance protection.

In addition, it should be noted that shunt reactors are usually connected to UHV transmission lines. In the case of shunt reactors installed at both ends of the line, the influence of the shunt reactor on distance protection can be eliminated by subtracting the ground current of the shunt reactor from the current flowing through the head of the line. In the case of a shunt reactor installed in the middle of the line, which side of the shunt reactor installation point contains the fault can be determined by calculation. If the fault occurs between the protection installation and shunt reactor, the calculation of distance protection may not be affected by the reactor. And if the fault occurs between the shunt reactor and the opposite bus, voltage and current at the installation point of the shunt reactor can be calculated using the long-line equation or Bergeron equation, and thus the influence of the reactor on the ground current can also be eliminated.

8.3.3 Case studies

(1) Simulation verification

The model of simulation verification is a 600 km long 1150 kV UHV transmission line, shown in Fig. 8.8:

The fault occurred at 40 ms. To clearly compare the relationship between the measured impedance and the fault distance in the following simulation results, the calculated values of the measured impedance and the fault distance for different functional relationships are listed in Table 8.1.

(2) Analysis of simulation results

(a) When three-phase metallic faults occur on UHV long transmission lines, it can be seen from Table 8.2 that the measured impedance at the head of the line is not directly proportional to the fault distance but rather a hyperbolic tangent function with the distance. If the distance protection still uses the conventional linearization setting, then the measured impedance may be greater than the fixed value

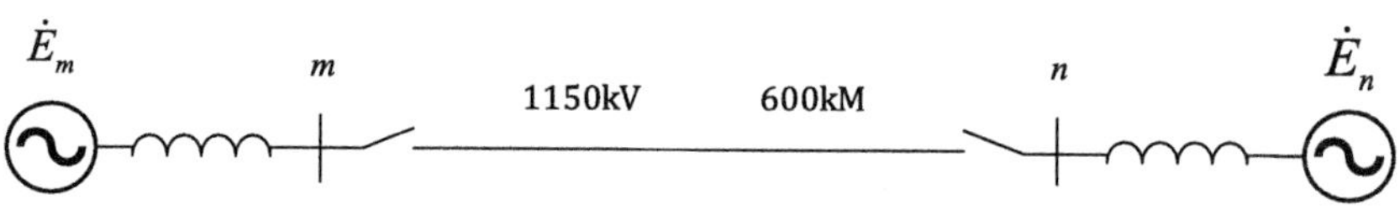

$$E_m = 1.027U_n,\ E_n = 1.087U_n,\ Z_m = 0.1Z_L,\ Z_n = 0.5Z_L$$
$$R_1 = 0.0049\Omega/km,\ L_1 = 0.930372mH/km,\ C_1 = 0.01764\mu F/km$$
$$R_0 = 0.2752\Omega/km,\ L_0 = 2.977197mH/km,\ C_0 = 0.0159\mu F/km$$

Figure 8.8 Simulation model of 1150 kV transmission line.

Table 8.1 Relationship between measured impedance and fault distance (linear function and hyperbolic tangent function).

Functional relationship	Measured impedance Z_m (Ω)				
	$l_f = 5\%\ l$ (degrees)	$l_f = 50\%\ l$ (degrees)	$l_f = 70\%\ l$ (degrees)	$l_f = 80\%\ l$ (degrees)	$l_f = 95\%\ l$ (degrees)
$Z_1 l$	$8.77\angle 89.0$	$87.65\angle 89.0$	$122.7\angle 89.0$	$140.3\angle 89.0$	$166.5\angle 89.0$
$z_c th\gamma l$	$8.78\angle 89.0$	$92.17\angle 88.99$	$135.9\angle 88.9$	$160.75\angle 88.9$	$203.55\angle 88.8$

Table 8.2 Measured impedance of distance protection when a three-phase metallic ground fault occurs along the transmission line.

Moment (ms)	Measured impedance Z_a (Ω)				
	$l_f = 5\%\ l$ (degrees)	$l_f = 50\%\ l$ (degrees)	$l_f = 70\%\ l$ (degrees)	$l_f = 80\%\ l$ (degrees)	$l_f = 95\%\ l$ (degrees)
60	$8.80\angle 88.9$	$92.98\angle 88.1$ degree	$151.23\angle 89.8$	$156.87\angle 89.8$	$194.53\angle 90.1$ degree
65	$8.58\angle 88.5$	$89.90\angle 88.3$	$148.06\angle 91.9$	$158.67\angle 91.8$	$194.71\angle 87.7$
70	$8.69\angle 90.3$	$92.18\angle 89.2$	$148.71\angle 92.6$	$169.25\angle 90.6$	$207.05\angle 87.4$
75	$8.98\angle 89.3$	$94.40\angle 89.0$	$144.74\angle 91.4$	$168.36\angle 92.1$ degree	$208.29\angle 88.6$
80	$8.84\angle 87.7$	$93.23\angle 89.2$	$142.71\angle 91.0$	$171.07\angle 89.1$	$206.67\angle 88.8$

when the internal fault occurs, so as to make the protection refuse to operate. Therefore, the setting of distance protection should be calculated by a hyperbolic function.

(b) When two-phase metallic faults occur on UHV long transmission lines, it can be seen from Table 8.3 that the measured impedance of the interphase impedance elements is a hyperbolic tangent function of distance, which is the same as when three-phase faults occur.

(c) Table 8.4 lists the calculation results of Eq. (8.18) when a single-phase ground fault occurs on a UHV long transmission line in which the value of zero-sequence voltage and current compensation coefficients K_u and K_i correspond to a fault occurring at 85% of total line length. As can be seen from Table 8.4, the measured impedance near the end of the protection range calculated by Eq. (8.18) is a hyperbolic tangent function of the fault distance, consistent with theoretical analysis. When a near fault occurs along the line, the calculation error is relatively large, which is caused by the values of K_u and K_i. However, when a near fault occurs, distance protection has enough margin to ensure that protection can operate reliably.

(d) As can be seen from Table 8.5, when conventional grounding impedance elements represented by Eq. (8.2) are applied to a UHV long transmission line, measured impedance cannot accurately reflect the functional relationship between impedance and fault distance. Moreover, when the fault occurs near the end of the protection range, measured impedance falls into the second quadrant with a larger angle, which may affect the selection of the distance protection action boundary. In addition, a large number of simulation tests show that when the line runs in different ways, such as unloaded or overloaded lines when a single-phase ground fault occurs, calculated measured impedances from Eq. (8.2) are very different for the same fault point running in different ways. If the value is set according to a hyperbolic function, it may happen that the protection overreach in unloaded line fault will refuse to operate in an overloaded line fault. The reason for this error is that the measured impedance of Eq. (8.2) for conventional grounding impedance elements cannot objectively and accurately reflect the functional relationship with the fault distance.

Table 8.3 Measured impedance of distance protection when an A–B interphase metallic fault occurs along the transmission line.

Moment(ms)	Measured impedance Z_{ab} (Ω)				
	$l_f = 5\%\ l$ (degrees)	$l_f = 50\%\ l$ (degrees)	$l_f = 70\%\ l$ (degrees)	$l_f = 80\%\ l$ (degrees)	$l_f = 95\%\ l$ (degrees)
60	$8.77\angle 87.0$	$91.64\angle 87.5$	$145.33\angle 87.4$	$157.57\angle 89.1$	$197.75\angle 89.7$
65	$8.51\angle 89.0$	$88.96\angle 88.9$	$143.16\angle 89.5$	$159.83\angle 91.5$	$192.47\angle 89.2$
70	$8.76\angle 90.7$	$92.29\angle 90.0$	$147.77\angle 89.9$	$166.00\angle 90.9$	$203.62\angle 88.5$
75	$9.02\angle 88.7$	$94.62\angle 88.5$	$149.40\angle 88.6$	$170.42\angle 88.0$	$209.07\angle 87.7$
80	$8.77\angle 87.4$	$92.60\angle 87.1$	$144.90\angle 88.0$	$163.09\angle 86.5$	$206.88\angle 87.3$

Table 8.4 Measured impedance calculated by Eq. (8.18) when an A phase metallic ground fault occurs along the transmission line.

Moment (ms)	**Measured impedance Z_a (Ω)**				
	$l_f = 5\%$ l (degrees)	**$l_f = 50\%$ l (degrees)**	**$l_f = 70\%$ l (degrees)**	**$l_f = 80\%$ l (degrees)**	**$l_f = 95\%$ l (degrees)**
60	$13.15\angle 65.6$	$93.79\angle 76.5$	$136.95\angle 81.1$	$169.22\angle 81.9$	$208.24\angle 88.2$
65	$12.39\angle 73.9$	$88.29\angle 81.6$	$130.78\angle 82.3$	$160.46\angle 86.4$	$196.99\angle 89.5$
70	$14.99\angle 76.8$	$94.25\angle 87.0$	$132.37\angle 87.6$	$164.35\angle 88.5$	$200.81\angle 91.5$
75	$13.87\angle 69.7$	$97.45\angle 80.7$	$141.57\angle 83.3$	$169.00\angle 87.4$	$209.81\angle 89.5$
80	$13.61\angle 72.6$	$94.74\angle 81.7$	$135.51\angle 84.3$	$164.51\angle 85.9$	$205.34\angle 89.7$

Table 8.5 Measured impedance calculated by Eq. (8.2) when an A phase metallic ground fault occurs on transmission line.

	Measured impedance Z_a (Ω)				
Moment (ms)	$I_f = 5\%\ I$ (degrees)	$I_f = 50\%\ I$ (degrees)	$I_f = 70\%\ I$ (degrees)	$I_f = 80\%\ I$ (degrees)	$I_f = 95\%\ I$ (degrees)
60	9.1∠89.2	90.3∠91.9	123.2∠94.4	140.7∠95.6	157.8∠97.1
65	8.7∠87.4	88.2∠90.9	117∠93.5	135.7∠95.4	155∠96.7
70	8.3∠89.1	86.7∠90.6	119.9∠92.9	137.3∠93.5	150.2∠94.2
75	8.7∠89.8	87.5∠90.9	121.9∠93.7	137.9∠95.5	156.6∠97.5
80	8.9∠89.1	87.9∠90.9	121.5∠94.1	140∠96.2	156.2∠96.6

References

[1] Que B-V, Khodabakhchian B, Landry M, Mahseredjian J, Mainville J. Performance of series-compensated line circuit breakers under delayed current-zero condition. IEEE Transactions on Power Delivery 1997;12(1):227–33.

[2] Li B, He J, Yang H, Huang S, Qing Y, Xu Z. Improvement of distance protection algorithm of UHV long transmission line. Automation of Electric Power Systems January 2007;31(1):43–6 [In Chinese].

CHAPTER 9

Overvoltage

Contents

9.1 Overview

9.1.1 Causes and effects of overvoltage

Power system overvoltage can be divided into two categories, internal and external. External overvoltage, also known as atmospheric overvoltage, is caused by lightning. Depending on the cause, internal overvoltage can be divided into switching and temporary overvoltage. Switching overvoltage is overvoltage during the electromagnetic transition process caused by fault tripping, reclosing, or normal operation. Temporary overvoltage includes resonance overvoltage and power-frequency overvoltage. In general, the overvoltage with a duration of less than 0.1 s is called switching overvoltage, while that with a longer duration is called temporary overvoltage. However, this classification is not clear, because temporary and transient are easily confusing. In fact, judging from the harmfulness of overvoltage to insulation, one aspect is the size of its value and the other is the duration of its action. Switching overvoltage generally has a high value and short acting duration, which is an electromagnetic transient process (including wave process) caused by a sudden change in system energy, so it should be called

Protection Technologies of Ultra-High-Voltage AC Transmission Systems
ISBN 978-0-12-816205-7
https://doi.org/10.1016/B978-0-12-816205-7.00009-1

transient overvoltage and measured by its maximum peak value. As for power-frequency and resonance overvoltage, they generally operate for a longer time and are the overvoltages when electromagnetic energy conversion reaches or approaches the steady state. They should be called steady-state or short-time overvoltage, measured by their amplitudes and acting durations. No matter how they are named, the difference depends mainly on the size of the value and the time of its action. The effect of lightning overvoltage has no difference in either UHV or EHV transmission lines, but internal overvoltage of UHV transmission lines is much more serious than that of EHV transmission lines, which is the key problem in the design and operation of UHV transmission lines.

Switching overvoltage is a transient process caused by the operation of circuit breakers and disconnecting switches, or by system faults. It includes both the normal operation of circuit breakers, such as overvoltage of the transient process caused by the closing of lines, transformers, and reactors, and overvoltage caused by removing and reclosing tripping or faults for various reasons. Switching overvoltage has the characteristics of high peak value, high frequency oscillation, and short duration. The influence of switching overvoltage on insulation and protection devices of electrical equipment mainly depends on its peak value, waveform, steepness, and duration. The head steepness of switching overvoltage is generally lower than that of lightning overvoltage.

Overvoltage is the decisive factor in the design of UHV transmission lines. Compared with a 500 kV line, a UHV transmission line has larger distributed capacitance and smaller wave impedance as well as a relatively smaller ratio of system short-circuit capacity to the natural power of the line (i.e., system capacity is relatively small while system impedance is relatively large). These characteristics, mentioned above, make the switching overvoltage of UHV lines more serious than that of 500 kV lines. According to the results of the Soviet study, if the maximum allowable overvoltage of a UHV line is set to 2.5 times as that of a 500 kV line, the cost of line construction may be greatly increased. If allowable overvoltage time is reduced by 0.1%, the construction cost of a UHV transmission line may be reduced by 0.5%—1%, and the cost of the substation (including equipment cost) may be reduced by 0.5%—2.5%. The range of variation depends on the allowed minimum distance from the conductor to the ground and the allowed maximum electric field intensity on the ground. In addition, the decrease of the minimum distance contributes to reducing the height of the pole and tower, however, an increase in the cross-sectional area of the phase bundled conductor and the bus of the outdoor substation is necessary to reduce surface electric field intensity and the corona, which may increase the cost again. Therefore, a comprehensive economic comparison must be conducted.

According to technical and economic analysis, under the conditions of social and economic development (1975–85), the Soviet Union took the maximum allowable transient overvoltage time for 750 kV transmission lines as 2.1 and the maximum allowable transient overvoltage time of 1150 kV transmission lines as 1.8.

9.1.2 Characteristics of overvoltage

Overvoltage is a form of electromagnetic process generated by structural changes of the power system. The development of electromagnetic processes with long-distance transmission lines can be divided into three stages as follows.

(1) Wave process. Wave process is the process of a travelling wave propagating and reflecting along transmission lines. This process is very short, just a few milliseconds. When a line is under conditions such as no-load closing from one end, three phases suddenly broken on one end, in fault, fault isolation, or reclosing, the sudden change of energy on the transmission line may produce a wave process resulting in high transient overvoltage that is called a switching surge. Switching surges are usually represented by the ratio of the maximum spike voltage to the rated voltage amplitude or per-unit (p.u.) value k, or characterized by the impact overvoltage coefficient K_{imp} defined as the ratio of the spike voltage to the steady-state voltage amplitude of the operating process and the time of reaching the spike voltage.

(2) Electromagnetic transient. The electromagnetic transient is the period from the decay of the travelling wave to the arrival of the steady state, which is about 100 ms to 1 s. This period is mainly the gradual decay process of the nonperiodic components produced by the release of magnetic energy in inductance and the high-frequency components produced by capacitance charging and discharging. The low-frequency signals below the fundamental frequency, the second and third harmonics generated during this period may produce very high transient overvoltage after superimposing with the fundamental wave. This overvoltage is also characterized by overvoltage multiple K.

(3) Steady-state process. After completion of the above process, the steady-state process is entered, which is mainly reflected as the increase in power-frequency voltage, but there are also steady-state low-order harmonics caused by transformer saturation or ferroresonance. The overvoltage multiple during this period is generally small, however the damage may still be heavy due to the long duration.

Theoretically, various operations, faults, protection trippings, and reclosings on the line may lead to the above three processes, all of which should be considered to ensure line insulation and equipment safety.

Figs. 9.1–9.3 show the measured waveform of the no-load closing overvoltage of a 765 kV transmission line 328 km long between Dumont substation and Jefferson substation of the AEP power company [1]. It can be clearly seen from Fig. 9.1 that the period of the step-shaped waveform caused by the traveling-wave reflection is 2.2 ms, which is approximately equal to the time from the departure of the traveling wave to its reflection back to the starting point [(328 × 2)/(3 × 105) = 2.14 ms]. The influence of harmonics in the second period, i.e., the electromagnetic transient can be seen from Fig. 9.2. Due to the superposition of fundamental and harmonic waves, the peak value of voltage reaches 1.15 ~ 1.30 p.u. From Fig. 9.3, we can see the waveforms of voltage and current in no-load closing, where the persistent third harmonics and fundamental wave superposition caused by transformer saturation result in permanent steady-state overvoltage. Due to saturation, a large zero-sequence current occurs in the field current. Therefore, the overvoltage of UHV lines may be even more serious in no-load closing.

Switching overvoltage with steep wavefront and high frequency may occur in a sulfur hexafluoride (SF_6) gas-insulated substation (GIS). When disconnectors are used to operate short buses in a GIS, overvoltage of very high frequency may occur due to multiple breakdowns and extinctions of switches, whose initial front is generally between 3 and 200 ns; this is called very fast transient overvoltage (VFTO). Moreover, its frequency and steepness are much higher than that of lightning overvoltage, and zinc oxide arrester (MOA) cannot limit this overvoltage. VFTO may threaten the security of a GIS and its adjacent equipment, especially interturn insulation of the transformer, and may also cause high-frequency oscillation

Figure 9.1 Measured wave process of no-load line closing.

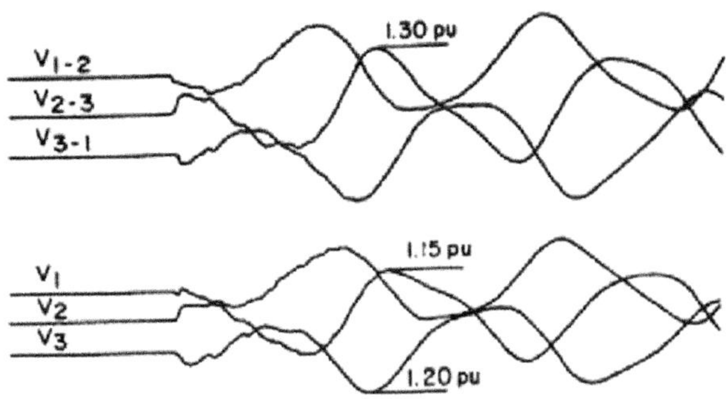

Figure 9.2 Measured electromagnetic transient process of no-load line closing.

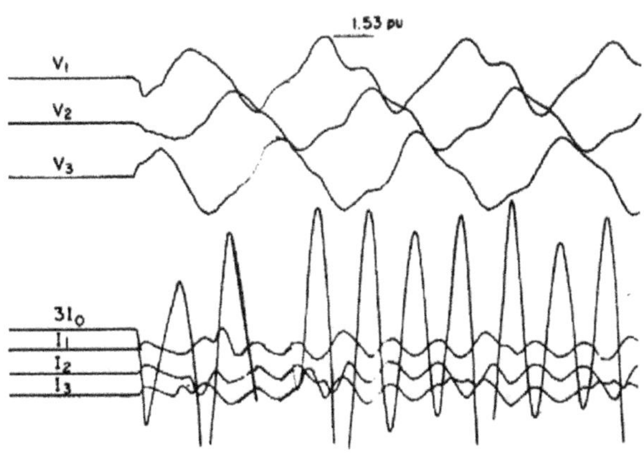

Figure 9.3 Measured steady-state overvoltage.

in the transformer. The accident of a VFTO damaging a large transformer has occurred in China's 500 kV system [2].

Overvoltage generated by series or parallel resonance in a power system is called resonance overvoltage, and system inductance includes linear inductance and nonlinear inductance (e.g., the shunt reactor and the excitation reactance of transformers). In the case of normal operation or fault, these inductances may resonate with capacitors in series or parallel (e.g., line capacitance and series/shunt compensation capacitances) to cause resonance overvoltage.

The frequency of power-frequency overvoltage is equal or close to power frequency. It is caused by the capacitance effect of the no-load long line, the non-fault phase voltage increase caused by asymmetric grounding fault, and sudden load change, and is related to the structure, capacity, parameters, and operating mode of the system. Generally, the amplitude of power-frequency overvoltage is not high, but its duration is long. Power-frequency overvoltage plays an important role in UHV systems:

(1) Its peak amplitude and the harmonic peak may directly increase the peak value of switching overvoltage.
(2) It is an important basis for determining the rated voltage of arresters.
(3) It may last for a long time and may endanger the safe operation of the equipment and system.

9.1.3 Measures to prevent internal overvoltage

Measures to prevent internal overvoltage are as follows:

(1) Reasonably arrange MOA.
(2) Reasonably arrange high-voltage shunt reactors or controllable shunt reactors.
(3) Install both closing auxiliary and tripping auxiliary resistance for the circuit breaker.
(4) Install a high-speed grounding switch.
(5) Adopt synchronous closing.
(6) Adjust the operating mode of the power grid.
(7) Comprehensively construct a unified overvoltage control system by changing the primary system operating procedure or network structure and setting up an emergency control device in addition to adopting relay protection, an automatic device, a shunt reactor, and a lightning arrester, which may play a protective role and prevent overvoltage when the system fails or some equipment is damaged.

9.2 Overvoltage characteristics of ultra-high-voltage transmission systems

9.2.1 Overvoltage of ultra-high-voltage transmission systems

The problems of electromagnetic transients and overvoltage in UHV transmission system are similar to those of the UHV system. However, due to the large transmission capacity and longer distance of a UHV transmission line, its reactive power is very large; the reactive power of a 1000 kV transmission line per 100 km can reach about 530 MVar, which brings up problems as follows:

- It may cause severe transient overvoltage during load shedding.
- When the load changes during normal operation, it may cause difficulties in reactive power regulation, voltage control, and secondary arcs extinguishing during single-phase reclosing. In UHV transmission

system design, an important problem is limiting all kinds of overvoltage to a reasonable level.

From the perspective of electromagnetic transients, the UHV transmission system in China has three characteristics as follows:

(1) Some UHV transmission lines have long distances and large transmission capacities.

(2) Great differences of power grids exist among different regions, including areas with concentrated loads and dense grid structures as well as newly built isolated power plants in transitional periods.

(3) Some UHV lines may pass through high altitudes or heavily polluted areas, which makes overvoltage an important issue to consider in UHV system design.

9.2.2 Overvoltage standards

One of the most important problems in designing and constructing UHV transmission lines is to establish the overvoltage standard. Tower structure, insulators, equipment selection, measures to suppress overvoltage, and so on should all be based on this standard. Table 9.1 gives the overvoltage standards for UHV systems in various countries, in which the p.u. value of switching overvoltage is the ratio of the peak value of overvoltage to the rated voltage amplitude.

The power-frequency overvoltage level of China's EHV power system is stipulated as follows: power-frequency overvoltage on the substation side of the circuit breaker should not be more than 1.3 p.u. and on the line side of the circuit breaker should not be more than 1.4 p.u. The allowable instantaneous overvoltage of switching overvoltage is related to the rated voltage of the system. The instantaneous switching overvoltage time of a 330 kV system should not exceed 2.2 p.u., that of a 500 kV system should not exceed 2.0 p.u., and that of a 750 kV system should not exceed 1.8 p.u. That is to say, the requirement for limiting the switching overvoltage is stricter under higher rated voltage of the grid.

At present, there is no overvoltage standard for UHV systems in China. After repeated calculation and comparison as well as drawing on the experience of other countries, the limit value of UHV internal overvoltage is temporarily taken as the following [2] as a reference for further research:

(1) Power-frequency overvoltage is limited to below 1.3 p.u.; in some cases the line side can be allowed to be below 1.4 p.u. for a short time, the duration of which is no more than 0.3 s.

Table 9.1 Overvoltage levels of ultra-high-voltage transmission systems in various countries.

Country item	Soviet union	Japan	Italy
Maximum operating voltage (kV)	1200	1100	1050
Power-frequency overvoltage (per unit)	1.3 allowed to continue for 0.1 s 1.4 allowed to continue for 0.05 s	1.4	1.35
Transient overvoltage (per unit)	1.8 (striv to reduce to 1.6)	1.6 −1.7	1.7

(2) Phase-to-ground statistical instantaneous switching overvoltage, whose occurrence probability is 2%, should be limited to less than 1.6 p.u. for substation switching equipment and less than 1.7 p.u. for the pole and tower parts of the long line.

(3) Statistical switching overvoltage between phases for substation switching equipment should be limited to less than 2.6 p.u., and for the pole and tower parts of a long line, it should be limited to less than 2.8 p.u.

In the foregoing, the reference voltage $= 1.0p.u. = 1100\sqrt{2}/\sqrt{3}kV$ —i.e., the peak value of phase voltage.

9.2.3 Various operations that generate overvoltage

Various operations that generate overvoltage are as follows:

(1) Line switch according to the plan: the closing the line from one end; the operation of disconnecting the no-load line from one end

(2) Sudden load shedding: three-phase tripping from one end of the line under natural power or 1.1−1.2 times of natural power

(3) When asymmetric fault occurs at the end of a long line, the circuit breaker disconnects the line from one end within 0.04−0.05 s

(4) Rapid three-phase automatic reclosing

(5) Single-phase automatic reclosing

(6) Secondary resection of asymmetric fault when automatic reclosing is not successful

(7) Disconnecting the line during the oscillation process

When calculating or testing the aforementioned overvoltage, the system impedance, magnitudes of electromotive forces, angle between electromotive forces, and number of shunt reactors should all be changed accordingly. For each operating condition of the circuit breaker, both the steady-state and the transient overvoltage should be calculated. The allowable steady-state overvoltage multiple is determined by both the duration of the line in a single-ended closed state under each operation and the overvoltage multiple that the line and equipment can withstand. Because it takes time for the overvoltage to reach its peak value, it is better to keep the line in a single-end closed state as short as possible.

9.3 Study on overvoltage of ultra-high-voltage transmission lines

The Soviet Union carried out a computer simulation calculation in the design of a 750 kV transmission line from Donbas to Dnipro to Venezia to West Ukraine. The results are explained below [3].

The allowable overvoltage stipulated by the Soviet Union is as follows:

- Power-frequency overvoltage: 1.3 times for 0.1 s, 1.4 times for 0.05 s.
- Maximum allowable switching surge: K = 2.1 times for 750 kV line, K = 1.8 for 1150 kV line, desire to reduce to 1.6 in the future.
- Power-frequency test voltage for equipment: 1.3 times/h, 1.6 times/min for transformer; 1.65 times/min for other equipment.

9.3.1 Overvoltage in the case of energizing the line

In this case, it may take a long time for the line to be in the single-ended closed state, but because a series of measures can be taken to reduce overvoltage during the scheduled closing time, the overvoltage may not be very high. Steady-state overvoltage for 750 kV lines can reach 1.05 ~ 1.1 p.u., while transient overvoltage has a 50% probability of reaching 1.5 ~ 1.6 p.u. and a probability of 2%–3% to reach 1.8 ~ 2.0 p.u.

9.3.2 Overvoltage in the case of asymmetric fault

The study shows that when the asymmetric fault is isolated, if the time difference between the disconnections of the two circuit breakers on both ends is more than 0.05–0.06 s, the overvoltage may endanger the safety of the equipment. The effect of this overvoltage is related to the duration of the single-ended state of the line. Without any measures to limit overvoltage when isolating the asymmetric short circuit at the end of the line, steady-state power-frequency overvoltage of 750 kV lines can reach 1.55 ~ 1.7 p.u. and for 1150 kV lines can reach 1.35 ~ 1.55 p.u., while the peak value of transient overvoltage can reach 1.8 ~ 2.2 p.u. In the case of short circuit under maximum load, due to the high electromotive force of the system generator, in order to balance reactive power, some of the shunt reactors are disconnected, and transient overvoltage may also increase. The influence of the angle between the electromotive forces on both sides is small because this angle generally does not exceed 30–50 degrees.

Table 9.2 shows the calculation results of overvoltage times when asymmetric short circuits and oscillations are isolated from the transmission lines [3]. The disconnected line load is large, the asymmetric short circuit occurs at end, the same circuit breaker is used to disconnect the line at the beginning, and the overvoltage of six different places is calculated. The overvoltages at six different breaking times are also calculated for unsynchronized cutoff lines, and calculation results under the condition of limiting overvoltage are given.

The magnitudes of switching and fault-isolation overvoltages are related to many factors, such as system operating condition, fault moment, type and location of the fault, operating time, switch-off time, and so on. Generally, they should be calculated or tested many times to obtain a probability distribution.

As can be seen from Table 9.2, when the reactor and discharger are not connected at the operating end (No. 1), the overvoltage may significantly exceed the allowable time by 2.1 p.u. when the asymmetric fault and nonsynchronous operation are removed. The overvoltage of isolating the asymmetric fault is not more than 2.65 p.u., while the overvoltage of isolating the nonsynchronous operation may reach 3.04 p.u. Under the same conditions, maximum overvoltage times can be reduced to 2.26 and 2.62 p.u. respectively by preswitching the reactor before the operation (No. 5) but 2.34 p.u. and 2.7 p.u. respectively by switching the reactor on simultaneously (No. 2). If the reactor is switched on simultaneously, the discharger discharges normally (No. 3), the overvoltage generally does not exceed 2.1 p.u. and only reaches 2.15 p.u. once. In order to reduce overvoltage further, the accelerating discharge mode (No. 4—i.e., reducing the discharge voltage) can be adopted, in which case the overvoltage does not exceed 2.04 p.u.

9.4 Field tests of operating overvoltage

In order to verify the calculated results of the computers, the Soviet Union had carried out nearly 900 field tests on 750 kV lines since its first 750 kV line being put into operation in 1967. The purpose was to verify the correctness of the line and equipment design, to actually test the ability of line and equipment to withstand overvoltage, and to discover the practical

Table 9.2 Maximum overvoltage time for isolation of asymmetric short circuits or oscillations on the Donbas—Western Ukraine 750 kV line.

No.	Operating type and working condition of protection devices	Overvoltage time when tripping from one end											
		Fault						Nonsynchronous operation					
		1	2	3	4	5	6	1	2	3	4	5	6
1	When tripping, the protection at the other end doesn't operate, the reactor is not connected, and the discharger is not discharged.	2,52	2.6	2.65	2.41	2.17	2.42	2.50	2.67	3.04	2.87	2.37	2.56
2	The reactor is connected at the same time and the discharger is not discharged.	2.26	2.34	2.17	2.10	—	—	2.29	2.40	2.7	2.51	—	—
3	The reactor is connected at the same time and the discharger discharges normally.	2.01	2.06	2.05	2.0	2.05	2.08	2.04	2.07	2.15	2.12	2.06	2.09
4	The reactor is connected at the same time and the discharger accelerate the discharging.	—	—	1.86	1.56	1.75	1.63	—	—	2.04	1.93	1.75	1.67
5	The reactor is connected beforehand and the discharger is not discharged.	2.18	2.26	—	—	1.98	2.12	2.28	2.31	2.62	2.41	2.12	2.27

problem that was not foreseen or considered in the design. Table 9.3 shows parts of the statistical result of the tests.

The value of switching overvoltage on the line is related to many random factors, such as switching angle, voltage level, quantity of charge remains on the line, and the number of lines and transformers connected to the bus. Moreover, it is also related to climate. Therefore, switching overvoltage has statistical properties and can be given in statistical form. If the value of overvoltage is given according to the actual conditions of various factors in the test, it may be very complex and not easy to apply. Because the operating condition of newly built lines cannot be completely consistent with the environmental conditions of these lines under testing, it may be convenient to give them in statistical form for reference in the design of newly built line projects.

The items measured in Table 9.3 and calculated in Table 9.2 are for different operating states. The data shown in Table 9.2 is in the case of line fault, while that shown in Table 9.3 is in the case of non-fault. Table 9.2 is illustrated by p.u. value, while Table 9.3 is illustrated by the impact coefficient, so it is not easy to compare the two directly. However, the following points can be made in Table 9.3:

(1) Table 9.3 shows two ways of line closing and reclosing: one is uncontrolled random closing where the closing angle is random; the other is so-called "synchronous closing," which is a method proposed by E. Mori to reduce line-closing overvoltage [4]. In this method, a mechanism is used to control the closing time of the breaker of each phase, which makes the three-phase contacts sequentially being closed 60° late in the order of A—B—C or C—B—A so that the absolute value of instantaneous voltage of the three-phase reclosing occurs at the same time. It is then controlled to close when the bus voltage crosses zero or at the minimum voltage of the voltage enveloping line between the two ends of the breaker contacts. In this way, overvoltage time can be significantly reduced. For reclosing, this method is called "synchronized reclosing." As can be seen from the table, the overvoltage time produced by synchronous reclosing and segregated synchronized reclosing are much smaller than those produced by uncontrolled closing and reclosing. The maximum overvoltage can be up to 1.9 times in uncontrolled closing but only 1.28 times in synchronous closing. The maximum overvoltage can reach 2.45 p.u. for uncontrolled three-phase fast reclosing, while for synchronized reclosing it is only 1.28 p.u.

Table 9.3 Field test statistics of switching overvoltage of Soviet 750 kV system.

Operating type	Test condition	Measurement unit	Number of operations	Statistics of overvoltage time	
				Average (probability of 50%)	Maximum (probability of 2%–3%)
Closing with no load	Uncontrolled	K_{imp}	168	1.55–1.65	1.8–1.9
	Synchronous closing	K_{imp}	58	1.15	1.28
Tripping	No reactor	K_{imp}	96	0.87	1.22
	Reactor and overcompensation	K_{imp}	36	1.15	1.35
	No reactor, quick reclosing after tripping, residual voltage 0.3–0.4 s	K_{imp}	96	0.77	1.14
	No reactor, quick reclosing after tripping, residual voltage 0.3–0.4 s, shunt reactor, and overcompensation	K_{imp}	36	0.7	1.05
Three-phase quick reclosing (with end disconnected)	No shunt reactor, uncontrolled reclosing	K_{imp}	81	1.55–1.65	2.45
	No shunt reactor, synchronized reclosing	K_{imp}	47	1.08	1.28
	Circuit breaker recloses when voltage between the two contacts reaches maximum value	K_{imp}	55	1.9	2.25* (2.7–2.9)**
	Circuit breaker recloses when the voltage between the two contacts reaches minimum value	K_{imp}	92	1.12	1.3–1.42
Line - transformer closing	Transient overvoltage of three-phase closing	K	60		2.2–2.35*
	Short-time overvoltage for open-phase closing	K	44		2.2–2.54*
	Short-time overvoltage for symmetric-state closing	K	6		1.55

*Represents that the arrester operation limits overvoltage.
**Represents that the condition has been reduced to that with no arrester.

Although synchronized reclosing is an effective measure for reducing overvoltage, it is complicated to realize. If it is not well controlled or closed at the maximum beat voltage time due to error, the maximum overvoltage can reach 2.7–2.9 times, which may cause serious consequences.

(2) After the tripping due to fault, the residual electric charge on the line discharges through the shunt reactor, which may cause resonance and produce beat voltages of different frequencies. Meanwhile, the voltage difference on both sides of the breaker contacts is also a beat voltage in reclosing. Hence, all three phases of the circuit breaker should be controlled to close when this beat voltage is minimum—that is, the so-called synchronized reclosing. As can be seen from Table 9.3, with synchronized reclosing, if the breaker is closed when the envelope line of the beat voltage is minimum, the impact coefficient $K_{imp} = 1.12$, and if the breaker is closed when the envelope line is maximum, $K_{imp} = 1.9$. Therefore, by adapting synchronized reclosing and the adaptive reclosing principle, reclosing overvoltage can be greatly reduced by searching for the time at which the envelope line of the beat voltage is at minimum and closing at exactly that time.

(3) The transient line overvoltage with transformer in no-load closing is larger than that of the line-only in no-load closing. This is caused by the superposition of harmonic voltage and power-frequency voltage generated by transformer inrush current. Therefore, this operation should be avoided.

(4) The value of overvoltage produced by open-phase no-load closing is similar in size to that produced by full-phase no-load closing, hence a too high overvoltage is not likely to occur under slow closing or rejecting closure of the one-phase circuit breaker for a no-load closing.

References

[1] Vassell Gregory S, Maliszewski Raymond M, Johnsen norman B. Experience with the AEP 765kV system: part II – system performance. IEEE PES Winter Meeting; 1972.

[2] Ultra-high voltage grid. Zhenya Liu. Beijing: China Economic Publishing House; 2005.

[3] Vinogradova AD, Lyskov YuI. Major problem in switching overvoltage protection of extra high voltage transmissions. US-USSR Symposium held by Joint Committee On scientific And Technical Cooperation, Washington D.C.;1975.

[4] Kazmicheva II, et al. Limiting of internal overvoltage in 750kV transmission systems with synchronous connection of series production air breaker. Elektrichestro No. 7 1975 [In Russian].

CHAPTER 10

Protection for transformers and shunt reactors

Contents

10.1 Structural features of ultra-high-voltage transformers

The capacity, volume, and weight of an ultra-high-voltage (UHV) transformer are all very large. Therefore, in order to facilitate the transportation and maintenance, three single-phase autotransformers with three weddings are used to form a three-phase transformer group. At the same time, the voltage regulator is separated from the main transformer, and the voltage regulating transformer and compensation transformer are set up independent of the main transformer [1].

The 1000 kV main transformer is composed of two parts, the main part (single-phase autotransformer without voltage regulation and low-voltage side voltage compensation) and regulating transformer (including

Protection Technologies of Ultra-High-Voltage AC Transmission Systems
ISBN 978-0-12-816205-7
https://doi.org/10.1016/B978-0-12-816205-7.00010-8

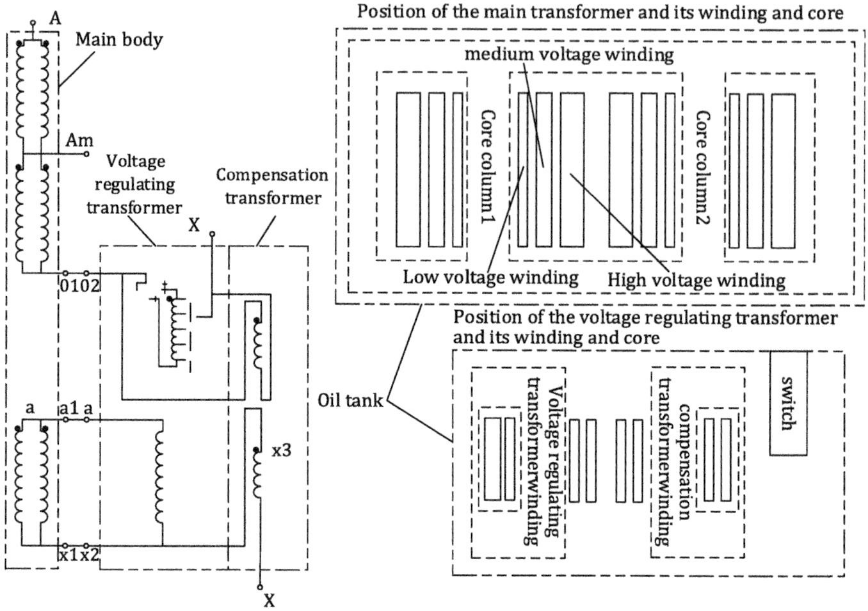

Figure 10.1 Transformer internal connection diagram provided by Shenyang Transformer Factory in China.

low-voltage compensator), such as shown in Fig. 10.1. The regulating transformer and main transformer are not in the same oil tank but are connected of rigid busbar. The two parts are not simply connected in series. The regulating transformer (including low-voltage compensation winding) consists of two parts, a voltage regulator and a low-voltage compensator that both use the same oil tank. The function of the regulating transformer is to stabilize the voltage of the medium-voltage side, and the role of the compensation transformer is to compensate for the influence imposed on the low-voltage side when adjusting the voltage of the medium-voltage side. The field winding (the primary winding of the power supply) of the regulating transformer is parallel to the low-voltage winding of the autotransformer, and the field winding (the primary winding of the power supply) of the low-voltage-side voltage compensator is parallel to the regulating winding. The winding of the regulating transformer is connected in series to the neutral point of the main transformer's high-voltage side and the medium-voltage side, and the low-voltage compensating winding is connected with the low-voltage winding of the main transformer in series. In order to adjust (stabilize) the voltage, the connection mode of negative feedback should be adopted—that is, the regulating winding and main transformer high-voltage winding connected in series, and the low-voltage compensating winding and main transformer low-voltage winding connected in series, should be connected by the mode of negative feedback respectively. This connection mode achieves that the voltage can be controlled in a stable state.

The core of the UHV transformer is a single-phase four-column type, and the middle two core columns are twined with windings. The arrangement of the windings of each core column is (from iron core to outer) core, low-voltage winding, common winding, and series winding. The high-voltage, medium-voltage, and low-voltage windings of core columns 1 and 2 are connected in parallel and winded in opposite directions (because the flux directions in the two columns are opposite). The two parts of voltage regulation and voltage compensation are separated independently, which is not only for the convenience of transportation but also for the reliability of the main transformer and convenience of maintenance. When the voltage regulation part fails, it can be untied from the main part of the main transformer without affecting normal operation. The disadvantage of this regulating mode is that when regulating the voltage on the medium-voltage side of the main transformer may affect the voltage on the low-voltage side. Therefore, in order to stable the voltage on the low-voltage side stable during the regulating process, the voltage on the low-voltage side must be increased by adding low-voltage compensating winding and introducing negative feedback voltage.

Figs. 10.1 to 10.3 represent the connection diagrams of phase A. The series winding, SV, of the autotransformer and the common winding, CV, are in series in a forward direction. A–X represents A-phase winding on the

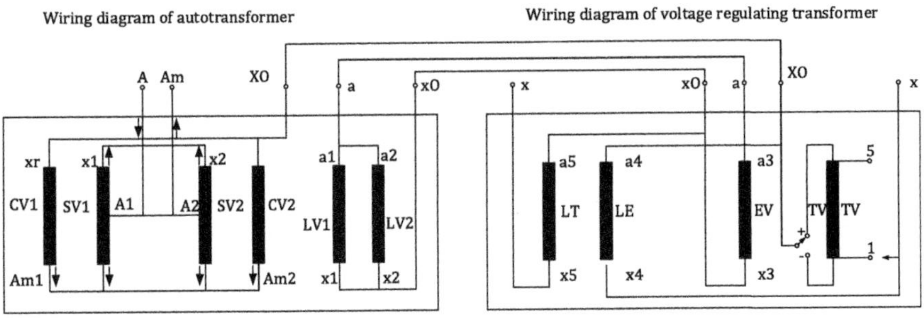

Figure 10.2 Transformer internal connection mode provided by Baoding Transformer Factory in China.
CV—common winding, *EV*—voltage-regulating field winding, *LE*—low-voltage field winding, *LT*—low-voltage compensation winding, *LV*—low-voltage winding, *SV*—series winding, *TV*—voltage-regulating winding.

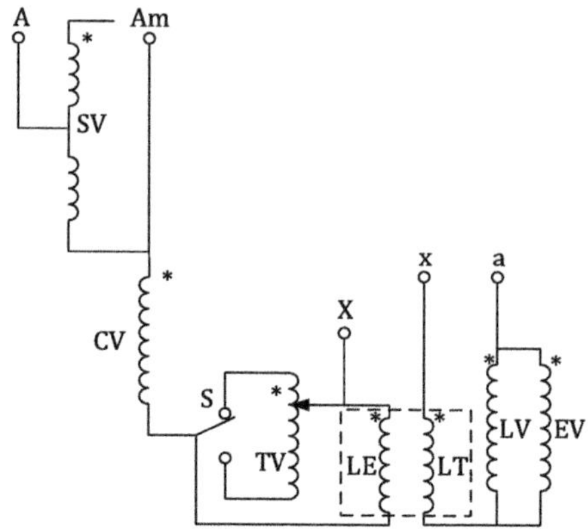

Figure 10.3 Elementary wiring diagram of an ultra-high-voltage three-winding autotransformer.

high-voltage side, Am—X represents A-phase winding on the medium-voltage-voltage side, and a—x represents A-phase winding on the low-voltage side. The field winding, EV (the primary winding of the power source), of the voltage-regulating transformer is parallel to the low-voltage winding, LV, of the autotransformer. The field winding, LE (the primary winding of the power source), of the low-voltage compensator is parallel to the regulating winding, TV. The regulating winding, TV, is connected in series to the neutral point of the main transformer (it can be connected in series in either a forward or a reverse direction, which is selected by selector switch S), and the connection should meet the requirement of negative feedback. The series between the low-voltage compensating winding, LT, and the low-voltage winding, LV, of the main transformer (autotransformer) should also be connected according to the requirements of negative feedback. The voltage in the low-voltage compensating winding, LT, is proportional to the voltage in the regulating winding, TV, and is also proportional to the voltage of the low-voltage winding, LV—that is, the voltage of both the medium-voltage and the low-voltage sides should be regulated according to the same primary voltage variation.

10.2 Protection configuration of ultra-high-voltage transformers

According to the structural characteristics of the UHV transformer and its requirements for protection sensitivity, the new principle of distributed transformer protection was put forward by NR Electric Co., Ltd. and a brief introduction is as follows [2].

From the structural characteristics and simulation study of the UHV transformer, it is found that the proportion of the turn number of the regulating transformer and compensation transformer to the total turn number of the whole transformer is relatively small, and the interturn voltage is also very small compared with that of the main transformer. When the regulating transformer or compensation transformer has a slight interturn fault, its influence on the whole transformer may be slight. Therefore, it is difficult in this case for transformer differential protection to take action when its protection range is the whole transformer.

According to the fault waveform (not shown here) of the 25% interturn short circuit of the regulating transformer in the 1000 kV transformer protection dynamic model test carried out by the China Electric Power Research Institute, even if there is a 25% interturn fault (the interturn fault of this turn ratio is already a very serious internal fault for the regulating transformer) at the regulating transformer, the differential current amplitude detected by the main transformer differential protection is only 0.47 times that of the rated current, which has just exceeded the setting value of the differential protection (0.41 times the rated current). When the short-circuit turns of the regulating transformer continue to decrease, the differential protection of the main transformer may not even be able to start.

Therefore, for the connection mode of *main transformer + voltage-regulating transformer + compensation transformer*, differential protection must be separately configured for the regulating transformer and compensation transformer to improve its sensitivity when an internal interturn fault occurs. In addition, differential protection separately configured for the regulating transformer and compensation transformer is mainly used to improve the sensitivity in small faults, so there is no need to configure differential instantaneous trip protection for it.

Therefore, the distributed protection configuration scheme is proposed and developed—that is, to add differential protections of the regulating transformer and compensation transformer on the basis of the original protection scheme—to improve the sensitivity of detecting interturn faults of the regulating transformer and compensation transformer. The protection configuration is shown in Figs. 10.4 and 10.5. TA1 and TA2 are used for differential protection of the main transformer, TA3 and TA4 are used for other protection, and QF1, QF2, and QF3 are three circuit breakers for the 3/2 connection on the HV side.

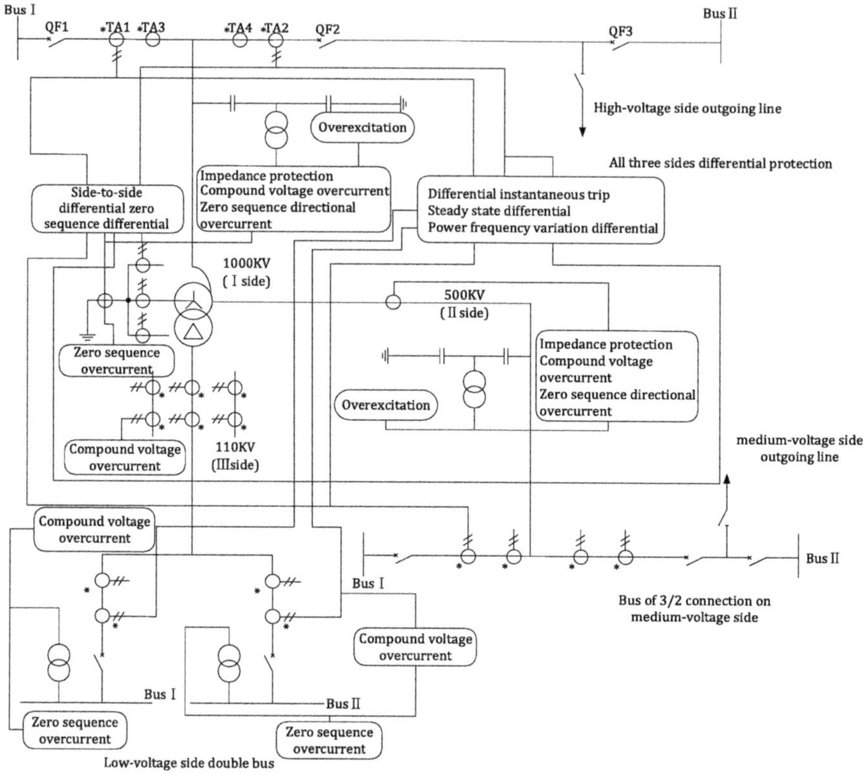

Figure 10.4 1000 kV transformer protection configuration schematic.

10.3 Distributed protection scheme for ultra-high-voltage transformers

In order to facilitate the withdrawal of the regulating and compensation transformers, NR Electric Co., Ltd. suggested that differential protection of the regulating transformers and compensation transformers is implemented by a single device, configuration of both protection and TA are dual, and secondary winding of TA and the main transformer protection are independent of each other. The protection configuration is shown in Table 10.1.

10.3.1 Current transformers

The specific parameters of the UHV transformer are as follows:

- capacity: 3 × 1000 MVA
- capacity ratio: 1:1:1/3

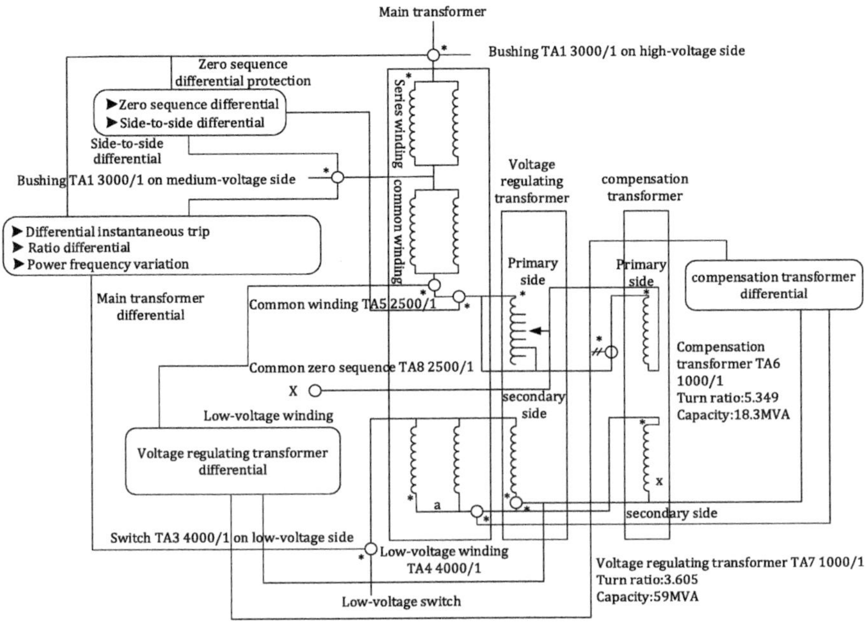

Figure 10.5 1000 kV transformer differential protection and TA configuration schematic diagram (including main, voltage-regulating, and compensation transformers).

- voltage ratio: $\frac{1050}{\sqrt{3}}/\frac{525}{\sqrt{3}} \pm 5\%\ \text{kV}/110\ \text{kV}$
- short-circuit reactance x_k: high-middle (H-M), 18%; high-low (H-L), 62%; middle-low (M-L), 40%
- mode of connection: $Y_{0\ \text{self}}/\ \Delta$-12-11

The rated primary current at each side is as follows:

- high-voltage side：1649.6 A
- medium-voltage side：3299.2 A
- low-voltage side：15,746.4 A (full capacity), 5248.8 A (the winding capacity)

The maximum short-circuit current passing through the current transformers at each side is as follows.

High-voltage side: when a three-phase metallic fault occurs, short-circuit current flowing through the high-voltage side is at maximum, but due to the influence of the short-circuit impedance of the 1000 kV system, this maximum short-circuit current is small, even smaller than that of some 500 kV systems.

Medium voltage side: when a three-phase metallic fault occurs, short-circuit current flowing through the medium voltage side is at maximum, which is determined by the short-circuit level of the 500 kV system.

Table 10.1 1000 kV transformer protection configuration scheme.

	Protection type	Zone number	Time delay setting each zone	Remarks
Voltage-regulating transformer	Ratio differential			
Compensation transformer	Ratio differential			
Main protection (main transformer)	Differential instantaneous trip	–	–	
	Ratio differential	–	–	
	Power frequency variation ratio differential	–	–	
	Zero-sequence and split ratio differential	–	–	
(Backup protection, auxiliary protection and signal) I side of high-voltage side	Interphase impedance	1	2/I	Zone I has 2 time delay settings. Forward and backward impedance can be set.
	Grounding impedance	1	2/I	Zone I has two time-delay settings. Forward and backward impedance can be set.
	Overcurrent	1	2/I	Zone I has two time-delay settings. Zone I can be locked by low-voltage relay or negative-sequence overvoltage relay. Zone I can also be locked by directional relay.
	Zero-sequence overcurrent	2	2/I,1/II	Zone I has two time-delay settings. Zone II has one time-delay setting. Zone I can be locked by directional relay.

	Overexcitation	2	1	With inverse time limit overexcitation function; the zone number of its fixed time limit is in the table.
	Overload	1	1	
	Starting cooler	1	1	
	Locking voltage regulation	1	1	
	Overexcitation alarm	1	1	With inverse time limit overexcitation alarming function; the zone number of its fixed time limit is in the table.
(Backup protection, auxiliary protection and signal) II side of middle-voltage side	Interphase impedance	1	2/I	Forward and backward impedance can be set.
	Grounding impedance	1	2/I	Forward and backward impedance can be set.
	Overcurrent	1	2/I	It can be locked by low-voltage relay or negative-sequence overvoltage relay and can also be locked by directional relay.
	Zero-sequence overcurrent	2	2/I, 1/II	Zone I has two time limits, zone II has one time limit, and zone I can be locked by directional relay
	Overexcitation	2	1	With inverse time limit overexcitation function; the zone number of its fixed time limit is in the table.

Continued

Table 10.1 1000 kV transformer protection configuration scheme.—cont'd

	Protection type	Zone number	Time delay setting each zone	Remarks
	Overload	1	1	
	Starting cooler	1	1	
	Locking voltage regulation	1	1	
	Overexcitation alarm	1	1	With inverse time limit overexcitation alarming function; the zone number of its fixed time limit is in the table.
(Backup protection, auxiliary protection and signal) low-voltage winding	Overcurrent	2	2/I, 1/II	Zone I has two time limits, zone II has one time limit, and zone II can be locked by low-voltage relay or negative-sequence overvoltage relay.
	Overload	1	1	
Common winding	Zero-sequence overcurrent	1	1	
	Overload	1	1	
	Starting cooler	1	1	

Low-voltage side: when a three-phase metallic fault occurs, short-circuit current flowing through the low-voltage side is at maximum, but its maximum value is only about two times the rated current of the transformer due to the influence of the large short-circuit impedance ratio of the transformer.

Common winding: in the case of neglecting the low-voltage winding, when a three-phase metallic fault occurs on the high- or medium-voltage side, short-circuit current flowing through the common winding is at maximum, and this maximum current is only (3299.2−1649.6)/ 0.18 = 9155.3 A (for infinity system). In actual application, the maximum short-circuit current of the common winding is lower than this current value.

From the above data, it can be seen that because of the limitation of the short-circuit level and transformer short-circuit impedance of the 1000 kV system, the maximum short-circuit current flowing through TA on each side of the 1000 kV transformer is not too large, and the demand for thermal stability of TA is even lower than that of the general 500 kV transformer.

It can be seen that when selecting the ratio of the TA on each side of 1000 kV transformer protection, we can select the TA with a relatively small variable ratio so that the requirements of dynamic and thermal stability of TA can be fully met. Especially for bushing TA of the low-voltage side, common winding, regulating transformer, and compensation transformer, bushing TA itself may not flow over a large short-circuit current, while reducing the TA ratio means that the volume of the device decreases, and the related cost of the bushing may also decrease.

10.3.2 The scheme of distributed transformer protection

With continuous improvement of the voltage grade of power systems (especially the UHV voltage grade), the electrical distance of the switches on each side of the transformer is getting farther and farther (above 400−500 m). This leads to extension of the secondary circuit wiring of the differential protection and increase the burden of TA's secondary circuit, which brings some influences on the differential protection and increases the cost of the secondary circuit. At the same time, the increased length of secondary cable in the trip circuit leads to an increase in the distributed capacitance of the circuit, which reducing protection reliability. The distributed transformer protection configuration can solve this problem better.

Distributed transformer protection adopts the structure of main station + substation, as shown in Fig. 10.6.

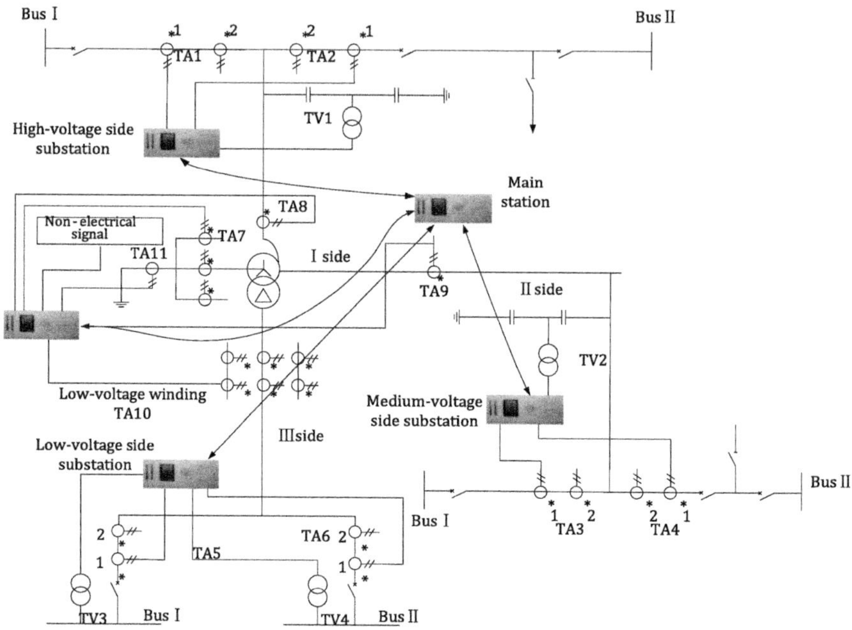

Figure 10.6 Schematic diagram of distributed transformer protection.

Substations collect the analog data needed for protection and transmit the data to the main station through fiber, meanwhile receiving and executing the tripping command sent by the main station through the fiber.

The main station receives analog data sent from substation though fiber and performs protection calculation, sending the protection tripping command to the substation through fiber.

The advantage of distributed transformer protection is that it can collect analog signals directly and send them to the main station through fiber, and thus it is easy to realize data sharing, simplify the secondary circuit, and reduce the possibility of the equipment burnt by overvoltage produced by a connection breakage of the TA secondary circuit. At the same time, using optical fiber to transmit data has a strong validation function, which improves secondary circuit reliability. Sending the tripping signal through optical fiber avoids the possibility of a protection misoperation.

10.3.3 Key technical points

10.3.3.1 Data transmission problem

The number of bay units in the distributed transformer protection communication network is large, which makes the volume of data transmission on the communication network larger and the burden of the communication network heavier. How to ensure real-time and reliability of data transmission are key technical problems in the realization of distributed transformer protection.

10.3.3.2 Synchronized sampling problem

The bay unit of distributed transformer protection corresponds to the primary devices connected to the transformer and is distributed in the vicinity of the corresponding equipment, thus producing distance to one another in space. There is no electrical connection between the sampling units, thus creating the problem of synchronization for each unit. The quality of the sampling clock affects differential protection performance directly.

It is feasible to adopt GPS technology to synchronize the time for each bay unit technically. However, this scheme increases complexity of hardware. More importantly, the bay unit is located in the switch field with serious electromagnetic interference, and its reliability is worrisome. The sampling synchronization clock of the bay unit requires that the relative clock should be accurate, while there is no requirement for absolute time. Without increasing the burden of the hardware and communication network, sampling synchronization of the bay unit is one of the key technical problems for the distributed transformer.

The synchronous sampling solution is like this: the time difference between the sampling time of each substation and the time when the sampling data is sent to the main station should be the same. The master control board of the main station releases the time reference through the internal bus, the data processing board records the time of receiving the sampling data, and the sampling data is interpolated in real time according to the receiving time and time reference. The accuracy of this synchronization scheme can be within 5 μs (0.09°).

With continuous improvements in computer technology, network communication, and hardware, the technical difficulties of configuring distributed protection have been solved, and the economic benefits of the distributed protection scheme are high. Therefore, new types of distribution transformer protection should be adopted by UHV transformers.

10.3.4 Identification of transformer magnetizing inrush current

The magnetizing inrush current of high-voltage and large-capacity transformers is small, and the decay time is long. In addition, according to the operational experience of the transformer, when the magnetizing inrush current is small in no-load input, secondary harmonic content is generally low, which is more likely to cause the misoperation of differential protection in the no-load inputting process of the transformer. In addition, when there are internal faults in the transformer due to the influence of line distribution capacitance and other parameters, the current and differential current in each side may contain large harmonic components, which may slow the speed of differential protection.

The identification method of magnetizing inrush current is based on second harmonic braking and waveform symmetry; both mathematical expressions are based on the characteristic detection of the magnetizing inrush current waveform. However, there are certain differences between the two methods. The second harmonic braking method only reflects the ratio of the value of the secondary harmonic and the fundamental amplitude of the differential current, while the waveform symmetry method reflects both the amplitude ratio and the phase relation of the even harmonic components and fundamental wave components. As in the process of no-load inputting, some even harmonic components other than the second harmonic are often included, which may make the waveform symmetry identification method better than the secondary harmonic method in some cases.

Compared with the choice of the second harmonic and waveform symmetry detection methods, the choice between "or brake" and "split-phase brake" methods has a greater influence on differential protection. The opening mode of "or brake"magnetizing inrush current is stronger in locking ability when inputting no-load to the transformer without any fault. When inputting no-load to the transformer with fault, especially to a transformer with small interturn fault, operating speed is slow. In cases of slight external fault occurring in the transfer zone with the saturation of TA, there are large harmonic components in the difference current caused by saturation, which leads to a decrease in protection operating speed. On the contrary, the mode of "split-phase brake" can guarantee the operating speed of differential protection in the case of failure, but there may be a lack of locking ability when no-load inputting to a transformer without fault. Therefore, a better method to identify magnetizing inrush current is proposed by NR Electric Co., Ltd., which introduces floating threshold technology into magnetizing inrush current identification, combining it with "split-phase brake" and then taking the variation trend of the secondary harmonic components in the differential current into consideration to identify the magnetizing inrush current.

When the device has detected the transformer input with no load, it may firstly reduce the locking fixed value of the magnetizing inrush current automatically so that in the initial stage, the device can get a very strong magnetizing inrush current locking capacity. As time goes on, it may continuously improve the locking fixed value of the magnetizing inrush current, thus ensuring that protection can also be quickly tripped to remove the fault in the case of a no-load inputting transformer. At the same time, the changing trend of secondary harmonic components of the differential current in the process of no-load input is also used in auxiliary judgment. This method of magnetizing inrush current identification method makes the protection device improve its ability to avoid the magnetizing inrush current and does not affect the operating speed when no-load input to the fault transformer and when there are internal faults occurring in the transformer under normal circumstances. Fig. 10.7 shows the waveform diagrams of a no-load inputting to transformer without any fault.

From the above waveform, it can be seen that if just the split-phase locking magnetizing inrush current identification method is adopted, the A-phase proportional differential protection element may misoperate in this no-load input case. However, after introducing the criteria of floating magnetizing inrush current fixed value and the second harmonic component trend, this case of no-load inputting to the transformer without fault can be locked correctly.

10.3.5 Settings of transformer protection

A 1000 kV transformer is obviously different from a conventional transformer, so its fixed value setting also has some particularities, which is mainly embodied in the following two aspects.

(1) Fixed value for transformer differential protection. Because the system impedance and short-circuit impedance of the transformer are both large, the short-circuit current is relatively small in cases of internal and external fault, which brings a higher requirement for protection sensitivity. The maximum adjusting range of the 1000 kV transformer tap is 5%, and the starting fixed value of transformer differential protection needs to avoid the maximum adjusting range of the tapping point, so the starting fixed value of the differential protection of the transformer can be set to 0.2 I_N (I_N is the rated current), and the proportional brake coefficient can be set to about 0.4. In this way, higher sensitivity is obtained, and meanwhile the reliability of protection under external fault can not be sacrificed. In addition, because the magnetizing inrush current of the 1000 kV transformer is small, the fixed value of the differential instantaneous trip can be set to 4 I_N. The setting value of the magnetizing inrush current can be set according

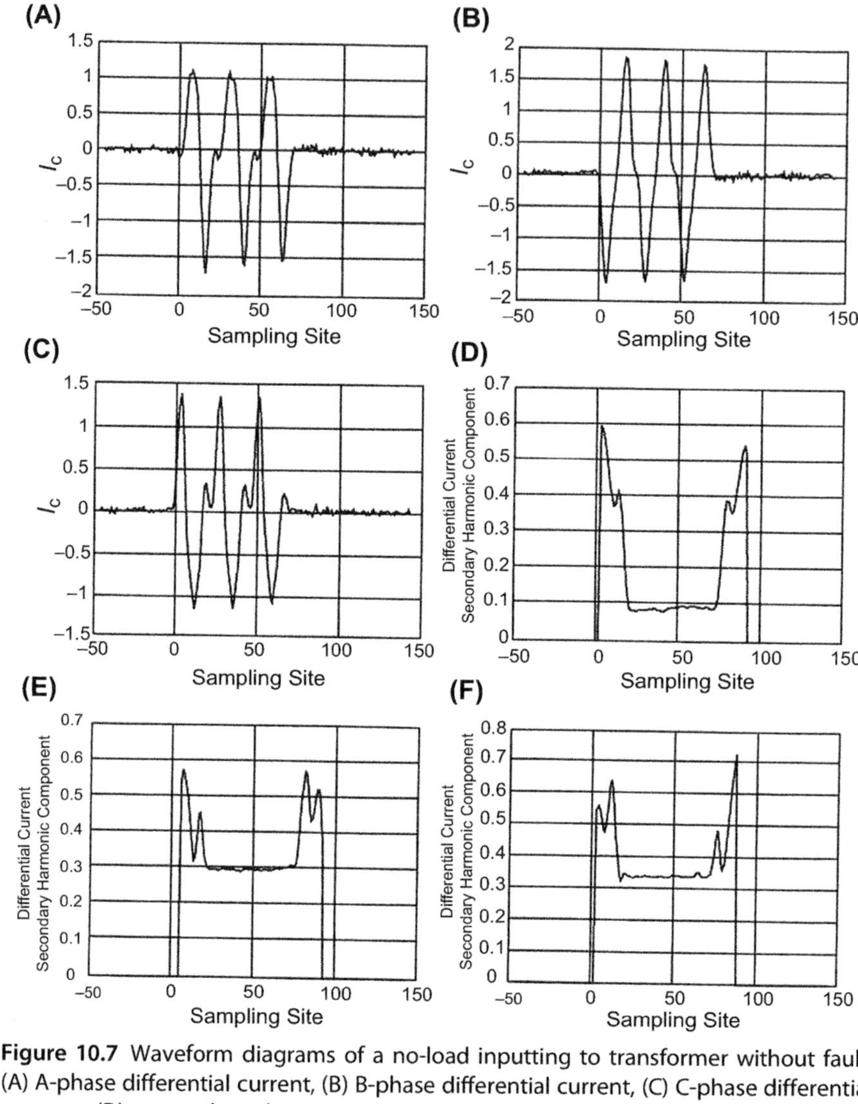

Figure 10.7 Waveform diagrams of a no-load inputting to transformer without fault. (A) A-phase differential current, (B) B-phase differential current, (C) C-phase differential current, (D) secondary harmonic components of A-phase differential current, (E) secondary harmonic components of B-phase differential current, (F) secondary harmonic components of C-phase differential current.

to the traditional method; generally, the second harmonic ratio is set to 15%, and the protection device can automatically take measures to improve the ability to avoid the magnetizing inrush current and improve reliability in the case of no-load input to the fault transformer.

(2) Setting of differential protection fixed values for regulating transformers and compensation transformers. The differential protection needs to be configured separately due to the differential sensitivity requirements of the regulating transformers and compensation transformers, but differential instantaneous trip protection is not required. Differential instantaneous trip protection is mainly set up to prevent differential protection from being unable to operate as a result of TA saturation in a serious internal fault, while in a fault serious enough to make TA badly saturated, transformer differential protection may surely operate. The fixed value setting principle of differential protection for the regulating transformers and compensation transformers is basically the same as that of transformer differential protection. As the short-circuit current flowing through the regulating transformers and compensation transformers is not large in the case of external fault, the differential fixed value can be reduced properly, which is suggested at 0.2—0.3 times its rated current, and the braking coefficient can be set to about 0.4 simultaneously.

10.4 Interturn short-circuit protection for ultra-high-voltage shunt reactors

10.4.1 Protection configuration [3]

UHV reactors are similar to 750 and 500 kV reactors in terms of structure and connection mode. The protection configuration of NR Electric Co., Ltd. is shown in Fig. 10.8. The protection configuration of a 1000 kV reactor is shown in Table 10.2.

10.4.2 Interturn short-circuit protection [4]

In the previous design, in order to limit short-circuit current, the short-circuit impedance of the 1000 kV system transformer has been greatly improved relative to that of the 500 kV system. At the same time, due to the longer line, the zero-sequence impedance of the system is larger. However, in the interturn protection of reactors, the zero-sequence impedance elements are often used as a criterion for protection. In the case of external fault, the zero-sequence impedance measured at the reactor port is the zero-sequence impedance of the reactor itself, and the impedance measured in the case of internal fault is the zero-sequence impedance of the system. Therefore, whether the fault is an interturn fault can be judged according to whether the zero-sequence impedance measured is

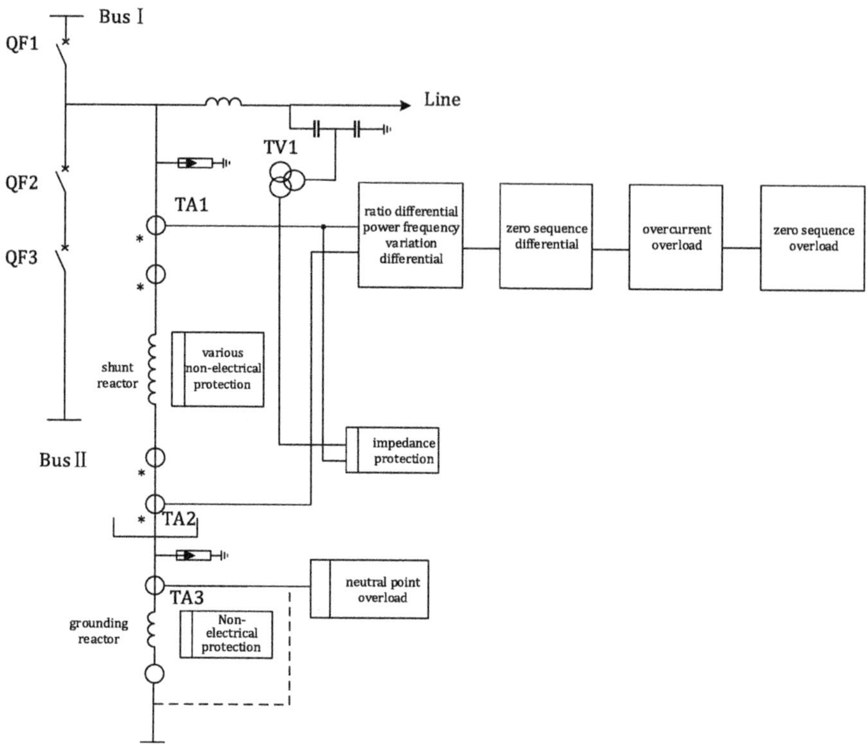

Figure 10.8 Schematic diagram of protection configuration for a 1000 kV ultra-high-voltage shunt reactor.

much larger than the zero-sequence impedance of the reactor. However, due to the large zero-sequence impedance of a 1000 kV system, especially the reactor installed in the intermediate switching station, excessive reliance on the zero-sequence impedance criterion may lead to misoperation of the interturn protection in UHV reactor protection. So the method of combining the two interturn protection principles of passing through the zero-sequence element and without passing through the zero-sequence element are adopted in the protection device.

(1) There may be a serious transient process in the operation of no-load closing, internal fault, external fault and fault removal, and the operation of line protection and reclosing. This transient process may bring some significantly attenuated DC components and harmonic components appearing in the reactor's voltage (especially in the voltage) and current. These components may lead to incorrect judgments of zero-sequence power directional components. To avoid this problem, harmonic

Table 10.2 Protection configuration of a 1000 kV reactor.

	Protection type	Zone numbers	Time delay settings each zone	Remarks
Main protection	Differential instantaneous trip	—	—	
	Percentage differential	—	—	
	Zero-sequence percentage differential	—	—	
	Interturn short-circuit protection	—	—	
	Frequency variation ratio differential	—	—	
Backup protection	Phase overcurrent	2	1	
	Zero-sequence overcurrent	2	1	
	Overload*	2	1	Zone I can be used for overload alarm, zone II can be used to start air cooling, etc.
	Neutral point overload*	1	1	Can be used for overloading alarm

* Representation of abnormal alarm function.

voltage detection elements are added to the device. Harmonic voltage detection elements can improve the operating threshold of interturn protection adaptively when having detected serious harmonic voltage to ensure interturn protection reliability in these transient processes. At the same time, whether the harmonic voltage element should be put into use is judged by the size of the zero-sequence current to ensure protection operating speed in an internal interturn fault. Meanwhile, the existence of line-distributed capacitance also affects the measurement of zero-sequence impedance at the reactor port.

(2) The compensation coefficient of the reactor in a 1000 kV transmission system is high, and therefore it is easy to produce LC resonance after the removal of a single-phase switch inline fault, which may lead to resonance voltage and current on the removed line. It also may directly affect the reactor's zero-sequence voltage and zero-sequence current and lead to misjudgment of the zero-sequence impedance component and zero-sequence power directional element. This phenomenon has appeared in the process of the 1000 kV reactor protection dynamic model test, so some measures such as the weakening voltage criterion under the condition of a large margin of current criterion, and the interturn fault criterion independent of voltage, have been adopted in the device to ensure protection reliability.

(3) The higher the protection's sensitivity, the more advantageous to the safe and stable operation of equipment. In order to improve the sensitivity of the interturn protection, the adaptive zero-sequence voltage compensation element is introduced into the protection. In the case of a small interturn fault, the sensitivity and judgment speed of zero-sequence power directional component detection can be improved by voltage compensation, it also improves the capability of protection: when an interturn short-circuit fault occurs inside the reactor, the phase of zero-sequence current goes ahead of that of zero-sequence voltage by about 90°; when a single-phase grounding fault occurs inside the reactor, the phase of zero-sequence current also goes ahead of that of zero-sequence voltage; when a single-phase grounding fault occurs out of the reactor, the phase of zero-sequence current goes behind that of zero-sequence voltage. Therefore, the phase relationship between the zero-sequence current and zero-sequence voltage on the side of the reactor can be used to distinguish the interturn short circuit, internal and external grounding short circuits for an interturn short-circuit fault inside the reactor, a single-phase grounding fault inside the reactor, and single-phase grounding fault out of the reactor occurring. As the zero-sequence impedance of the system is very small compared with the zero-sequence impedance of the reactor, when an interturn short circuit occurs, the zero-sequence source is inside the reactor and the voltage drop (zero-sequence voltage) of the zero-sequence current on the system zero-sequence impedance is very small. Hence in order to improve the sensitivity of interturn short-circuit protection, the zero-sequence voltage needs to be compensated.

(4) A study on TA ratio of reactor protection. The reactor parameters of a 1000 kV system are shown in Table 10.3.

The rated primary current of the above four sets of reactors is shown in Table 10.4.

It can be seen from Table 10.4 that the rated primary current of the reactor is not very large, and there may not be any traversing short-circuit current flowing through the reactor itself. In the selection of the TA transformation ratio, instead of selecting according to the short-circuit level of the system, the load current under normal conditions should be considered to some extent. Therefore, it is suggested that for a 1000 kV shunt reactor, the TA transformation ratio should be selected as 1000/1, which is basically equivalent to the secondary load current of the TA transformation ratio being selected as 300/1 for a traditional 500 kV shunt reactor.

(5) The study of the setting of reactor protection. There is no obvious difference between 1000 kV reactor protection configuration and 500 kV reactor protection configuration, and its fixed value is similar to that of 500 kV reactor protection.

When the TA transformation ratio is selected as 1000/1, the fixed value of the differential protection can be set to 0.2 I_N, and if the TA transformation ratio is selected as 2000/1, the fixed value of the differential protection needs to be properly improved, which needs to be set to (0.3—0.4) I_N. Its braking coefficient can be set to 0.5 by the traditional way, and its differential fast tripping fixed value can be set to 2—3 I_N.

10.5 Fast impedance protection for large-capacity transformers [5]

State Grid in China has made a provision of duality for main equipment protection, and the "double main and double back up" protection scheme becomes the criterion for the configuration of main equipment protection. Since there may be performance defects in protection based on any principle, the duality should be two main protections based on different principles and two backup protections based on different principles.

At present, for a large-capacity and multibranched generator, the main protections based on different principles, such as complete pilot differential protection, incomplete pilot differential protection, single-element transverse, split-phase transverse, and pilot zero-sequence differential

Table 10.3 Reactor parameters of a 1000 kV system.

	# 1 reactor	# 2 reactor	# 3 reactor	# 4 reactor
Rated capacity	960 Mvar	720 Mvar	720 Mvar	600 Mvar
Rated reactance	1260 Ω	1680 Ω	1680 Ω	2016 Ω
Rated neutral reactance	280 Ω	327 Ω	334 Ω	440 Ω

Table 10.4 The rated primary current of the above four sets of reactors.

	# 1 reactor	# 2 reactor	# 3 reactor	# 4 reactor
Rated primary current	528 A	396 A	396 A	330 A

protections, can be set up. For the interphase short-circuiting of a large capacity transformer, there has been only main protection based on the pilot differential principle for quite a long time, which is greatly influenced by magnetizing inrush current. Although several methods can identify magnetizing inrush current, there is no guarantee of 100% reliability. Especially, extra-high-voltage (EHV) and UHV large-capacity transformers have small magnetizing inrush current, slow attenuation, fewer secondary harmonics, and poor locking ability. Furthermore, EHV and UHV transformers are connected by long-distance EHV and UHV transmission lines whose distributed capacitance and saturation reactance resonance enlarge harmonic contents in cases of fault. Thus it makes it more difficult to distinguish magnetizing inrush current and short-circuit current accurately and makes the protection not reliable. The interphase short circuit is the most harmful to large-capacity transformers. Although EHV and UHV transformers usually consist of three single-phase transformers with no direct interphase short circuit occurring, there is still a possibility of interphase short circuit between the leads of three single-phase transformers. In addition, during thunderstorm season, there is also the possibility of two-phase grounding short circuits, which are actually interphase short circuit. For this kind of short circuit, the only main protection at present is the pilot differential method, and the requirement of dual main protection based on different principles thus cannot be realized. This is a hidden danger for EHV and UHV transformers that cannot be ignored.

As we all know, impedance protection can be used not only as backup protection in the case of an interphase fault on one side of the transformer but also in detection of interphase short circuit on the opposite side of the Y/△ connection transformer. It only needs to consider the phase change of the fault current and the voltage at the fault point passing through the Y/△ connection in the settings. In this way, it is entirely feasible to transplant the impedance pilot protection of the transmission line to the transformer. When the transformer is energized without power load, magnetizing inrush current may be produced. In this case, the power frequency component of the current increases quite a bit, while the power frequency of the voltage is reduced a little, so the measuring impedance is not reduced very much and may not make the impedance protection misoperate. This is the greatest advantage of impedance protection.

Impedance pilot protection of the transmission line mainly consists of four types: underrange setting direct trip, underrange setting permissive trip, overrange setting permissive, and overrange setting locking. Due to the low sensitivity of impedance protection when there is an interphase short circuit near the neutral point, the underrange direct trip scheme cannot be applied. Both overrange setting locking and permissive types can be applied.

10.5.1 Overrange locking type

The overrange locking type can be applied to a double or multiple winding transformers, which requires two sets of impedance relays (elements) installed on each side of the transformer, of which one points to the inner (i.e. the forward direction) and the other to the external (or the reverse direction) parts of the transformer. Every impedance relay in forward direction is set in an overrange way so that both whole-transformer and lead wire faults can be reflected. The impedance relay pointing to the reverse direction should be set according to the action range which exceeds that of the forward action impedance relay on the other sides. Because the reverse phase impedance relay only reflects the short circuit on the reverse direction line, as long as this principle is set when the transformer is externally short-circuited, the reverse direction impedance relay on the open circuit side acts, sending out locking signals by the "or" door to the protection on each side and reliably closing the protection on all sides, as shown in Fig. 10.9. In setting the forward impedance relay, when the opposite two phases are short-circuited, the conversion of current and voltage passing through Y/△ should be considered; the setting method is omitted here. In Fig. 10.9, 21 represents interphase impedance protection, and 52 represents circuit breaker, which is the international general equipment number.

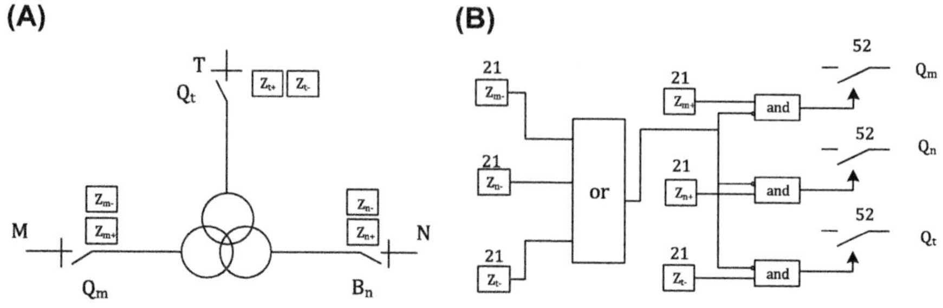

Figure 10.9 Logic block diagram for overrange locking type impedance protection of transformer:
(A) protection configuration, (B) logic block diagram of protection action.

10.5.2 Overrange setting permissive type

In this way, only one set of the interphase impedance relay is installed on each side, and setting is made according to being greater than the transformer impedance and accounting for the maximum auxiliary increasing coefficient. Its action directions all point to the transformer interior. When a short circuit occurs within the transformer or in the lead, the impedance relays on each side act, sending a permissive signal to the other sides and tripping the circuit breakers on each side through the "and" door, as shown in Fig. 10.10. When a short circuit occurs outside the transformer, the protection on the near side of the short-circuit point may not act and the "and" door may not be opened, and neither side may trip. This scheme is simpler than the overrange locking scheme, as only one forward impedance relay needs to be installed on each side. However, when there is no power supply and some internal fault occurs in the protection on either side, the

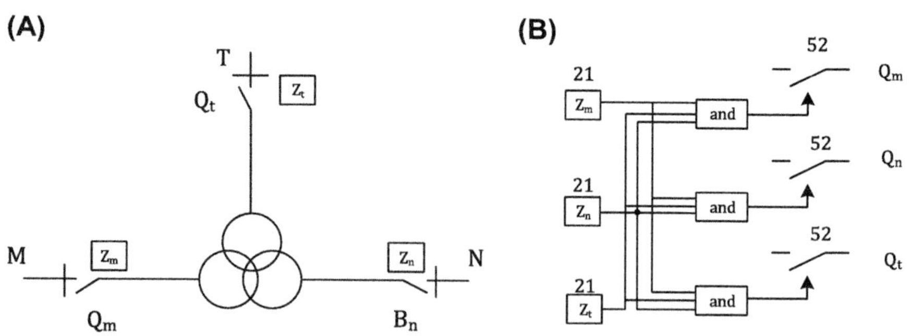

Figure 10.10 Logic block diagram for overrange permissive type impedance protection of transformer:
(A) protection configuration, (B) logic block diagram of protection action.

protection on this side may not act, and the protection on other sides may not trip either. For this reason, it is necessary to install low-voltage terminal protection on each side because the power supply may be lost—i.e. when the voltage and current at the terminal are less than the minimum, the protection should act and send permissive signals to other terminals. When the transformer is no-load input from one side, function of the "and" door can be relieved by a manual closing signal. The logic block diagram for overrange permissive type impedance protection of a transformer is shown in Fig. 10.10. Simulation results show the correctness and reliability of the two schemes.

References

[1] Okuyama K, Hosi M, Mori E, Kashima Y, Kamata Y. Development of a prototype UHV transformer and its application to 500 kV transformers. IEEE Transactions on Power Apparatus and Systems 1984;(9):2545−52.
[2] Li Y, He J. New principle relay for protecting power transformers. Automation of Electric Power Systems 1995;(7):15−9.
[3] Li B, Li Y, He J. Study on the performance of protecting relaying and stunt reactors of 750kV transmission Line 2005 (11):40−44.
[4] Li B, Li Y, Chen J, Zhang Y. Turn-to-Turn faults protection of EHV shunt reactors. Journal of Tianjin University 2005;(8):717−21.
[5] He J, He J, Wang G, Xue S. Impedance backup protection for UHV large capacity transformers. Power System Protection and Control 2014;42(22):1−8.

CHAPTER 11

Multi-circuit transmission lines

Contents

With the increase of power demand and the shortage of transmission corridors, the multi-circuit transmission lines on the same tower are widely adopted in extra-high-voltage (EHV)/ultra-high-voltage (UHV) transmission systems. Considering the demand of the transmission system, multi-circuit lines in the form of completely on same tower, partially on same tower, multi-voltage level, and phase conductor untransposition are also common.

11.1 Tower structure and conductor arrangement

11.1.1 Tower structure

Due to increasingly tight land resources, the construction cost of transmission networks is growing progressively higher. As a result, multicircuit transmission lines have become an inevitable trend in the development of UHV power grids. The research and construction of multicircuit transmission lines are very extensive. In the United States, the ratios of double-circuit lines at voltage levels of 220 and 345 kV are 47% and 52% respectively. All British 400 kV transmission lines use same-tower double-circuit lines. More than 90% of 500 kV lines in Japan are double-circuit lines, while the four 1000 kV UHV lines that have been built are

Protection Technologies of Ultra-High-Voltage AC Transmission Systems
ISBN 978-0-12-816205-7
https://doi.org/10.1016/B978-0-12-816205-7.00011-X

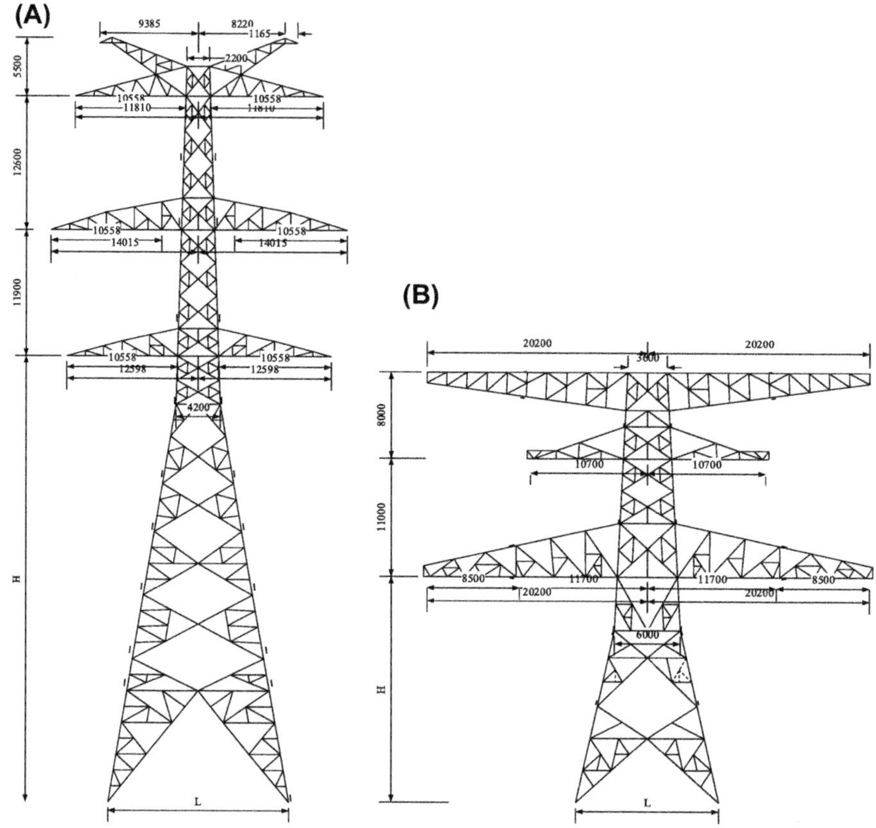

Figure 11.1 Tower structure of typical 500 kV double-circuit lines (unit: mm). (A) Conventional drum-type tower. (B) Compact double-triangle-type tower.

all double-circuit lines. In addition, as pioneers of compact transmission technology, the Soviet Union and Japan have built UHV double-circuit compact transmission lines on the same tower. The double-circuit transmission line is widely used in China's EHV/UHV transmission system. The typical 500 kV double-circuit line tower structure is shown in Fig. 11.1.

Operating experience of Soviet and Japanese UHV lines shows that lightning strikes are the main causes of common faults or cross-line faults. The conventional drum-type tower structure shown in Fig. 11.1A is advantageous for preventing the ice-shedding jump, while the umbrella-type tower is more advantageous in lightning protection performance. The compact tower shown in Fig. 11.1B has been successfully applied in China's 500 kV system. The spacing between the wires of compact double-circuit lines is closer. Especially under UHV electric field conditions, the skin effect caused by alternating current flowing through the wire and the earth makes the parameter frequency characteristic more obvious. China proposed a 750 kV compact double-circuit tower structure and further proposed a symmetrical inverted-triangle arrangement with a more compact structure and superior electromagnetic environment for vertical arrangement and double-triangle arrangement.

With the development of same-tower transmission technology, the number of multicircuit transmission lines is increasing. Quadruple-circuit lines are the most common, and triple-circuit lines have also been adopted in some transmission networks. The tower structures of typical quadruple- and triple-circuit lines are respectively shown in Figs. 11.2A and B.

Considering the complexity of tower structure, the arrangement of the wires varies, the electromagnetic and electrostatic coupling of multicircuit lines between phases and circuits that are asymmetrical, so the parameter asymmetry and current unbalance of the UHV multicircuit transmission lines need to be taken seriously.

11.1.2 Phase conductor arrangement

It can be seen from the tower structure that the conductor arrangement for each tower circuit can be generally divided into four arrangement categories: vertical, erected triangular, horizontal, and inverted triangular, as shown in Fig. 11.3. The phase conductor spacing of 1000 kV UHV transmission lines can reach 20 m, while the phase conductor spacing of 500 kV EHV lines is generally 10—15 m.

For double-circuit lines, the wires of two circuits are mostly vertically arranged. As the primary factor affecting the electrical unbalance of the double-circuit line, the phase arrangement is quite diverse. The wire typical arrangement and phase arrangement of double-circuit lines is shown in Fig. 11.4.

The research shows that the phase arrangement, height to the ground, spacing between circuits, and conductor property are the main factors that determine the electrical unbalance of multicircuit transmission lines. For conventional UHV double-circuit lines, the negative-sequence unbalanced current of the abc—abc arrangement is largest, followed by the abc—bca arrangement, and the negative-sequence unbalanced current of the abc—cba arrangement is smallest. For the same tower structure and wire arrangement, the circulating and electrostatic unbalances of UHV double-circuit lines are more significant than those of EHV lines.

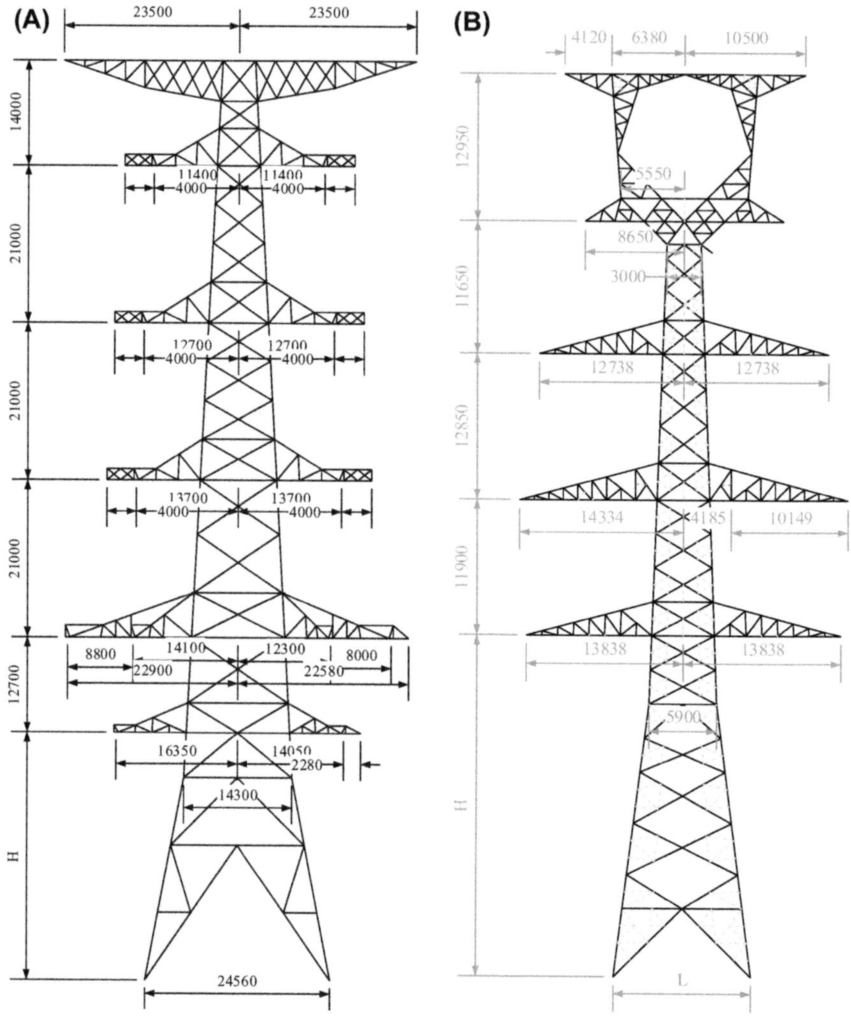

Figure 11.2 Tower structure of typical multicircuit lines (unit: mm). (A) Tower structure of typical multi-voltage quadruple-circuit lines. (B) Tower structure of typical triple-circuit lines.

In order to limit the unbalanced current caused by electromagnetic coupling, transmission lines with a length exceeding 100 km should be transposed, and the length of transposition circulation should not exceed 200 km. In real transmission system, it is often difficult to achieve full transposition of phase conductor. The dense substation distribution and the shortage of transmission corridors make the length of each transmission line shorter than the specified transposition length. The variety of erection

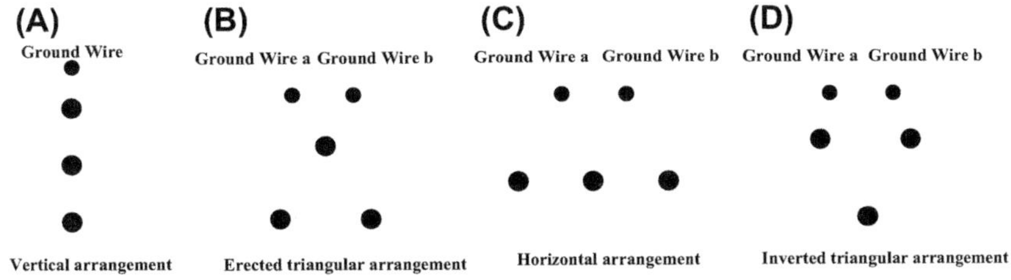

Figure 11.3 Wire arrangement of a single-circuit line.

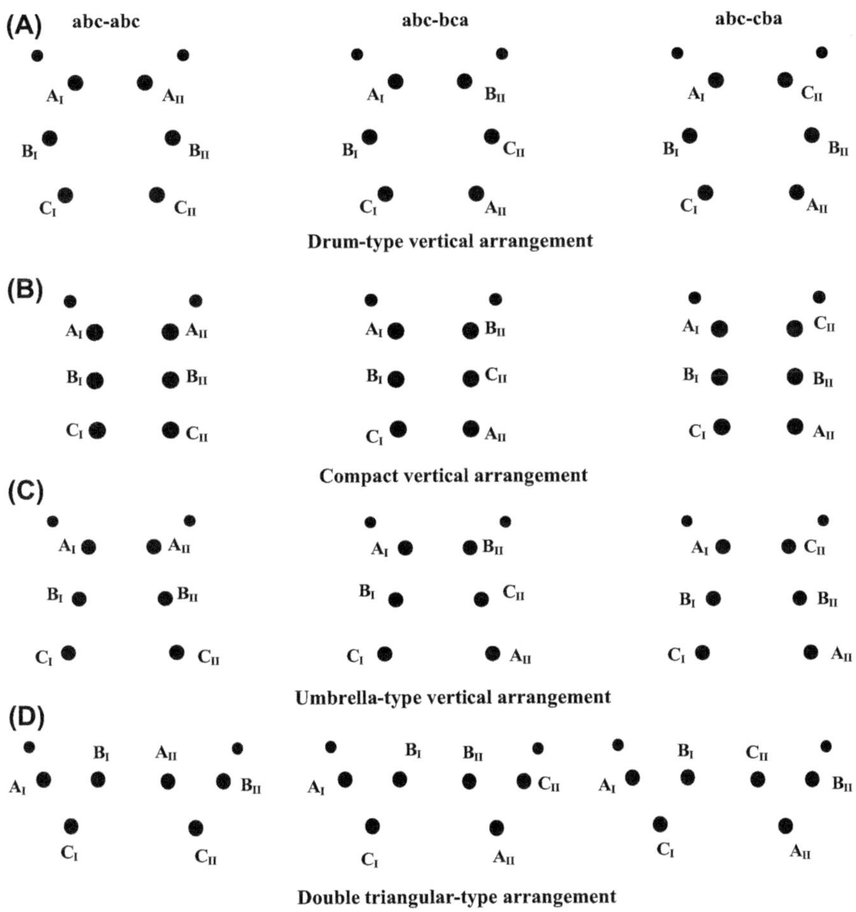

Figure 11.4 Wire arrangement of a double-circuit line.

forms and topologies result in more significant and complex electromagnetic coupling. The fault characteristics of multicircuit lines and their impact on protection have naturally been the focus of attention.

11.2 Unbalanced current of double-circuit lines

11.2.1 Unbalanced factors of double-circuit lines

The distance between the wires of a double-circuit transmission line on the same tower is relatively close, and there is electrostatic and electromagnetic coupling between the phases and between the wires. In addition, UHV tower double-circuit transmission line wires may not be completely transposed, and incompletely transposed or even untransposed lines may aggravate the unbalanced current problem of the UHV double-circuit transmission line.

Two types of current unbalance in double-circuit lines are formulated by Hesse [1,2]: net through unbalance factors (m_{t2}, m_{t0}) and circulating unbalance factors (m_{c2}, m_{c0}):

$$\begin{cases} m_{t2} = \left|\dfrac{\dot{I}_2 + \dot{I}'_2}{\dot{I}_1 + \dot{I}'_1}\right| \times 100\%, \quad m_{t0} = \left|\dfrac{\dot{I}_0 + \dot{I}'_0}{\dot{I}_1 + \dot{I}'_1}\right| \times 100\% \\ m_{c2} = \left|\dfrac{\dot{I}_2 - \dot{I}'_2}{\dot{I}_1 + \dot{I}'_1}\right| \times 100\%, \quad m_{c0} = \left|\dfrac{\dot{I}_0 - \dot{I}'_0}{\dot{I}_1 + \dot{I}'_1}\right| \times 100\% \end{cases} \tag{11.1}$$

where

$\dot{I}_1$, $\dot{I}_2$, $\dot{I}_0$ are the respective positive-, negative-, and zero-sequence current of double-circuit line I.

$\dot{I}'_1$, $\dot{I}'_2$, $\dot{I}'_0$ are the respective positive-, negative-, and zero-sequence current of double-circuit line II.

In order to clearly explain the negative- and zero-sequence current of each circuit, current unbalances m_2, m_0 and m'_2, m'_0 of individual circuit I or II are also observed:

$$\begin{cases} m_{I2} = \left|\dfrac{\dot{I}_2}{\dot{I}_1}\right| \times 100\%, \quad m_{I0} = \left|\dfrac{\dot{I}_0}{\dot{I}_1}\right| \times 100\% \\ m_{II2} = \left|\dfrac{\dot{I}'_2}{\dot{I}'_1}\right| \times 100\%, \quad m_{II0} = \left|\dfrac{\dot{I}'_0}{\dot{I}'_1}\right| \times 100\% \end{cases} \tag{11.2}$$

11.2.2 Unbalanced current under normal operation

Double-circuit lines are represented by circuits I and II, respectively. Suppose that $[Z_I]$ and $[Z_{II}]$ are the three-phase impedance matrix of circuits I and II, respectively; $[Z_M]$ is the three-phase mutual-impedance matrix of circuit II to circuit I; and $[Z_M]^T$ is the three-phase mutual-impedance matrix of circuit I to circuit II. Assuming that the earth is a plane with infinite conductivity, each element in the impedance matrix can be determined by the following equation:

$$\begin{cases} Z_{\mathrm{ii}} = R + \mathrm{j}\dfrac{\omega\mu_0}{2\pi}\ln\dfrac{2h_{\mathrm{i}}}{r_{\mathrm{i}}} \\ Z_{\mathrm{ij}} = \mathrm{j}\dfrac{\omega\mu_0}{2\pi}\ln\dfrac{D'_{\mathrm{ij}}}{D_{\mathrm{ij}}} \end{cases} \tag{11.3}$$

where h_{i} is the average height of wire i to ground; r_{i} is the radius of wire I; D'_{ij} is the distance between wires i and j; D_{ij} is the distance between the mirror of wire i and the mirror image of j; Z_{ii} is corresponding-phase self-impedance; and Z_{ij} corresponds to the mutual impedance between two phases.

Double-circuit transmission lines with common two-terminal buses are shown in Fig. 11.5.

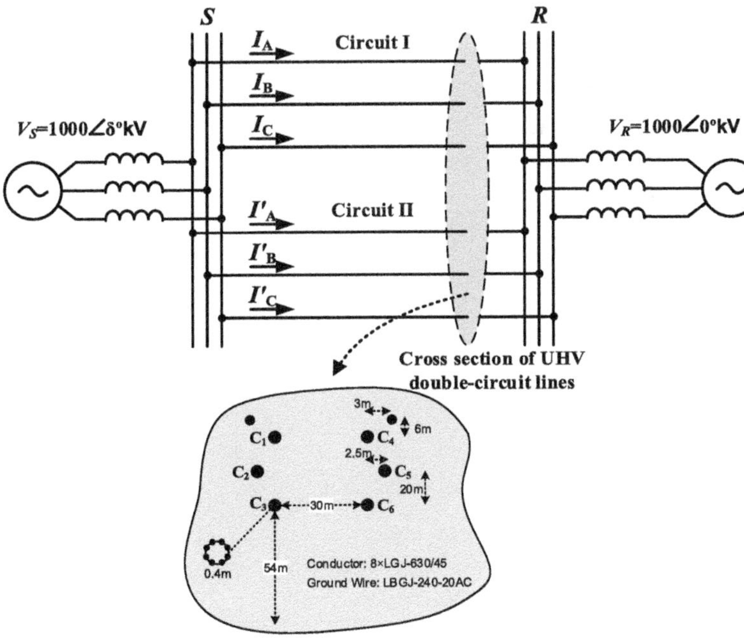

Figure 11.5 Diagram of an ultra-high-voltage double-circuit transmission line.

The current vectors of two circuits can be represented as $[\dot{I}] = [\dot{I}_A \ \dot{I}_B \ \dot{I}_C]^T$, $[\dot{I}'] = [\dot{I}'_A \ \dot{I}'_B \ \dot{I}'_C]^T$. The voltage vectors at buses are $[\dot{U}_S] = [\dot{U}_{AS} \ \ \dot{U}_{BS} \ \ \dot{U}_{CS}]^T$, $[\dot{U}_R] = [\dot{U}_{AR} \ \ \dot{U}_{BR} \ \ \dot{U}_{CR}]^T$.

Generally, the electrostatic contribution to current unbalances for untransposed lines is negligible for short lines [1]. The following equation can be obtained:

$$\begin{bmatrix} [\dot{U}_{SR}] \\ [\dot{U}_{SR}] \end{bmatrix} = \begin{bmatrix} [\dot{U}_S] - [\dot{U}_R] \\ [\dot{U}_S] - [\dot{U}_R] \end{bmatrix} = \begin{bmatrix} [Z_S] & [Z_M] \\ [Z_M]^T & [Z'_S] \end{bmatrix} \begin{bmatrix} [\dot{I}] \\ [\dot{I}'] \end{bmatrix} \tag{11.4}$$

Here $[Z_S]$, $[Z'_S]$ and $[Z_M]$, $[Z_M]^T$ are impedance matrices of individual circuits and their mutual-impedance matrices. If ground wires are included, a matrix with 6×6 dimensions can be obtained by corresponding reduction methods [3].

The relationship among sequence voltages and currents of any circuit line can be obtained when three-phase symmetrical component transformation matrix $\boldsymbol{T}$ is applied to (11.4):

$$\begin{bmatrix} [\dot{U}_{SR}]_{012} \\ [\dot{U}_{SR}]_{012} \end{bmatrix} = \begin{bmatrix} T[Z_S]T^{-1} & T[Z_M]T^{-1} \\ T[Z_M]^T T^{-1} & T[Z'_S]T^{-1} \end{bmatrix} \begin{bmatrix} [\dot{I}]_{012} \\ [\dot{I}']_{012} \end{bmatrix} \tag{11.5}$$

where $T = \begin{bmatrix} 1 & 1 & 1 \\ 1 & \alpha & \alpha^2 \\ 1 & \alpha^2 & \alpha \end{bmatrix}$, $\alpha = e^{j\frac{2\pi}{3}}$. By matrix inversion, Eq. (11.6) can be obtained from (11.5):

$$\begin{bmatrix} \dot{I}_0 \\ \dot{I}_1 \\ \dot{I}_2 \\ \dot{I}'_0 \\ \dot{I}'_1 \\ \dot{I}'_2 \end{bmatrix} = \begin{bmatrix} y_{00} & y_{01} & y_{02} & y_{00'} & y_{01'} & y_{02'} \\ y_{10} & y_{11} & y_{12} & y_{10'} & y_{11'} & y_{12'} \\ y_{20} & y_{20} & y_{20} & y_{20} & y_{20} & y_{20} \\ y_{0'0} & y_{0'1} & y_{0'2} & y_{0'0'} & y_{0'1'} & y_{0'2'} \\ y_{1'0} & y_{1'1} & y_{1'2} & y_{1'0'} & y_{1'1'} & y_{1'2'} \\ y_{2'0} & y_{2'1} & y_{2'2} & y_{2'0'} & y_{2'1'} & y_{2'2'} \end{bmatrix} \begin{bmatrix} \dot{U}_0 \\ \dot{U}_1 \\ \dot{U}_2 \\ \dot{U}_0 \\ \dot{U}_1 \\ \dot{U}_2 \end{bmatrix} \tag{11.6}$$

where $U_{0,1,2}$, $I_{0,1,2}$, and $I'_{0,1,2}$ are zero-, positive-, and negative-sequence voltages and currents. y_{ii} and y_{ij} (i, j = 0, 1, 2, 0′, 1′, 2′) represent self-admittance and mutual admittance among different sequence components.

Obviously, line-to-line mutual influence and conductor asymmetrical spacing may result in different self-impedance values and different mutual-impedance values [2,4]. Therefore, the sequence components of any circuit line in (11.5) or (11.6) are not decoupled completely. That is to say that even if the voltage between two buses has perfect symmetry, zero- and negative-sequence unbalanced current may emerge.

Zero- and negative-sequence unbalanced current components in one circuit may not be in phase, respectively, with those in other circuits. The inphase portions lead to the overall net through unbalance. The out-of-phase portions, however, lead to circulating current components flowing down one circuit and returning through others. Previous research and current unbalance calculations in the following context show that the net through unbalance in comparatively short lines is not significant, whereas the circulating current unbalance can be quite large [1,4].

Based on the parameters shown in Fig. 11.5, a frequency-dependent model of UHV double-circuit lines is adopted using EMTP simulation. Simulation results are listed in Table 11.1.

For untransposed double-circuit transmission lines, net through unbalances m_{t2} and m_{t0} exhibit an apparently increasing trend with line length increasing, whereas circulating unbalances m_{c2} and m_{c0} are basically unchanged with line length for a certain phase arrangement. The current unbalances m_0, m_2, m'_0, and m'_2 of an individual circuit are determined by overall net through and circulating current unbalances. Obviously, phase arrangements have a great impact on unbalance characteristics. For example, abc—abc phasing shows the highest net through unbalance and lowest circulating unbalance among phase arrangements. Although phase arrangements abc—bca and abc—acb lead to low net through unbalances, they have high circulating unbalances. In real applications, many phase arrangements are possible, so circulating and net through unbalanced currents should always be watched.

The values in Table 11.1 indicate that m_{c2} and m_{c0} may reach 8%—9% and 4%—5% respectively. However, it should be emphasized that the values of m_{c0} here are not the most significant. Due to different tower structures and phase arrangements, m_{c0} may reach 8% or even more, as discussed in Ref. [1]. Since the length of untransposed lines is comparatively short, the influence of capacitive current is limited. Plenty of simulation results show that the characteristics of the unbalanced current of EHV and UHV untransposed lines are basically the same.

Table 11.1 Zero- and negative-sequence unbalancing factors for untransposed double-circuit lines in normal operation (30-degree load angle).

Phase arrangement	Line length (km)	Circuit I			Circuit II			Double circuit				
		$\lvert i_1 \rvert$(kA)	m_0(%)	m_2(%)	$\lvert i_1' \rvert$(kA)	m_0'(%)	m_2'(%)	$\lvert i_1 + i_1' \rvert$(kA)	m_{t0}(%)	m_{c0}(%)	m_{t2}(%)	m_{c2}(%)
abc—*abc*	25	1.48	0.52	0.45	1.48	0.52	0.45	2.95	0.52	0.00	0.45	0.00
	50	1.42	1.13	1.10	1.42	1.13	1.10	2.84	1.12	0.00	1.09	0.00
	100	1.34	1.82	2.12	1.34	1.82	2.12	2.69	1.82	0.00	2.12	0.00
	150	1.28	2.26	3.11	1.28	2.26	3.11	2.56	2.26	0.00	3.11	0.00
	200	1.23	2.56	4.09	1.23	2.56	4.09	2.45	2.56	0.00	4.08	0.00
abc—*bca*	25	1.47	4.16	8.70	1.47	4.43	8.68	2.94	0.27	4.28	0.05	8.70
	50	1.43	4.10	8.66	1.43	4.55	8.72	2.86	0.48	4.30	0.10	8.74
	100	1.36	4.06	8.59	1.36	4.73	8.82	2.72	0.79	4.34	0.19	8.71
	150	1.30	4.08	8.90	1.30	4.83	8.92	2.60	0.99	4.38	0.30	8.72
	200	1.26	4.13	8.51	1.25	4.91	8.94	2.51	1.14	4.38	0.40	8.69
abc—*cba*	25	1.47	1.24	0.58	1.47	1.10	0.58	2.94	0.30	1.13	0.15	0.56
	50	1.43	1.37	0.67	1.43	1.18	0.69	2.86	0.56	1.15	0.30	0.61
	100	1.36	1.63	0.89	1.36	1.46	0.95	2.72	0.99	1.19	0.60	0.70
	150	1.30	1.86	1.14	1.30	1.79	1.23	2.61	1.36	1.22	0.91	0.76
	200	1.26	2.11	1.41	1.26	2.12	1.53	2.51	1.69	1.26	1.22	0.82
abc—*acb*	25	1.47	3.40	9.47	1.47	2.98	9.78	2.94	0.66	3.13	0.36	9.60
	50	1.43	3.71	9.33	1.43	2.96	9.93	2.85	1.18	3.14	0.70	9.61
	100	1.35	4.27	9.04	1.36	3.08	10.25	2.70	1.98	3.16	1.36	9.55
	150	1.29	4.71	8.71	1.30	3.31	10.58	2.58	2.55	3.17	1.99	9.50
	200	1.24	5.05	8.50	1.25	3.56	10.91	2.48	2.99	3.19	2.61	9.44

11.2.3 Unbalanced current under external faults

Very large positive-sequence current passes through double-circuit lines when external symmetrical faults occur [5]. Consequently, circulating and net through unbalanced currents with significant values may arise in double-circuit lines.

The two terminal systems shown in Fig. 11.5 are assumed to be ideal infinite power sources. In these systems, $\dot{U}_{AR} = U\angle 0°$, $\dot{U}_{AS} = U\cdot e^{j\delta}$, and δ is the power angle difference. For normal operation, the voltage drop along the line is $\left[\dot{U}_0 \;\; \dot{U}_1 \;\; \dot{U}_2\right]_{SR}^T = \left[0, \;\left(e^{j\delta}-1\right)U, \;0\right]^T$. Obviously, the sequence voltage drop in the case of an external symmetrical fault has a similar format. According to Eq. (11.6), the zero-sequence unbalanced current factors for normal operation or external symmetrical fault can be estimated:

$$\begin{cases} m_{t0} = \left|(y_{01}+y_{01'}+y_{0'1}+y_{0'1'})/(y_{11}+y_{11'}+y_{1'1}+y_{1'1'})\right| \times 100\% \\ m_{c0} = \left|(y_{01}+y_{01'}-y_{0'1}-y_{0'1'})/(y_{11}+y_{11}+y_{1'1}+y_{1'1'})\right| \times 100\% \end{cases} \tag{11.7}$$

Negative-sequence unbalanced current factors can be deduced in the same way. Undoubtedly, negative- and zero-sequence unbalanced currents have significant values under the condition of external symmetrical faults due to large positive-sequence current passing through double-circuit lines.

For an external phase-to-phase fault, assume the voltages at bus S are symmetrical when an external fault occurs at side R—that is, $\left[\dot{U}_S\right] = \left[\dot{U}_{AS} \;\; \dot{U}_{BS} \;\; \dot{U}_{CS}\right]^T$. First, for a BC phase-to-phase fault, voltages at bus R can be given by $\left[\dot{U}_R\right] = \left[\dot{U}_{AR} \;\; -\dot{U}_{AR}/2 \;\; -\dot{U}_{AR}/2\right]^T$. Consequently, the sequence voltage drop between buses S and R in the case of a BC phase-to-phase fault can be calculated as

$$\left[\dot{U}_0 \;\; \dot{U}_1 \;\; \dot{U}_2\right]_{SR}^T = [T]\left[\dot{U}_S - \dot{U}_R\right] = \left[0, \;2e^{j\delta}-1, \;-1\right]^T \cdot \frac{U}{2} \tag{11.8}$$

Similarly, for the case of a CA phase-to-phase fault at bus R,

$$\left[\dot{U}_0 \;\; \dot{U}_1 \;\; \dot{U}_2\right]_{SR}^T = \left[0, \;2e^{j\delta}-1, \;-\alpha\right]^T \cdot \frac{U}{2} \tag{11.9}$$

For the case of an AB phase-to-phase fault at bus R,

$$\left[\dot{U}_0 \;\; \dot{U}_1 \;\; \dot{U}_2\right]_{SR}^T = \left[0, \;2e^{j\delta}-1, \;-\alpha^2\right]^T \cdot \frac{U}{2} \tag{11.10}$$

Substituting the respective sequence voltages of (11.8)–(11.10) into (11.6), the zero-sequence unbalanced current factors for external phase-to-phase faults can be estimated:

$$\begin{cases} m_{t0} = \dfrac{\left|(y_{01}+y_{01'}+y_{0'1}+y_{0'1'})(2e^{j\delta}-1)+K(y_{02}+y_{02'}+y_{0'2}+y_{0'2'})\right|}{\left|(y_{11}+y_{11'}+y_{1'1}+y_{1'1'})(2e^{j\delta}-1)+K(y_{12}+y_{12'}+y_{1'2}+y_{1'2'})\right|}\times 100\% \\ m_{c0} = \dfrac{\left|(y_{01}+y_{01'}+y_{0'1}+y_{0'1'})(2e^{j\delta}-1)-K(y_{02}+y_{02'}+y_{0'2}+y_{0'2'})\right|}{\left|(y_{11}+y_{11'}+y_{1'1}+y_{1'1'})(2e^{j\delta}-1)+K(y_{12}+y_{12'}+y_{1'2}+y_{1'2'})\right|}\times 100\% \end{cases} \quad (11.11)$$

Here K is -1, $-\alpha$, or $-\alpha^2$, respectively, for a BC, a CA, or an AB phase-to-phase fault. Obviously, different external phase-to-phase faults lead to different current unbalances. Still, as shown in Fig. 11.5, line length is assumed to be 100 km. Zero-sequence current unbalances for external phase-to-phase faults at bus R are presented in Table 11.2.

There is no circulating unbalance in double-circuit lines with an abc—abc arrangement due to the structural symmetry of this arrangement. It can be seen from Table 11.2 that the magnitudes of unbalanced currents differ for the different external phase-to-phase faults. Different from the results for normal operation, the unbalanced current in the case of external fault, and its influence, cannot be ignored because there is a large positive-sequence current passing through double-circuit lines.

Table 11.2 Zero-sequence current unbalances for external phase-to-phase faults (30-degree load angle).

Phase arrangement	Fault type	Circuit I I_1/kA	Circuit II m_0	Double circuit I'_1/kA	m'_0	m_{t0}	m_{c0}
abc—abc	AB	2.53	3.42	2.53	3.42	3.42	0.00
	BC	2.57	0.81	2.57	0.81	0.81	0.00
	CA	2.44	2.69	2.44	2.69	2.69	0.00
abc—bca	AB	2.65	3.77	2.46	3.07	0.53	3.44
	BC	2.67	6.06	2.48	8.53	1.00	7.30
	CA	2.37	3.13	2.75	5.24	1.50	4.20
abc—cba	AB	2.58	2.55	2.55	0.78	1.04	1.58
	BC	2.57	1.18	2.59	1.98	1.55	0.51
	CA	2.54	1.59	2.54	2.60	0.53	2.08
abc—ach	AB	2.61	4.79	2.42	2.16	3.55	1.49
	BC	2.70	6.49	2.47	4.99	1.03	5.76
	CA	2.31	2.53	2.76	6.13	2.47	4.35

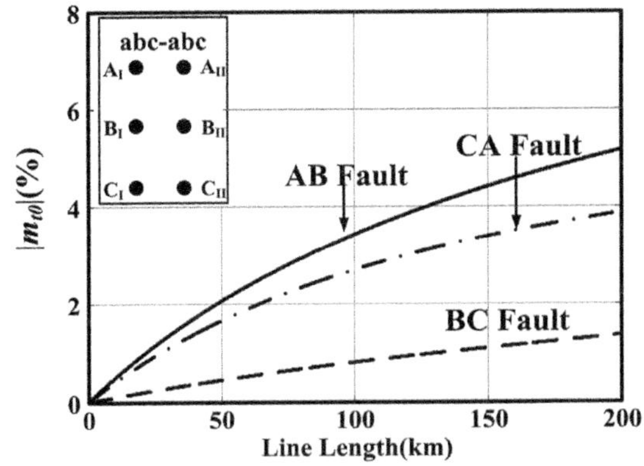

Figure 11.6 Net through zero-sequence unbalances of double-circuit lines in cases of external phase-to-phase faults (abc—abc phase arrangement).

The relationships between zero-sequence current unbalances and line lengths under the condition of external phase-to-phase faults are shown in Figs. 11.6—11.9. Since there is no circulating unbalanced current for the abc—abc phase arrangement, only the trends of net through zero-sequence unbalances with increases in line length are included in Fig. 11.6. For other phase arrangements, net through and circulating current unbalances m_{t0} and m_{c0}, are shown in Figs. 11.7—11.9.

Compared with current unbalances under normal operating conditions, there are similarities and differences for current unbalances in cases of external fault. On the one hand, similar to that under normal operating conditions, net through zero-sequence unbalances for external faults grow

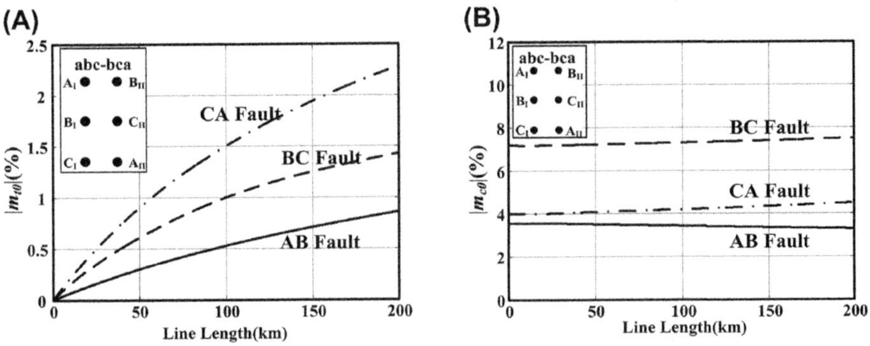

Figure 11.7 Zero-sequence current unbalances of double-circuit line in cases of external phase-to-phase faults (abc—bca phase arrangement). (A) Net-through current unbalance m_{t0} versus line length. (B) Circulating current unbalance m_{c0} versus line length.

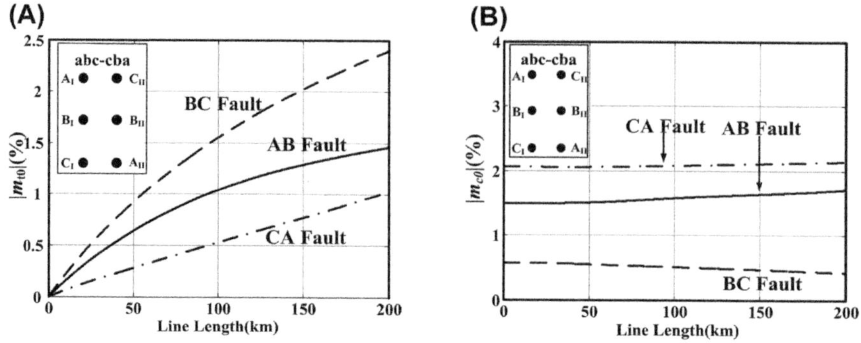

Figure 11.8 Zero-sequence current unbalances of double-circuit line in cases of external phase-to-phase faults (abc—cba phase arrangement). (A) Net-through current unbalance m_{t0} versus line length. (B) Circulating current unbalance m_{c0} versus line length.

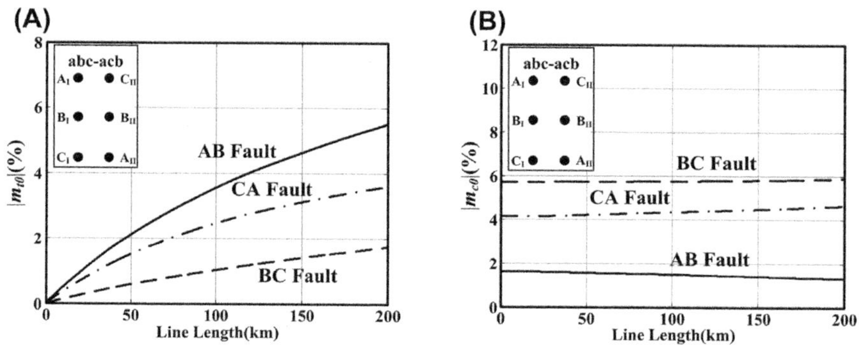

Figure 11.9 Zero-sequence current unbalances of double-circuit line in cases of external phase-to-phase faults (abc—acb phase arrangement). (A) Net through current unbalance m_{t0} versus line length. (B) Circulating current unbalance m_{c0} versus line length.

with increased line length, whereas circulating zero-sequence unbalances basically do not vary with line length. On the other hand, unlike under normal operating conditions, net through and circulating zero-sequence unbalances are different for cases of different external phase-to-phase faults, even under the same phase arrangement.

As shown in Figs. 11.6—11.9, external AB phase-to-phase faults under the abc—abc arrangement show the biggest net through zero-sequence unbalance, and external CA phase-to-phase faults with abc—cba phasing show the lowest net through zero-sequence unbalance. The external BC phase-to-phase fault with abc—bca phasing exhibits the largest circulating zero-sequence unbalance. From a quantitative view, zero-sequence

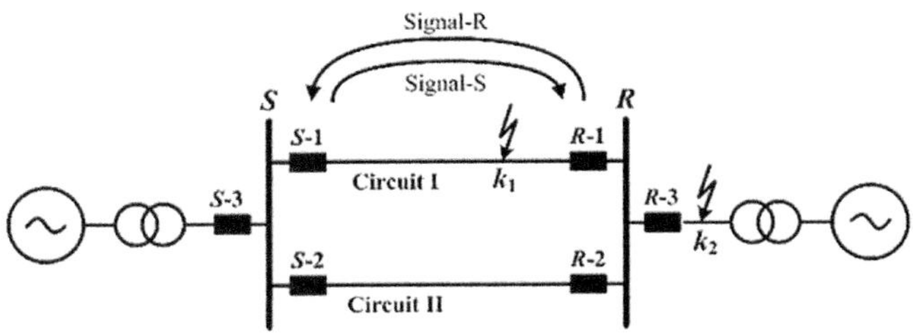

Figure 11.10 Double-circuit transmission line and its pilot directional protection.

unbalances can reach 6%–8%, which are larger than those under normal operating conditions. Since there is large positive-sequence current passing through the line in the case of external fault, the values of unbalanced currents may be large enough to have a great influence on protective relaying.

11.2.4 Influence on directional pilot protection

Fig. 11.10 shows EHV/UHV double-circuit transmission lines where pilot directional protection is applied as one of the main protections.

The specific criterion of a zero-sequence directional relay can be expressed as follows. Generally, $\varphi_{sen.max}$ = approximately −90 to −110 degrees:

$$90^\circ \geq \arg\left(3\dot{U}_0 \cdot e^{-j\phi_{sen.max}} / 3\dot{I}_0\right) \geq -90^\circ \tag{11.12}$$

It is well recognized that zero- and negative-sequence-based protections may not respond to any symmetrical three-phase or phase-to-phase faults. However, circulating and net through unbalanced zero-sequence currents may exist in double-circuit transmission lines when an ungrounded fault k_2 occurs, as shown in Fig. 11.11. I_0 and I'_0 of double-circuit lines are reverse in phase. In addition, some zero-sequence unbalanced voltages always exist at both ends due to electromagnetic and electrostatic coupling.

For circuit I, it can be seen that zero-sequence directional protection at R-1 may incorrectly determine the fault as a forward fault because zero-sequence current is passing from the line to bus. At the same time, zero-sequence directional protection at S-1 may also determine the fault as a forward fault if the phase of U_{0S} at bus S is reversed from U_{0R} at bus R. In fact, the same problem may exist for protections at S-2 and R-2. Consequently, the directional pilot protections may mal-trip when external fault k_2 occurs due to the influence of circulating current unbalance. This fact

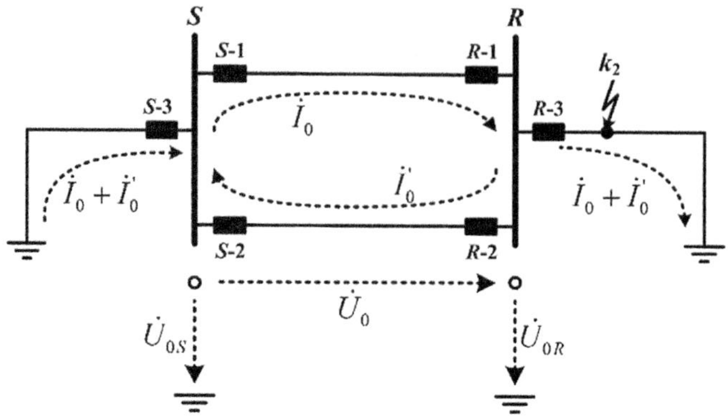

Figure 11.11 Zero-sequence unbalanced current in double-circuit transmission lines with external three-phase symmetrical fault or phase-to-phase fault at k_2.

was verified by the following real fault case on an EHV power system as well as simulation tests.

The real EHV double-circuit line shown in Fig. 11.10 is located in Northwest China. Side S is a huge-capacity power plant, and side R is the power grid. The phase arrangement of double-circuit lines is cab—acb. The zero-sequence component-based directional pilot protections of circuit I mal-tripped when a phase-to-phase fault occurred at point k_2. Fig. 11.12 shows the real fault waveforms provided by the power grid utility (CT ratio = 2500: 5). The external fault occurs at 0.04 s (t_1), and the breakers are disconnected at about 0.1 s (t_2).

It can be clearly seen from real fault record data that significant circulating currents $3I_0$ and $3I'_0$ occur on the double-circuit line when an external phase-to-phase fault occurs. As shown in Fig. 11.12A, $3I_0 \approx 3I'_0 \approx 3.4$ A. As for the secondary side, its value is large enough to enable protection pickup. For side R of circuit I, it can be seen from Fig. 11.12B that the relationship of $\arg(3\dot{U}_0/3\dot{I}_0) \approx -90°$ is satisfied so that directional protection at R-1 sends a permission signal to S-1 at remote side S. For this real fault, the zero-sequence directional relay at S-1 is not a pickup because the zero-sequence voltage at bus S is too small. However, the k_2 fault is within the protected zone of the overreaching pilot distance relay at S-1. Finally, the k_2 fault is determined by the directional pilot protection of line I as an internal fault, and then line I is isolated.

For the untransposed double-circuit lines shown in Fig. 11.5, line length is assumed to be 100 km. A fault occurs at 0.3 s. An external CA

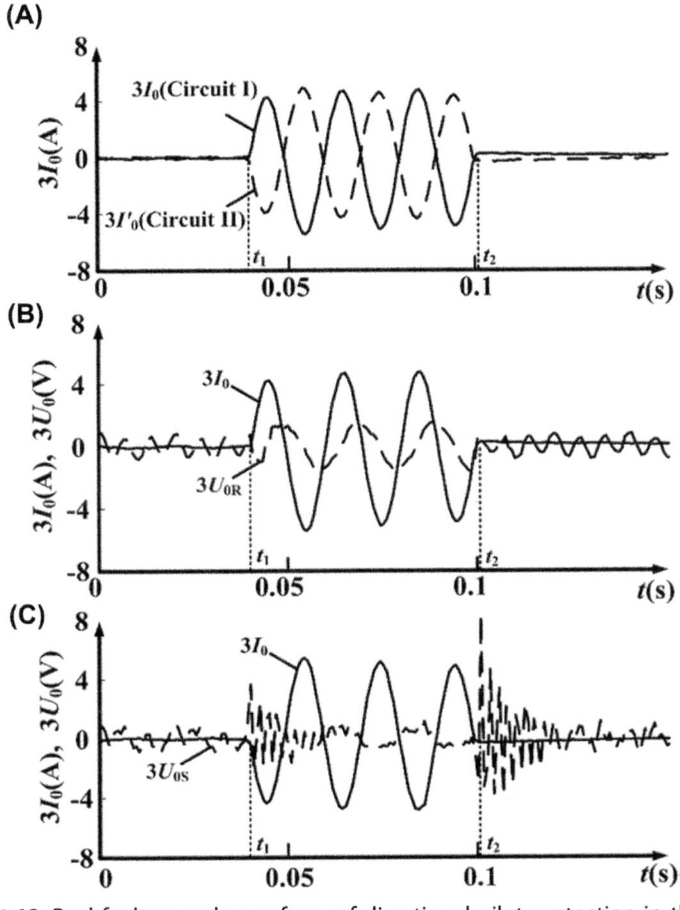

Figure 11.12 Real fault record waveform of directional pilot protection in the case of an external phase-to-phase fault. (A) Zero-sequence unbalanced current of two circuits. (B) Zero-sequence voltage and current at R-1 of receiving side. (C) Zero-sequence voltage and current at S-1 of sending side.

phase-to-phase fault at point k_2 is simulated for a double-circuit line with an abc—acb phase arrangement. The simulation results are demonstrated in Fig. 11.13.

The amplitude of positive-sequence fault current is approximately equal to that in case I. So the ratio of $(3I'_0/I_1)$ for circuit II reaches 15%. The unbalanced zero-sequence currents in the two lines are distinctly different in phase and magnitude. For circuit II, $3I'_{0S}$, $3I'_{0R}$ measured at two terminals lead $3U_{0S}$, $3U_{0R}$ by about 80.5 degrees and 77.6 degrees respectively. Therefore, zero-sequence directional pilot protection of circuit II may maloperate for this external CA phase-to-phase fault.

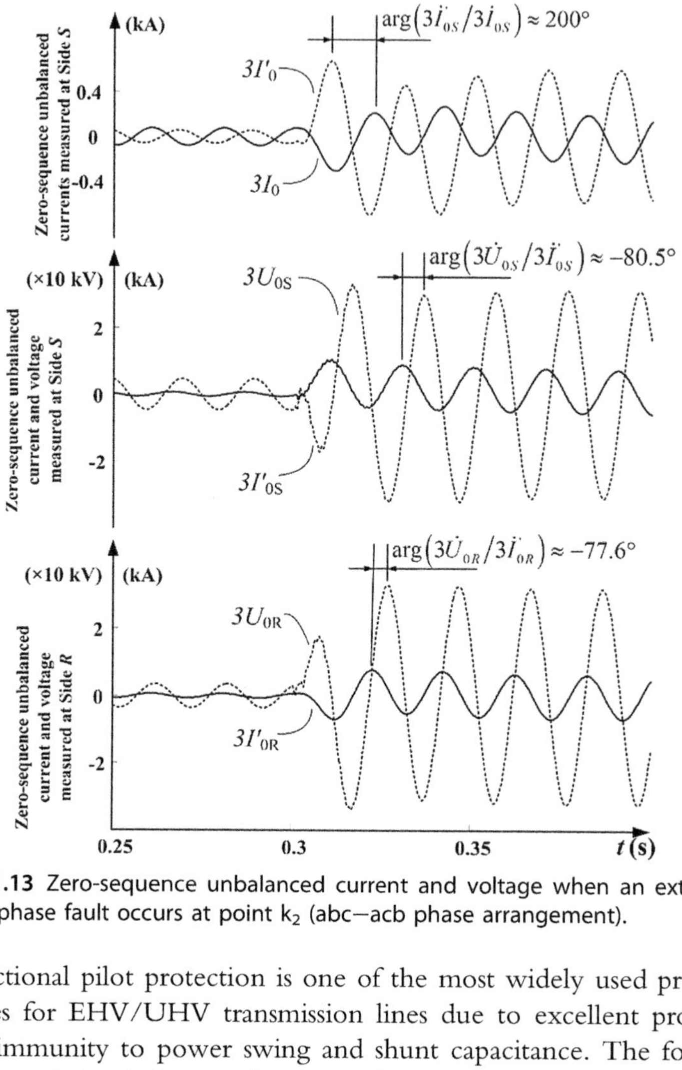

Figure 11.13 Zero-sequence unbalanced current and voltage when an external CA phase-to-phase fault occurs at point k_2 (abc—acb phase arrangement).

Directional pilot protection is one of the most widely used protective principles for EHV/UHV transmission lines due to excellent properties, such as immunity to power swing and shunt capacitance. The foregoing real fault and simulation results prove that conventional directional protection may be influenced by circulating unbalanced current to maloperate in the case of an external fault. Since the values of unbalanced currents may be large under external fault, maloperation cannot be avoided by simply increasing the threshold.

As we know, the superimposed energy directional principle has been used successfully to realize directional protection with ultrahigh speed. A forward fault decision is obtained when Eq. (11.13) is satisfied:

$$S = \int (\Delta u \cdot \Delta i)dt < 0 \tag{11.13}$$

Either the superimposed energy directional element in the time domain or the conventional directional element in the frequency domain is based on the comparison of polarity between voltage and current signals. According to Figs. 11.12 and 11.13, the zero-sequence directional element maloperates because the polarity relationship between unbalanced $3U_0$ and $3I_0$ under the condition of an external ungrounded fault is nearly in accord with that of $3U_0$ and $3I_0$ under the condition of an internal grounded fault. Therefore, it is inevitable for both directional principles to provide wrong direction determination under the influence of circulating unbalanced currents. However, by utilizing the rapidity of the energy directional element and the stability of the conventional directional element, an improvement protection scheme is proposed based on conventional pilot directional protection on the transmission line, as shown in Fig. 11.14.

As shown in Fig. 11.14, a forward fault decision of local protection may be made when any directional element gives a forward fault decision, and no backward fault decision is made by other directional elements.

(1) For external phase-to-phase faults, a negative-sequence-based directional element gives a definite backward fault decision because distinct negative-sequence current and voltage exist. Therefore, even if the zero-sequence directional element maloperates due to the influence of circulating unbalanced current, local protection may not trip with high security.

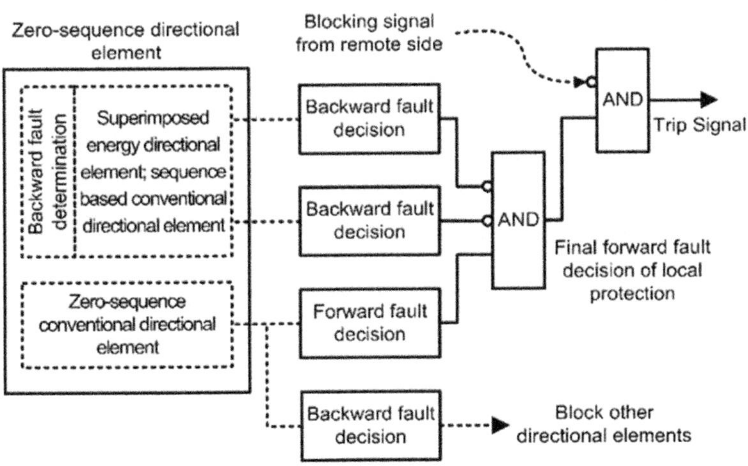

Figure 11.14 Improvement scheme of pilot direction protection (zero-sequence directional element).

(2) The same conclusion can be obtained for the case of external symmetrical faults, because a superimposed positive-sequence directional element can reliably block protection.
(3) The superimposed energy directional element has ultrahigh speed and less reliability, whereas the conventional directional element has better stability. Therefore, both principles are mutually complementary in performance. In order to avoid maloperation during the initial fault stage, a backward fault decision is made by a superimposed energy directional element with high speed.

The operational results of directional elements for the external CA phase-to-phase fault at point k_2 discussed in Case II are shown in Fig. 11.15. Obviously, a superimposed energy directional element gives a backward fault decision with high speed during the initial fault stage. Furthermore, the currents I_{2R}, I'_{2R} of double circuits measured at side R have nearly the same phase. The voltage U_{2R} at bus R leads I_{2R} or I'_{2R} by about 90 degrees. Therefore, the backward fault decision is made reliably in the case of an external k_2 fault. Referring to Fig. 11.13, although the zero-sequence

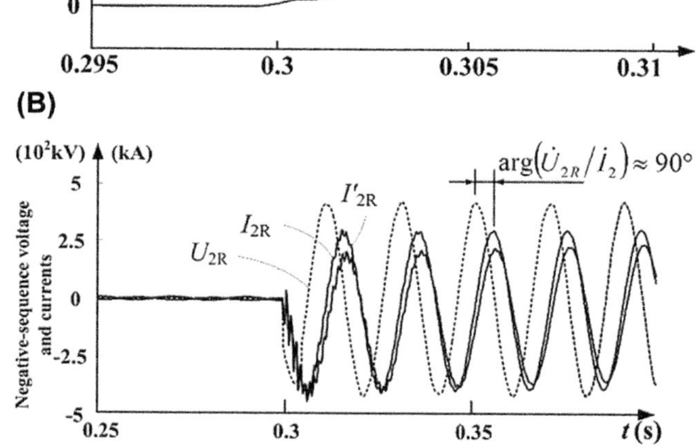

Figure 11.15 Operational results of directional elements. (A) Results of a superimposed energy directional element. (B) Results of conventional negative-sequence directional elements at side R.

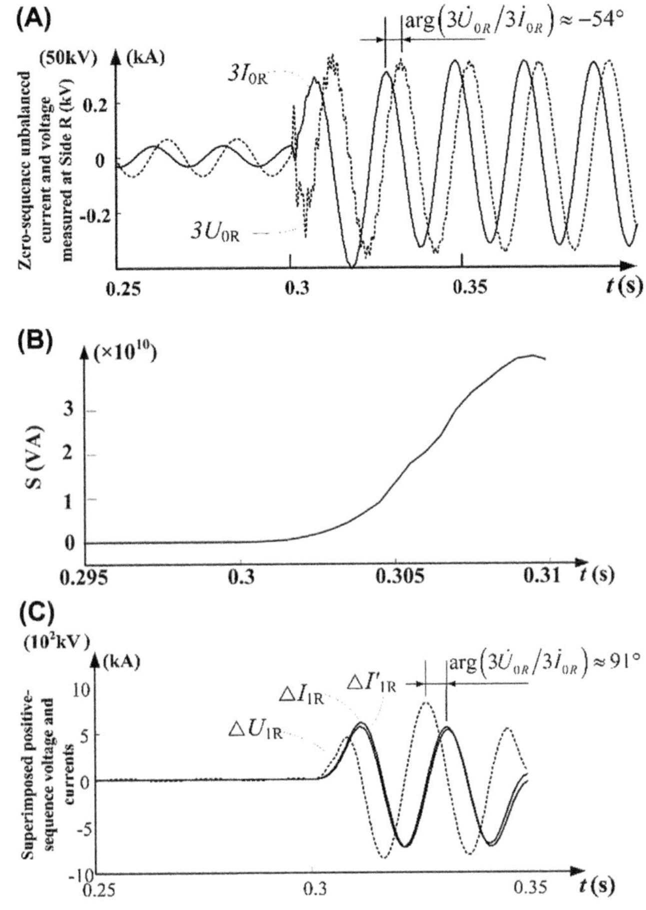

Figure 11.16 Operational results of directional elements. (A) Unbalanced zero-sequence voltage and current at side R. (B) Results of superimposed energy directional element. (C) Results of conventional superimposed positive-sequence directional elements at side R.

directional element may maloperate, the proposed directional protection scheme can be blocked with high security.

Similarly, for an external symmetrical fault at point k_2, the operation results of directional elements are shown in Fig. 11.16. Influenced by circulating unbalanced current and voltage, the zero-sequence directional element of circuit II may maloperate because $3U_{0R}$ lags $3I_{0R}$ by about 54 degrees. On the contrary, neither the superimposed energy directional element nor positive-sequence directional element may determine the forward fault decision, as shown in Fig. 11.16B and C.

11.3 Unbalanced current of multicircuit lines

11.3.1 Electromagnetic coupling characteristics of multicircuit lines

With the development of multicircuit transmission technology, the application of triple- and quadruple-circuit lines is increasingly widespread in EHV/UHV transmission systems. In a real EHV/UHV transmission project, a set of parallel double-circuit lines with common buses at both ends is often erected on the same tower with another single-circuit line or parallel double-circuit lines forming partial triple- and quadruple-circuit lines. Typical triple- and quadruple-circuit lines in 500 kV transmission systems are shown in Figs. 11.17A and B, respectively.

The asymmetry of three-phase parameters of circuits leads to an unbalanced current, so long-distance EHV/UHV transmission lines are often subjected to wire transposition. According to the regulations of the China State Grid Corporation, EHV transmission lines with a length of

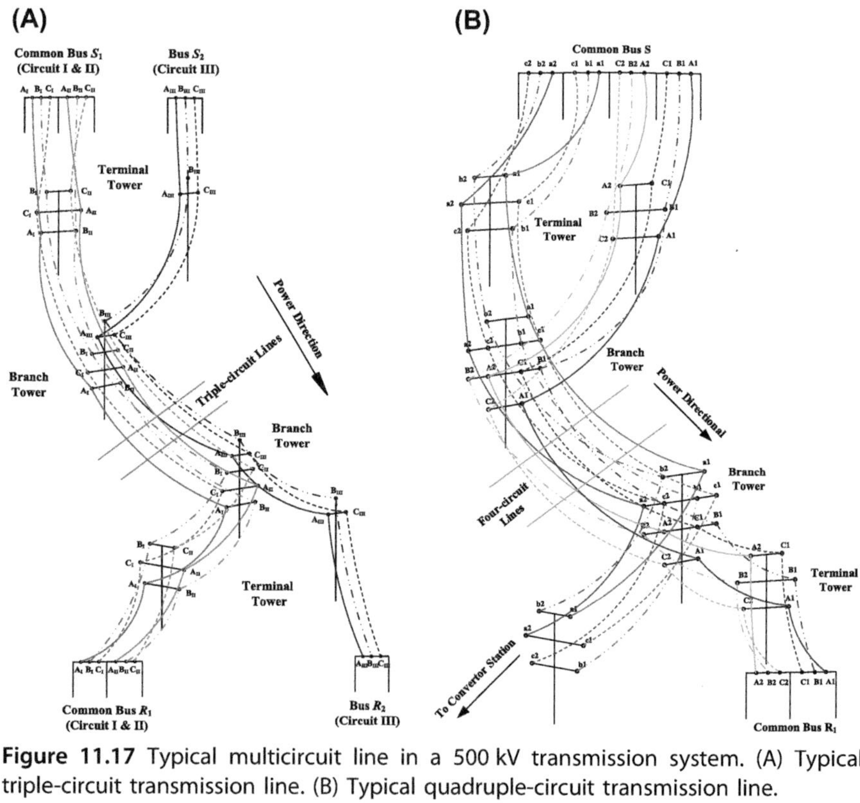

Figure 11.17 Typical multicircuit line in a 500 kV transmission system. (A) Typical triple-circuit transmission line. (B) Typical quadruple-circuit transmission line.

more than 100 km and UHV transmission lines with a length of more than 150 km should be considered for transposition. However, due to the complicated tower structure and wire arrangement of multicircuit lines, ideal transposition of phase conductors is very difficult. For partial multicircuit lines, it is difficult to form a complete transposition cycle in the part erected on the same tower. In addition, the length of each segment of multicircuit lines in real EHV/UHV transmission systems is often shorter than the length required to be transposed. Therefore, EHV/UHV multicircuit transmission lines on the same tower are mostly untransposed lines. Since the spacing between circuits on the same tower is close to the spacing of phase conductors, mutual inductance between circuits is significantly enhanced, and the strong magnetic coupling characteristics between multicircuit lines cannot be ignored. The value of the unbalanced current of multicircuit lines is more significant not only in normal operation but also under faults.

Taking the untransposed triple-circuit line shown in Fig. 11.17A as an example, it consists of a set of parallel double-circuit lines I and II with common buses at both ends as well as a circuit III. The phase conductors of parallel double-circuit lines I and II are vertically arranged in drum type. Circuit III is erected above parallel double-circuit lines I and II, and its phase conductors are arranged in erected triangular type. Due to the asymmetry of wire space position and line parameters, unbalanced current flows through each circuit during normal operation. The unbalanced current of parallel double-circuit lines I and II is analyzed in Chapter 11.2. In the triple-circuit line, although circuit III has no direct electrical connection with parallel double-circuit lines I and II, the mutual inductance coupling affects the unbalanced currents of parallel double-circuit lines I and II due to the close spacing between circuits. The diagram of a triple-circuit line is shown in Fig. 11.18.

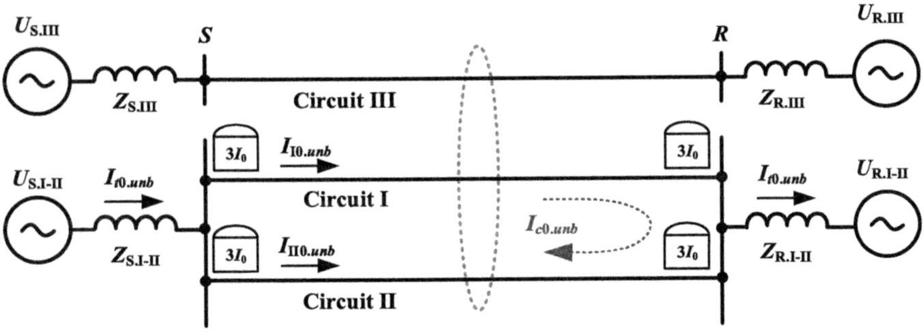

Figure 11.18 Diagram of a triple-circuit line.

11.3.2 Unbalanced current of triple-circuit lines and its impact

In the triple-circuit line shown in Fig. 11.18, when the load current of circuit III is large, the zero-sequence unbalanced current of parallel double-circuit lines I and II can be considerable. In a 500 kV transmission system, the active power delivered by one circuit can reach 2000 MW, and the load current can reach 2200 A. When circuit III is operating on heavy load, the zero-sequence unbalanced current of circuits I and II can be very large even if their own load current is small.

According to real operational data of the triple-circuit line of China's 500 kV transmission system, when circuit III was operating on heavy load, zero-sequence unbalanced currents $3\dot{I}_0$ of parallel circuits I and II erected on the same tower reached 275 and 250 A respectively, which is close to the setting value of zero-sequence overcurrent protection, 300 A. The three-phase currents of circuits I and II are shown in Figs. 11.19A and B, respectively, and their three-phase currents are significantly asymmetrical. Obviously, when zero-sequence short-circuit current flows through circuit III, its impact on parallel double-circuit lines I and II may be more significant even though there is no direct electrical connection between circuit III and parallel circuits I/II, which highlights the "strong magnetic weak current" coupling characteristics of multicircuit transmission lines.

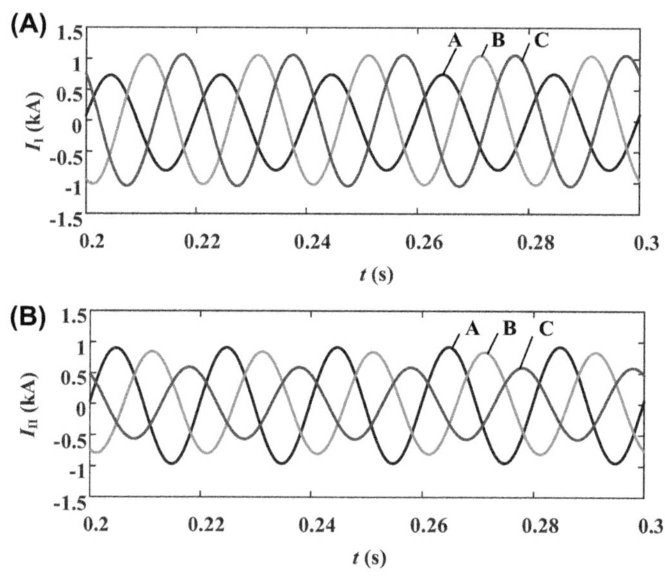

Figure 11.19 Three-phase currents of circuits I and II when circuit III is operating on heavy load. (A) Three-phase current of circuit I. (B) Three-phase current of circuit II.

Since EHV/UHV transmission lines are mostly equipped with zero-sequence overcurrent protection for high-resistance ground faults, the setting value is generally 300 A. The impact of strong magnetic coupling characteristics on zero-sequence overcurrent protection of multicircuit lines needs to be taken seriously.

References

[1] Harry Hesse M. Circulating currents in paralleled untransposed multicircuit lines: I — numerical evaluations. IEEE Transactions on Power Apparatus and Systems July 1966;PAS-85(7):802—11.
[2] Harry Hesse M. Circulating currents in paralleled untransposed multicircuit lines: II — methods for estimating current unbalance. IEEE Transactions on Power Apparatus and Systems July 1966;PAS-85(7):812—20.
[3] Kurokawa S, Pissolato Filho J, Tavares MC, Portela CM, Prado AJ. Behavior of overhead transmission line parameters on the presence of ground wires. IEEE Transactions on Power Apparatus and Systems April 2005;20(2):1669—76.
[4] Kalyuzhny A, Kushnir G. Analysis of current unbalance in transmission systems with short lines. IEEE Transactions on Power Delivery April 2007;19(3):1040—8.
[5] Li B, Guo F, Li X, Bo Z. Circulating unbalanced currents of EHV/UHV untransposed double-circuit lines and their influence on pilot protection. IEEE Transactions on Power Delivery April 2014;29(2):825—33.

CHAPTER 12

Half-wavelength transmission lines

Contents

12.1 Basic characteristics

12.1.1 Half-wavelength transmission lines

Half-wavelength transmission is an ultra-long-distance three-phase AC transmission technology whose transmission distance is close to a half-wave of power frequency—i.e., 3000 km (50 Hz) or 2500 km (60 Hz). As a kind of ultra-long-distance transmission method, half-wavelength transmission technology has received extensive attention [1]. The characteristics of half-wavelength transmission lines are as follows:

(1) The half-wavelength transmission line actually shortens the electrical distance between two points where the electrical distance is far apart. The half-wavelength transmission line can be approximated as an ideal

Protection Technologies of Ultra-High-Voltage AC Transmission Systems
ISBN 978-0-12-816205-7
https://doi.org/10.1016/B978-0-12-816205-7.00012-1

transformer whose variable ratio is 1 and the voltage of the beginning and ending point are at the same value and opposite phase. The voltage at both ends is not affected by the line's transmission power and remains stable when the line power factor changes.

(2) Theoretically, the power limit of a half-wavelength transmission line could reach positive infinity. In fact, due to the limitation of the line voltage withstand level, insulation strength, and other factors, the transmission capacity of an actual half-wavelength transmission line is generally about 1.0−1.2 times the natural power of the transmission line.

(3) Because of system stability and reactive power compensation, the conventional ultra-high-voltage (UHV) long line needs to set the switch station; the half-wavelength transmission line is different from this, and no reactive power compensation is needed. It can be seen that there is no need to set up a switch station in the middle of a half-wavelength transmission line.

(4) There is no need to install reactive power compensation equipment on the entire line. When the half-wavelength transmission line is in normal operation, half the line emits reactive power, half absorbs reactive power, and the entire line maintains reactive power self-balancing. Therefore, in normal operation, no reactive power compensation device is required at each point of the half-wavelength line.

(5) The economic efficiency of the UHV half-wavelength line is much higher than that of the conventional UHV AC transmission line and even better than the long-distance UHV DC transmission line.

12.1.2 Steady-state characteristics

Since the transmission distance of the half-wavelength line has reached half the wavelength of the electromagnetic wave transmitted on it, the simple concentrated parameter equivalent of the line cannot be used in the analysis, and an analysis method based on the distribution parameter must be adopted.

The parameters of the UHV AC half-wavelength transmission line are shown in Table 12.1 [2]. From Table 12.1, the wave impedance Z_c of the half-wavelength line is calculated as $(246.11 - j3.7455)\ \Omega$, and the natural power P_n of the line at 1000 kV is $(4062.3 + j61.8)$ MVA. The exact length of the half-wavelength line is 2938.36 km.

The voltage and current characteristics along the half-wavelength transmission line are quite different from those of the conventional AC

Table 12.1 Half-wavelength transmission line unit parameter.

Phase sequence	**R_0/(Ω/km)**	**$j\omega L_0$/(Ω/km)**	**C_0/(μF/km)**
Positive-sequence	0.00801	0.8379	0.013830
Zero-sequence	0.15630	2.4908	0.008955

transmission line. Eqs. (12.1) and (12.2) are the relationships between both voltage and current along the half-wavelength line and the voltage at the head end:

$$\dot{U}_x = \left(ch\gamma x - \frac{k_g ch\gamma l + sh\gamma l}{ch\gamma l + k_g sh\gamma l} sh\gamma x \right) \dot{U}_1 \tag{12.1}$$

$$\dot{I}_x = \left(\frac{ch\gamma l + \frac{1}{k_g} sh\gamma l}{\frac{1}{k_g} ch\gamma l + sh\gamma l} ch\gamma x - sh\gamma x \right) \frac{\dot{U}_1}{Z_c} \tag{12.2}$$

In the equations, $\dot{U}_1$ is the voltage at the head of the line, and k_g is the ratio of the line transmission power to natural power. According to Eqs. (12.1) and (12.2), the voltage and current amplitude along the half-wavelength transmission line can be obtained as shown in Figs. 12.1 and 12.2.

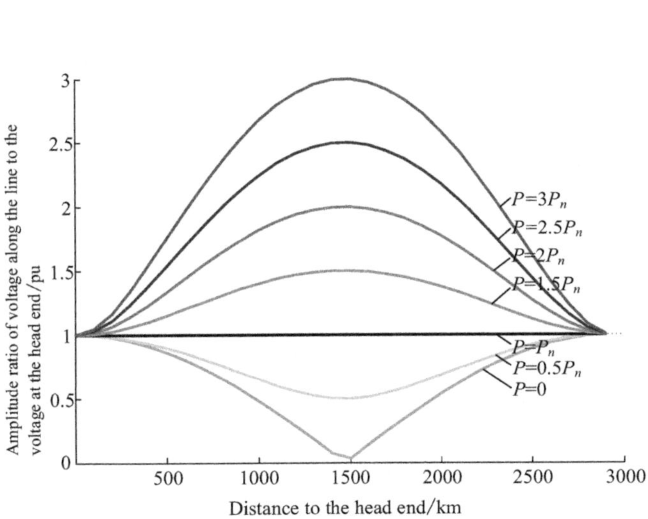

Figure 12.1 Voltage along the half-wavelength line under different transmission powers.

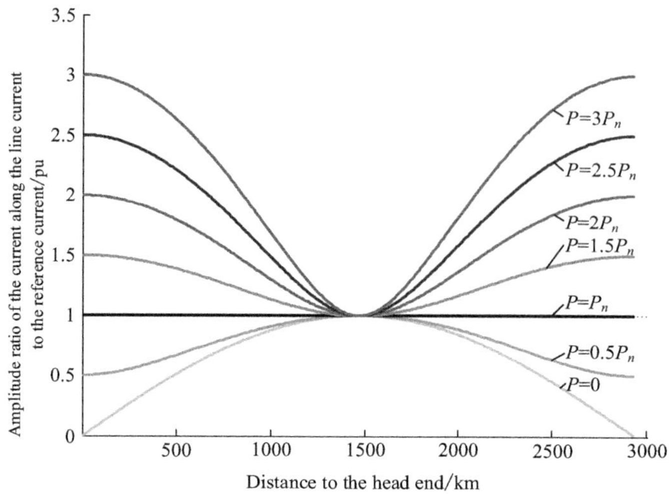

Figure 12.2 Current along the half-wavelength line under different transmission powers.

It can be seen from Fig. 12.1 that when the half-wavelength transmission line delivers natural power, voltage along the line is always equal to the voltage at the head end. When transmission power is greater than natural power, voltage along the line increases, the midpoint voltage is highest, and its per-unit value is the ratio of delivered power to natural power; When transmission power is less than natural power, voltage along the line decreases, the midpoint voltage is lowest, and the lowest voltage-per-unit value is the ratio of transmitted power to natural power. When the line is no-load, midpoint voltage is zero.

12.2 New principle of current differential protection

12.2.1 Deficiency of conventional current differential protection

12.2.1.1 Adaptability of conventional current differential protection

For the conventional UHV AC long line, differential current is not zero during normal operation because of the large distributed capacitance current. It is necessary to compensate the differential current according to the capacitor current, and the compensation current on both sides of the line should satisfy [3].

$$\dot{I}_c = \frac{\dot{U}_\varphi - \dot{U}_0}{-2jX_{c_1}} + \frac{\dot{U}_0}{-2jX_{c_0}} \tag{12.3}$$

where X_{c_1} and X_{c_0} are the positive-sequence and zero-sequence impedances of the line, and $\dot{U}_\varphi$ and $\dot{U}_0$ are the phase and zero-sequence voltages of the line.

When the half-wavelength line is in normal operation or under a fault condition, voltage along the line changes nonlinearly, and there is no monotonous change relationship between the line voltage and the distance from the measurement point to the line side. So there is no means, using the voltage at the protection installation point, to precisely calculate distributed capacitor current along the line. Therefore, the compensation algorithm shown in Eq. (12.3) cannot accurately compensate for the capacitor current of the line.

In addition, the electromagnetic wave traveling from one end of the half-wavelength line to the other costs 10 ms, and the communication delay between the two ends is longer than 20 ms because of fiber channel delay.

12.2.1.2 Current differential protection based on the Bergeron model

Current differential protection based on the Bergeron model was successfully applied in the Jindongnan–Nanyang–Jingmen UHV AC demonstration project in 2009 [4]. Some scholars have proposed the principle of split-phase current differential protection based on the Bergeron model as the main protection principle of half-wavelength transmission lines [5]. Values for $|F_1|$ and $|F_2|$ are calculated using the half-wavelength transmission line parameters. Referring to Chapter 7, the reference point is set as the terminal of the line here. Fig. 12.3 shows the value of differential current changes when fault occurs on different positions of the line. Here L is the ratio of the distance from the fault point to the reference point and the full line length.

According to Fig. 12.3, it can be seen that $|F_1| < |F_2|$, especially when the distance from the fault point to the reference point is about 40%–50% of line length. If the half-wavelength line still adopts the conventional Bergeron's differential protection principle in which the reference point is fixed, the following equation can be concluded:

$$I_{da} = F_1 I_{fa} < F_2 I_{fa} = I_{db} = I_{dc} \tag{12.4}$$

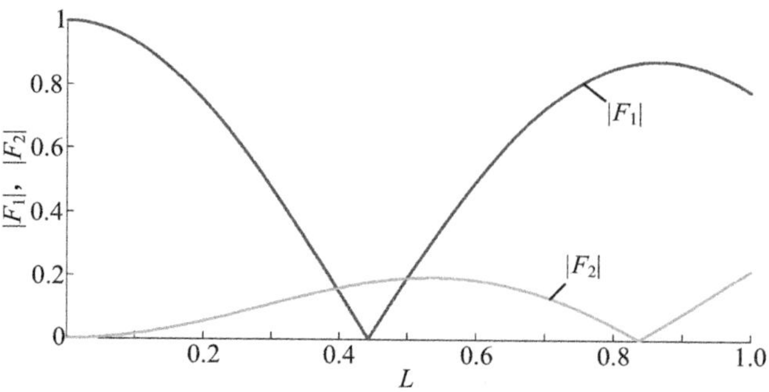

Figure 12.3 $|F_1|$ and $|F_2|$ curves when a single-phase ground fault occurs.

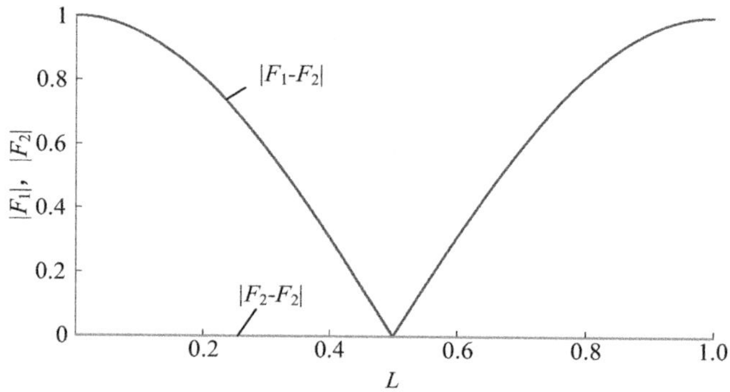

Figure 12.4 $|F_1|$ and $|F_2|$ curves when a three-phase short-circuit fault occurs.

Under this condition, the fault-phase differential current is smaller than the non-fault-phase differential current. It not only can make the fault-phase protection fail to operate, but also may cause non-fault-phase protection maloperation. Therefore, if the reference point is fixed at a certain point of the long transmission line, current differential protection based on the Bergeron model cannot protect the full line. The closer the distance from the reference point to the fault point, the higher the sensitivity and reliability of the protection.

12.2.2 Current differential protection for half-wavelength transmission lines

12.2.2.1 Differential current

The F_1 and F_2 curves above are only related to the line parameters and the distance between the fault point and the reference point independent of the actual position of the fault point. Therefore, in the case where the line parameters are determined, enough reference points are taken on the line for the Bergeron differential calculation, and the calculation results are added (equivalent to integrating the curves of the absolute values of F_1 and F_2). There is a certain quantitative relationship between the result of the fault phase and the current value of the fault point, as shown in the following equation:

$$\lim_{n \to \infty} \frac{(F_{11} + F_{12} + \ldots + F_{1n})}{n} = pI_f \tag{12.5}$$

where n is the number of reference points in the line, F_1 to F_{1n} are the values of $|F_1|$ in the reference points, and p is the ratio of the calculation result to the fault current.

Although the fault current can be calculated accurately according to the above calculation process, the method of selecting infinite reference points on the line for the calculation of the Bergeron differential calculation will lead to extremely large computational complexity, which cannot be completed by the existing protection device and cannot be applied to the actual project. Therefore, on the basis of fully retaining the above integral characteristics, the number of reference points should be reduced as much as possible to achieve the purpose of reducing computational complexity.

12.2.2.2 Reference point selection

$|F_1 - F_2|$ is a cosine curve and only determined by the positive-sequence parameters of the line. When the single phase-to-ground fault occurs, F_1 is the superposition of the cosine curves of two different angular frequencies, which is also very similar to the cosine curve. The aberration rate depends on the ratio of the positive- and zero-sequence propagation constants of the line. Therefore, the Bergeron differential protection method can be improved by using the cosine curve feature of $|F_1 - F_2|$ and then observe the calculation error of the $|F_1|$.

It can be seen from the analysis in Section 12.2.2 that the Bergeron differential calculation using a single reference point cannot meet the requirements of the half-wavelength line. Therefore, in order to make the calculation result error and the fault current error small enough, the number of reference points are added to the line by adding one point each time to find the minimum number of reference points n.

First, set $n = 2$ namely, add one reference point. In order to preserve the integral characteristics of the calculation results, the distribution of the reference points in the line should be as uniform as possible. If the half-wavelength line is divided into two equal-length segments by using the reference point, there are two ways to arrange the reference points:

That is, place the reference point at the head and midpoint of the line, as shown in Fig. 12.5A, or place the reference point at the one-quarter and three-quarter points of the line, as shown in Fig. 12.5B. Among them, the black point is the end of the line connected end to end, and the red points are the reference points. Both reference point arrangements divide the end-to-end line into two equal segments. However, when a fault occurs at the end of the line, the reference point at the head of the line is far from the fault point, and the calculation result has a large error with the fault current. Therefore, the reference point should be placed with reference to Fig. 12.5B—that is, the reference point divides the line equally, and its position should also be symmetric about the midpoint of the line.

In summary, the reference point selection principle is as follows:

(1) Determine the number of reference points n;
(2) Ensure that the reference points divide the end-to-end line into n segments of equal length, and each segment length is l/n;
(3) Guarantee the central symmetry of the reference points on the line.

Taking $n = 3$ as an example, according to the above-mentioned reference point selection principle, set the reference points at places that are $l/6$, $l/2$, and $5l/6$ from the head of the line, as shown in Fig. 12.5C.

12.2.2.3 Determination of the number of reference points

Taking $n = 3$ as an example, the theoretical values shown in Figs. 12.3 and 12.4 can be calculated with the method mentioned above, to obtain the relationship curve between the calculated value of the theoretically derived new method and the actual fault current, as shown in Fig. 12.6. Fig. 12.6 also reveals the theoretical relationship between the location of the fault point and the error between the calculated result and the magnitude of the

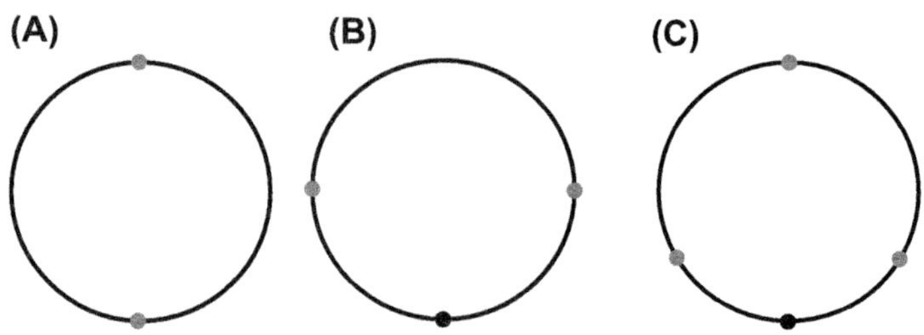

Figure 12.5 Schematic diagram of reference points.

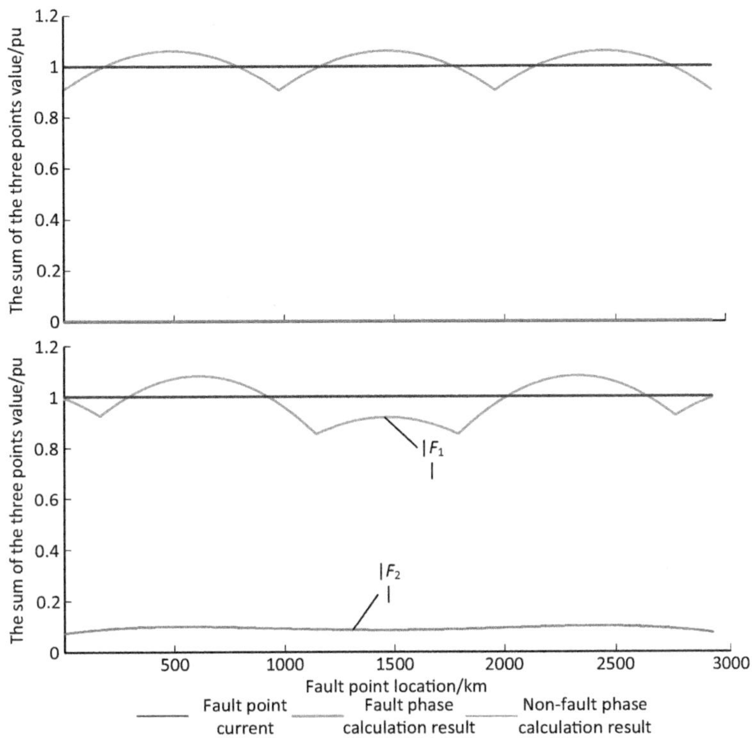

Figure 12.6 Relationship between calculated value and fault current in single-phase-to-ground fault and two-phase short circuit.

fault current. In contrast, the two curves plotted in Fig. 12.6 are the $|F_1|$ and $|F_2|$ curves obtained with the new method. When the number of reference points increases, the maximum error between the calculation result and the actual fault current is as shown in Table 12.2.

It can be seen from Fig. 12.6 and Table 12.2 that for the two-phase short circuit, the larger the number of reference points, the smaller the maximum error, and the more the calculation result can characterize the fault current. For a single-phase-to-ground fault, the maximum error position of the fault-phase calculation is achieved as the number of reference points moves from 39% to 61% of the line to 50% of the line, and finally to the midpoint of the line. When $n = 3$, the maximum error of the Bergeron differential calculation result is slightly less than 15%, and it hardly changes with an increase in the number of reference points. Therefore, this calculation result can be used to characterize the fault current more accurately. When $n = 2$, the error of the Bergeron differential calculation exceeds 20%,

Table 12.2 Difference between multipoint differential calculation result and fault current.

Fault type	Two-phase short circuit		Single-phase-to-ground fault	
n	Maximum positive direction difference of $\lvert F_1 - F_2 \rvert$ (%)	Maximum negative direction difference of $\lvert F_1 - F_2 \rvert$ (%)	Maximum positive direction difference of $\lvert F_1 \rvert$ (%)	Maximum negative direction difference of $\lvert F_1 \rvert$ (%)
2	7.62	22.58	17.35	22.98
3	5.29	8.82	8.19	14.64
4	3.04	4.80	5.34	12.79
5	2.30	2.70	4.73	12.30
6	1.75	1.72	3.33	12.22
7	1.48	1.06	3.62	12.28

which is not sufficient accuracy. Therefore, the number of reference points of multipoint split-phase current differential protection applied to the half-wavelength line is at least 3. When $n > 3$, the maximum negative direction difference of the single-phase-to-ground fault is reduced but is still greater than 10%, so the number of reference points should be taken as 3. If $n = 3$, the reference point arrangement is shown in Fig. 12.5C.

In summary, the specific method of the split-phase current differential protection method based on multiple reference points proposed in this paper is as follows: set the reference points at places that are $l/6$, $l/2$, and $5l/6$ from the head of the line, calculate the differential current amplitude at three points with the Bergeron differential method, and then sum the three differential current amplitudes and divide by pn. The reference points are shown in Fig. 12.7, where l indicates the full length of the line.

Since the new method proposed in this book can accurately calculate the fault current with the fault-phase action quantity on the half-wavelength line, the above protection criteria can still be used, and

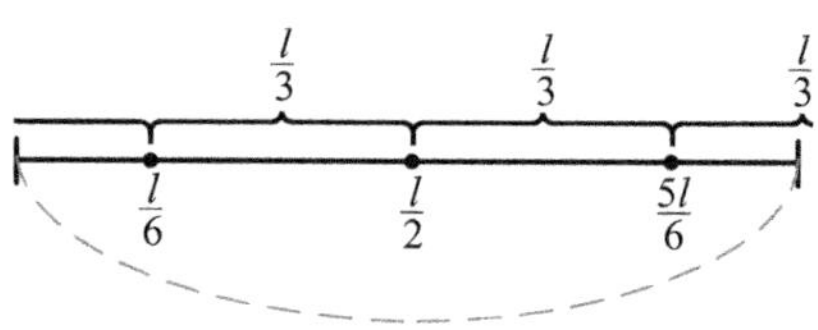

Figure 12.7 Schematic diagram of reference points of the three reference points method.

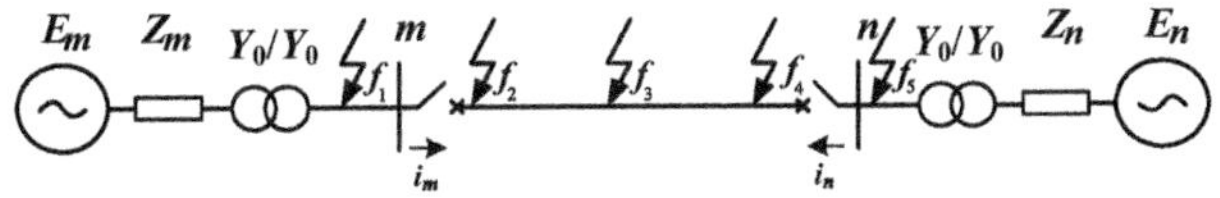

Figure 12.8 Schematic diagram of fault locations.

$|F_1|$ and $|F_2|$ are replaced with the value of the fault and nonfault phases in Fig. 12.6, where I_s is set as 300 A. Moreover, as with the split-phase current differential protection based on the Bergeron model, the new method itself has the ability to select the phase.

12.2.2.4 Case studies

The simulation model of a half-wavelength 1000 kV transmission line is built with PSCAD/EMTDC. The simulation parameters are shown in Table 12.1. When the half-wavelength line transmits natural power, the voltage at each point along the line is equal to the voltage at the head, and there is no midpoint overvoltage problem, so in the simulation, the active power transmitted on the line is 1 times natural power. Since the oscillation time of a half-wavelength line is long in simulation, the fault occurrence time is selected as 2 s after the start of the simulation. Various faults that may occur in a half-wavelength line are simulated in the model and calculated using MATLAB.

12.2.2.4.1 External faults:

When a three-phase short circuit occurs at the location of f_1, the voltage and current at the protection installation are as shown in Fig. 12.9.

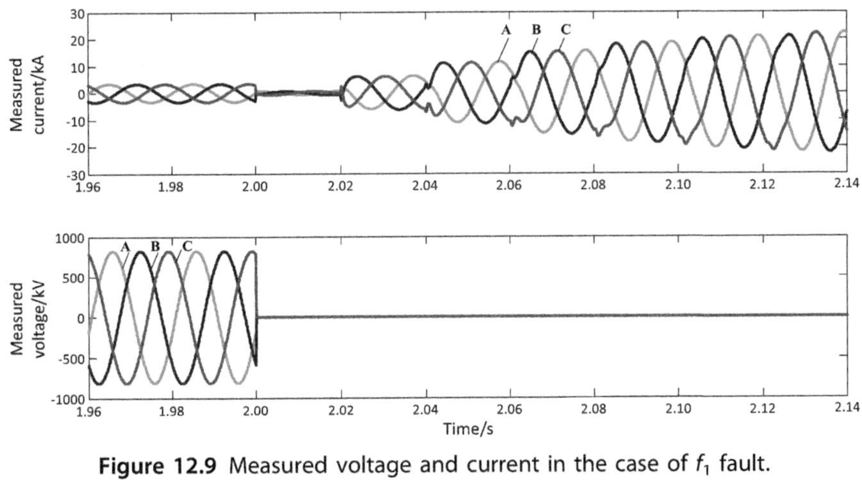

Figure 12.9 Measured voltage and current in the case of f_1 fault.

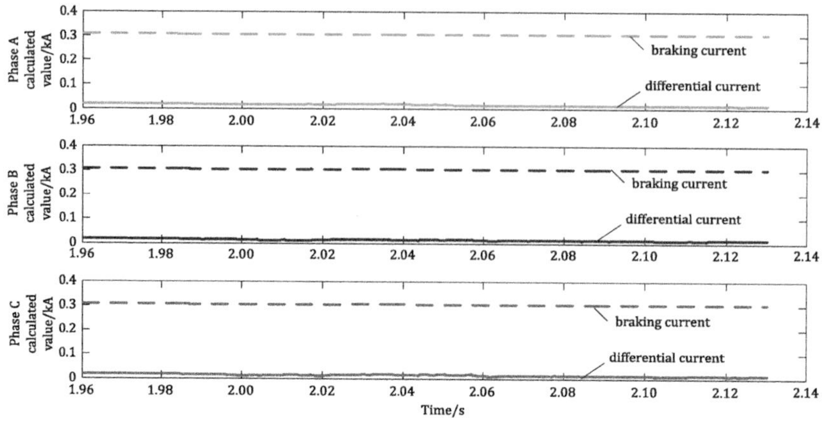

Figure 12.10 Differential current and braking current in the case of f_1 fault.

The differential and braking currents of phase a are shown in Fig. 12.10. The other two phases are similar. Obviously, in this case, the braking current is always much larger than the differential current. Therefore, the new method does not work reliably when there is an external fault on the line, which is in agreement with the theoretical analysis.

12.2.2.4.2 Internal faults:

Different types of internal faults are simulated at the fault points f_2, f_3, and f_4 to evaluate the performance of the new method. Figs. 12.11 and 12.12 show the simulation results for the b and c phase short circuit occurring at f_2. The fault occurs when $t = 2$ s. The voltage and current at the protection

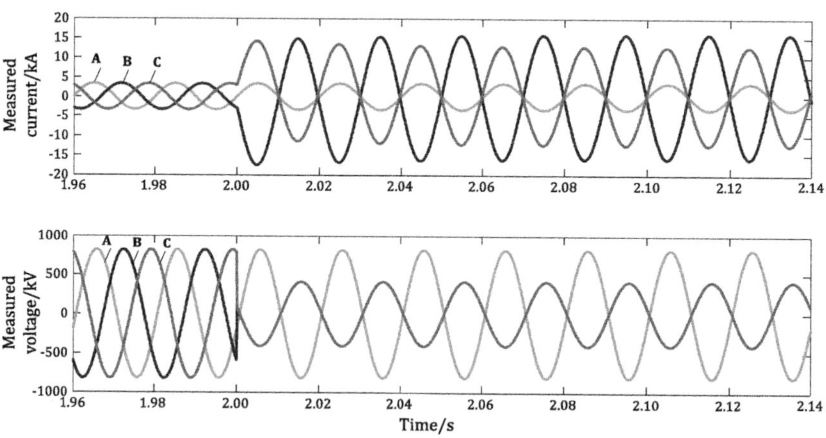

Figure 12.11 Measured voltage and current in the case of f_2 fault.

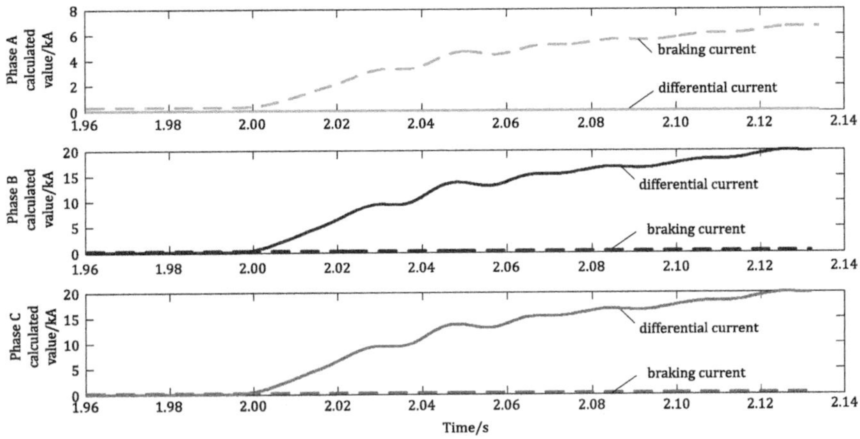

Figure 12.12 Differential current and braking current in the case of f_2 fault.

installation are shown in Fig. 12.11. As shown in Fig. 12.12, the differential current quickly exceeds the braking current in phase b and phase c, thus the protection can identify the fault accurately.

Figs. 12.13 and 12.14 show the simulation results of the phase ground fault at f_3. The fault current measured at the receiving end is large because of the internal fault. The differential current of the fault phase is greater than the brake current, therefore, the protection of the fault phase operate reliably, while the protections of non-fault phases not maloperate.

When a three-phase short circuit occurs at f_4, the simulation results are shown in Figs. 12.15 and 12.16. The three-phase differential current exceeds the braking current, and the protection can accurately identify the fault.

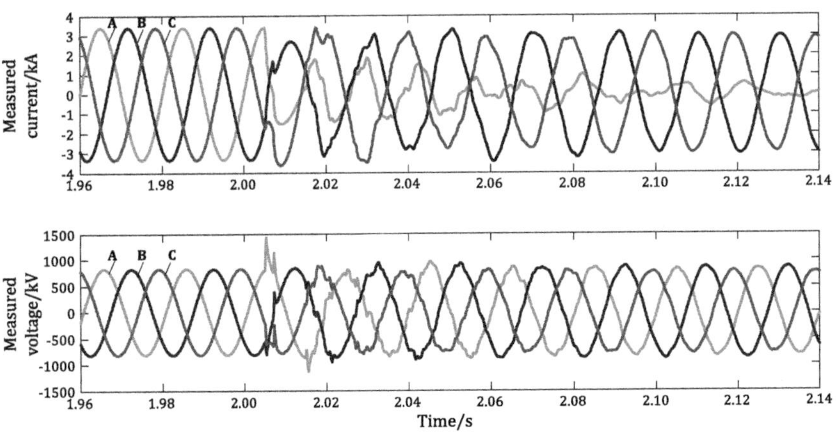

Figure 12.13 Measured voltage and current in the case of f_3 fault.

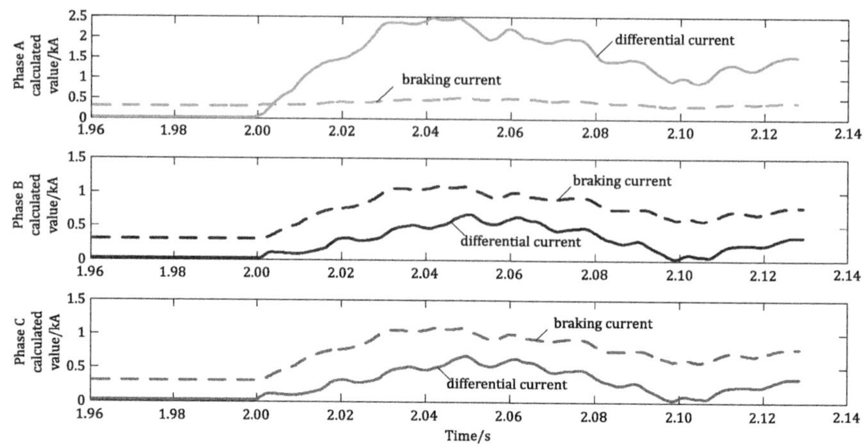

Figure 12.14 Differential current and braking current in the case of f_3 fault.

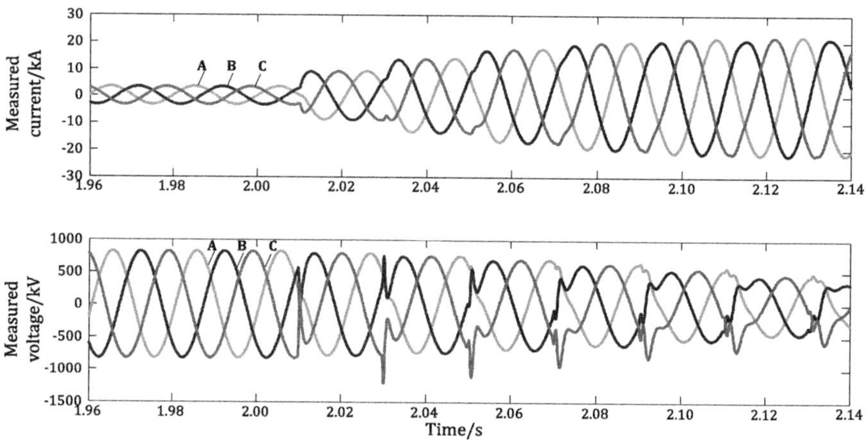

Figure 12.15 Measured voltage and current in the case of f_4 fault.

12.3 New principle of directional pilot protection

12.3.1 Conventional directional protection

Conventional directional protections include superimposed component directional protections and sequence component directional protections [6]. Operating characteristics of superimposed component direction protection are shown in Fig. 12.17.

When a fault occurs at f_1 of a half-wavelength line as shown in Fig. 12.18, the superimposed component of the interphase current and voltage of the relay location at the m side of the line are as shown in

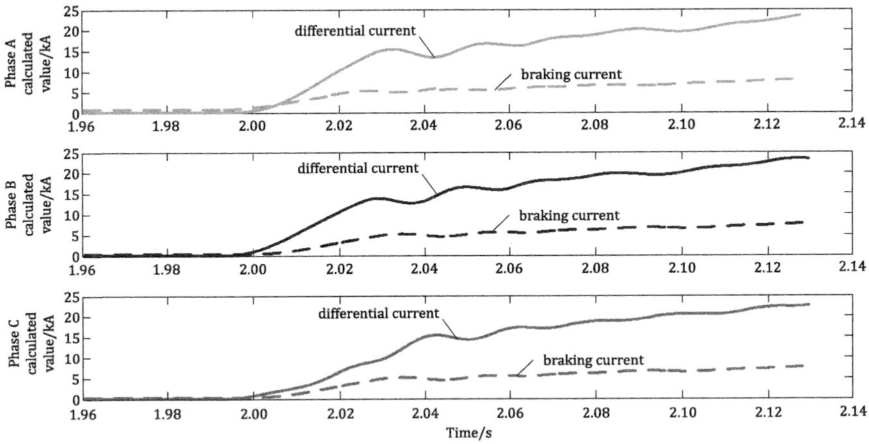

Figure 12.16 Differential current and braking current in the case of f_4 fault.

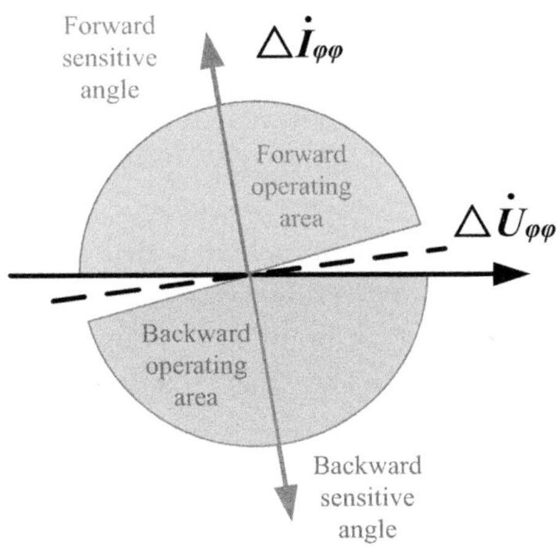

Figure 12.17 Operating characteristic of directional protection.

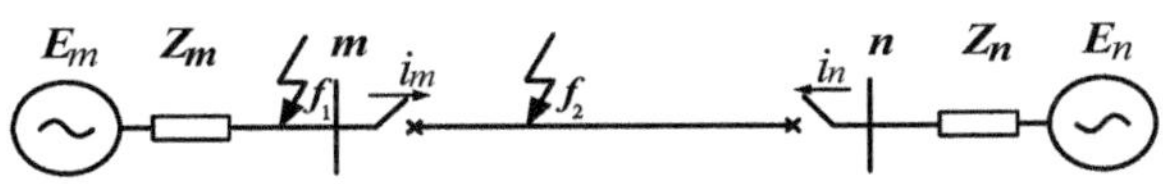

Figure 12.18 Schematic diagram of half-wavelength transmission line.

Fig. 12.19A, where Z_c is the positive-sequence wave impedance of the transmission line, and the time corresponding to zero on the time axis is the time of fault. The impedance angle of the n side system is set as δ_n, Fig. 12.19A shows that phase difference δ of the superimposed component of interphase current and voltage is 0, and the amplitude ratio is wave impedance at the initial stage of backward fault on the m side of the half-wavelength line. After 20 ms, δ gradually changes from 0 to $-\delta_n$.

When the fault occurs in the area of the half-wavelength line as shown in Fig. 12.18 (for example, the distance between f_2 and the m side of the line is 20% of the line's total length), the interphase current and voltage superimposed component of the relay location at the m side of the line are as shown in Fig. 12.19B. From Fig. 12.19B, it can be seen that there is no clear amplitude—phase relationship between the superimposed component of interphase current and voltage in the initial fault stage.

It can be seen that the phase difference of the superimposed component of interphase current and voltage at the relay location is quite different from that of the conventional line, when the backward fault occurs in the half-wavelength line. If directional pilot protection with the operating characteristics shown in Fig. 12.17 is still installed on the half-wavelength line, phase difference δ of the superimposed component of interphase current and voltage will fall on the boundary of the protection backward operating area at the initial stage of the backward fault, which seriously affects the judgment on the backward fault of the line of the directional pilot protection.

The reason for the above phenomenon is that the backward direction component reflects the impedance observed at the back of the relay location when looking from the fault point to the relaying. When a backward fault occurs on a conventional line (the fault point is f_3), the schematic

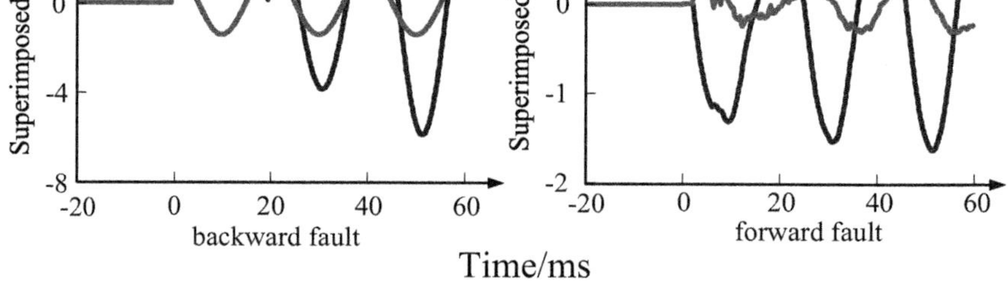

Figure 12.19 Superimposed voltage and current in the case of a phase-to-phase fault.

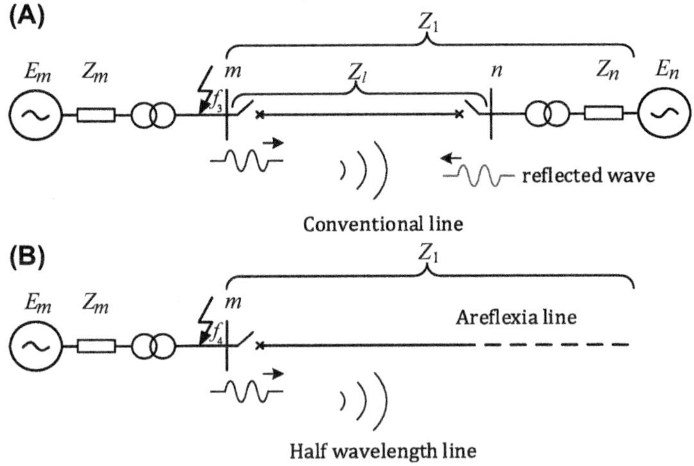

Figure 12.20 The m side backward fault diagram.

diagram of the line measured by the m side protection is shown in Fig. 12.20A, where Z_m and Z_n are the system impedance of the m side and n side respectively, and Z_l is the total impedance of the line. Because the time of wave process on conventional lines is short, in the early stage of fault, the m side protection can receive the backward traveling wave of fault current and the voltage wave reflected on the n side. Then impedance Z_l measured by the m side protection is the sum of line impedance and system impedance of the n side—that is to say, for Fig. 12.20A, there is

$$Z_1 = -(Z_l + Z_n) \tag{12.6}$$

Because the impedance angle of the conventional line is approximately equal to that of system impedance, the backward operating area for conventional directional protection of superimposed components is shown in Fig. 12.17.

Compared with conventional lines, the wave process of half-wavelength lines is longer. The turnback time of positive- and negative-sequence electromagnetic waves in a half-wavelength line is about 20 ms, which is several times that of a conventional UHV long line. Simulation results show that the wave process of the half-wave long line usually lasts more than 1 s.

When a backward fault occurs on the half-wavelength line (the fault point is f_4), the schematic diagram of the line measured by m side protection is shown in Fig. 12.20B. Because the half-wavelength line wave process takes a long time, the reflection wave of fault voltage and current on the n side of the line cannot be received by the protection on the m side at the early stage of the fault (20 ms after the fault occurs). Then, the electrical quantity measured by m side protection shows the characteristics of a terminal non-reflective circuit—that is, the phase of the voltage and current superimposed components measured by the m side protection are the same, and the amplitude ratio is the line wave impedance. Only when the half-wavelength line wave process is over can impedance Z_1 measured by m side protection satisfy Eq. (12.6), which has the same characteristics as the conventional line. Therefore, the impedance angle of Z_1 changes from $-Z_c$ to $-Z_n$ during the continuous wave process of the line—i.e., the phenomenon that δ changes gradually from 0 to $-\delta_n$.

From the above analysis, before the first turnback of the fault voltage and current wave is completed—that is, within 20 ms after the half-wavelength line backward fault—the phase of the voltage and current superimposed components measured by m side protection is the same, and the amplitude ratio is the line wave impedance. This special phenomenon, which is determined by the half-wavelength line wave process, is an important feature of the fault outside the half-wavelength line area. Based on this, a criterion can be designed for fault identification.

12.3.2 New principle of directional pilot protection

12.3.2.1 Waveform identification coefficient

It is assumed that fault occurs at f_1 as shown in Fig. 12.18. During the initial stage of the fault, the superimposed component of interphase current and voltage measured at the m side relay location should satisfy the following requirements:

$$\Delta u_{\varphi\varphi}(k) - Z_c \Delta i_{\varphi\varphi}(k) = 0 \tag{12.7}$$

where $\Delta u_{\varphi\varphi}(k)$ is the sampling value of the voltage phase difference superimposed component, and $\Delta i_{\varphi\varphi}(k)$ is the sampling value of current phase difference superimposed component.

Therefore, if the sampling value is accurate and the calculation result of $\Delta u_{\varphi\varphi}(k) - Z_c \Delta i_{\varphi\varphi}(k)$ on the line side protection is 0 in the early stage of startup, the backward fault of the line can be identified. Because the calculation applies the sampling value of the voltage and current phase difference superimposed component, no filter is needed, and the fault direction can be judged by protection in a very short time.

However, considering that the sampling values may be inaccurate due to CT, PT errors, and other factors, $\Delta u_{\varphi\varphi}(k) - Z_c \Delta i_{\varphi\varphi}(k)$ should be calculated within a time window to eliminate accidental error caused by single calculation. Therefore, the waveform identification coefficient is defined as follows:

$$K_{af} = \frac{\sum_{1}^{N_1} \left| \Delta u_{\varphi\varphi}(k) - Z_c \Delta i_{\varphi\varphi}(k) \right|}{\sum_{1}^{N_1} \left| \Delta u_{\varphi\varphi}(k) \right|} \tag{12.8}$$

where N_1 is the number of sampling points used to calculate the waveform identification coefficient.

At the beginning of the backward fault, the waveform identification coefficient calculated at the relay location should meet

$$K_{af} = 0 \tag{12.9}$$

Therefore, the criterion adopted by protection for backward fault on one side of the line is that the waveform identification coefficient calculated at the initial startup stage of the protection on that side satisfies

$$K_{af} < 0.20 \tag{12.10}$$

Setting the sampling point of the protection device in a cycle as N, in order to account for protection rapidity and reliability, the waveform identification coefficient calculation in this paper adopts a 5 ms data window—that is, $N_1 = N/4$.

According to the analysis in Section 12.3.1, Eq. (12.10) can only discriminate backward faults and is only valid in the initial stage of line faults. Therefore, besides calculating the waveform identification coefficient, the new method also uses the directional components of power frequency faults and negative-sequence directional elements as auxiliary criteria to determine the forward faults of lines.

12.3.2.2 Case studies

A half-wavelength 1000 kV transmission line simulation model is built on the PSCAD/EMTDC platform. And the simulation parameters are shown in Table 12.1. Simulation is carried out under the condition that active power delivered is the same as the natural power, and the angle of the power supply on both sides of the line is 205.15 degrees. The system impedance on both sides of the half-wavelength line is $Z_m = (1.972 + j19.728)\ \Omega$, $Z_n = (1.840 + j19.408)\ \Omega$.

Because the simulation oscillation time of the half-wavelength line is long, the fault occurring time is selected as 2 s after the simulation start, and this time is taken as the zero of the simulation time axis. In the simulation model, various faults of the half-wavelength line are simulated and analyzed.

When the single-phase-to-ground fault occurs near bus m on the left side of the half-wavelength line, the superimposed components of phase difference current and voltage measured by m side protection and the calculated waveform identification coefficient are shown by the double longitudinal axes in Fig. 12.21A. At this time, the measured voltage phase

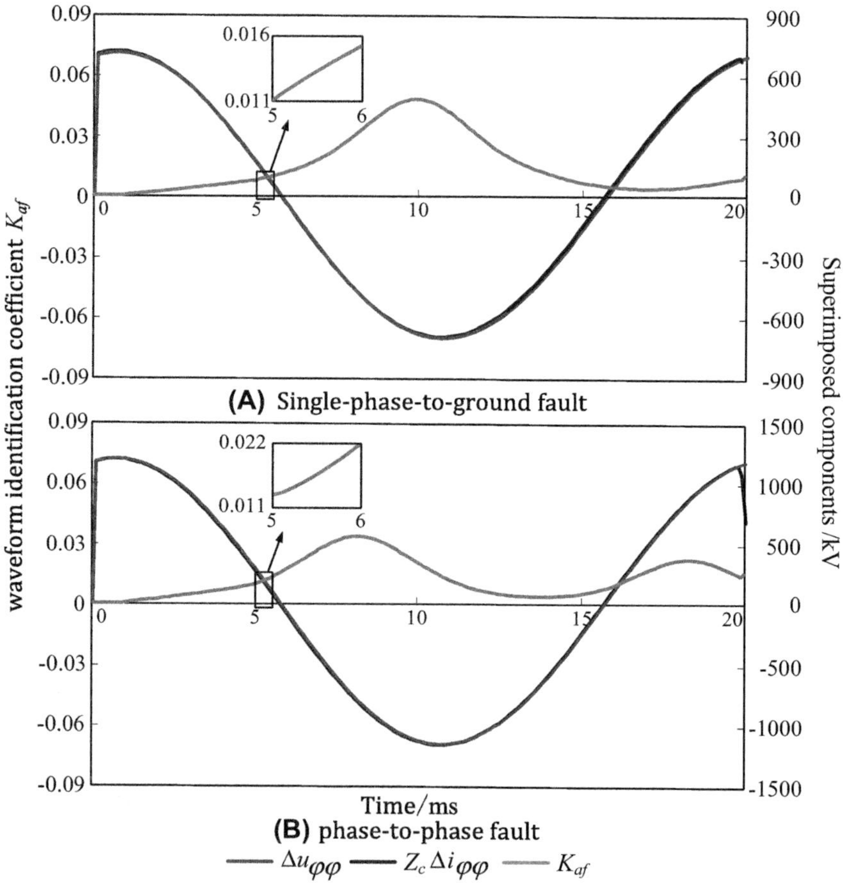

Figure 12.21 Coefficient K_{af} of m side protection in the case of a backward fault.

difference superimposed component is almost the same, as the waveforms reflect the product of current phase difference superimposed component and the positive sequence impedance of the line, which coincides with the analysis in Section 12.3.1. When the waveform identification coefficient of m side protection is calculated—that is, when the protection is started from 0 to 20 ms—the waveform identification coefficient calculated by the protection is close to 0, which basically coincides with the analysis in Section 12.3.2.1. At 5 ms after protecting startup, the waveform identification coefficient calculated by m side protection is about 0.011, much less than 0.2. At this time, m side protection can quickly distinguish the backward fault according to Eq. (12.10) and issue backward fault markers to the opposite side to ensure that protection on both sides is reliable and will not mal-operate.

When a external phase-to-phase fault occurs near the bus m on the left side of the half-wavelength line, the superimposed components of phase difference current and voltage measured by m side protection and the calculated waveform identification coefficient are shown by the double longitudinal axes in Fig. 12.21B. Similar to Fig. 12.21A, the measured voltage phase difference superimposed component is almost the same, as the waveforms reflect the product of the current phase difference superimposed component and the positive- sequence impedance of the line. At 5 ms after protecting startup, the waveform identification coefficient calculated by m side protection is about 0.013, much less than 0.2. At this time, the m side protection can quickly distinguish the backward fault according to Eq. (12.10) and issue backward fault markers to the opposite side. From Fig. 12.21, it can be seen that determination of the backward fault is not affected by fault type because the calculation of the waveform identification coefficient is based on the component of current and voltage phase difference superimposed component.

When single-phase-to-ground fault occurs 20% away from the m side of the half-wavelength line, the waveform identification coefficient calculated by the protection on both sides of the line is as shown in Fig. 12.22A. Due to the long transmission distance of the half-wavelength line, the protection located on the two sides of the line takes a long time to react to the fault. Fig. 12.22A shows that the starting time of m side protection is about 2 ms after the fault, and that of *n* side protection is about 8 ms after the fault. In fact, calculation of the waveform identification coefficient for both sides of the protection starts at their respective startup times. After the protection is started, the waveform identification coefficient calculated by both sides of

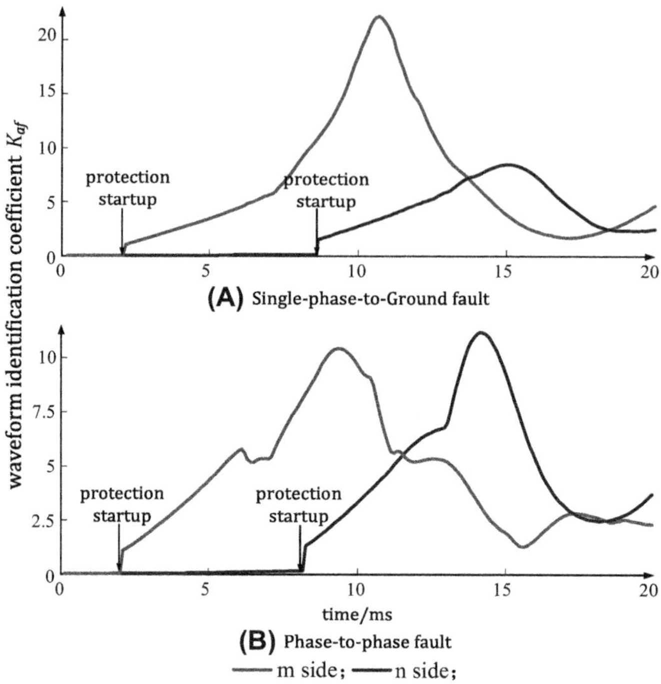

Figure 12.22 K_{af} in the case of internal fault at 20% of the line.

the protection is far greater than 1, which is basically consistent with the analysis in Section 12.3.1. Therefore, in case of internal fault, the waveform identification coefficient calculated by the protection on both sides of the line does not satisfy Eq. (12.10). The protection will not judge that the backward fault occurs on the line based on the backward identification coefficients, nor will it issue backward fault signs to the opposite side.

When a phase-to-phase short circuit occurs 20% away from the m side of a half-wavelength line, the waveform identification coefficient calculated by the protection on both sides of the line is as shown in Fig. 12.22B. Due to the same fault location, the start time of both sides of the protection is the same as in Fig. 12.22A. Similar to Fig. 12.22A, the waveform identification coefficient calculated by both sides of the protection is far greater than 1, which is not satisfied with Eq. (12.10). Similarly, the protection will not identify the backward fault on the basis of the waveform identification coefficient.

In this case, the auxiliary criterion will be used to identify the forward faults. Since the protection will first use the directional component of the superimposed component to determine the forward fault of the line from

20 to 40 ms after startup and then put the negative- sequence directional component into operation, only the results of the fault determination of the directional component of the superimposed component when the internal fault occurs are given here.

When a single-phase-to-ground fault occurs 20% away from the m side of a half-wavelength line, the phase difference of the voltage and current superimposed components calculated by m side and *n* side protection is as shown in Fig. 12.23A. Since the fault direction determination at this time matches the calculation of the waveform identification coefficient of Fig. 12.22A, the starting time of the protection on both sides of the line is the same as that of Fig. 12.22A. Fig. 12.23A shows that when the protection is started 20 ms later, the phase difference of the voltage and current superimposed components can be obtained exactly by full-wave Fourier algorithm. At this time, the phase differences of the voltage and current superimposed components on both sides of the line are both about 90°, which is basically consistent with the theoretical

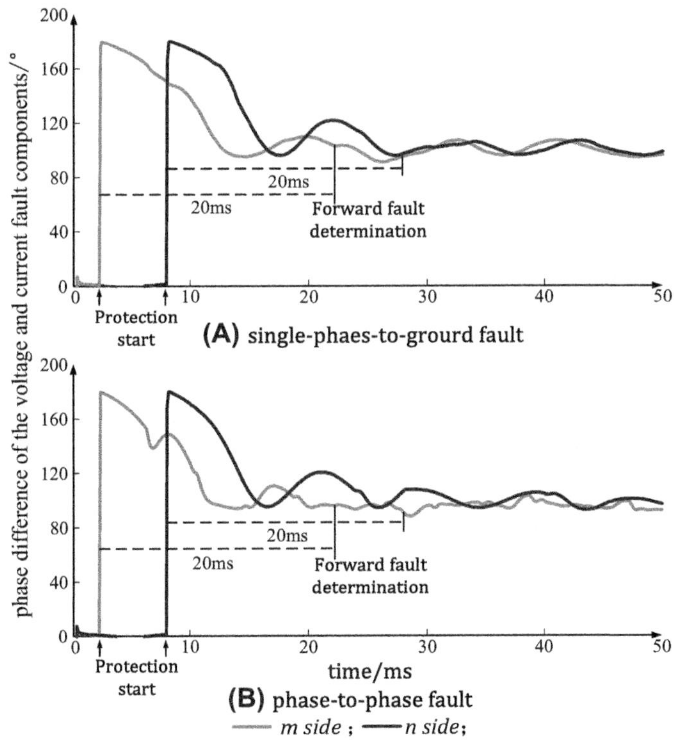

Figure 12.23 Phase difference of superimposed component of interphase current and voltage in the case of internal fault at 20% of the line.

analysis. Therefore, the protection on both sides can identify the forward fault after the start of 20 ms.

When a phase-to-phase fault occurs 20% away from the m side of a half-wavelength line, the phase difference of voltage and current power-frequency fault components calculated by m side and *n* side protection is as shown in Fig. 12.23B. At this time, the identification of fault direction matches the calculation of the waveform identification coefficient of Fig. 12.22B. Fig. 12.23B is similar to Fig. 12.23A for forward fault.

According to the high-speed operating logic adopted by the protection, once the forward fault is judged by the protection and the external fault marker is not emitted by the opposite side before the forward fault is judged, it can operate on tripping after a time delay of 5 ms.

For the half-wavelength transmission line, the proposed method can quickly identify the backward fault 5 ms after the start; the forward fault is determined 20 ms after the start, and the full line high-speed operation is achieved 25 ms after the start. The simulation results verify the correctness and reliability of the new method.

References

[1] Renzo G, Maria C. Faulted phase selection for half-wavelength power transmission lines. IEEE Transactions on Power Delivery 2018;33(2):992−1001.

[2] Ferreira Gomes LC, Da Silva LCP, Tavares MC. Half-wavelength transmission lines for connecting power plants in Amazon region to the Brazilian system. In: Proceedings of 2013 IEEE Grenoble PowerTech (POWERTECH). Grenoble, France: IEEE; 2013.

[3] Wu Y, Zou D. The influence of capacitance current to differential protection and its compensation scheme. Relay 1997;25(4):4−8 [in Chinese].

[4] Li B, He J, Guo Z, et al. Differential protection in transmission line with shunt reactor in stalled in the middle segment. Automation of Electric Power Systems 2006;30(7):31−6.

[5] Xiao S, Cheng Y, Wang Y. A Bergeron model based current differential protection principle for UHV half wavelength AC transmission line. Power System Technology 2011;35(9):46−50.

[6] He J, Song C. Power system relay protection. Beijing: China Electric Power Press; 2004.

Index

'*Note*: Page numbers followed by "f" indicate figures and "t" indicates tables'

P

Q

R

S

T

U

W

Z

$165

24

CPI Antony Rowe
Eastbourne, UK
February 05, 2020

Yuliya A. Komarova, Associate Professor of Marketing, Fordham School of Business, Fordham University, USA.

Manfred Krafft, Director and Professor of Marketing, Institute of Marketing, University of Münster, Germany.

Nandini Krishnamoorthy, Research Associate, J. Mack Robinson College of Business, Georgia State University, USA.

V. Kumar, PhD, Regents Professor, Chang Jiang Scholar, Huazhong University of Science and Technology, Richard and Susan Lenny Distinguished Chair and Professor of Marketing, Executive Director, Center for Excellence in Brand and Customer Management, and Director of the PhD Program in Marketing, J. Mack Robinson College of Business, Georgia State University, USA.

Seoungwoo Lee, PhD candidate, Robert H. Smith School of Business, University of Maryland, USA.

Donald R. Lehmann, George E. Warren Professor of Business, Chair of the Marketing Division, Columbia Business School, Columbia University, USA.

Robert P. Leone, Wilson Chair and Professor of Marketing, Neeley School of Business, Texas Christian University, USA.

Michael Lewis, Associate Professor of Marketing, Goizueta Business School, Emory University, USA.

Anita Luo, Assistant Professor of Marketing, J. Mack Robinson College of Business, Georgia State University, USA.

Mohammad Nejad, Associate Professor of Marketing, Fordham School of Business, Fordham University, USA.

Scott A. Neslin, Albert Wesley Frey Professor of Marketing, Tuck School of Business, Dartmouth College, USA.

Anita Pansari, PhD student, J. Mack Robinson College of Business, Georgia State University, USA.

Kay Peters, Professor of Marketing, Institute of Marketing, Hamburg, Germany.

J. Andrew Petersen, Assistant Professor of Marketing, Assistant Director of the Center for Integrated Marketing and Sales (CIMS), Kenan-Flagler Business School, University of North Carolina at Chapel Hill, USA.

Girish Ramani, formerly Assistant Professor of Marketing, Lebow College of Business, Drexel University, USA.

Werner Reinartz, Professor of Marketing, University of Cologne, Germany.

Roland T. Rust, Distinguished University Professor, David Bruce Smith Chair in Marketing, Executive Director of Center for Excellence in Service, Executive Director of Center for Complexity in Business, Robert H. Smith School of Business, University of Maryland, USA.

Donald E. Sexton, Professor of Marketing and of Decisions, Risk, and Operations, Faculty Director, Center for International Business Education and Research, Columbia Business School, Columbia University, USA.

Denish Shah, PhD, Assistant Professor of Marketing, Assistant Director, Center for Excellence in Brand and Customer Management, J. Mack Robinson College of Business, Georgia State University, USA.

Gayatri Shukla, Research Associate, J. Mack Robinson College of Business, Georgia State University, USA.

Bernd Skiera, Chaired Professor of Electronic Commerce, Department of Marketing, Faculty of Business and Economics, University of Frankfurt, Germany.

Raji Srinivasan, Professor of Marketing, Department of Marketing, McCombs School of Business, University of Texas at Austin, USA.

Shuba Srinivasan, Professor of Marketing, Dean's Research Fellow, Boston University School of Management, Boston University, USA.

Crina O. Tarasi, Assistant Professor of Marketing, College of Business Administration, Central Michigan University, USA.

Rajkumar Venkatesan, Bank of America Research Associate Professor of Business Administration, Darden Graduate School of Business, University of Virginia, USA.

Peter C. Verhoef, Professor of Marketing, Faculty of Economics and Business, Department of Marketing, University of Groningen, the Netherlands.

Julian Villanueva, Professor, Head of the Marketing Department, IESE Business School, University of Navarra, Spain.

Thorsten Wiesel, Professor of Value-Based Marketing, Marketing Center Münster, University of Münster, Germany.

Shijin Yoo, Associate Professor of Marketing, Business School, Korea University, South Korea.

Introduction: the evolution of customer equity

V. Kumar and Denish Shah

Customer equity is defined as the sum of the lifetime value of all customers of the firm. The customer lifetime value (CLV) may be defined as the net present value of future profits from a customer. The important keywords embedded in the definition are 'future' and 'profits'. Computing *future profits* of a customer makes the customer equity metric inherently forward looking, thereby enabling marketers to invest the right amount of marketing resources in a customer *today* to get the expected value from the customer in the *future*. In contrast, traditional backward-looking metrics such as past customer value and past customer profits provide limited (if any) guidance to managers in terms of the expected revenue stream from a customer in the foreseeable future.

Today, customer equity has established itself as one of the most dominant metrics for managers to measure and manage firm performance. In fact, it has evolved into a way of thinking and doing business as evinced by numerous research and business case studies. Consistent with the developments in the marketplace, this *Handbook* covers a broad range of tactical and strategic issues related to understanding, measuring, managing, and implementing the customer equity metric for maximizing firm performance.

The content of the *Handbook* is divided into 21 chapters drawn from a global pool of leading scholars in the field and offers different dimensions of insights with respect to the customer equity metric. Figure I.1 offers a synthesis of the broad range of topics included in this *Handbook* relevant to understanding, measuring, managing, and implementing the customer equity metric for enhancing firm performance.

As illustrated in Figure I.1, both customer acquisition efforts (or how a firm acquires a customer) as well as customer retention efforts (how a firm retains a customer) influence customer equity. Given the importance of the customer equity metric, it is imperative for firms to understand the drivers of the customer equity metric and the appropriate measurement and methodological approach to apply correctly the lifetime value of each customer and hence the customer equity. The drivers of customer equity may be impacted through a firm's marketing or how a firm acquires and

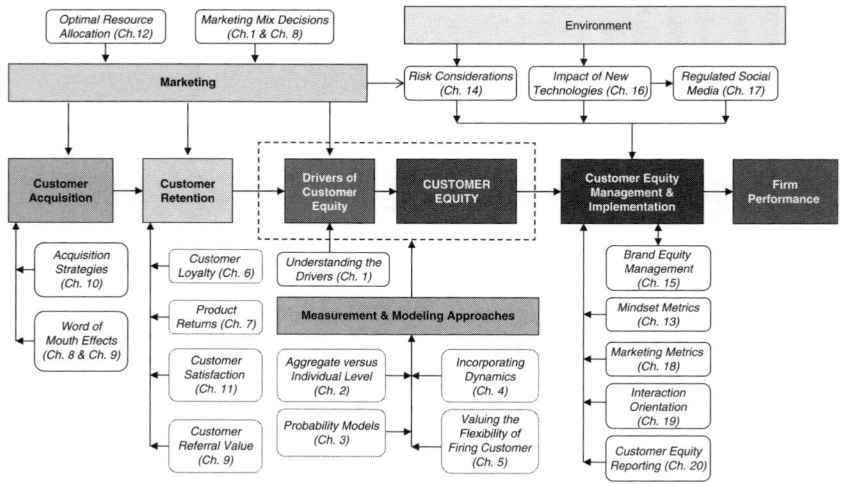

Figure I.1 *An integrated framework for understanding, measuring, managing, and implementing customer equity for enhancing firm performance*

subsequently manages a customer in the firm. A thorough understanding of these factors is critical for managers to identify and understand the levers that can lead to an increase (or decrease) of customer equity.

Finally, several issues arise related to the proper implementation and strategic management of customer equity in a firm. For example, environment-related factors in the form of new technologies, competitive changes, proliferation of new media, as well as internal factors such as the brand equity of the firm bear important implications for the strategic management of customer equity. In addition, firms need to have a clear understanding of the role of the right metrics, orientation, and reporting practices in enabling the customer equity metric to enhance firm performance upon implementation. Overall, the *Handbook* serves as a comprehensive and authoritative guide to customer equity for marketing scholars, practitioners, and students.

HIGHLIGHTS OF THE BOOK

The *Handbook* is organized in five parts. Each part contains several chapters, deals with a particular theme related to customer equity, and addresses a set of key questions, as illustrated in Figure I.2.

Part 1: Understanding and Measuring Customer Equity

The first part of the book delves into topics related to understanding and measuring customer equity. Chapter 1, aptly titled 'Drivers of customer equity' by Rust, J. Kim, Dong, T.J. Kim and Lee, addresses the most fundamental question: what are the core building blocks of customer equity? The chapter discusses in detail three different types of drivers of customer equity: value equity, brand equity, and relationship equity. Although all three drivers of customer equity are important, the relative strength of influence of these drivers on the customer equity of a firm can vary across firms and different industries. For example, brand equity is important for low-involvement goods such as toilet paper and soda, while relationship equity is important in industries such as financial services or airlines, where repeat purchase is important and the evaluation of quality for a good or service is difficult. The chapter further breaks up the three main drivers into their respective sub-drivers and discusses when a firm should invest most in which driver (or sub-driver) of customer equity. Each of the sub-drivers is conceptualized as either a means or the end to achieving customer equity. All of the sub-drivers correspond to both a marketing program driven by management and a customer response, thereby

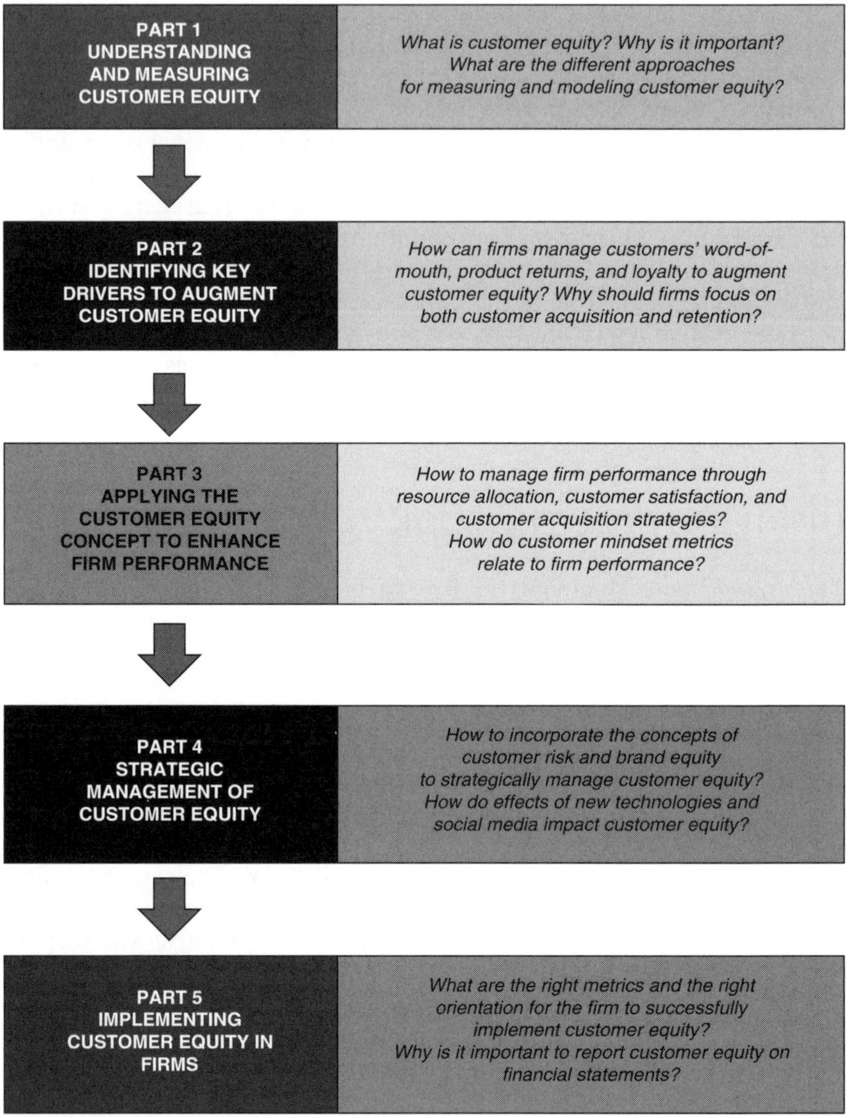

Figure I.2 Overview of the Handbook

rendering each sub-driver measurable. The authors conclude the chapter with a discussion on how a return on marketing model can be adopted for implementing an actionable customer equity program based on the aforementioned drivers of customer equity.

Customer equity measurement can be performed at the aggregate (i.e., at a customer segment or firm level) or the individual customer level. In Chapter 2, Kumar and Pansari offer a comprehensive discussion on *when* and *how* firms can measure customer equity at the aggregate versus individual customer level. Both approaches have their share of benefits and drawbacks. For example, an aggregate-level measurement of customer equity is relatively easy to compute with data related to marketing activities of the firm at the overall firm level or customer segment level. The relevant insights can enable firms to evaluate the overall effectiveness of their marketing programs and their impact on the overall profitability of the customer base of the firm. In contrast, customer lifetime value measurement at the individual customer level is relatively complex and requires marketing and revenue data specific to each customer of the firm. Individual customer lifetime value measures offer richer insights that can enable the respective firm to identify profitable customers and hence differentiate marketing resource allocations and strategies based on the profitability of each customer. Given the relative pros and cons of the two approaches, it is imperative for firms to choose the right approach in the light of their overall marketing objectives, data constraints, and ability to implement differentiated marketing programs. The chapter concludes with a framework in the form of a flowchart to help managers decide which customer equity measurement approach would be the most optimal to implement in their respective firms.

Forecasting buyer behavior over a multi-period time horizon is central to estimating the future revenue stream from a customer and hence the customer equity of the firm. In Chapter 3, Fader and Hardie review stochastic probability models that can be applied to predict future buyer behavior in different business scenarios. The chapter classifies the models on the basis of two important criteria: (1) whether the buyer behavior is occurring in a contractual versus non-contractual setting and (2) whether the opportunities for transactions are restricted to discrete points in time or whether they can occur at any point in time. For example, participation in academic conferences represents a contractual setting where the revenue from the participants can only occur at discrete points in time (at the time of registering or attending the conference). In contrast, customers shopping from say an online retailer represents a non-contractual setting where the revenue from the customers can occur at any time, as there are no constraints as to when the customer can purchase from the retailer.

The chapter recommends application of different models to measure the CLV (and hence the customer equity) given the business context of the firm and underscores the need for the managers to distinguish between (1) the expected lifetime value of an as-yet-to-be-acquired customer, (2) the expected lifetime value of a just-acquired customer, or (3) the expected residual lifetime value of an existing customer.

In Chapter 4, Lewis discusses the need for incorporating the dynamics of customer behavior in CLV models. Customers are intelligent entities that not only learn from their experiences transacting with the firm but also develop expectations of the future. As a consequence, when firms implement different marketing programs, consumer preferences and expectations are likely to evolve over time. For example, firms' promotional discounts can decrease future revenues by changing reference prices or by training consumers to wait for future promotions. Standard CLV calculations typically do not account for the dynamics of customer behavior as they assume that customers' revenues, costs, and retention rates stay constant. Such assumptions lead to incorrect estimation of the CLV and hence sub-optimal managerial decisions based on the CLV. However, modeling customers as intelligent and dynamically evolving actors greatly complicates the modeling of customer decisions. In addition, the creation of optimal relationship marketing strategies by itself is a challenging endeavor. Drawing upon the dynamic programming approach employed by the marketing and economics literature, this chapter discusses in detail how the dynamics of customer behavior can be accounted for in the CLV computation. The methodology entails treating customer relationship marketing (CRM) of the firm as a stochastic dynamic optimization problem. The solution of an appropriate dynamic programming model is potentially of great value to firms as it yields policies that map marketing actions to customer states in a manner that maximizes CLV and hence the customer equity of the firm.

In Chapter 5, Haenlein argues that standard CLV calculations assume that, once a customer has been acquired, a company will continue to retain the customer, regardless of whether that customer turns out to be profitable. In reality, this assumption is likely to be violated as firms typically have the flexibility to directly or indirectly abandon (or fire) their unprofitable (or problem) customers. For example, the US telecommunications firm Sprint Nextel directly fired customers in July 2007 by terminating the contracts of 1000 clients, while the brokerage house Charles Schwab is claimed to have implemented indirect abandonment by making low-profitability accounts wait longer in call center queues. Consequently, a more realistic computation of customer equity should entail incorporating the firm's *option* to fire unprofitable customers in the CLV computation.

The chapter elaborates on how this may be accomplished by applying the concept of the valuation of (real) options from the interface of the strategic management and finance literature. A related issue is whether some abandonment strategies of the firm (for example: direct versus indirect approach) are associated with more severe negative consequences and higher abandonment costs than other ones. Interestingly, the negative effects of unprofitable customer abandonment in response to direct as well as indirect strategies result in approximately the same levels of exit intention among remaining customers, purchase intention among potential customers, and boycott likelihood for both customer types. The chapter concludes with a discussion of future avenues of research with respect to this relatively unexplored area from both the methodological as well as the managerial perspective.

Part 2: Identifying Key Drivers to Augment Customer Equity

The second part of the book discusses important aspects of customer behavior that managers can leverage to augment customer equity. For example, loyal customers can have a substantial impact on customer equity through their extended lifetime duration with the firm. However, loyal customers are not necessarily profitable. In Chapter 6, Reinartz and Eisenbeiss highlight the complex processes involved in generating customer loyalty. The complexity arises from the fact that the satisfaction–loyalty–profit chain links are almost always non-linear, asymmetric, and industry and customer specific. Thus, the current industry practice of using customer satisfaction or customer loyalty as a simple proxy measure for customer profits is not sufficient. Consequently, the chapter discusses why and how the CLV metric offers an elegant way to measure and quantify customer loyalty in the context of the satisfaction–loyalty–profit chain and directly connects loyalty with profitability. In addition, this forward-looking metric supports long-term decision-making. As a flexible measure, it can adapt to the specific business context of any industry and evaluate each individual customer according to his or her expected contributions to the company. The chapter concludes with a discussion on how firms can profitably manage customer loyalty through (1) a cross-functional integration of processes, people, operations, and marketing capabilities, enabled through information, technology, and applications, (2) adoption of a distinctive customer orientation, and (3) a proper alignment of the management processes and activities with the different (and changing) phases of a customer–firm relationship.

One of the interesting aspects of customer behavior (specific to product-based industries) is returning previously purchased products. Firms have often seen customer product returns as just a cost of doing business and a drain on profitability. However, recent research has shown that customer product return behavior plays a more integral role in the firm–customer exchange process than just a cost to the firm. In Chapter 7, Petersen and Anderson find that firm decisions on managing product return behavior through return policies and marketing mix variables impact the way the customer thinks about the products and the firm, what the customer does in terms of purchase and product return behavior, and in turn the profitability of the firm. This raises two important questions: (1) Can firms manage customers' product return behavior? (2) If so, given the known and potential costs along with the benefits of customer product return behavior, how can firms optimally manage customer product returns to maximize customer equity? The chapter explains how this can be accomplished by linking relevant marketing metrics to the product return behavior of the customer. Consequently, the product return behavior of the customer is integrated in a customer value framework. With such a set up, firms can optimally plan their return policies and marketing mix elements with the objective of leveraging customers' product return behavior to maximize customer equity.

In Chapter 8, Hanssens, Villanueva, and Yoo discuss the word-of-mouth and marketing effects on customer equity. The authors argue that contrary to the popular belief, effective customer equity management should not only focus on customer retention efforts but also on customer acquisition efforts. This is because a firm acquiring the wrong customers will have difficulties in trying to retain and grow the value of the potentially problem customers subsequently. Consequently, in the first part of the chapter, the authors first discuss different customer acquisition channels and underscore the strategic importance of word-of-mouth vis-à-vis traditional marketing channels for acquiring customers. In the second part of the chapter, the authors discuss different methodological approaches for measuring the contribution of marketing efforts on the evolution of customer equity. Broadly speaking, these methods fall under three categories: judgment-based methods, switching matrices based on stated preferences, and econometric methods based on transactions databases. The authors illustrate the application of these methods to measure, manage, and maximize customer equity in the context of relationship businesses (e.g., insurance) as well as product-centric sectors (e.g., automobiles and consumer packaged goods).

In Chapter 9, Leone and Christodoulopoulou argue that a customer's value to the firm should not only be based on the respective customer's purchase behavior but also on their non-purchase behavior such as customer referrals that indirectly create value for the firm. This chapter builds upon the discussion of Chapter 8 concerning the importance of customer referrals as an important customer acquisition tool, characterized by lower costs and a relatively higher value of newly acquired customers, thereby offering the potential for significantly increasing the customer equity of the firm. This chapter conducts a detailed examination of the customer referral process, explains how firms use referral programs, discusses the value of referred customers, presents strategies for initiating the referral process by pursuing different seeding strategies and defines the referral rewards in detail. The chapter also discusses how to measure customer referral value (CRV), highlights the strategic importance of this metric, and explains how firms can enhance customer equity by using CRV to evaluate and manage customers based on their referral activity, and in tandem with their CLV. Finally, although the concept of customer referrals is mostly used in consumer markets, its application in the context of business markets is also discussed in this chapter.

Part 3: Applying the Customer Equity Concept to Enhance Firm Performance

The third part of the book broadly focuses on how firms can implement the customer equity metric to manage the overall performance of the firm. In Chapter 10, Peters, Verhoef, and Krafft demonstrate and discuss the important role of customer acquisition in customer equity management. This is because (1) success in cross-buying, retention, and win-back strategies is to some extent conditional on what the firm does for acquisition in the first place, and (2) even with generally successful retention, natural attrition requires continuous replacement with new customers through effective customer acquisition efforts. The authors discuss the fundamental aspects of customer acquisition and illustrate the complex nature as well as the relevant challenges related to proper implementation of customer acquisition processes in firms. The authors conduct a detailed literature review and discuss the state-of-the-art research findings related to customer acquisition. The authors conclude with a set of recommendations for managers. The overarching message of this chapter is that 'customer equity management should govern acquisition strategies as they predetermine later outcomes to a substantial degree' (p. 228).

Customer acquisition and retention efforts of the firm go hand in hand. Customer satisfaction is a necessary condition to retain customers. However, managers often struggle to consistently achieve a positive financial outcome (or customer equity) from their customer satisfaction and retention efforts. In Chapter 11, Keiningham, Aksoy, Komarova, and Nejad identify how and why the links in the chain of effects from satisfaction to firm performance break down. The authors identify three primary issues in connecting satisfaction to customer behavior and to financial performance: (1) share of category spending – not retention – the most relevant behavioral metric, (2) satisfaction's impact can only be understood relative to competitive alternatives, and (3) not all customers can be profitably satisfied. The authors propose that firms should attempt to maximize share-of-wallet, revenue, and profitability simultaneously to fix the links in the chain of effects from satisfaction to firm performance.

Any customer acquisition or retention efforts of a firm typically entail investment of marketing resources. The question then is, how to determine the right amount of marketing resources to allocate across different customers of the firm? The complexity of the task is often compounded by the fact that firms typically encounter marketing budget constraints. In Chapter 12, Venkatesan proposes a five-step process for managers to optimally allocate marketing resources at the individual customer level based on their lifetime value. The first step requires developing the appropriate CLV formulation for a business considering its targeting capability and contractual versus non-contractual relationships. The second step is to break CLV down into its components. The third step in the resource allocation process is to determine the unknown CLV components that must be predicted (for example, a customer's margin, retention cost, retention rate and so on). The fourth step involves developing a predictive model for each of the unknown components. The final step is to find the optimal level of marketing resources to maximize CLV while considering factors such as budget constraints. The chapter offers managerial guidance on implementing these steps. Consequently, managers can maximize the overall firm performance given the marketing budget constraints of the firm.

In Chapter 13, Shuba Srinivasan develops a conceptual framework that investigates the effects of customer mindset metrics on firm performance. The objective here is to illustrate the two routes through which marketing impacts firm performance: (1) a transaction route where 'customers-as-assets' directly impact firm performance through increase in cash flows and (2) a mindset route where 'customers-as-information' indirectly impact firm performance through changes in customers' mindset metrics and customer equity. Focusing on the link between customer mindset metrics and firm performance, the chapter provides a metrics overview and offers extant findings on the customer-mindset-metrics/financial-performance interface. The chapter concludes with an agenda for future studies that addresses gaps and challenges in this important and growing area of research.

Part 4: Strategic Management of Customer Equity

The fourth part of the book comprises an eclectic set of issues concerning strategic management of the customer equity metric in a firm. For example, one of the most important determinants of customer equity is the expected stream of cash flow from the customers. In Chapter 14, Bolton and Tarasi discuss why it is critical for firms to understand the risks associated with the unexpected variability in customers' cash flows in order to effectively forecast, allocate resources, manage customer equity, and prepare for the future. The variability or unpredictability of future cash flows arise from underlying customer interactions, consumption patterns and purchase behavior – as well as from organizational processes that support them. This chapter summarizes alternative approaches to reducing variability in future cash flows without adversely affecting cash flow levels. First, firms can target customers with low cash flow variability or low risk. Second, since customers do not have fixed cash flow characteristics, the variability of individual customers' future cash flows can be decreased through appropriate management actions. Third, firms can allocate resources to specific customers and market segments (through their marketing plan) to yield future cash flows that complement each other and thereby decrease aggregate variability of future cash flows. The chapter offers a practical roadmap for firms to strategically manage customer equity through optimization of their customer base by balancing return and risk.

Another important aspect of strategically managing customer equity is its linkage to brand equity. Brand equity and customer equity are marketing assets that enhance short-term profit and long-term firm value and are essential to the success of marketing management. Most of the literature views brand equity as a determinant of customer equity. In Chapter 15, Luo, Lehmann, and Neslin argue that customer equity feeds back to brand equity. The managerial upshot is that brand equity and customer equity need to be managed strategically, taking into account both the more immediate-term impact of brand equity on customer equity, and the longer-term feedback of customer equity on brand equity. The chapter refers to this as 'co-management' of brand and customer equity. The chapter demonstrates how brand equity and customer equity can be co-managed in a firm using a matrix with customers as the rows and brands as columns.

Rapid advances in technologies such as Internet, mobile telephony, and related technologies can have a significant impact on how firms manage customer relationships, with important implications for customer equity management. In Chapter 16, Srinivasan's insightful review of emerging technology developments suggests many dramatic effects on businesses in general, but customer relationship management in particular, resulting in a double-edged effect presenting opportunities and threats for customer equity management practice. The chapter identifies the challenges and threats posed by new technologies to markets, customers, basis of competition, customer relationships, marketing program effectiveness, and customer value propositions – all key elements of the customer equity management framework. The chapter discusses the important managerial implications for customer equity management arising from the effects of new technologies.

Technological developments have contributed to the proliferation of social media, which have revolutionized the way consumers go about their purchase decisions, as well as the way consumers and businesses stay connected with each other. In Chapter 17, Kumar, Krishnamoorthy, and Shukla discuss the latest trends in social media as well as the inherent flaws associated with taking a short-term approach to social media marketing (SMM) by offering discount coupons, standalone offers/promotions and one-dimensional advertisements in the light of findings from a recent business case study of the topic. Consequently, the chapter recommends a novel 'regulated social media marketing' (regulated SMM) approach that integrates group purchases, social networks, and mobile marketing. This approach empowers businesses to identify and target the most influential customers on the basis of their customer influence effect (CIE) and customer influence value (CIV), so that they serve as word-of-mouth (WOM) agents of the business and channelize its message across their social networks in a regulated, measurable, and effective way. Regulated SMM can enhance not only customer acquisition, but also customer equity, shareholder value, and the overall profits of firms.

Part 5: Implementing Customer Equity in Firms

The fifth part of the book discusses important issues related to the implementation of the customer equity metric. Given the current competitive environment, marketing managers are under intense pressure to make sound decisions. In general, besides the customer equity metric, several other performance metrics exist that are widely employed by managers across firms. However, several surveys of managers in the past decade indicate a general disconnect with the marketers' ability to improve marketing decisions and the performance metric being implemented by the firms. In Chapter 18, Sexton argues that the relative success of implementing a marketing return metric such as the customer equity metric is linked to the effort invested by an organization in developing and utilizing metrics, the data and skills available, and the overall environment created by the organization's leadership.

Since customer equity represents an inherently customer-centric way of dealing with customers, successful implementation of the metric is likely to result in firms adopting a customer-centric orientation with respect to their marketing practices. In Chapter 19, Ramani and Knox talk about how an 'interaction orientation' can enhance a firm's ability to interact with individual customers and take advantage of information obtained from them through successive dyadic interactions in order to achieve profitable customer relationships. The chapter reviews the concept of interaction orientation and traces the adoption of interaction orientation by firms. The chapter also extends the interaction orientation framework by discussing the usefulness of the framework in the context of customer complaints and the company's superior capability to recover from product and service failures.

In Chapter 20, Wiesel and Skiera present a compelling argument for why firms should report customer equity on their financial documents. Current financial reporting practices are primarily backward oriented or based on reporting business that was realized in the past. However, managers and investors typically need more information about the future performance of a company, in particular the long-term value of the company's customers. Customer equity reporting presents a mean to focus on the long-term value of the company's customers. It measures the long-term value of a company's customer base, which is the discounted profit that a company will make with its current customers. Its reporting supports decisions that are long-term rather than short-term value oriented. As such, it avoids the risk that short-term profits are increased at the expense of long-term value creation and its central metric, customer equity, and allows for monitoring companies' performance with respect to the value of their current customers.

Collectively, the different chapters of the *Handbook* offer the following important takeaways with respect to customer equity:

- Both customer acquisition as well as retention efforts of the firm are important to manage the customer equity of the firm.
- To compute customer equity, different measurement and modeling approaches exist. It is imperative for firms to fully understand the nuances of the different approaches in order to apply the best approach – given the specific marketing objectives, data availability, and resource constraints of the firm.
- Customer equity can be augmented in different ways. The differ-

ent drivers of customer equity may be impacted through marketing interventions and prudent customer management.

- Strategic management of customer equity can enable firms to regard customer equity as a means to enhance the overall performance of the firm.
- Adoption of the right orientation, performance metrics, and reporting practices enable successful implementation of the customer equity metric in firms.
- Customer equity is more than a metric. Proper measurement, management, and implementation of the metric renders customer equity as a way of thinking and doing business.

In the concluding section (Chapter 21) of the book, Kumar and Shukla discuss how business practices based on the customer equity metric will evolve further due to the following three major forces of change: (1) explosion and availability of data, (2) development of more sophisticated modeling techniques aided by an increase in computation prowess, and (3) spatial diffusion of the implementation of the customer equity metric across markets (developed versus emerging) and different industries.

PART I

UNDERSTANDING AND MEASURING CUSTOMER EQUITY

1. Drivers of customer equity

Roland T. Rust, James Kim, Yue Dong, Tom J. Kim and Seoungwoo Lee

INTRODUCTION

Making Customer Equity Strategic

The 'service revolution' has shifted the core of exchange between firms and customers from products to the provision of service (Vargo and Lusch, 2004). Long-term one-to-one customer relationships have come to the fore while customer transactions, the hallmark of a product economy, are being increasingly de-emphasized (Peppers and Rogers, 1993). While firms can't ignore the need for attracting new customers, the retention of existing customers has become more crucial to long-term profitability. Innovations in information technology further accelerate such structural changes in the economy as true interactivity between firms and customers, customer specific personalization, and real-time adjustments to a firm's offerings are becoming reality (Rust and Lemon, 2001). Such trends that define today's economy necessitate the fundamental rethinking of business and strategy around the customer.

The notion that customer assets are becoming more and more important is not new. For example, Procter & Gamble renamed its sales department as 'Customer Business Development' more than a decade ago to integrate such thinking into its operations (Sheth and Parvatiyar, 2002). However, the increasing pace of economic and technological change forces firms to move beyond simply recognizing that the customer lies at the heart of business – to evolve into truly customer-centric companies (Rust, Zeithaml, and Lemon, 2000, 2004). This calls for marketing, and potentially the entire firm, to organize and deploy all tangible and intangible assets into maximizing customer equity. Customer equity should lie at the heart of any firm's philosophy and vision. The decision criterion for all business functions and operations should be whether an action contributes to growing customer equity. In this chapter we lay the groundwork for firms to adapt to become true customer-centered companies by providing an overview of the drivers of customer equity and introducing a method for implementing a customer equity program.

The Three Drivers

In order to evolve into true customer-centered companies, firms need to know the specific drivers of customer equity. Following the framework of Rust et al. (2000), we present value equity, brand equity, and relationship equity as three main drivers of customer equity. We adhere to the definitions of Rust, Zeithaml, and Lemon (2004). Value equity stems from the perceptions of objective value, based on attributes such as the quality, price, and convenience of a good or service. Value equity is created when customers choose a good or service that provides greater value than its cost. Brand equity originates from subjective attributes, such as customers' subjective and intangible assessments of the brand and emotional attachments to the brand. Brand equity helps the differentiation of a product and the formation of positive associations in customers' minds that simplify customer choice. Relationship equity (originally called 'retention equity') is created from customers' intentions to maintain a relationship with a brand, over and above the value customers derive from their objective and subjective assessments of the brand. Strong relationships between a firm and its customers increase long-term profitability through increased repeated purchase. Thus, each driver of customer equity increases customer lifetime value (CLV).

Although all three drivers of customer equity are important, the strength of influence that any one of the drivers exerts on CLV varies across industries (Lemon, Rust, and Zeithaml, 2001). For example, brand equity is important for low-involvement goods such as a toilet paper and soda, where customers do not want to spend too much time on decision-making, and are therefore likely to be influenced heavily by their prior attitudes toward brands (Keller, 1993). However, brand equity is less important in industries where the objective value of a good or service is important, and customers are motivated to process information more carefully. Relationship equity is important in industries such as financial services or airlines, where repeat purchase is important and the evaluation of quality for a good or service is difficult.

The customer equity framework provides solid ground for firms to develop actionable initiatives to maximize CLV given budget constraints. Firms need to explore which driver, or combination of drivers, has the maximum impact on customer equity and CLV. Throughout the chapter we describe each driver of customer equity and its respective sub-drivers in detail, and provide a method for implementing a customer equity program. We conceptualize the sub-drivers as both the means and the ends to achieving customer equity. All of the sub-drivers can correspond to both a marketing program driven by management (e.g., a new ad campaign to increase brand awareness) and a customer response (e.g., a customer's perception of brand awareness following a new ad campaign). In any case, the sub-drivers of customer equity are measureable in terms of budget or associated expenditures or in terms of customer-measured items (e.g., 'To what degree are you aware of the brand?' on a Likert scale).

VALUE EQUITY

The Nature of Value Equity

Defining the term 'value' is difficult since the value that an individual customer derives from a good or service is based on his or her own underlying needs and perceptions. We adopt the definition put forth by Rust et al. (2000), following Zeithaml (1988). From the customer's perspective, value means one of the following four categories: low price, what the customer wants in a good or service, the quality the customer receives for the price the customer pays, and what the customer receives for what the customer gives up (e.g., time and effort). These categories indicate that some customers concentrate primarily on what they receive from the good or service, whereas others focus mainly on what they give up or sacrifice in return for the good or service. Thus, value can be defined as the consumer's objective assessment of the utility of a good or service based on perceptions of what is given up for what is received (Rust et al., 2000).

Value equity represents the objective appraisal of the utility of a good or service. Value often lies at the core of customers' relationships with a firm (Lemon et al., 2001). The firm's ability to offer goods or services that satisfy customers' needs and meet their expectations is a necessary condition that precedes brand-building and relationship-building strategies in the successful implementation of a customer equity program. Advertising campaigns and loyalty programs can only create synergy with goods and services when customers perceive that the firm's offerings provide value. For example, car manufacturers' extensive investments in TV commercials are ineffective if the quality of their cars does not live up to customers' standards. Membership programs offered by airlines, hotel chains, and retailers also need to be built upon the core value of their goods and services in order to have an effect in increasing CLV.

Drivers of Value Equity

The key sub-drivers of value equity are quality, price, and convenience (Rust et al., 2000). Quality can simply be considered in terms of how well the customer is served. Quality includes tangible and intangible aspects of a good or service that are under a firm's control. Quality is driven by physical goods, service products, service delivery, and the service environment (Lemon et al., 2001). Firms can enhance value equity by introducing additional variety to their physical goods, by differentiating the features and the benefits of its service products, by improving the service delivery process, or by making its service environment more interesting for the customer. Introducing variety to a product assortment can increase a customer's value by accounting for customer heterogeneity in individual preferences. Although Henry Ford remarked that any customer could buy a Model T 'painted any color that he wants so long as it is black', customers could pick one out of nine body styles that best fitted their preferences. Toyota's mission statement, 'Through our commitment to quality, constant innovation and respect for the planet, we aim to exceed expectations and be rewarded with a smile' emphasizes the best customer experience and dealer support, exemplifying firms' efforts to drive value equity by providing top quality.

The price of a good or service is another important driver of value equity. Firms must exercise caution when adopting pricing strategies. While price discounts may contribute to the growth of value equity, they have the potential to simultaneously deteriorate customers' perceptions of the firm's overall quality and value in the long run. Products purchased on an infrequent basis for which the price–quality relationship is stronger than for frequently purchased items, are more susceptible to such potential negative effects of price promotions (Gerstner, 1985). There are four practical ways that a firm can compete on price. A price discount strategy offers temporary bargains. An everyday low pricing strategy gives customers the image that the firm consistently provides good prices, reducing the likelihood that customers will wait for a price discount to make their next purchase. A complex pricing strategy allows the customer to exercise flexibility in the overall price paid, the timing of payment, or the overall product received for a given period. A firm can also apply a situation-based pricing strategy, which offers different pricing based on business circumstances.

Providing convenience to customers strengthens value equity by reducing customers' search cost, time, and effort. Providing additional physical locations can create a value for customers. Opening an online or mobile website enables customers to access the firm's goods and services conveniently. Gilt City, a high-end website promoting local deals, recently demonstrated the impact of convenient access through mobile channels on value equity. The firm's transactions that were settled through mobile devices doubled within a year to 20 percent of total sales in 2012 (Indvik, 2012). Ease of use is a particularly important sub-driver for web-based services and software products. Automating routine operations on the firm's website based on customers' behavior and emphasizing the intuitive interface during the process of software development can increase value equity. Availability is another sub-driver of value equity that emphasizes the 'when' more than the 'where' (i.e., location). For example, AT&T's Global Customer Service Center offers 24-hour access, seven days a week, allowing customers to make immediate contact with the firm whenever they want.

When Value Equity Matters Most

Value equity is a crucial driver of customer equity since value precedes customers' relationships with the firm. Under specific circumstances, value equity may be the most important customer equity driver in maximizing CLV (Rust et al., 2000).

First, value equity plays a pivotal role when the customer has the ability to discern differences between competing products. Driving value equity is generally very challenging for firms whose offerings do not provide clearly differentiated benefits, and in markets where little differentiation exists among competing products. In such conditions, firms may need to find other venues of opportunity for outpacing competitors. However, value equity can be built in environments where there are clear differences among competing products, or if a firm can add distinct features to its products.

Second, a firm can employ a value equity strategy when launching innovative goods and services. Customers tend to scrutinize new goods or services, comparing them to existing offerings in deciding whether they provide enough benefits that warrant purchase. Positive signals of quality help lower customers' perceptions of the risk of consuming a new good or service that does not deliver (Kirmani and Rao, 2000), which in turn leads to the growth of value equity.

Third, value equity matters most when customers face complex purchase decisions. Large expenditures or business-to-business purchases, which generally involve a long-term relationship between the buyer and the provider, are examples that involve complex decision-making. Companies that successfully minimize their customer's time and efforts are well equipped to build value equity.

Finally, value equity is important when firms want to rejuvenate mature goods or services. By adding new features or benefits to the goods or services in the maturity stage, firms can vitalize their goods and services, adding extra life to the life cycle. The successful execution of such efforts leads to value equity.

The Future of Value Equity

As the business environment continues to evolve from mass marketing to one-to-one marketing (Peppers and Rogers, 1997) building adhesive customer relationships has been emphasized while cost reduction through customer relationship management has been relatively de-emphasized. Thus, value equity has been receiving less attention.

However, value equity is still the fundamental element that precedes all other drivers in building and maintaining long-term relationships with customers (Rust et al., 2000). Rather than questioning the importance of value equity, the challenge going forward lies in how firms can create value equity in the changing business environment.

Advances in technology and the widespread application of sophisticated customer relationship management tools give firms unprecedented access to more information about their customers. The Internet also enables firms to explore and access more information on their potential customers. With such broad and individualized data, firms may implement value equity strategies by fine-tuning the three levers of quality, price, and convenience. Richards and Jones (2008) provide some examples of how firms can use three levers to drive value equity. First, firms can classify their customers in terms of profitability (Thomas, Reinartz, and Kumar, 2004). Furthermore, technology allows firms to identify the impact of each sub-driver at the individual customer level. Targeting the most profitable customers on a one-to-one basis will open the door to maximizing value equity. Second, access to individual data greatly increases the allocation of costs to individual customers (Rust, Lemon, and Zeithaml, 2001). Price strategies can be implemented with surgical precision. Third, customer relationship management streamlines the firm's information flow in a customer-centric direction (Buttle, 2004), enabling the firm to provide convenient services to customers by building adhesive relationships, improving customer responsiveness, and sharing more information with customers.

Advances in technology are culminating toward firms providing goods and services that are perfectly individualized and change with customer needs (Oliver, Rust, and Varki, 1998). Location-based mobile applications such as Google Now and Google Field Trip provide highly individualized services based on users' current locations and/or their historical Google website activity. For example, Chung, Rust, and Wedel (2009) developed an adaptive personalization system for digital audio-players that enables service providers to dynamically individualize and fine-tune their service based on changing customer needs. Coupled with the changing business environment, value equity will remain a critical component in reshaping the firm to create customer equity.

BRAND EQUITY

The Nature of Brand Equity

Brand equity is defined as the marketing effects uniquely attributable to a brand (Keller, 1993). Brand equity can be considered the added value of a good or service with a brand name in comparison to the same good or service without the specific brand name attached (Aaker, 1991; Ailawadi, Neslin, and Lehmann, 2003). Brand equity can be measured using the customer mindset approach, the product market approach, or the financial market approach, each with its strengths and weaknesses (Keller and Lehmann, 2003). We focus on the customer mindset approach to present the drivers of customer equity from the viewpoint of the customer. From the customer's perspective, brand equity represents the customer's subjective and intangible assessment of the brand beyond an objective value (Rust et al., 2000; Rust, Lemon, and Narayandas, 2005). The value of a brand is ultimately derived from 'what customers have learned, felt, seen, and heard about the brand' (Keller, 2008).

More specifically, a brand adds value to firms by performing three vital roles (Rust et al., 2000). First, it attracts new potential customers through brand awareness and recognition. Second, a brand reminds existing customers of the firm's goods and services. Third, a brand helps customers form emotional ties to the firm, which leads to increased sales and purchase share (Park et al., 2010).

Brand equity has a positive impact on a variety of firm performance metrics. Brand names are positively related to stock prices (Kerin and Sethuraman, 1998; Madden, Fehle, and Fournier, 2006; Mizik and Jacobson, 2008). Brand equity positively impacts customer loyalty (Chaudhuri and Holbrook, 2001), willingness to pay (Park et al., 2010), and firm revenue (Ailawadi et al., 2003). Brand equity also has positive impacts on customer acquisition, retention, and profitability, even after controlling for the influence of firms' marketing activities (Stahl et al., 2012).

Drivers of Brand Equity

Brand awareness, attitudes toward the brand, brand attachment, and corporate ethics are actionable drivers that firms have at their disposal to drive brand equity (Rust et al., 2000, 2005; Park et al., 2010).

Brand equity stems from customers' knowledge about the brand, which consists of brand-related thoughts, feelings, perceptions, images, and experiences (Keller, 2003). Therefore, customers must be aware of the brand to build strong brand equity. In the absence of sufficient brand awareness, it is highly unlikely that customers will form any attitudes toward the brand or emotional ties with the brand. To increase brand awareness, firms need to deliver a consistent message to customers using an integrated marketing communication mix in the most effective media channels.

Strong brand equity comes from favorable, strong, and unique associations with a brand (Keller, 1993). Therefore, brand equity increases when customers evaluate a brand favorably, with confidence and certainty. Celebrity endorsements, special events, product placement, brand partners, and brand extensions are some examples of firm initiatives that aim to create positive and strong brand attitudes.

Brand attachment, the strength of the bond between the brand and the self (Park et al., 2010), increases brand equity by forming self–brand connections and building brand prominence. Customers are more likely to use their resources to maintain their relationship with a brand when they have formed a cognitive and emotional connection between the self and the brand. Firms need to form brand attachment by gratifying customers' self through aesthetic experiences with the brand, by offering a chance to enrich the self through brand concept, and by providing consistent and reliable performance (Park, MacInnis, and Priester, 2008).

In addition to brand awareness, positive brand attitudes, and brand attachment, brand equity is influenced by customers' perceptions of the firm organization. When considering a relationship with a brand, customers examine whether the values that a brand represents (i.e., in the sense of principles and beliefs, not as in value equity) coincide with their own values. For example, environmentally conscious customers will prefer brands or firms that promote green marketing and eco-friendliness. Firms can express their strong corporate ethics through means such as community sponsorships, donations, environmentally friendly initiatives, the promoting of privacy policies, and the fostering of firm–employee relations.

When Brand Equity Matters Most

Brand equity is most critical under the following circumstances. First, brand equity is important for purchases of low-involvement products. For certain products, such as frequently purchased low-price consumer packaged goods, customers apply routinized decision processes with little consideration. In this case, the role of brand equity is critical, because customers will be easily influenced by brand awareness and emotional attachment to the brand (Rust et al., 2000; Lemon et al., 2001). Brand equity from positive post-purchase experiences influences customers' opinions for their subsequent decisions. Therefore, developing a 'loyalty loop' is important for creating superior marketing performance (Court et al., 2009).

Second, brand equity is vital when the use of the good or service is publicly visible to others. Customers often regard brands as an extension of the self (Belk, 1988) and use brands as a method for expressing self-identity to others (Berger and Heath, 2007). Therefore, customers are more likely to consider the fit between the brand and the self-concept when their own brand usage or consumption activity is of a public nature.

Third, brand equity matters when customers evaluate a new or unfamiliar product based on their knowledge of pre-existing brands. Brand extension, a strategy of using established brand names to launch new products (Völckner and Sattler, 2006), leverages the brand equity of existing brands to enhance the awareness and positive perception of new products. For example, Arm & Hammer leveraged its brand equity built upon its baking soda, to extend into toothpaste, laundry detergent, the underarm deodorant, and the pet litter deodorizer industry.

Fourth, brand equity is important when experiences of the good or service can be shared with others or passed from one to another. For instance, parents' brand preference of toothpaste or soap can pass down to their children (Rust et al., 2000; Lemon et al., 2001). The sharing of consumption experiences among customers can also lead to positive word-of-mouth effects that increase brand equity. Word-of-mouth and assistance from personal acquaintances are especially effective in driving sales for products that signify a high financial or psychological risk (Gershoff and Johar, 2006). The potential of word-of-mouth in surpassing traditional media channels in effective persuasion (Herr, Kardes, and Kim, 1991; Godes and Mayzlin, 2004), and the proliferation of online reviews, recommendations, and forums for sharing product usage or service experiences provide firms ample opportunity to build brand equity.

Fifth, the role of the brand is important for credence goods (Rust et al., 2000; Lemon et al., 2001). When customers face difficulty in evaluating the quality of a good or service prior to consumption, they need some type of proxy to 'anticipate' quality. Since positive brand associations and brand images can be used as the proxy, brand equity is important in decisions of credence goods such as the choice of lawyers, doctors, and investment consultants.

The Future of Brand Equity

The marketing environment has changed rapidly during the past decade. First, consumers have become increasingly diversified. Second, competition between firms has increased. Third, technology has advanced rapidly.

Fourth, there is a growing interest in environment, community, and social concerns (Keller, 2011).

The diversification of customers and the evolution of their needs over time necessitate the customization of goods and services. In order to increase brand equity firms need to transcend traditional mass customization and relationship marketing and focus on real-time marketing, not only providing customized offerings but also adapting to the changing needs of customers over time (Oliver et al., 1998; Chung et al., 2009).

Technology has allowed firms to greatly diversify the means of communication with customers. Internet technology has enabled interactive communication between firms and customers, and also increased the influence of user-generated contents (Fader and Winer, 2012) on brand equity through new forms of social media such as social networking services, YouTube, and Internet user forums. Customers now contribute to firms' value chains by providing reviews and feedback that firms can capitalize on as sources of ideas, improvements, innovations, and customer knowledge (Kumar et al., 2010). Furthermore, technology facilitates the co-creation of products through venues such as self-design, self-production, and customer participation in the new product development process (Hennig-Thurau et al., 2010; Hoyer et al., 2010). Such customer engagement (Van Doorn et al., 2010; Brodie et al., 2011) often leads to commitment, emotional attachment, self–brand connection, and brand loyalty (Moreau and Herd, 2010; Troye and Supphellen, 2012). Therefore, the integration of new interactive communication channels with traditional one-way communication channels has become an important issue for managing brand equity.

As media coverage of business practices and the speed of information diffusion have increased, customers have become more sensitive to firm activities. As a result, the importance of corporate social responsibility and ethics on brand equity has increased during recent decades. Firms should adopt a long-term view in building and managing their brand equity and prioritize corporate ethics as a key driver of brand equity.

RELATIONSHIP EQUITY

The Nature of Relationship Equity

Providing great value through a strong brand may be insufficient to keep customers 'glued' to a firm and preventing them from switching to competitors' offerings since customers now have the ability to interact directly with companies (Rust, Moorman, and Bhalla, 2010). Firms may need to provide relational benefits outside their key offering itself, which reward customers for staying with the firm or impose a cost for breaking off from the firm to switch to a competitor.

Relationship equity is defined as 'the tendency of the customer to stick with the brand, above and beyond the customer's objective and subjective assessments of the brand' (Rust et al., 2005, p. 25). It represents the strength of the relationship between a firm and its customers, based upon the firm and the customer's efforts to establish, build, and maintain a relationship. The stronger the relationship equity, customers perceive that they have more to lose from breaking the relationship with the firm. A firm's retention programs and relationship-building efforts can build relationship equity by conferring tangible (e.g., loyalty program points) and/or intangible benefits to its customers (Rust et al., 2000). Customers can reduce choice by maintaining ongoing relationships (Ganesan, 1994; Kalwani and Narayandas, 1995; Sheth and Parvatiyar, 1995), which lightens the burden of having to make decisions (Gwinner, Gremier, and Bitner, 1998). Relationships can provide emotional benefits such as feelings of familiarity, personal recognition, and social support (Berry, 1995). Firms are also rewarded for building strong relationship equity since customers who perceive themselves as a firm's valuable customer tend to purchase more and stay with the firm longer (Berry and Parasuraman, 1991). Building relationship equity also increases the likelihood that the most profitable customers return for future purchases (Rust et al., 2000).

Drivers of Relationship Equity

Firm initiatives for strengthening relationship equity should maximize the likelihood of retaining customers and the size of their future purchases while minimizing the likelihood of customers switching to competitors (Rust et al., 2000). The key levers that firms may implement to achieve such objectives are loyalty programs, special recognition and treatment, affinity programs, community-building programs, and knowledge-building programs (Lemon et al., 2001).

Loyalty programs are actions that reward customers' specific behavior such as frequent purchase with tangible benefits. The loyalty program has become a key element of marketing for many firms such as airlines, liquor stores, supermarket chains, and banks. In order for loyalty programs to be effective, three criteria must be met. First, customers need to earn points frequently enough to accumulate sufficient rewards. Second, the rewards must be perceived as meaningful and worth receiving to customers. Third, the firm's margins of its goods or services need to be high enough to justify a loyalty program. If these conditions are not met, loyalty programs may deteriorate into an industry-wide prisoner's dilemma game where firms suffer from being forced to fight loyalty wars and from eroding margins (Rust et al., 2000; Lemon et al., 2001).

Special recognition and treatment are accomplished through recognizing customers for specific behavior with intangible benefits rather than monetary rewards. For example, many airlines grant early boarding to their 'platinum' level members. US Airways' 'Chairman Preferred' customers receive perks such as guaranteed seating and reservations assistance. Whereas customers' perceptions of a reward program may degrade from a firm's relationship-building effort into an alternative form of price discount or rebate over time (thereby leading to value equity rather than relationship equity), special recognition and treatment may become more valuable than monetary awards to a firm's best customers (Rust et al., 2000).

Affinity programs also seek to provide non-monetary benefits by creating strong emotional connections between a firm and its customers. Affinity programs are effective for brands that represent strong lifestyles (Johnson, 1998) that facilitate a link between the customer's relationship with the firm and other important aspects of the customer's life. The GOLF Magazine Rewards Visa Card 'designed with serious golfers in mind' rewards its customers with points toward golf equipment and apparel, preferred tee times, personalized golf balls, and other golf experiences. Affinity programs create a switching cost in the form of customers investing a sense of self into the firm, or into the firm's good or service.

Community-building programs seek to cement relationships by creating a customer community that links like-minded customers. Saturn, General Motors' automobile brand that was discontinued in 2010, fostered customer community through activities such as hosting 'homecoming' celebrations at its Spring Hill plant, attracting crowds of 30000 and 44000 to Tennessee in 1994 and 1999 respectively (Rust et al., 2000). Rosenbaum, Ostrom, and Kuntze (2005) find that members of loyalty programs who perceive a sense of community from the brand tend to be more loyal customers than members of programs that rely only on financial incentives without offering this relationship.

Knowledge-building programs seek to create learning relationships or structural bonds between the customer and the firm that provide a disincentive for customers to switch. Firms such as Amazon.com provide suggestions tailored to customers' preferences learned from purchase behavior over time. British Airways tracks customer food and drink choices, saving time for the customer in future interactions while simultaneously increasing the customer's switching costs in terms of having to 'train' competing airlines on their preferences (Lemon et al., 2001).

When Relationship Equity Matters Most

First, relationship equity is crucial when customers associate greater benefits with the firm's loyalty program than the actual monetary value of the benefits they receive. This 'aspiration value' of a loyalty program creates a strong incentive for customers to return to the firm in the future. The frequent flyer program is an example that capitalizes on the difference between the true value (approximately three to four cents per mile) and the customer's aspiration value of a frequent flyer mile – the perception of getting that much closer to a free trip (Rust et al., 2000; Lemon et al., 2001).

Second, relationship equity is critical when the community associated with the good or service is as important as the good or service itself. Customers will often continue to purchase from firms whose goods or services attract a strong community of enthusiasts in order to maintain 'membership' status in the community. For example, members of a HOG (Harley-Davidson Owners Group) are fiercely loyal to the Harley-Davidson brand (Lemon et al., 2001).

Third, relationship equity is critical when firms have the opportunity to create learning relationships with their customers. The learning relationship between the firm and the customer may become as important as the provision of the good or service itself. Collecting, tracking, and utilizing information on customers' purchase behavior reveals their preferences and buying habits, allowing firms to pay personal attention to customers, a benefit that customers can only receive from alternative providers by 'training' the new provider (ibid.).

Finally, relationship equity is critical when customer action is required to discontinue the service. Book clubs, insurance, Internet service providers, and negative-option services require customers to actively stop consuming the goods or services. In such cases, inertia can help customers stay in a relationship, providing firms with a unique opportunity to strengthen the bonds with their customers (Rust et al., 2000; Lemon et al., 2001).

The Future of Relationship Equity

The Internet's potential to be utilized as the 'ultimate customer equity tool' was recognized more than a decade ago (Rust et al., 2000). Information technology enables consumers to directly take care of numerous functions that previously required the firm's input (e.g., automated teller machines [ATMs], automated hotel checkout, mobile banking, and Internet services such as package tracking and online brokerage services) (Meuter et al., 2000). Furthermore, as a platform of customer-to-firm communication, the Internet will continue to expand the role of customers as co-creators of goods and services, pricing, and distribution (Sheth, Sisodia, and Sharma, 2000). This presents opportunities for firms to strengthen the relationship 'glue' that sticks customers to the firm while reducing the internal costs of serving the customer (Rust and Lemon, 2001).

With technological innovation, firms now have the unprecedented ability to amass and to analyze customer data, and thereby build customer knowledge and strengthen learning relationships. With increased customer knowledge, firms can better implement co-creation marketing (Sheth et al., 2000). Customer co-creation enabled through technology creates not only new opportunities to drive brand equity, but also presents new opportunities for relationship equity. The truly personalized customer experiences from the co-creation of value (Prahalad and Ramaswamy, 2004) provide firms with the potential to add layers of depth to learning relationships. However, technological innovation also raises challenges to firms in driving relationship equity. How will customers' perceptions and expectations of their relationships with firms evolve? Will customers expect full personalization in not only a firm's goods and service offerings, but also in their relationships? Firms may need to optimize each driver of relationship equity building on an individual basis, providing different levels of programs and relational benefits that maximize the probability of retention and future spending for each individual customer. In an era of real-time marketing and segments of one, successfully offering fully personalized goods and services will strengthen the stickiness of customers to the firm, increasing the importance of relationship equity in driving a firm's profitability.

The Internet increases the connectivity not only between a firm and its customers, but also between a firm and its partners, suppliers, and employees. The wealth-creating potential of a firm's relationships with such parties, and the interactions among these parties and customers will become more important. For example, highly motivated employees may engender tighter relationships between the firm and its customers, and between the firm and its partners, who in turn reciprocate by strengthening the relationship with increased loyalty and larger future purchases (Heskett, Sasser, and Schleslinger, 1997; Reichheld, 2001). A broader perspective of relationship equity that incorporates the relationships between a firm and its stakeholders may be necessary to successfully utilize a firm's relational assets in driving value outside of that created by customers (Sawhney and Zabin, 2002).

Customer-to-customer relationships also warrant more attention. As the Internet makes it easier for firms to learn about their customers, customers are better able to learn about firms than ever before. The proliferation of social media facilitates the sharing of knowledge among diverse customer groups and strengthens the potential for relationships between individual customers. Such customer-to-customer relationships may become more relevant as a reason for staying with a firm. Instead of certain goods and services attracting a strong community, in time customers may demand that all firms engage in more online community-building activities.

RETURN ON MARKETING

Managers are constantly faced with dilemmas such as whether they should improve service quality or decrease price, or whether they should invest in a loyalty program or in increasing advertising. Customer equity can help answer these questions. In this section, we describe a unified and executable strategic framework for trading off marketing strategy options based on the projected financial return of each action. We present the 'return on marketing' approach proposed by Rust, Lemon, and Zeithaml (2004). The procedures and equations outlined henceforth are taken directly from their model. We believe this approach provides a solid groundwork for maximizing customer equity by driving value equity, brand equity, relationship equity, or by developing a joint strategy of the three drivers.

Modeling the Chain of Effects

Departing from the methods used in direct marketing/customer relationship management (CRM) models (e.g., Venkatesan and Kumar, 2004; Rust and Verhoef, 2005), acquisition and retention models (Blattberg and Deighton, 1996), and customer-retention–based models (e.g., Rust, Zahorik, and Keiningham, 1995; Bolton, 1998; Gupta, Lehmann, and Stuart, 2004), the return on marketing model is centered on the brand-switching pattern of individual customers (Rust, Lemon, and Zeithaml, 2004).

Rust, Lemon, and Zeithaml's (2004) return on marketing model uses Markov switching matrices to capture brand-switching patterns. This method examines the chain of effects resulting from how customer equity originates from driver perceptions, that is:

Driver perceptions ➔ *Switching matrix* ➔ *Customer lifetime value* ➔ *Customer equity*

Modeling this chain of effects starts from the utility function. Consider that the utility an individual i derives from buying brand k is the sum of inertia utility from the previous purchase of brand j, and the conjoint impacts of the customer equity drivers of brand k. Statistically, this utility function can be modeled as:

$$U_{ijk} = \beta_{0k}\, LAST_{ijk} + X_{ik}\, \beta_{1k} + \varepsilon_{ijk} \tag{1.1}$$

where U_{ijk} is the utility of brand k to individual i, who most recently purchased brand j. The dummy variable $LAST_{ijk}$ equals one if $j = k$ and equals zero otherwise. X_{ik} is a row vector of the drivers. The coefficient β_{0k} corresponds to the coefficient for inertia, and β_{1k} to the coefficients for the drivers.

Based on this utility function, we model the choice of individual i as:

$$P_{ijk*} = Pr\ [\textit{individual } i \textit{ chooses brand } k^*\textit{, given that brand } j \textit{ was most recently chosen}] = \exp(U_{ijk*})/\Sigma_k \exp(U_{ijk}). \tag{1.2}$$

The Rust–Lemon–Zeithaml approach specifies a brand-switching matrix to model customer acquisition and retention effects. Assuming that each customer i has an associated J × J Markov switching matrix M_i is:

$$\underset{\mathrm{J}\times\mathrm{J}}{M_i} = \begin{bmatrix} P_{i11} & P_{i12} & \cdots & P_{i1J} \\ P_{i21} & P_{i22} & \cdots & P_{i2J} \\ \vdots & \vdots & \ddots & \vdots \\ P_{iJ1} & P_{iJ2} & \cdots & P_{iJJ} \end{bmatrix} \tag{1.3}$$

where J is the number of brands, and p_{ijk} is the switching probability that customer i will choose brand k in the next purchase conditional on having purchased brand j in the most recent purchase. Therefore, the retention is modeled by the diagonal element while the off-diagonal elements represent acquisition from competing brands. By the same token the off-diagonal elements simultaneously represent customer attrition to competition for brand j. Furthermore, to calculate the purchase probability of a specific brand, we denote a 1 × J row vector A_i with its elements representing the probabilities of purchase for customer i's current transaction. Multiplying A_i with M_i, we obtain the 1 × J row vector B_{it}, with elements B_{ijt} representing the probability that customer i buys brand j in purchase t:

$$B_{it} = A_i M_i t \tag{1.4}$$

The lifetime value, CLV_{ij}, of customer i to brand j is:

$$CLV_{ij} = \sum_{t=0}^{T_{ij}} (1 + d_j)^{-t/f_i} v_{ijt} \pi_{ijt} B_{ijt} \tag{1.5}$$

where d_j is firm j's discount rate, f_i is customer i's average purchase rate per unit time (e.g., three purchases per year), v_{ijt} is customer i's expected purchase volume in a purchase of brand j in purchase t, and π_{ijt} is the expected contribution margin per unit of firm j from customer i in purchase t. T_{ij} is the number of purchases customer i is expected to make before firm j's time horizon H_j (e.g., a typical time horizon ranges from three to five years) and B_{ijt} is a firm-specific element of B_{it}. Therefore T_i is the integer part of H_{jfj}. Summing up the discounted CLV of all the firm's current and potential customers, we express firm j's customer equity, CE_j, as:

$$CE_j = \text{mean}_i\,(CLV_{ij}) \times POP \tag{1.6}$$

where mean $_i(CLV_{ij})$ is the average lifetime value for firm j's customers across the sample, and POP is the total number of customers in the market across all brands.

Data for Customer Equity Analysis

The key of the model is to obtain the brand-switching matrix, which depends on customers' perceptions of drivers. Therefore, a firm needs to collect cross-sectional survey data that include customer ratings of each competing brand on each driver.

The first step is to examine the competing and internal environment of a specific company. Typically, researchers outside of the firm should conduct manager interviews and exploratory research to acquire such information. Researchers should extend effort to acquire internal financial information such as the discount rate and relevant time horizon if possible. Based on such information, a firm or researcher can identify competing firms and customer segments, choose drivers that correspond to the current or potential management initiatives, and obtain the size of the market (i.e., the total number of customers across all brands). In addition, the contribution margins for all competitors can be estimated. Any predictable trends in any competitors' contribution margins should be incorporated into the model. Utilizing secondary sources and focus group interviews, additional drivers can be identified, which can then be reviewed with management to ensure that they are managerially actionable items.

Driver items should not be selected based solely on exploratory research with customers (e.g., focus groups). Preserving balance between importance of the driver and implementation of the improvement process is important.

If data on actual purchase behavior is available, it should be combined with survey data to improve the accuracy of estimation. If a customer database exists, firms have access to panel data on customers' subsequent purchases (i.e., next product purchased). Otherwise, purchase intentions from surveys can be used as a proxy for the probability of each brand being chosen in the next purchase. Since self-reported data may suffer from bias, it is important to calibrate survey responses, which is often performed by matching the aggregate survey responses with the aggregate sales or market share figures. For example, survey responses suggest that the three firms in a market have the following market shares: Alpha 50 percent, Beta 25 percent, and Gamma 25 percent. However, the actual market shares may be: Alpha 40 percent, Beta 30 percent, and Gamma 30 percent. Then the survey responses should be re-weighted accordingly to form a close match to observed aggregate behavior (Rust, Lemon, and Zeithaml, 2004).

Focusing Marketing Efforts

Once survey and longitudinal data on consumer purchase information are collected, the 'chain of effects' outlined previously can be modeled and estimated. Because the drivers of customer equity are likely to be highly multi-collinear, principal components analysis needs to be conducted to reduce the dimensionality of the independent variables. Then a multinomial logit regression can be conducted to evaluate the importance of each driver (ibid.).

After the important drivers and sub-drivers have been identified, they can be projected into an 'importance–performance map' that allows comparisons of performance on the items with key competitors (Rust et al., 2000). To uncover the key areas for improvement, a firm should focus on two elements: (1) understanding which drivers will have the most impact on customer equity for the organization, and (2) understanding where the firm's relative standing is to its competitors on each of the drivers of customer equity identified as most important. The firm can then decide which areas require the most marketing efforts, and determine the strategic direction of whether to invest in a brand strategy, value strategy, relationship strategy, or in a joint strategy.

Projecting Return on Investment

A firm's strategic expenditure to improve a driver needs to be justified by the profitability of such an investment. The impact on customer equity and CLV given the targeted amount of improvement can be projected as discussed in the 'chain-of-effects' section. Considering the cost associated with improving the drivers, the projected financial impact or ROI (return on investment) can be defined as:

$$ROI = \frac{Change\ in\ customer\ equity - marketing\ expenditure}{marketing\ expenditure} \quad (1.7)$$

Operationally, ROI can be simulated by using a simple spreadsheet program or a dedicated software package. The advantage of using a simulator is that the firm can test many possible expenditures, and then choose only the projects warranted by the top ROI figures. Since market dynamics and the competitive environment change continuously, a firm must track customer equity and ROI over time, to capture competitive trends, capitalize on emerging opportunities, overcome potential threats, and evaluate the progress of marketing initiatives designed to maximize customer equity.

Caution should be exercised in selecting the appropriate time horizon and discount rate for projecting ROI. Shorter time horizons or larger discount rates give less credit to future profits and therefore result in a lower ROI. Even so, a shorter time horizon or larger discount rate may be justified if the future is more uncertain (ibid.).

Measuring Marketing Effectiveness

The goal of a successful customer management strategy is to develop and implement marketing strategies that result in profitable long-term ROI (e.g., Zeithaml, Berry, and Parasuraman, 1996; Zeithaml, 2000; Berger et al., 2002). Given changes in market dynamics and the competitive landscape, it is necessary to measure and monitor changes in customer equity to evaluate the effectiveness of marketing actions.

A firm should conduct a customer equity analysis at least annually, although it may be useful to survey customers more frequently if the market is particularly turbulent (e.g., really new goods and services) (Rust et al., 2000). This allows the firm to (1) assess the effectiveness of a marketing investment, (2) track changes in customer needs, and (3) track competitive strategies and actions.

If possible, implementing a marketing investment and evaluating the changes in customer equity in a test market relative to a control market in which the marketing program is not implemented may prove effective. Comparing the effectiveness of a marketing investment in the test market and the control market allows firms to gain a strong sense of the impact of a specific marketing intervention in driving customer equity. However, this approach runs the risk that the firm's competitors will observe the firm's strategy in action during the test, and then formulate responses or retaliation tactics that greatly undermine the effectiveness of the strategy when it is fully implemented.

In summary, the return on marketing approach can be used for two purposes: (1) to project the ROI of a potential customer equity program before making any expenditures to determine whether to implement the program or not (or to project the ROIs of multiple customer equity programs and make an budget allocation decision); (2) to assess the actual ROI that was derived from investing in a particular customer equity program (or particular sub-driver of customer equity).

As a hypothetical illustration, suppose that Amazon wants to make an investment to improve relationship equity. Amazon may subsidizing the $79 annual fee it charges for its Amazon Prime membership program so that customers pay less membership fees. On the other hand, Amazon may consider hosting offline Amazon nights out to encourage community building. Following the steps outlined above, the company can select the project that yields the highest projected ROI. In another hypothetical scenario, the Scott Brand could determine whether its $35 million TV ad campaign in the first quarter of 2013 was successful in improving the brand image of Scott Paper Towels in a way that is profitable for the company by calculating the ROI from the $35 million expenditure.

DISCUSSION AND CONCLUSIONS

Focusing Strategy

In today's competitive landscape, customer equity provides a framework and 'roadmap for effective strategy' for firms to fully evolve into customer-centered organizations. At the most basic level, customers stay with a brand because the firm offers better value than competitors, the firm has a stronger brand than competitors, and switching costs are too high. An understanding of the drivers of customer equity allows firms to identify which specific drivers of value equity, brand equity, and relationship equity customers perceive as most important or crucial to a certain industry. Furthermore firms can determine their 'true' competitive standing in the industry. Such an analysis grounded in customer equity provides insight into key opportunities and unforeseen threats, laying the groundwork for firms to strategically focus their resources on the drivers of customer equity with the greatest marketing return on the long-term profitability of their customer base (Rust, Zeithaml, and Lemon, 2000, 2004).

Competitive Analysis

Before developing specific strategies to drive customer equity, firms need to understand the market in which they do business. Successful implementation of marketing strategies depends on the firm understanding the market, its current performance, and its competitors (Rust et al., 2005).

The first step is to analyze the market at the industry level. Firms need to understand which factors most influence customers' purchase decisions in the specific industry. Firms must first identify which driver of customer equity is most important in its competitive arena and to its customers, then dig deeper to understand which specific sub-driver is most critical given business circumstances.

Once the key drivers of customer equity have been identified, firms should analyze their current performance. Careful monitoring of firm performance and the importance of each driver in building customer equity helps identify areas for future improvement. Firms need to optimize the balance between cultivating strength in important drivers and de-emphasizing less important drivers, given their budget constraints.

After analyzing their own performance, firms need to analyze competition in order to implement their strategies successfully. Understanding key competitors aids in measuring customer equity, because the future behavior of customers depends not only on the firm's behaviors but also on competitors' behaviors (Rust et al., 2000, 2005). Comparing the firm's current performance to that of competitors provides information on which specific areas the firm is competent or incompetent. Analyzing value equity, brand equity, and relationship equity will help identify the strongest competitor that outperforms on the drivers, and the competitor with the largest growth potential that may become the next big player in the industry

Return on Marketing

Marketing managers constantly face decisions that involve trade-offs between competing initiatives, and the need to make their decisions financially accountable. We described a framework for evaluating return on marketing by calculating customer equity, summing up the customer lifetime values of each individual customer (Rust, Lemon, and Zeithaml, 2004). Adopting the return on marketing model, we can model CLV on the basis of the brand-switching matrix, which is useful for capturing both the brand-switching pattern of individual customers and the competitive effects between firms. The brand-switching approach employs customer surveys to identify the underlying reasons that customers choose to remain loyal to their current brand or to switch to a competitor. Manager interviews and exploratory research is necessary to identify potential drivers that affect customers' utility perception. Then the customer survey can be developed, balancing between the importance of the driver and the feasibility of the improvement process. Using principal components and multinomial logit regression, firms can identify the underlying reasons for customer acquisition or retention. Taking into account the importance of the drivers, the competitive standing of performance on such drivers, and the expenditures for improving performance, firms can identify the key areas of improvement. We encourage firms to examine ROI and measure marketing effectiveness over time. If feasible, firms may also implement a test market or partial roll-out to arrive at a more precise estimate.

Directions for the Future

In the past, the full-fledged implementation of the customer equity framework was limited. Information technology (or lack thereof) and resultant firm infrastructure impeded the realization of real-time marketing on a completely one-to-one basis. Now, firms have access to massive amounts of individual-level customer data. Technology and fundamental shifts in the economy now provide opportunities for firms to fully capitalize on this information to become fully customer-centric. However, the very same forces that enable such, also empower customers to become a firm's partners or co-creators of firm value. As customers, we now have more access to information on the firm, and on the firm's goods and services than ever before. Community among customers is more tightly knit than ever before as virtually all barriers to the flow of communication are eliminated. We see this bi-directional bond between the customer and the firm as a rich, promising area for driving customer equity. A wealth of information on changing consumer preferences and value can be captured on Internet forums and communities. For some time, firms have utilized the Internet and mobile applications for various brand- and community-building activities. We see room for further exploitation. For example, downloadable content such as music can now be fully customized according to user preferences at low cost (Chung et al., 2009). Technology is at a point where mobile applications can uncover user preferences that may not even be fully known by the users themselves. The branding and commercialization of such applications will truly provide customers irreplaceable value, a brand to stay loyal to, and a reason to stay with the brand from the compelling experience of learning, discovering, and sharing preferences.

With such developments, the drivers of customer equity may become inextricably linked. For example, over 650 000 customers acted as beta-testers for Microsoft Windows 2000 in its development phase. This collective R&D effort on behalf of Microsoft's customers (estimated to exceed $500 million in time, effort, and fees) not only helped to eliminate bugs from the final product, but also provided the customers the opportunity to learn how the software could create value in their own businesses or customer environments (Prahalad and Ramaswamy, 2000). This effort contributed to value equity through the quality improvement of the final product. At the same time, the beta-testers' customer engagement would have led to improvements in brand equity drivers such as commitment and self–brand connection. In addition to increasing value and brand equity, such engagement can also strengthen relationships between customers and the firm by increasing the number of connection points at various stages of value creation (Hoyer et al., 2010).

As the traditional customer role of consumption becomes further blurred by increased participation in the value creation process, the drivers of customer equity may represent a feedback loop. Achieving an improvement in one driver may lead to improvements in another driver. It may even be impossible to achieve an improvement in one driver without achieving an improvement in another. For example, an improvement in brand equity may be impossible to implement without achieving relationship equity, or vice versa. Identifying the greatest sources of synergy in the feedback loop of customer equity may become an important priority in the future.

However, the increased connectivity between customers and the firm also provides challenges to driving customer equity. Firms will need to monitor the abundant information and content that is customer created, which is by definition not under the direct control of the firm. The sheer amount of information may impose costs to the firm in data collection and analysis. Negative word-of-mouth or incorrect information on goods and services may spread like wildfire even before the firm is aware of the existence of a customer problem. Additionally, customers may become more and more concerned with privacy issues as firms' methods of collecting purchase data and precision in uncovering customers' unobserved preferences continue to become more sophisticated.

Regardless, we believe the potential for the customer equity framework in driving firm profitability and, ultimately, fundamental changes in the firm itself, is immense. If anything, we believe that it is not yet utilized enough in business practice. Although the continuous innovations in technology will create challenges to implementing a customer equity program, we believe that the customer equity framework is, and will remain, an important tool for overcoming these challenges and leading the firm to success.

REFERENCES

Aaker, D.A. (1991), *Managing Brand Equity*, New York: Free Press.

Ailawadi, K.L., S.A. Neslin and D.R. Lehmann (2003), 'Revenue premium as an outcome measure of brand equity', *Journal of Marketing*, **67**(4), 1–17.

Belk, R.W. (1988), 'Possessions and the extended self', *Journal of Consumer Research*, **15**(2), 139–68.

Berger, J. and C. Heath (2007), 'Where consumers diverge from others: identity signaling and product domains', *Journal of Consumer Research*, **34**(2), 121–34.

Berger, P.D., R.N. Bolton, D. Bowman, E. Briggs, V. Kumar, A. Parasuraman and C. Terry (2002), 'Marketing actions and the value of customer assets: a framework for customer asset management', *Journal of Service Research*, **5**(1), 39–54.

Berry, L.L. (1995), 'Relationship marketing of services – growing interest, emerging perspectives', *Journal of the Academy of Marketing Science*, **23**(4), 236–45.

Berry, L.L. and A. Parasuraman (1991), *Marketing Services: Competing Through Quality*, New York: Free Press.

Blattberg, R.C. and J. Deighton (1996), 'Manage marketing by the customer equity test', *Harvard Business Review*, **74**(4), 136–44.

Bolton, R.N. (1998), 'A dynamic model of the duration of the customer's relationship with a continuous service provider: the role of satisfaction', *Marketing Science*, **17**(1), 45–65.

Brodie, R.J., L.D. Hollebeek, B. Jurić and A. Ilić (2011), 'Customer engagement conceptual domain, fundamental propositions, and implications for research', *Journal of Service Research*, **14**(3), 252–71.

Buttle, F. (2004), *Customer Relationship Management: Concept and Tools*, Burlington, MA: Elsevier.

Chaudhuri, A. and M.B. Holbrook (2001), 'The chain of effects from brand trust and brand affect to brand performance: the role of brand loyalty', *Journal of Marketing*, **65**(2), 81–93.

Chung, T.S., R.T. Rust and M. Wedel (2009), 'My mobile music: an adaptive personalization system for digital audio players', *Marketing Science*, **28**(1), 52–68.

Court, D., D. Elzinga, S. Mulder and O.J. Vetvik (2009), 'The consumer decision journey', *McKinsey Quarterly*, June 2009.

Fader, P.S. and R.S. Winer (2012), 'Introduction to the special issue on the emergence and impact of user-generated content', *Marketing Science*, **31**(3), 369–71.

Ganesan, S. (1994), 'Determinants of long-term orientation in buyer–seller relationships', *Journal of Marketing*, **58**(2), 1–19.

Gershoff, A.D. and G.V. Johar (2006), 'Do you know me? Consumer calibration of friends' knowledge', *Journal of Consumer Research*, **32**(4), 496–503.

Gerstner, E. (1985), 'Do higher prices signal higher quality?', *Journal of Marketing Research*, **22**(2), 209–15.

Godes, D. and D. Mayzlin (2004), 'Using online conversations to study word-of-mouth communication', *Marketing Science*, **23**(4), 545–60.

Gupta, S., D.R. Lehmann and J.A. Stuart (2004), 'Valuing customers', *Journal of Marketing Research*, **41**(1), 7–18.

Gwinner, K.P., D.D. Gremler and M.J. Bitner (1998), 'Relational benefits in services industries: the customer's perspective', *Journal of the Academy of Marketing Science*, **26**(2), 101–14.

Hennig-Thurau, T., E.C. Malthouse, C. Friege, S. Gensler, L. Lobschat, A. Rangaswamy and B. Skiera (2010), 'The impact of new media on customer relationships', *Journal of Service Research*, **13**(3), 311–30.

Herr, P.M., F.R. Kardes and J. Kim (1991), 'Effects of word-of-mouth and product-attribute information on persuasion: an accessibility-diagnosticity perspective', *Journal of Consumer Research*, **17**(4), 454–62.

Heskett, J., E. Sasser and L.A. Schleslinger (1997), *The Service Profit Chain*, New York: Free Press.

Hoyer, W.D., R. Chandy, M. Dorotic, M. Krafft and S.S. Singh (2010), 'Consumer cocreation in new product development', *Journal of Service Research*, **13**(3), 283–96.

Indvik, L. (2012), 'Mobile set to surpass desktop sales at some flash sales sites', accessed 5 September 2014 at http://mashable.com/2012/05/08/mobile-commerce-flash-sales-rue-la-la-gilt/.

Johnson, K. (1998), 'Choosing the right program', *Direct Marketing*, **61**(2), 36–8.

Kalwani, M.U. and N. Narayandas (1995), 'Long-term manufacturing–supplier relationships: do they pay off for supplier firms?', *Journal of Marketing*, **59**(1), 1–16.

Keller, K.L. (1993), 'Conceptualizing, measuring, managing customer-based brand equity', *Journal of Marketing*, **57**(1), 1–22.

Keller, K.L. (2003), 'How do brands create value?', *Marketing Management*, **12**(3), 26–31.

Keller, K.L. (2008), *Strategic Brand Management: Building, Measuring, and Managing Brand Equity*, 3rd edition, Upper Saddle River, NJ: Prentice-Hall.

Keller, K.L. (2011), 'How to navigate the future of brand management', *Marketing Management*, **20**(1), 36–43.

Keller, K.L. and D.R. Lehmann (2003), 'How do brands create value?', *Marketing Management*, **12**(3), 26–31.

Kerin, R.A. and R. Sethuraman (1998), 'Exploring the brand value–shareholder value nexus for consumer goods companies', *Journal of the Academy of Marketing Science*, **26**(4), 260–73.

Kirmani, A. and A.R. Rao (2000), 'No pain, no gain: a critical review of the literature on signaling unobservable product quality', *Journal of Marketing*, **64**(2), 66–79.

Kumar, V., L. Aksoy, B. Donkers, R. Venkatesan, T. Wiesel and S. Tillmanns (2010), 'Undervalued or overvalued customers: capturing total customer engagement value', *Journal of Service Research*, **13**(3), 297–310.

Lemon, K.N., R.T. Rust and V.A. Zeithaml (2001), 'What drives customer equity?', *Marketing Management*, **10**(1), 20–25.

Madden, T.J., F. Fehle and S. Fournier (2006), 'Brands matter: an empirical demonstration of the creation of shareholder value through branding', *Journal of the Academy of Marketing Science*, **34**(2), 224–35.

Meuter, M.L., A.L. Ostrom, R.I. Roundtree and M.J. Bitner (2000), 'Self-service technologies: understanding customer satisfaction with technology-based service encounters', *Journal of Marketing*, **64**(3), 50–64.

Mizik, N. and R. Jacobson (2008), 'The financial value impact of perceptual brand attributes', *Journal of Marketing Research*, **45**(1), 15–32.

Moreau, C.P. and K.B. Herd (2010), 'To each his own? How comparisons with others influence consumers' evaluations of their self-designed products', *Journal of Consumer Research*, **36**(5), 806–19.

Oliver, R.W., R.T. Rust and S. Varki (1998), 'Real-time marketing', *Marketing Management*, **7**(2), 28–37.

Park, C.W., D.J. MacInnis and J. Priester (2008), 'Brand attachment and a strategic brand exemplar', in B.H. Schmitt and D.L. Rogers (eds), *Handbook on Brand and Experience Management*, Cheltenham, UK and Northampton, MA, USA: Edward Elgar Publishing, pp. 3–17.

Park, C.W., D.J. MacInnis, J. Priester, A.B. Eisingerich and D. Iacobucci (2010), 'Brand attachment and brand attitude strength: conceptual and empirical differentiation of two critical brand equity drivers', *Journal of Marketing*, **74**(6), 1–17.

Peppers, D. and M. Rogers (1993), *The One to One Future: Building Relationships One Customer at a Time*, New York: Currency Doubleday.

Peppers, D. and M. Rogers (1997), *Enterprise One to One: Tools for Competing in the Interactive Age*, New York: Currency Doubleday.
Prahalad, C.K. and V. Ramaswamy (2000), 'Co-opting customer competence', *Harvard Business Review*, **78**(1), 79–88.
Prahalad, C.K. and V. Ramaswamy (2004), 'Co-creation experiences: the next practice in value creation, opting customer competence', *Journal of Interactive Marketing*, **18**(3), 5–14.
Reichheld, F. (2001), *Loyalty Rules! How Today's Leaders Build Lasting Relationships*, Cambridge, MA: Harvard Business School Press.
Richards, K.A. and E. Jones (2008), 'Customer relationship management: finding value drivers', *Industrial Marketing Management*, **37**(2), 120–30.
Rosenbaum, M.S., A.L. Ostrom and R. Kuntze (2005), 'Loyalty programs and a sense of community', *Journal of Services Marketing*, **19**(4), 222–33.
Rust, R.T. and K.N. Lemon (2001), 'E-service and the consumer', *International Journal of Electronic Commerce*, **5**(3), 83–99.
Rust, R.T. and P.C. Verhoef (2005), 'Optimizing the marketing interventions mix in intermediate-term CRM', *Marketing Science*, **24**(3), 477–89.
Rust, R.T., K.N. Lemon and D. Narayandas (2005), *Customer Equity Management*, Upper Saddle River, NJ: Prentice-Hall.
Rust, R.T., K.N. Lemon and V.A. Zeithaml (2001), 'Where should the next marketing dollar go?', *Marketing Management*, **10**(3), 24–8.
Rust, R.T., K.N. Lemon and V.A. Zeithaml (2004), 'Return on marketing: using customer equity to focus marketing strategy', *Journal of Marketing*, **68**(1), 109–27.
Rust, R.T., C. Moorman and G. Bhalla (2010), 'Rethinking marketing', *Harvard Business Review*, **88**(11), 94–101.
Rust, R.T., A.J. Zahorik and T.L. Keiningham (1995), 'Return on quality (ROQ): making service quality financially accountable', *Journal of Marketing*, **59**(2), 58–70.
Rust, R.T., V.A. Zeithaml and K.N. Lemon (2000), *Driving Customer Equity*, New York: Free Press.
Rust, R.T., V.A. Zeithaml and K.N. Lemon (2004), 'Customer-centered brand management', *Harvard Business Review*, **82**(9), 110–18.
Sawhney, M. and J. Zabin (2002), 'Managing and measuring relational equity in the network economy', *Journal of the Academy of Marketing Science*, **30**(4), 313–32.
Sheth, J.N. and A. Parvatiyar (1995), 'Relationship marketing in consumer markets: antecedents and consequences', *Journal of the Academy of Marketing Science*, **23**(4), 255–71.
Sheth, J.N. and A. Parvatiyar (2002), 'Evolving relationship marketing into a discipline', *Journal of Relationship Marketing*, **1**(1), 3–16.
Sheth, J.N., R.S. Sisodia and A. Sharma (2000), 'The antecedents and consequences of customer-centric marketing', *Journal of the Academy of Marketing Science*, **28**(1), 55–66.
Stahl, F., M. Heitmann, D.R. Lehmann and S.A. Neslin (2012), 'The impact of brand equity on customer acquisition, retention, and profit margin', *Journal of Marketing*, **76**(4), 44–63.
Thomas, J.S., W. Reinartz and V. Kumar (2004), 'Getting the most out of all of your customers', *Harvard Business Review*, **82**(7/8), 116–23.
Troye, S.V. and M. Supphellen (2012), 'Consumer participation in coproduction: "I made it myself" effects on consumers' sensory perceptions and evaluations of outcome and input product', *Journal of Marketing*, **76**(2), 33–46.
Van Doorn, J., K.N. Lemon, V. Mittal, S. Nass, D. Pick, P. Pirner and P.C. Verhoef (2010), 'Customer engagement behavior: theoretical foundations and research directions', *Journal of Service Research*, **13**(3), 253–66.
Vargo, S.L. and R.F. Lusch (2004), 'Evolving to a new dominant logic for marketing', *Journal of Marketing*, **68**(1), 1–17.
Venkatesan, R. and V. Kumar (2004), 'A customer lifetime value framework for customer selection and resource allocation strategy', *Journal of Marketing*, **68**(5), 106–25.

Völckner, F. and H. Sattler (2006), 'Drivers of brand extension success', *Journal of Marketing*, **70**(2), 18–34.
Zeithaml, V.A. (1988), 'Consumer perceptions of price, quality and value: a means-end model and synthesis of evidence', *Journal of Marketing*, **52**(3), 2–22.
Zeithaml, V.A. (2000), 'Service quality, profitability, and the economic worth of customers', *Journal of the Academy of Marketing Science*, **28**(1), 67–85.
Zeithaml, V.A., L.L. Berry and A. Parasuraman (1996), 'The behavioral consequences of service quality', *Journal of Marketing*, **60**(2), 31–46.

2. Aggregate- and individual-level customer lifetime value

*V. Kumar and Anita Pansari**

INTRODUCTION

Over the years, companies have focused on building and maintaining profitable customer relationships for maximizing firm profitability (Kumar and Rajan, 2012). Tracking individual customers and measuring their profitability is challenging and requires innovative customer relationship management strategies. This becomes even more challenging in a non-contractual setting, where the customer is not tied to the company by any bond or contract. However, there are various ways to measure individual customer profitability. Some of the earlier metrics were recency-frequency-monetary (RFM) value, share-of-wallet (SOW), and past customer value (PCV). However, all these measures use past customer behavior under the assumption that it will be demonstrated in the future too (Kumar, 2008). Yet, this may not be true in several situations, especially in a non-contractual environment. Further, these metrics do not provide any customer segmentation strategies.

In view of all the shortcomings of these metrics, a forward-looking metric that measures customer lifetime value (CLV) has been advocated. CLV is defined as 'the net present value of future profits from a customer' (ibid.). It is a forward-looking metric, which can be used as a basis for formulating and implementing customer-specific strategies (Kumar and Rajan, 2012). This metric is based on the cumulated cash flow and the value accruing directly or indirectly from a customer over his or her lifetime with the company. By quantifying this value, firms are able to determine how much they should invest in a customer to achieve a positive return on investment (ROI) or maximized profitability. CLV can also be used to identify the most and least profitable customers and direct the marketing efforts at the most valuable ones by classifying customers into high-, medium-, and low-value segments. This classification not only helps managers allocate their resources optimally but also helps them to design appropriate marketing strategies, which would maximize their profits. It also helps managers to focus on long-term strategies, rather than concentrate their efforts on a per-transaction basis. There are two levels of CLV based on the data and the resources a firm is willing to invest. The first is the individual-level CLV, which requires the transaction data of customers along with the firm–customer interaction data; and the second is the aggregate-level CLV, which requires data on a firm's overall marketing activities and the aggregate data of a segment of customers or all customers of the firm. Individual-level CLV helps firms to personalize strategies based on customer needs and the profitability of the individual customer. On the other hand, the aggregate-level CLV does not help in customizing strategies for customers; it only helps firms in evaluating the overall effectiveness of the marketing plan. There are various approaches through which CLV can be computed. Some of the popular approaches have been discussed in Table 2.1. Although the underlying objective is the same (to identify profitable customers, to optimize resource allocation, etc.) in all of these approaches it is the different analytical models that set them apart from each other.

This chapter can be reviewed either with the modeling approaches discussed in Table 2.1 or with the unit of analysis (individual or aggregate). A detailed discussion of all the models is out of scope for this *Handbook*; therefore we only focus on the different facets of individual-level and aggregate-level CLV in this chapter.

Aggregate-level and the individual-level CLVs have their pros and cons; for instance, although the aggregate-level CLV is easier to compute, it does not provide much information on the customer-level strategies, whereas the individual-level CLV provides guidance on many customer-level strategies like acquisition, retention, targeting, and so on, but uses complex models. These two levels of CLVs are discussed extensively in this chapter. The chapter begins with a basic overall discussion of the two levels of CLVs in different business settings and then proceeds to an extensive review of select literature, highlighting the kind of data used and the outcomes of the studies. We further throw light on the five key approaches of customer equity (the total sum of all individual-level and aggregate-level CLVs). Then, we move on to the strategies that can be used to maximize CLV and finally conclude with thought-provoking ideas for future research.

AGGREGATE-LEVEL AND INDIVIDUAL-LEVEL MEASURES OF CLV

Aggregate-level CLV is a measure of computing the lifetime of a customer segment or the average lifetime of all the customers of the firm. It can be maximized by the drivers of customer equity (Rust, Lemon, and Zeithaml,

Table 2.1 Popular modeling approaches for CLV

Modeling Approach	Salient Features	Unit of Analysis		Select Publications
		Aggregate	Individual	
Probabilistic	Explains observed behavior based on data and empirics. Imposes the assumption that customer behavior is based on probability distribution, e.g., Pareto/NBD model		X	Reinartz and Kumar (2000, 2003); Fader, Hardie, and Lee (2005)
Econometric	Similar underlying philosophy as the probabilistic model, except that researchers use economic theory to explain customer behavior, e.g., logit/probit models, hazard models, etc.		X	Thomas, Reinartz, and Kumar (2004); Venkatesan and Kumar (2004)
Persistence	Projects the long-run or equilibrium behavior of a variable/group of variable. Useful when long time series data are available. Examines the evolution of variables of interest, e.g., VAR model	X	X	Dekimpe and Hanssens (2004); Yoo and Hanssens (2005)
Diffusion Growth	Can be used only when aggregate data is available. Helps in forecasting the number of new customers. Can help in assessing the value of a lost customer	X		Hogan, Lemon, and Libai (2003); Gupta, Lehmann, and Stuart (2004)

2004). Individual-level CLV is a measure to compute the profitability of a customer over his or her entire lifetime with the company. It can be maximized by customer-level strategies like balancing acquisition and retention, optimal resource allocation, and so on (Kumar and George, 2007).

CLV can be computed for all firms whether they operate in a contractual setting or a non-contractual setting. A contractual setting is one in which the customer is contractually obligated to be with the firm (mobile carriers like AT&T, Verizon, etc.) and a non-contractual setting is when the customer has the freedom to leave the relationship at any point in time. The CLV metric in a contractual setting is fairly straightforward as there is a fixed time frame available for the customer's relationship with the firm. Conversely, in a non-contractual setting, due to the absence of an explicit expiry date, it may be difficult to determine how long a customer has been associated with a firm. In such a scenario, predicting the lifetime duration of a customer by observing buying patterns and other explanatory factors assumes importance. This section discusses the different approaches in contractual and non-contractual settings for estimating CLV at different levels.

Individual-level CLV

Individual-level CLV is calculated as the sum of the cumulated cash flows – discounted using the weighted average cost of capital (WACC) of a customer over his or her entire lifetime with the company (Kumar, 2008). It is a function of the predicted contribution margin, the propensity for a customer to continue in the relationship, and the marketing resources allocated to the customer. Individual-level CLV is useful in industries that require formulation of customer-level strategies, firm-level strategies or financial valuation of the firm. Individual-level CLV can be executed only in firms that have end user transaction data. Therefore, it is challenging for large firms with millions of customers to estimate individual-level CLV. This is an econometric approach, which generally combines customer acquisition, retention, and expansion to estimate CLV. A simple way in which CLV can be estimated is:

$$CLV_i = \sum_{t=1}^{T} \frac{(Future\ Contribution\ Margin_{it} - Future\ Cost_{it})}{(1+d)^t} \tag{2.1}$$

Where:
i = customer index;
t = time index;
T = the number of time periods considered for estimating CLV; and
d = discount rate.

The future contribution margin and the future costs are adjusted for time value. The future contributions of a customer can be calculated in a non-contractual setting, only if the probability that the customer continues to do business with the firm in future time periods or stays active, *P* (*Alive*) is available. The net present value (NPV) of expected gross contribution (EGC) can be calculated by using *P* (*Alive*) in the following manner (Reinartz and Kumar, 2003):

$$NPV\ of\ EGC_{it} = \sum_{n=t+1}^{t+x} P(Alive)_{in} * \frac{AMGC_{it}}{(1+d)^n} \tag{2.2}$$

Where:
$AMGC_{it}$ = average gross contribution margin in period t based on all prior purchases;
i = customer index;
t = the period for which NPV is being estimated;
x = the future time period;
n = the number of periods beyond t;
d = discount rate;
$P(Alive)_{in}$ = the probability that customer i is active in period n.

CLV is derived by subtracting the acquisition cost and the marketing cost (discounted to present value) from the NPV. However, if the marketing costs are accounted for at the beginning of a given time period and the gross contribution at the end of that time period, the CLV can be computed as:

$$CLV\ of\ Customer\ i = \sum_{n=t+1}^{t+x} P(Alive)_{in} * \frac{AMGC_{it}}{(1+d)^n} - \sum_{n=1}^{x} M_{in} * \left(\frac{1}{1+d}\right)^{n-1} - A \tag{2.3}$$

The weakness of the above method is that *P* (*Alive*) assumes that, when a customer ends a relationship once, he or she never comes back to the firm. This method is called 'lost-for-good' and is questionable because it undervalues CLV (Rust et al., 2004). This limitation has been taken care of in the 'always-a-share' approach, which takes into consideration the possibility of a customer returning (Venkatesan and Kumar, 2004). This method is most appropriate for a non-contractual business setting. Here, predicting the customers' purchase frequency is a better way of projecting his or her activity. This can be computed as:

$$CLV_i = \sum_{y=1}^{T_i} \frac{CM_{i,y}}{(1+r)^{\frac{y}{frequency_i}}} - \sum_{l=1}^{n} \frac{\sum_m C_{i,m,l} * X_{i,m,l}}{(1+d)^{l-1}} \tag{2.4}$$

Where:
CLV_i = lifetime value of customer i;
$CM_{i,y}$ = predicted contribution margin from customer i in purchase occasion y;
d = discount rate;
$C_{i,m,l}$ = unit marketing cost for customer i in channel m in year l;
$X_{i,m,l}$ = number of contacts to customer i in channel m in year l;
$frequency_i$ = predicted purchase frequency for customer i;
n = number of years to forecast; and
T_i = predicted number of purchases made by customer i until the end of planning period.

The customers of a firm could either be in a contract with a firm or they may be non-contractual consumers who have no bonds/contracts with the firm. Computing the CLV of both these sets of customers is also essential, as this would help to distinguish the amount of investment a firm should make in their contractual and non-contractual customers. In a contractual setting, computing CLV is straightforward, and is calculated as shown in equation (2.5):

$$CLV_i = \sum_{t=1}^{T} \frac{Base\ GC}{(1+r)^t} + \sum_{t=1}^{T} \frac{\hat{p}(Buy_{it} = 1) * \hat{G}C_{it}}{(1+r)^t} - \frac{\hat{M}C_{it}}{(1+r)^t} \tag{2.5}$$

Where:
CLV_i = lifetime value for customer i;
$\hat{p}(Buy_{it})$ = predicted probability that customer i will purchase in time period t;
$\hat{G}C_{it}$= predicted gross contribution margin provided by customer i in time period t;
$\hat{M}C_{it}$= predicted marketing costs directed toward customer i in time period t;
t = index for time periods, such as months, quarters, years, etc.;
T = marks the end of the calibration or observation time frame;
r = monthly discount factor;
$Base\ GC$ = predicted base monthly gross contribution margin.

In a contractual setting, the first term $\sum_{t=1}^{T} Base\ GC/(1+r)^t$ corresponds to the constant gross contribution that the customer is going to give to the firm. This could be either on a monthly, quarterly or annual basis. The second term $\sum_{t=1}^{T} \hat{P}(Buy_{it} = 1) * \hat{G}C_{it}/(1+r)^t$ corresponds to the predicted net present value of future purchases by the customer in a particular time period. The third term $\hat{M}C_{it}/(1+r)^t$ corresponds to the

Table 2.2 Transaction details for Sara

	May	June	July	August
$\hat{p}$ (*Buy*)	0.55	0.50	0.40	0.20
Monthly purchase amount (in $)	20	10	10	15
Profit margin (in %)	20	20	20	20
Marketing cost (in $)	5	7	7	10

net present value of future purchases attributed to the customer in a particular time period. The final value of these three components would yield the CLV in a contractual setting. The following example demonstrates how a company uses a customer's data in a contractual setting to compute the CLV.

Assume that a mobile phone company XYZ studies the CLV of Sara, who is one of its customers, and the monthly subscription or the Base GC provided by Sara is $40 and the annual discount rate is 12 percent. At the beginning of May, company XYZ wants to know the additional value Sara is likely to provide to it in the next four months (May, June, July, and August). Table 2.2 shows Sara's probability of buying additional services or $\hat{p}$ (*Buy*) (such as data, applications, text messaging, etc.) for the next four months, her monthly purchase amount, the percentage of margin for each purchase, and the marketing cost incurred by the company in contacting Sara. Keeping this example, we calculate the CLV of Sara:

$$\text{CLV of Sara on 1 May} = \left(\frac{40}{(1.01)^1} + \frac{(0.55)*(20*20\%)}{(1.01)^1} - \frac{5}{(1.01)^1}\right)$$

$$+ \left(\frac{40}{(1.01)^2} + \frac{(0.50)*(10*20\%)}{(1.01)^2} - \frac{7}{(1.01)^2}\right)$$

$$+ \left(\frac{40}{(1.01)^3} + \frac{(0.40)*(10*20\%)}{(1.01)^3} - \frac{7}{(1.01)^3}\right)$$

$$+ \left(\frac{40}{(1.01)^3} + \frac{(0.20)*(15*20\%)}{(1.01)^3} - \frac{10}{(1.01)^3}\right) = 132.87$$

Unlike in the contractual setting, there is no assurance of the customer being tied to the firm for a fixed period of time in the non-contractual setting, because of which the computation of CLV would differ. In a non-contractual setting, the first term $\sum_{t=1}^{T} BaseGC/(1 + r)^t$ would not be valid as there is no continual flow of base income from a customer on a periodic basis. In short, there is no guaranteed income due to subscriptions. As with the previous case, the second term $\sum_{t=1}^{T} \hat{P}(Buy_{it} = 1) * \hat{G}C_{it}/(1 + r)^t$ corresponds to the predicted net present value of future purchases by the customer in a particular time period. The third term $\hat{M}C_{it}/(1 + r)^t$ corresponds to the net present value of future purchases attributed to the customer in a particular time period. The final value of these three components would yield the CLV in a non-contractual setting. The formula for this is:

$$CLV_i = \sum_{t=1}^{T} \frac{\hat{p}(Buy_{it} = 1) * \hat{G}C_{it}}{(1 + r)^t} - \frac{\hat{M}C_{it}}{(1 + r)^t} \tag{2.6}$$

Let us assume that Good Food Limited, a grocery store, is interested in estimating the CLV of one of its customers, Charles, at the beginning of October. Since it's a grocery store, there is no contractual agreement between Charles and Good Food Limited. Therefore, while computing the CLV of Charles, there would be no base contribution margin. We will assume that the probability of Charles buying in the month of October is 0.55 and the predicted dollar value of the purchase is \$20, the predicted gross contribution margin is 20 percent, the marketing cost that the firm would incur would be \$5, and the monthly discount factor is 1 percent:

$$\text{Estimation for the month of October} = \frac{(0.55)*(20*20\%)}{(1.01)^1} - \frac{5}{(1.01)^1}$$
$$= -2.78$$

Similarly, we could compute the same for the month of November and December with the data shown in Table 2.3.

We would get the estimated prediction for November as −\$6.01 and December as −\$9.03. The total sum of all three months would give us the CLV of Charles for the beginning of October. Therefore, the CLV of Charles on 1 October = −2.78 + −6.01 + −9.03 = −17.82. The CLV of Charles indicates that Charles is not a profitable customer for Good Food Limited, as the revenues earned from Charles are less than the amount spent on marketing activities for Charles.

There are various modeling approaches through which individual-level CLV can be calculated and different perspectives that researchers use. Some researchers use an arbitrary time horizon (Reinartz and Kumar, 2003), whereas some use a boundless time horizon (Gupta, Lehmann, and Stuart, 2004; Fader, Hardie, and Lee, 2005). Gupta and Lehmann (2003) note that if the contribution margins and retention rates are constant over a period of time then an infinite/unbound time horizon can be used. Fader, Hardie, and Berger (2004) use a Pareto/negative binomial distribu-

Table 2.3 Transaction data for Charles

	November	December
$\hat{p}$ *(Buy)*	0.50	0.20
Monthly purchase amount (in $)	10	15
Profit margin (in %)	20	20
Marketing cost (in $)	7	10

tion (NBD) framework in non-contractual settings. Their model can be used only when the value of the transactions varies randomly around the mean transaction value and the mean transaction varies across customers but does not vary over time. Fader et al. (2005) demonstrate that data on recency, frequency, and monetary value is adequate to compute customers' purchase history. They further showed that the iso-value curves display different values for the three variables, recency, frequency, and monetary value, but produce the same CLV for a customer.

Most modeling approaches ignore competition because of the lack of access to competitive data. Researchers either build separate models for customer acquisition, retention, and margin or sometimes combine two elements. Some researchers simultaneously captured customer acquisition and retention (Thomas, 2001; Reinartz, Thomas, and Kumar, 2005), whereas some built one model for recency and frequency and a separate model for monetary value (Fader et al., 2005).

The computation of CLV in a B2C setting is a great challenge, as it requires transaction-level data. This is very expensive and in some cases it is also difficult to obtain data for all the customers, as there may not be any direct contact between company and customer. In such cases, the computation and application of CLV needs to be modified to make the maximum use of the framework. Classifying the consumers as high and low according to the value they provide and treating them differentially may lead to repercussions if the process is not managed in a proper manner. Not all firms would have the required skill set or goal to estimate the CLV for individual customers. Nonetheless, they would want to know the overall performance of the firm's marketing actions, understand if their customers are profitable and, if yes, to what extent. All these questions of the firm can be answered by aggregate-level CLV.

Aggregate-level CLV

In the aggregate-level measure of calculating CLV, firms use segment- or the firm-level data to compute the average lifetime value of the customer and multiply the same with the number of customers to arrive at the CLV (Kumar and George, 2007). This is known as the top-down approach. The estimation of CLV through this approach can be accomplished by identifying and measuring the factors that drive CLV.

These factors are measured based on the customer's objective and subjective assessment of value equity, brand equity, and relationship equity (Kumar, 2008). They differ depending on the contexts in which the CLV is being studied. Prior literature introduced CLV calculations that are proponents of Kumar's (2008) approach (Blattberg and Deighton, 1996; Berger and Nasr, 1998; Blattberg, Getz, and Thomas, 2001; Gupta and Lehmann, 2003; Rust et al., 2004). This measure of calculating CLV can be used in industries where it is difficult to collect transaction data or for large retailers where there is no loyalty program (Kumar and George, 2007). There are various ways of computing aggregate-level CLV; a few of them have been discussed here. Table 2.4 provides a detailed analysis of the academic literature, which has used either the aggregate-level measure or the individual-level measure of CLV. The three different methods of computing aggregate-level CLV are given below.

1 In the customer equity approach of calculating CLV, the sum of the lifetime values of all customers (customer equity – CE) is calculated, which gives the economic value of the firm. Average CLV is calculated by dividing the CE by the number of customers (Kumar, 2008):

$$CE = \sum_{i=1}^{I} \sum_{t=1}^{T} CM_{it} \left(\frac{1}{1 + \delta} \right)^{t} \tag{2.7}$$

Where:
CE = customer equity of customer base in \$ (sum of individual lifetime values);
CM = contribution margin in time period t;
δ = discount rate;
i = customer index;
t = time period;
T = the number of time periods for which CE is being estimated.

2 Another approach of calculating aggregate-level CLV is when the average CLV of a customer is calculated based on the lifetime value of the customer segment (Berger and Nasr, 1998). The average CLV can be measured as:

$$CLV_1 = \sum_{t=0}^{T} \left[\frac{(GC - M)}{(1 + d)^t} r^t \right] - A \tag{2.8}$$

Where:
r = rate of retention;
d = discount rate or the cost of capital for the firm;
t = time period;
T = the number of time periods considered for estimating CE;
GC = the average gross contribution;
M = marketing cost per customer;
A = the average acquisition cost per customer.

This approach takes the retention rate r as a constant, which is a drawback of this method as many customers may leave the firm at different points and the retention probabilities of each of them may vary.

Assume that a segment of 500 customers of Bank ABC are observed for a quarter (three months) starting January 2013. Assume the average contribution margin for these customers in the month of April is $1000, the marketing cost $50 and the acquisition cost $150, respectively, with a discount rate of 15 percent per year and a retention rate of 0.75. The aggregate-level CLV is calculated as:

$$CLV = \left\{ \frac{950}{(1.0125)^1}(0.75)^1 + \frac{950}{(1.0125)^2} * (0.75)^2 + \frac{950}{(1.0125)^3} * (0.75)^3 \right\} - 150$$

$$= \$1461 \text{ (approx.)}$$

3 In Blattberg et al.'s (2001) approach, the customer equity of the firm is first calculated as the sum of the return on acquisition, return on retention, and return on add-on selling. This is expressed in a mathematical equation as follows:

$$CE(t) = \sum_{i=0}^{I} \Big[N_{i,t} a_{i,t} (S_{i,t} - C_{i,t}) - N_{i,t} B_{i,a,t}$$

$$+ \sum_{k=1}^{\infty} N_{i,t} a_{i,t} \{ \pi_{j=1}^{k} p_{j,t+k} \} (S_{i,t+k} - C_{i,t+k} - B_{i,r,t+k} - B_{i,Ao,t+k}) \left\{ \frac{1}{1+d} \right\}^k \Big] \tag{2.9}$$

Where:
$CE(t)$ = the customer equity value for customers acquired at time t;
$N_{i,t}$ = the number of potential customers at time t for segment i;
$a_{i,t}$ = the acquisition probabilities at time t for segment i;
i,t = the retention probability at time t for a customer in segment i;
$B_{i,a,t}$ = the marketing cost per prospect (N) for acquiring customers at time t for segment i;

$B_{i,r,t}$ = the marketing in time period t for retained customers for segment i;
$B_{i,AO,t}$ = the marketing costs in time period t for add-on selling for segment i;
d = discount rate;
$S_{i,t}$ = sales of the product/services offered by the firm at time t for segment i;
$C_{i,t}$ = cost of goods at time t for segment i;
I = the number of segments;
i = the segment designation;
t_0 = the initial time period.

In simple terms, the above equation shows that customer equity is the summation of returns on acquisition, retention, and add-on selling across a segment of customers over a period of time. This indicates that the focus of the company should be on maximizing the sum of the total customer equity than any of its individual parts.

The published financial reports on the gross contribution margins, and the marketing/advertising spending of competitors could be used to estimate the aggregate-level CLV of competitors (Gupta and Lehmann, 2003). Although aggregate-level CLV has simplicity, it has its own set of challenges. The aggregate-level CLV is only for a group of customers, therefore it is done on the basis of sampling. The sampling method of data has its own challenges; the sample has to be representative of the population to obtain an accurate picture of reality. In the aggregate-level measure of CLV, since we use the average CLV of the selected sample to compute the customer equity of a firm, the customer equity relies on the selected sample, its revenue distribution and its size. This fails to differentiate between the customers of different segments and different levels of revenue, and the firm uses it as a general measure for all its customers. In spite of this drawback, aggregate-level CLV has been used and experimented upon in the domain of marketing. This can be seen in Table 2.4, which also provides an overview of the select literature from the extensive field of CLV.

DRIVERS OF CLV

There are various drivers that affect CLV. The drivers define the nature of the relationship between the firm and the customers and help assess the level of profitability and finally maximize CLV. A clear understanding of these drivers is vital for building profitable customer loyalty, as this forms the basis of customer management strategies. The drivers are typically

Table 2.4 Summary of select CLV literature

Citation	Purpose of the Study	Modeling Type	Level of CLV	Model Setting	Data Used	Sample Size (No. of Customers)	Findings
Rust, Kumar, and Venkatesan (2011)	Propose adoption of Monte Carlo simulation algorithm that provides predictions on customer purchase propensity, profit, and firm marketing actions	Stochastic	Individual	Non-contractual	Firm, which sells high-technology products	191 customers over a period 1997–2003	The proposed model provides large improvements in prediction over the simpler models shown in the literature
Pfeifer (2011)	To use company-reported summary data to estimate the total CLV of a firm's current customers and how to estimate that model using periodically reported company, customer and financial summary results	Deterministic	Aggregate	Contractual	Netflix company data	10 310	Summary data are used to estimate both retention and revenue per renewal

Fader and Hardie (2010)	To develop a model of relationship duration that can be used to predict the retention rates beyond the observed retention rates. This can further be used to calculate a customer's expected residual lifetime	Stochastic	Aggregate	Contractual	Hypothetical data	26 569	Failing to account for cohort-level retention-rate dynamics leads to biased estimates of the residual value of a customer
Kumar (2010)	To study a multimedia multi-channel communication framework based on CLV	NA	NA	B2C/B2B	NA	NA	Maximizing firm's profitability is critical to understand both drivers of CLV and 'customer referral value (CRV)' and manage customers accordingly
Kumar and Shah (2009)	Linking customer equity (determined by the CLV metric) to market capitalization	Stochastic	Individual and aggregate	B2B and B2C	Two Fortune 1000 firms	Monthly transaction data of all customers from January 2000 to 2007	A CLV-based framework can reliably predict firm value and

Table 2.4 (continued)

Citation	Purpose of the Study	Modeling Type	Level of CLV	Model Setting	Data Used	Sample Size (No. of Customers)	Findings
							marketing strategies targeted at maximizing CLV can increase firm value and thus ultimately stock price
Kumar, Venkatesan et al. (2008)	To show the implementation of CLV on IBM and its effects on the profitability and resource allocation	Stochastic	Individual	Non-contractual, B2B	IBM	35 000	CLV-based reallocation of marketing resources yielded a $20 million revenue increase without any additional resource investment
Benoit and Van den Poel (2009)	To analyze CLV by means of quantile regression and compare the effects of the covariates that are missed in linear regression	Deterministic	Individual	Contractual	Transaction data over a period of four years	22 665	Quantile regression provides insights into the effects of the covariates on the

							conditional CLV distribution that may be missed by traditional least squares
Donkers, Verhoef, and de Jong (2008)	Prediction of CLV in multi-service industry. Insurance industry	Stochastic	Individual	B2C, Contractual	Transaction data over six years	30000	Simple models perform well. Focusing on customer retention is not enough, cross-buying also needs to be accounted for
Fader, Hardie, and Jerath (2007)	To demonstrate the use of Pareto/model or negative binomial distribution models of repeat buying in a non-contractual setting for computing CLV	Stochastic	Individual	Non-contractual	Transaction data over a period of five years	7953 customers over a period of five years	Different kind of economic efficiencies for different acquisition strategies can be made through this model and customer segmentation can be done on the same basis

Table 2.4 (continued)

Citation	Purpose of the Study	Modeling Type	Level of CLV	Model Setting	Data Used	Sample Size (No. of Customers)	Findings
Fader, Hardie, and Lee (2005)	Propose a new model that links RFM metric with CLV using iso-value curves	Stochastic	Individual	B2C, non-contractual	CDNOW – online music site	23 570 customers for a period of one year	Iso-value curves are highly non-linear, as customers with low recency values but high frequency present lower CLV than customers who have lower frequency
Thomas, Blattberg, and Fox (2004)	Examine an optimal pricing strategy, when the customer reinitiates the relationship	Stochastic	Individual	B2C, contractual	Newspaper subscription database	566	An optimal pricing strategy involves low re-acquisition price and higher prices after the customer has been re-acquired

Rust, Lemon, and Zeithaml (2004)	Present a framework that enables competing marketing strategy options to be traded off on the basis of projected financial returns	Stochastic (Markov)	Aggregate	B2C non-contractual	Survey data	355	The framework enables a 'what-if' evaluation of marketing ROI, given a particular shift in customer perceptions
Venkatesan and Kumar (2004)	Develop a dynamic framework to maintain/ improve customer relationships through optimal allocation of marketing resources and maximize CLV, simultaneously	Stochastic	Individual	B2B, non-contractual	Computer hardware and software firm	2189	Marketing contacts over various channels influence CLV non-linearly. Customers who are selected on the basis of their CLV provide higher profits in the future compared to customers selected by other metrics
Reinartz and Kumar (2003)	Develop a framework to compute CLV and identify factors that explain the variation in the profitable lifetime duration	Deterministic	Individual	B2B and B2C, non-contractual	Merchandise catalog retailer – three-year daily data	11 992	Demonstrated the superiority of CLV framework by comparing it with other traditional frameworks

Table 2.4 (continued)

Citation	Purpose of the Study	Modeling Type	Level of CLV	Model Setting	Data Used	Sample Size (No. of Customers)	Findings
Dreze and Bonfrer (2003)	To understand the impact of frequency of communication, customer retention and CLV	Stochastic	Aggregate	B2C, non-contractual	Entertainment products company	2 490 164	Inter-communication time affects both customer attrition and surplus. Therefore, it has a critical impact on the value of the consumers
Libai, Narayandas, and Humby (2002)	Introduces a segment-based approach for customer probability analysis	Stochastic	Customer segment (alternative to individual)	B2C, non-contractual	NA	NA	Segment-based approach to estimate long-run customer equity probability using switching probabilities between segments
Reinartz and Kumar (2000)	Proposes a method of computing CLV to test the	Deterministic	Individual	B2C, non-contractual	Catalog retailer	967	Although loyal consumers demand

	propositions between customer loyalty and profitability						premium service and believe they deserve lower prices, they are not always profitable nor do they contribute more to profits in the long run
Pfeifer and Carraway (2000)	Introduces the use of Markov chain models for modeling customer relationships and CLV	Stochastic	Individual	B2C, non-contractual	Hypothetical data		Markov chain models can handle complicated customer relationship situations. It can also handle some complicated situations that algebraic solutions cannot
Berger and Nasr (1998)	Presents mathematical models for determination of CLV	Deterministic	Both aggregate and individual	Both B2B and B2C, contractual	Numerical examples provided		Introduces the first five general classes of CLV models

classified into 'exchange characteristics' and 'customer heterogeneity'. The exchange characteristics include all the variables that affect and influence the customer–firm relationship and customer heterogeneity refers to the demographic variables of the customers, for example age, income levels, and location. As shown in Table 2.5, while the drivers between B2B and B2C firms do not differ, the relative influence of the drivers can vary between and within the B2B and B2C firms.

In the aggregate-level CLV approach, the drivers are firm-level strategies for enhancing customer equity. The drivers could be value equity, brand equity or relationship equity, as discussed by Kumar and George (2007). In the individual-level CLV approach, the customer-specific variables like purchase frequency, cross-buying, product returns, and so on, are the drivers of CLV.

KEY APPROACHES OF CUSTOMER EQUITY

Customer lifetime value when aggregated at the firm level leads to customer equity. There are five different approaches of maximizing and measuring customer equity, which have been extensively discussed in the literature. These approaches differ in the kind of aggregation, underlying assumptions, data requirement and the metrics computed (Kumar and George, 2007). Figure 2.1 provides an overview of these approaches, their assumptions, the benefits of each of the approaches, the cost involved, ease of implementation, time to implement, and the metrics. In Figure 2.1, 'VK approach' stands for the CLV approach developed by Venkatesan and Kumar (2004); it is an individual-level CLV measure in which the gamma distribution is used to model inter-purchase time and panel data regression methodologies are employed to model the contribution margin. The 'BGT approach' stands for Blattberg, Getz, and Thomas's (2001) approach. In this approach customer equity is computed as the sum of return on acquisition, return on retention, and return on add-on selling across a firm's entire portfolio. The 'BN approach' stands for Berger and Nasr's (1998) approach, in which the basic CLV model relies on the three assumptions mentioned in the figure; it is a firm-level approach. The 'RLZ approach' is the Rust, Lemon, and Zeithaml (2004) approach in which the information from the focal brand and the competing brand are used to model the acquisition and retention of customers in the context of brand switching. The last approach is the Gupta and Lehmann (2003) approach (GL approach), which simplifies the CLV computation by assuming constant average margins and retention rate and an infinite projection period.

Although various methods of estimating CLV in different business situ-

Table 2.5 Drivers of CLV

Drivers for Both B2B and B2C Firms	Definition	Effect on CLV
Higher the past customer purchase[a]	The level of purchases made by the customer in the past	Higher
Higher the share-of-wallet[b]	The degree to which the customer meets his/her needs in the category with a focal brand	Higher
Higher the cross-buy[c]	The degree to which the customer buys products from a large number of available products/categories offered by the firm	Higher
Higher the focused buying[d]	Level of purchase made by the customer within a single category	Lower
Higher the customer-based brand equity[e]	Total value provided by the customer to a focal brand	Higher
Moderate the product returns[f]	Number of products the customer returns between the two purchase periods	Higher
Moderate the marketing efforts[g]	Number of times the customer is contacted through the various communication channels between the two purchase periods	Higher
Belonging to the loyalty program[h]	The customer belongs to the loyalty program of the firm	Higher
Shopping in multiple channels[i]	The customer shops in multiple channels of the firm	Higher
Longer duration[j]	The customer's tenure with the firm	Higher

Notes:

a. Kumar (2008).
b. Ibid.
c. Kumar, George, and Pancras (2008).
d. Kumar (2008)
e. Ibid.
f. Petersen and Kumar (2009).
g. Ventatesan and Kumar (2004).
h. Kumar and Shah (2004).
i. Venkatesan, Kumar, and Ravishanker (2007).
j. Kumar, Ramani, and Bohling (2004).

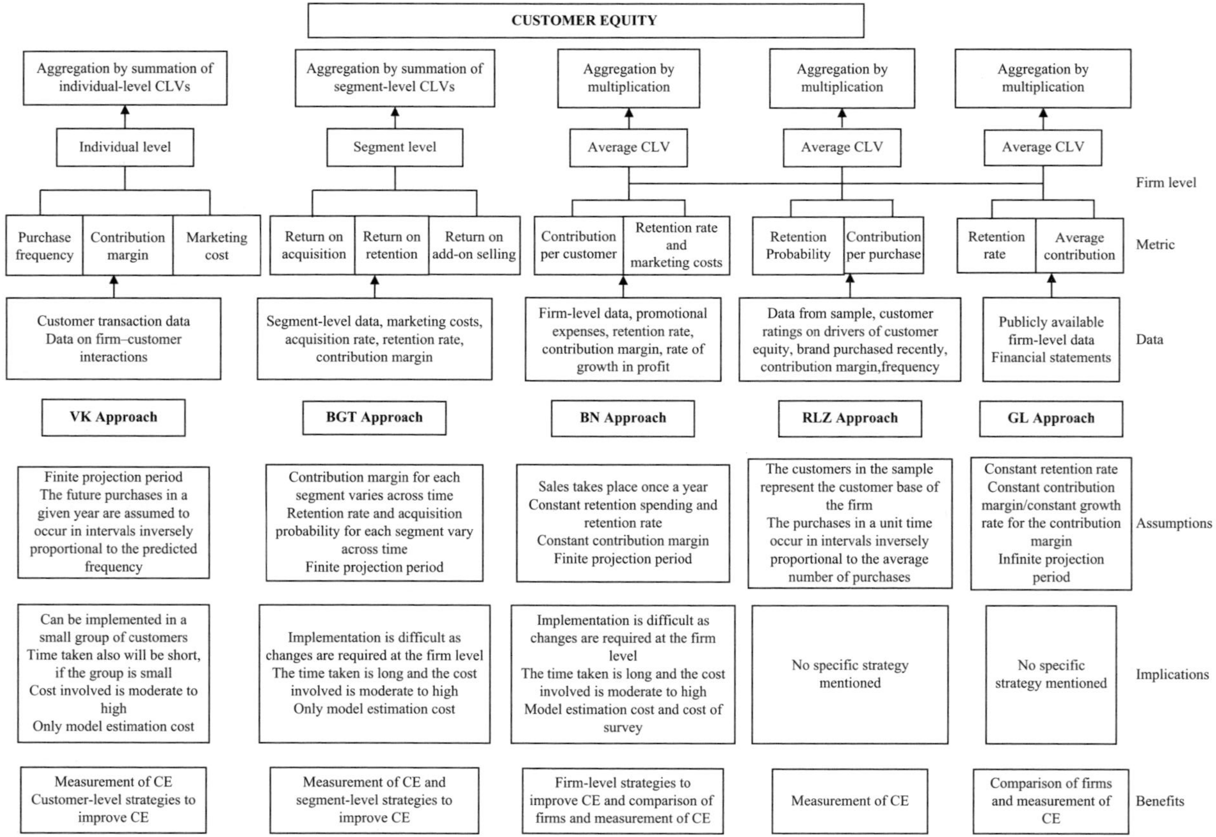

Source: Adapted from Kumar and George (2007).

Figure 2.1 Key approaches of CLV

Table 2.6 Should a firm choose individual-level or aggregate-level CLV?

	Individual-level CLV	Aggregate-level CLV
Objective	To design customer-level strategies for profit maximization	To measure the effectiveness of the marketing strategies and programs
Industry	Financial services, telecom, retail, airlines. Any other industry where customer transaction data are available from loyalty cards, credit cards, account numbers and contractual agreements	All industries
Size of the firm	Large firm with record of transaction data and firm interaction data	Small or large firm with a record of marketing expenditure of the firm
Data	Transaction data, competitors' customer-level data (if possible), data on acquisition and retention per customer, marketing data per customer, any other operational expense	Data on companies overall marketing expenses Data on a group of consumers (a representative sample of all customers)
Resources	Firms would require a competent and skilled\analytical team, along with the required technical expertise. Further, it would also require ongoing research on the consumer database for modifying strategies, as buying behavior of consumers is dynamic	Simple statistical software and a competent analytics employee would be required. The time lapse would be a year after the implementation of the marketing programs

ations have been discussed, deciding which method to use based on the constraints the firm has can be challenging. Table 2.6 provides guidelines to ease the decision-making process of firms. While determining a marketing strategy, every firm has to decide its overall objective of undergoing any change in strategies or developing a new one. If the firm's objective is only to evaluate the effect of its marketing strategies on its overall profitability, then it should opt for the aggregate-level CLV method as it would help the firm clearly analyze the return on retention, acquisition, and other marketing costs. The firm could use the method suggested by Blattberg et al. (2001). However, if the firm is interested in designing customer-specific strategies for each of its unique customers or for its customers who belong to a similar group, then the firm should opt for the individual-level CLV method, as this method estimates the lifetime of each individual customer and helps in segmenting them in groups as validated by Venkatesan and Kumar (2004).

Once the firm has determined its objective, the industry in which the firm operates would be the next important criterion to consider. The industry in which the firm operates is essential, as that would determine the data that the firm possesses. Although aggregate-level CLV can be estimated in all industries that have marketing activities, it is the individual-level CLV that requires customer transaction data along with firm-level data. Therefore, if the firm is operating in a retail industry that has access to customer transaction data, then the firm can accrue the benefits of estimating individual-level CLV. Sometimes it may not be the industry barrier but budgetary constraints, size of the firm, lack of technology, lack of skill set, and so on that may restrain the firm from opting for the individual-level CLV method, and instead choose the aggregate-level CLV. Individual-level CLV can be computed in large firms with a huge customer database, as large firms would possess the various resources required for modeling CLV. This would be a very expensive affair for small firms; hence they would prefer the aggregate-level CLV method.

We have discussed when a firm should use the individual-level CLV and the aggregate-level CLV, but sometimes decision-making is a challenge when resource constraints exist or when alternative modeling approaches/tools are available. Clarity on which modeling approach to use with the given constraints is essential. Figure 2.2 is a decision-making guide for managers as it provides a clear flowchart to implement CLV; it shows the tools to be used when data are available and the actions to be taken if the firm does not have any data. The flowchart is a complete process from the time of data collection to the final effect of customer equity on market capitalization.

STRATEGIES FOR MAXIMIZING CLV

Most analytical activities are designed by a firm to modify or design strategies. Similarly, there are various strategies that can be implemented for maximizing CLV.

Individual-level CLV

The strategies that can be used for maximizing individual-level CLV have been discussed in the 'wheel of fortune' framework given by Kumar (2008). We look at an overview of all the wheel of fortune strategies in this section:

- *Selecting the right customer.* It is important to have a profitable customer base, as the marketing budget is limited and firms want to maximize their profits within the limited budget. Acquiring

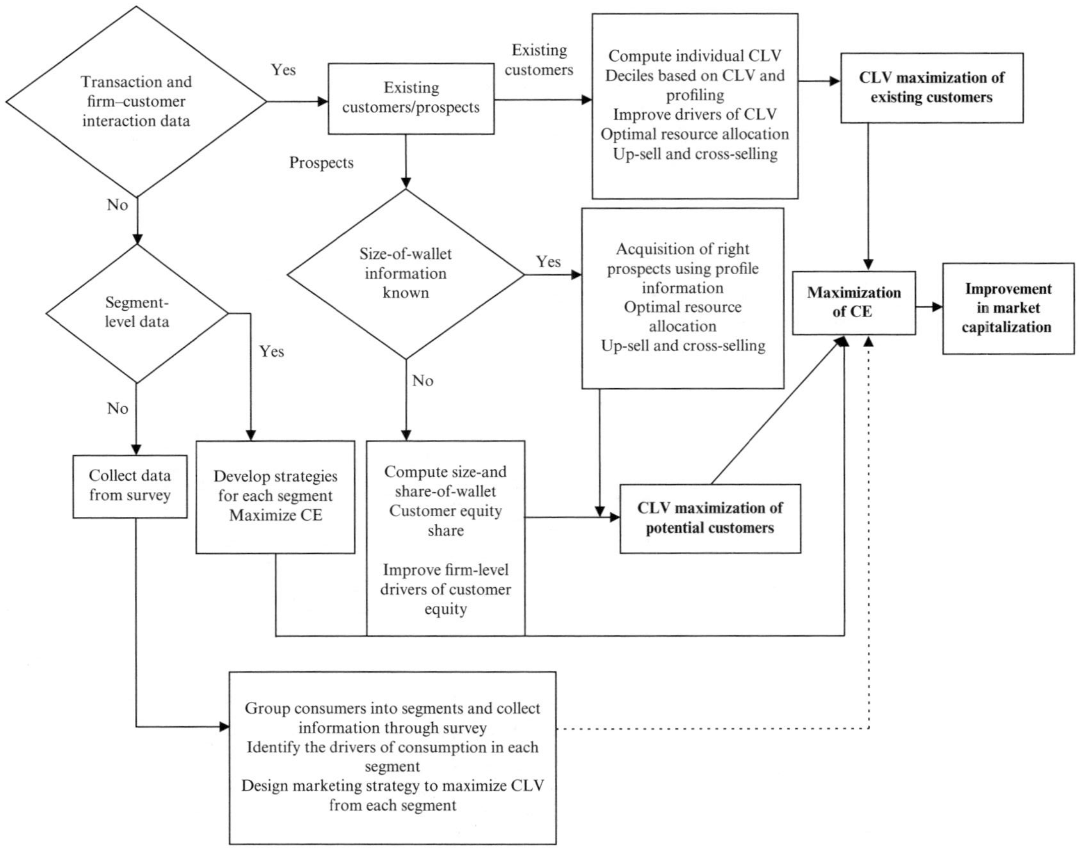

Source: Adapted from Kumar and George (2007).

Figure 2.2 Implementation of CLV

the right customers is possible by using CLV because CLV is a forward-looking metric that can project the future profitability of a customer. Once the right customer base is established, the firm can strategize its future activities for enhancing customer profitability.

- *Manage loyalty and profitability.* A popular notion has been that loyal customers are profitable. However, this relationship does not hold well in many instances, a fact that has been tested in a US high-tech company by Reinartz and Kumar (2002). Some customers may be loyal and not profitable, whereas some may be both loyal and profitable. Distinguishing between the various levels of customer loyalty and profitability is essential, as the objective of the marketing program for each segment of customers will be different. Therefore, firms should define the different levels of loyalty and profitability among customers and segment them accordingly. A segmentation classification has been provided by Reinartz and Kumar (2002) and Kumar and Shah (2004).
- *Optimum allocation of resources.* Since firms have limited resources and would want to make the best possible use of these resources, as mentioned earlier they should begin with choosing the right customers. Once the firm has established a customer base it should understand the right communication mix for each of its customers and use only that channel(s) to which the customer is most responsive. The firm should also be aware when to cross-sell, what product to cross-sell, and so forth. This is possible only if the firm regularly tracks the purchase frequency of the customer, the inter-purchase time, and the contribution to profit by the customer (Kumar, 2008).
- *Pitching the right product to the right customer at the right time.* Cross-selling and up-selling are important for a firm. The more product categories the customer buys in, the longer he or she will remain with the company. Therefore, firms will pitch multiple products to a customer. Knowing which product to pitch, at what time, and how much the customer is willing to invest are the three important things a firm should know while offering an additional product/service to its customers. A Bayesian estimation of the purchase sequence and purchase timing would help the firm achieve this (Kumar and Rajan, 2012).
- *Preventing attrition of customers.* Retaining profitable customers is one of the important challenges that firms face. The dynamic churn models can be used to predict which customers are likely to leave the company. The firms should design strategies based on when to approach the customer, the channel through which to approach the customer, and how to approach the customer. These strategies

should be based on the CLV of the customer to ensure that the firm doesn't spend more than the CLV of the customer. This would help ensure that the customer is still profitable to the firm.

- *Managing multi-channel shoppers.* Kumar and Venkatesan (2005) conducted a study on the multi-channel shoppers of a B2B firm. The results of the study indicated that multi-channel shoppers initiate more contact with the firm, have longer tenure, purchase more frequently, and are more receptive to communication through multiple channels. Managers can determine the profitability of multiple channels and design strategies accordingly by using various metrics along with CLV, such as share-of-wallet, customers' past profitability, how much a customer spends, and so on.
- *Linking investments in branding to customer profitability.* Linking the brand value of the firm to CLV maximizes CLV. The measurements of the components of individual brand value suggest what customers think about the brand. A framework for the same has been designed by Kumar, Luo, and Rao (2008) establishing the link between individual brand value and CLV. Their framework enables optimization of CLV, which in turn allows simultaneous growth in brand equity and customer equity.
- *Acquiring profitable customers.* Acquiring customers is generally the primary objective of the firm, because of the belief that the higher the number of customers of a firm the higher the profitability. This, however, may not always be true, because after a certain point the cost of acquisition is higher than the profits accrued from the customer. Firms should be mindful of this fact and of their limited resources. Similarly, retaining all of the customers is not always a good idea, because not all existing customers are profitable. Firms should know which customers to let go and which to retain. Analyzing the customers' actions, competitors' actions, and considering acquisition and retention simultaneously with profitability would help the firm decide the right amount of resources to be spent on the right customer (Reinartz et al., 2005).
- *Managing customer referral behavior.* Although CLV cannot provide much insight into the referral behavior of the customers, it can act as a tool along with the customer referral value (CRV) score to help firms decide the most effective and promising customers to target (Kumar, Petersen, and Leone, 2010). CRV is the value of the referral behavior of a customer; it is useful in calculating the indirect impact of customers on the firm's profitability.
- *Implementation and organizational changes.* Firms should focus on timely interactions with the customers and effective management of

this interaction. This interaction between firms and customers helps in developing organizational resources for successful management strategies such as interaction orientation (Ramani and Kumar, 2008). Continuous interaction with the customers, empowering them, continuously monitoring the CLV, and allocating resources to the CLV would lead to increased customer satisfaction, positive word-of-mouth, and so on.

Aggregate-level CLV

Customization of strategies, specialization of focus and targeting of customers specifically per their needs may not be the focus of all firms. Instead, the firm's objective could to be evaluate the overall effectiveness of its marketing objectives. The aggregate-level CLV helps firms evaluate their overall effectiveness, by considering the rate of retention, acquisition, and overall profits. This metric can also be used to evaluate the firm with respect to its competition. Competitive firms' data can be acquired from published financial reports and firms can estimate the competitors' profitability. This would help the firm to estimate if the customers of the competition are profitable or not (Gupta and Lehmann, 2003).

All the above strategies tend towards maximizing firm profitability and aligning the goals of the marketing department with the overall objectives of the firm. Finally, CLV (individual-level or aggregate-level) can also be used to demonstrate the effect of the firm's marketing activities on the market capitalization of the firm as shown in Figure 2.2. Gupta et al. (2004), demonstrated the effect of aggregate CLV on firm value, with data from five companies (four Internet firms and one traditional firm). They found that for three out of five firms, customer value was very close to firm value. Their study provides various acquisition and retention strategies that the firm should follow to increase its firm value. On the other hand, Kumar and Shah (2009, 2011) demonstrate how individual-level CLV can be used in both B2B and B2C firms to establish the relationship between customer equity and firm value (measured by the stock price of the firm). Their study not only demonstrates this link but also discusses the customer-specific strategies that are responsible for driving future profits. They further show that the relationship between the CLV and market capitalization is moderated by risk factors in the form of volatility and vulnerability of cash flows from customers.

SCOPE FOR FUTURE RESEARCH

We have discussed various methods of calculating CLV at both the aggregate and individual level in this chapter, both in contractual and non-contractual settings. However, it would be interesting to explore this field of study from a few different viewpoints.

It would be interesting to understand the effect of the customer's behavioral aspects (like customer attitudes, satisfaction) on his or her CLV. There have been studies that demonstrate that customer attitudes affect customer behavior (Hogan, Lemon, and Libai, 2003) and customer behavior influences CLV (Reinartz and Kumar, 2003), but the question remains: do customer attitudes affect CLV directly?

Although CLV has been studied in the service industry, it would be interesting to understand if the CLV of consumers in the consumer packaged goods industry would follow the same pattern as that of customers in the service industry. This would help firms design their sales and promotional strategies. This may also provide a new direction to the consumer packaged goods industry.

The changes in the economy affect all firms and individuals. Since CLV is a predictive and forecasting technique, it would be interesting to account for the macroeconomic factors that indirectly influence firm strategies and customer behavior. For instance, employment rates, GDP, the international trade deficit situation, and so on, would contribute to a more accurate CLV that would reflect market realities. None of the current CLV measures considers the impact of competitors. It would be thought-provoking to understand the effect of the CLV of the customers of focal firms with respect to their competitor's strategies and new product innovations by competitors. This would help in designing more holistic and effective marketing strategies.

Finally, it is not only the firm's activities that affect the firm's customers and profitability, but also the other partners of the firm like the employees, stakeholders, distributors and so on that influence its customers and profitability. It would be exciting to study the lifetime value of suppliers (dealers) of the firm, employees of the firm and other stakeholders. The dealers' lifetime value would help in allocating resources and charting out distribution strategies. The employee lifetime value would help in allocating resources for soft skills like motivating employees, training them, and ensuring that their lifetime value is high, as the longer the employee would stay with a company, the greater would be his or her knowledge of the firm's policies and procedures and the better he or she would serve the customers. This too would contribute toward ensuring a higher CLV.

CONCLUSION

Firms can design their marketing strategies at either the individual level or by categorizing similar customers at the segment level. Assessing the profitability of these customers with the various resources available should help firms design strategies to maximize customer profitability. Firms now want a 360-degree approach to measuring customer profitability. Such an approach has been designed by Kumar (2014), by assessing the customer brand value (CBV), customer lifetime value (CLV), customer referral value (CRV), customer influence value (CIV), and customer knowledge value (CKV), and would further help firms in designing strategies that would maximize their profits and in determining what strategy would work across different communication channels and customers segments.

NOTE

* The authors thank Renu and Gayatri Shukla for copy-editing the original manuscript of this chapter.

REFERENCES

Benoit, D.F. and D. van den Poel (2009), 'Benefits of quantile regression for the analysis of customer lifetime value in a contractual setting: an application in financial services', *Expert Systems with Applications*, **36**(7), 10475–84.

Berger, P.D. and N.I. Nasr (1998), 'Customer lifetime value: marketing models and applications', *Journal of Interactive Marketing*, **12**(1), 17–30.

Blattberg, R.C. and J. Deighton (1996), 'Manage marketing by the customer equity test', *Harvard Business Review*, **74**(4), 136–44.

Blattberg, R.C., G. Getz and J.S. Thomas (2001), *Customer Equity: Building and Managing Relationships as Valuable Assets*, Boston, MA: Harvard Business Press.

Dekimpe, M. and D.M. Hanssens (2004), 'Persistence modeling for assessing marketing strategy performance', in C. Moorman and D. Lehmann (eds), *Assessing Marketing Strategy Performance*, Cambridge, MA: Marketing Science Institute.

Donkers, B., P.C. Verhoef and M. de Jong (2007), 'Modeling CLV: a test of competing models in the insurance industry', *Quantitative Marketing and Economics*, **5**(2), 163–90.

Dreze, X. and A. Bonfrer (2002), 'To pester or leave alone: lifetime value maximization through optimal communication timing', working paper, Marketing Department, University of California, Los Angeles, CA.

Fader, P. and B.G.S. Hardie (2010), 'Customer-base valuation in a contractual setting: the perils of ignoring heterogeneity', *Marketing Science*, **29**(1), 85–93.

Fader, P., B. Hardie and P.D. Berger (2004), 'Customer-base analysis with discrete-time transaction data', accessed 30 May 2013 at http://ssrn.com/abstract=596801.

Fader, P., B.G.S. Hardie and K. Jerath (2007), 'Estimating CLV using aggregated data: the Tuscan Lifestyles case revisited', *Journal of Interactive Marketing*, **21**(3), 55–71.

Fader, P., B.G.S. Hardie and K.L. Lee (2005), 'RFM and CLV: using iso-value curves for customer base analysis', *Journal of Marketing Research*, **42**(4), 415–30.

Gupta, S. and D. Lehmann (2003), 'Customers as assets', *Journal of Interactive Marketing*, **17**(1), 9–24.

Gupta, S., D. Lehmann and J.A. Stuart (2004), 'Valuing customers', *Journal of Marketing Research*, **41**(1), 7–18.

Hogan, J.E., K.N. Lemon and B. Libai (2003), 'What is the true value of a lost customer?', *Journal of Service Research*, **5**(3), 196–208.

Kumar, V. (2008), *Managing Customers for Profit: Strategies to Increase Profits and Build Loyalty*, Upper Saddle River, NJ: Prentice Hall Professional.

Kumar, V. (2010), 'A customer lifetime value-based approach to marketing in the multichannel, multimedia retailing environment', *Journal of Interactive Marketing*, **24**(2), 71–85.

Kumar V. (2014), *Profitable Customer Engagement: Concepts, Metrics and Strategies*, Thousand Oaks, CA: Sage Publications.

Kumar, V. and M. George (2007), 'Measuring and maximizing customer equity: a critical analysis', *Journal of the Academy of Marketing Science*, **35**(2), 157–71.

Kumar, V. and B. Rajan (2012), 'Customer lifetime value management: strategies to measure and maximize customer profitability', in V. Shankar, G.S. Carpenter, and J. Farley (eds), *Handbook of Marketing Strategy*, Cheltenham, UK and Northampton, MA, USA: Edward Elgar Publishing, pp. 107–34.

Kumar, V. and D. Shah (2004), 'Building and sustaining profitable customer loyalty for the 21st century', *Journal of Retailing*, **80**(4), 317–29.

Kumar, V. and D. Shah (2009), 'Expanding the role of marketing: from customer equity to market capitalization', *Journal of Marketing*, **73**(6), 119–36.

Kumar, V. and D. Shah (2011), 'Can marketing lift stock prices', *Sloan Management Review*, **52**(4), 24–6.

Kumar, V. and R. Venkatesan (2005), 'Who are the multi-channel shoppers and how do they perform? Correlates of multi-channel shopping behavior', *Journal of Interactive Marketing*, **19**(2), 44–62.

Kumar, V., M. George and J. Pancras (2008), 'Cross-buying in retailing: drivers and consequences', *Journal of Retailing*, **84**(1), 15–27.

Kumar, V., M. Luo and V. Rao (2008), 'Linking brand value to customer lifetime value', working paper, J. Mack Robinson College of Business, Georgia State University.

Kumar, V., J.A. Petersen and R.P. Leone (2010), 'Driving profitability by encouraging customer referrals: who, when, and how', *Journal of Marketing*, **74**(5), 1–17.

Kumar, V., G. Ramani and T. Bohling (2004), 'Customer lifetime value approaches and best practice applications', *Journal of Interactive Marketing*, **18**(3), 60–72.

Kumar, V., R. Venkatesan, T. Bohling and D. Beckmann (2008), 'Practice prize report – the power of CLV: managing customer lifetime value at IBM', *Marketing Science*, **27**(4), 585–99.

Libai, B., D. Narayandas and C. Humby (2002), 'Toward an individual customer profitability model: a segment-based approach', *Journal of Service Research*, **5**(1), 69–76.

Petersen, J.A. and V. Kumar (2009), 'Are product returns a necessary evil? Antecedents and consequences', *Journal of Marketing*, **73**(3), 35–51.

Pfeifer, P.E. (2011), 'On estimating current-customer equity using company summary data', *Journal of Interactive Marketing*, **25**(1), 1–14.

Pfeifer, P.E. and R.L. Carraway (2000), 'Modeling customer relationships as Markov chains', *Journal of Interactive Marketing*, **14**(2), 43–55.

Ramani, G. and V. Kumar (2008), 'Interaction orientation and firm performance', *Journal of Marketing*, **72**(1), 27–45.

Reinartz, W. and V. Kumar (2000), 'On the profitability of long-life customers in a noncontractual setting: an empirical investigation and implications for marketing', *Journal of Marketing*, **64**(4), 17–35.

Reinartz, W. and V. Kumar (2002), 'The mismanagement of customer loyalty', *Harvard Business Review*, **80**(7), 86–95.

Reinartz, W. and V. Kumar (2003), 'The impact of customer relationship characteristics on profitable lifetime duration', *Journal of Marketing*, **67**(1), 77–99.

Reinartz, W., J.S. Thomas and V. Kumar (2005), 'Balancing acquisition and retention resources to maximize customer profitability', *Journal of Marketing*, **69**(1), 63–79.
Rust, R.T., V. Kumar and R. Venkatesan (2011), 'Will the frog change into a prince? Predicting future customer profitability', *International Journal of Research in Marketing*, **28**(4), 281–94.
Rust, R.T., K.N. Lemon and V.A. Zeithaml (2004), 'Return on marketing: using customer equity to focus marketing strategy', *Journal of Marketing*, **68**(1), 109–27.
Thomas, J. (2001), 'A methodology for linking customer acquisition to customer retention', *Journal of Marketing Research*, **38**(2), 262–8.
Thomas, J., R.C. Blattberg and E.J. Fox (2004), 'Recapturing lost customers', *Journal of Marketing Research*, **41**(1), 31–45.
Thomas, J., W. Reinartz and V. Kumar (2004),'Getting the most out of all your customers', *Harvard Business Review*, **82**(7–8), 116–23.
Venkatesan, R. and V. Kumar (2004), 'A customer lifetime value framework for customer selection and resource allocation strategy', *Journal of Marketing*, **68**(4), 106–25.
Venkatesan, R., V. Kumar and N. Ravishanker (2007), 'Multichannel shopping: causes and consequences', *Journal of Marketing*, **71**(2), 114–32.
Yoo, S. and D.M. Hanssens (2005), 'Modeling the sales and customer equity effects of the marketing mix', working paper, Anderson School of Management, University of California, Los Angeles, CA.

3. Simple probability models for computing CLV and CE

Peter S. Fader and Bruce G.S. Hardie

INTRODUCTION

The past two decades have seen an evolution from transaction-oriented to customer-centric marketing strategies (Fader, 2012). As part of this shift, it is now common to think of a firm's customers as assets that generate cash flow, not just this period but in future periods as well. This is formalized in the notion of customer lifetime value (CLV), which Pfeifer, Haskins, and Conroy (2005) define as 'the present value of the future cash flows attributed to the customer relationship' (p. 17). Using the term 'customer equity' (CE) to denote the sum of the lifetime values of a firm's customers, Blattberg and Deighton (1996) suggested that a central goal for the marketing executive (and for the firm as a whole) should be to maximize CE.

At the heart of any attempt to measure CE is the calculation of CLV, which requires us to answer two questions: (1) How long will the customer remain 'alive'? and (2) What is the net cash flow per period while 'alive'? This requires the analyst to make multi-period forecasts of key behaviors (i.e., for each period, is the customer still 'alive' and, if so, how much do they spend?). In other words, this is a time-series modeling problem.

A popular classification scheme for time-series models is that proposed by Cox (1981), who distinguishes between observation-driven and parameter-driven models. Let Y_t denote the random variable(s) associated with the behavior(s) of interest at time t; this is assumed to be a function of parameter(s) θ_t. In an observation-driven model, θ_t is modeled as some function of past observed behavior: $\theta_t = f(y_{t-1}, y_{t-2}, \ldots)$. In a parameter-driven model, θ_t is a function of past values of the latent variable(s): $\theta_t = g(\theta_{t-1}, \theta_{t-2}, \ldots)$.

While observation-driven models are well suited for the task of making one-period-ahead forecasts, there are problems when using them to generate the multi-period forecasts required for the estimation of CLV and CE. Suppose we specify a model in which $\theta_t = f(y_{t-1}, y_{t-2}, \ldots)$ and have calibrated the parameters of $f(\cdot)$ using data on y_1, y_2, y_3. Standing at the end of period 3, we can predict period 4 behavior using y_2 and y_3. However, in order to predict period 5 behavior, we need y_3 and y_4, yet period 4 is currently in the future. We therefore have to use the model to simulate period 4 behavior, and so on. As we move further away from the last period for which we have realizations of Y_t, we expect the forecasts of the behavior of interest to become increasingly unreliable as any prediction error propagates through the forecasts.[1]

Parameter-driven models do not suffer from this problem. The stochastic process underpinning $g(\cdot)$ is estimated using the observed data y_1, y_2, y_3 and no additional observations (actual or simulated) are required to generate forecasts of behavior regardless of the length of the forecast horizon.

The majority of models developed by researchers interested in CE are observation-driven models (e.g., Kumar and Petersen, 2012). In this chapter, we review the key parameter-driven models that have been developed by marketing scientists that can be used to generate the multi-period forecasts of key behaviors required for the calculation of CLV and CE. These models go under the label of 'probability models for customer-base analysis', and are specified, tested, and communicated following the well-established traditions of stochastic models of buyer behavior, which have been part of the marketing science literature from its very beginning (Fader, Hardie, and Sen, 2014).

Rather than try to tease out the effects of all the factors that may influence buyer behavior, a probability model completely embraces the notion of stochasticity, viewing the behavior of interest as the outcome of one or more probabilistic process(es). This does not imply that we think the individual is truly purchasing 'at random'. Instead it simply reflects our uncertainty regarding the factors that influence buying behavior; the probabilistic process captures the net effect of all the influences not explicitly considered in the model.

A probability model of buyer behavior typically has two components. First, the individual behavior of interest (e.g., length of time they remain a customer) is characterized in terms of a probability distribution (or several distributions in the case of more complex models), with the parameters of this distribution reflecting the individual's underlying behavioral propensities. Second, differences in these underlying behavioral propensities are captured by an additional probability distribution. These two components result in a mixture distribution, which characterizes the behavior of a randomly chosen individual. While such models do not give us any direct insights as to 'how' or 'why' consumers behave the way they do, they have proven capable of generating the accurate multi-period forecasts of buyer behavior needed for the calculation of CLV and CE.

CLASSIFYING ANALYSIS SETTINGS

Before reviewing the key models, we must first highlight some of the key types of firm–customer relationships, which will drive model formulations.

Consider, as a starting point, the following two statements regarding the size of a company's customer base:

- Based on numbers presented in a news release that reported Vodafone Group Plc's results for the six months ended 30 September 2013, we see that Vodafone UK had 11.3 million 'pay monthly' customers at the end of that period.
- In his 'Q3 2013 Earnings Conference Call', the CFO of Amazon made the comment that '[a]ctive customer accounts exceeded 224 million', where customers are considered active if they have placed an order during the preceding 12-month period.

At first glance, both statements seem like perfectly logical characterizations of the size of each firm's customer base. However, after careful consideration, we realize that only one of them (the first one) is a valid description. We can be confident about the number of 'pay monthly' customers that Vodafone UK has at any single point in time. As these are contract customers (as opposed to customers on 'pay as you talk'/prepaid tariffs), the customer must tell Vodafone when they wish to switch their mobile phone provider. Therefore, Vodafone knows for sure the time at which any customer 'dies'. Such a business is viewed as having a subscription-based or 'contractual' relationship with its customers.

In contrast, the Amazon statement defines an 'active' customer as someone who has placed an order in the past year. Can we assume that someone is no longer a customer just because their last order was placed 366 days ago? Furthermore, can we necessarily assume that, just because someone placed an order 364 days ago, they are still a customer? This 12-month cut-off point is completely arbitrary. If we were to change the cut-off point to six months, the apparent size of Amazon's customer base would be smaller, even though the true size would remain unchanged. In such a business setting, the 'death' of customers is unobserved by the firm (Schmittlein, Morrison, and Colombo, 1987). Customers do not formally end their relationship with the firm; instead they just 'silently attrite' (Mason, 2003). As such, the firm can never know for sure how many customers it has at any point in time. Reinartz and Kumar (2000) introduced the label 'non-contractual' to characterize such a business setting.

While ignored by a number of researchers, this contractual/non-contractual distinction is of fundamental importance for any researcher developing models for estimating CLV and CE.[2] The key challenge of non-contractual settings is how to differentiate those customers who have ended their relationship with the firm (without informing the firm) from those who are simply in the midst of a long hiatus between transactions.

A secondary distinction concerns the treatment of time. Consider the following four specific business settings: (1) airline lounges, (2) electrical utilities, (3) academic conferences, and (4) mail-order clothing companies. The first two are clearly contractual settings, while the last two are clearly non-contractual settings. Reflecting on the first and third settings, we see that the transactions can only occur at discrete points in time. The conference occurs at a specific point in time in any given year, and one either attends or does not attend. Similarly, the airline lounge membership lapses at a specific point in time and the member either renews or does not renew. On the other hand, a common characteristic of the second and fourth settings is that there are no constraints as to when the customer can purchase clothing or end his or her relationship with the firm. Thus, we can talk of a second distinction: are the opportunities for transactions restricted to discrete points in time or can they occur at any point in time?

These two dimensions lead to a classification of customer bases, adapted from Schmittlein et al. (1987), as illustrated in Figure 3.1.[3] When developing a model for CLV and CE estimation, we must ask ourselves in which quadrant we are operating. While in certain circumstances a model developed for a discrete setting can be applied in a continuous setting (and

Opportinities for transactions	Non-contractual	Contractual
Continuous	Grocery purchasing Doctor visits Hotel stays	Credit cards Utilities Continuity programs
Discrete	Conf. attendance Prescription refills Charity fund drives	Magazine subs Insurance policies 'Friends' schemes

Type of relationship with customers

Figure 3.1 Classifying customer bases

vice versa), the contractual/non-contractual divide is fundamental and the boundary cannot be crossed. We use this framework to guide our review of the literature.

CLV REVISITED

As noted in the introduction, CLV is the present value of the future cash flows attributed to the customer relationship. Following Rosset et al. (2003), we can express this mathematically as:

$$E(CLV) = \int_0^\infty E[v(t)]S(t)d(t)dt, \tag{3.1}$$

where, for $t \geqslant 0$ (with $t = 0$ representing 'now'), $E[v(t)]$ is the expected value (or net cash flow) of the customer at time t, $S(t)$ is the probability that the customer has remained alive to at least time t, and $d(t)$ is a discount factor that reflects the present value of money received at time t.

It is important to distinguish between (1) the expected lifetime value of an as-yet-to-be-acquired customer, (2) the expected lifetime value of a just-acquired customer, and (3) the expected *residual* lifetime value, $E(RLV)$, of an existing customer. (The difference between the first two quantities is simply the value of the first transaction that signals the start of the relationship.) We can express the notion of RLV mathematically as:

$$E(RLV) = \int_{t'}^\infty E[v(t)]S(t|t > t')d(t - t')dt, \tag{3.2}$$

where t' is the 'age' of the customer at which time their residual lifetime value is computed.

Consider a contractual setting. Following Fader and Hardie (2010), if we assume a constant expected net cash flow per contract period (i.e., $E[v(t)] = \bar{v}$), we can express equation (3.1) as:

$$E(CLV) = \bar{v}\, DEL,$$

where the 'discounted expected lifetime' (DEL) of the customer is defined as:

$$DEL = \int_0^\infty S(t)d(t)dt. \tag{3.3}$$

(With no discounting, we have $\int_0^\infty S(t)dt$, which is simply $E(T)$, the customer's expected lifetime.) Similarly, we can express equation (3.2) as:

$$E(RLV) = \bar{v}\, DERL,$$

where the 'discounted expected residual lifetime' (DERL) of the customer is defined as:

$$DERL = \int_{t'}^{\infty} S(t|t > t')d(t - t')dt$$

$$= \int_{t'}^{\infty} \frac{S(t)}{S(t')} d(t - t')dt. \tag{3.4}$$

At the heart of a probability model for a contractual setting is an expression for $S(t)$. For a given model specification, we can derive the corresponding expressions for DEL and DERL. We review the basic results in the next section.

Now consider a non-contractual setting. Following Fader, Hardie, and Lee (2005b), if we assume a constant net cash flow per transaction of $\bar{v}$, we have $v(t) = \bar{v}\, t(t)$, where $t(t)$ is the transaction rate at time t, and can express equation (3.1) as:

$$E(CLV) = \bar{v}\, DET,$$

where the 'discounted expected transactions' (DET) of the customer is:

$$DET = \int_{0}^{\infty} t(t) S(t) d(t) dt. \tag{3.5}$$

(For a just-acquired customer, DET measures the present value to the firm of all future transactions by the customer, with transactions at a future time point appropriately discounted to obtain their present values.) Similarly, we can express equation (3.2) as:

$$E(RLV) = \bar{v}\, DERT,$$

where the 'discounted expected residual transactions' (DERT) of the customer is:

$$DERT = \int_{t'}^{\infty} t(t) S(t|t > t') d(t - t') dt. \tag{3.6}$$

At the heart of a probability model for a non-contractual setting are expressions for $t(t)$ and $S(t)$. For a given model specification, we can derive the corresponding expressions for DET and DERT. We review the basic results in the 'Non-contractual Settings' section below.

CONTRACTUAL SETTINGS

Numerous researchers working in the areas of marketing, applied statistics, and data mining have developed a number of models that attempt to identify those customers with the greatest risk of churning in the next time period (Parr Rud, 2001; Lemmens and Croux, 2006; Neslin et al., 2006; Risselada, Verhoef, and Bijmolt, 2010; Berry and Linoff, 2011). A typical analysis exercise may use a logit model to predict churn where the independent variables include behavioral variables such as usage in the previous period, changes in usage over the last two periods, customer-initiated contacts with the company (e.g., contacting the call center), and marketing activity by the firm. As such analyses are observation-driven models, they are of limited use to the analyst interested in estimating CLV and CE. For example, in order to predict churn in t period's time from now, we need to know the number of calls made to the call center over the next $t-1$ periods, which is an unknown.

If we consider a cohort of customers acquired at the same time, we typically observe that the cohort-level retention rates increase over time (e.g., Kumar and Reinartz, 2012, Figure 5.2). While it is tempting to tell a story of individual-level time dynamics (e.g., increasing loyalty as the customer gains more experience with the firm, and/or increasing switching costs with the passage of time), a far simpler story – and one consistent with the fundamental marketing concept of segmentation – is that of a sorting effect in a heterogeneous population.

A very simple parameter-driven model for discrete-time duration data is the beta-geometric (BG) model (Potter and Parker, 1964). In a contractual setting, it is formulated by assuming that (1) at the end of each contract period, an individual remains a customer of the firm with constant retention probability $1-\theta$ (which is equivalent to assuming that the duration of the customer's relationship with the firm is characterized by a geometric distribution), and (2) individual differences in θ are characterized by a beta distribution with parameters γ and δ. The resulting survivor function is:

$$S(t|\gamma,\delta) = \frac{B(\gamma,\delta+t)}{B(\gamma,\delta)}, \quad t = 0, 1, 2, \ldots \tag{3.7}$$

The corresponding cohort-level retention rate is:

$$r(t) = \frac{\delta + t - 1}{\gamma + \delta + t - 1},$$

which increases over time, even though each individual's unobserved (and unobservable) churn probability is assumed to be constant over time. Despite what may seem to be overly simplistic assumptions, the analyses presented in Fader and Hardie (2007a) demonstrate that this two-parameter model (which can easily be implemented in a basic Excel spreadsheet) generates very accurate forecasts of retention. As such, we can confidently use it as the basis of attempts to compute CLV and CE in contractual settings, something explored in Fader and Hardie (2010) and extended in Braun and Schweidel (2011).

Substituting equation (3.7) in equation (3.3) and solving the integral (which becomes a sum in a discrete-time setting) gives us the following expression for the discounted expected lifetime of an as-yet-to-be-acquired customer:

$$DEL(\gamma, \delta, d) = {}_2F_1\left(1, \delta; \gamma + \delta; \frac{1}{1 + d}\right), \tag{3.8}$$

where ${}_2F_1(\cdot)$ is the Gaussian hypergeometric function and d is the appropriate discount factor used to represent the time value of money. (When there are k contract periods per year, an annual discount rate of $(100 \times r)$ percent maps to a period discount rate of $d = (1 + r)^{1/k} - 1$.) To compute DEL for a just-acquired customer, we simply subtract 1 from this quantity.

Substituting equation (3.7) in equation (3.4) and solving the integral gives the following expression for the discounted expected residual lifetime of a customer who has renewed his or her contract $n - 1$ times, evaluated *just before* we find out whether they will make an nth contract renewal:

$$DERL(\gamma, \delta, d; n-1 \text{ renewals})$$

$$= \left(\frac{\delta + n - 1}{\gamma + \delta + n - 1}\right) {}_2F_1\left(1, \delta + n; \gamma + \delta + n; \frac{1}{1+d}\right). \tag{3.9}$$

The continuous-time analog of the BG is the Pareto distribution of the second kind (Pareto Type II), one characterization of which is a gamma mixture of exponentials. In other words, it is assumed that (1) the duration of the customer's relationship with the firm is characterized by an exponential distribution with rate parameter λ, and (2) individual differences in λ are characterized by a gamma distribution with shape parameter s and scale parameter β. The associated survivor function is:

$$S(t|s, \beta) = \left(\frac{\beta}{\beta + t}\right)^s, \quad t \geq 0. \tag{3.10}$$

The continuous-time analog of the churn rate is the hazard function. The hazard function of the Pareto Type II is:

$$h(t) = \frac{s}{\beta + t},$$

which decreases as a function of time (i.e., it exhibits negative duration dependence), even though the hazard function associated with the individual-level exponential distribution is constant over time. Thus, once again, the aggregate duration dependence is simply the result of a sorting effect in a heterogeneous population, as is the case with the BG model in discrete-time settings.

Substituting equation (3.10) in equations (3.3) and (3.4), and solving the integrals gives us:

$$DEL(s, \beta, d_c) = \beta^s d_c^{s-1} \Psi(s, s; \beta d_c) \tag{3.11}$$

$$DERL(s, \beta, d_c; \text{tenure of } t')$$
$$= (\beta + t')^s d_c^{s-1} \Psi(s, s; (\beta + t')d_c) \tag{3.12}$$

where $\Psi(\cdot)$ is the confluent hypergeometric function of the second kind and d_c is the continuously compounded discount rate. (A discrete-time discount rate of d is equivalent to a continuously compounded rate of $d_c = \ln(1+d)$. If the data are recorded in time units such that there are k periods per year ($k = 52$ if the data are recorded in weekly units of time), an annual discount rate of $(100 \times r)$ percent maps to a continuously compounded rate of $d_c = \ln(1+r)/k$.)

One stream of CE research has focused on the problem of valuing a customer base in contractual settings (e.g., Bauer, Hammerschmidt, and Braehler, 2003; Gupta and Lehmann, 2003; Gupta, Lehmann, and Stuart, 2004; Wiesel, Skiera, and Villaneuva, 2008; Libai, Muller, and Peres, 2009; Pfeifer, 2011). CLV and subsequent CE calculations are performed using a given constant retention rate (which implies the parameter-driven model underpinning this work is the geometric distribution for contract durations). The justification for such an assumption, if ever given, is that published retention rate numbers appear to be fairly constant – see, for example, Figure 3.2, which plots two years of quarterly annualized churn numbers from Vodafone Germany.

How do we reconcile such data with the above comments about increasing retention rates? The following comment on a professional society's membership patterns hints at the answer: 'I am happy to report that 41% of new members who joined in 2011 renewed their membership in 2012,

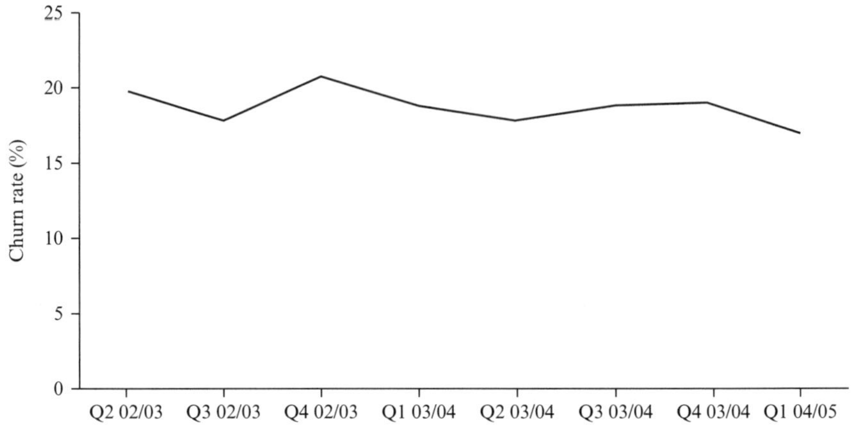

Source: Vodafone Germany, 'Vodafone analyst & investor day', presentation, 27 September 2004.

Figure 3.2 Annualized churn numbers for Vodafone Germany

and that ION has an overall retention of 78%' (Walter, 2013, p. 2). The aggregate retention rate numbers are a weighted average of the retention rates across all cohorts at any given time (i.e., a mix of 'young' and 'old' customers), and this weighted average will mask (and moderate) the within-cohort retention patterns, giving the analyst the impression that the retention dynamics are mild (and therefore possibly ignorable). This is illustrated in Figure 3.3, where the retention rate curve for each cohort is that associated with a beta-geometric distribution with parameters $\gamma = 0.764$ and $\delta = 1.296$.[4] Assuming the cohorts are of equal size and that the first cohort was acquired in period 0, the aggregate retention rate for any given period is simply the weighted average of each cohort-specific retention rate where the weights are the number of surviving cohort members at that point in time.

Does this mean we can 'assume away' the within-cohort retention dynamics as a minor source of noise when computing CE in contractual settings? The answer is a definite no. To illustrate, taking the aggregate retention rate in Figure 3.3 as a constant 0.833, Figure 3.4 compares the geometric survival curve given this constant retention rate with that implied by the common cohort-level retention curves plotted in Figure 3.3. Clearly any estimates of customer survival computed using the aggregate retention rate will be severely biased compared to those from a heterogeneous model, such as the BG, which allows for increasing cohort-level

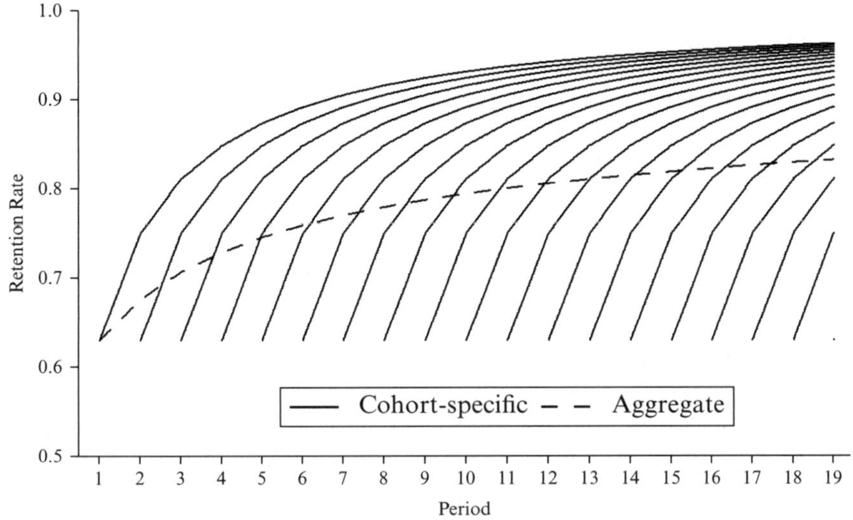

Figure 3.3 Aggregate vs cohort-level retention rates

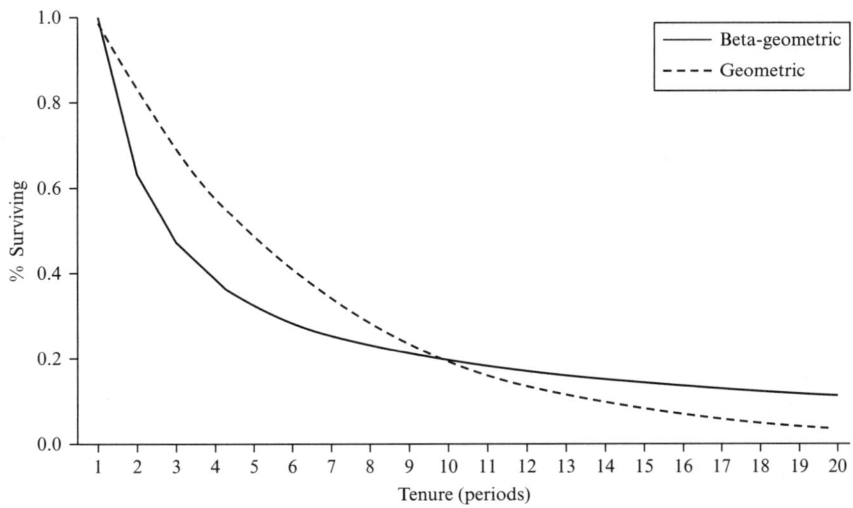

Figure 3.4 Comparing the geometric survival curve (computed using the constant aggregate retention rate) with the cohort-level curve (from the BG model)

retention rates. This has obvious implications for estimates of CLV and CE. As we extend into the future (e.g., beyond period 20 in Figure 3.4), the gap between the two curves will persist, leading to a substantial underestimation of CLV (and CE) for the homogeneous model.

Consider a setting in which the net cash flow per period is \$100 and the annual discount rate is 10 percent. Taking the aggregate retention rate as a constant 0.833, the expected residual lifetime value of a customer (computed at the end of the annual contract period) would be a constant \$343, regardless of how many periods the individual has been a customer. In contrast, any calculation of the expected residual lifetime value of a customer that acknowledges the phenomenon of increasing cohort-level retention rates will see the estimate increase as we lengthen in tenure of the customer. Using equation (3.9) we find that the expected residual lifetime value is \$288 for a customer approaching the end of their first contract period with the firm, \$394 for a customer approaching the end of their second contract period with the firm, and \$568 for a customer approaching the end of their fifth contract period with the firm. Fader and Hardie (2010) show that failing to account for cohort-level retention-rate dynamics will lead to systematically biased estimates of the residual value of the customer base (i.e., CE). Their analysis indicated that valuations performed using an aggregate retention rate will underestimate the true value of the customer base by a magnitude of 25–50 percent in standard settings.

Why then do we continue to see researchers performing CE calculations in contractual settings using a constant aggregate retention rate? Perhaps one reason is the belief that individual-level data or longitudinal data for each cohort are required in order to estimate the parameters of the model use to characterize contract durations (e.g., the BG distribution). It turns out that this need not be the case; Fader and Hardie (2007b), show that, under certain assumptions, all we require are data on the number of new subscribers and the total number of subscribers for each period. However, it is assumed that we have these data from the time the service of interest was launched on the market and that the reporting period is the same as the contract period. Fader, Hardie, and Liu (2012) show how to deal with the case where the reporting period spans multiple renewal periods and the data series is left censored (i.e., we do not have the data from the time the service was launched).

NON-CONTRACTUAL SETTINGS

The retention rate is defined as 'the ratio of customers retained to the number at risk' (Farris et al., 2010, p. 156). As noted above, the defining characteristic of a non-contractual setting is that customer 'death' is unobserved (and unobservable). As such, we cannot compute a retention rate directly from customer purchasing data (i.e., without an explicit model of the latent attrition process). Two metrics we can compute in non-contractual settings are the repeat-buying rate and repurchase rate, defined as 'the percentage of brand customers in a given period who are also brand customers in the subsequent period' and 'the percentage of brand customers for a brand who repurchase that brand on their next purchase occasion' respectively (ibid., p. 48). The measures are frequently incorrectly referred to as retention rates; for example, the annual J.D. Power Customer Retention Study for automobiles talks of retention when they are actually computing a repurchase rate. Similarly, we can find a number of references to 'retention rates' in settings where the measure being discussed is clearly a repeat-buying rate (e.g., Ofek, 2002; Dev and Stroock, 2007). This has seen researchers unwittingly using these 'retention rate' numbers as if they were actually retention rates, using them in the basic formulas we associate with textbook discussions of CLV (Fader and Hardie, 2012). This leads to seriously biased estimates of CLV and CE.

If we track over time the repeat purchasing of a cohort of customers (acquired at the same time) in a non-contractual setting, we see the general pattern illustrated in Figure 3.5. Some individuals make frequent purchases; for others, there are long time intervals between their pur-

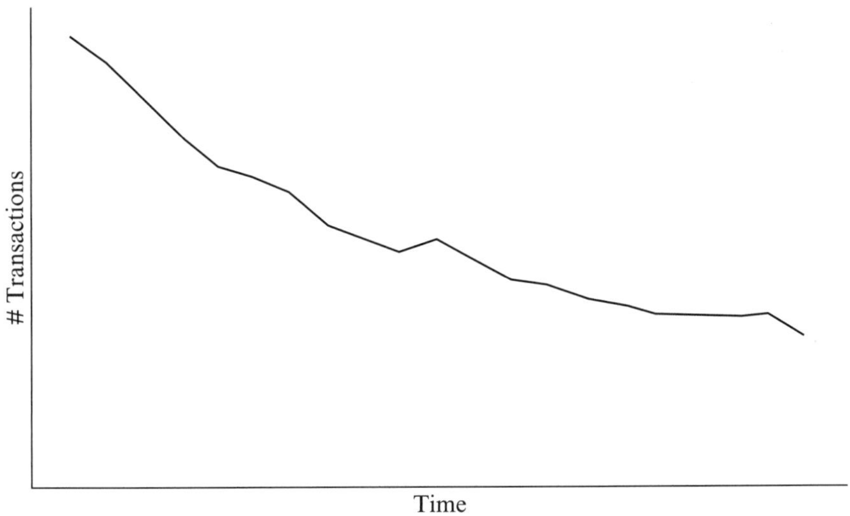

Figure 3.5 Stylized pattern of purchasing over time by a cohort of customers in a non-contractual setting

chases. (In contrast to a contractual setting, the fact that a customer has not made a purchase does not mean that they are 'dead'.)

While there are many non-stationary processes that could give rise to such a decline in repeat purchasing at the cohort level (e.g., Fader, Hardie, and Huang, 2004; Moe and Fader, 2004), one approach that has proven to be very popular is the latent attrition (or 'buy till you die') framework put forward by Schmittlein et al. (1987). A customer's relationship with a firm has two phases: they are 'alive' for an unobserved period of time, then 'dead'. Ignoring the effect of random purchasing around their means, individual customers purchase the product at steady but different underlying rates. At different unobservable points in time they 'die'.[5] This is illustrated in Figure 3.6, where each individual's purchasing while alive is represented by a dashed line, and the unobserved point at which they die by a diamond. The observed decline in repeat purchasing (the solid line) is due to the unobserved death of customers A, C, and D.

Schmittlein et al.'s (1987) operationalization of this framework is for a continuous-time setting and assumes that (1) while 'alive' the customer's purchasing is characterized by the negative binomial distribution (NBD) model (i.e., a gamma mixture of Poissons) with parameters r and α, and (2) the unobserved customer 'lifetimes' are treated as if random and are characterized by the Pareto Type II distribution (i.e., a gamma mixture of exponentials) with parameters s and β. Thus, the resulting four-parameter model of buyer behavior for non-contractual settings is called the Pareto/NBD.

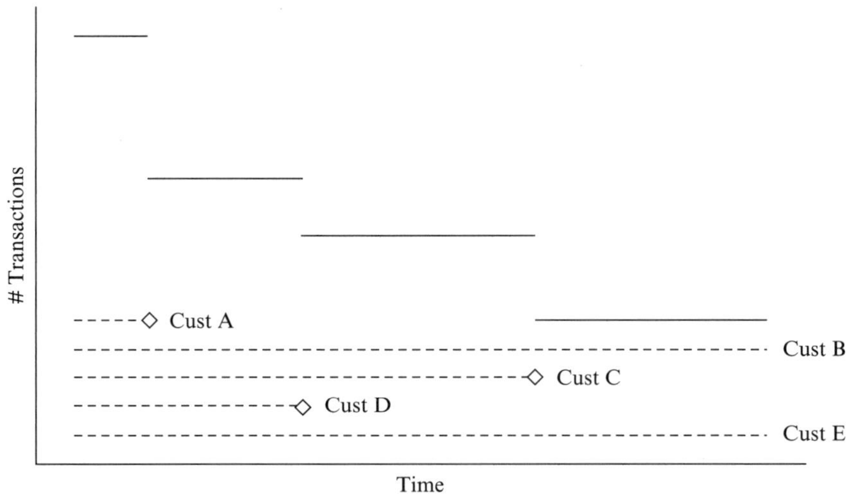

Figure 3.6 Visualizing the latent-attrition explanation for the decline in cohort-level purchasing in a non-contractual setting

Empirical validations of the model are presented in Schmittlein and Peterson (1994) and Fader et al. (2005b), amongst others. We find that its predictive performance is impressive, able to closely track the sales data in a hold-out period (at both the aggregate and the individual level). This gives us confidence in the Pareto/NBD model as the most appropriate starting point for CLV and CE calculations in a non-contractual setting. Two important early applications of the model are given by Reinartz and Kumar (2000, 2003), who used the Pareto/NBD model in their analysis of customer profitability.

The assumptions of the Pareto/NBD model mean that we do not need to know customers' entire purchase history; the only customer-level information required to estimate the four model parameters and then make individual-level predictions are 'recency' and 'frequency'. The notation used to represent this recency and frequency information is (x, t_x, T), where x is the number of transactions observed in the time interval $(0, T]$ and t_x $(0 < t_x \leq T)$ is the time of the last transaction.

One of the key expressions derived by Schmittlein et al. (1987) is $P(\text{alive} \mid x, t_x, T)$, the probability that an individual with observed behavior (x, t_x, T) is still 'alive' at time T. This quantity is of limited value when computing CLV or CE, as we need to know the probability that the customer is alive at all future points. Of greater use is the expression they derive for the conditional expectation, $E(X(T, T + t) \mid x, t_x, T)$, which is the expected number of transactions in the future period $(T, T + t]$ for an individual with observed behavior (x, t_x, T). This quantity can be used to compute DERT in the following manner:

$$DERT = \sum_{i=1}^{\infty}\left(\frac{1}{1+d}\right)^{i-0.5}\{E(X(T, T+i)|x, t_x, T) - E(X(T, T+i-1)|x, t_x, T)\}. \tag{3.13}$$

A more elegant solution is to substitute the Pareto/NBD model expressions for $t(t)$ and $S(t)$ in equations (3.5) and (3.6) and solve the integrals, giving us (Fader et al., 2005b; Jerath, Fader, and Hardie, 2014):

$$DET(r, \alpha, s, \beta, d_c) = \frac{r}{\alpha}\beta^s d_c^{s-1}\Psi(s, s; \beta d_c), \tag{3.14}$$

$$DERT(r,\alpha,s,\beta,d_c;x,t_x,T) = \frac{\Gamma(r+x+1)}{\Gamma(r)}\frac{\alpha^r\beta^s d_c^{s-1}}{(\alpha+T)^{r+x+1}} \times \frac{\Psi(s,s;(\beta+T)d_c)}{L(r,\alpha,s,\beta|x,t_x,T)}, \tag{3.15}$$

where $L(\cdot \mid \cdot)$ is the Pareto/NBD model likelihood function. (Note that the DET quantity is that for a just-acquired customer. To compute DET for an as-yet-to-be-acquired customer, we need to add 1 to this quantity (i.e., the purchase at time $t = 0$ that corresponds to the customer's first-ever purchase with the firm, and which starts the repeat purchase clock).)

To illustrate the importance of distinguishing between DET and DERT, let us consider an example where $r = 0.553$, $\alpha = 10.578$, $s = 0.606$, and $\beta = 11.669$.[6] Assuming an annual discount rate of 10 percent, DET = 4.8. The value of DERT obviously depends on the customer's exact purchase history. For the case of customers acquired one year ago, we evaluate equation (3.15) for all combinations of recency ranging from $t_x = 0$ to 52 weeks and frequency ranging from 0 to 10 transactions. For the case of customers acquired two years ago, we evaluate equation (3.15) for all combinations of recency ranging from $t_x = 0$ to 104 weeks and frequency ranging from 0 to 20 transactions. The corresponding contour (or 'iso-value') plots are presented in Figure 3.7. Each curve in the plots links customers with equivalent future value (in terms of discounted transactions) despite differences in their prior behavior. Depending on the age of the customer and their past purchasing behavior, their residual value ranges from almost zero to an order of magnitude greater than DET.

In order to compute CLV or RLV using the Pareto/NBD model, we need to augment equations (3.14) and (3.15) with a model for spend per

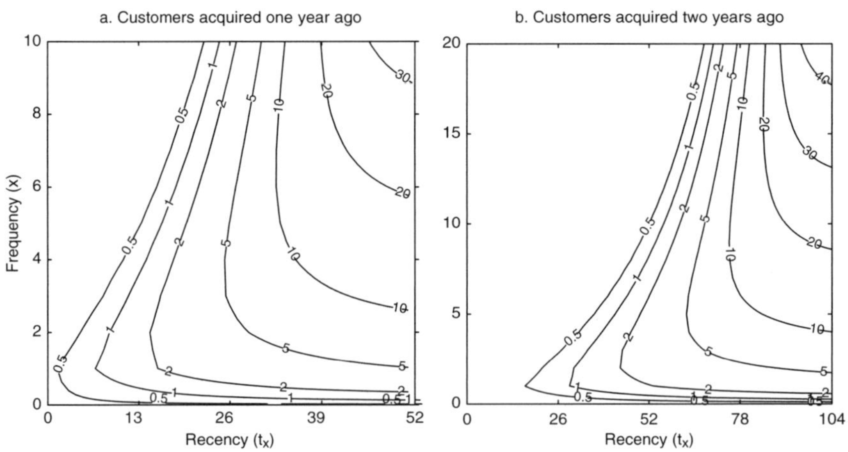

Figure 3.7a,b DERT as a function of recency and frequency for customers acquired one and two years ago

transaction. Fader et al. (2005b) proposed the three-parameter 'gamma mixture of gammas' model for spend per transaction – also see Colombo and Jiang (1999) and Fader and Hardie (2013) – that can be used to estimate an individual customer's underlying mean transaction value given their average transaction value over the x observed transactions. This means we have closed-form expressions for the residual lifetime value of a customer conditional on the so-called RFM summary measures of their behavior: recency (the time of their last transaction), frequency (the number of transactions observed in the calibration period), and monetary value (average spend per transaction in the calibration period).

In some situations we do not have access to individual-customer-level recency and frequency data; for example, we may only have a series of cross-sectional summaries, such as those reported in the Tuscan Lifestyles case (Mason, 2003). By only relying on data in histogram form, any individual-level information about each customer (e.g., recency and frequency) is lost, which may lead some to think that we cannot apply the Pareto/NBD model. Fader, Hardie, and Jerath (2007) and Jerath et al. (2014) show how the model parameters can still be estimated using such 'repeated cross-sectional summary' data, despite this limitation. As such, we are able to compute CE in the absence of the raw individual-customer-level data (or the associated recency and frequency summaries).

Despite being published in 1987, the Pareto/NBD saw relatively limited 'real-world action' for a number of years, the major problem being perceived challenges with respect to parameter estimation. To address this problem, Fader, Hardie, and Lee (2005a) developed a variant of the Pareto/NBD model that they called the BG/NBD. Changing the 'death' story to one where a customer could only 'die' immediately after a transaction (as opposed to any point in continuous time as for the Pareto/NBD) and assuming that this time-to-death in 'transaction time' is characterized by the BG distribution with parameters a and b, resulted in a model that is vastly easier to implement. In fact, its key selling point is that its parameters can be obtained quite easily using Microsoft Excel. As the two models yield very similar results in a wide variety of purchasing environments, the BG/NBD can be viewed as an attractive alternative to the Pareto/NBD in most applications, including as the basis for CLV and CE calculations (e.g., Lerner, Fader, and Hardie, 2012; Song, Kim, and Kim, 2013).

Note that the BG/NBD model expressions for DET and DERT are very messy. It makes more sense to compute DERT using the model-based conditional expectations, computed using:

$$E(X(T,T+t)|r,\alpha,a,b;x,t_x,T,) = \left(\frac{a+b+x-1}{a-1}\right)$$

$$\times\left[1-\left(\frac{\alpha+T}{\alpha+T+t}\right)^{r+x}{}_2F_1\left(r+x,b+x;a+b+x-1;\frac{t}{\alpha+T+t}\right)\right]$$

$$\div\left[1+\delta_{x>0}\frac{a}{b+x-1}\left(\frac{\alpha+T}{\alpha+t_x}\right)^{r+x}\right], \qquad (3.16)$$

in conjunction with equation (3.13). DET can be computed in a similar manner using the unconditional expectation formula:

$$DET = \sum_{t=1}^{\infty}\left(\frac{1}{1+d}\right)^{i-0.5}\{E(X(t)|r,\alpha,a,b)$$

$$-E(X(t-1)|r,\alpha,a,b)\}, \qquad (3.17)$$

where the expected number of purchases in the interval (0, t] is:

$$E(X(t)|r,\alpha,a,b) = \frac{a+b-1}{a-1}\left[1-\left(\frac{\alpha}{\alpha+t}\right)^{r}{}_2F_1\left(r,b;a+b-1;\frac{t}{\alpha+t}\right)\right]. \qquad (3.18)$$

The Pareto/NBD and its extensions are for non-contractual settings where transactions can occur at any point in time. As reflected in Figure 3.1, such an assumption does not always hold (e.g., an annual conference can only be attended at a discrete point in time). In other situations, the transaction can perhaps occur at any point in time but is treated as discrete by management. For example, a charity may record the behavior of each member of its supporter base in terms of whether or not they responded to the year i fund drive, even though the check could be received at any point in calendar time. Even for business settings that are truly non-contractual/continuous-time in nature, management may wish to discretize them for ease of summarization or data storage; this is particularly appropriate for very rare events. (See, for example, Berger, Weinberg, and Hanna's (2003) characterization of the 'repeat cruising' behavior of the customers of a cruise ship company in terms of whether or not they made a repeat cruise in each of the four years following the year of their first-ever cruise with the company.)

In these discrete-/discretized-time settings, a customer's transaction history can be expressed as a binary string, where 1 indicates that a transaction took place at the discrete point in time (or during the specified time interval), 0 otherwise. As in the case of the continuous-time setting, the challenge facing the modeler is to determine whether a trailing sequence of 0s reflects a 'dead' customer or simply one that is in the middle of a long hiatus since the last transaction. Assuming the latent-attrition/'buy till you die' framework illustrated in Figure 3.6, Fader, Hardie, and Shang (2010) develop a model in which, while 'alive', the customer's purchasing is characterized by a Bernoulli process with beta heterogeneity (i.e., the BB model with parameters α and β), and the unobserved customer 'lifetimes' are treated as if random and characterized by the beta-geometric distribution (i.e., a beta mixture of geometrics with parameters γ and δ). The BB is the discrete-time analogue of the NBD, and the BG is the discrete-time analogue of the Pareto Type II. Thus, the resulting BG/BB model of buyer behavior is the discrete-time analogue of the Pareto/NBD, which tends to the Pareto/NBD when the length of the discrete time period tends to zero. As with the Pareto/NBD and BG/NBD models, it turns out that the BG/BB model does not require information on when each of the x transactions occurred (i.e., the complete binary string of purchases); the only customer-level information required to estimate the four model parameters and then make individual-level predictions are recency and frequency, denoted by (x, t_x, n) for a calibration period of length n discrete/discretized time-periods. As with the other models discussed in this review, the predictive performance of the BG/BB model is impressive, able to closely track the sales data in a holdout period (at both the aggregate and the individual level). As such, we have confidence in using it as the basis for CLV and CE calculations in discrete-time non-contractual settings. Such calculations would be based off the following expressions:

$$DET(\alpha, \beta, \gamma, \delta, d) = \left(\frac{\alpha}{\alpha+\beta}\right)\left(\frac{\delta}{\gamma+\delta}\right)\left(\frac{1}{1+d}\right) \times {}_2F_1(1, \delta+1; \gamma+\delta+1; \tfrac{1}{1+d}) \tag{3.19}$$

$$DERT(\alpha, \beta, \gamma, \delta, d; x, t_x, n) = \frac{B(\alpha+x+1, \beta+n-x)}{B(\alpha, \beta)} \frac{B(\gamma, \delta+n+1)}{B(\gamma, \delta)(1+d)} \times \frac{{}_2F_1\left(1, \delta+n+1; \gamma+\delta+n+1; \frac{1}{1+d}\right)}{L(\alpha, \beta, \gamma, \delta \mid x, t_x, n)} \tag{3.20}$$

where $L(\cdot|\cdot)$ is the BG/BB model likelihood function. (Note that the DET quantity is that for a just-acquired customer. To compute DET for an as-yet-to-be-acquired customer, we need to add 1 to this quantity (i.e., the purchase at time $t = 0$ that corresponds to the customer's first-ever purchase with the firm, and which starts the 'transaction opportunity' clock).)

REFLECTIONS

At the heart of any attempt to calculate CLV and CE is a model that can be used to make 'multi-period forecasts' of buyer behavior. We have identified a core set of simple probability models of buyer behavior well suited to this task; these models can be classified according to (1) whether they are designed to characterize buyer behavior in contractual versus non-contractual settings, and (2) whether the behavior of interest is best characterized as occurring in discrete or continuous time – see Figure 3.8 for a summary. We have deliberately limited our discussion to the core set of models for each quadrant. A discussion of the various extensions to these basic models can be found in Fader and Hardie (2009).

Whenever we are computing CLV (and therefore CE), it is crucially important that we distinguish between the expected lifetime value of a new customer and the expected residual lifetime value of an existing customer. Failure to make this distinction leads to severely biased estimates in the value of the firm's customer base. Making the (common) simplifying assumption of constant net cash flow per contract period or transaction,[7] the key quantities of interest to the analyst are DEL and DERL in a contractual setting, and DET and DERT in a non-contractual setting. We have presented the equations for computing these quantities for the five core models presented in Figure 3.8.

With all of these models, observed behavior is modeled as a function of an individual's latent behavioral characteristics using simple probability distributions (or combinations thereof), with other probability distributions used to characterize heterogeneity in these latent characteristics. The resulting mixture models are 'simple' (Fader and Hardie, 2005): they are parsimonious (i.e., have very few parameters) and have limited data requirements (e.g., very simple data structures). With the exception of the Pareto/NBD, all the models can be implemented from scratch in an Excel spreadsheet; Matlab code[8] and an R package[9] are available for the Pareto/NBD model. Most importantly, the multi-period forecasting performance of these models has been established, which is fundamental to their use as the basis for any CLV and CE calculations.

Another thing these models have in common is the absence of any covariates that reflect the marketing activities of the firm. Any analyst using a with-covariates model will need to forecast the values of these

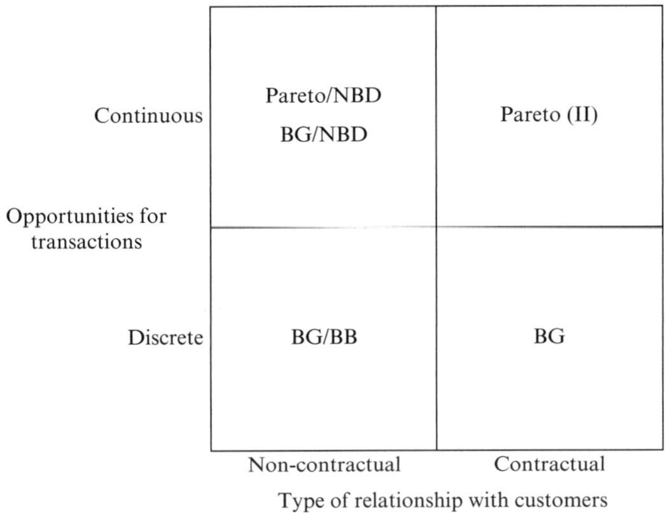

Figure 3.8 *The core set of probability models for computing CLV by type of customer base*

covariates far into the future for any true CLV or CE calculations. This clearly introduces a lot of additional noise into the exercise. The documented multi-period forecasting performance of these models indicates that the absence of marketing covariate effects is not a fundamental flaw. In fact, we view it as a positive characteristic. The only real downside is that the models cannot be used for any CE-based resource allocation exercise. It is worth noting that as covariates are incorporated, data structures and model estimation issues become more complex; if customers have been targeted with different marketing activities on the basis of their past behavior, we must also account for endogeneity (Schweidel and Knox, 2013).

Reflecting on the gap between what marketing academics create and what managers use, Urban and Karash (1971, p. 62) proposed an evolutionary approach to the development of marketing models: 'The introduction of models as an evolutionary development from simple to more complex but a related one would foster managerial acceptance, encourage an orderly development of data and analysis systems, and reduce risks of failure'. The models considered in the chapter represent an easy first step into the world of developing the models that can be used to make the multi-period forecasts of buyer behavior required for the calculation of CLV and CE.

NOTES

1. See the discussion of scoring models in Fader, Hardie and Lee (2005b) for further issues concerning the use of observation-driven model specifications.
2. This distinction is not the same as the 'lost-for-good' versus 'always-a-share' classification used elsewhere in the literature. See Fader and Hardie (2014) for a discussion of this matter.
3. Further dimensions of possible interest in contractual settings include whether usage while under contract is observed or unobserved, and whether the revenue associated with the contract is known in advance or not known in advance. These factors may be determined by technology and the firm's pricing policies (Ascarza, 2009).
4. These parameters come from fitting the BG model to the first five observations of the 'regular' dataset used in Fader and Hardie (2007a).
5. What lies behind this death? It could be a change in customer tastes, financial circumstances, and/or geographical location, the outcome of bad customer service experiences, or even physical death, to name but a few possible causes. But given the modeling objectives, *why* this death occurs is of little interest to the researcher focusing on CLV and CE, whose primary goal is to ensure that the phenomenon is captured in the model in order to make accurate forecasts.
6. These are the parameters associated with the CDNOW dataset as reported in Fader, Hardie, and Lee (2005a).
7. This assumption is not as simple as it first may seem. First, we are not assuming that $\bar{v}$ is constant across individuals. For example, the gamma-gamma spend model discussed above explicitly allows for heterogeneity between customers. Second, the assumption of stationarity can be relaxed without too much difficulty. For example, if we assume that $v(t)$ grows over time at a constant rate of $(100 \times g)$ percent per period, we can still use $\bar{v}$ and replace the discount rate d with $d' = (d - g)/(1 + g)$.
8. See http://brucehardie.com/notes/008/; last accessed 9 September 2014.
9. See http://cran.r-project.org/package=BTYD; last accessed 9 September 2014.

REFERENCES

Ascarza, E. (2009), 'Modeling customer behavior in contractual settings', unpublished PhD thesis, University of London.

Bauer, H.H., M. Hammerschmidt and M. Braehler (2003), 'The customer lifetime value concept and its contribution to corporate valuation', *Yearbook of Marketing and Consumer Research, Vol. 1*, Berlin: GfK-Nürnberg, e.V., Duncker & Humblot, pp. 47–67.

Berger, P.D., B. Weinberg and R.C. Hanna (2003), 'Customer lifetime value determination and strategic implications for a cruise-ship company', *Journal of Database Marketing & Customer Strategy Management*, **11**(1), 40–52.

Berry, M.J.A. and G.S. Linoff (2011), *Data Mining Techniques*, 3rd edition, Indianapolis, IN: Wiley Publishing, Inc.

Blattberg, R.C. and J. Deighton (1996), 'Manage marketing by the customer equity test', *Harvard Business Review*, **74**(4), 136–44.

Braun, M. and D.A. Schweidel (2011), 'Modeling customer lifetimes with multiple causes of churn', *Marketing Science*, **30**(5), 881–902.

Colombo, R. and W. Jiang (1999), 'A stochastic RFM model', *Journal of Interactive Marketing*, **13**(3), 2–12.

Cox, D.R. (1981), 'Statistical analysis of time series: some recent developments', *Scandinavian Journal of Statistics*, **8**(2), 93–108.

Dev, C.S. and L.M. Stroock (2007), 'Rosewood Hotels & Resorts: branding to increase customer profitability and lifetime value', *Harvard Business Publishing HBP No. 2088*, 15 June 2007.

Fader, P. (2012), *Customer Centricity: Focus on the Right Customers for Strategic Advantage*, 2nd edition, Philadelphia, PA: Wharton Digital Press.
Fader, P.S. and B.G.S. Hardie (2005), 'The value of simple models in new product forecasting and customer-base analysis', *Applied Stochastic Models in Business and Industry*, **21**(4–5), 461–73.
Fader, P.S. and B.G.S. Hardie (2007a), 'How to project customer retention', *Journal of Interactive Marketing*, **21**(1), 76–90.
Fader, P.S. and B.G.S. Hardie (2007b), 'Fitting the sBG model to multi-cohort data', accessed 9 September 2014 at http://brucehardie.com/notes/017/.
Fader, P.S. and B.G.S. Hardie (2009), 'Probability models for customer-base analysis', *Journal of Interactive Marketing*, **23**(1), 61–9.
Fader, P.S. and B.G.S. Hardie (2010), 'Customer-base valuation in a contractual setting: the perils of ignoring heterogeneity', *Marketing Science*, **29**(1), 85–93.
Fader, P.S. and B.G.S. Hardie (2012), 'Reconciling and clarifying CLV formulas', accessed 9 September 2014 at http://brucehardie.com/notes/024/.
Fader, P.S. and B.G.S. Hardie (2013), 'The gamma-gamma model of monetary value', accessed 9 September 2014 at http://brucehardie.com/notes/025/.
Fader, P.S. and B.G.S. Hardie (2014), 'The Pareto/NBD is not a "lost-for-good" model', accessed 7 September 2014 at http://brucehardie.com/notes/031/.
Fader, P.S., B.G.S. Hardie and C.-Y. Huang (2004), 'A dynamic changepoint model for new product sales forecasting', *Marketing Science*, **23**(1), 50–65.
Fader, P.S., B.G.S. Hardie and K. Jerath (2007), 'Estimating CLV using aggregated data: the Tuscan Lifestyles case revisited', *Journal of Interactive Marketing*, **21**(3), 55–71.
Fader, P.S., B.G.S. Hardie and K.L. Lee (2005a), '"Counting your customers" the easy way: an alternative to the Pareto/NBD model', *Marketing Science*, **24**(2), 275–84.
Fader, P.S., B.G.S. Hardie and K.L. Lee (2005b), 'RFM and CLV: using iso-value curves for customer base analysis', *Journal of Marketing Research*, **42**(4), 415–30.
Fader, P.S., B.G.S. Hardie and Y. Liu (2012), 'Bringing heterogeneity into customer equity calculations', presentation at the 34th ISMS Marketing Science Conference, 7–9 June.
Fader, P.S., B.G.S. Hardie and S. Sen (2014), 'Stochastic models of buyer behavior', in R. Winer and S. Neslin (eds), *The History of Marketing Science*, Singapore: World Scientific Publishing, pp. 165–205.
Fader, P.S., B.G.S. Hardie and J. Shang (2010), 'Customer-base analysis in a discrete-time non-contractual setting', *Marketing Science*, **29**(6), 1086–108.
Farris, P.W., N.T. Bendle, P.E. Pfeifer and D.J. Reibstein (2010), *Marketing Metrics*, 2nd edition, Upper Saddle River, NJ: Wharton School Publishing.
Gupta, S. and D.R. Lehmann (2003), 'Customers as assets', *Journal of Interactive Marketing*, **17**(1), 9–24.
Gupta, S., D.R. Lehmann and J.A. Stuart (2004), 'Valuing customers', *Journal of Marketing Research*, **41**(1), 7–18.
Jerath, K., P.S. Fader and B.G.S. Hardie (2014), 'Customer-base analysis using repeated cross-sectional summary (RCSS) data', accessed 9 September 2014 at http://brucehardie.com/papers/025/.
Kumar, V. and J. Andrew Petersen (2012), *Statistical Methods in Customer Relationship Management*, Chichester, UK: John Wiley & Sons, Ltd.
Kumar, V. and W. Reinartz (2012), *Customer Relationship Management*, 2nd edition, Heidelberg: Springer.
Lemmens, A. and C. Croux (2006), 'Bagging and boosting classification trees to predict churn', *Journal of Marketing Research*, **43**(2), 276–86.
Lerner, S.P., P.S. Fader and B.G.S. Hardie (2012), '"Buy till you die" to improve operational efficiency: realigning the sales force using the BG/NBD', presentation at the 23rd Advanced Research Techniques Forum, 24–27 June.
Libai, B., E. Muller and R. Peres (2009), 'The diffusion of services', *Journal of Marketing Research*, **46**(2), 163–75.

Mason, C.H. (2003), 'Tuscan Lifestyles: assessing customer lifetime value', *Journal of Interactive Marketing*, **17**(4), 54–60.
Moe, W.W. and P.S. Fader (2004), 'Capturing evolving visit behavior in clickstream data', *Journal of Interactive Marketing*, **18**(1), 5–19.
Neslin, S., S. Gupta, W. Kamakura, J. Lu and C.H. Mason (2006), 'Defection detection: measuring and understanding the predictive accuracy of customer churn models', *Journal of Marketing Research*, **46**(2), 204–11.
Ofek, E. (2002), 'Customer profitability and lifetime value', *Harvard Business School Note 9–503–01*, rev. 7 August 2002.
Parr Rud, O. (2001), *Data Mining Cookbook*, New York: John Wiley & Sons.
Pfeifer, P.E. (2011), 'On estimating current-customer equity using company summary data', *Journal of Interactive Marketing*, **25**(1), 1–14.
Pfeifer, P.E., M.E. Haskins and R.M. Conroy (2005), 'Customer lifetime value, customer profitability, and the treatment of acquisition spending', *Journal of Managerial Issues*, **17**(1), 11–25.
Potter, R.G. and M.P. Parker (1964), 'Predicting the time required to conceive', *Population Studies*, **18**(1), 99–116.
Reinartz, W. and V. Kumar (2000), 'On the profitability of long-life customers in a non-contractual setting: an empirical investigation and implications for marketing', *Journal of Marketing*, **64**(4), 17–35.
Reinartz, W. and V. Kumar (2003), 'The impact of customer relationship characteristics on profitable lifetime duration', *Journal of Marketing*, **67**(1), 77–99.
Risselada, H., P.C. Verhoef and T.H.A. Bijmolt (2010), 'Staying power of churn prediction models', *Journal of Interactive Marketing*, **24**(3), 198–208.
Rosset, S., E. Neumann, U. Eick and N. Vatnik (2003), 'Customer lifetime value models for decision support', *Data Mining and Knowledge Discovery*, **7**(3), 321–39.
Schmittlein, D.C. and R.A. Peterson (1994), 'Customer base analysis: an industrial purchase process application', *Marketing Science*, **13**(1), 41–67.
Schmittlein, D.C., D.G. Morrison and R. Colombo (1987), 'Counting your customers: who are they and what will they do next?', *Management Science*, **33**(1), 1–24.
Schweidel, D.A. and G. Knox (2013), 'Incorporating direct marketing activity into latent attrition models', *Marketing Science*, **32**(3), 471–87.
Song, T.H., J.Y. Kim and S.Y. Kim (2013), 'Optimality of customer relationship management: does profitability really matter?', *Asia Marketing Journal*, **15**(3), 141–57.
Urban G.L. and R. Karash (1971), 'Evolutionary model building', *Journal of Marketing Research*, **8**(1), 62–6.
Walter, T. (2013), 'Changing of the guard and thank you', *ION Newsletter*, Winter 2012–13, 2.
Wiesel, T., B. Skiera and J. Villanueva (2008), 'Customer equity: an integral part of financial reporting', *Journal of Marketing*, **72**(2), 1–14.

4. Incorporating dynamics in customer lifetime value models

Michael Lewis

INTRODUCTION

Over the past two decades the marketing literature has increasingly adopted the perspective that customer relationships should be viewed and managed as economically valuable assets (Blattberg and Deighton, 1996; Gupta, Lehmann, and Stuart, 2004; Rust, Lemon, and Zeithaml, 2004; Lewis, 2005a). In particular, customer equity (CE) and customer lifetime value (CLV) have been advanced as key CRM metrics related to the long-term value of customers. However, the most common CLV formulas contain significant limitations. For instance, many CLV models include assumptions of constant retention and revenue rates. Given that these inputs (retention and expected revenue) are likely to change based on consumer experiences and the firm's marketing tactics, the lack of explicit consumer dynamics in standard CLV models is likely to result in biased predictions. CLV projections based on these traditional models also limit the usefulness of the metric. If the goal is segmentation of customers based on expected CLV, then exclusion of marketing mix variables may be acceptable. In contrast, if CLV is a viewed as a goal to be maximized by the selection of marketing tactics then CLV calculations need to be executed within a dynamic optimization model rather than via the standard net-present-value–based formulas. In this chapter I focus on the challenges of modeling dynamic aspects of customer behavior and incorporating managerial decision-making in customer lifetime value modeling and CRM efforts.

The central fact that motivates the development of sophisticated dynamics in CLV and customer equity models is that consumer behavior is fundamentally dynamic. Dynamic aspects of customer behavior are important because if customers are evolving or changing over time then relationship marketing policies are essentially trying to appeal to a moving target. In other words, a challenge in CRM is that customers are intelligent entities who both learn from experience and develop expectations of the future. For example, common tactics such as promotional discounts can have significant long-term consequences. Promotional discounts may decrease future revenues by changing reference prices or by training consumers to wait for future promotions. However, modeling customers as intelligent and dynamically evolving actors greatly complicates the modeling of customer decisions.

In addition to the need to consider customer dynamics, the creation of optimal relationship marketing strategies is itself a challenging endeavor. If CLV, and by extension customer equity, is viewed as an objective to be maximized via the selection of marketing actions over time then relationship management becomes a dynamic optimization problem. The challenge in CRM applications is that dynamic optimization techniques suffer from the 'curse of dimensionality'. Simply, the curse of dimensionality means that as problems grow linearly in size the required computations grow multiplicatively. This is a significant problem in CRM applications because firms typically have a large number of marketing options and consumers may be best characterized using a large number of dimensions. For example, suppose that a cable TV firm characterizes customers based on transaction histories. This firm might possess data on the number of months as a subscriber, the current and previous packages of channels purchased, previous prices paid, previous promotional efforts, service complaints, whether other products are bundled with the service and other factors. The state space for even this simple illustration is the product of the number of levels for each dimension. The firm's action space might also be multidimensional and include decisions related to pricing, service quality, and equipment upgrades. Given the computational demands of dynamic programming models, the marketer's dynamic optimization problem often becomes intractable.

The purpose of the current chapter is to review the existing literature focused on incorporating dynamic elements into customer lifetime value and customer equity models. The chapter is structured as follows. I begin with a brief review of customer economics models. This review highlights the problematic nature of not including explicit dynamic structures in these calculations. The third section of the chapter then considers issues related to modeling the decision processes of forward-looking consumers. The fourth section of the chapter discusses the methods and challenges associated with computing firms' optimal dynamic marketing policies. These sections each include reviews of existing applications. The final section of the chapter discusses future areas for research. In that section, I consider questions relating to non-quantifiable elements of marketing such as branding, relaxations of assumptions about consumer rationality, and real-world implementation issues.

CLV AND CE BASICS

The last two decades have witnessed a significant change in the practice of marketing. As information and communications technologies have improved, firms have adopted a greater focus on relationship marketing. With this transition new metrics such as CLV have gained prominence. Gupta and Lehmann (2005, p. 15) define CLV as 'the present value of all current future profits generated from a customer over the life of his or her business with a firm'. CLV is therefore a measure of the profitability of individual customers. Berger and Nasr (1998) give the basic structural model of CLV as:

$$CLV = \sum_{t=0}^{T} \frac{\alpha^t (R_t - C_t)}{(1 + d)^t} \tag{4.1}$$

where t indexes time periods, R_t is the revenue from the customer in period t, C_t is the cost of serving a customer in period t, T is the total number of time periods used in the analysis, d is a single-period discount factor and α is a retention rate. Berger and Nasr (1998) also discuss procedures for adapting the calculations to account for variations in purchase frequency and revenue patterns. It should be noted, however, that the variations in purchase timing and revenue are incorporated by assuming specific deterministic patterns. A detailed summary of the customer lifetime value literature is provided by Jain and Singh (2002).

The primary value of this type of net present value approach to measuring customer relationship value is as a segmentation scheme. The logic is that the higher-valued customer should be given priority. This approach provides benefits as the firm is directed to link marketing investments to higher-value customer assets (Zeithaml, Rust, and Lemon, 2001). Where this approach falls short, however, is that it does not consider the marginal costs and benefits of investing in different types of customers. For example, a firm's most valuable customers may have very strong preferences for the firm's brands and products. Additional investments in this segment may earn limited rewards. In contrast, newer customers may have relatively low expected CLVs but may be more responsive to a firm's marketing and service efforts.

The basic CLV formulations suffer from several limitations related to consumer dynamics (Bolton, Lemon, and Verhoef, 2004). First, the expression does not account for the differences between individual customers. In particular, these expressions do not include customer transaction history variables. This is problematic since expected revenue and retention rates are likely to be strongly correlated with transaction history measures. This means the models do not account for the tendency of consumer preferences and loyalty to change with experience. It should be noted that the literature does include a number of CLV models that have begun to address this first issue. Schweidel, Fader, and Bradlow (2008) consider segment-level heterogeneity by using cohort-level information to predict CLV. Reinartz and Kumar (2000) use individual-level historical gross contribution and costs to account for individual differences. Fader, Hardie, and Lee (2005) use individual-level RFM (recency, frequency, and monetary value) measures to predict customer lifetime length and CLV.

Second, the formulas lack mechanisms for evaluating alternative marketing policies since retention and revenue rates are not treated as functions of marketing decisions. This is also a dynamic issue since static marketing policies are unlikely to maximize customer lifetime value. For example, Rust, Kumar, and Venkatesan (2011) and Kumar et al. (2008) calculate CLV using an assumption that future marketing contacts can be modeled as a function of lagged values of individual-level contacts and the customer's past purchase behavior. While this approach has the benefit of considering variations in individual-level marketing activities and costs, it does include an implicit assumption that marketing policies are not treated as decision variables of the firm.

The extension from customer lifetime value to customer equity is straightforward. Rust et al. (2004, p. 110) define customer equity as 'the total of the discounted lifetime values summed over all the firm's current and potential customers'. The preceding definition of customer equity with its emphasis on the discounted profitability of the firm's present and future customers highlights that a firm's value is the sum of the economic value of its customer relationships. Gupta and Lehmann (2005) emphasize this point and discuss how customer equity can be used to value firms and for merger and acquisition analyses. Other researchers have considered how customer equity can be used to guide marketing decision-making. Blattberg and Deighton (1996) use customer equity calculations to balance customer acquisition and retention spending levels while Rust et al. (2004) provide models for linking marketing investments in advertising, loyalty programs or service quality, to customer equity. Other work has examined the link between customer equity and firm value (Wiesel, Skiera, and Villaneuva, 2008).

CUSTOMER DYNAMICS

The limitations of the basic customer economics calculations discussed in the previous section suggest a need to include more sophisticated models of consumer behavior in the CLV equation. In particular, there is a need to include structures that include elements of consumer learning and evolution. Consumer learning, in particular, may be especially challenging to model as past experience may impact consumers' expectations of the future. For example, when brands frequently promote, consumers may develop expectations of future prices. This presents analysis challenges since consumer response is not based only on current market conditions but also on consumers' expectations of future prices and promotions.

Modeling consumers as dynamic or forward-looking agents is both logically appealing and analytically challenging. The basic premise of these models is to build a structure that replicates the decision-making process of a consumer when dynamic incentives exist. As such, the model needs to include the factors that affect the attractiveness of a consumer's options at a given time and a structure that captures dynamic considerations. A basic model begins with a customer (customer-specific indices are dropped for clarity) making repeated choices among J mutually exclusive alternatives in each of T discrete periods. The choice set is defined based on the specific application and may include order incidence, order size, brand choice, contract length, and other types of consumer decisions.

The customer's decision is likely to be based on marketing factors such as prices and individual-level factors such as inventories or previous experience. A myopic decision-maker would select from the options 1,..,J, by considering the relative benefits associated with each option, denoted as R_j. The standard assumption in choice models is that consumers select the alternative that yields the greatest benefits for the purchase occasion. However, when dynamic incentives exist consumers may balance current utility with expectations of future rewards. For example, consumers may choose unfamiliar brands in order to learn about alternative brands or pay a premium to make a purchase from a firm offering a loyalty program in the hopes of earning a future reward. If rewards are earned over time, a multiple-period objective function is proper for the customer's decision problem. The objective function of a dynamically oriented customer is given as:

$$\max E\left[\sum_{t=1}^{T} \alpha^{t-1} \sum_{j=1}^{J} R_j(t) d_j(t) \mid S(t)\right] \tag{4.2}$$

where $R_j(t)$ is the single-period reward associated with alternative j at time t, $d_j(t)$ is an indicator variable set equal to 1 if option j is chosen at time t, and zero otherwise, and α is a one-period discount factor. An important element of this formulation is that the reward functions $R_j(t)$

are conditional on the state of the environment, $S(t)$. The state space, $S(t)$, is a vector of information about the environment that is relevant to the customer's forward-looking optimization problem. The state space may consist of elements of the marketing mix, such as the pricing environment in a given week, and customer-specific information such as cumulative purchases.

The customer's decision problem involves selecting the option in each period that maximizes the expected discounted utility for the remainder of the time horizon. This decision is represented by the value function, V, which is defined as the maximum value of discounted expected utility over the horizon:

$$V(S(t)) = \max_{d(t)} E\left[\sum_{\tau=t}^{T} \alpha^{\tau-t} \sum_{j=1}^{J} R_j(\tau) d_j(\tau) | S(\tau)\right] \tag{4.3}$$

The value function is the utility that can be achieved over the time horizon if the individual selects the optimal sequence of quantities. The value function may also be written in terms of alternative specific expected value functions as $V(S(t)) = \max_{j \in J} V_j(S(t))$, where the alternative specific value function $V_j(S(t))$ is the expected value of selecting option j at time t and then selecting optimal actions subsequently. Furthermore, the alternative specific value functions satisfy the Bellman equation (Bellman, 1957) given in equation (4.4):

$$V_j(S(t)) = E[R_j(t)|S(t)] + \alpha E[V(S(t+1)|S(t), d_j(t) = 1],$$

$$\text{when } t = 1, \ldots, T-1 \text{ and } V_j(S(T)) = E[R_j(T)|S(T)], \text{ at } t = T. \tag{4.4}$$

The form of the alternative specific value functions implies that decisions are based on both the immediate utility provided by an alternative and the expected total future utility from subsequent purchases and loyalty rewards. A key detail is that the expected future benefits from period t + 1 forward may depend on the alternative selected. In general, the evolution of the state space will be a function of both the current environment $S(t)$ and the choices made by the customer $d_j(t)$ over time. Estimation of these types of dynamic structure models requires the definition of the consumer's single-period reward function and the specification of the consumer's expectations. Estimation is conducted using an approach analogous to that used in standard choice models. The significant point of departure is that in contrast to the simple comparisons of utilities used in a choice model, each comparison of alternatives requires the solution of the consumer's dynamic optimization problem. Typically, this means that estimation requires the nesting of a dynamic programming model within a maximum likelihood routine (Rust, 1994).

Applications

Dynamic programming-based choice models are increasingly common in the marketing literature as they are applicable to situations where consumers make sequential decisions under uncertainty. This literature includes a number of applications focused on different aspects of the marketing mix and across a range of industries. The following describes several applications of discrete choice dynamic programming to modeling consumer behavior.

Learning Erdem and Keane (1996) used a discrete choice dynamic programing model to study product trial and consumer learning. In this model consumers make current-period decisions based on estimates of immediate utility and also the value of learning about alternative brands for future consumption. The model is implemented using data on liquid detergents. The key to this market is that frequent product introductions result in consumer uncertainty. The authors find that adding a forward-looking component produces significantly better in-sample fit.

Price expectations The literature includes a number of studies that focus on consumer expectations of price and promotions. These chapters all focus on contexts where consumers attempt to minimize long-term costs by anticipating future deals or long-term price trends. Gönül and Srinivasan (1996) used a dynamic programming structure to model consumer expectations of promotions and stocking up behavior in the disposable diaper category. The key findings relate to how consumers use expectations of promotions and stockpiling strategies in order to minimize long-term expenditures. Gönül (1999) examined the role of price expectations in an over-the-counter medicine market. She found that the dynamic model improved fit relative to a myopic model and that price expectations influence consumers' current-period choice of brands. Lewis (2005b) modeled consumers' expectations about promotions in the context of newspaper subscription discounts. In this model, consumers may act strategically by opportunistically cancelling or allowing subscriptions to expire, based on beliefs about the availability of future discounts.

Loyalty programs Loyalty programs are unique among marketing promotions in that they are often set up to act as dynamic incentive schemes. This is a critical factor for modeling response to these programs because consumers may be motivated by loyalty program rewards over an extended time period. As such loyalty programs may cause consumers to shift from single-period to sequential decision-making; Lewis (2004) examined consumer response to an online grocer's loyalty program. The key to this formulation is that expectations of reward program participation are a function of the changing values of the offered rewards, cumulative purchasing (points), and the program time constraints. Kopalle et al. (2012) used a similar dynamic structure to model response to an airline loyalty program that includes rewards and status tiers based on cumulative purchases.

Replacement decisions In many high-technology categories, replacement decisions are an important element of overall demand. These replacement decisions often include an element of forward-looking behavior since consumers may have expectations of future price changes and technology improvements. Gordon (2009) estimated a model of computer replacement decisions. This model involves consumers' expectations of both future quality and price levels.

CLV and Dynamic Structural Models

The preceding dynamic programming-based models of consumer behavior do not directly yield optimal marketing policies. From the firm's perspective the creation of dynamic marketing policies that take into account customer evolution is a challenging technical problem. However, the models do provide estimates of consumer primitives (utility function, subjective beliefs about uncertain future states) and so they may be used to evaluate policy changes (Lucas, 1976).

Simulation to evaluate the impact of alternative marketing policies on customer lifetime value has been used in the context of loyalty programs (Lewis, 2004), high-tech product sales, and maintenance (Venkatesan, Kumar, and Bohling, 2007), stock-outs (Jing and Lewis, 2011), and sales promotions (Lewis, 2005b). While these papers have used simulations to explicitly consider the impact of marketing actions on customer lifetime value, other researchers have used the results of structural models to evaluate the impact of marketing variables on other marketing objectives. For example, Erdem and Keane (1996) use simulations to predict the effect of advertising policies on brand equity. In this and other papers, it would be straightforward to use simulation studies to also evaluate the impact of alternative marketing policies on CLV.

OPTIMAL DYNAMIC MARKETING POLICIES

The idea of using optimization techniques to maximize the value of customers is intuitive and has been contemplated since at least the 1950s (Howard, 2002). However, the computational challenges associated with dynamic programming and the costs of maintaining large-scale, detailed customer databases have left these types of techniques infeasible until only recently. A first step in relating marketing decisions to CLV is to acknowledge that the key elements of the CLV equations are dependent on marketing tactics. In equation (4.5), equation (4.1) is rewritten to make explicit that retention rates, revenues, and costs may all be functions of marketing activity, M:

$$CLV = \sum_{t=0}^{T} (\alpha(M_t)^t) \times (R_t(M_t) - C_t(M_t)) \tag{4.5}$$

Writing the equation for CLV as a function of marketing actions alters how CLV is viewed. Rather than as a static quantity that may be used to segment customers, CLV is now an objective to be maximized by selecting marketing actions. For example, pricing may increase per period revenues but simultaneously decrease retention. The proper pricing strategy therefore involves balancing short-term revenue with long-term retention.

In addition to including marketing actions in the CLV calculation it is also useful to include observable customer data. For example, Dwyer's (1989) migration model uses recency or time since last purchase to better account for differences in retention rates across customers. Other salient transaction history variables include a customer's past frequency of buying and the monetary value of past transactions. These three measures are collectively referred to as RFM and are frequently used in database marketing applications (Hughes, 2001).

Furthermore, as transaction history measures may be reflective of underlying customer preferences it may be useful to select marketing actions based on past customer behavior. Specifically, marketing tactics may have differential effects across transaction-history–based customer segments. This is salient as it suggests that current behavior may be determined by interactions between past buying activity and marketing actions. For instance, there is some belief that long-term customers may be less price sensitive than newer customers (Reichheld and Teal, 1996). Equation (4.6) makes explicit that customer behavior may be a function of a set of state variables, X_t, that describe a customer at time t:

$$CLV = \sum_{t=0}^{T} (\alpha(M_t,X_t)^t) \times (R_t(M_t,X_t) - C_t(M_t,X_t)) \tag{4.6}$$

Maximizing Customer Lifetime Value

Equation 4.6 provides a starting point for the idea that the goal of CRM is to determine the sequence of marketing activities that maximize customer lifetime value. The multi-period time horizon of the problem combined with the probabilistic nature of consumer behavior mean that the development of optimal CRM marketing policies requires some form of stochastic dynamic optimization. Equation (4.6) provides an objective function for CRM applications.

In this section, I briefly describe how CRM can be formulated as a dynamic programming problem. The emphasis is on the general form of the problem rather than on solution techniques. For the general case, customers are classified into one of $x_t \in X$ possible states based on transaction history at each point in time, t. The firm is assumed to be able to select a marketing action or controls, m_t, from a set of feasible actions, M. Defining the single-period profit function as $\pi(x_t, m_t)$, a single-period discount factor as d, and the decision-horizon length as T, the firm's CRM objective function may be written as $\max_{m_t \in M} E\left\{\sum_{t=0}^{T} d^t * \pi(x_t, m_t)\right\}$. This objective function is the discounted sum of single-period rewards over a T period horizon. In this expression, the single-period profit, π, is equivalent to the revenue minus cost term $(R_t - C_t)$, in the previous CLV expressions.

The customer management goal is to determine the mapping of marketing policies to observed customer states $m(x_t)$ that maximizes CLV. The solution to the dynamic optimization problem is defined in terms of the value function, denoted as V. By definition the value function is the greatest feasible expected payoff from time t forward, given that the customer is in state x_t at time t. The value function is equivalent to the maximum expected CLV based on optimal selection of marketing policies:

$$V = CLV(x_0) = \max_{m_t \in M} E\left\{\sum_{t=0}^{T} d^t * \pi(x_t, m_t)\right\} \tag{4.7}$$

Thus far the objective function does not include a direct statement of how customers evolve over time. In equation (4.8) optimal customer value (CLV) is rewritten to make the relationship dynamics more explicit. In this equation $\Pr(x_{t+1}|x_t, m(x_t))$ is the probability a customer in state x_t transitions to state x_{t+1} given the marketing actions specified by $m(x_t)$:

$$CLV(x_t, m(x_t)) = E[\pi(x_t, m(x_t))] + \sum_{\tau=t+1}^{T} \sum_{x' \in X} \alpha^{\tau} * \pi(x'_\tau, m(x_\tau)) * Pr(x'_{\tau+1}|x_\tau, m(x_\tau)) \tag{4.8}$$

The first term on the right side is the expected one-period profit from applying the selected marketing tactics in the current time period.

The term involving the double summations represents the sum of the expected profits for future periods from $t + 1$ to T. This term includes the transition probabilities through the inner summation over the customer's potential future states (x'). The outer summation is over future time periods.

The term for the current period profits, $E[\pi(x_t, m_t)]$, is an expectation as customers' decisions are stochastic from the firm's perspective. Purchasing probabilities also define the state-to-state transition rates, $Pr(x_{t+1}|x_t, m_t)$, if the state variables include transaction history measures such as recency or frequency. Customer buying decisions are embedded in both the single-period profit and in the state-to-state transition probabilities. The solution of the dynamic programming problem may be accomplished using a variety of algorithms that vary based on whether the problem is formulated as having an infinite (e.g., policy evaluation or value iteration) or finite time horizon (e.g., backwards recursion) (see Bertsekas, 1995).

Applications

The preceding material describes the general approach to formulating CLV as a dynamic optimization problem. These types of models generate two related managerially relevant outputs. First, the models generate dynamically oriented marketing policies. Specifically, the models generate the sequence of marketing actions that maximize the profitability of a customer over time. The second output is an estimate of CLV. Unlike the estimates provided by equation (4.1) these estimates include the value of managerial flexibility in terms of marketing actions. Previous research has implemented dynamic optimal CRM policies for a variety of marketing mix elements:

Contact frequency The first applications of dynamic optimization to customer management were in the area of contact management (Bitran and Mondschein, 1996; Gönül and Shi, 1998; Simester, Sun, and Tsitsiklis, 2006). These models seek to determine the mailing policies that maximize customer lifetime value. For example, Simester et al. (2006) compute and then implement dynamic mailing strategies in a large-scale field test. They find that the dynamic policies improve profits by between 22 percent and 29 percent when the model assumes a 3 percent per month discount rate. Rust and Verhoef (2005) also optimize CLV by manipulating marketing communications. In this case they extend the contact frequency model and also consider whether communications should be relationship oriented versus action oriented.

Pricing Lewis (2005a) developed a dynamic programming for computing optimal relationship pricing policies in the context of subscriptions. The logic of this model is that product usage results in positive purchase feedback effects that result in changing consumer response parameters. If this is the case then the firm can use a dynamic programming model to determine the optimal sequence of prices to charge a customer. Lewis finds that the optimal relationship pricing policies consist of a deep acquisition discount followed by a shallower discount for first-time renewals. The logic of this policy is that a deep discount is needed to overcome the risks of trial and then a shallow discount is useful for reinforcing the relationship.

Promotion policy Khan, Lewis, and Sing (2009) investigate the value of implementing a dynamic one-to-one promotion policy relative to mass market, segment- and individual-level marketing policies. The marketing variables considered in this study include loyalty program points, short-term discounts and free shipping. Khan et al. use dynamic programming to determine the optimal sequence of these promotions based on models that include transaction history measures, latent class specific response parameters and individual response parameters. They find that the majority of gains can be achieved by optimizing based on transaction history measures.

DISCUSSION

The development of extensive customer data warehouses combined with advances in computer technology have made it increasingly possible for firms to consider dynamic factors in the valuation of customer assets and the creation of relationship-oriented marketing policies. There are, however, a number of remaining challenges. In this section I comment on three areas that offer promising opportunities for future research.

Studies of Qualitative Marketing Mix Elements

The preceding discussion tends to focus on quantifiable aspects of the marketing mix such as pricing or contact frequency. While these topics are important, there are a number of broad marketing issues that have not been considered in the literature. For example, branding is an area where marketing researchers have studied key aspects of brand–consumer relationships using psychological and anthropological methods (Fornier, 1998). The idea of a brand–consumer relationship has very obvious dynamic implications as most theories of relationships involve evolutionary stages such as initiation, experimenting, bonding, stagnating, and terminating (Knapp, 1978). Branding strategies may therefore benefit by being customized based on relationship stage. However, a complication of treating brand–consumer relationship development as a dynamic optimization problem is that it is difficult to identify a closed set of branding strategies.

The literature does include streams of research that may become the foundation for studies into CLV-maximizing branding policies. On the behavioral side, Aaker (1997) has proposed that brand personalities can be placed into five categories: ruggedness, sincerity, excitement, competence, and sophistication. This framework has dynamic implications if certain brand personalities are more resilient or more conducive to creating loyalty. For example, Aaker, Fournier, and Brasel (2004) find that 'exciting' brands were better able to recover following a transgression than 'sincere' brands.

One recent example of applying quantitative techniques to relationship building is the work on website morphing (Hauser et al., 2009). In website morphing, clickstream data is used to infer consumers' cognitive traits and to customize the presentation of information. In this work the authors design a variety of 'morphs'. These include morphs that emphasize detailed comparisons, different levels of information, or whether information should be graphical or verbal. While the website morphing applications have not used CLV as an objective, this type of approach has potential value as a CRM tool.

Consumer Rationality Assumptions

The dynamic models of customer behavior discussed above necessarily involve strong assumptions about customers' abilities to optimize utility over extended time periods. The assumptions that consumers are capable of and willing to devote the needed effort to optimize long-run utility are needed to formulate and solve the dynamic programming models of consumer decision-making. Much of the consumer behavior literature, of course, casts doubt on the notion that consumers act as dynamic optimizers. The economics literature also contains a significant body of research concerned with the bounded nature of consumer rationality (Conlisk, 1996; Kahneman, 2003). In terms of the marketing literature, Shugan (2006) discusses the role of rationality assumptions in models of consumer behavior.

The issue of consumer rationality is best viewed as a model specification issue. In the case of dynamic programming models, a frequent assumption is that consumers optimize over an infinite horizon. This assumption is usually justified based on the relative fit of the dynamic model relative to a myopic specification where future rewards are not considered. Similarly, constraints may operate in terms of the range of possible consumer decisions. It remains an open research question as to how notions of bounded rationality may be used to inform the development of dynamic models of consumer behavior.

Implementation Issues

Incorporation of dynamic models into customer economics metrics or CRM systems involves substantial challenges. These challenges are particularly daunting in the context of one-to-one marketing (Peppers and Rogers, 1993). The basic idea in one-to-one marketing is that the firm should customize marketing efforts based on individual-level data. The paper by Khan et al. (2009) is an example of this approach, as individual-level parameters are estimated using a hierarchical model and these parameters are fed into a dynamic programming model that calculates individually optimal promotions policies. This paper notes several significant implementation challenges.

For example, new customers are a challenge because the firm has little information available about the individual. Efforts to optimize CLV are therefore likely to be less effective since firms will not possess detailed individual data. The 'website morphing' paper from Hauser et al. (2009) addresses this same point by noting that customization needs to done based on few clicks. The development of methods that optimize the value of new and prospective customers is straightforward but may be of limited value since the firm has limited information about each customer's state. In these instances it is likely that the ideal approach would focus on simultaneously managing CLV while generating customer information. Methods for simultaneously managing rewards while learning are well established in operations research (Bertsekas, 1995; Sutton and Barto, 1998).

Another implementation issue is basic computation feasibility. In addition to dynamic programming models being time consuming to solve, firms may also need to solve the models for a large number of customer segments. The marketing literature has devoted a considerable effort to incorporating heterogeneity in choice models. At the most fundamental level firms can estimate individual-level response parameters (Rossi, McCulloch, and Alenby, 1996) and then optimize over individual customers. For firms with millions of customers, this type of individual-level marketing approach is unlikely to be feasible in the near term. For these types of problems it will be necessary to develop heuristic procedures that balance consumer heterogeneity with dynamic considerations.

CONCLUSIONS

This chapter focuses on the interaction of dynamic consumer behavior and customer economics management. The models and methods described above all begin from the premise that the past and current treatment of consumers impacts future customer behavior. The mechanisms by which past consumption and exposure to marketing instruments influence future behavior may differ across categories. In some cases product usage may create habits or other switching costs. In other categories consumers may learn from exposure to marketing activities such as short-term promotions and may begin to plan purchases to coincide with future promotions offers. When these types of dynamics effects exist, traditional CLV methods that do not include marketing activities and assume unchanging retention and revenue rates can lead to significant problems. First, these methods may result in biased CLV projections. Second, and perhaps more importantly, these simple net-present-value–type calculations do not provide an explicit link between marketing and the value of a firm's customer relationship assets. A primary benefit of viewing CLV as an objective to be maximized by selecting the ideal sequence of marketing actions is that the marketing function becomes the unambiguous manager of the firm's success.

REFERENCES

Aaker, J. (1997), 'Dimensions of brand personality', *Journal of Marketing Research*, **34**(3), 347–56.

Aaker, J., S. Fournier and A. Brasel (2004), 'When good brands do bad', *Journal of Consumer Research*, **31**(1), 458–69.

Bellman, R. (1957), *Dynamic Programming*, Princeton, NJ: Princeton University Press.

Berger, P. and N. Nasr (1998), 'Customer lifetime value: marketing models and applications', *Journal of Interactive Marketing*, **12**(1), 17–30.

Bertsekas, D.P. (1995), *Dynamic Programming and Optimal Control, Vol. 1*, 2nd edition, Belmont, MA: Athena Scientific.

Bitran, G.R. and S.V. Mondschein (1995), 'Mailing decisions in the catalog sales industry', *Management Science*, **42**(9), 1364–81.

Blattberg, R. and J. Deighton (1996), 'Manage marketing by the customer equity', *Harvard Business Review*, **74**(4), 136–44.

Bolton, R., K. Lemon and P. Verhoef (2004), 'The theoretical underpinnings of customer asset management: a framework and propositions for future research', *Journal of the Academy of Marketing Science*, **32**(3), 1–20.

Conlisk, J. (1996), 'Why bounded rationality?', *Journal of Economic Literature*, **34**(2), 669–700.

Dwyer, F. (1989), 'Customer lifetime value to support marketing decision making', *Journal of Direct Marketing*, **8**(2), 73–81.

Erdem, T. and M. Keane (1996), 'Decision-making under uncertainty: capturing dynamic brand choice processes in turbulent consumer markets', *Marketing Science*, **15**(1), 1–20.

Fader, P., B. Hardie and K.L. Lee (2005), 'RFM and CLV: using iso-value curves for customer base analysis', *Journal of Marketing Research*, **42**(6), 415–30.
Fornier, S. (1998), 'Consumers and their brands: developing relationship theory in consumer research', *Journal of Consumer Research*, **24**(4), 343–73.
Gönül, F. (1999), 'Estimating price expectations in the OTC medicine market: an application of dynamic stochastic discrete models to scanner panel data', *Journal of Econometrics*, **89**(1–2), 41–56.
Gönül, F. and M.Z. Shi (1998), 'Optimal mailing of catalogs: a methodology using estimable structural dynamic programming models', *Management Science*, **44**(9), 1249–62.
Gönül, F. and K. Srinivasan (1996), 'Estimating the impact of consumer expectations of coupons on purchase behavior: a dynamic structural model', *Marketing Science*, **15**(3), 262–79.
Gordon, B.R. (2009), 'A dynamic model of consumer replacement cycles in the PC processor industry', *Marketing Science*, **28**(5), 846–67.
Gupta, S. and D. Lehmann (2005), *Managing Customers as Investments: The Strategic Value of Customers in the Long Run*, Philadelphia, PA: Wharton School Press.
Gupta, S., D. Lehmann and J. Stuart (2004), 'Valuing customers', *Journal of Marketing Research*, **41**(1), 1–17.
Hauser, J., G. Urban, G. Liberali and M. Braun (2009), 'Website morphing', *Marketing Science*, **28**(2), 202–23.
Howard, R. (2002), 'Comments on the origin and application of Markov decision processes', *Operations Research*, **50**(1), 100–102.
Hughes, A. (2000), *Strategic Database Marketing*, 2nd edition, New York: McGraw-Hill.
Jain, D. and S. Singh (2002), 'Customer lifetime value research in marketing: a review and future directions', *Journal of Interactive Marketing*, **16**(2), 34–46.
Jing, X. and M. Lewis (2011), 'Stockouts in online retailing', *Journal of Marketing Research*, **48**(2), 342–54.
Kahneman, D. (2003), 'Maps of bounded rationality: psychology for behavioral economics', *The American Economic Review*, **93**(5), 1449–75.
Khan, R., M. Lewis and V. Sing (2009), 'Dynamic customer management and the value of one-to-one marketing', *Marketing Science*, **28**(6), 1063–79.
Knapp, M. (1978), *Social Intercourse: From Greeting to Goodbye*, Needham Heights, MA: Allyn and Bacon.
Kopalle, P.K., Y. Sun, S.A. Neslin, B. Sun and V. Swaminathan (2012), 'The joint sales impact of frequency reward and customer tier components of loyalty programs', *Marketing Science*, **31**(2), 216–35.
Kumar, V., R. Venkatesan, T. Bohling and D. Beckmann (2008), 'The power of CLV: managing customer lifetime value at IBM', *Marketing Science*, **27**(4), 585–99.
Lewis, M. (2004), 'The influence of loyalty programs and short-term promotions on customer retention', *Journal of Marketing Research*, **41**(3), 281–92.
Lewis, M. (2005a), 'Research note: a dynamic programming approach to customer relationship pricing', *Management Science*, **51**(6), 986–94.
Lewis, M. (2005b), 'Incorporating strategic consumer behavior into customer valuation', *Journal of Marketing* (Special Issue on Customer Relationship Management), **69**(3), 230–38.
Lucas, R.E. Jr. (1976), 'Econometric policy evaluation: a critique', in K. Brunner and A.K. Meltzer (eds), *The Phillips Curve and Labor Markets*, Carnegie-Rochester Conference on Public Policy, North-Holland: Amsterdam.
Peppers, D. and M. Rogers (1993), *The One to One Future*, New York: Doubleday.
Reichheld, F. and T. Teal (1996), *The Loyalty Effect*, Boston, MA: Harvard Business School Press.
Reinartz, W.J. and V. Kumar (2000), 'On the profitability of long-life customers in a noncontractual setting: an empirical investigation and implications for marketing', *Journal of Marketing*, **64**(4), 17–35.
Rossi, P., R. McCulloch and G. Alenby (1996), 'The value of purchase history data in target marketing', *Marketing Science*, **15**(4), 321–40.

Rust, J. (1994), 'Structural estimation of Markov decision processes', in D. McFadden and R. Engle (eds), *Handbook of Econometrics, Vol. 4*, Amsterdam: North-Holland, pp. 3081–143.
Rust, R. and P. Verhoef (2005), 'Optimizing the marketing interventions mix in intermediate-term CRM', *Marketing Science*, **24**(3), 477–89.
Rust, R., V. Kumar and R. Venkatesan (2011), 'Will the frog change into a prince? Predicting future customer profitability', *International Journal of Research in Marketing*, **28**(4), 281–94.
Rust, R., K. Lemon and V. Zeithaml (2004), 'Return on marketing: using customer equity to focus marketing strategy', *Journal of Marketing*, **68**(1), 109–27.
Schweidel, D., P. Fader and E. Bradlow (2008), 'Understanding service retention within and across cohorts using limited information', *Journal of Marketing*, **72**(1), 82–94.
Shugan, S. (2006), 'Are consumers rational? Experimental evidence', *Marketing Science*, **25**(1), 1–7.
Simester, D., P. Sun and J. Tsitsiklis (2006), 'Dynamic catalog mailing policies', *Management Science*, **52**(5), 683–96.
Sutton, R. and A. Barto (1998), *Reinforcement Learning: An Introduction*, Cambridge, MA: MIT Press.
Venkatesan, R., V. Kumar and T. Bohling (2007), 'Optimal customer relationship management using Bayesian decision theory: an application for customer selection', *Journal of Marketing Research*, **44**(4), 579–94.
Wiesel, T., B. Skiera and J. Villanueva (2008), 'Customer equity: an integral part of financial reporting', *Journal of Marketing*, **72**(2), 1–14.
Zeithaml, V., R. Rust and K. Lemon (2001), 'The customer pyramid: creating and serving profitable customers', *California Management Review*, **43**(4), 118–42.

5. The value of flexibility: real options and customer lifetime value

Michael Haenlein

INTRODUCTION

It is a cornerstone of relationship marketing and customer relationship management (CRM) that creating and maintaining customer relationships drives profitability and growth. Managers have claimed that firms benefit from long-term customer relationships, as customers with greater tenure make more purchases, show reduced operating costs and are more likely to engage in referral behavior, and are associated with lower operating costs (Reichheld and Sasser, 1990) – at least in the case of contractual relationships (Reinartz and Kumar, 2000). Likewise, academic research has shown a positive relationship between satisfaction and share-of-wallet (Cooil et al., 2007), analyzed the benefits of allocating scarce marketing resources based on past behavior and customer value (Kumar et al., 2008), and provided an indication that referred customers are more profitable and more loyal (Schmitt, Skiera, and Van den Bulte, 2011). It is therefore not surprising that relationship maintenance is the dimension of the CRM process that is most strongly related to perceived and objective performance (Reinartz, Krafft, and Hoyer, 2004).

Given the high importance attributed to loyalty and retention, academics and managers sometimes seem to forget that not all customer relationships are actually beneficial for the firm. Unreasonable customers, for example, who insist on 'always being right' and demand unwarranted privileges and compensation, can represent a burden for firms, employees, and other clients (Berry and Seiders, 2008). Unfriendly customers may generate unpleasant work experiences for front-line employees and lead to lower job satisfaction and higher quitting intentions (Walsh, 2011). And unprofitable customers can cause firms to incur substantial financial costs, which continue to increase the longer such relationships are maintained (Haenlein and Kaplan, 2009). In recent years it has therefore been argued that companies should consider not retaining certain customers (Mittal, Sarkees, and Murshed, 2008), either by no longer investing in the relationship or by actively refusing to continue to provide service to these customers. This idea has been discussed in the literature under the labels of customer abandonment (Haenlein, Kaplan, and Schoder, 2006) and customer divestment (Mittal and Sarkees, 2006).

A firm's decisions on which customers to keep and which ones to dispense with are likely to be based on the long-term profitability of the client relationship, measured in terms of customer lifetime value (CLV). This perspective is consistent with the managerial practice of allocating scarce marketing resources across customers according to their CLV (Rust, Lemon, and Zeithaml, 2004; Venkatesan and Kumar, 2004; Reinartz, Thomas, and Kumar, 2005). Frequently, a firm uses CLV to develop a rank order of customers, with the intention of subsequently devoting more resources to customers with higher ranks. Within such a framework it is sensible to assume that customers with low ranks (and particularly those with negative CLV) should be abandoned to increase overall profitability of the customer base (Zeithaml, Rust, and Lemon, 2001).

Interestingly, for over two decades researchers have suggested that the simple net present value (NPV) analysis that forms the foundation of most traditional CLV calculations does not reflect the value of the flexibility to make such decisions. Rather, the use of real options analysis may be more appropriate in certain situations (Pindyck, 1991; Dixit and Pindyck, 1995; Amram and Kulatilaka, 1999). For example, standard CLV calculations assume that once a customer has been acquired a company will continue to deal with him or her, regardless of whether that customer turns out to be profitable. Indeed, a company may decide to do so for a variety of reasons, in which case regular CLV calculations are appropriate. However, when a firm decides to discontinue an unprofitable customer relationship because it believes that the relationship cannot become profitable in future periods, a modification of standard CLV calculations is required.

Take the mobile phone industry as an example. Because of the absence of indicators of customer quality at the time of acquisition, mobile phone operators usually attribute the same customer acquisition cost (mainly consisting of handset subsidies) to all customers acquired through the same acquisition channel. To prioritize acquisition channels in the allocation of a limited total budget, the company calculates the CLV of the 'average' customer acquired through each channel. Because customers expect the company to provide them with upgraded handsets at regular time intervals, this CLV includes the expected cost of future upgrades. In reality, however, a company upgrades a customer's handset at the end of his or her acquisition contract only if the revenues the customer has generated until that point justify a reinvestment in the relationship. At this point, revenue that the customer generated in the past serves as a much better predictor of future customer quality than does information about the acquisition channel.

By following this practice, mobile phone operators keep the freedom to exercise their option not to upgrade a customer whose prior revenues are unfavorable. This usually equates to an abandonment decision, because a customer who does not receive a desired upgrade is likely to leave. However, CLV calculated using traditional formulas does not account for this flexibility, since it assumes a fixed expected cost of future upgrades for all customers acquired through the same channel. Thus, the value of the flexibility not to upgrade low-value customers is not included in the calculation of CLV. This makes the traditional CLV measure inconsistent with actual behavior.

This is particularly problematic because, as shown in the following sections, there may be a substantial divergence between CLV calculations that take option value into account and those that do not, and the magnitude of this divergence may vary across customers. It is therefore possible that inclusion of option value in CLV calculations might influence customer CLV rankings. Using CLV to allocate a limited total customer acquisition budget across acquisition channels, without incorporating the value of the option to make future decisions that may alter customer relationships, can lead to an overall biased result. The following section provides a more detailed overview of real options theory, showing how this valuation technique can overcome some limitations of simple NPV analysis and how it can be applied in a marketing context.

REAL OPTIONS THEORY

Authors in the area of strategic investment decision-making have noted for a long time that simple NPV analysis may be associated with several implementation and conceptual problems. Several modifications of the NPV analysis approach have been proposed to remedy some of these problems (for a detailed list and explanation, see Adler, 2000). Yet one shortcoming remains (Dixit and Pindyck, 1995; Amram and Kulatilaka, 1999): simple NPV analysis by definition assumes that one specific value for some kind of metric (e.g., cash flow) can be defined for each period within the time horizon of the analysis. Although it is possible to account for future uncertainty by specifying more than one state of nature per period, different values of this metric must be combined into one expected value before NPV can be calculated.

This methodological constraint translates into a conceptual one, namely, that simple NPV analysis builds on the assumption that a project or customer relationship can be completely planned beforehand. It therefore ignores the project's potential to create future options that are not foreseeable based on the current stock of knowledge. Such options could include the freedom to abandon a project in the case of unfavorable conditions or to extend its scope in the case of favorable ones. By ignoring such options, NPV analysis often underestimates the true value of an investment (Benninga and Tolkowsky, 2002).

To address this weakness, it is necessary to use other valuation methods. Extensive literature on the valuation of (real) options has emerged in fields situated at the interface between strategic management and finance. Yet very few studies in the marketing discipline have explored applications of such options (Finch, Becherer, and Casavant, 1998; Slater, Reddy, and Zwirlein, 1998; Levett et al., 1999; Hogan and Hibbard, 2002). This is especially surprising given that the use of real options seems to be particularly appropriate in the context of CRM and specifically in the analysis of customer acquisition/retention strategies based on CLV. Acquisition and retention activities usually create future options because the decision to acquire or reinvest in a customer always creates the choice of further investment in the relationship in subsequent periods.

Three different approaches can be used to evaluate real options and to incorporate their value into the value of a project: the Black–Scholes model (Black and Scholes, 1973), contingent claims valuation (Cox, Ross, and Rubinstein, 1979), and dynamic programming (Copeland and Tufano, 2004). The Black–Scholes model has the advantage of providing a closed-form solution for option value. It has previously been applied in the context of real options analysis (Luehrman, 1998) but has the drawback of being relatively computationally demanding. Contingent claims valuation is substantially easier to apply and uses less stringent assumptions, though it includes the Black–Scholes formula as a limiting case. It has also been used to value real options (Sachdeva and Vandenberg, 1993), but builds on the assumption of arbitrage-free markets and thus assumes that the real option is tradable. This assumption may be questionable in many situations, especially when applied to a marketing context.

Dynamic programming is a class of solution methods for solving sequential decision problems. It builds on the Bellman principle of optimality, which states that the optimal way of getting from a particular state to an end point is not a function of the path taken to that state (Elton and Gruber, 1971). The optimal decision at any time t can therefore be identified by maximizing the sum of the paybacks from following a particular policy at time t, plus the expected paybacks from following an optimal policy from time $t + 1$ onward. In the context of real options analysis, dynamic programming helps examine options from a decision theory point of view. The following section uses an example to illustrate how dynamic programming can be applied in order to calculate the value of a seller's 'abandonment option', that is, the option to terminate business with a customer who has proved to be unprofitable.

HOW TO INCORPORATE OPTION VALUE INTO CLV

To better understand the mechanics behind real options, imagine a marketing manager who must choose between implementing one of two alternative direct marketing campaigns (A and B). With Campaign A, the company can attract 'good' customers with a probability of 25 percent and 'bad' ones with a probability of 75 percent, where a 'good' client generates a net profit of $1500 in the period following the campaign (Period 1), and a 'bad' one causes a net loss of −$500 in the same period. In the subsequent period (Period 2), each client's net contribution either improves or deteriorates by 50 percent, where 'improvement' refers to either an increase in the profit the company gains from a client or a decrease in the loss the company incurs from the client (Figure 5.1). The likelihood of improvement is 25 percent. With Campaign B, only one type of customer can be acquired: one who generates a constant net profit of $46 in each period (Periods 1 and 2). For both campaigns, the average cost of recruiting the customer and setting up a new account is assumed to be $50, and the discount rate is 15 percent. Assuming an average lifetime of two periods, a simple NPV analysis indicates that the CLV of a newly acquired customer is approximately −$140 for Campaign A and $75 for Campaign B. Because the CLV obtained under Campaign B is greater than that obtained under Campaign A and greater than the acquisition cost, the decision would be in favor of Campaign B.

However, as noted previously, such a simple NPV analysis does not account for the flexibility of the marketing manager after the end of Period 1. At that point, the marketing manager knows (e.g., on the basis of experience from comparable campaigns already carried out by the company) that a customer who proved to be bad in Period 1 will generate an expected loss in Period 2. Given this knowledge, the marketing manager has the option to behave asymmetrically in Period 2; that is, in the case of a good customer, the client relationship will continue, and in the case of a bad one, it will be abandoned. This asymmetrical behavior protects the manager from the downside risk of realizing loss as a result of a bad customer in Period 2. If the marketing manager decides to terminate the client relationship with a bad customer at the end of Period 1, and if this termination can be conducted without additional cost, the lifetime value of a bad customer increases from −$907 to −$435, which increases the expected CLV from −$140 to $212. On the basis of this increase of

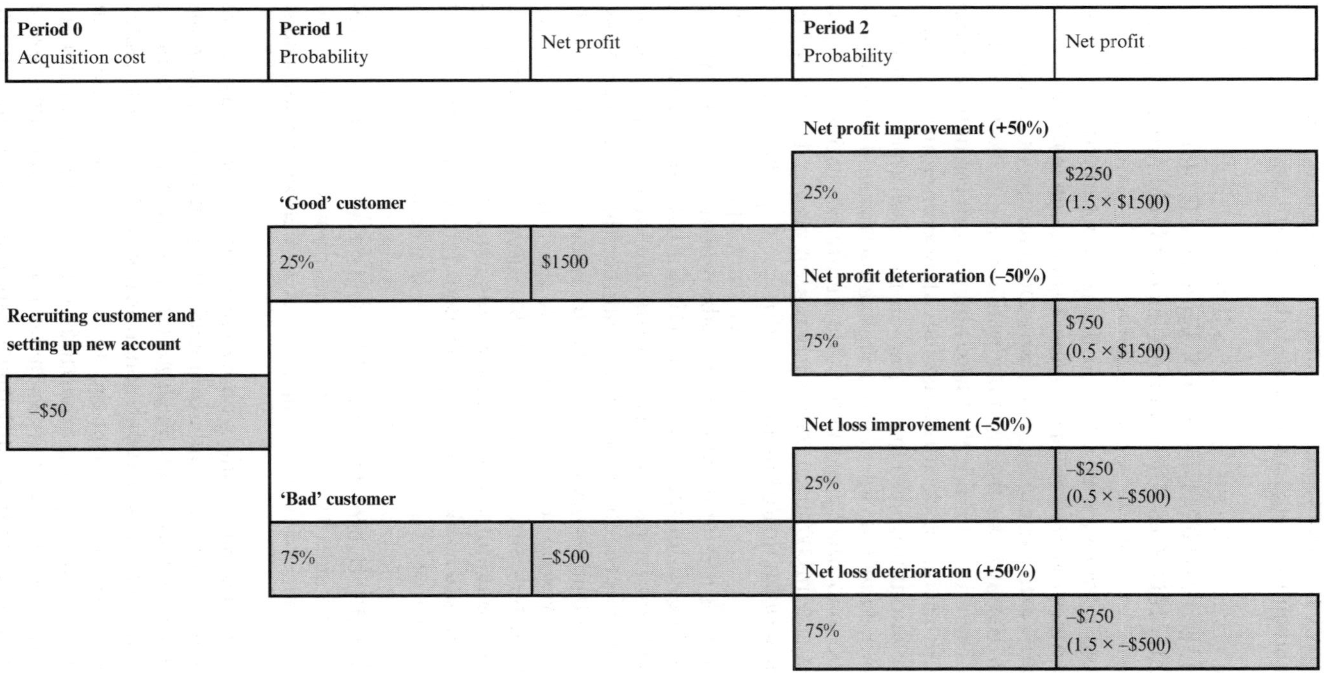

Figure 5.1 Expected net contribution of customers acquired through Campaign A (Example 1)

$352, Campaign A now seems far more attractive than Campaign B. This increase can be interpreted as the value of the abandonment option.

WHAT DIFFERENCE DOES IT MAKE? CLV INCLUDING VS EXTENDING OPTION VALUE

The previous example, which is built on hypothetical figures, provides a basic illustration of how real options can be included in CLV calculations. It does not, however, provide an empirical assessment of the extent to which CLV calculations that include option value differ from those that exclude it. Therefore, it is not clear how much bias is introduced by neglecting the value of future decision-making when computing the value of a customer relationship. The current section describes an empirical analysis that was carried out in order to investigate this issue. The analysis relied on a dataset provided by the Direct Marketing Educational Foundation (DMEF, Academic Dataset 3), consisting of sanitized purchasing data from a specialty catalog company.

To determine CLV including vs excluding option value within this dataset, we used the following four-step approach (see Haenlein et al., 2006 for additional methodological details). First, the total dataset was split into 36 distinct cells on the basis of recency (i.e., the time elapsed since the previous purchase, operationalized as the number of years since the last order) and frequency (i.e., the number of purchases made in a specified time period, operationalized as the total number of years in which at least one purchase occurred). Subsequently, the past purchase behavior of the customers in each cell was determined according to the first eight periods of data available in the dataset. This past purchase pattern was then used as a proxy for purchase behavior (i.e., probability of purchase and dollar amount expected to be spent) in the remaining four periods in the respective cell. This widely used approach for forecasting customer activity – referred to as the recency-frequency-monetary (RFM) value framework – was selected because of its simplicity and ease of use.

Second, we estimated the costs generated per customer and year, consisting of the cost of goods sold and direct marketing expenditures. We computed the cost of goods sold in a given year as 55 percent of the average dollar amount spent by the company in that year, estimated on the basis of data available in the dataset. Due to a lack of information on marketing cost, and because actual marketing costs may vary considerably across different companies, we carried out computations for 100 different scenarios, implementing a different value for marketing expenditure in each scenario. Specifically, we assumed that marketing expenditure per customer per year was in the range of \$0–\$10 and varied the value in increments of \$0.10.

Third, the inputs generated in the previous two steps were used to estimate the expected future revenue and profit contribution of an average customer in each of the 36 recency-frequency cells. Expected future revenue was calculated as the average amount spent given a certain purchasing pattern multiplied by the probability that this purchasing pattern would occur. Profit contribution was subsequently calculated as the expected revenue, minus the cost of goods sold, minus direct marketing expenditures.

Finally, we generated two estimates of CLV, one including option value and one excluding it. CLV excluding option value was calculated as the sum of all discounted profit contributions in year t using a discount rate of 20 percent. To calculate CLV including option value, it was first necessary to decide when a client relationship should be terminated, which in turn would determine the activity status (active = 1; inactive = 0) of each customer in each period. To this end, we made two assumptions. First, a customer should be abandoned in year T, whenever the sum of expected future profit contributions from T onward discounted back to T is less than the cost associated with the abandonment. Second, an abandoned customer becomes inactive for all years after abandonment (i.e., a customer, once abandoned, cannot be reactivated later). On the basis of these rules it was straightforward to determine the activity status of each customer, contingent on his or her purchasing pattern. CLV including option value was then calculated for a given customer as the sum of all discounted profit contributions in year t using a discount rate of 20 percent multiplied by the corresponding activity status.

An examination of the divergence between CLV including vs excluding real option value across all recency-frequency cells showed that the value estimates differed substantially. For direct marketing expenditures equal to \$5 per customer per year, the divergence was ~1.5 percent on average, and for direct marketing expenditures greater than \$6.50, the divergence consistently exceeded 5 percent. Interestingly, for any given direct marketing expenditure level, the divergence at the customer segment level differed greatly across customer segments. For example, in the case of marketing expenditures of \$7.50 per customer per year, segment-specific divergence varied from 0 percent to 70 percent. This latter point shows that there may be factors other than cost that determine the divergence. This question is addressed in more detail in the following section.

WHICH FACTORS DRIVE REAL OPTION VALUE?

To gain a better understanding of the factors other than cost that influence the divergence between CLV including option value and CLV excluding option value, we performed a Monte Carlo simulation. In it, we focused on the potential impact of a change in expected revenues, driven by a change in the distribution (i.e., variation and mean) of the probability of purchase and dollar amount expected to be spent. Analytically, it is possible to show that the value of the real option of abandonment increases with increasing uncertainty and decreasing present value of expected cash flows (Haenlein et al., 2006). This suggests that the difference between CLV including vs excluding option value should increase with increasing variation and decreasing mean values of future purchase probabilities and dollar amounts expected to be spent.

We built our simulation on the always-a-share CLV example illustrated in Dwyer (1989). This example is similar to the empirical analysis above. It assumes that purchase probabilities, dollar amounts expected to be spent, and marketing expenditures are dependent on the recency cell of each customer (Table 5.1) and that the cost of goods sold represents 60 percent of revenue. Dwyer's example is more realistic than our empirical analysis in some dimensions (e.g., by assuming that marketing expenditures are dependent on each customer's recency rather than constant) and less in others (e.g., by assuming that purchase probabilities and dollar amounts expected to be spent depend only on recency rather than on recency as well as frequency). However, since the focus of the Monte Carlo simulation was to investigate how the divergence between CLV including vs exclud-

Table 5.1 Assumptions of always-a-share CLV example as proposed by Dwyer (1989)

Recency Cell	Description	Probability of Purchase in Period t (%)	Amount Expected to be Spent in Period t ($)	Marketing Expenditures in Period t ($)
1	Customer last purchased in t–1	30	100	3.60
2	Customer last purchased in t–2	20	80	3.10
3	Customer last purchased in t–3	15	60	1.80
4	Customer last purchased in t–4	5	40	1.30

ing option value changes with a variation in model parameters, it was not essential for the simulation to be completely identical to the empirical validation.

Our final simulation analyzed the difference between CLV including vs excluding option value across $3 \times 3 \times 10 = 90$ scenarios with 20000 runs each, resulting in a total of 1.8 million runs. For the average of the two main variables, that is, the probability of purchase in period t and the amount expected to be spent in period t, we used either the values given in Table 5.1 or, alternatively, values equal to double or half these amounts. This resulted in three different sets of average values for both variables and a total of $3 \times 3 = 9$ different settings. Within each setting, we assume that the purchase probability and dollar amount expected to be spent varies according to a normal distribution (or truncated normal in case of probabilities) around these averages, with a standard deviation ranging between 5 percent and 50 percent of these average values in 5 percent increments. This results in a total of nine (settings) × 10 (standard deviations) = 90 scenarios. For each scenario we subsequently simulate 20000 customers by drawing individual purchase probabilities and dollar amounts expected to be spent from the underlying distributions and determined the divergence between CLV including vs excluding option value on the individual level. The final output is the median of this divergence across all 20000 customers.

Three important findings emerge from the outcome of this simulation. First, there appears to be a quadratic relationship between the average divergence and the standard deviation. This is intuitive, since increasing variance is associated with increasing probability of unprofitable customer relationships and, thus, increasing occurrence (and value) of abandonment decisions. Second, the absolute impact of standard deviation on average divergence seems to be marginal. Although the standard deviation increases tenfold from 5 percent to 50 percent, the average divergence increases only slightly (from 0.85 percent to 1.21 percent). Thus, although our simulation confirms the analytical result that average divergence increases with increasing standard deviation, this increase seems to be of limited size. Finally, we observe that the overall average divergence increases with decreasing purchase probabilities and dollar amounts spent, which is equally in line with the theoretical expectations. For the range of values investigated in this simulation, this effect seems to be substantially larger than the impact of standard deviation on average divergence.

THE IMPORTANCE OF CONSIDERING OPTION VALUE WHEN CALCULATING CLV

The discussion until this point has shown how real options can contribute to research in the area of CLV by proposing a methodology that can be used to incorporate the value of the seller's flexibility into CLV calculations – a point that has been raised by Rust et al. (2004), among others. On the basis of the findings discussed herein, it seems recommendable to systematically include option value in the calculation of CLV.

In particular, results of our empirical study and simulation show that neglecting option value can lead to a substantial underestimation of true customer value. This underestimation is not uniform across the whole customer base: customers with low probabilities of future purchase and low dollar amount expected to be spent are more likely to be underestimated compared with customers with more favorable values along these two dimensions. Thus, a rank order of customers based on CLV may be different when CLV is calculated including option value than when it is calculated excluding option value. Most recommended methods for the allocation of scarce marketing resources (e.g., Rust et al., 2004; Venkatesan and Kumar, 2004; Reinartz et al., 2005) build on such a rank order of customers and recommend the allocation of more resources to customers with higher ranks. Thus, our results suggest that the incorporation of option value into CLV calculations could substantially influence such allocation decisions.

STRATEGIES FOR THE ABANDONMENT OF UNPROFITABLE CUSTOMER RELATIONSHIPS

Approaches that incorporate the value of real options for future decision-making into CLV calculations implicitly assume that a firm has the ability to change its course of action. In our specific context, this implies that abandoning unprofitable customer relationships is an option that can actually be implemented by companies. Approaches to 'firing' certain groups of customers have been discussed in literature for a long time under the label 'selective demarketing' (Kotler and Levy, 1971). With the increasing importance of CRM and profitability-based customer management, the topic has recently received the attention of a broader range of researchers (Mittal and Sarkees, 2006; Mittal et al., 2008; Haenlein and Kaplan, 2009).

Several techniques for abandoning unprofitable customers have been proposed. Generally, firms are faced with a choice between direct and indirect abandonment approaches. Direct abandonment strategies refer to cases in which the company explicitly states to the customer its desire to exit the relationship, whereas indirect strategies describe cases in which the firm tries to accomplish the break up without an explicit statement of the goal (Baxter, 1985). Both techniques are implemented in practice: the US telecommunications firm Sprint Nextel, for example, directly abandoned customers in July 2007 by terminating the contracts of 1000 clients, while the brokerage house Charles Schwab is claimed to implement indirect abandonment by making low-profitability accounts wait longer in call center queues.

From the (unprofitable) customer's perspective, it seems likely that being abandoned in a direct or indirect way will be perceived as negative and create a certain level of dissatisfaction, ultimately leading to negative word-of-mouth that may generate attention among the general public. From the abandoning firm's point of view, such publicity is likely to have two negative consequences. First, it could lead to the (involuntary) loss of remaining customers that the abandoning firm would like to retain, and, second, it could influence potential customers not to join the focal firm but rather to join one of its competitors. Both consequences are likely to result in abandonment costs, which the firm should consider prior to making a divestment decision.

A relevant question in this context is whether some abandonment strategies are associated with more severe negative consequences and higher abandonment costs than other ones. In fact, a formal investigation of the negative effects of unprofitable customer abandonment on the abandoning firm's remaining and potential customer base suggests that such differences are relatively minor (Haenlein and Kaplan, 2011). Direct as well as indirect strategies result in approximately the same levels of exit intention among remaining customers, purchase intention among potential customers, and boycott likelihood for both customer types.

It is interesting to observe that customers do not seem to perceive indirect abandonment (e.g., raising prices, decreasing service levels) as less severe than direct abandonment (i.e., a direct refusal to serve unprofitable customers). One explanation for this result could be that in abandoning an unprofitable customer, the firm conveys a message that is so strong in terms of content that the recipients see no need to differentiate between the means by which the message is delivered. In an environment where the customer is used to being treated as 'king', the news of customer abandonment may be interpreted as a violation of implicitly assumed social norms (see Bolton, Warlop, and Alba, 2003 for a similar argument in the context of price fairness perceptions; and Xia, Monroe, and Cox, 2004).

It is therefore possible that customers focus their attention only on the presence of abandonment itself and not the way in which it is carried out.

CUSTOMER-SPECIFIC RESPONSE TO ABANDONMENT

Prior research (Haenlein and Kaplan, 2012) indicates that the market is likely to penalize companies that abandon unprofitable client relationships. In particular, other customers of the firm, whom the company would like to retain, are more likely to respond *actively* to unprofitable customer abandonment – by either deciding to leave the abandoning firm or speaking out against such behavior – than *passively* through silence and loyalty.

Notably, a given customer's (negative) reaction toward an instance of abandonment appears to be independent of the strength of that customer's relationship (tie strength) with the abandoned customer. Specifically, other customers are *not* likely to perceive the abandonment of a customer with whom they have a weak-tie relationship as less severe than that of a strong-tie contact. Therefore, negative collateral effects might not be limited to the immediate social networks of abandoned customers. This suggests that firms should closely monitor online news forums and other types of social media to assess negative reactions to an abandonment decision (Kaplan and Haenlein, 2010), especially at a time in which customer engagement is of increasing importance (Verhoef, Reinartz, and Krafft, 2010).

Not all customers react negatively to a firm's abandonment of (other) unprofitable customers. Most customers maintain a minimum level of loyalty intentions. This can result from superior product quality, inertia ('because I have been with them so long'), or personal utility maximization ('because it did not happen to me and I only take action with programs that affect me'). Surprisingly, some customers may even react positively to unprofitable customer abandonment. These customers expect that the money saved from no longer serving unprofitable customers will be reinvested into a firm's existing client relationships. This mirrors prior research in the area of price fairness that has shown that consumers have a certain understanding of the constraints faced by companies and accept that firms need to make profit in order to survive (Kahneman, Knetsch, and Thaler, 1986). It appears that at least some clients may equally understand that firms that decide to provide above-standard service to their best customers also need to cut costs among their unprofitable customers in order to remain profitable overall.

Although existing customers do tend to react negatively to customer abandonment, the most common reactions are verbal statements of negative emotions and feelings (e.g., anger, disappointment) and beliefs (e.g., the perception of a lack of appropriate customer service) instead of intended actions (Haenlein and Kaplan, 2010). Furthermore, in some cases a firm can compensate for the negative consequences of unprofitable customer abandonment by improving its core service quality. Generally, it seems advisable for companies that decide to abandon unprofitable customers to pay special attention to making their offers as unique as possible. This can be achieved, for example, by choosing a positioning strategy that highlights the exclusive characteristics of the product or service. Such a strategy is likely to motivate current customers to perceive the abandoning firm as less comparable to potential competitors, which ultimately decreases the attractiveness of competitors in the marketplace.

LIMITATIONS AND AVENUES FOR FUTURE RESEARCH

Both the methodological focus of this chapter (i.e., incorporating real options into CLV calculations) as well as the managerial focus (i.e., whether and how to abandon unprofitable customer relationships) are still two relatively unexplored areas, and more research is needed in both domains.

Regarding the methodological focus, the findings reported here apply only to the specific type of option analyzed, namely, abandonment options. A better understanding of such options is of value in itself, particularly since some authors have suggested that the flexibility underlying the real options approach is dependent on the possibility of abandoning investment initiatives. This indicates that abandonment is essential for limiting downside risk (Adner and Levinthal, 2004). Nevertheless, the empirical conclusions outlined herein – that is, the degree of divergence between CLV incorporating real options and CLV excluding them, and the factors that influence this divergence – should not be generalized to other types of real options. It would be interesting to analyze the consequences of considering other option types that have been discussed in the literature, for example, deferment, learning, growth, and switching options. A relevant next step would then be to formally include real options into other approaches that have been proposed to calculate CLV (Fader, Hardie, and Lee, 2005), probably by relying on the Black–Scholes model, which allows for a closed-form expression of option value.

Furthermore, there is a rich literature in marketing that uses dynamic programming models to optimally develop relationship marketing policies. This body of research dates back to the 1950s (Howard, 2002) and includes, among others, work on catalog mailing decisions (Bitran and Mondschein, 1996; Gönül and Shi, 1998; Simester et al., 2006) and the analysis of managerial flexibility in terms of promotions (Lewis, 2005; Khan, Lewis, and Singh, 2009). Many of these decisions can also be interpreted as real options (e.g., deciding to stop mailing catalogs or delivering promotions could be considered as indirect abandonment strategies), and an extension of our approach to these areas could provide interesting results.

Looking at the managerial focus, it appears that some customers or customer groups react positively to firms' abandonment of unprofitable customers. This implies that unprofitable customer divestment could generate benefits that go beyond the option value of terminating the client relationship itself. Future studies could investigate whether the abandoning firm can implement specific strategies to foster such positive associations. Also, it should be highlighted that there might be reasons to keep a customer, even if the relationship itself is unprofitable, for example, if the customer is able to refer new profitable customers to the company.

A relevant question in the context of customer divestment is to what extent unprofitable customers are likely to possess social networks that include profitable prospects. Research in the area of social network analysis and referral behavior has shown that customers who belong to the same social environment are likely to be similar, share the same behaviors and prefer the same brands (Ward and Reingen, 1990; McPherson, Smith-Lovin, and Cook, 2001). If this is the case, it seems probable that such customers also share similar profitability patterns from the company's perspective. Indeed it has been shown that low- (high-) revenue customers tend to be connected to other low- (high-) revenue clients (Haenlein, 2011) and that retention rates and customer lifetimes are likely to be correlated within a social network (Nitzan and Libai, 2011; Haenlein, 2013). Further research is needed to better understand these social network effects in customer-level profitability and their consequences for managerial decision-making. If, for example, it turns out that an unprofitable customer's friends are likely to be unprofitable themselves, their word-of-mouth might be less powerful than that of the 'average' customer (Haenlein and Libai, 2013).

CONCLUSION

Previous research in the area of customer profitability measurement has indicated that unprofitable customers can represent a substantial share of a company's client base. Niraj, Gupta, and Narasimhan (2001) analyzed customer profitability for a grocery distributor in a business-to-business setting and found that when cost was allocated using activity-based costing, about 32 percent of all customers were unprofitable. Bowman and Narayandas (2004) investigated the customer base of a vendor in the processed metal business industry and showed that 31 percent of all relationships recorded a negative operating profit. Haenlein, Kaplan, and Beeser (2007) analyzed the customer base of a retail bank in a business-to-consumer environment and identified five customer segments (accounting for 27 percent of the total customer base) that had a negative contribution margin.

Several factors contributing to negative customer profitability have been discussed in the literature. Some customers may become unprofitable because their needs and requirements are incompatible with the company's business model (Rosenblum, Tomlinson, and Scott, 2003; Bhargava and Feng, 2005). Furthermore, differences in cost-to-serve and customer situations can contribute to significant differences in customer profitability (Shapiro et al., 1987). Therefore, clients requiring complex and non-standard processes stand a high chance of being unprofitable (Kaplan and Anderson, 2004). For example, customers with a disproportionate share of last-minute requests often have negative profitability (Van Hoek and Evans, 2005).

Due to the increasing importance of profitability-based customer management, the idea of proactively terminating customer relationships is far from uncommon and has received increasing interest in managerial practice in recent years. For example, it has become common business practice to route call center calls based on customer profitability in order to minimize waiting time for the most profitable clients – an approach that can be considered an indirect abandonment strategy. In a similar spirit, the 'customer divestment continuum' proposed by Mittal et al. (2008) includes relationship termination as a (last) resort in case other strategies such as relationship reassessment and customer education or migration do not prove to be successful. Similarly, the consumer electronics retailer Best Buy announced a new strategic initiative called 'Devil and Angel' in 2004, consisting of modifications in store layout and product offerings to discourage unprofitable customers from patronizing their retail outlets (McWilliams, 2004).

As discussed above, strategic implementation of customer abandonment requires that firms modify the standard methods by which they determine CLV; simple NPV analysis does not account for the value of (real) abandonment options. Failure to consider the value of such options can lead to biased computations of CLV. The purpose of this chapter has been to outline a method for incorporating option value when calculating CLV, to describe the problems that can arise when such options are not taken into account, and to provide insight into the strategies that can be used for customer abandonment and the customer-specific costs associated with them.

REFERENCES

Adler, R.W. (2000), 'Strategic investment decision appraisal techniques: the old and the new', *Business Horizons*, **43**(6), 15–22.

Adner, R. and D.A. Levinthal (2004), 'Real options and real tradeoffs', *Academy of Management Review*, **24**(1), 120–26.

Amram, M. and N. Kulatilaka (1999), *Real Options: Managing Strategic Investment in an Uncertain World*, Boston, MA: Harvard Business School Press.

Baxter, L.A. (1985), 'Accomplishing relationship disengagement', in S. Duck and D. Perlman (eds), *Understanding Personal Relationships: An Interdisciplinary Approach*, London: Sage Publications.

Benninga, S. and E. Tolkowsky (2002), 'Real options: an introduction and an application to R&D valuation', *The Engineering Economist*, **47**(2), 151–68.

Berry, L.L. and K. Seiders (2008), 'Serving unfair customers', *Business Horizons*, **51**(1), 29–37.

Bhargava, H.K. and J. Feng (2005), 'America online's internet access service: how to deter unwanted customers', *Electronic Commerce Research and Applications*, **4**(1), 35–48.

Bitran, G.R. and S.V. Mondschein (1996), 'Mailing decisions in the catalog sales industry', *Management Science*, **42**(9), 1364–81.

Black, F. and M. Scholes (1973), 'The pricing of options and corporate liabilities', *Journal of Political Economy*, **81**(3), 637–54.

Bolton, L.E., L. Warlop and J.W. Alba (2003), 'Consumer perceptions of price (un)fairness', *Journal of Consumer Research*, **29**(4), 474–91.

Bowman, D. and D. Narayandas (2004), 'Linking customer management effort to customer profitability in business markets', *Journal of Marketing Research*, **41**(4), 433–47.

Cooil, B., T.L. Keiningham, L. Aksoy and M. Hsu (2007), 'A longitudinal analysis of customer satisfaction and share of wallet: investigating the moderating effect of customer characteristics', *Journal of Marketing*, **71**(1), 67–83.

Copeland, T. and P. Tufano (2004), 'A real-world way to manage real options', *Harvard Business Review*, **82**(3), 90–99.

Cox, J.C., S.A. Ross and M. Rubinstein (1979), 'Option pricing: a simplified approach', *Journal of Financial Economics*, **7**(3), 229–63.

Dixit, A.K. and R.S. Pindyck (1995), 'The options approach to capital investment', *Harvard Business Review*, **73**(3), 105–15.

Dwyer, F.R. (1989), 'Customer lifetime valuation to support marketing decision making', *Journal of Direct Marketing*, **3**(4), 8–15.

Elton, E.J. and M.J. Gruber (1971), 'Dynamic programming applications in finance', *Journal of Finance*, **26**(2), 473–506.

Fader, P.S., B.G.S. Hardie and K.L. Lee (2005), 'RFM and CLV: using iso-value curves for customer base analysis', *Journal of Marketing Research*, **42**(4), 415–30.

Finch, J.H., R.C. Becherer and R. Casavant (1998), 'An option-based approach for pricing perishable services assets', *Journal of Services Marketing*, **12**(6), 473–83.

Gönül, F. and M.Z. Shi (1998), 'Optimal mailing of catalogs: a new methodology using estimable structural dynamic programming models', *Management Science*, **44**(9), 1249–62.

Haenlein, M. (2011), 'A social network analysis of customer-level revenue distribution', *Marketing letters*, **22**(1), 15–29.

Haenlein, M. (2013), 'Social interactions in customer churn decisions: the impact of relationship directionality', *International Journal of Research in Marketing*, **30**(3), 236–48.

Haenlein, M. and A.M. Kaplan (2009), 'Unprofitable customers and their management', *Business Horizons*, **52**(1), 89–97.

Haenlein, M. and A.M. Kaplan (2010), 'An empirical analysis of attitudinal and behavioral reactions toward the abandonment of unprofitable customer relationships', *Journal of Relationship Marketing*, **9**(4), 200–28.

Haenlein, M. and A.M. Kaplan (2011), 'Evaluating the consequences of abandoning unprofitable customers: a comparison of direct and indirect abandonment strategies', *Zeitschrift fuer Betriebswirtschaft*, **81**(2), 77–94.

Haenlein, M. and A.M. Kaplan (2012), 'The impact of unprofitable customer abandonment on current customers' exit, voice, and loyalty intentions: an empirical analysis', *Journal of Services Marketing*, **26**(6), 458–70.

Haenlein, M. and B. Libai (2013), 'Targeting revenue leaders for a new product', *Journal of Marketing*, **77**(3), 65–80.

Haenlein, M., A.M. Kaplan and A.J. Beeser (2007), 'A model to determine customer lifetime value in a retail banking context', *European Management Journal*, **25**(3), 221–34.

Haenlein, M., A.M. Kaplan and D. Schoder (2006), 'Valuing the real option of abandoning unprofitable customers when calculating customer lifetime value', *Journal of Marketing*, **70**(3), 5–20.

Van Hoek, R. and D. Evans (2005), 'When good customers are bad', *Harvard Business Review*, accessed 9 September 2014 at http://hbr.org/2005/09/when-good-customers-are-bad/ar/1.

Hogan, J.E. and J.D. Hibbard (2002), 'A real options-based framework for valuing business relationships as strategic assets', Boston, MA: Boston University School of Management.

Howard, R.D. (2002), 'Comments on the origin and application of Markov decision processes', *Operations Research*, **50**(1), 100–102.

Kahneman, D., J.L. Knetsch and R. Thaler (1986), 'Fairness as a constraint on profit seeking: entitlements in the market', *The American Economic Review*, **76**(4), 728–41.

Kaplan, A.M. and M. Haenlein (2010), 'Users of the world, unite! The challenges and opportunities of social media', *Business Horizons*, **53**(1), 59–68.

Kaplan, R.S. and S.R. Anderson (2004), 'Time-driven activity-based costing', *Harvard Business Review*, **82**(11), 131–8.

Khan, R., M. Lewis and V. Singh (2009), 'Dynamic customer management and the value of one-to-one marketing', *Marketing Science*, **28**(6), 1063–79.

Kotler, P. and S.J. Levy (1971), 'Demarketing, yes, demarketing', *Harvard Business Review*, **49**(6), 74–80.

Kumar, V., R. Venkatesan, T. Bohling and D. Beckmann (2008), 'The power of CLV: managing customer lifetime value at IBM', *Marketing Science*, **27**(4), 585–99.

Levett, P., M. Page, D. Nel, L. Pitt, P. Berthon and A. Money (1999), 'Towards an application of option pricing theory in the valuation of customer relationships', *Journal of Strategic Marketing*, **7**(4), 275–84.

Lewis, M. (2005), 'Research note: a dynamic programming approach to customer relationship pricing', *Management Science*, **51**(6), 986–94.

Luehrman, T.A. (1998), 'Investment opportunities as real options: getting started on the numbers', *Harvard Business Review*, **76**(4), 51–67.

McPherson, M., L. Smith-Lovin and J.M. Cook (2001), 'Birds of a feather: homophily in social networks', *Annual Review of Sociology*, **27**(1), 415–44.

McWilliams, G. (2004), 'Minding the store: analyzing customers, Best Buy decides not all are welcome', *The Wall Street Journal*, accessed 15 September 2014 at http://online.wsj.com/articles/SB109986994931767086.

Mittal, V. and M. Sarkees (2006), 'Customer divestment', *Journal of Relationship Marketing*, **5**(2/3), 71–85.

Mittal, V., M. Sarkees and F. Murshed (2008), 'The right way to manage unprofitable customers', *Harvard Business Review*, **86**(4), 95–102.
Niraj, R., M. Gupta and C. Narasimhan (2001), 'Customer profitability in a supply chain', *Journal of Marketing*, **65**(3), 1–16.
Nitzan, I. and B. Libai (2011), 'Social effects on customer retention', *Journal of Marketing*, **75**(6), 24–38.
Pindyck, R.S. (1991), 'Irreversibility, uncertainty, and investment', *Journal of Economic Literature*, **29**(3), 1110–48.
Reichheld, F.F. and W.E. Sasser (1990), 'Zero defections: quality comes to service', *Harvard Business Review*, **68**(5), 105–11.
Reinartz, W.J. and V. Kumar (2000), 'On the profitability of long-life customers in a non-contractual setting: an empirical investigation and implications for marketing', *Journal of Marketing*, **64**(4), 17–35.
Reinartz, W., M. Krafft and W.D. Hoyer (2004), 'The customer relationship management process: its measurement and impact on performance', *Journal of Marketing Research*, **41**(3), 293–305.
Reinartz, W., J.S. Thomas and V. Kumar (2005), 'Balancing acquisition and retention resources to maximize customer profitability', *Journal of Marketing*, **69**(1), 63–79.
Rosenblum, D., D. Tomlinson and L. Scott (2003), 'Bottom-feeding for blockbuster businesses', *Harvard Business Review*, **81**(3), 52–9.
Rust, R.T., K.N. Lemon and V.A. Zeithaml (2004), 'Return on marketing: using customer equity to focus marketing strategy', *Journal of Marketing*, **68**(1), 109–27.
Sachdeva, K. and P.A. Vandenberg (1993), 'Valuing the abandonment option in capital budgeting: an option pricing approach', *Financial Practice and Education*, **3**(2), 57–65.
Schmitt, P., B. Skiera and C. Van den Bulte (2011), 'Referral programs and customer value', *Journal of Marketing*, **75**(1), 46–59.
Shapiro, B.P., V. Rangan, R.T. Moriarty and E.B. Ross (1987), 'Manage customers for profits (not just sales)', *Harvard Business Review*, **65**(5), 101–8.
Simester, D.I., P. Sun and J.N. Tsitsiklis (2006), 'Dynamic catalog mailing policies', *Management Science*, **52**(5), 683–96.
Slater, S.F., V.K. Reddy and T.J. Zwirlein (1998), 'Evaluating strategic investments: complementing discounted cash flow analysis with options analysis', *Industrial Marketing Management*, **27**(5), 447–58.
Venkatesan, R. and V. Kumar (2004), 'A customer lifetime value framework for customer selection and resource allocation strategy', *Journal of Marketing*, **68**(4), 106–25.
Verhoef, P.C., W.J. Reinartz and M. Krafft (2010), 'Customer engagement as a new perspective in customer management', *Journal of Service Research*, **13**(3), 274–52.
Walsh, G. (2011), 'Unfriendly customers as a social stressor: an indirect antecedent of service employees' quitting intention', *European Management Journal*, **29**(1), 67–78.
Ward, J.C. and P.H. Reingen (1990), 'Sociocognitive analysis of group decision making among consumers', *Journal of Consumer Research*, **17**(3), 245–62.
Xia, L., K.B. Monroe and J.L. Cox (2004), 'The price is unfair! A conceptual framework of price fairness perceptions', *Journal of Marketing*, **68**(4), 1–15.
Zeithaml, V.A., R.T. Rust and K.N. Lemon (2001), 'The customer pyramid: creating and serving profitable customers', *California Management Review*, **43**(4), 118–42.

PART 2

IDENTIFYING KEY DRIVERS TO AUGMENT CUSTOMER EQUITY

6. Managing customer loyalty to maximize customer equity

Werner Reinartz and Maik Eisenbeiss

INTRODUCTION

In the past two decades, customer loyalty has become a top priority for many companies. Motivated by the expectation that loyalty contributes positively to firm performance (Reichheld, 1996), companies have invested considerable resources in the measurement and management of customer loyalty. Despite its intuitive appeal, translating this conceptual link into managerial practice remains a challenge; the process of generating and then financially benefitting from customer loyalty is far more complex than initially believed (Anderson and Mittal, 2000), and not every loyal customer is a profitable customer (Reinartz and Kumar, 2002). Field evidence from various industries also indicates that even when returns to customer loyalty are positive, they exhibit substantial variation – ranging between 25 percent and 95 percent for a 5 percent increase in loyalty (Reichheld, 1996). A key implication of these findings is that when managing customer loyalty, care is necessary on several fronts.

First, the mechanisms required to build customer loyalty and benefit economically from it are highly complex. As we now know, customer satisfaction is almost always non-linearly and asymmetrically connected to customer loyalty (Anderson and Sullivan, 1993; Mittal, Ross, and Baldasare, 1998). Ignoring possible asymmetries and non-linearities can substantially affect the efficiency of loyalty investment decisions. Second, the nature and extent of these patterns can differ across industries, firms, or even customers (Anderson and Mittal, 2000). Third, generating revenue from loyal customers comes at a cost; some customers may become profitable to serve only over time (ibid.; Kumar and Reinartz, 2012). Thus, customer loyalty is a net present value proposition that needs to be evaluated and monitored from a dynamic perspective.

Building on these aspects, we first introduce the concept of customer loyalty and link it conceptually to customer equity and firm value. We then summarize existing research dealing with specific behavioral facets of customer loyalty (positive word-of-mouth, repeat purchase behavior, cross-buying behavior, and willingness-to-pay) to highlight the complexity associated with generating loyalty and translating it into value for the firm (i.e., 'Understanding Customer Loyalty'). Next, we discuss several measurement approaches to customer loyalty. Although the customer lifetime value (CLV) metric emerges as the most sophisticated approach, we also highlight various value levers of the CLV formula to reveal its conceptual links to customer loyalty ('Measuring Customer Loyalty'). Finally, we briefly introduce the most important strategic and tactical aspects for any firm to consider to form a basis for successfully building and managing a loyal and profitable customer base ('Managing Customer Loyalty').

UNDERSTANDING CUSTOMER LOYALTY

Customer loyalty has a long tradition in marketing literature. Most early studies conceptualized loyalty from a mere behavioral perspective (e.g., Churchill, 1942; Brown, 1952), but later studies began to treat it as a two-dimensional construct, consisting of a behavioral (e.g., repurchase behavior) and an attitudinal (e.g., deeply held commitment to repurchase) dimension (e.g., Dick and Basu, 1994; Oliver, 1999). This extended perspective emerged because repurchase alone, as a single example of loyal behavior, does not necessarily represent a deeply held commitment to repurchase a particular brand; rather, the mere act of repurchase could be the result of situational factors, such as happenstance or a preference for convenience (Jacoby and Chestnut, 1978; Zeithaml, Berry, and Parasuraman, 1996). Without any deeper commitment to the focal brand, these customers can easily switch to another brand if it offers a better deal. In contrast, in other situations customers might lack the opportunity or the need to repurchase but, from a psychological perspective, still remain loyal. Such customers might engage in positive word-of-mouth communication, for example, and thus contribute indirectly to firm performance (Zeithaml et al., 1996; Oliver, 1999).

In addition to underlining the importance of treating customer loyalty as a behavioral and attitudinal disposition of customers, extant discussions have revealed that repurchase is not the only quantifiable benefit of a psychological state of loyalty (Oliver, 1999). Besides exhibiting higher repurchase probabilities, loyal customers likely buy additional products and services (i.e., cross-buying) and are willing to pay higher prices (Reichheld and Sasser, 1990). Moreover, the value a loyal customer grants to a firm consists of more than the revenue stream that results from his or her purchases of products and services. In the Internet era, with the rapid growth of social communities (e.g., Facebook), product-rating websites,

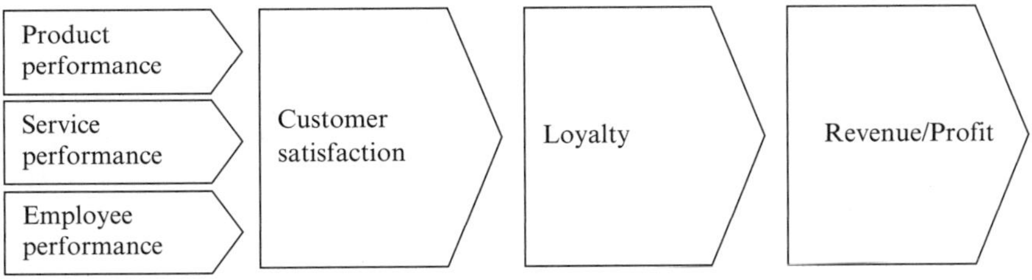

Source: Adapted from Anderson and Mittal (2000, p. 107).

Figure 6.1 Satisfaction–loyalty–profit chain

and weblogs, sharing personal opinions about a product or firm can contribute substantially to the firm's value. For example, customer referrals can result in new customer acquisitions; negative product reviews might deter potential customers from buying. If we combine these activities under the umbrella term 'word-of-mouth', we might predict that loyal customers engage more heavily in word-of-mouth than less loyal customers do (ibid.; Kumar and Reinartz, 2012). When customers recommend a firm, they put their reputation on the line; only loyal customers are willing to take such a risk (Reichheld, 2003). Finally, recent research indicates that loyal customers are more forgiving, such that they attribute occasional product or service failures to external causes rather than to the focal firm (Mittal and Frennea, 2012).

The satisfaction–loyalty–profit chain (Figure 6.1) offers a more structured and detailed understanding of how customer loyalty builds and ultimately translates into value for the firm (Anderson and Mittal, 2000; Kumar and Reinartz, 2012).

The underlying idea is that improving product/service attributes leads, first and foremost, to more satisfied customers. More satisfied customers seemingly become more loyal over time, which in turn should lead to greater profitability. Despite the almost self-evident nature of these positive links, empirical evidence from years of research offers only partial support for them (Oliver, 1999; Zeithaml, 2000). Translating the conceptual framework into practical reality has also been challenging for many firms (Anderson and Mittal, 2000; Kumar and Reinartz, 2012). For example, a firm might improve its performance on several key attributes, only to find that its overall satisfaction score remains unaffected. Or changes in overall satisfaction scores might not exert any demonstrable impact on retention behavior (Ittner and Larcker, 2003). Reichheld (1996) notes that the defection rates of customers in the automobile industry claiming to be satisfied or very satisfied are surprisingly high – between

60 percent and 70 percent. At first glance, these observations seem to stand in sharp contrast with the satisfaction–loyalty–profit chain. However, what they really document is that the links are simply more complex than initially assumed. We discuss these individual linkages in greater detail, to highlight those aspects that contribute primarily to the complexity of the satisfaction–loyalty–profit chain.

Link between Attribute-level Performance and Customer Satisfaction

On the attribute level, the satisfaction–loyalty–profit chain suggests that improving product/service attributes leads to more satisfied customers. Thus, Zappos Inc., an online shoe and apparel retailer, goes to great lengths to satisfy its customers with excellent service. In addition to offering free return shipping on all purchases, the company concentrates on building and maintaining a qualified and satisfied workforce. To sort out less committed employees, Zappos adopts unconventional approaches, such as offering new employees $2000 to quit after one week (McFarland, 2008). The underlying idea is that qualified employees should be the primary touchpoint for customers, so they constitute a key attribute driving customer satisfaction.

However, as indicated before, increasing attribute performance does not always result in more satisfied customers. This is mainly because customers do not perceive each product/service attribute as equally important. In most cases, these attributes can be classified as either satisfaction maintaining or enhancing (Anderson and Mittal, 2000). Satisfaction-maintaining attributes are generally core attributes that are taken for granted by customers, such that a failure to perform adequately on these attributes harms customer satisfaction, but their marginal impact on customer satisfaction above an adequate performance level is subject to diminishing returns. Consider available seating in an airline's business lounge. Offering too few seats triggers dissatisfaction, but offering too many seats has, if any influence at all, only a marginal impact on customer satisfaction. In contrast, satisfaction-enhancing attributes are unexpected or unusual, relative to what customers normally encounter during their product/service consumption. Because they offer a much larger 'surprise' potential, customers respond more strongly and favorably to these attributes, and their marginal impact on customer satisfaction is subject to increasing returns. Singapore Airlines offers airport lounges with separate seating areas, including some with sofas for catching a nap. Lounge staff provide blankets and pillows, then note the guests' flight number, to ensure that they waken the travelers before their flights depart. This additional service represents a satisfaction-enhancing attribute and will likely trigger customer satisfaction to a greater extent than the offer of more seating space (i.e., satisfaction-maintaining attribute).

Beyond distinguishing these different types of attributes, we need a situation-specific approach to fully understand the influence of attribute performance on customer satisfaction. Customer attributions about the cause of high or low performance and emotions such as disappointment, anger, elation, or delight may influence satisfaction outcomes (Homburg, Koschate, and Hoyer, 2005; Mittal and Frennea, 2012). Similarly, contingency theory suggests that the satisfaction experience generally depends on contextual factors (Fournier and Mick, 1999), which alter the nature of the satisfaction experience in two ways: (1) comparison standards applied in the satisfaction formation process may differ by type (e.g., expectations versus desires) or significance, and (2) cognition and emotion have different roles to play in the satisfaction experience (ibid.).

Link between Customer Satisfaction and Loyalty

The second link in the satisfaction–loyalty–profit chain suggests that when customers experience greater satisfaction with a firm's offering, they become more loyal to it. In addition to higher repurchase probabilities, these loyal customers likely buy additional products and services (i.e., cross-buying), spread favorable word-of-mouth communication, and pay a price premium or exhibit less sensitivity to price increases (Reichheld and Sasser, 1990). However, customer satisfaction does not always translate into greater loyalty. In general, a satisfied customer considers multiple factors when deciding what and where to purchase; in modern competitive markets, customers usually have several alternatives to choose from. Even if the level of satisfaction is high, repurchase or any other type of loyal behavior is not guaranteed, because another offering from a competitor might attract the customer to a similar extent. Conversely, if the customer is dissatisfied, the alternative offer is nearly always more appealing. This non-linear, asymmetric relationship between customer satisfaction and loyalty appears frequently in recent studies that address various satisfaction–loyalty links. We illustrate some typical shapes in Figure 6.2.

Repurchase is perhaps the most heavily studied loyalty response (e.g., Fornell, 1992; Anderson and Sullivan, 1993; Mittal et al., 1998; Bolton and Lemon, 1999; Mittal and Kamakura, 2001; Seiders et al., 2005). Among studies that account for higher-order complexities, a consistent finding shows that the impact of satisfaction on repurchase is greater at the extremes, implying an inverse S-shaped relationship, as in the first panel in Figure 6.2. The flat part of the curve reflects the 'zone of indifference' or 'zone of tolerance' (e.g., Woodruff, Cadotte, and Jenkins, 1983),

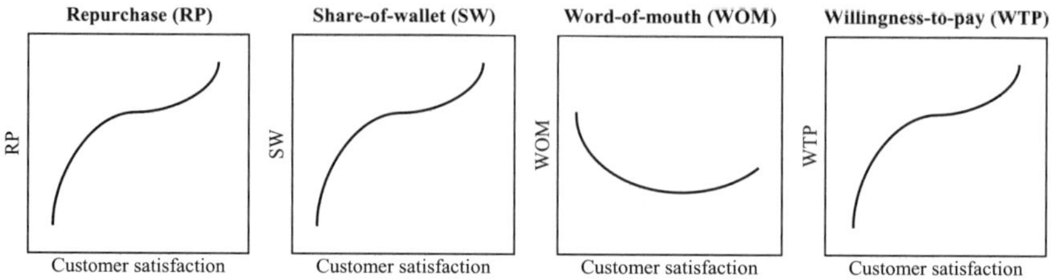

Study support

- Anderson and Mittal (2000)
- Mittal and Kamakura (2001)
- Mittal, Ross, and Baldasare (1998)

- Keiningham, Munn, and Evans (2003)

- Anderson (1998)
- Bowman and Narayandas (2001)

- Homburg, Koschate, and Hoyer (2005)
- Eisenbeiss et al. (2014)

Note: Study support – studies where the functional form of the respective relationship was found.

Figure 6.2 Typical functional shapes for links of customer satisfaction and various loyalty dimensions

within which the behavioral impact of changes in satisfaction is negligibly small. Outside the zone though (i.e., beyond upper and lower thresholds), strong behavioral reactions result. On the right-hand side, the functional form exhibits increasing returns, in line with the concept of customer delight (e.g., Jones and Sasser, 1995; Oliver, Rust and Varki, 1997). That is, because of the many existing alternatives in the marketplace, it may not be sufficient for firms to satisfy their customers; rather, they need to delight them, which then might produce more loyal customers. On the left-hand side, the function becomes much steeper again, suggesting that dissatisfied customers more likely refuse to repatronize a firm. Moreover, the functional form is asymmetric: dissatisfaction exhibits a stronger effect on repurchase than does satisfaction (e.g., Mittal et al., 1998). Theoretically, this finding is in line with the concept of loss aversion (prospect theory; Kahneman and Tversky, 1979), such that realized losses loom larger than realized gains.

Similar, asymmetric, inverse S-shaped patterns also emerge for the influences of customer satisfaction on share-of-wallet (second panel) and willingness-to-pay (fourth panel). Concepts such as the zone of indifference, customer delight, and loss aversion, which are inherent to an asymmetric, inverse, S-shaped function, hold not only for loyalty dimensions that lead to more stable customer relationships (i.e., repurchase) but also for dimensions that lead to growing relationships (i.e., share-of-wallet and willingness-to-pay). Only for word-of-mouth communication (third panel), a different function following an asymmetric inverse U-shaped form has been identified. Thus, the amount of word-of-mouth communication is particularly high for very satisfied and very dissatisfied customers, though very dissatisfied customers engage in even greater word-of-mouth than very satisfied ones. This aspect again reflects the behavioral phenomenon of loss aversion (i.e., dissatisfying experiences evoke stronger behavioral reactions than satisfying ones), similar to the effect on the other loyalty dimensions.

Although the functional shapes in Figure 6.2 reflect consistent functional patterns for various satisfaction–loyalty links identified in prior studies, considerable departures from these shapes also are possible in practice. Contextual factors such as competitive intensity, degree of switching costs, or a consumer's product involvement may alter the functional form of the underlying relationship (Jones and Sasser, 1995; Eisenbeiss et al., 2014). Figure 6.3 illustrates an example of the variability of the link between satisfaction and repurchase, depending on the underlying industry (Jones and Sasser, 1995).

In the highly competitive automotive industry for example, very high levels of customer satisfaction are necessary for a customer to repurchase, mainly because the customer can choose from a vast number of alternative car brands. In the airline industry though, the number of available alternatives is restricted because a limited number of airlines serve a particular airport. Consumers may incur considerable switching costs in this industry too, such as through their membership in frequent flyer programs and accumulated reward points. Thus, in contrast with the automotive industry, even with moderate satisfaction consumers tend to repatronize airlines.

Figure 6.4 provides a similar example of the variability of the link between satisfaction and willingness-to-pay, depending on a customer's product involvement (Eisenbeiss et al., 2014).

In this case, highly involved customers react with greater intensity to extreme satisfaction changes than do customers who exhibit less involvement. Increasing returns to customer satisfaction, as suggested in the fourth panel of Figure 6.2, only hold for extremely involved customers. Interestingly, with decreasing involvement, this moderation is so strong that we even observe a reversing of increasing into decreasing returns to customer satisfaction. Hence, whether delighting a customer (Oliver et al., 1997) actually leads to disproportionately small or high willingness-to-pay reactions depends strongly on the involvement of a particular customer. When striving for very high satisfaction levels, the expected average willingness-to-pay of highly involved participants can increase to a level more than twice as high as that for low-involvement customers.

Although the functional patterns for the various satisfaction–loyalty

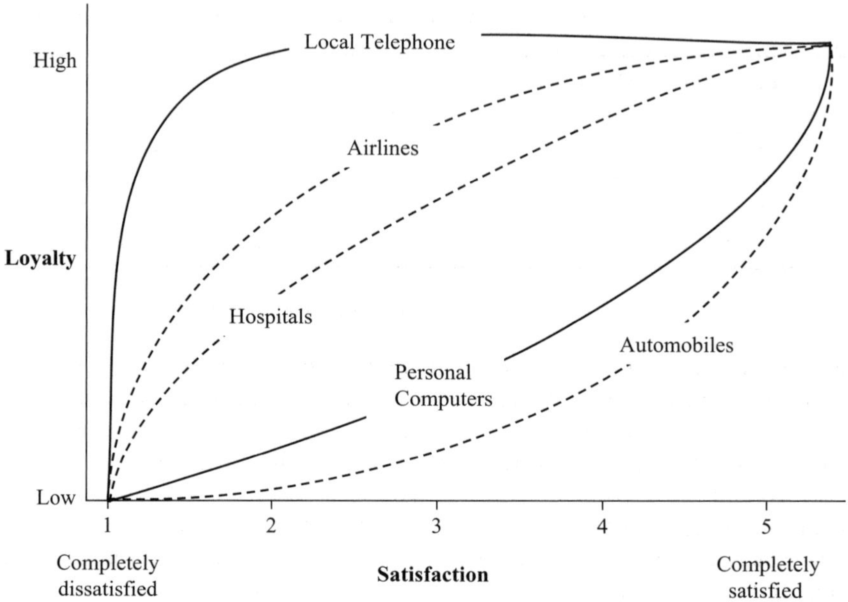

Non-competitive Zone
- Regulated monopoly or few substitutes
- Dominant brand equity
- High cost of switching
- Powerful loyalty program
- Proprietary technology

Highly Competitive Zone
- Commoditization or low differentiation
- Consumer indifference
- Many substitutes
- Low cost of switching

Note: Loyalty was measured as the customer's stated intent to repurchase.

Source: Adapted from Jones and Sasser (1995, p. 5).

Figure 6.3 Moderating influence of industry type on satisfaction–repurchase link

links in Figure 6.2 seem to hold in many situations, they represent baseline links, whose functional form can be altered by the underlying context. Therefore, firms should always thoroughly investigate the nature of a specific satisfaction–loyalty link for each industry, category, or customer segment.

Link between Loyalty and Profits

The final link in the satisfaction–loyalty–profit chain suggests that loyal customers generate higher profits for the firm than their less loyal

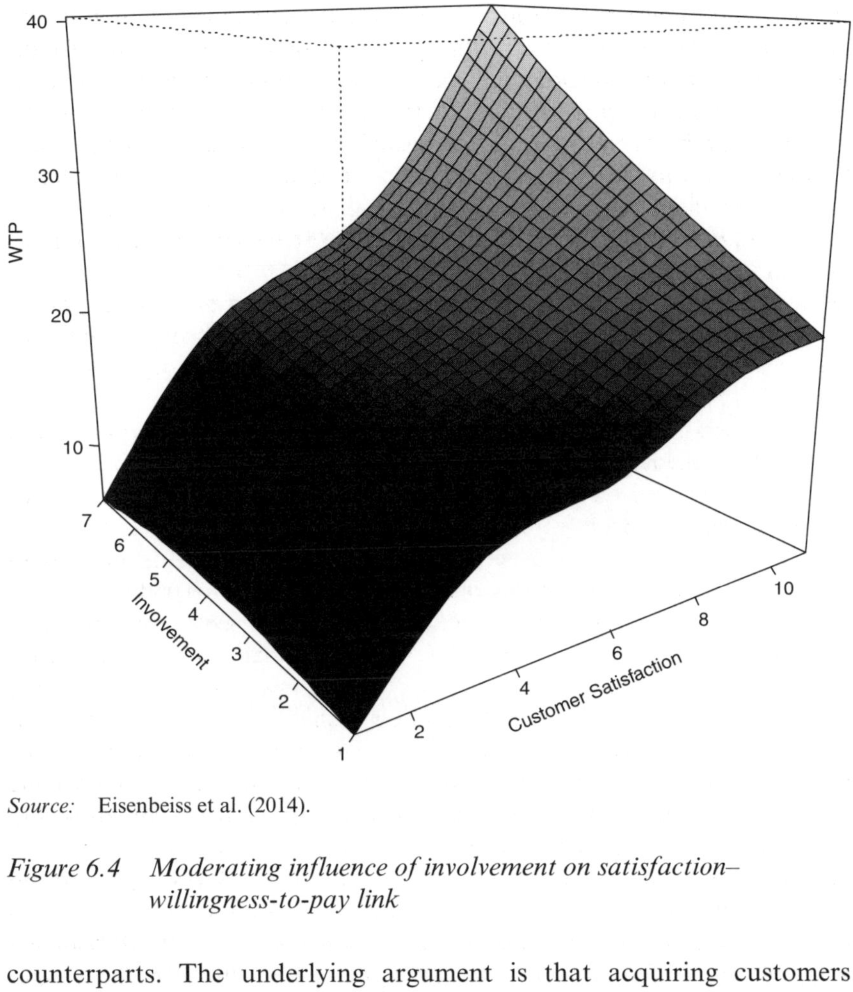

Source: Eisenbeiss et al. (2014).

Figure 6.4 Moderating influence of involvement on satisfaction–willingness-to-pay link

counterparts. The underlying argument is that acquiring customers involves costs, and this investment needs to be recovered over time. Considering the favorable characteristics of loyal customers (i.e., stable repurchase behavior, high cross-buying, positive word-of-mouth, and high willingness-to-pay), they should pay off the acquisition investment and become profitable earlier in their relationship with the firm than do less loyal customers. Such a development likely holds true in a contractual setting (e.g., mobile phone contract, magazine subscription), but it does not automatically apply to non-contractual settings. In such relationships, the revenue stream must be balanced by the cost of constantly sustaining the relationship and preventing the customer from switching to competitive offerings. However, managing these constant investments

	Short-term customers	Percentage of customers	Long-term customers	Percentage of customers
High profitability	Corporate service provider	20%	Corporate service provider	30%
	Grocery retail	15%	Grocery retail	36%
	Mail-order	19%	Mail-order	31%
	Direct brokerage	18%	Direct brokerage	32%
Low profitability	Corporate service provider	29%	Corporate service provider	21%
	Grocery retail	34%	Grocery retail	15%
	Mail-order	29%	Mail-order	21%
	Direct brokerage	33%	Direct brokerage	17%

Source: Reinartz and Kumar (2002, p. 86).

Figure 6.5 Lifetime duration–profitability association

can be tricky. Achieving and maintaining a loyal customer base is highly desirable, yet firms need to know that increasing marginal loyalty demands increasing costs. Efforts to increase customer satisfaction and loyalty not only consume the firm's resources but are typically subject to diminishing returns too.

Against this backdrop, Reinartz and Kumar (2000, 2002) investigate a sample of 16000 individual customers across four industries to analyze whether the loyalty–profitability link, as suggested by the satisfaction–loyalty–profit chain, holds in non-contractual settings. Their results offer a decidedly mixed picture (Figure 6.5).

To begin, the link between customer loyalty (e.g., retention) and customer profits is not as strong as anticipated. Across all four industries they study, a substantial segment of customers is loyal (long-term relationships) but not very profitable (mainly due to excessive resource allocation); another segment generates very high profits, despite its short tenure. Because these short-term customers can also be very profitable, it is clear that loyalty is not the only path to profitability. Figure 6.5 also reveals that, again across all industries, the two largest segments are loyal customers (long-term relationships) who are highly profitable and less loyal customers (short-term relationships) who offer low profitability. That is, even in non-contractual settings, there is a positive correlation between loyalty and profitability. Yet the non-negligible number of outliers – loyal (less loyal) customers who are low (high) in profitability – makes the link less strong or obvious than it might be in most contractual settings. Therefore, caution should be exercised before equating customer loyalty with customer profitability. Firms cannot take relationships for granted but rather must undertake efforts to monitor revenues and relationship maintenance costs continuously at least at the segment level, if not the individual customer level.

In summary, the satisfaction–loyalty–profit chain is conceptually sound, but the linkages are far more complex than initially assumed, especially in practice. Failing to understand the exact nature (strength, non-linearity, asymmetry) and underlying mechanisms of the individual links has led many firms to seriously misallocate their resources. To manage customer loyalty, every firm must develop a detailed understanding of the unique links in its own satisfaction–loyalty–profit chain. As we now know, these links are almost always non-linear, asymmetric, and industry and customer specific. Thus, the current industry practice of using customer satisfaction or customer loyalty as a simple proxy measure for customer profits is not sufficient. Only with a detailed understanding of the whole chain, managers can optimally allocate resources and make good loyalty investment decisions.

MEASURING CUSTOMER LOYALTY

Measurement Level

When monitoring and implementing the satisfaction–loyalty–profit chain, it is worthwhile for the firm to consider the level of analysis. Most academic studies adopt an aggregate-level view, pooling observations across different firms or different customers of the same firm. For example, one series of studies (Anderson, Fornell, and Lehmann, 1994; Anderson, Fornell, and Rust, 1997) analyzes how firm-level customer satisfaction indices relate to firm-level performance. Because customers are increasingly managed at the individual level though, it seems critical to understand how customers respond at this level, or at least at the segment level (e.g., in terms of loyalty behavior or profitability), when they become more or less satisfied. Kumar and Reinartz (2012, p. 158) emphasize this focus by stating:

> [I]n a world where resources are allocated on the individual customer level, the [satisfaction–loyalty–profit] chain needs to be implemented at a disaggregate level. There is much less hard evidence of the nature of the [satisfaction–loyalty–profit chain] on the individual customer level. Although one would expect a correlation between firm-level and individual-level results, it is not clear how strong this correlation really is.

In other words, if different customers have heterogeneous expectations and perceptions, a firm-level analysis cannot appropriately reflect the individual-level performance implications of constructs in the satisfaction–loyalty–profit chain.

Loyalty Measurement Metrics

Over the years, the increased availability of customer-level data has resulted in the development of a new set of metrics that reflect the need to evaluate constructs and links in the satisfaction–loyalty–profit chain at the individual level. The psychological state of loyalty includes an attitudinal and a behavioral dimension, with different behavioral manifestations, so a suitable approach for categorizing these different metrics is in reference to their dimensionality. On the one hand, some metrics focus on the attitudinal or a particular behavioral manifestation of loyalty, such as repurchase behavior (retention), cross-buying, willingness-to-pay, or word-of-mouth. In a sense, these metrics are uni-dimensional. On the other hand, a set of multivariate metrics combine various loyalty manifestations into a single number. We briefly discuss the most popular metrics for each group (Table 6.1; for an overview, see Kumar and Reinartz, 2012).

The retention rate relates to customer repurchase behavior. It is defined as the average likelihood that a customer purchases from the focal firm in a period (t), given that this customer has also purchased in the previous period ($t - 1$). Theoretically, the retention rate differs for each customer but can be approximated by the average retention rate of a (homogeneous) customer group or segment. Rarely any distinction is made between the individual level and the average retention rate. A key assumption of the retention rate concept is that once customers leave the relationship, they are gone forever, so it does not allow for temporary dormancies. Managers need to assess whether the dormancy phenomenon plays a major or minor role in their business.

Conceptually, the retention rate is also limited to an activity-based perspective; it only relates to the act of repurchase. However, loyal customers might not only buy more regularly, but also buy more offerings from the firm (i.e., cross-buying). This action increases the customer's value contribution to the firm, beyond the considerations of the retention rate.

Table 6.1 Selected loyalty measurement metrics/methods

Uni-dimensional Perspective		Multi-dimensional Perspective
Metric/method	Associated loyalty behavior	
Retention rate	Repurchase/retention	Customer lifetime value (CLV)
Size-of-wallet	Cross-buying	
Share-of-wallet		
Contingent valuation	Willingness-to-pay	
Conjoint analysis		
Becker-DeGroot-Marschak (BDM) lottery		
Customer referral value (CRV)	Word-of-mouth	

Popular surrogate measures of a customer's value contribution include the size-of-wallet or share-of-wallet. The size-of-wallet refers to the amount of a customer's total spending in a given category; share-of-wallet captures the proportion of a customer's wallet accounted for by a specific firm. The share-of-wallet thus indicates the degree to which a customer satisfies his or her needs in the category with that focal firm. Generally speaking, share-of-wallet is a value-based loyalty measure that clarifies how important the firm is for an individual customer in terms of his or her expenditures in the category. In addition to using share-of-wallet to track the development of an individual customer's value contribution over time, this measure can help separate profitable from unprofitable customers. This latter purpose requires the combination of share- and size-of-wallet. Even if two customers have the same share-of-wallet, they could differ in their overall profitability, depending on their total spending in a category (i.e., size-of-wallet).

In contrast with repurchase or cross-buying, a customer's willingness-to-pay reflects the maximum amount of money he or she is willing to pay for a certain good, so it cannot be observed directly (Kalish and Nelson, 1991). Each method for measuring willingness-to-pay represents only an estimate of the true value, though several procedures have been proposed for eliciting more accurate estimates. These methods mainly differ in whether they require real economic commitments from the respondents or instead elicit hypothetical willingness-to-pay (Voelckner, 2006), such as through contingent valuation (e.g., Mitchell and Carson, 1989) or conjoint analysis (e.g., Kohli and Mahajan, 1991). In contingent valuation studies, respondents directly state the maximum amount of money they are willing to pay for a product (Mitchell and Carson, 1989) or decide whether they would buy a product at a given price (Wertenbroch and Skiera, 2002). Conjoint analysis instead is indirect and derives a respondent's willingness-to-pay from rankings or ratings of various product choice decisions, each of which is defined by a set of attributes (including price). Methods to elicit real willingness-to-pay instead require respondents to pay a stated price or price of the chosen product. One of the most prominent methods is the BDM lottery (ibid.), which requires a customer to state the maximum price he or she is willing to pay for the product. However, unlike contingent valuation, the respondent has to buy the product if an issued price, randomly determined in a lottery setting, is lower than his or her initially stated price, whereas if the randomly determined price is higher, the product becomes unavailable for purchase. Because this method avoids requiring customers to determine the purchase price directly, it ensures that customers more accurately reveal their willingness-to-pay (ibid.). Despite their higher measurement accuracy, managers often refrain from using real willingness-to-pay measures, which can be very costly to implement (Voelckner, 2006). However, if they use a hypothetical measure, managers need to consider that the resultant willingness-to-pay is likely overestimated, so to avoid overpricing they must adjust their final estimates accordingly (Wertenbroch and Skiera, 2002).

A sophisticated metric to measure word-of-mouth communication at the individual customer level is customer referral value (CRV; Kumar, Petersen, and Leone, 2007; Kumar and Reinartz, 2012). The CRV quantifies the value contribution that a customer makes to a firm through his or her referral behavior. A customer's referral can reduce acquisition costs while also enhancing the future buying behavior of the newly acquired customer, so the CRV distinguishes between 'Type 1' and 'Type 2' referrals. For example, if a new customer would not have joined a firm without the referral, the referral is Type 1, and the savings in acquisition costs and business earned from the newly acquired customer both get included in the calculation of the CRV of the customer who offered the referral. If the new customer would have joined the firm anyway, even without the referral, this Type 2 referral requires a CRV measure that relies only on the present value of savings in acquisition costs.

Finally, CLV not only translates various behavioral characteristics of a loyal customer into a single value for that customer (i.e., multivariate perspective), but it also is a forward-looking metric. Evaluating the long-term economic value of a customer to the firm is a critical goal, as a direct consequence of the shift from transactional to relational marketing. Because customer loyalty does not translate automatically into customer profitability (as we discussed in the previous section), CLV can focus managers' attention on the actual profits a customer generates, rather than on proxies for profits such as loyalty. Unlike the preceding (univariate) measures, CLV also accounts not only for a customer's revenue stream but also for the costs associated with initializing, maintaining, and growing the relationship with the customer.

Formally, CLV refers to the net present value of future profits a customer provides over his or her entire life with the firm. In its simplest definition, the CLV of customer *i* is the sum of his or her discounted gross contribution over the time horizon *T* (Kumar and Reinartz, 2012):

$$CLV_i = \sum_{t=1}^{T} GC_{it}\left(\frac{1}{1+\delta}\right)^t \tag{6.1}$$

where:
i = customer;
t = time period;
T = observation time horizon;
δ = interest (or discount) rate; and
GC_{it} = gross contribution of customer i at time t.

This formulation assumes that gross contribution and retention rates remain constant over the customer's lifetime. Moreover, the simple CLV formulation does not include acquisition costs. A more comprehensive formulation that incorporates both retention probabilities and acquisition costs is as follows (Kumar and Reinartz, 2012):

$$CLV_i = \left(\sum_{t=1}^{T}\left(\prod_{k=1}^{t} Rr_k\right) GC_{it}\left(\frac{1}{1+\delta}\right)^t\right) - AC_i \tag{6.2}$$

where:
i = customer;
t = time period;
T = observation time horizon;
δ = interest (or discount) rate;
Rr_t = average retention rate at time t;
GC_{it} = gross contribution of customer i at time t; and
AC_i = costs of acquiring customer i (acquisition costs).

In this formulation, the CLV concept integrates several univariate metrics and thus can account for various behavioral facets of a loyal customer simultaneously. For example, the retention rate relates directly to the customer's repurchase behavior, but the development of the gross contribution may capture expected changes in a customer's cross-buying behavior or willingness-to-pay.

In the same manner, additional loyal customer benefits, such as word-of-mouth communication, could be incorporated into the CLV calculation. To do so, we would need to determine the savings in acquisition costs due to customer *i*'s referrals (AC savings = ACS_i), as well as the number of new acquisitions at time t (n_{it}) that arose through the referrals issued by customer i. Then, the CLV formula, accounting for a customer's referral value, extends to:

$$CLV_i = \left(\sum_{t=1}^{T}\left(\prod_{k=1}^{t} Rr_k\right)(GC_{it} + n_{it}ACS_i)\left(\frac{1}{1+\delta}\right)^t\right) - AC_i \quad (6.3)$$

By including n_{it}, this extended CLV model acknowledges that a customer's referral value depends on two essential factors: his or her *willingness* to engage in positive word-of-mouth and his or her *capability* to do so (Kiss and Bichler, 2008). The latter factor acknowledges that some opinion leaders have a stronger influence on their peers and that customers differ with respect to the size of their social network. Thus, the stronger a customer's willingness or capability to recommend a firm, the higher is the resulting n_{it} in the CLV model, and the more important is the customer's referral behavior in determining his or her total lifetime value. Kumar et al. (2007) also demonstrate that in many situations, referral value adds considerably to a customer's overall CLV, such that many low-CLV customers (measured by their own transactions with the firm) become almost as valuable as those with the highest CLV when the calculation includes their referral value. Yet a customer's willingness and capability to make a referral is difficult to quantify. Surveys and questionnaires may offer some insights, but they are also costly, time consuming, and often not reliable. Recent research recommends network metrics though, such as centrality or betweenness degrees, to approximate these aspects (e.g., Kiss and Bichler, 2008; Lee, Cotte, and Noseworthy, 2010).

In summary, using the CLV metric offers an elegant way to measure and quantify customer loyalty in the context of the satisfaction–loyalty–profit chain and directly connects loyalty with profitability. In addition, this forward-looking metric supports long-term decision-making. As a flexible measure, it can adapt to the specific business context of any industry and evaluate each individual customer according to his or her expected contributions to the company. Yet all forecasts are inherently subject to uncertainty. Being aware of the assumptions (especially for the prediction of the retention rate, GC, and the referral capability) is important for achieving an accurate interpretation of CLV and implementing appropriate actions.

MANAGING CUSTOMER LOYALTY

When firms want to generate and maintain a loyal and profitable customer base, their organizational set up likely has a catalyzing or detracting influence (Kumar and Reinartz, 2012; Venkatesan, Kumar, and Reinartz, 2012). A prerequisite for an appropriate organizational set up is a cross-functional integration of processes, people, operations, and marketing capabilities, enabled through information, technology, and applications (Reinartz, Krafft, and Hoyer, 2004; Becker, Greve, and Albers, 2009). The still prevalent notion that managing customer relationships affects only activities and processes in sales and customer services is, quite simply, false. Rather, best-in-class organizations adopt cross-functional processes instead of functional silos (Kumar and Reinartz, 2012). With a process perspective, managers must think more deeply about the purpose of their activities and the expected outcomes. Likewise, this approach better aligns employee incentives with customer management objectives. In general, it is important that 'marketing thought' penetrates other functions, so that it is included in the mission of generating and maintaining a base of loyal customers.

A second organizational prerequisite is that the firm adopts a distinctive customer orientation, which may require a shift in focus from a product-centric to a customer-centric organization. A product-centric perspective implies selling the firm's offerings to anyone willing to buy them; a customer-centric approach makes the customer the focal point of all business decisions. A customer orientation also requires top management's commitment to the notion that the customer is the center of all activity. Without a corresponding structural design and reward system, the outcomes could be highly disappointing. A customer orientation recognizes that customers are heterogeneous in their needs and in their individual value contributions to the firm. Thus, the firm is ready to treat different customers differently.

Finally, managing a loyal customer base at the operational level can involve multiple forms; which is good, because in practice, processes and activities must align with the specific context (e.g., industry, customer base). However, to ensure that all implemented activities and processes remain in compliance with the overarching objective, it is helpful to impose some structure. The vast scope of potential activities and processes thus can be described according to two dimensions (Reinartz et al., 2004): (1) a customer dimension, which pertains to the changing phases of a customer–firm relationship (i.e., customer acquisition, growth, retention, and exit), and (2) a management dimension, involving managerial processes and activities. The management dimension might be analytical (i.e.,

		Customer Dimension		
		Acquisition Stage	Growth and Retention Stage	Decline and Exit Stage
Management Dimension	Analytical CRM			
	Operational CRM			

Source: Kumar and Reinartz (2012, p. 40).

Figure 6.6 CRM implementation matrix

activities and processes to obtain a good understanding of customer needs, behaviors, and expectations and their respective value) or operational (i.e., activities and processes to roll out and manage interactions with customers across demands) in nature. The resulting CRM implementation matrix (Figure 6.6) allows managers to map a set of relevant managerial activities and processes onto various phases of the customer–firm relationship.

CONCLUSION

This review of literature related to customer loyalty reveals that a loyal customer base is by no means a sufficient condition, because it does not automatically translate into higher profitability. That is, using loyalty as a proxy for customer profits is not viable. Instead, accurate measures of loyalty and a rigorous, firm-specific understanding of both its antecedents and its consequences are critical for managing loyalty effectively. Only with the appropriate organizational set up can loyalty ultimately generate the desired results.

As we look ahead, the major challenges facing academics and practitioners for loyalty management include the following:

- Developing a more fine-grained understanding of the context-specific linkages in the satisfaction–loyalty–profit chain at the individual customer level.
- Evaluating the implications of the growing popularity of social media for the loyalty management process. Social media not only increase the engagement level of consumers but also reduce the communication gap between consumers and companies in various (i.e., business-to-business, business-to-consumer, and consumer-to-consumer) markets.
- Understanding and outlining how loyalty management might harness the potential and cope with the challenges of unstructured

data generated by websites, social media platforms, and mobile devices (i.e., 'big data').

We hope this chapter, by providing a summary of the status quo in loyalty management research, also offers a helpful starting point for stimulating additional research in these areas.

REFERENCES

Anderson, E.W. (1998), 'Customer satisfaction and word-of-mouth', *Journal of Service Research*, **1**(1), 5–17.

Anderson, E.W. and V. Mittal (2000), 'Strengthening the satisfaction–profit chain', *Journal of Service Research*, **3**(2), 107–20.

Anderson, E.W. and M.W. Sullivan (1993), 'The antecedents and consequences of customer satisfaction for firms', *Marketing Science*, **12**(2), 125–43.

Anderson, E.W., C. Fornell and D.R. Lehmann (1994), 'Customer satisfaction, market share, and profitability: findings from Sweden', *Journal of Marketing*, **58**(3), 53–66.

Anderson, E.W., C. Fornell and R.T. Rust (1997), 'Customer satisfaction, productivity, and profitability: differences between goods and services', *Marketing Science*, **16**(2), 129–45.

Becker, J.U., G. Greve and S. Albers (2009), 'The impact of technological and organizational implementation of CRM on customer acquisition, maintenance, and retention', *International Journal of Research in Marketing*, **26**(3), 207–15.

Bolton, R.N. and K.N. Lemon (1999), 'A dynamic model of customers' usage of services: usage as an antecedent and consequence of satisfaction', *Journal of Marketing Research*, **36**(2), 171–86.

Bowman, D. and D. Narayandas (2001), 'Managing customer-initiated contacts with manufacturers: the impact on share of category requirements and word-of-mouth behavior', *Journal of Marketing Research*, **38**(3), 281–97.

Brown, G.H. (1952), 'Brand loyalty – fact or fiction?', *Advertising Age*, **23**(25), 53–5.

Churchill, H. (1942), 'How to measure brand loyalty', *Advertising and Selling*, **35**(24), 11–16.

Dick, A.S. and K. Basu (1994), 'Customer loyalty: toward an integrated conceptual framework', *Journal of the Academy of Marketing Science*, **22**(2), 99–113.

Eisenbeiss, M., M. Cornelissen, K. Backhaus and W.D. Hoyer (2014), 'Nonlinear and asymmetric returns on customer satisfaction: do they vary across situations and consumers?', *Journal of the Academy of Marketing Science*, **42**(3), 242–63.

Fornell, C. (1992), 'A national customer satisfaction barometer: the Swedish experience', *Journal of Marketing*, **56**(1), 6–21.

Fournier, S. and D.G. Mick (1999), 'Rediscovering satisfaction', *Journal of Marketing*, **63**(4), 5–23.

Homburg, C., N. Koschate and W.D. Hoyer (2005), 'Do satisfied customers really pay more? A study of the relationship between customer satisfaction and willingness to pay', *Journal of Marketing*, **69**(2), 84–96.

Ittner, C.D. and D.F. Larcker (2003), 'Coming up short on nonfinancial performance measurement', *Harvard Business Review*, **81**(11), 88–95.

Jacoby, J. and R.W. Chestnut (1978), *Brand Loyalty: Measurement and Management*, New York: John Wiley & Sons.

Jones, T.O. and W.E. Sasser Jr. (1995), 'Why satisfied customers defect', *Harvard Business Review*, **73**(6), 88–91.

Kahneman, D. and A. Tversky (1979), 'Prospect theory: an analysis of decision under risk', *Econometrica*, **47**(2), 263–91.

Kalish, S. and P. Nelson (1991), 'A comparison of ranking, rating and reservation price measurement in conjoint analysis', *Marketing Letters*, **2**(4), 327–35.
Keiningham, T.L., T. Perkins-Munn and H. Evans (2003), 'The impact of customer satisfaction on share-of-wallet in a business-to-business environment', *Journal of Service Research*, **6**(1), 37–50.
Kiss, C. and M. Bichler (2008), 'Identification of influencers – measuring influence in customer networks', *Decision Support Systems*, **46**(1), 233–53.
Kohli, R. and V. Mahajan (1991), 'A reservation-price model for optimal pricing of multiattribute products in conjoint analysis', *Journal of Marketing Research*, **28**(3), 347–54.
Kumar, V. and W. Reinartz (2012), *Customer Relationship Management*, 2nd edition, Heidelberg: Springer.
Kumar, V., A. Petersen and R.P. Leone (2007), 'How valuable is word-of-mouth?', *Harvard Business Review*, **85**(10), 139–46.
Lee, S.H., J. Cotte and T.J. Noseworthy (2010), 'The role of network centrality in the flow of consumer influence', *Journal of Consumer Psychology*, **20**(1), 66–77.
McFarland, K. (2008), 'Why Zappos offers new hires $2,000 to quit', *Businessweek.com*, 16 September.
Mitchell, R.C. and R.T. Carson (1989), *Using Surveys to Value Public Goods: The Contingent Valuation Method*, Washington, DC: Resources for the Future.
Mittal, V. and C. Frennea (2012), 'Managing customer satisfaction', in V. Shankar and G.S. Carpenter (eds), *Handbook of Marketing Strategy*, Cheltenham, UK and Northampton, MA, USA: Edward Elgar Publishing.
Mittal, V. and W.A. Kamakura (2001), 'Satisfaction, repurchase intent, and repurchase behavior: investigating the moderating effect of customer characteristics', *Journal of Marketing Research*, **38**(1), 131–42.
Mittal, V., W.T. Ross Jr. and P.M. Baldasare (1998), 'The asymmetric impact of negative and positive attribute-level performance on overall satisfaction and repurchase intentions', *Journal of Marketing*, **62**(1), 33–47.
Oliver, R.L. (1999), 'Whence consumer loyalty?', *Journal of Marketing*, **63**(Special Issue), 33–44.
Oliver, R.L., R.T. Rust and S. Varki (1997), 'Customer delight: foundations, findings, and managerial insight', *Journal of Retailing*, **73**(3), 311–36.
Reichheld, F.F. (1996), *The Loyalty Effect*, Boston, MA: Harvard Business School Press.
Reichheld, F.F. (2003), 'The one number you need to grow', *Harvard Business Review*, **81**(12), 46–54.
Reichheld, F.F. and W.E. Sasser (1990), 'Zero defections: quality comes to services', *Harvard Business Review*, **68**(5), 105–11.
Reinartz, W.J. and V. Kumar (2000), 'On the profitability of long-life customers in a noncontractual setting: an empirical investigation and implications for marketing', *Journal of Marketing*, **64**(4), 17–35.
Reinartz, W.J. and V. Kumar (2002), 'The mismanagement of customer loyalty', *Harvard Business Review*, **80**(7), 86–94.
Reinartz, W.J., M. Krafft and W.D. Hoyer (2004), 'The customer relationship management process: its measurement and impact on performance', *Journal of Marketing Research*, **41**(3), 293–305.
Seiders, K., G.B. Voss, D. Grewal and A.L. Godfrey (2005), 'Do satisfied customers buy more? Examining moderating influences in a retailing context', *Journal of Marketing*, **69**(4), 26–43.
Venkatesan, R., V. Kumar and W.J. Reinartz (2012), 'Customer relationship management in business markets', in G.L. Lilien and R. Grewal (eds), *Handbook of Business-to-Business Marketing*, Cheltenham, UK and Northampton, MA, USA: Edward Elgar Publishing.
Voelckner, F. (2006), 'An empirical comparison of methods for measuring consumers' willingness-to-pay', *Marketing Letters*, **17**(2), 137–49.
Wertenbroch, K. and B. Skiera (2002), 'Measuring consumers' willingness-to-pay at the point of purchase', *Journal of Marketing Research*, **39**(2), 228–41.

Woodruff, R.B., E.R. Cadotte and R.L. Jenkins (1983), 'Modeling consumer satisfaction processes using experience-based norms', *Journal of Marketing Research*, **20**(3), 296–304.

Zeithaml, V.A. (2000), 'Service quality, profitability, and the economic worth of customers: what we know and what we need to learn', *Journal of the Academy of Marketing Science*, **28**(1), 67–85.

Zeithaml, V.A., L.L. Berry and A. Parasuraman (1996), 'The behavioral consequences of service quality', *Journal of Marketing*, **60**(2), 31–46.

7. Leveraging product returns to maximize customer equity

J. Andrew Petersen and Eric T. Anderson

INTRODUCTION

Product returns occur when a customer brings back a previously purchased product with the intent to receive a repair, refund, or replacement. Product returns provide a significant expense to businesses, with some studies showing that product returns cost US manufacturers and retailers an estimated $100 billion annually, reducing profits by around 3.8 percent on average per retailer or manufacturer (Blanchard, 2007). And, this cost is only likely to rise as retailers increasingly engage in multi-channel strategies. The increasing flexibility of retailers to allow customers to shop and purchase both online and offline has offered numerous benefits to consumers such as increases in value and convenience. However, it has also impacted customer product return behavior. Product returns are potentially more costly in this environment as customers cannot 'touch and feel' the product before purchase online and purchases online are often returned to the retail store (Ofek, Katona, and Sarvary, 2011). This is likely to decrease profit for the retailer through both increases in product returns and increases in product return management costs.

One solution to this product return cost problem can come in the form of increases in reverse supply chain efficiency, a problem that has been addressed at length by research in logistics and supply chain management (see Dekker et al., 2004 for a thorough review of the role of product returns in the supply chain and reverse logistics literature). By increasing the efficiency of the reverse supply chain the firm can effectively lower the costs of managing product returns without changing the firm–customer exchange process. Alternatively, another solution to this product return cost problem can come in the form of changes in customer management strategies (e.g., product return policies). Increasing product return policy strictness would likely lead to reductions in the number of, and likely the total cost, of product returns for the firm. However, it is not clear if making it more difficult for customers to return products is ideal. Research has shown that increasing the 'hassle' for customers to return products can negatively affect a customer's willingness to purchase (Nasr

Bechwati and Siegal, 2005), decrease total customer demand for products by reducing a customer's option value to purchase (Anderson, Hansen, and Simester, 2009a), and adversely impact long-term customer value through reductions in future purchases (Petersen and Kumar, 2009). This raises a key question. Given the known and potential costs along with the benefits of customer product return behavior, how can firms optimally manage customer product returns to maximize customer equity?

The purpose of this chapter is to address this key research question. We do this in the following manner. First, we develop a broad conceptual framework that describes how product returns have an impact throughout all aspects of the firm–customer exchange process. Specifically, we identify the ways research has shown that product return policies and customer product return behavior play a role in the way firms act, how customers think, how customers behave, and how firms can become more profitable. Next, we discuss how firms can approach setting optimal product return policies and potentially leverage information about customer product return behavior to maximize customer equity. Finally, we discuss opportunities to explore new research to further enhance knowledge on the role of product returns in the firm–customer exchange process.

THE ROLE OF PRODUCT RETURNS IN THE FIRM–CUSTOMER EXCHANGE PROCESS

We can generally think of the firm–customer exchange process as being a series of four interrelated steps that help to describe the relationship between the customer and the firm dynamically over time. These steps include: (1) what firms do, (2) what customers think, (3) what customers do, and (4) what firms get (Gupta and Zeithaml, 2006). We adapt this conceptual model to specifically reflect the role of product return policies and customer product return behavior in the firm–customer exchange process (Figure 7.1).

In Figure 7.1 we focus on the key relationships between each of the steps of the firm–customer exchange process and decisions by the customer and firm with regard to product return policies and product return behavior. We expand on this discussion in the following sections.

What Firms Do

When evaluating how to handle product return issues firms have two key decisions to make. First, the firm is able to set product return policies that affect each customer's ease or difficulty of returning a product. Second,

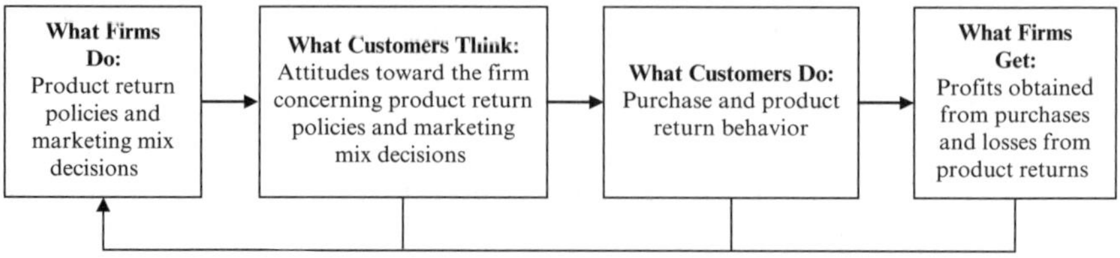

Source: Adapted from Gupta and Zeithaml (2006).

Figure 7.1 Product returns and the firm–customer exchange process

the firm also can make strategic and targeted marketing mix decisions that serve not only to influence customer purchase behavior, but also to directly or indirectly influence customer product return behavior.

Product return policies

There is a high degree of variance in product return policy strictness across different retailers and even within different departments of the same retailers (Anderson et al., 2009a). A recent survey of retailers[1] found that many stores are actually continuing to either keep the same return policy (83 percent) or even increase the return policy strictness (7 percent). Often many of these increases in strictness come in the form of setting different and often complicated rules across different channels (online vs offline), categories, or even items. The high variation in product return policies across firms could be a result of several things. It might result from managers not knowing the true impact of product returns on customer value. However, product return policies may also serve a strategic purpose for a given firm. For instance, a lenient product return policy may be an effort by a firm to signal product quality or allow customers to try several sizes of a given product and return those that do not fit.

In general, from a regulatory perspective, firms can make decisions on whether to allow product returns in the first place (i.e., all sales final), whether to charge a non-refundable fee for the return of a product (e.g., restocking fees or return shipping fees) (Bower and Maxham, 2012), or whether to put a temporal constraint on the product return decision (e.g., 30-day return window). Many customers value more lenient product return policies and this has been shown to increase overall demand (Wood, 2001). However, firms are often wary of lenient product return policies because increases in overall demand are also often correlated with increases in the total number of products returned (Davis, Hagerty, and Gerstner, 1998). In fact, usually as the purchase rate increases, the product return rate stays the same or increases at an incrementally faster rate.

Additionally, recent research has shown that firms often allocate fewer marketing efforts to customers who return more products, even if it is inevitably detrimental to the firm's profitability (Petersen and Kumar, 2009). This makes it unclear whether it is ideal for a firm to have a lenient versus strict return policy and whether this policy should even vary across departments within the firm or across customers. Given this trade-off between the benefits of product return policies between customers and firms, firms have to make key decisions on the optimal level of 'hassle' (Davis et al., 1998) to put in a product return policy to maximize net demand, or profit from purchases net of the cost of product returns, rather than just trying to maximize purchase behavior or minimize product return behavior.

Marketing mix decisions

There are a number of marketing mix decisions (product, price, place, and promotion) that firms can make that have been shown to directly (or indirectly) affect the customer's decision and ability to return products and the firm's ability to handle product returns profitably.

From a *product* perspective, firms often find product return rates in different product categories to vary dramatically, with some product categories seeing almost no product returns (e.g., socks, gloves) and other product categories like shoes and swimwear having return rates of over 25 percent (Hess and Mayhew, 1997). Thus, it is critical to first understand how the category of the product might affect the product return rate. Additionally, firms cannot necessarily assume that decisions to augment products with services will lead to differences in customer behavior (i.e., reductions in product return behavior). For instance, research suggests that the strategic implications of managing customer product return behavior and managing customer complaint behavior is quite similar, where customer complaints should be allowed to a certain extent to maximize profitability (Fornell and Wernerfelt, 1988; Petersen and Kumar, 2009). This suggests that it is important for firms to understand the characteristics of products being sold when trying to simultaneously understand customer purchase and product return behavior.

From a *price* perspective, firms often set price levels in order to affect demand for products, where lowering prices is often done to increase demand for products. For instance, Anderson et al. (2009b) found that perceived value hypothesis explained why lower-priced products had lower return rates compared to higher-priced products all else being equal. Similarly, Petersen and Kumar (2009) found that products purchased on sale are less likely to be returned. This finding suggests that pricing decisions by firms not only impact the revenue (and gross margin) generated per sale, but can also have an impact on the propensity of a customer to return the product. It also suggests that lowering the price on a product might hurt the margins when the product is sold, but it might also decrease the costs incurred by the firm due to a decrease in products returned. Thus, decisions on pricing need to reflect both the probability of purchase and the corresponding probability of return so the firm has the entire picture of the potential profitability of the product being sold.

From a *place* (distribution) perspective, firms have to make decisions about how to handle both forward and reverse logistics. This means that if the firm chooses a product return policy or makes another marketing mix decision that leads toward greater (or fewer) product returns, the firm not only has to consider how the customer will purchase the product, but also how the customer will potentially return the product and how the firm can potentially salvage value from the product (Shulman and Coughlan, 2007; Shulman, Coughlan, and Savaskan, 2010). This includes decisions both in terms of optimal channel structure and on how to potentially realize any salvage value for the returned product. Additionally, multi-channel retailers also have to consider the cross-channel impact of purchases that occur in one channel with returns that may occur in a different channel (Ofek et al., 2011). For instance, the decision for a brick-and-mortar store to open an online store may impact the customer's ability to 'touch and feel' the product and increase returns to the store of items that were purchased online. It may also encourage customers to purchase the product online and later return the product to the brick-and-mortar store – which may or may not be equipped to handle a potential increase in the volume of product returns.

From a *promotion* perspective, firms often allocate resources to customers based on past exchange characteristics (e.g., product return behavior) and customer characteristics (e.g., demographics). For instance, Petersen and Kumar (2009) found that a catalog retailer was sending catalogs to customers where customers who had purchased more *and* returned less received the most marketing efforts. However, the authors also found that customers who returned more products to a threshold (about 10–15 percent of what they purchased) tended to have the highest future profitability. In this case the catalog retailer was likely leaving money on the table by overinvesting in customers who were less profitable by treating product return behavior only as a negative. Recent research has shown through an extensive field study with a catalog retailer that the profit maximizing strategy to a set of customers can be realized through optimal resource allocation when both purchase and product return behavior are taken into account (Petersen and Kumar, 2014). This suggests that strategic decisions on marketing efforts directed at customers should not ignore the potential implications of setting a marketing allocation algorithm that focuses only on customer purchase behavior and ignores customer product return behavior.

What Customers Think

From a customer's perspective, the firm's product return policy and marketing mix decisions can have a direct impact on the customer's perceptions about the firm and about the products the firm sells. For instance, product return policies can signal product quality, reduce perceived risk of purchase, increase/decrease perceived hassle, post-purchase evaluation, and intentions to complain. And, these perceptions of how product returns impact customer perceptions can happen before, during, and even after purchase occurs – making the need for managing customer perceptions throughout the entire firm–customer exchange process critical.

Pre-purchase

Often product return policies and marketing mix decisions by firms are used by customers as signals. For instance, research suggests customers often perceive a strict return policy as a potential sign of lower product quality (Wood, 2001). Customers tend to believe that firms who stand behind their products with an extremely lenient product return policy are likely signaling that they believe that the product will not need to be returned. Thus, while increasing leniency is likely to proportionally increase product return rates (Davis et al., 1998), customers may choose not to purchase in the first place if they think product quality is low due to the strict product return policy. This is especially important to consider for firms in remote purchase environments (e.g., online), where perceived purchase risk is higher due to the lack of the customer's ability to experience the product before purchase.

Purchase

Customers also consider product return policies when they evaluate products at the time of purchase. When faced with a purchase decision, research has shown that customers create an option value of returning the product that is embedded in the price they are willing to pay (Anderson et al., 2009a). This suggests that as return policies become stricter the customer's willingness to pay for a given product decreases, all else being equal, because the customer fears that the product may not fit his or her needs or the product may be of lower quality. So even during the purchase process, customers are often simultaneously considering the potential of returning the product as they are making the purchase – even if the product is not likely to be returned in the end.

Post-purchase

After the purchase occurs customers are often faced with information that may or may not validate their purchase decision in their minds. For instance, when customers receive disconfirming information post-purchase it may cause the customer to return the product (Nasr Bechwati and Siegal, 2005). This research suggests that managers can impact the customer's post-purchase attitudes towards returning the product by properly designing the presentation of product alternatives at the time of purchase, for example, not allowing customers to simultaneously compare product alternatives. When a customer is unable to compare alternatives appropriately it is likely to lead to a situation where the customer experiences some purchase regret when the alternatives are evaluated after purchase. Additionally, this research suggests that limiting the time that customers can make purchase decisions (e.g., flash sales) will lead to a higher likelihood of disconfirming information and thus the desire to return products, or in the case where customers cannot return the product it might also cause the customer to lower their future purchase probability.

What Customers Do

When firms draw inferences on expected customer demand, it is important to consider both customer purchase behavior and customer product return behavior, otherwise the estimate for the actual demand will be biased (Anderson et al., 2009a). A model that is able to account for both the purchase and product return behavior of customers begins to identify the potential gaps between products and customers' preferences, which were not observable by the firm or the customer at the time of the sale. However, research has shown that it is possible to begin to explain which customer behaviors describe the likelihood of a customer to return a product.

For instance, research (Anderson et al., 2009b; Petersen and Kumar, 2009; Shah et al., 2012) has empirically shown the following:

- Products purchased as gifts are less likely to be returned.
- Products purchased during the holiday season are more likely to be returned.
- Products purchased in new product categories (i.e., cross-buying behavior) are more likely to be returned.
- Similar products purchased in new channels (i.e., multi-channel shopping) are less likely to be returned.
- Products purchased in new categories and new channels are more likely to be returned.
- Products purchased on sale are less likely to be returned.

This suggests that even behaviors that have been linked to increases in customer profitability from purchase behavior, such as through cross-buying, can also have a positive impact on customer product return behavior (Shah et al., 2012). Thus, it is important that firms leverage data they already collect on other exchange characteristics when trying to explain customer product return behavior, especially those metrics that have already been linked to a customer's purchase behavior.

It is also the case that product return behavior can have an impact on future customer purchase behaviors. Research has shown that customers who return more products to a threshold (about 10–15 percent) are more likely to purchase incrementally more in the future (Petersen and Kumar, 2009) and provide more referrals on behalf of the firm (Petersen and Kumar, 2010). So, even if it is the case that the product return behavior does have a negative short-term effect on the firm's profit through increases in costs associated with the reverse logistics of product returns, it is possible that future purchase behavior or customer referrals can create additional value in the long term for the firm through more effective customer acquisition and retention.

What Firms Get

In the short term, firms do incur costs from managing customer product returns through both reverse logistics and profits lost from returned products. The key question is whether this customer product return behavior also has an impact on customer and firm profit into the future. Evidence suggests that firms can receive benefits from the proper management of product return behavior in both the short and long term.

Short-term profitability

First, a firm must understand how the customers interact with the firm across the different channels the firm provides. The optimal channel structure allows firms to lower the cost of product returns by understanding the flow of products out and potentially back into the firm and potentially provide an opportunity for the firm to receive some salvage value from the products that are returned (Shulman and Coughlan, 2007; Shulman et al., 2010). Also, a model that understands the customer's option value of returning products can help firms optimize return policies across products and categories and can potentially lead to increases of average purchase rates of more than 50 percent (Anderson et al., 2009a). Finally, understanding past purchase and product return behavior allows firms to optimally allocate resources to individual customers to maximize profitability (Petersen and Kumar, 2009). Firms can focus on allocating more resources to customers based on both their purchase and product return behavior to maximize profitability. A recent field study even showed that optimal resource allocation when accounting for customer product return behavior over a six-month time period led to increases in the short-term profitability from customers when compared to several other resource allocation strategies that either ignored or punished customers through reductions in marketing efforts for increases in product return behavior (Petersen and Kumar, 2014).

Long-term profitability

If a firm is able to obtain profitability in the short term by better managing customer product return behavior, then repeating this strategy over time should lead to higher profitability in the long term as well. However, research has also found additional long-term benefits for better short-term customer product return management strategies. For instance, Petersen and Kumar (2014) ran a study in which five random samples of 5200 customers each were allocated resources using four different resource allocation algorithms (and one control group that received no marketing efforts). The authors found that the group that was allocated resources taking into account both purchase and product return behavior had a higher post-study customer lifetime value (CLV) than the other four groups at the end of the six-month field study – where all groups had the same pre-study CLV level. This suggests that better product return management not only provides short-term gains in profit, but can also upwardly change the trajectory of a customer's long-term profitability as well.

LEVERAGING CUSTOMER PRODUCT RETURNS

The research in marketing to this point has provided evidence of the following two things: (1) ignoring customer product return behavior when managing customers is *bad* and (2) punishing customers through increasing strictness in return policies and reductions in marketing efforts is *worse*. The purpose of this section of the chapter is to begin to outline a strategy to leverage knowledge about product return policies and customer product return behavior to maximize customer equity. We see four key areas that when addressed properly will help firms better understand how customer product returns will potentially drive what customers think, what customers will do, and eventually the benefits the firms will reap. These areas are:

- setting an optimal product return policy;
- understanding how marketing mix decisions impact product returns;

- linking metrics with product return behavior;
- understanding the customer's value to the firm.

In the following subsections our goal is to walk through examples in the literature that have started to address these key issues and give a potential roadmap to managers on how to use information about product return policies and customer product return behavior to maximize customer equity.

Setting an Optimal Product Return Policy

Research has shown that a lenient product return policy is likely better than a strict return policy – at least in the many contexts where it was empirically tested. However, a key challenge in understanding the optimal product return policy for a firm is how changes in the current product return policy to a new product return policy will affect customer purchase and return behavior. In addition, it is critical to understand how a new return policy will affect the cost of managing reverse logistics due to increases or decreases in product returns. From a reduced-form modeling perspective the only way to understand how customers will respond to changes in a product return policy is to create sufficient variance in product return policy over time. This is often neither feasible nor desirable for a firm since frequent changes in product return policies are likely to confuse and frustrate customers and can even have unintended effects on purchase and product return behavior that are unrelated to the return policy itself and only related to the frequency of return policy change.

So how can a firm determine an optimal return policy? It is possible to create an econometric structural model of purchase incidence and return behavior that can help firms determine optimal product return policies without necessitating frequent changes in product return policies over time. Anderson et al. (2009a) provide a great example to consider. The authors develop an econometric structural model of product returns, allowing them to construct counterfactuals that provide product-category–specific optimal product return policies.

The model in Anderson et al. (2009a) describes a customer's decision to either return or keep a product post-purchase. Consider a customer i who makes a purchase at a given time period t. After purchase the customer has to decide whether to keep or return the product, where the decision to keep the product takes the following utility:

$$U(return)_{it} = -R_i$$

And, the decision to keep the product takes the following utility:

$$U(keep)_{it} = \mu_{it} + \psi_{it} + \varepsilon_{it}$$

Where:

$$\mu_{it} \equiv \beta'_i X_t$$

Here the utility of returning the item is represented by the cost of the return to the customer, which is always considered to be negative and can vary across customers but not over time. This cost is assumed to include both the hassle of the return as well as the financial return cost. The utility of keeping the product is made up of deterministic utility (μ_{it}), which is known by the customer and firm at the time of purchase, the fit of the transaction (ψ_{it}), which is known only by the customer and not the firm after the transaction has been completed, and the standard econometric error component (ε_{it}). Thus, the decision to keep or return the product purchased is determined by the net utility of keeping the item versus returning the item, which is represented by the following equation:

$$U^K_{it} = \mu_{it} + \psi_{it} + R_i + \varepsilon_{it}$$

The customer will choose to keep the product as long as the utility for keeping the product is greater than 0 ($U^K_{it} > 0$). Otherwise the customer will choose to return the product. Using this formulation of the problem it is now possible to derive expressions for the three possible outcomes: (1) no order, (2) order and return, and (3) order and keep. Additionally, it is possible to determine the option value that the customer places on returning the product and in turn how the net demand (i.e., purchases – product returns) is affected as R_i is finite (lower return cost) to when R approaches ∞ (infinite return cost), which is equivalent to the firm prohibiting product returns. For complete details on the set up and estimation, the reader is encouraged to refer to Anderson et al. (2009a).

The authors test their modeling framework on data from a mail-order catalog retailer that primarily sells private label men's and women's clothing on three categories of products: (1) women's tops, (2) men's tops, and (c) women's footwear. With regard to optimal return policies the authors ran counterfactuals that allowed the return cost to the customer (R_i) and determined at which point net demand was maximized. An interesting finding from this analysis is the fact that the optimal return policy varies by category and by customer. For instance, the authors found that it was optimal for the firm to have a stricter return policy on men's tops than on women's tops. In fact, it was optimal to not allow any returns on men's tops.

What this example shows is that it is possible for a firm to determine its optimal product return policy without making frequent changes to its current policy. Using a model such as this, a firm can determine how to vary its return policy by product category or by customer type (e.g., elite loyalty customers may have more return privileges) to maximize gross profit.

Understanding How Marketing Mix Decisions Impact Product Returns

Often when projecting inventory needs, many firms only take into account the outward flow of purchased merchandise and not the reverse flow of returned merchandise. For instance, if a firm changes the price of a product, how does that affect both the demand for purchases and the change in customer product return behavior? A current study by Anderson et al. (2009b) looks at how changes to the marketing mix, in this case price, have an effect on both the demand for purchases and the change in product return behavior.

In order to isolate the effect that price has on purchase and product return behavior, the authors run a field experiment with a catalog retailer that sells women's apparel. The field experiment separates a group of 90000 randomly selected customers into three experimental conditions. For each group of customers, each receives a catalog with the same products and layout, but the only variation happens on price. For the control group the prices stayed unchanged from the original prices (average price is about $58). For group A the prices of 55 items were increased by $1 to $6 and the prices of 53 items were lowered by $0.50 to $4.00. For group B the prices of 58 items were increased by $1.00 to $4.50 and the prices of 52 items were lowered by $0.50 to $10.00.

In general the findings across-item found similar numbers of items were purchased (16.9) and returned (3.9) across the three conditions. However, when the authors compared the within-item results they found that the product return rates were lowest for the conditions where the prices on the specific products were lowest. This suggests that lowering the price not only has a positive effect on demand, but also a positive effect on product returns (i.e., lowering the return rate).

This study provides an example of a controlled experiment that allows firms to test how changes in the marketing mix drive purchase and product return behavior for customers in general. Managers can run experiments like this with other marketing mix variables, which will allow them to determine optimal marketing mix decisions.

Linking Metrics with Product Return Behavior

There are numerous marketing metrics that firms measure (Petersen et al., 2009). Often these metrics such as multi-channel shopping or cross-buying behavior are directly linked with desirable outcomes such as CLV or customer equity (CE). However, it is often not the case that these metrics are also linked to customer product return behavior, usually because it is assumed that these marketing metrics are always positively correlated with customer profitability. This is not always the case.

So how can a manager link relevant marketing metrics to product return behavior and manage those metrics for profitability? Here we discuss a recent example from Shah et al. (2012) that uses data from five different firms (both B2C and B2B) and describes the perils of ignoring negative behavioral traits such as product return behavior when firms engage in cross-selling activities.

Cross-buying behavior, in this case operationalized as the number of categories purchased, is often positively linked to customer profitability. And, the authors do find that at the aggregate level cross-buying is positively linked to customer profitability. However, the authors show that at the customer level, increasing cross-buying behavior does not always lead to increases in customer profitability. In fact, the most unprofitable customers for all firms in the study are the customers who cross-buy the most (it is also the case that the most profitable customers also cross-buy the most). These unprofitable customers who score highly on cross-buying were found to engage in several adverse behavioral traits that include high numbers of service requests, low revenue growth, buying always on promotion, and a large number of revenue reversals (e.g., product returns). This raises the question of how a firm can determine when it is ideal to cross-sell to a given customer since not all cross-selling activities by the firm are profitable.

The authors propose a general framework that can help managers determine when each customer should receive cross-selling efforts by focusing on answering three specific questions:

1 What is the customer's willingness to purchase?
2 What is the customer's ability to purchase?
3 What is the customer's need to purchase?

Each of these questions can be answered with the firm's internal database. For example, in another study, several of the same authors give some examples of how managers can answer these three questions (Shah and Kumar, 2008). The authors suggest that a customer satisfaction score may be a good proxy to answer the question on willingness to purchase, a credit score or measure of income may be a good proxy of a customer's ability to purchase, and a next-product-to-buy model (Knott, Hayes, and Neslin, 2002) can be used to predict a customer's need to purchase.

The study we highlighted on the potential downsides of cross-buying shows how a firm can link marketing metrics, even those that have already been gathered and measured such as cross-buying, and link them to customer product return behavior. Then, it also shows how managers can further manage those metrics to maximize net purchase behavior (purchases – product returns) rather than just driving more purchase behavior and ignoring product returns.

Understand the Customer's Value to the Firm

To this point many of the customer value frameworks have focused on measuring CLV as the sum of expected discounted future cash flows from purchases minus the sum of future expected marketing costs for a given customer. What's missing from this measure of CLV is a customer's product return behavior. If it were the case that very few customers returned products or that most customers returned about the same number or percentage of products the current measures of CLV that do not consider product returns would still over-predict the actual value of the customer. But, the general rank order of highest to lowest CLV would still generally reflect the rank order of customers by actual value.

So how much value is generated by integrating customer product returns in a customer value framework? Here we discuss a current research paper by Petersen and Kumar (2014), which first creates a new measure of CLV that includes product returns and then runs a field test with a firm to quantify the value of product returns in CLV.

CLV, when measured in a discrete time framework, is generally represented by the following equation:

$$CLV_{i,t=0} = P(Relationship_{i,t=1}) * \sum_{t=1}^{T} \frac{\pi(Purchases_{it}) - Marketing_{it}}{(1+r)^t} \quad (7.1)$$

where:
$CLV_{i,t=0}$ = customer lifetime value for customer i at time $t = 0$;
$\pi(Purchases_{it})$ = gross margin from purchases by customer i in time t;
$Marketing_{it}$ = marketing costs spent on customer i in time t;
r = discount rate (approximately 3.56 percent by quarter – or 15 percent annually);
n = all customers in the sample for equation (7.1);

T = number of time periods in the prediction horizon;
$P(Relationship_{i,t=1})$ = the probability customer i is active in the relationship at $t = 1$.

As you can see equation (7.1) does not include product returns directly in the objective function. Usually this type of framework only includes past product returns as an independent variable to predict the likelihood and amount of purchase in a given time period. The next equation in the study (7.2) does include product returns in the objective function, but only within a measure that combines both purchases and product returns (*NetPurchases*). The equation takes the following format:

$$CLV_{i,t=0} = P(Relationship_{i,t=1}) * \sum_{t=1}^{T} \frac{\pi(NetPurchases_{it}) - Marketing_{it}}{(1+r)^t} \quad (7.2)$$

where:
$CLV_{i,t=0}$ = customer lifetime value for customer i at time $t = 0$;
$\pi(NetPurchases_{it})$ = gross margin from purchases minus gross margin lost and costs incurred by returns by customer i in time t;
$Marketing_{it}$ = marketing costs spent on customer i in time t;
r = discount rate (approximately 3.56 percent by quarter – or 15 percent annually);
n = all customers in the sample for equation (7.2);
T = number of time periods in the prediction horizon;
$P(Relationship_{i,t=1})$ = the probability customer i is active in the relationship at $t = 1$.

At a minimum this equation is able to account for product returns directly, although not independently from customer purchase behavior. The newly proposed model equation (7.3) from the study includes product returns directly in the objective function for a given CLV and independently from a customer's purchase behavior. The equation takes the following format:

$$CLV_{i,t=0} = P(Relationship_{i,t=1}) * \sum_{t=1}^{T} \frac{\pi(Purchases_{it}) - \pi(Returns_{it}) - Marketing_{it}}{(1+r)^t} \quad (7.3)$$

where:
$CLV_{i,t=0}$ = customer lifetime value for customer i at time $t = 0$;
$\pi(Purchases_{it})$ = gross margin from purchases by customer i in time t;
$\pi(Returns_{it})$ = gross margin lost and costs incurred from returns by customer i in time t;
$Marketing_{it}$ = marketing costs spent on customer i in time t;

r = discount rate (approximately 3.56 percent by quarter – or 15 percent annually);
n = all customers in the sample for equation (7.3);
T = number of time periods in the prediction horizon;
$P(Relationship_{i,t=1})$ = the probability customer i is active in the relationship at $t = 1$.

Then the question arises of how much benefit a firm obtains from integrating customer product return behavior directly into a measure of CLV. To answer this question the authors of the study ran a field study with a B2C company that sells footwear, apparel, and other accessories through the Internet and mail-order catalogs. The authors took a random sample of customers and allocated them to each of the different equations, in this case 5200 customers were allocated to each segment. Then for a three-month time period the authors determined the optimal marketing resource allocation to each customer in each segment using the CLV objective functions (equations 7.1, 7.2, and 7.3) and then continued to observe purchase and product return behavior for another three months.

The authors found that the sample associated with segment equation (7.1) outperformed the other two samples in both the short and long term. In the short term, customers who were allocated resources using equation (7.3) as the objective function to maximize purchased more, returned fewer products, and required less marketing efforts over the six-month time period than customers in the equation (7.1) and (7.2) samples. In fact the average profit of \$352.77 for the sample in equation (7.3) was significantly higher than the average profit for equation (7.1) (\$273.48) or equation (7.2) (\$294.21). In the long term, the authors re-estimated CLV for the customers in each of the samples after the end of the study. The authors found that the post-study CLV was significantly higher for the sample using equation (7.3) (\$1402.96) when compared to the sample using equation (7.1) (\$1172.17) or equation (7.2) (\$1223.16), where the CLV pre-study was similar for all groups.

This study showed that it is possible to integrate customer product returns into a CLV framework and that using traditional CLV models that generally ignore customer product return behavior significantly lowers the potential profitability of the firm.

OPPORTUNITIES FOR FUTURE RESEARCH

While there has been interest recently in research related to customer product return behavior, there is still relatively little known about the role of product returns in the firm–customer exchange process. There needs to be continued theoretical and empirical research to help both academics and practitioners understand not only how product returns directly impact customer and firm value, but also how customer product returns impact other customer behaviors (e.g., cross-buying, multi-channel shopping, etc.). Here are some potential avenues for future research.

There are several similarities between customer product return behavior in a product environment and customer complaint behavior in a service environment. Can we build a framework that helps to broadly explain how undesirable customer-initiated contacts (UCICs) can be managed by firms profitably? This is especially important as firms that have focused mainly on product offerings begin augmenting their products with service offerings.

There has been a recent push to increase the amount of research in areas such as shopper marketing. But just as in much of the firm–customer exchange process research of past, much of it has focused on how customer experiences in stores affect customer purchase behavior. Does the way that customers experience either offline (e.g., brick-and-mortar) or online shopping also affect the desire of the customers to return products? Again, by ignoring customer product return behavior any strategy to understand customer purchase behavior is likely to be sub-optimal for the firm.

SUMMARY

What seems to be clear is that product returns are not a necessary evil of the firm–customer exchange process. Additionally, the relationship between product return behavior and the firm–customer exchange process is more complex than previously thought. This makes it critical to further understand how actions that firms take influence customer product return behavior and how customer product return behavior influences other customer behaviors. Only then will firms be able to manage customer product return behavior to maximize customer equity.

NOTE

1. See 'Three major chains shorten return windows' (2013), *Consumer World*, 18 December 2013, accessed 12 September 2014 at http://www.consumerworld.org/pages/returns.htm.

REFERENCES

Anderson, E.T., K. Hansen and D. Simester (2009a), 'The option value of returns: theory and empirical evidence', *Marketing Science*, **28**(3), 405–23.

Anderson, E.T., K. Hansen, D. Simester and L.K. Wang (2009b), 'How price affects returns: the perceived value and incremental customer effects', Working Paper, Kellogg School of Management.

Blanchard, D. (2007), 'Supply chains also work in reverse', *Industry Week*, 10 April 2007, accessed 10 September 2014 at http://www.industryweek.com/planning-amp-forecasting/supply-chains-also-work-reverse.

Bower, A.B. and J.G. Maxham III (2012), 'Return shipping policies of online retailers: normative assumptions and the long-term consequences of fee and free returns', *Journal of Marketing*, **76**(5), 110–24.

Davis, S., M. Hagerty and E. Gerstner (1998), 'Return policies and the optimal level of "hassle"', *Journal of Economics and Business*, **50**(5), 445–60.

Dekker, R., M. Fleischmann, K. Inderfurth and L.N. van Wassenhove (2004), *Reverse Logistics: Quantitative Models for Closed-loop Supply Chains*, Berlin/New York: Springer.

Fornell, C. and B. Wernerfelt (1988), 'A model for customer complaint management', *Marketing Science*, **7**(3), 287–98.

Gupta, S. and V.A. Zeithaml (2006), 'Customer metrics and their impact on financial performance', *Marketing Science*, **25**(6), 718–39.

Hess, J.D. and G.E. Mayhew (1997), 'Modeling merchandise returns in direct marketing', *Journal of Direct Marketing*, **11**(2), 20–35.

Knott, A.I., A. Hayes and S.A. Neslin (2002), 'Next-product-to-buy models for cross-selling applications', *Journal of Interactive Marketing*, **16**(3), 59–75.

Nasr Bechwati, N. and W.S. Siegal (2005), 'The impact of the prechoice process on product returns', *Journal of Marketing Research*, **42**(3), 358–67.

Ofek, E., Z. Katona and M. Sarvary (2011), '"Bricks and clicks": the impact of product returns on the strategies of multichannel retailers', *Marketing Science*, **30**(1), 42–60.

Petersen, J.A. and V. Kumar (2009), 'Are product returns a necessary evil? Antecedents and consequences', *Journal of Marketing*, **73**(3), 35–51.

Petersen, J.A. and V. Kumar (2010), 'Can product returns make you money?', *MIT Sloan Management Review*, **51**(3), 85–9.

Petersen, J.A. and V. Kumar (2014), 'Perceived risk, product returns, and optimal resource allocation: evidence from a field experiment', forthcoming, *Journal of Marketing Research*.

Petersen, J.A., L. McAlister, D.J. Reibstein, R.S. Winer, V. Kumar and Geoff Atkinson (2009), 'Choosing the right metrics to maximize profitability and shareholder value', *Journal of Retailing*, **85**(1), 95–111.

Shah, D. and V. Kumar (2008), 'Research before you leap: does cross-sell always lead to higher profit?', *Marketing Research: A Magazine of Management and Applications*, **20**(3), 26–32.

Shah, D., V. Kumar, Y. Qu and S. Chen (2012), 'Unprofitable cross-buying: evidence from consumer and business markets', *Journal of Marketing*, **76**(3), 78–95.

Shulman, J. and A. Coughlan (2007), 'Used goods, not used bads: profitable secondary market sales for a durable goods channel', *Quantitative Marketing and Economics*, **5**(2), 191–210.

Shulman, J.D., A.T. Coughlan and R.C. Savaskan (2010), 'Optimal reverse channel structure for consumer product returns', *Marketing Science*, **29**(6), 1071–85.

Wood, S.L. (2001), 'Remote purchase environments: the influence of return policy leniency on two-stage decision processes', *Journal of Marketing Research*, **38**(2), 157–69.

8. Word-of-mouth and marketing effects on customer equity

Dominique M. Hanssens, Julian Villanueva and Shijin Yoo

INTRODUCTION

Companies can acquire and grow customers in many ways, and different strategies will bring very different results to the firm. Many chief marketing officers (CMOs) rely on short-term metrics such as quarterly sales and earnings to assess their return on marketing. Inasfar as good marketing creates and cements customer relationships with the firm, these short-term metrics risk underestimating the true impact of their efforts. Customer equity has emerged as a powerful paradigm to monitor and measure the long-term financial impact of marketing spending and of word-of-mouth, not only in inherently relationship-oriented sectors (such as insurance) but also in transactions-dominated sectors (such as consumer packaged goods).

We start this chapter by proposing that managers improve the measurement of their customer acquisition efforts. Some believe that customer equity management should focus on already-acquired customers. This reasoning is flawed, because a firm acquiring the wrong customers will have serious difficulties in trying to retain and grow them later on. The literature has shown that long-term customer equity will depend on these acquisition efforts. On the other hand, some research has shown the superiority of word-of-mouth over marketing-induced customer acquisition methods. Hence, understanding the different short- and long-run dynamics of these two efforts will help CMOs do a better acquisition resource allocation.

Once acquired, customer's lifetime value should be increased through retention and share-of-wallet efforts. An important stream of literature aims at developing marketing-mix models that connect spending on current customers to customer equity. However, that literature has been primarily focused on relationship businesses, where there is a contractual setting. We will show that customer equity models can also be built in transactions-focused settings such as those of consumer packaged goods. In such cases, the use of customer equity as a long-term ('stock') performance metric greatly reduces the short-term volatility that characterizes the behavior of standard ('flow') metrics such as sales and market share. We conclude the chapter by offering several ideas for further research.

CUSTOMER ACQUISITION AND WORD-OF-MOUTH

Customer Acquisition and its Impact on Customer Equity

A large percentage of the customer equity and customer lifetime value literature (CE and CLV, henceforth) has focused on models of retention and cross-selling efforts. These models help managers extract as much value as possible from existing customers. Despite the need and merits of this literature stream, we argue that the quality of a customer base is heavily dependent on the acquisition effort: first, because customer acquisition represents an important share of marketing spending for many firms, thus having an important influence on its value; second, because a firm should not acquire any customer, but the 'right' kind of customers – those bringing in the highest CE contribution. We therefore think that there is an opportunity for further research on this area.

Linking customer acquisition efforts to long-term value creation is particularly important in the following cases: when customers exhibit high switching costs, or low switching costs with undifferentiated products, when the market is in its early life-cycle stage, when the product is infrequently purchased and when the firm is a new entrant (Villanueva and Hanssens, 2007). Additionally, for any firm losing a share of its current customer base, acquisition becomes most critical, because: (1) it will have a huge impact on the firm's ability to maintain or grow its market share, and (2) depending on the quality of the customers acquired, it will be easier or tougher to retain and grow them later on.

We review current and potential research in the following sections: optimal acquisition spending, prospect profiling, and acquisition channel allocation.

Optimal acquisition spending

Deciding the optimal acquisition spending is one of the first steps in maximizing CE. This decision will not only impact the number of customers acquired, but also their CE. Since it is a well-known fact that the relationship between CE and acquisition spending will exhibit decreasing returns to scale (Blattberg and Deighton, 1996), deciding the optimal spending level is critical.

We find in the literature two alternative approaches to solving this problem. The first is the 'breakeven method'. This approach is a simple heuristic in which it is suggested that managers invest in customer acquisition as long as the cost is lower than CLV (e.g., Jackson, 1989a, 1989b, 1989c; Hansotia and Wang, 1997). The main limitation of this approach is that it helps managers decide the maximum they could spend, not the optimal. The second is the 'acquisition response function approach'. This is a more rigorous approach, in which managers should maximize an objective profit function that depends on the acquisition response function (see Blattberg and Deighton, 1996). However, in order to optimize the acquisition spending, the model should also include the retention response function, given that this latter function will most probably depend on previous acquisition efforts (Thomas, 2001; Reinartz, Thomas, and Kumar, 2005).

Optimizing acquisition spending with the goal of maximizing CE can present some problems in competitive situations. Villanueva et al. (2007) show that, in contrast with the commonly held view, profits in competitive situations may be maximized when firms ignore the long term and, instead of trying to maximize CE, focus on period-by-period profits. The rationale is that maximizing the long term will lead to less switching, but higher profit erosion through increased price competition.

Identifying the best prospects

In situations where prospects are addressable, firms not only have to decide how much to spend, they also have to decide which ones are the most interesting potential customers to target first. Direct marketers and firms using direct sales forces typically profile their prospect lists to identify those customers with the highest propensity to buy (Blattberg, Getz, and Thomas, 2001). In a CE paradigm, however, firms should rank-order prospects based on their CE contribution. CLV does not necessarily equal CE contribution, as the latter includes a customer's role in attracting other customers to the brand. For example, Kumar et al. (2007), using data from financial and telecommunication services, showed that customers with the highest CLV are not necessarily the ones generating the highest word-of-mouth (WOM). Hence, a firm may prefer to acquire customers that will possibly generate a lower CLV, if they will generate a larger number of WOM acquisitions. Similarly, Gupta and Mela (2008) showed that buyers on eBay, even though they generate no direct revenue, are very valuable because they attract fee-paying sellers.

Acquisition budget allocation across different channels

Firms acquire customers using a myriad of channels (Neslin and Shankar, 2009). Each of these channels will bring different qualities of customers with respect to their CE to the firm (Villanueva, Yoo, and Hanssens, 2008). Hence, we believe that, in order to maximize CE, a firm should ask itself: (1) which channels can be used to acquire customers (2) how much CE contribution is generated from an average customer acquired through each of these channels, (3) what should be the acquisition budget allocation across all channels?

A taxonomy of acquisition channels We propose a classification of different acquisition channels that incorporates direct as well as indirect means of acquiring customers, such as word-of-mouth and public relations. The Internet and the more recent advent of social media have demonstrated the increasing power of these non-traditional channels. Therefore, comparing the short- and long-term effectiveness among the different acquisition channels is of extreme importance for the acquisition spending allocation. Our classification uses two concepts developed in previous research: contact level and level of intrusiveness.

The first concept is the 'level of contact' a given acquisition channel has with customers, which can be personal or broadcast. For example, if customers learn about the firm from a friend or from a personalized email, they are getting a more personalized contact than if they acquire information from TV advertising or a newspaper article. In the former, someone has decided to send a message to that specific customer, while in the latter, the message is available for anyone who could turn on the TV or read the newspaper. This is similar to the concept of 'audience addressability', as explained in Blattberg and Deighton (1996). In general, we would expect personal contacts to have high addressability and broadcast contacts to have low.

The second concept is the 'level of intrusiveness' of the acquisition channel, which can be low or high. This concept derives from the persuasion knowledge model (Friestad and Wright, 1994, 1995), which proposes that acquisition channels that are perceived by the customer as more intrusive will be less effective. Consumers interpret and cope with marketers' communication attempts (e.g., advertising) based on contingent persuasion knowledge (e.g., schemer schema, folk wisdom). Consumers understand that the main goal of marketing communications is influencing their own beliefs and/or attitudes about the firm's products or services. There is a difference between marketers' persuasion goals and the consumers' coping goals. We argue that acquisition channels such as direct marketing or mass advertising will be perceived as more intrusive than channels such as social media and word-of-mouth. An excellent illustration of this phenomenon is provided in a field experiment by Goldfarb and Tucker (2011): online display ads that are highly targeted *or* that are viewed as intrusive can increase purchase intent. However, when the ads are highly targeted *and* intrusive, their effectiveness decreases.

Level of Intrusiveness \ Level of Contact	Personal	Broadcast
Low	**WOM** • Referral from friends or colleagues • SEO	**PR** • Mentioned or linked from other websites • Magazine or newspaper articles • Influencers (e.g., blog posts, tweets, etc.)
High	**DM** • Personalized email • Personalized mail • SEM • Online ad banners • Sales force • Telemarketing	**AD** • Non-personalized email or mail • Magazine or newspaper advertisement • Radio advertisement • Television advertisement • Outdoor advertisement

Note: SEO = Search Engine Optimization; SEM = Search Engine Marketing.

Figure 8.1 Classification of acquisition channels

Our classification of acquisition channels using the concepts of level of contact and level of intrusiveness, leads us to four different categories: word-of-mouth (WOM), direct marketing (DM), advertising (AD), and public relations (PR). A wide array of acquisition tactics can be incorporated into one of these categories and we present some of them in Figure 8.1.

The CE contribution of a customer acquired through a specific channel In previous work (Villanueva et al., 2008), we stated that different acquisition strategies bring to the firm different qualities of customers. We differentiated between two broad channels – marketing-induced vs word-of-mouth – but the model could be extended to a larger number of channels. We used a vector-autoregressive (VAR) model to represent the inherent endogeneity that exists between customer acquisition and long-run profitability. This VAR model allowed us to quantify the following effects: (1) 'direct effects' of acquisition on the performance of the firm, (2) 'cross-effects' between two types of acquisition channels, (3) 'feedback effects' (i.e., current firm performance impacting future acquisitions), and (4) 'reinforcement effects' (i.e., momentum in firm performance and customer acquisitions). An advantage of this model is that it is well suited to capturing long-run dynamics among the variables of study.

For example, an increase in the rate of customer acquisitions impacts firm performance, which, in turn, triggers an increase in subsequent acquisition spending.

The three-variable VAR system to capture the above-mentioned dynamic interrelationships was expressed as:

$$\begin{pmatrix} MKT_t \\ WOM_t \\ VALUE_t \end{pmatrix} = \begin{pmatrix} a_{10} \\ a_{20} \\ a_{30} \end{pmatrix} + \sum_{l=1}^{p} \begin{pmatrix} a_{11}^l & a_{12}^l & a_{13}^l \\ a_{21}^l & a_{22}^l & a_{23}^l \\ a_{31}^l & a_{32}^l & a_{33}^l \end{pmatrix} \begin{pmatrix} MKT_{t-l} \\ WOM_{t-l} \\ VALUE_{t-l} \end{pmatrix} + \begin{pmatrix} e_{1t} \\ e_{2t} \\ e_{3t} \end{pmatrix} \quad (8.1)$$

where *MKT* stands for the number of customers acquired through the firm's marketing actions, *WOM* stands for the number of customers acquired through word-of-mouth, and *VALUE* is the firm's performance. The subscript t stands for time, and p is the lag order of the model. For this VAR model, where white-noise disturbances are distributed as $N(\mathbf{0},\Sigma)$, the direct effects are captured by a_{31}, a_{32}, cross-effects by a_{12}, a_{21}, feedback effects by a_{13}, a_{23}, and, finally, reinforcement effects by a_{11}, a_{22}, a_{33}. The dynamic impact of interest is captured by impulse response functions (IRFs) that trace the present and future responses of a variable to an unexpected shock in another variable.

Allocation across channels Direct effects shown in the previous model could eventually rank-order the different acquisition channels in terms of benefit per dollar. This approach, while managerially insightful, has two limitations. First, it assumes that, for each channel, the acquisition cost per customer does not change with the number of customers acquired. This is equivalent to assuming that the acquisition response function is linear with no intercept. Second, it does not address the question of how much to spend on acquisition, nor does it reveal how the budget should be allocated among the different acquisition channels.

A managerial decision support system (MDSS) similar to that of Mantrala, Sinha, and Zoltners (1992) could be proposed, using sub-market (in this case, channel) acquisition response functions to derive optimal spending and allocation across sub-markets. For example, assuming four acquisition channels $k=\{AD,DM,PR,WOM\}$, we could define a concave acquisition response function for each in the form:

$$c_k = \alpha_k + (s_k - \alpha_k)(1 - \exp(-\beta_k x_k)) \quad (8.2)$$

where c_k is the number of customers acquired, x_k is the amount of money spent, s_k is the maximum number of customers that can be acquired (saturation level), β_k represents the rate at which the number of customers approaches the saturation level, and α_k is an intercept that captures the number of customers acquired when no investment is made. This specification assumes that there are no cross-effects of acquisition responses among channels. This assumption could be relaxed by incorporating in each acquisition response function the effect that a certain spending in another channel will have on the number of customers acquired through that specific channel (see Rangaswamy, Sinha, and Zoltners, 1990).

These acquisition response functions can be parameterized for each channel using decision calculus (see Blattberg and Deighton, 1996 for a similar approach). The optimal resource allocation finds the best investment for each acquisition channel $x_k^*(B)$, given a fixed budget B. It is also possible to determine the optimal acquisition budget and then derive $x_k^*(B^*)$. The allocation problem can be expressed as:

$$\underset{x_k(B)}{Max}\,\Pi(B) = \sum_k m_k(\alpha_k + (s_k - \alpha_k)(1 - \exp(-\beta_k x_k))) - B$$

$$s.t. \sum_k x_k \leq B,\ x_k \geq 0,\ k = \{AD, DM, PR, WOM\} \tag{8.3}$$

where m_k is the contribution margin for each customer acquired from a specific acquisition channel, and B is the acquisition budget.

An estimation alternative to decision calculus is to use a statistical model on historical data. This may present several challenges. First, it may prove difficult to collect data on some channels such as public relations and word-of-mouth. Second, a sufficient number of data points with enough variability is required. Third, the data generation process should be able to predict future behavior. If these requirements are not met, decision calculus may be superior to statistical modeling. An example of a successful implementation of decision calculus can be found in Lodish et al. (1988), and an example using statistical modeling is reported in Hanssens, Thorpe, and Finkbeiner (2008).

The Emerging Importance of WOM

Positive word-of-mouth has always been important for the viability of firms and brands. However, it is not until recently that word-of-mouth has sparked increasing interest among academics and managers alike. Armelini and Villanueva (2010) explain the reasons for this

Table 8.1 Articles using 'word-of-mouth' and 'WOM' as keywords

	'Word-of-mouth'	'WOM'
2012[a]	13	3
2011	15	3
2010	15	2
2009	6	2
2008	10	2
2007	7	1
2006	5	1
2005	0	0
2004	2	0
2003	1	0
2002	3	0
2001	3	0
2000	2	0
Total*	82	14

Note: a. 2012 data until October.

phenomenon. First, consumers exhibit declining interest in advertising and traditional media. For example, it has been shown that advertising return on investment (ROI) has substantially decreased in the last decades (Sethuraman, Tellis, and Briesch, 2011). Second, the technology revolution of the last two decades has shortened the effective distance among consumers, making it easier for them to share information and opinions about products, services, brands, and firms. Hence, customer-to-customer interactions have dramatically increased (Nitzan and Libai, 2011). These two facts, combined with the recent economic crisis in much of the developed world, have spurred an unprecedented interest from marketers in word-of-mouth marketing. Scholars have followed this trend. As it can be seen in Table 8.1, the number of papers published in the *Journal of Marketing*, *Journal of Marketing Research*, *Marketing Science*, and *International Journal of Research in Marketing* using the keywords 'word-of-mouth' and 'WOM' has significantly increased in the last five years.

In a marketing context, word-of-mouth may be defined as social interactions '(a) taken place by an individual not actively engaged in selling a product or service that (b) impact others' expected utility for that product or service' (Godes et al., 2005, pp. 416–17). While these interactions have always existed, marketers now have new technologies and methods to elicit them.

WOM vs marketing-induced acquisition

Advertising is 'a non-personal form of communication offered through paid media with clear sponsorship' (Kotler, in Lilien et al., 1992), while a referral is the opposite: personal and non-paid. However, when we are to compare WOM with marketing-induced acquisition, the distinction becomes more blurred. For example, what is a referral reward program – where the customer is enticed to recommend a product or service in exchange for a benefit? That is certainly WOM, but it is also marketing induced. Or what about an influencer who writes a positive blog post about a firm or service he or she has some relationship with? Is that marketing-induced or WOM? What is a viral campaign that spurs lots of buzz? It may have started as marketing-induced but results in WOM. In order to deal with such ambiguities, a more recent distinction has been made between 'owned', 'paid', and 'earned' media (Stephen and Galak, 2012). A brand's website, for example, is an owned medium, TV advertising is paid, and blogs are earned.

Companies attempt to ignite WOM in different ways, often by starting a consumer conversation through paid media. They do so because of the inherent advantages of WOM with respect to advertising: (1) WOM has higher *credibility* than advertising (Day, 1971; Sheth, 1971; Loda and Coleman, 2005), due in part to the fact that recommendations come through stronger ties (Granovetter, 1973; Brown and Reingen, 1987), and (2) advertising is the most expensive communication medium (Vakratsas and Ambler, 1999; Vakratsas, 2005), while WOM is *free or inexpensive*. However, as Armelini and Villanueva (2010) point out, there are also complementarities between WOM and advertising. Hence, studies using different marketing channels will probably face endogeneity problems that should be accounted for (Villanueva et al., 2008; Trusov, Bucklin, and Pauwels, 2009). For example, higher advertising can lead to an increase in WOM, but a surge in WOM can also entice a brand to leverage the effect through advertising.

Measuring the return of WOM

Marketers are therefore interested in measuring the impact and the dynamics of WOM in comparison to other marketing-induced acquisition channels (MKTG henceforth). In previous work (Villanueva et al., 2008), we were able to estimate the short- and long-term impact of WOM and MKTG for an Internet web-hosting firm. Our results showed that, while MKTG had a stronger positive effect than WOM in the short term (first week of impact), the long-term impact of WOM was twice as much as that of MKTG. Moreover, the effect of MKTG acquisition settled down after only three weeks, while the effect of WOM lasted for six weeks.

Thus, MKTG acts as a 'loud, immediate-impact and short-lived' driver of customer acquisition, while WOM is a 'quiet, gradual-impact and long-lasting' driver.

Other scholars have studied the return of WOM in terms of sales or customer acquisition. For example, Trusov et al. (2009) use a VAR model with data from an Internet social site and find that WOM referrals have an impact on customer acquisition that is approximately 20 times higher than that of marketing events and 30 times that of media appearances. Chevalier and Mayzlin (2006) studied the effect that customer reviews have on sales and found that positive reviews had a positive effect on total sales at the bookstore site. Furthermore, the impact of negative reviews is even larger, and in the expected direction.

A highly relevant research area is that of referral programs, as they involve brand investments. In a study using data from a German bank, Schmitt, Skiera, and Van den Bulte (2011) find that referred customers have a higher margin, retention rate, and average value than non-referred customers. This study suggests that firm-induced WOM can also have a more positive effect than traditional means of firm-induced acquisition methods.

Negative WOM has been studied in the literature as well. For example, Hogan, Lemon, and Libai (2003), using the Bass diffusion model, show that the lost value of customer attrition depends not only on the customer's foregone revenue, but also on the lost WOM. Additionally, the impact of negative WOM on short-term stock prices has been studied by Luo (2007), while its long-term impact was investigated by Luo (2009) using a VAR model. Based on data from the airline industry, these studies are the first to demonstrate that negative WOM can cause a profound destruction of shareholder value, especially when competition is taken into consideration.

MARKETING EFFECTS ON CUSTOMER EQUITY

Importance and Challenges

The ultimate goal of any business executive should be to maximize intrinsic firm value. Marketing managers are no exception. They strive to design and implement various marketing activities to effectively maximize firm value. However, key performance indicators for marketers in many organizations are short term and/or product oriented, such as quarterly sales or market share.[1] Such a short-term and product orientation can lead to sub-optimal allocation of marketing resources when marketers do not fully appreciate how their marketing actions create or diminish long-term firm value. For example, short-term sales lifts due to temporary discounting may create an anticipation of future discounts and thus erode long-term profit margins. Similarly, advertising aimed at the upper funnel of prospects may not result in meaningful sales lifts, but may set the stage for future performance increases as more prospects place the advertised brand in their evoked set.

By contrast, customer equity (CE) is a meaningful long-term metric that marketers should measure, manage, and maximize since it is a good proxy for firm value (Kumar and Shah, 2009). CE is a stock variable that can indicate the long-term health of a brand/company, and CE may change over time due to the firm's marketing actions and/or changes in consumer preferences or the competitive environment. For example, a firm's CE may be significantly increasing after a successful new-product launch or decreasing due to a failed sales promotion. Therefore, understanding the impact of marketing on CE is a crucial task for managers and academics who view marketing as an important value-driving function of the firm. Firms may optimally allocate their marketing resources across tools/customers/brands/time to maximize their CE and thus intrinsic firm value (e.g., Reinartz et al., 2005).

However, linking marketing efforts and CE is a challenging task for several reasons. First, the metric CE itself is not easy to measure (Rust, Lemon, and Zeithaml, 2001). Unlike conventional flow metrics such as sales or market share based on revealed performance, CE heavily involves projecting future firm actions as well as consumer responses – that is, it is a forward-looking metric. Unlike other stock metrics such as market capitalization, CE does not have the benefit of an established system to collect the necessary data, at least for publicly traded firms. Even the simple task of counting a firm's number of customers is challenging for non-contractual product categories (Schmittlein, Morrison, and Colombo, 1987). Consequently, measurement has become an important issue in the CE literature (e.g., Kumar and George, 2007).

Second, there exists a dynamic – possibly simultaneous – relationship between marketing actions and CE (Berger et al., 2002). For example, customer retention rates may not only be influenced by cumulative past marketing activities, but may also drive future marketing decisions as well. Price discounts may deteriorate CE if there exists significant decision inertia, that is, prices do not quickly return to normal after the promotion campaign has ended. Third, as illustrated earlier, CE is indirectly influenced by marketing actions. For example, a newly acquired customer through advertising may indirectly contribute to the firm value by inviting other customers via referrals. Such spontaneous value contribution has gained in importance as ubiquitous social networks make consumers well connected to each other. Failure to control this organic growth of CE when assessing marketing effects on CE results in under- or overestimation.

In light of these challenges, the next section will summarize selected methodological approaches to measuring the contribution of marketing efforts on the evolution of CE. Broadly speaking, these methods fall under three categories: judgment-based methods, switching matrices based on stated preferences, and econometric methods based on transactions databases, which we will discuss in turn.

Measuring Marketing Effects on CE: Methods and Approaches

Judgment-based methods

In their seminal article, Blattberg and Deighton (1996) coined the term 'customer equity' and proposed a framework for maximizing CE by allocating marketing resources between retention and acquisition efforts. They measure the impact of marketing on CE by decision calculus (Little, 1979), assuming a one-to-one correspondence between acquisition/retention spending and acquisition/retention rates, respectively. For example, the acquisition rate a can be modeled as a function of the acquisition expenditure per prospect A as follows:

$$a = \textit{ceiling rate} \times [1 - \exp(-kA)] \tag{8.4}$$

The critical parameters, *ceiling rate* and k are estimated by asking a few simple questions to managers such as 'How much did you spend last period?' and 'What is the maximum acquisition rate you could reasonably expect, given market conditions?' The linkage between marketing spending and CE is found by the two important components of CE: acquisition rate and retention rate. As a result, an optimal allocation rule between acquisition and retention spending can be derived to maximize the firm's customer equity.

Although it can be applied to various industries relatively quickly without heavy data requirements, this approach is subject to two limitations. First, it does not recognize the dynamic relationship between marketing spending and consumer response. As indicated by the economics and marketing literature (e.g., Heckman, 1991; Seetharaman and Chintagunta, 1998), customer purchase behavior is not zero order, but instead exhibits an important dynamic component. In addition, marketing spending levels may also depend on past decisions and market performance (Dekimpe and Hanssens, 1995). Second, this approach ignores other factors that may influence CE. For example, if higher retention spending leads to increased customer satisfaction and thus higher retention, acquisition rates may increase as well due to enhanced word-of-mouth from satisfied retaining customers. Ignoring these forces will likely result in an underestimation of the CE impact of marketing. Despite these limitations, Blattberg and Deighton's work (1996) opens the avenue of investigating the relationship between marketing actions and customer equity.

Switching matrix models based on stated preferences

Rust and his colleagues proposed a multinomial logit model based on a purchase intention survey to estimate marketing effects on CE (Rust, Lemon, and Zeithaml, 2004). They derive each probability component of a Markov switching matrix under the utility maximization framework. More specifically, the probability of individual i choosing brand k^* given that brand j was most recently chosen can be expressed as:

$$P_{ijk^*} = \exp(U_{ijk^*}) / \sum_k \exp(U_{ijk}) \tag{8.5}$$

where:

$$U_{ijk} = \beta_{0k} LAST_{ijk} + \mathbf{X}_{ik}\boldsymbol{\beta}_{1k} + \varepsilon_{ijk}. \tag{8.6}$$

The dummy variable *LAST* is equal to one if $j = k$ and zero otherwise; $\boldsymbol{X}$ is a row vector of marketing activities (e.g., price); and ε is a random error term following an extreme value distribution.

This model captures the impact of marketing activities on CE since CLV is a function of a series of purchase probabilities over time. The authors estimate the model parameters based on stated responses in consumer surveys conducted for various companies. They successfully show that return on marketing investments can be measured by marketing effects on CE. For example, in their illustrative application, American Airlines' $70 million investment in improving passenger comfort on its flights was found to lift its customer equity by 1.39 percent, or equivalently $101.3 million. One of the advantages this approach offers is that a wide range of marketing activities (e.g., launching loyalty programs) can be considered. However, it also has a few limitations. As shown in the literature, stated preferences are known to be only weakly associated with actual purchasing behavior (Morrison, 1979). Furthermore, this approach is necessarily static in nature.

Time-series econometrics

The marketing effects on CE can be measured by modeling complex and dynamic interactions between marketing variables and the components of CE such as acquisition/retention rates. In doing so, a vector-autoregressive

(VAR) model is a good choice for several reasons. First, the model can capture the long-run aspects of the relationship among the variables. Since the measurement of CE should consider future acquisition/retention processes, the model should be able to separate short-term and long-term effects. Moreover, the VAR model is capable of dealing with evolving variables that do not have fixed means and variances. Second, the VAR model can also capture possible indirect effects among the variables. For example, acquisition rates may be affected not only by advertising, but also by retention rates, inasfar as existing customers generate word-of-mouth to future prospects. Impulse response functions (IRFs) quantify such complex dynamic interactions among the variables in the VAR model. Third, due to parsimonious estimation and intuitive metrics (e.g., IRF elasticity), the VAR model can be easily understood and adapted by managers.

This econometric model requires different variable operationalization depending upon the product category. For consumer durables (e.g., automobiles) or service industries (e.g., insurance) where acquisition/retention can be observed by product ownership and contract renewal behavior, consumer response variables (e.g., acquisition rates) can be explicitly defined. For example, Yoo and Hanssens (2007) measured the acquisition rate (*a*) and retention rate (*r*) in the automobile industry based on the following formula:

$$a_t = n_t^{AP}/n_{t-1}^{P} \quad r_t = n_t^{RC}/n_{t-1}^{C} \tag{8.7}$$

where the superscript *P* stands for total prospects, *AP* for acquired prospects, *C* for existing customers, *RC* for retained customers. The subscript *t* stands for time. They measure the net impact of marketing activities on acquisition rates and retention rates (and thereby on CE) by constructing a VAR model such that:

$$\begin{pmatrix} a_t \\ r_t \\ x_t \\ z_t \end{pmatrix} = \begin{pmatrix} c^a \\ c^r \\ c^x \\ c^z \end{pmatrix} + \sum_{l=1}^{p} \begin{pmatrix} \phi_{11}^l & \phi_{12}^l & \phi_{13}^l & \phi_{14}^l \\ \phi_{21}^l & \phi_{22}^l & \phi_{23}^l & \phi_{24}^l \\ \phi_{31}^l & \phi_{32}^l & \phi_{33}^l & \phi_{34}^l \\ \phi_{41}^l & \phi_{42}^l & \phi_{43}^l & \phi_{44}^l \end{pmatrix} \begin{pmatrix} a_{t-l} \\ r_{t-l} \\ x_{t-l} \\ z_{t-l} \end{pmatrix} + \begin{pmatrix} \varepsilon_t^a \\ \varepsilon_t^r \\ \varepsilon_t^x \\ \varepsilon_t^z \end{pmatrix} \tag{8.8}$$

For example, the impact of own marketing efforts *x* (e.g., price discounting) on the acquisition rate *a* – and thereby CE – is investigated by the impulse response of acquisition rate to a shock in own marketing efforts. Other endogenous variables *z* (such as competitive marketing efforts) can be included in the model to control for additional drivers of the marketing–CE relationship.

The authors apply this model to weekly transactions data in the luxury passenger car product category. They find that luxury automobile brands fail to increase CE by additional mass marketing activities such as price discounting or advertising. Some brands have positive marketing effects on CE components such as acquisition rates, but these effects are insufficient to increase their CE, mainly due to the sustained nature of their marketing spending. For example, Cadillac's price discounting shows strong inertia such that 95 percent of an initial price shock (discount) remains in place during and even after a 26-week period. This is a good example of the 'escalation' scenario, discussed in Dekimpe and Hanssens (1999), that is, the phenomenon that changes in marketing variables (e.g., price) are sustained even though there is no persistence in response (e.g., acquisition rate).

For consumer packaged goods (CPG) categories where customer acquisition or retention is not directly observable by product ownership, one cannot directly estimate the impact of marketing on the components of CE. Due to the non-contractual setting and heterogeneity of purchase frequency, measuring acquisition and retention rates is challenging and requires more sophisticated statistical methods. For instance, in CPG categories, a customer who has not repeat-purchased a product in a given recent time period is not necessarily 'lost for good'. Perhaps this consumer made his or her last purchase for an unusually large quantity because the product was on sale, and is therefore 'out of the market' in the current time period. In the same vein, a manager should not de facto treat all customers in the firm's database as existing customers, since silent attrition may have occurred that cannot be measured without using sophisticated statistical models (Schmittlein et al., 1987; Schmittlein and Peterson, 1994). Using Dwyer's (1997) terminology, measuring CE in CPG industries should be based on always-a-share rather than lost-for-good buyer–seller relationships.

Yoo, Hanssens, and Kim (2012) proposed a novel approach to overcome this hurdle. They recursively project the CLV of individual customers for a CPG brand using individual-level stochastic models at different points in time. These individual values are aggregated and used to estimate the brand's CE at a given point in time. Then the authors link variations of these equity estimates to the brand's marketing activities such as price promotions. In this model more specifically, they first evaluate customer *i*'s lifetime value at time *t* by the number of expected transactions (*ET*), purchase quantity per transaction (*PQ*), price per purchased unit (*PR*), contribution margin (*CM*), and any discretionary marketing investments (*FC*) such that:

$$CLV_{it} = \sum_{j=t}^{\infty} (1 + d)^{-(j-t)} [ET_{it} PQ_{it} PR_{it} CM_{it} - FC_{it}] \tag{8.9}$$

where d is a time discount factor. CE can then be obtained by summing up all customers' CLV at a certain point in time.

Based on the beta-geometric/beta-Bernoulli (BG/BB) model (Fader, Hardie, and Shang 2010) for the number of expected transactions and the Normal/Normal model (Schmittlein and Petersen, 1994) for purchase quantity per transaction, the authors derive the discounted sum of purchase quantities (DPQ), which is a multiplication of the discounted expected number of transactions (DET) by the expected purchase quantity per transaction (PQ). In this model, DPQ plays a critical role in investigating the linkage between marketing actions and CE. In their empirical illustration, they obtain a time series of DPQ for each brand by recursively estimating the BG/BB and Normal/Normal model parameters, and calculating a new value of each brand's DPQ for each individual. Finally, a VAR model on marketing actions (MKT), short-term sales lift (SLS), and long-term expectation on customer value (DPQ) is specified as follows:

$$\begin{pmatrix} MKT_t \\ SLS_t \\ DPQ_t \end{pmatrix} = \begin{pmatrix} \mathbf{c}_{1t} \\ \mathbf{c}_{2t} \\ \mathbf{c}_{3t} \end{pmatrix} + \sum_{l=1}^{L} \begin{pmatrix} \varphi_{11}^{l} & \varphi_{12}^{l} & \varphi_{13}^{l} \\ \varphi_{21}^{l} & \varphi_{22}^{l} & \varphi_{23}^{l} \\ \varphi_{31}^{l} & \varphi_{32}^{l} & \varphi_{33}^{l} \end{pmatrix} \begin{pmatrix} MKT_{t-l} \\ SLS_{t-l} \\ DPQ_{t-l} \end{pmatrix} + \begin{pmatrix} \varepsilon_{1t} \\ \varepsilon_{2t} \\ \varepsilon_{3t} \end{pmatrix} \tag{8.10}$$

where the **c** vectors contain the impact of other exogenous variables on the system, such as seasonality, competitive activity, and macroeconomic conditions; L is the lag order of the system; and ε is a vector of white-noise processes with zero mean and covariance matrix Σ. Once the model parameters of the VAR system are obtained, one can investigate how the firm's marketing actions affect current and future DPQs and thereby CE.

An empirical illustration for the leading national and private-label brands in two CPG categories shows that the evolution of CE and the impact of marketing actions are different across brands. For example, in the yogurt category, all major brands' CE declined over time – that is, the category is becoming financially less attractive – but some deteriorated much faster than others. In the ketchup category, on the other hand, one brand faced declining CE while two others were able to reverse such negative trends. In conclusion, CE estimates may be evolving in different directions that are not readily captured from the brands' conventional business metrics such as revenue or market share.

Additional Issues in CE Effects of Marketing

The literature to date has focused more on the relationship between marketing actions and two critical consumer response metrics, acquisition and retention rates, to shed light on the marketing effects on CE. However, there exist other components that influence CE estimates such as contribution margin and time discount factor. For example, Venkatesan and Kumar (2004) regressed contribution margins on several covariates including firm size, industry dummies, and total number of contacts made to a customer to capture the dynamics of contribution margin. The time discount factor has been assumed as fixed in most of CE-related studies, but this assumption may not hold under turbulent economic conditions where the firm's cost of capital dramatically changes.

Another promising research avenue is the role of social media and social network services in creating CE by marketing efforts. As the role of these non-traditional earned and less-intrusive media becomes more important among the firm's marketing efforts, incorporating social aspects of consumer behavior will be a crucial task in revealing the relationship between marketing and CE. For example, a recent study by Kim and Ko (2012) investigated how social media marketing activities enhance customer equity for luxury fashion brands.

Finally, since CE is an important mediator for intrinsic firm value, investigating the marketing effects on CE will help us understand the role of marketing in increasing firm value (Hanssens, Rust, and Srivastava, 2009; Srinivasan and Hanssens, 2009). Although researchers have attempted to examine the relationship between CE and the firm's market value of equity (e.g., Gupta, Lehmann, and Stuart, 2004; Kumar and Shah, 2009), there still exist unresolved issues that provide further research opportunities. For example, Choi et al. (2011) tested the relevance of customer equity in explaining a firm's market value using data from the mobile telecommunications industry.

CONCLUSIONS

This chapter has focused on two important drivers of customer equity, marketing and word-of-mouth, which are at least partially under the control of brand management. Broadly speaking, customer equity is created or enhanced by profitable increases in either customer acquisition, customer retention or customer value creation (consumption and/or word-of-mouth).

We started, logically, with a discussion of customer acquisition. After quantifying the value of acquisition, we described a taxonomy of acquisition channels and discussed various budget allocations across channels. Next, we examined the role of word-of-mouth in customer acquisition and proposed methods for quantifying its importance. The discussion paid particular attention to new measurement opportunities offered by ever-expanding Internet data sources.

Next, we addressed the broader question of marketing and overall customer equity. We propose that customer equity should be *the* ultimate long-term value metric for a brand or a firm, and describe some unique challenges in assessing customer equity. Fortunately, three different measurement methods may be identified, each with unique strengths and limitations. Our illustrations of these methods include relationship businesses such as insurance, as well as product-centric sectors such as automobiles and consumer packaged goods. Thus, we provide evidence that customer equity may be viewed as a *universal* performance metric. Our chapter concluded with a discussion of several remaining challenges and directions for future research.

NOTE

1. 'Product' in this context means 'offering', and can be either a physical product or a service.

REFERENCES

Armelini, G. and J. Villanueva (2010), 'Marketing expenditures and word-of-mouth communication: complements or substitutes?', *Foundations & Trends in Marketing*, **5**(1), 5–20.

Berger, P.D., R.N. Bolton, D. Bowman, E. Briggs, V. Kumar, A. Parasuraman and C. Terry (2002), 'Marketing actions and the value of customer assets: a framework for customer asset management', *Journal of Service Research*, **5**(1), 39–54.

Blattberg, R. and J. Deighton (1996), 'Managing marketing by the customer equity test', *Harvard Business Review*, **75**(4), 136–44.

Blattberg, R.C., G. Getz and J.S. Thomas (2001), *Customer Equity: Building and Managing Relationships as Valuable Assets*, Boston, MA: Harvard Business School Press.

Brown, J.J. and P.H. Reingen (1987), 'Social ties and word-of-mouth referral behavior', *Journal of Consumer Research*, **14**(3), 350–62.

Chevalier, J.A. and D. Mayzlin (2006), 'The effect of word of mouth on sales: online book reviews', *Journal of Marketing Research*, **43**(3), 345–54.

Choi, Y., J. Lee, S. Yoo and Y.K. Yoo (2011), 'Relevance of customer equity in explaining the market value of the firm's equity: evidence from the mobile telecommunications industry', Working Paper, Korea University Business School.

Day, G.S. (1971), 'Attitude change, media and word of mouth', *Journal of Advertising Research*, **11**(6), 31–40.

Dekimpe, M.G. and D.M. Hanssens (1995), 'The persistence of marketing effects on sales', *Marketing Science*, **14**(1), 1–21.

Dekimpe, M.G. and D.M. Hanssens (1999), 'Sustained spending and persistent response: a new look at long-term marketing profitability', *Journal of Marketing Research*, **36**(4), 397–412.
Dwyer, R.F. (1997), 'Customer lifetime valuation to support marketing decision making', *Journal of Direct Marketing*, **11**(4), 6–13.
Fader, P.S., B.G.S. Hardie and J. Shang (2010), 'Customer-base analysis in a discrete-time noncontractual setting', *Marketing Science*, **29**(6), 1086–108.
Friestad, M. and P. Wright (1994), 'The persuasion knowledge model: how people cope with persuasion attempts', *Journal of Consumer Research*, **21**(1), 1–31.
Friestad, M. and P. Wright (1995), 'Persuasion knowledge: lay people's and researchers' beliefs about the psychology of advertising', *Journal of Consumer Research*, **22**(1), 62–74.
Godes, D., D. Mayzlin, Y. Chen, S. Das, C. Dellarocas, B. Pfeiffer, B. Libai, S. Sen, M. Shi and P. Verlegh (2005), 'The firm's management of social interactions', *Marketing Letters*, **16**(3–4), 415–28.
Goldfarb, A. and C. Tucker (2011), 'Online display advertising: targeting and intrusiveness', *Marketing Science*, **30**(3), 389–404.
Granovetter, M. (1973), 'The strength of weak ties', *The American Journal of Sociology*, **78**(6), 1360–80.
Gupta, S. and C.F. Mela (2008), 'What is a free customer worth?', *Harvard Business Review*, **86**(11), 102–9.
Gupta, S., D.R. Lehmann and J.A. Stuart (2004), 'Valuing customers', *Journal of Marketing Research*, **41**(1), 7–18.
Hansotia, B.J. and P. Wang (1997), 'Analytical challenges in customer acquisition', *Journal of Direct Marketing*, **11**(2), 7–19.
Hanssens, D.M., D. Thorpe and C. Finkbeiner (2008), 'Marketing when customer equity matters', *Harvard Business Review*, **86**(5), 117–23.
Hanssens, D.M., R.T. Rust and R.K. Srivastava (2009), 'Marketing strategy and Wall Street: nailing down marketing's impact', *Journal of Marketing*, **73**(6), 115–18.
Heckman, J.J. (1991), 'Identifying the hand of the past: distinguishing state dependence from heterogeneity', *American Economic Review*, **81**(2), 75–9.
Hogan, J.E., K.N. Lemon and B. Libai (2003), 'What is the true value of a lost customer?', *Journal of Service Research*, **5**(3), 196–208.
Jackson, D. (1989a), 'Determining a customer's lifetime value', *Direct Marketing*, **51**(11), 60–62, 123.
Jackson, D. (1989b), 'Determining a customer's lifetime value (Part 2)', *Direct Marketing*, **52**(1), 24–32.
Jackson, D. (1989c), 'Insurance marketing: determining a customer's lifetime value (Part 3)', *Direct Marketing*, **52**(4), 28, 30.
Kim, A.J. and E. Ko (2012), 'Do social media marketing activities enhance customer equity? An empirical study of luxury fashion brand', *Journal of Business Research*, **65**(10), 1480–86.
Kumar, V. and M. George (2007), 'Measuring and maximizing customer equity: a critical analysis', *Journal of the Academy of Marketing Science*, **35**(2), 157–71.
Kumar, V. and D. Shah (2009), 'Expanding the role of marketing: from customer equity to market capitalization', *Journal of Marketing*, **73**(6), 119–36.
Kumar, V., A.J. Petersen and R.P. Leone (2007), 'How valuable is word of mouth?', *Harvard Business Review*, **85**(10), 139–46.
Lilien, G.L., P. Kotler and S. Moorthy (1992), *Marketing Models*, Upper Saddle River, NJ: Prentice-Hall International Editions.
Little, J.D.C. (1979), 'Decision support systems for marketing managers', *Journal of Marketing*, **43**(3), 9–26.
Loda, M.D. and B.C. Coleman (2005), 'Sequence matters: a more effective way to use advertising and publicity', *Journal of Advertising Research*, **45**(4), 362–72.
Lodish, L.M., E. Curtis, M. Ness and M.K. Simpson (1988), 'Sales force sizing and deployment using a decision calculus model at Syntex Laboratories', *Interfaces*, **18**(1), 5–20.

Luo, X. (2007), 'Consumer negative voice and firm-idiosyncratic stock returns', *Journal of Marketing*, **71**(3), 75–88.
Luo, X. (2009), 'Quantifying the long-term impact of negative word of mouth on cash flows and stock prices', *Marketing Science*, **28**(1), 148–65.
Mantrala, M.K., P. Sinha and A.A. Zoltners (1992), 'Impact of resource allocation rules on marketing investment-level decisions and profitability', *Journal of Marketing Research*, **29**(2), 162–75.
Morrison, D.G. (1979), 'Purchase intentions and purchase behavior', *Journal of Marketing*, **43**(2), 65–74.
Neslin, S.A. and V. Shankar (2009), 'Key issues in multichannel customer management: current knowledge and future directions', *Journal of Interactive Marketing*, **23**(1), 70–81.
Nitzan, I. and B. Libai (2011), 'Social effects on customer retention', *Journal of Marketing*, **75**(6), 24–38.
Rangaswamy, A., P.A. Sinha and A. Zoltners, 'An integrated model-based approach for sales force structuring', *Marketing Science*, **9**(4), 279–99.
Reinartz, W., J.S. Thomas and V. Kumar (2005), 'Balancing acquisition and retention resources to maximize customer profitability', *Journal of Marketing*, **69**(1), 63–79.
Rust, R.T., K.N. Lemon and V.A. Zeithaml (2001), 'Driving customer equity: how customer lifetime value is reshaping corporate strategy', *Journal of Business Research*, **57**(6), 679–80.
Rust, R.T., K.N. Lemon and V.A. Zeithaml (2004), 'Return on marketing: using customer equity to focus marketing strategy', *Journal of Marketing*, **68**(1), 109–27.
Schmitt, P., B. Skiera and C. Van den Bulte (2011), 'Referral programs and customer value', *Journal of Marketing*, **75**(1), 46–59.
Schmittlein, D.C. and R.A. Peterson (1994), 'Customer base analysis: an industrial purchase process application', *Marketing Science*, **13**(1), 41–67.
Schmittlein, D.C., D.G. Morrison and R. Colombo (1987), 'Counting your customers: who are they and what will they do next?', *Management Science*, **33**(1), 1–24.
Seetharaman, P.B. and P. Chintagunta (1998), 'A model of inertia and variety-seeking with marketing variables', *International Journal of Research in Marketing*, **15**(1), 1–17.
Sethuraman R., G.J. Tellis and R.A. Briesch (2011), 'How well does advertising work? Generalizations from meta-analysis of brand advertising elasticities', *Journal of Marketing Research*, **48**(3), 457–71.
Sheth, J.N. (1971), 'Word-of-mouth in low-risk innovations', *Journal of Advertising Research*, **11**(3), 15–18.
Srinivasan, S. and D.M. Hanssens (2009), 'Marketing and firm value: metrics, methods, findings, and future directions', *Journal of Marketing Research*, **46**(3), 293–312.
Stephen, A.T. and J. Galak (2012), 'The effects of traditional and social earned media on sales: a study of a microlending marketplace', *Journal of Marketing Research*, **49**(5), 624–39.
Thomas, J.S. (2001), 'A methodology for linking customer acquisition to customer retention', *Journal of Marketing Research*, **38**(2), 262–8.
Trusov, M., R.E. Bucklin and K. Pauwels (2009), 'Effects of word-of-mouth versus traditional marketing: findings from an internet social networking site', *Journal of Marketing*, **73**(5), 90–102.
Vakratsas, D. (2005), 'Advertising response models with managerial impact: an agenda for the future', *Applied Stochastic Models in Business & Industry*, **21**(4/5), 351–61.
Vakratsas, D. and T. Ambler (1999), 'How advertising works: what do we really know?', *Journal of Marketing*, **63**(1), 26–43.
Venkatesan, R. and V. Kumar (2004), 'A customer lifetime value framework for customer selection and resource allocation strategy', *Journal of Marketing*, **68**(4), 106–25.
Villanueva, J. and D.M. Hanssens (2007), 'Customer equity: measurement, management and research opportunities', *Foundations & Trends in Marketing*, **1**(1), 1–95.
Villanueva, J., S. Yoo and D.M. Hanssens (2008), 'The impact of marketing-induced versus word-of-mouth customer acquisition on customer equity', *Journal of Marketing Research*, **45**(1), 48–59.

Villanueva, J., P. Bhardwaj, S. Balasubramanian and Y. Chen (2007), 'Customer relationship management in competitive environments: the positive implications of a short-term focus', *Quantitative Marketing & Economics*, **5**(2), 99–129.
Yoo, S. and D.M. Hanssens (2007), 'Modeling the sales and customer equity effects of the marketing mix', Working Paper, UCLA Anderson School of Management.
Yoo, S., D.M. Hanssens and H. Kim (2012), 'Marketing and the evolution of customer equity of frequently purchased brands', Working Paper, UCLA Anderson School of Management.

9. The power of customer referrals

Robert P. Leone and
Angeliki Christodoulopoulou

INTRODUCTION

A customer's contribution to the firm is not limited only to the profits generated from his or her purchase behavior, it also encompasses the benefits that result from his or her referral behavior. Research has clearly shown that a customer's referrals can be quite valuable to a firm since they can generate new customers in various situations, including during new-product diffusion (Mahajan, Muller, and Bass, 1995). Word-of-mouth (WOM) can also be quite important, since even when consumers are making general purchase decisions they often seek opinions from others who have specific product or market knowledge (Feick and Price, 1987). Furthermore, they tend to trust other consumers considerably more than they trust marketer-generated information. For example, Nielsen reports that 92 percent of consumers say who they trust recommendations from people they know,[1] compared to just 47 percent who trust TV advertising. Moreover, 77 percent of consumers are more likely to buy a new product that was introduced to them through word-of-mouth originating from friends or family.[2] Given the trust and importance that consumers place on product and service information provided by their social networks, many firms have now incorporated a referral program in their customer acquisition strategy. In addition, research has shown that firms who acquire customers via referrals are acquiring more profitable customers, making it desirable to run referral marketing campaigns (Villanueva, Yoo, and Hanssens, 2008).

This chapter examines the power of customer referrals and their ability to generate value for the firm. After introducing the concepts of word-of-mouth and the referral process, the way referral programs work and some relevant definitions are briefly described. Then, the value of referred customers is discussed in more detail, in addition to some strategies for initiating the referral process and defining the referral rewards. Next, the limitations of solely measuring referral intentions are discussed, and the value of customers who generate referrals is outlined. The chapter explains how to use customer referral value (CRV) to evaluate and manage customers based on their referral activity, and in tandem with their customer lifetime value (CLV). Finally, customer referrals in the context of B2B markets are discussed, and the chapter concludes with some implementation considerations.

WOM AND REFERRALS

In the present context, word-of-mouth refers to the exchange of information about products and services between consumers. It may involve descriptions, evaluations, and/or recommendations about new or existing products and brands. The natural occurrence of word-of-mouth has led marketers to develop different techniques for inducing and facilitating it, which can be summarized as word-of-mouth marketing. Such practices aim to educate customers about the product offerings of the firm and encourage them to share this information with their social network. Certainly, this requires that customers are satisfied with the firm (Anderson, 1998) and are willing to share their opinions. Ultimately, these efforts are aimed at drawing attention to the brand and attracting potential customers to the firm's clientele. Word-of-mouth may not always be positive, so additional care should be taken when consumers are sharing negative word-of-mouth in order to counter it constructively.

Customer referrals can be described as a type of positive word-of-mouth that originates from a firm's customers and focuses on recommending the purchase of a product/service and, sometimes, facilitating it. Customer referrals are increasingly relevant today, as numerous online and mobile social tools make word-of-mouth communication and referrals very easy. This availability of information has permeated the customer purchase decision in a unique way. In addition, even strangers' opinions matter. As reported by Google's Wildfire,[3] 83 percent of consumers report that online user reviews 'often or sometimes impact their purchase decisions'. These changes have led many companies to adopt referral programs as part of their marketing efforts. The following section provides a detailed description of the structure of referral programs and a brief illustration of the referral process.

REFERRAL PROGRAMS AND THE REFERRAL PROCESS

As a type of positive word-of-mouth, customer referrals are commonly a natural outcome of customer satisfaction with the firm. Given the

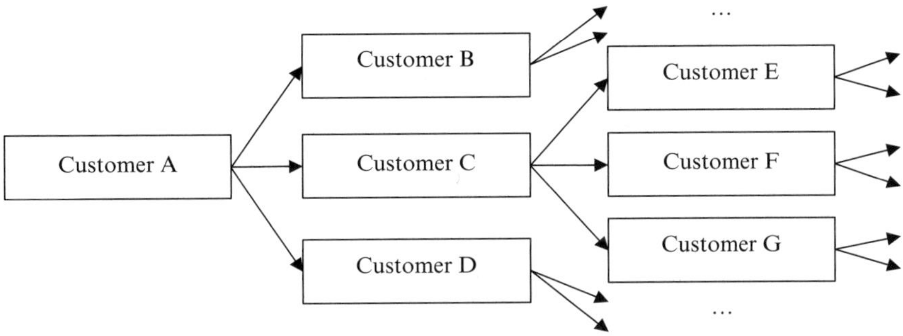

Figure 9.1 *The customer referral process*

value of customer referrals and in order to increase them, firms develop specialized programs to encourage and manage referrals, termed as customer referral programs, or referral reward/bonus programs. In the context of these programs, a referral is a recommendation from a current customer (referring customer or referrer) to a prospective customer (referred customer) for purchasing the product offerings of the firm, and it is accompanied by an incentive. The main difference between traditional word-of-mouth referrals and customer referrals that are stimulated by referral programs is that the latter are associated with rewarding customers. For example, Dish[4] offers a $50 credit to their customers' accounts for each successful referral and another $50 credit to their referrals' accounts when they subscribe to its 24-month satellite television service.

Figure 9.1 illustrates how the typical referral process works in a firm's referral program. This process starts from an existing customer of the firm (referrer; customer A), who recommends a product/service to his or her family, friends, coworkers, or other social contacts (referred prospects). This interaction can take place via a physical conversation, online conversation, email or mail communication, or via any other channel of communication. In many cases, such interactions also happen via the firm's website, where some of these prospects eventually become customers (customers B, C, and D). As a result, the referring customer (customer A) and/or the referred customers (customers B, C, and D), depending on the particular program, receive the associated rewards. This process can continue as the newly acquired customers become referrers by participating in the referral program as well. Another succession of referrals that generate from customers B, C, and D to their own social network (customers E, F, G, etc.) can lead to additional referrals, which are called second-generation referrals. This process can go on to include additional tiers of customers, so customer referral programs have the potential to increase the firm's customer base significantly.

The main mechanism through which customer referral programs encourage referral behavior is the use of incentives, or rewards. These incentives, sometimes called referral bonuses, vary greatly across firms, though they largely depend on the nature of the firm's products and services. Among commonly used referral incentives are free products, free service, program points, monetary credit, gifts, vouchers, discounts, or cash. For companies that also deploy loyalty programs, it is common to combine the two, at least in terms of the incentive; that is, bonus points towards the loyalty program can be offered as a reward for customer referrals. Another important aspect of referral rewards is the receiver. That is, rewards may be offered to the referring customers, the referred customers, or both, though most programs adopt a referrer-only reward. For instance, customers of the Discover credit card receive a $50 cash bonus for each referred friend that applies and are granted a credit card as part of the credit card provider's referral program.[5] This offer has a limit of ten referrals per year for each customer. Limiting the number of extended referral rewards to a specified maximum is a common practice in order to avoid customers exploiting such offers.

Clearly, referral programs entail certain investments including the management and promotion of the program through referral marketing campaigns in addition to the cost of the promised incentives. However, they are still less expensive than traditional acquisition techniques, such as direct mail and advertising, as the rewards are only extended when a purchase is made. The relatively low investment is one of the main reasons behind the rapid expansion of customer referral programs as a means to affordably attract new customers. This is especially true among smaller companies, who usually have limited disposable funds. The lower acquisition costs, together with the potential for future profits from referred customers, demonstrate how investing in customer referral programs creates prospects for enhancing a firm's customer equity.

The benefits associated with customer referral programs have made them particularly popular among companies operating in a variety of industries, among which is the education industry. An example of a higher education institute that employs a referral program for recruiting new students is Ashworth College,[6] a distance learning institute in Georgia, USA, with high school and college degree programs. As part of a customer referral program, current students receive $50 in the form of account credit or a check for each new student that enrolls in the college because of their referral (limited to $100 per student per year).

REFERRED CUSTOMERS' VALUE

According to research, customers that emerge through referrals are not only less expensive to acquire but also more valuable for the firm, as compared to the customers that are acquired through traditional methods (Villanueva et al., 2008; Schmitt, Skiera, and Van den Bulte, 2011). One such study examined the difference in value derived from each customer depending on their acquisition method, using customer data from a financial services firm (Schmitt et al., 2011). This data included the acquisition and profitability information of two groups of customers over 33 months; one group was acquired in a one-year period through the firm's referral program (5181 new referred customers), while the other was a random sample that joined the firm's customer base during the same period through other acquisition methods (4633 new non-referred customers). The customer referral program encouraged referrals by offering referrers a voucher worth 25 euros for each new referred customer.

Overall, Schmitt et al. (2011) found that referred customers are more profitable than non-referred customers. In particular, referred customers had a daily contribution margin of 7.6 cents more than non-referred customers. In other words, customers that joined due to the referral program contributed 27.74 euros per year more than those who joined through traditional methods, which corresponded to a 25 percent increase in profitability. Moreover, the researchers found that this contribution difference was higher at first, but it eroded and eventually disappeared 29 months after the acquisition. Therefore, referred customers are more profitable in the short term than non-referred customers, and they remain more profitable for almost 2.5 years after their acquisition. Besides, even after accounting for the costs of the referral program, referred customers still incurred an acquisition cost that was 20 euros less than that for non-referred customers. The study also found evidence that referred customers are more loyal than non-referred customers. Specifically, customers acquired through a referral showed a higher retention probability than their counterparts, which corresponded to an 18 percent lower likelihood of defection. In other words, referred customers were 18 percent less likely to terminate their relationship with the firm, as compared to non-referred customers. Moreover, this retention difference was found to persist over time, which demonstrates that customers that emerge through the referral process are more loyal. Finally, the researchers found that the added value of referred customers persists throughout their lifetime with the firm. In particular, their CLV for a total of six years after their acquisition was calculated as 40 euros or 16 percent more than that for non-referred customers. When considering the cost savings over traditional acquisition methods, referred customers' CLV was higher by 25 percent than that of non-referred customers.

Overall, the value of customers that join due to referrals is considerably higher than that of other customers, as they are more profitable and more loyal. This is expressed through their higher contribution margins, lower acquisition costs, lower defection rates, and higher CLV. Based on the purchase behavior of the participants in the previous study and on their calculated CLV, the value of referred customers is 25 percent higher than that of non-referred customers with similar demographic information.

Another study found that customers acquired via traditional techniques are more valuable in the short term to some extent, but referred customers are twice as valuable in the long term (Villanueva et al., 2008). Specifically, the researchers found that short-term earnings can increase by 28 percent due to customers acquired through referrals, and by almost 34 percent due to customers acquired through other methods. In contrast, the value of the referral efforts is highlighted in the long run, as long-term earnings accumulate to an almost 10 percent increase for referred customers, compared to only 5 percent for non-referred customers. This difference can be explained by referred customers' longer tenure with the firm, as their dropout rates were found to be consistently lower than those of non-referred customers. In addition, the value differential becomes even larger when considering the disparity in acquisition costs, due to the considerably lower costs of referrals. The above findings highlight the popularity of referral reward programs across a plethora of companies. They also explain why some firms offer incentives of considerably high value. For instance, DIRECTTV[7] offers a $100 credit to both referred and referring customers in ten monthly credit installments of $10 toward their respective accounts.

SEEDING STRATEGIES

It is clear at this point that the acquisition costs associated with referred customers are lower, and their overall value is higher, when compared to customers recruited through other methods. These benefits render referral programs a superior acquisition method. However, these programs still require substantial investments in order to be successfully implemented and promoted. In particular, referral campaigns can be expensive, as they frequently involve personal communications such as direct mailings and phone calls. In order to control these costs, firms need to be selective in approaching their customers through such channels. Therefore, they should target those individuals from their customer base that have the highest potential to generate successful referrals. Initiating word-of-mouth, like referrals, by contacting specific individuals and providing them with information to be shared with others is called 'seeding'. Seeding strategies involve selecting certain groups of people to receive a word-of-mouth communication (or seed) over others. In the context of referral campaigns, firms need to identify which customers are more prone to refer the firm to others, thereby developing the optimum seeding strategy in order to maximize referral activity.

Before determining possible seeding strategies, however, customers should be categorized into groups. Given that word-of-mouth depends on social networks, customers may be assessed on the basis of some indicators of their social network presence, such as the extent and type of their connectivity with others, and this information can guide grouping. An individual's level of connectivity represents the number of his or her social connections, and it may be measured using the degree centrality scale. An individual's type of connectivity shows the number of his or her unique connections, and it may be measured using the betweenness centrality scale. Unique connections occur when social groups are connected through only one individual. Based on these two measures, customers can be divided into three groups, termed as 'hubs', 'fringes', and 'bridges' (Hinz et al., 2011). Hubs are people with relatively many social connections (high-degree centrality), fringes are those with relatively few connections (low-degree centrality), and bridges are those with relatively many unique connections (high-betweenness centrality). Figure 9.2 illustrates the position of these three groups in a simplified depiction of a social network.

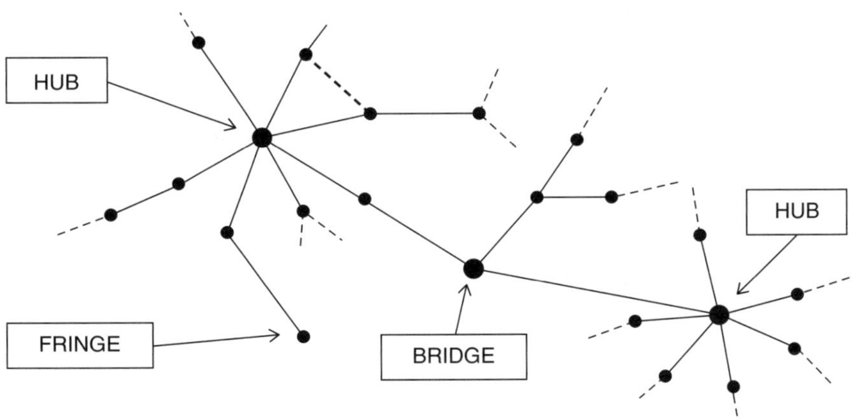

Figure 9.2 Location of hubs, fringes, and bridges in a social network

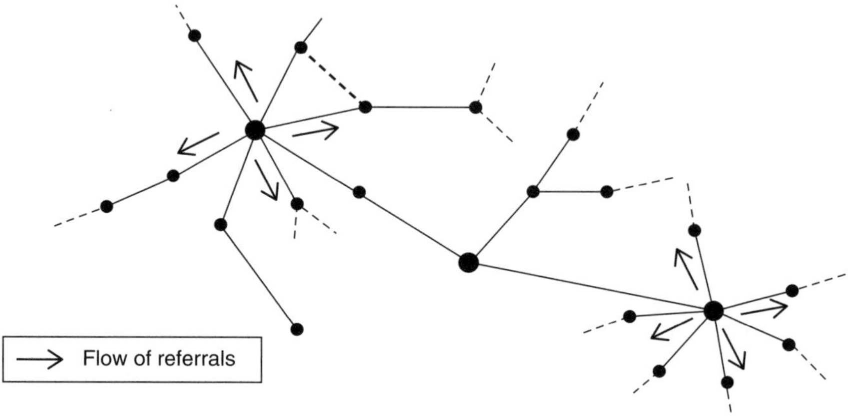

Figure 9.3 Flow of referrals in high-degree seeding strategy

Depending on the connectivity characteristics of the customers that receive the initial message (seed), we distinguish among four different seeding strategies, as described next.

High-degree seeding strategy

This strategy involves transmitting the information seeds to people with high-degree centrality, that is, hubs. Hubs are expected to be more influential than other individuals because of the central position they assume in their respective networks. Also, their numerous connections increase the chances of them informing, persuading, and/or successfully recruiting more prospects for the firm. The flow of referrals generated through the high-degree seeding strategy is depicted in Figure 9.3.

Low-degree seeding strategy

This involves transmitting the information seeds to people with low-degree centrality, that is, fringes. This strategy is based on the premise that hubs, due to their large number of connections, receive too much information, which in turn lowers the chance that they will actually process marketing information and be persuaded by it. On the contrary, fringes might be more receptive and influential, so they are selected to receive the seeds. The flow of referrals made through the low-degree seeding strategy is shown in Figure 9.4.

High-betweenness seeding strategy

This strategy involves transmitting the information seeds to people with high-betweenness centrality, that is, bridges. It attempts to maximize the

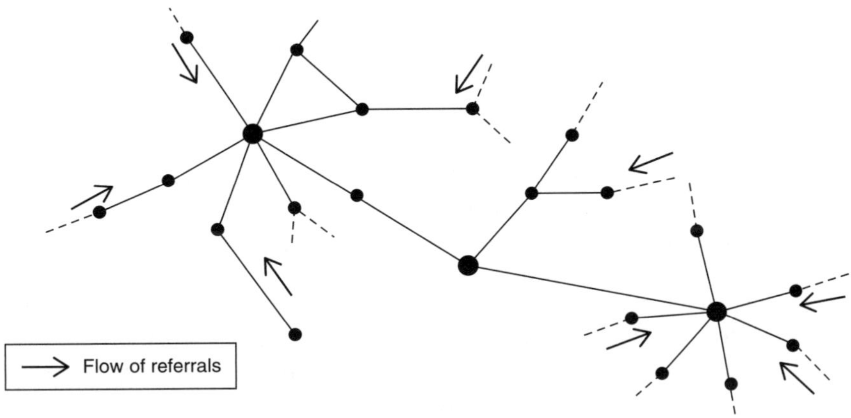

Figure 9.4 Flow of referrals in low-degree seeding strategy

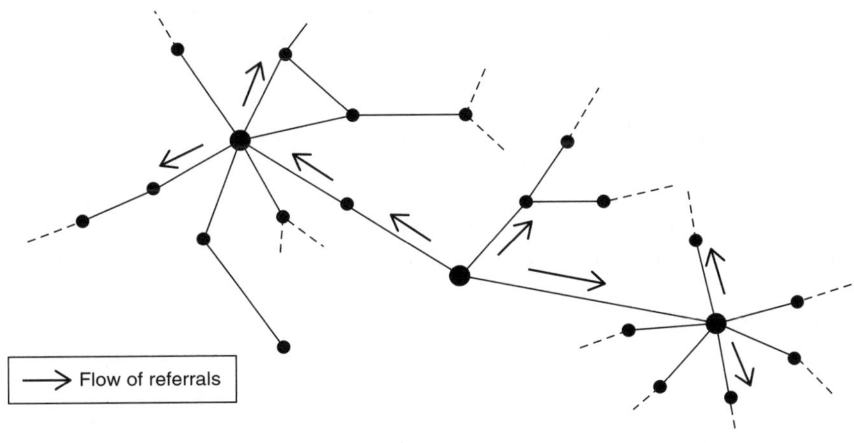

Figure 9.5 Flow of referrals in high-betweenness seeding strategy

total number of referrals and subsequent (second-generation) referrals, based on the position of bridges in their respective social networks. That is, because bridges are located between unconnected groups of people, the chances of the referral program information spreading are much higher. The referrals generated through the high-betweenness seeding strategy and their flow are depicted in Figure 9.5.

Random seeding strategy

This involves transmitting the information seeds to random people, regardless of their connectivity with others. This strategy is used either when there is no available information about the social network of customers or when there are no actual/expected differences between hubs, fringes, or bridges with regard to sharing information about the referral program, participating customers, or successfully recruiting referred customers.

Among these strategies, the high-degree seeding and high-betweenness seeding strategies generate the most referrals, as demonstrated in two different research settings (Hinz et al., 2011). Initially, the researchers tested the four seeding strategies by conducting an experiment through an online social networking website. The referral process was tested by using friend lists of the participants (n = 120) and categorizing them into three groups – hubs, fringes, and bridges. Accordingly, different seeding strategies were applied by sending unique seeds to hubs (high-degree seeding), fringes (low-degree seeding), bridges (high-betweenness seeding), or random people (random seeding) via personal messages on the social networking platform. The participants who received the message were asked to forward it to their friends, and these friends were asked to pass it along as well. They were also informed about one of two small cash rewards, either for both referrers and referred participants, or only for those referred. Those who accepted and acted on the referrals (referred) logged on a website using unique credentials, a secret word contained in the message they received, and the name of the person who sent them the message (referrer), thus allowing for tracking of successful referrals and the respective referrers. As a result, a total of 1155 referrals were made by 55 percent of the participants.

The results of this experiment indicate that seeding to hubs and bridges as well as extending rewards to referrers increases the chances of successful referrals. Specifically, when compared to random seeding, the likelihood of successful referrals was higher by 39 percent through high-betweenness seeding and by 53 percent through high-degree seeding, but it was lower by 81 percent through low-degree seeding. Therefore, low-degree seeding appears to be ineffective in inducing successful referrals. Even though high-degree seeding was more effective in producing successful referrals than high-betweenness seeding (53 percent vs 39 percent) when compared to the random seeding strategy, they were not that different when compared to each other. Therefore, both strategies are expected to perform well and quite similarly in practical applications, given that most companies are able to group their customers. However, acquiring information about degree centrality requires information on only the social connections of individuals, and hubs and fringes can be easily identified thereon. In contrast, information about a greater part of the social networks (such as friends of friends) of customers is needed for measuring individuals' betweenness centrality, so that they can be categorized as bridges. Therefore, a high-degree seeding strategy is considerably easier to implement for most firms. Finally, the study revealed the significance of providing incentives to referrers, which is a popular characteristic of current referral programs. In particular, participants exhibited a considerably higher rate of successful referrals (38.11) when incentivized by rewards for both themselves and their referred friends than when the rewards were only intended for their referred friends. In other words, offering an incentive to the referrer increases successful referrals by more than 37 times. The following paragraphs discuss the performance of seeding strategies in the context of a marketing campaign.

The second part of the study by Hinz et al. (2011) involved comparing the seeding strategies when applied in a campaign to promote an existing referral program of a telecommunications firm. Due to the availability of customers' communication data, it was possible to assess their social networks and measure their level of connectivity. However, information about non-customers' social networks was not accessible to the firm, so there were no measures of betweenness centrality. As part of the campaign, the mobile service provider increased its referral incentive to 15 euros (up from 10 euros) for each successful referral during one month, and notified its entire customer base (n = 208 829) through text messages. Consequently, referral activities increased by 50 percent during that month, as a total of 6392 referrals were made by 2.1 percent of the customers.

The results of this campaign indicate that seeding to highly connected individuals (hubs) increases the chances of successful referrals. In particular, customers with higher levels of connectivity are more likely to generate referrals, as those who participated in the referral program had roughly nine more social contacts than those who did not. Also, customers with higher levels of connectivity made more referrals to prospects overall. Additionally, some customer characteristics were related to participation in the referral program, such as gender (male), age (older customers), customer tenure (recently acquired), service usage (heavy), and acquisition channel (online). These characteristics can further guide selection for possible seeding targets. In order to directly compare the seeding strategies, the researchers estimated probabilities of participation in the referral program for all customers, and assigned them into ordered groups. Table 9.1 shows that the top group in terms of participation likelihood also had the highest average degree centrality (70.8), whereas the bottom group had the lowest average degree centrality (18.0). These groups are representative of hubs and fringes respectively, so targeting the first implies a high-degree seeding strategy while targeting the second implies a low-degree seeding strategy.

Table 9.1 Referral activity rates per connectivity level

	Top 5000 (Hubs)	Bottom 5000 (Fringes)	Average (Random)
Participation rate	4.4%	0.5%	2.2%
Referral rate	5.8%	0.5%	3.0%
Successful referral rate	3.8%	0.4%	2.0%
Average degree centrality	70.8	18.0	36.5

Source: Adapted from Hinz et al. (2011).

Again, a positive relation between connectivity levels and referral success was found, which indicates that the high-degree seeding strategy is much more effective than the low-degree seeding strategy. Furthermore, considering the average rates of the whole customer base to be equivalent to those of a random seeding strategy, Table 9.1 reveals that high-degree seeding is still more effective than random seeding. Overall, the findings of both studies illustrate that high-degree seeding greatly increases the chances of participation in a referral program, the referrals generated, and their success.

REFERRAL AWARDS

Referral incentives vary greatly across firms and industries in monetary value, form, and their recipient. With regard to the monetary value, past research has tried to determine the right amount of incentives. One study examined the optimal amount of customer referral rewards alongside the price of the product/service (Biyalogorsky, Gerstner, and Libai, 2001). Based on the concept of customer delight, customers make recommendations when they are delighted (very satisfied) by the product offering and its price, when they are offered referral rewards, or when they have a combination of both. Therefore, referrals may be encouraged by a low price and/or a referral reward, depending on the level of customer delight. Additional research has incorporated the concern of customers for the product outcomes of their friends and family as well (Kornish and Li, 2010). That is, when customers are highly concerned about the referred individuals' satisfaction, they require incentives of high value or a lower price. The researchers suggest that customers' greater concern possibly makes them more selective in providing referrals, so this should be balanced by extending a high-value reward. They also found that, despite the presence of a reward for the referrer, customers still make credible recommendations because the product/service is of value to their friends (ibid.). The above findings show that referral rewards are important, and defining a proper incentive value is necessary for boosting the number of customer referrals. In line with these findings, companies that offer low prices tend to extend low-value incentives. Groupon,[8] which is known for its deals, provides a $10 account credit to its customers for each successful referral. In contrast, firms that cannot charge low prices are more likely to extend high-value incentives. For example, AT&T[9] provides a $25 voucher to both referring and referred customers, for each referred customer who subscribes to one of the firm's services. These vouchers may be used towards any product/service provided by the firm.

Firms may also adjust the value of the referral rewards on an individual customer basis, if possible. This approach is implemented by the hotel chain Marriott.[10] The firm's referral program incentivizes referrals through loyalty points rewards, varying the reward as per the referred customers' usage. That is, the more nights the referred customers spend in the firm's hotels, the more points are offered to both them and their referrers, up to a certain limit.

As described earlier, referral rewards may be encountered in a variety of forms, including free service, loyalty points, vouchers, and cash. Indeed, the previous examples include a variety of incentives: Dish provides account credit and Marriott awards loyalty points as referral rewards. It is clear that most incentives are usually connected to the referred product/service, which makes them easy and cheap to manage. In some cases, however, determining the reward type is not straightforward. For example, offering discounts for additional purchases is not ideal for infrequently purchased products. In such instances, incentives may entail accompanying products/services. In other cases, firms may hand over control to their customers and allow them to choose from an array of referral reward options. For instance, we saw that AT&T offers vouchers redeemable for any of its services. The bottom line is, firms need to provide their customers with rewards that are valued and desired by them, since offering an assortment of different incentives is not a viable option for many referral programs.

Furthermore, regarding the recipient of the incentives, research by Ryu and Feick (2007) has shown that referrer rewards increase the chances of referrals towards relatively weaker social connections of customers, such as their distant friends and acquaintances. On the other hand, referred-customer rewards are important in incentivizing customers to provide referrals towards their stronger/closer social connections, such as their close friends and family. The same study revealed that providing an incentive to referring customers is especially beneficial for non-established brands, whereas rewarding prospects is effective for more well-known brands (ibid.). Therefore, offering the right type of rewards to the right individuals depends on a number of factors, and it can successfully result in increased customer acquisition.

REFERRAL INTENTIONS

Over the last few years, interest in word-of-mouth and customer referrals has increased given the fact that many managers have started adopting metrics such as the 'net promoter score' (Reichheld, 2003) to strategically manage firm revenue growth. This score is easily obtained by asking customers a simple question: 'How likely is it that you would recommend our company to a friend or colleague?' Using a response scale from 0 to 10, a firm can find the ratio of 'promoters' (9–10 ratings), 'passively satisfied' (7–8 ratings) and 'detractors' (0–6 ratings). The 'net promoter score' is given by subtracting the detractors' percentage from the promoters' percentage (ibid.). While easy to compute, Keiningham et al. (2007) empirically showed that the 'net promoter score' shows only a weak link to a firm's revenue growth. Moreover, this metric is not a reliable measure of the likelihood of recommendation or the results of this recommendation. Specifically, there is a lot of uncertainty from the beginning of the referral process to the point that it generates profits for the firm as it is dependent on multiple conditions: (1) customers who report willingness to recommend may not actually refer the firm to their social contacts, (2) referred customers may not be interested in hearing about the firm and its products, (3) referred customers may not be willing to purchase, or (4) they may not become profitable customers for the firm.

A field study examined the actual possibility of the above conditions occurring across the customers of two companies, a telecommunications firm and a financial services firm (Kumar, Petersen, and Leone, 2007). Using a sample of customers from each firm (9900 from the telecommunications firm and 6700 from the financial services firm), the researchers obtained measures of the customers' referral intentions. In addition, they monitored the consequent referral behavior of these customers as well as the purchase behavior of the referred customers that they recruited for the respective firms. The results of this study showed that a large number of customers reported an intention to recommend the firms to their friends (financial services: 68 percent, telecommunications: 81 percent), but only a few actually did make a referral (financial services: 33 percent, telecommunications: 30 percent). Furthermore, less than half of the referrals resulted in new customers for the firms (financial services: 14 percent, telecommunications: 12 percent), and not all of these newly acquired customers proved to be profitable (financial services: 12 percent, telecommunications: 8 percent). These findings illustrate that expressed intention to recommend does not necessarily translate into action, and it certainly does not lead to substantial increase in customer equity. This has led to an ongoing debate over the issues with measuring and managing metrics that predict firm and customer growth and the need for measures and frameworks that do generate growth in a firm's revenues and sales.

REFERRING CUSTOMERS' VALUE

While the existence of word-of-mouth and referral behavior has been studied within the marketing literature for decades, only recently have there been studies that have examined the monetary value of word-of-mouth and referral behavior, or studied how information on word-of-mouth or referral behavior can be used to effectively manage customers. This is important since, as noted by Rust, Zeithaml, and Lemon (2000, p. 46): 'the effect [of word-of-mouth] is notoriously hard to measure, but is frequently significantly large'.

Given that word-of-mouth and referral behavior can play a significant role in consumer decision-making and purchase behavior, the challenge faced by practitioners is in finding a way to harness word-of-mouth and referral behavior by managing their customers. There have been a number of studies published in the marketing literature that have shown that measuring the value of word-of-mouth or referrals can provide several key implications for managing customers and that referral marketing campaigns can be highly valuable. Recently, the findings from these studies have been extended further to aid in the practice of marketing by providing frameworks for the selection of the 'best' customers for such referral marketing campaigns. Table 9.2 summarizes the key contributions and implications from past studies.

CUSTOMER REFERRAL VALUE

Many firms in both the B2C and B2B space are now using referral programs to harness the power of word-of-mouth. In particular, firms are launching campaigns that attempt to increase referrals by existing customers in order to acquire new customers. The individuals that are successful in 'recruiting' prospects for the firm through their recommendations are highly valuable. Therefore, firms need a measure that represents customers' contribution through their referrals and aids in identifying their most

Table 9.2 Contributions of past studies

Criteria	Helm (2003)	Hogan et al. (2003)	Hogan et al. (2004)	Ryu and Feick (2007)	Kumar et al. (2007)	Kumar et al. (2010)	Kumar et al. (2013)
1. Computation of referral value	N/A	Value lost from customer churn	WOM multiplier from advertising	Referral likelihood based on rewards	Actual past referral behavior	Predicted future referral behavior	Predicted reference value
2. Key objective of the paper	Reviewing three approaches to computing referral value	A method to account for social effects in the management of customers	Quantifying the advertising ripple effect	Determine the effectiveness of rewards for referral marketing	Comparison of customer segmentation based on CLV and CRV	Optimal customer targeting for referral marketing campaigns	Determine the impact of references and the types of references that are most effective
3. Contribution to marketing	An extensive literature review	Customers cannot be valued in isolation due to WOM effects	Managers should account for customer WOM interactions when advertising	Shows how extrinsic motivations affect customer referral behavior	Computing past referral value of each customer	Determining behavioral drivers of referral value that can help in targeting customers for referral marketing campaigns	Determining the characteristics that lead to successful references

4. Field experiment(s) and/or empirical application	N/A	A simulation exercise using data collected about the online banking market	A survey of students based on their hair salon choice and the amount of WOM they engaged in following the last haircut	A series of four laboratory experiments that investigate the effects of different types of rewards on different types of customers	Simple field experiment with a telecom firm based on computed CLV and CRV where all customers were targeted with an incentive to make referrals	A series of field experiments that uncover 1) the most effective way to target customers for referral marketing and 2) the effectiveness of using the drivers of CRV as a method for customer selection and resource allocation	A series of field experiments to determine the most effective ways to design references
5. Key learning	The three models discussed offer insights into the determinants and effects of word-of-mouth	The value of a lost customer changes as a function of the time in the product life cycle	The WOM generated after an advertising-motivated purchase can be significant	Offering a reward does increase referral likelihood, but the size of the reward is not significant	Customers with high CLV are not the same as customers with high CRV	Firms can effectively select customers within segments of high and low CLV/CRV for referral marketing campaigns using the drivers of CRV (dynamic targeting)	Firms can effectively use business references to pull in new customers. Rich media and similar companies are the key drivers of success

referral-active customers. One such metric is customer referral value (CRV). The CRV of a customer is defined as 'the net present value of the future profits of new customers who purchased products/services due to the referral behavior of the current customer' (Kumar, 2013, p. 147). It represents the value of a customer's contribution to the firm indirectly, through his or her referral activity. This value differs according to the referred prospects' intentions to purchase from the firm prior to the referral. That is, there is a distinction between the customers that initiated a relationship with the firm solely because of the referral, and those that would have become customers even in the absence of a referral. This leads to the following basis for measuring CRV (Kumar et al., 2007):

$$CRV = \frac{\textit{Value of customers who join due to the referral}}{\textit{Discount rate}} + \frac{\textit{Value of customers who would join anyway}}{\textit{Discount rate}}$$

We distinguish between these two groups of referred customers because they contribute to a customer's CRV in distinct yet definite ways. Specifically, referrals recruit customers for lower costs than other methods, so they lead to savings in acquisition costs that apply to all referred customers. In addition, referrals are the main reason for attracting some of these customers (those who join due to the referral), so they lay the foundation for future purchases by those referred customers. Therefore, CRV includes both the acquisition savings and the profits of the future transactions of customers who join due to the referral. However, it includes only the acquisition savings of customers who would join anyway, as their future transactions would happen in spite of the referral, so they cannot be attributed to the referrer. Based on this, the expanded formula for computing CRV is the following:

$$CRV_i = \sum_{t=1}^{T}\sum_{y=1}^{n_1}\frac{A_{ty} - a_{ty} - M_{ty} + ACQ1_{ty}}{(1+r)^t} + \sum_{t=1}^{T}\sum_{y=n_1+1}^{n_2}\frac{ACQ2_{ty}}{(1+r)^t}$$

Where:
CRV_i = referral value of customer i;
T = number of periods predicted into the future (e.g., years);
n_1 = number of customers who join due to the referral per period;
n_2 = total number of referred customers per period;
$n_2 - n_1$ = number of customers who would join anyway per period;
A_{ty} = gross margin contributed by customer y (who would not purchase otherwise);

a_{ty} = cost of the referral to customer y;
M_{ty} = marketing costs needed to retain customer y;
$ACQ1_{ty}$ = savings in acquisition cost towards customers who join due to the referral;
$ACQ2_{ty}$ = savings in acquisition cost towards customers who would join anyway;
r = discount rate.

The preceding formula shows that CRV is a forward-looking metric that represents the value of a customer's referral activity using information about his or her past referral behavior, their recruits' buying behavior, and the firm's tactics. This information is used to predict the elements required for computing CRV, which include the number, timing, and cost of referrals, the profits that result from the referred customers' purchasing activities and the savings in their acquisition costs. The total CRV for each customer is the sum of his or her CRV for each referred customer per period and for each period that is examined in the future. The number of periods considered depends on the industry, the product cycle, and the firm, although it is usually confined to three years to limit long-term prediction inaccuracies.

Each of the elements used in the calculation of CRV are important determinants of the value that customers contribute to the firm through their referral activities. First, it is important to incorporate the cost of the referral, which represents the cost of the rewards extended to the participants of the referral (referring and/or referred customer). Based on the referral cost and the cost of alternative acquisition methods, firms can also measure the savings in acquisition costs that each referral generates, and include it in the computation. That is, the CRV of a customer increases by the amount of acquisition savings for each referred customer he or she brought to the firm. Furthermore, the profits generated from the purchases of referred customers are taken into consideration as well, by adding the gross contribution margin and subtracting the marketing costs per customer per period. Moreover, after establishing the number of periods predicted into the future, it is necessary to determine the number and timing of each customer's referrals. These two elements depend solely on the referral activity and referral success of the referring customer. Finding when each referral is made and assigning it to the correct period is important for an accurate discounting of future value to current rates. After determining referral timing, the number of referrals for each period is easily obtained, and this information can guide the summation for estimating individual CRV.

As explained earlier, referred customers are divided into two groups,

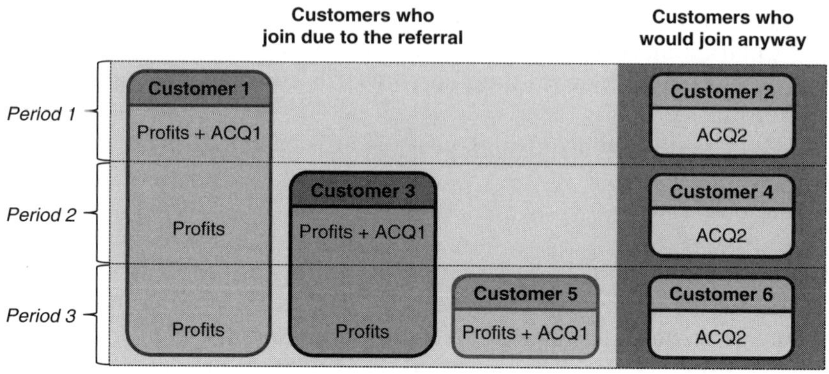

Figure 9.6 Contribution of referred customers to referrer's CRV

depending on whether the referral was the main reason for their purchase. Their distinct impact on their referrers' CRV is expressed in two ways. First, their respective contribution is calculated in two separate parts of the CRV formula, allowing for the inclusion of different elements for each group. Second, their contribution is considered along a different number of periods. That is, individuals who would become customers even without a referral are considered only once – during the period they join the firm's customer base. On the other hand, the other group of referred customers should be included in all time periods examined in CRV calculations. This happens because the reason they become customers is the referral, so their continued relationship with the firm should be credited to their referrers. Figure 9.6 illustrates a breakdown of the contribution of the two groups of referred customers to their referrer's CRV per period.

Calculating CRV

Early research focused on attempting to forecast an individual's referral behavior by developing models that used only a customer's actual past referral behavior to compute an individual customer's referral value (Kumar et al., 2007). Additional research by Kumar et al. (2010) developed a four-step approach that allows one to compute individual CRV, as illustrated in the example in Table 9.3.

Using information on customers' referral behavior and referred customers' purchasing behavior, we can easily calculate CRV. For example, we can use the information provided in Table 9.3 to estimate Mary's CRV through the following four steps. First, it is important to determine

Table 9.3 Mary's referral behavior information

Information	Value
Number of periods predicted (T)	2 years
Total number of referred customers per period (n_2)	6
Number of customers who join due to the referral (n_1)	3
Number of customers who would join anyway ($n_2 - n_1$)	3
Average gross margin (A_{ty})	\$120
Cost of referrals (a_{ty})	\$40
Average marketing costs (M_{ty})	\$30
Savings in acquisition costs ($ACQ1_{ty}$ & $ACQ2_{ty}$)	\$5
Discount rate (r)	15%

whether each referred customer would have bought anyway, in order to assign him or her to the correct group. This information can be obtained by simply asking new referred customers whether they would have joined in the absence of the referral. In this example, Mary successfully refers six customers to the firm per year, three of whom join solely due to her referral. Second, the future value of each referred customer should be predicted. This can be estimated through a number of approaches; for example, information from the transactions of previously referred customers can be used. Here, an average gross margin of \$120 and marketing costs of \$30 are assumed to be incurred towards Mary's referred customers. Including the cost of each referral for the firm, which is \$40 in this case, we have all the information needed for calculating the profit contribution of customers who join as a result of the referral. In the third step, the number of referrals generated by the customer is predicted, using information about his or her past referral activity in previous years. In this case, Mary is predicted to make a total of 12 successful referrals during the next two years. The final step entails predicting the timing of customer referrals, for which information on past referrals can also be used. Mary, the customer of this example, is predicted to make six referrals during the first year, and another six referrals during the second year of this estimation. In addition to this data, we have assumed that acquisition costs amount to \$40 through referrals and to \$45 through other methods, which results in \$5 acquisition savings per referral. Finally, a yearly discount rate of 15 percent is used to showcase the calculation of Mary's CRV below:

Period 1:

$$CRV_{t=1} = \sum_{y=1}^{n_1} \frac{A_{1y} - a_{1y} - M_{1y} + ACQ1_{1y}}{(1+r)^1} + \sum_{y=n_1+1}^{n_2} \frac{ACQ2_{1y}}{(1+r)^1}$$

$$CRV_{t=1} = \sum_{y=1}^{3} \frac{\$120 - \$40 - \$30 + \$5}{(1 + 0.15)^1} + \sum_{y=4}^{6} \frac{\$5}{(1 + 0.15)^1} \approx \$157$$

Period 2:

$$CRV_{t=2} = \sum_{y=1}^{n_1} \frac{A_{2y} - M_{2y}}{(1 + r)^2} + \sum_{y=1}^{n_1} \frac{A_{2y} - a_{2y} - M_{2y} + ACQ1_{2y}}{(1 + r)^2} + \sum_{y=n_1+1}^{n_2} \frac{ACQ2_{2y}}{(1 + r)^2}$$

$$CRV_{t=2} = \sum_{y=1}^{3} \frac{\$120 - \$30}{(1 + 0.15)^2} + \sum_{y=1}^{3} \frac{\$120 - \$40 - \$30 + \$5}{(1 + 0.15)^2} + \sum_{y=4}^{6} \frac{\$5}{(1 + 0.15)^2} \approx \$340$$

Total:

$$CRV_{Mary} = CRV_{t=1} + CRV_{t=2} \approx \$497$$

From an implementation standpoint, research by Kumar et al. (2010) shows how to determine the behavioral drivers of CRV and then how to identify the most effective methods of targeting the most promising customers for soliciting referrals based on each customer's CLV and CRV scores. They show results from four separate field experiments with firms from two different industries – financial services and retailing – providing strong support for how targeting the right customer segments with the right marketing campaigns can lead to significant increases in both a customer's CRV as well as their CLV. By sending the right message to the right segment (i.e., a campaign to increase the low-scoring metric), the return on marketing investment (ROMI) was found to be just over 10. Kumar et al. (2010) also found that identifying and using behavioral drivers significantly improves firms' ability to effectively target low-CRV customers via referral marketing campaigns. Their research showed that such targeted efforts have the potential to increase targeted customers' CRV by as much as 30 percent. Therefore, in order to maximize firm profitability, it is critical to manage customers based on both their CLV and CRV scores, and understand the behavioral drivers of CRV so as to better target the most profitable customers through referral marketing campaigns.

BUSINESS REFERENCE VALUE

Many firms have started programs using client referencing as part of their marketing and sales efforts to encourage new customer adoption. This is particularly true in the B2B space where the use of references is often the only way to leverage the value of current clients on new customer adoption. Kumar et al. (2013) define the concept of business reference value (BRV) as the ability of and degree to which a client's reference influences prospects to purchase. They develop a three-step method to compute BRV for a given client using a measure of reference value calculated on the basis of self-reported data from the prospect on (1) the degree that references impacted the prospect purchase decision, (2) the degree of influence the client reference had on converting the prospect to a customer, and (3) the CLV of the converted prospect.

Kumar et al. (2013) identify and empirically determine the key drivers of BRV using data from a telecommunications and financial services firm. The drivers capture four different areas: (1) length of client relationship, (2) client firm size, (3) reference media format, and (4) reference congruency. The researchers found that the source of the reference matters, both in terms of the client firm size and the length of the client relationship. Therefore, it is very important that managers select the 'right' clients to use as references. They also found that the reference media format is an important driver of BRV and that richer media formats were more influential. Kumar et al. (2013) also showed that reference congruency significantly increases BRV. Therefore, firms should attempt to match clients with prospects who are congruent based on product/service purchased (highest effect size), industry (second highest effect size), and/or having the same role/position within their firm (third highest effect size). Finally, they found that when reference clients are larger, this provides an incrementally larger degree of adoption influence on all other drivers, suggesting that larger client firms add direct and indirect value to social influence.

These findings suggest that a firm using a client referencing strategy needs to leverage a portfolio of client references including clients with varied characteristics in order to match the client to the prospect successfully. They also suggest that it might be more effective to bring in only a limited set of 'targeted' references in the sales process. In some situations, it may be possible to provide customized client references during a sales engagement and in those situations it is clear that a firm should identify those customers from its customer base that match the prospect firm as closely as possible.

IMPLEMENTATION CONSIDERATIONS

There are several important factors to consider before implementing marketing programs to increase CRV and/or use referral marketing programs to bring in new customers. First, CRV might not apply to all firms universally. For example, the value of some products makes it unsuitable to use rewards for referrals towards the sales of such products. In some such cases, research recommends increasing the satisfaction and, thus, willingness-to-recommend of current customers by lowering their price as a better strategy (Biyalogorsky et al., 2001). Second, not every firm has the ability to track referral behavior and even if it does, it may have neither the systems in place to do so nor a long history of tracking customer referral behavior to estimate the models required. Third, the methods that have been proposed only take into account the extrinsic motivation of a customer to make referrals (i.e., through incentives) and do not try to determine which customers generate referrals through an intrinsic motivation. Finally, these approaches work well in a fairly stable environment, but in cases where the market is more turbulent, it may be necessary to use a different approach to bring in new customers.

NOTES

1. Neilsen (2012), 'Consumer trust in online, social and mobile advertising grows', *Neilsen*, 4 December 2013, accessed 10 September 2014 at http://www.nielsen.com/us/en/newswire/2012/consumer-trust-in-online-social-and-mobile-advertising-grows.html.
2. Neilsen (2013), 'A multi-mix media approach drives new product awareness', *Neilsen*, 2 December 2013, accessed 10 September at http://www.nielsen.com/us/en/newswire/2013/a-multi-mix-media-approach-drives-new-product-awareness.html.
3. Google Wildfire (2013), 'Engaging consumers where it matters most', *Google Wildfire*, accessed 10 September 2014 at http://www.ufsa.ufl.edu/uploads/mcda/Engaging_Consumers_Where_It_Matters_Most.pdf.
4. DISH referral program (2013), *Dish*, accessed 10 September 2014 at http://www.dish.com/customerservice/referral/friendsandfamily.
5. Credit card referral bonus (2013), *Discover.com*, accessed 12 September 2014 at https://www.discover.com/credit-cards/exclusives/sharediscover.
6. Ashworth College (2014), 'Refer a friend make $50', *Ashworth College*, accessed 12 September 2014 at http://www.referashworth.com.
7. DIRECTV referral offer (n.d.), *DIRECTTV*, accessed 12 September at http://www.directv.com/DTVAPP/referral/referralProgram.jsp.
8. 'Refer a friend and earn $10 in Groupon Bucks!' (2013), *Groupon.com*, accessed 12 September 2014 at http://www.groupon.com/referral.
9. AT&T refer-a-friend (2013), *ATT.com*, accessed 12 September 2014 at http://referrals.att.com/common/overview.pg.
10. Marriott Rewards® Refer a Friend (2013), *Marriott*, accessed 12 September 2014 at http://www.marriott.com/marriott-rewards/refer-a-friend.mi.

REFERENCES

Anderson, E.W. (1998), 'Customer satisfaction and word-of-mouth', *Journal of Service Research*, **1**(1), 5–17.

Biyalogorsky, E., E. Gerstner and B. Libai (2001), 'Customer referral management: optimal reward programs', *Marketing Science*, **20**(1), 82–95.

Feick, L.F. and L.L. Price (1987), 'The market maven: a diffuser of marketplace information', *Journal of Marketing*, **51**(1), 83–97.

Helm, S. (2003), 'Calculating the value of customers' referrals', *Managing Service Quality*, **13**(2), 124–33.

Hinz, O., B. Skiera, C. Barrot and J.U. Becker (2011), 'Seeding strategies for viral marketing: an empirical comparison', *Journal of Marketing*, **75**(6), 55–71.

Hogan, J.E., K.N. Lemon and B. Libai (2003), 'What is the true value of a lost customer?', *Journal of Service Research*, **5**(3), 196–208.

Hogan, J.E., K.N. Lemon and B. Libai (2004), 'Quantifying the ripple: word-of-mouth and advertising effectiveness', *Journal of Advertising Research*, **44**(3), 271–80.

Keiningham, T.L., B. Cooil, T.W. Andreassen and L. Aksoy (2007), 'A longitudinal examination of net promoter and firm revenue growth', *Journal of Marketing*, **71**(3), 39–51.

Kornish, L.J. and Q. Li (2010), 'Optimal referral bonuses with asymmetric information: firm-offered and interpersonal incentives', *Marketing Science*, **29**(1), 108–21.

Kumar, V. (2013), *Profitable Customer Engagement: Concepts, Metrics & Strategies*, Thousand Oaks, CA: Sage Publications.

Kumar, V., J.A. Petersen and R.P. Leone (2007), 'How valuable is word-of-mouth?', *Harvard Business Review*, **85**(10), 139–46.

Kumar, V., J.A. Petersen and R.P. Leone (2010), 'Driving profitability by encouraging customer referrals: who, when, and how', *Journal of Marketing*, **74**(5), 1–17.

Kumar, V., J.A. Petersen and R.P. Leone (2013), 'Defining, measuring, and managing business reference value', *Journal of Marketing*, **77**(1), 68–86.

Mahajan, V., E. Muller and F.M. Bass (1995), 'Diffusion of new products: empirical generalizations and managerial uses', *Marketing Science*, **14**(3), G79–G88.

Reichheld, F.F. (2003), 'The one number you need to grow', *Harvard Business Review*, **81**(12), 46–54.

Rust, R.T., V.A. Zeithaml and K.N. Lemon (2000), *Driving Customer Equity*, New York: Free Press.

Ryu, G. and L. Feick (2007), 'A penny for your thoughts: referral reward programs and referral likelihood', *Journal of Marketing*, **71**(1), 84–94.

Schmitt, P., B. Skiera and C. van den Bulte (2011), 'Referral programs and customer value', *Journal of Marketing*, **75**(1), 46–59.

Villanueva, J., S. Yoo and D.M. Hanssens (2008), 'The impact of marketing-induced versus word-of-mouth customer acquisition on customer equity growth', *Journal of Marketing Research*, **45**(1), 48–59.

PART 3

APPLYING THE CUSTOMER EQUITY CONCEPT TO ENHANCE FIRM PERFORMANCE

10. Customer acquisition strategies: a customer equity management perspective

Kay Peters, Peter C. Verhoef and Manfred Krafft

INTRODUCTION

Firms can employ multiple strategies to increase customer equity (e.g., Rust, Lemon, and Zeithaml, 2004) and subsequently firm value (e.g., Gupta and Lehmann, 2003; Kumar and Shah, 2009). The most commonly considered strategies are to increase cross-buying, customer retention or to systematically win back lost customers. Since Reichheld's (1996) claim that loyal customers are much more profitable, increasing loyalty has been the dominant mantra in many industries. Also, cross-buying (e.g., Kumar, Venkatesan, and Reinartz, 2006; Kumar, George, and Pancras, 2008; Kumar, Venkatesan et al., 2008) and win-back have been shown to produce substantial returns (e.g., Tokman, Davis, and Lemon, 2007). Indeed, Gupta, Lehmann, and Stuart (2004) show that retention efforts have the strongest effect on customer equity (e.g., Gupta and Zeithaml, 2006). When evaluated in isolation, customer acquisition seems to be the lowest-yielding activity along the customer life cycle, while requiring substantial investment per new customer and specific campaign. Does this still hold when accounting for customer heterogeneity, long-term dynamics, employing new media like social networks, or cross-media efforts and related synergies? Specifically, employing one-fits-all or even to some degree targeted but isolated campaigns may result in higher cost per acquired customer due to the lack of appropriate customization with respect to media preferences, timing, and so on. Accounting for dynamics implies that when discussing customer acquisition, a firm should not only consider short-term outcomes, but also how acquisition strategies may drive long-term outcomes like customer lifetime value (e.g., Leeflang et al., 2009). At the same time, innovative approaches via new or more efficient social media may lead to reconsidering current efficacy assessments between the life cycle stages. Finally, evaluating acquisition efforts across all employed media and their respective synergies over time may allow better coordination and leverage of efforts.

Recent research has shown that customer acquisition, cross-buying, retention, and win-back are in fact intertwined processes (e.g., Blattberg and Deighton, 1996; Thomas, 2001; Reinartz, Krafft, and Hoyer, 2004; Reinartz, Thomas, and Kumar, 2005; Schweidel, Fader, and Bradlow, 2008). Hence, what firms do for customer acquisition actually defines the long-term value outcomes of their customer pool: from the composition of the customer database (e.g., Libai, Muller, and Peres, 2013) to the profitability of active customers (e.g., Lewis, 2006a, 2006b; Hanssens et al., 2008) to the retention and win-back efforts required later (e.g., Thomas, Blattberg, and Fox, 2004; Schweidel et al., 2008). For example, expansion strategies like cross-buying are not always successful when acquisition efforts are misguided: recently, Shah et al. (2012) show that cross-buying can be unprofitable for a considerable share of customers and that initial losses at the time of acquisition of these customers even tend to increase over time with higher levels of cross-buying (see also Shah and Kumar, 2012). Hence, what a firm does for acquisition is what it gets in the long term (e.g., Gupta and Zeithaml, 2006). In other words, success in cross-buying, retention, and win-back strategies is to some extent conditional on what the firm did for acquisition in the first place. A second reason for the importance of customer acquisition is that even with generally successful retention, natural attrition requires continuous replacement with new customers (e.g., Verhoef and Langerak, 2002).

Given the importance of customer acquisition, what can firms do to improve their acquisition performance? Among other tactics, they may improve their targeting (e.g., Bodapati and Gupta, 2004), focus on specific acquisition channels (e.g., Verhoef and Donkers, 2005), allocate their marketing resources better (e.g., Thomas, Reinartz, and Kumar, 2004; Venkatesan and Kumar, 2004), or improve specific tactics like their direct mailings (e.g., Van Diepen, Donkers, and Franses, 2009; Feld et al., 2013). Recent research offers various ways for firms to improve their customer acquisition process.

In this chapter, our overall contribution is to demonstrate and discuss the important role of customer acquisition in customer equity management. Our main claim is that customer equity management should govern acquisition strategies as they predetermine later outcomes to a substantial degree. These underlying dynamics in the customer portfolio are profound and need to be well understood in order to value, manage, and profit from optimal customer acquisition strategies. Second, we extract the main managerial insights from recent state-of-the-art research on customer acquisition: what is known and how can managers profit from this knowledge? Third, based on this collective knowledge we identify remaining research gaps and subsequently prioritize them into an agenda for future research.

We organize this chapter accordingly. In the next section we identify the fundamentals of customer acquisition and integrate them into a framework. This framework guides our subsequent literature review in the third section. From this review, we subsequently extract the main managerial insights and derive the research agenda.

FUNDAMENTALS OF CUSTOMER ACQUISITION

To ensure a proper understanding of the current research insights and remaining gaps we first need to define customer acquisition. It needs to be put into the context of the customer life cycle, which is a cornerstone of customer equity management. Based on this understanding, we need proper metrics to guide managerial efforts on customer acquisition. These metrics should reflect the important dimensions and dynamics that shape customer acquisition as well as its connections to the other activities along the customer life cycle. Next, we need to look at communication and sales channels through which a firm offers its products and services. Integrating these aspects leads to a process perspective of customer acquisition (e.g., Reinartz et al., 2004). Based on this process perspective, we conclude with challenges that practitioners frequently encounter when implementing customer acquisition as a value-driven process.

Defining Customer Acquisition

Customer acquisition is about acquiring new customers, that is, prospects (or lapsed customers) that buy for the first time (again) (e.g., Thomas, Blattberg, and Fox, 2004; Gupta and Zeithaml, 2006). New customers are generally persons (in B2C settings) or other firms (in B2B settings) that did not buy from the focal firm before.

A rather important issue for defining customer acquisition is that specifically in non-contractual relationships one might argue that firms constantly acquire customers. Firms have to compete with other firms as customers switch from one brand to the other over time. For example, in grocery markets national brands heavily compete with each other by constantly using deep price promotions to entice brand switching. In the same vein retailing firms in multiple markets (i.e., clothing, electronics, and groceries) rely strongly on sales promotions to attract customers to their stores. Berger and Nasr (1998) and Rust et al. (2004) refer to these specific forms of relationships as always-a-share customer relationships, whereas relationships in contractual settings are typically referred to as lost-for-good relationships.

Despite the fact that many firms constantly use acquisition strategies to also attract existing customers, it should not be considered as customer acquisition. Customers have entered into a relationship with a firm, although it is an always-a-share relationship. The objective of firms here is to gain a larger share-of-wallet per customer or to increase the purchase likelihood or purchase occasion. This requires strategies focusing on relationship development, despite the fact that acquisition tactics are used. Here, firms usually employ cross-buying or loyalty programs (Dowling and Uncles, 1997; Leenheer et al., 2007; Dorotic et al., 2011, 2012).

Related Strategic Approaches at the Aggregate Level

When a firm is trying to acquire new customers, it may distinguish between different target groups, that is, prospects that are from its *current* target segments featuring similar needs, preferences, and the like. Alternatively, it may pursue prospects from other or new segments that have *different* needs and preferences, requiring a distinct acquisition approach. The underlying strategic process is usually called STP – segmenting the market, targeting specific segments, and positioning the brand or offer in the competitive environment accordingly. The resulting prospect pool or target group may be characterized via demographics, attitudes, and behavior (e.g., Kotler and Armstrong, 1996).

Once the target groups have been selected and described, brand management frameworks suggest that customers move along stylized processes from awareness via interest and desire to an actual buying of the brand (AIDA, e.g., Lavidge and Steiner, 1961). Alternatively, customers may move along a hierarchy of intermediate effects of advertising, namely affect, cognition, and experience, to sales (e.g., Vaughn, 1986; Vakratsas and Ambler, 1999; Bruce, Peters, and Naik, 2012). These concepts focus on the intangibles of a brand and its effects on customers' perceptions. In the context of customer relationship management (CRM), these effects are reflected in metrics like brand equity (e.g., Keller, 1993). This metric and its components have been shown to exert a substantial influence along the customer life cycle (e.g., Leone et al., 2006; Stahl et al., 2012), especially in the early phase of the customer relationship (e.g., Bolton, Lemon, and Verhoef, 2004; Bügel, Verhoef, and Buunk, 2011).

Customer Acquisition Within the Customer Life Cycle

A prospect turns into a customer when he or she purchases from the firm for the first time. Once a customer, the firm tries to cross- and up-sell to the customer in order to retain the customer and maximize his or her value to the firm. When a customer lapses, win-back or customer re-acquisition activities are triggered (e.g., Thomas, Blattberg, and Fox, 2004; Tokman et al., 2007). This brings us back to customer acquisition as the CRM loop closes. In many instances, firms implementing customer relationship management need to define when re-acquisition efforts of lapsed customers stop and when these are subsequently reintegrated into the pool of prospects. Accordingly, the definition of customer acquisition may be extended to include previously lapsed customers for which re-acquisition efforts failed: prospects are then not only potential customers who did not yet buy from the firm, but also customers who bought from the firm only a long time ago.

One caveat though is that all lapsed customers should probably be targeted differently than prospects. This may especially hold for recently lapsed customers where re-acquisition efforts have failed so far. If lapsed customers are simply reintegrated into the pool of prospects an important characteristic is potentially being ignored, which may reduce the effectiveness of acquisition strategies. It is thus important to carefully establish when lapsed customers indeed become regular prospects again and are treated as such. So far, research on this issue is lacking (see also corresponding dotted box and line in Figure 10.1).

Prospect Value and Customer Lifetime Value

Customer lifetime value (CLV) is a forward-looking metric that sums up all discounted margins from future customer purchases and subtracts discounted future marketing expenditures required to generate those future purchases. CLV has been shown to be the best metric to inform resource allocation efforts at the individual customer level (e.g., Venkatesan and Kumar, 2004). In line with definitions of CLV, prospect lifetime value (PLV) is defined as the net present value of all expected generated profits of a prospect. This implies that for PLV, the CLV is usually extended to include the margin of the initial purchase, times the acquisition rate (i.e., probability of acquisition), while subtracting the acquisition spending required for turning the prospect into a customer:

$$PLV = Acquisition\ Rate^{*}\ [Initial\ Margin + CLV] - Acquisition\ Cost \tag{10.1}$$

The CLV component may be differentiated according to the prevailing type of relationship, that is, 'always-a-share' or 'lost-for-good'. We will discuss the implications of this value-based metric in later sections.

Communication Channels

An important element of the acquisition process are the employed communication channels. Importantly, these channels directly represent a major part of acquisition costs and thus drive PLV to a substantial degree (see Figure 10.1). Before turning into a customer, a prospect may be exposed to multiple contacts with the focal firm across a variety of advertising media (e.g., Smith, Gopalakrishna, and Chatterjee, 2006; Wiesel, Pauwels, and Arts, 2011). The prospect may also encounter the offerings of the firm when the customer actively looks for information, in the process even contacting the firm but without actually buying (e.g., Murray, 1991; Hoffman and Novak, 2000; Moe and Fader, 2004). From the perspective of the firm, these contacts may be attributed to distinct communication channels within the offline or online domain. Some of these channels may have a one-to-many character like TV, PR or online banner advertising, whereas others may be (personalized) one-to-one channels like calls, direct mailings, newsletters or emails (e.g., Ansari and Mela, 2003). Another even more important distinction between such media contacts is whether they are taking place with firm-controlled touch points or third-party–generated content like on social networks such as Facebook. Additionally, new hybrids like social coupons may be somewhere on the continuum between those poles (e.g., Kumar and Rajan, 2012). Jointly, combinations of these advertising channels may exhibit synergies that result in higher effectiveness than their single effects alone (e.g., Naik and Raman, 2003; Naik and Peters, 2009). Such distinctions and the implications for data collection, analysis, and interpretation have profound implications for calculating acquisition expenditure at the individual level.

Sales and Acquisition Channels

As the initial purchase turns a prospect into a newly acquired customer, the used sales channel is usually labeled 'acquisition channel' (e.g., online shop, branch, mail order). However, the immediate communication channel used prior to the initial buy is also often called 'acquisition channel' (e.g., search advertising, direct mail, referral, TV). Additionally, some studies describe the acquisition channel further by characteristics of the offer (e.g., a price promotion). Whatever a firm describes as acquisition channel for its purpose, it needs to distinguish properly between marketing instruments, communication, and sales channels in order not confuse marketing inputs and outputs within its customer relationship framework.

Customer Acquisition: A Process Perspective

Customer acquisition needs to be viewed as a process, often called lead management. In the B2C context, classic advertising models like AIDA depict a stylized process, from awareness via familiarity to initial purchase. In the B2B domain, sales funnel progressions are often employed to generate and qualify leads up to the initial purchase. For innovations, adoptions are the outcome of a combined process of external information by the firm and observational learning from previous adopters. This process perspective underlines that customer acquisition is governed by dynamics before the initial purchase, that is, a sequence of potentially multi-directional communication contacts (firm → customer, customer → firm, customer ↔ customer) across multiple stages.

Additionally, dynamics unfold after the initial purchase as the customer progresses along the life cycle, which is captured in the CLV. We will later present several studies that empirically show that these two processes are often interlinked rather than independent. If this is the case, then (1) what a firm does to acquire its customers partly defines the long-term value of its customer (base), and (2) an acquisition channel deemed cheap (expensive) when comparing cost per new customer may turn out to be rather expensive (cheap) when taking long-term implications into account. Hence, firms need to ensure that they learn for better acquisition policies not only from previous isolated campaigns, but also from general advertising policy, customer retention, and customer win-back.

Figure 10.1 connects all these aspects of customer acquisition.

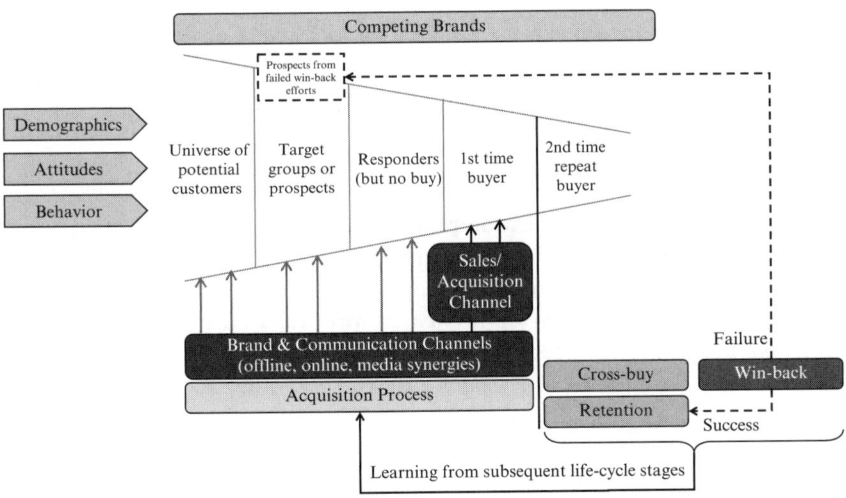

Figure 10.1 Integrated customer acquisition perspective

WHAT DO WE KNOW ABOUT CUSTOMER ACQUISITION?

In the previous section, we identify and briefly discuss the fundamentals of customer acquisition. We illustrate their complex nature and the resulting challenges in implementing proper acquisition processes in practice. Given the importance of customer acquisition for the long-term survival of firms, we continue to identify the relevant literature on customer acquisition and evaluate it along these dimensions. This serves three purposes: (1) we assess the current state-of-the-art on customer acquisition, (2) we extract the key learnings for practitioners, and (3) it enables us to identify important research gaps that need to be addressed. We first classify the studies in the literature based on these fundamentals. Next, we discuss the main findings arising from the literature

Classification of Literature

Our review discusses the following fundamentals, which we explicitly describe by the sub-criteria employed in the columns of Table 10.1:

- definition of customer acquisition process;
- target group definition;
- value orientation;
- communication, sales, and acquisition channels;
- study context.

Searching the major marketing research outlets we identify more than 70 studies that deal with customer acquisition in one way or another. From this pool of studies, we select a prominent and representative collection of 32 studies that demonstrate the current state of the art. Table 10.1 contains and characterizes all these studies.

Definition of customer acquisition process

Within this topic we need to account for the demarcation of customer acquisition within the customer life cycle as different specifications (e.g., the first buy with the firm or for a new product) may lead to different conclusions. We also need to assess whether and how feedback from subsequent CRM stages filters back into the customer acquisition process to ensure continuous improvement (e.g., learning from value patterns indicated by the acquisition process).

About half of all selected studies define customer acquisition in the context of the customer life cycle. But many times it is not explicitly defined

Table 10.1 Selected research studies on customer acquisition

Selected Studies	Focus of Study	Definition of Customer Acquisition Process		Target Group or Prospect Characterization					Value Orientation		
		Demarcation in Life Cycle	Learning from Feedback	New/Old Target Groups	Demo-graphics	Atti-tudes	Behavior	Hetero-geneity	Key Metric	Acquisition Cost	Value Dynamics
Dwyer (2012)	Acquisition and CLV	–	(Yes)	–	–	–	Yes	–	CLV	Partly	Yes
Gupta, Lehmann, and Stewart (2004)	Acquisition and CLV	–	–	Old	–	–	–	–	CLV	Yes	Yes
Coviello, Winklhofer, and Hamilton (2006)	Acquisition and CLV	Yes	–	–	Yes	–	–	–	Profitability	–	–
Pfeiffer and Ovchinnikov (2011)	Acquisition and CLV	Yes	Yes	–	–	–	Yes	Yes	CLV	Yes	Yes
Thomas (2001)	Acquisition vs retention	Yes	Yes	–	–	–	Yes	Yes	NPV	Yes	Yes
Reinartz, Thomas, and Kumar (2005)	Acquisition vs retention	Yes	Yes	Old	Yes	–	Yes	Yes	CLV	Yes	Yes

Table 10.1 (continued)

Selected Studies	Focus of Study	Definition of Customer Acquisition Process		Target Group or Prospect Characterization					Value Orientation		
		Demarcation in Life Cycle	Learning from Feedback	New/Old Target Groups	Demo-graphics	Atti-tudes	Behavior	Hetero-geneity	Key Metric	Acquisition Cost	Value Dynamics
Schweidel, Fader, and Bradlow (2008)	Acquisition vs retention	Yes	(Yes)	–	–	–	Yes	Yes	Acq.- & Ret.-Spells	–	Yes
Voss and Voss (2008)	Acquisition vs retention	–	–	–	–	–	Yes	–	Firm Performance	Yes	–
Musalem and Joshi (2009)	Acquisition vs retention	–	–	–	–	Yes	Yes	Yes	Value	Yes	Yes
Grewal, Chandra-shekaran, and Citrin (2010)	Acquisition vs retention	Yes	Yes	–	–	Yes	Yes	Yes	Profitability	Yes	Yes
Dong, Yao, and Cui (2011)	Acquisition vs retention	Yes	Yes	–	–	–	Yes	–	CLV	Yes	Yes
Stahl et al. (2012)	Acquisition vs retention	–	–	–	–	Yes		–	CLV	–	–
Martín-Herrera, McQuitty, and Sigué (2012)	Acquisition vs retention	Yes	Yes	–	–	–	–	–	Profitability	Yes	Yes

Hansotia and Wang (1997)	Prospect selection	–	(Yes)	–	–	–	Yes	Yes	CLV	Yes	Yes
Bodapati and Gupta (2004)	Prospect selection	–	–	Old	Yes	–	–	–	Sales	–	–
Cao and Gruca (2005)	Prospect selection	–	–	–	Yes	–	Yes	Yes	Profitability	Yes	Yes
Von Wangenheim and Bayōn (2007)	Acquisition channel: Referrals	–	(Yes)	–	–	Yes	–	–	Acquisition	–	–
Villanueva, Yoo, and Hanssens (2008)	Acquisition channel: Referrals vs. other marketing	Yes	Yes	–	–	–	Yes	–	Value Proxy	–	Yes
Kumar, Petersen, and Leone (2010)	Acquisition channel: Referrals	–	Yes	–	Yes	–	Yes	Yes	CLV CRV	Yes	Yes
Schmitt, Skiera, and van den Bulte (2011)	Acquisition channel: Referrals	Yes	Yes	–	Yes	–	Yes	(Yes)	CLV	Yes	Yes
Kumar, Petersen, and Leone (2013)	Acquisition channel: Referrals	Yes	Yes	–	Yes	–	Yes	Yes	CLV & BRV	Yes	Yes
Libai, Muller, and Peres (2013)	Acquisition channel: Referrals	Yes	Yes	–	–	–	Yes	–	Customer Equity	–	Yes

Table 10.1 (continued)

Selected Studies	Focus of Study	Definition of Customer Acquisition Process		Target Group or Prospect Characterization					Value Orientation		
		Demarcation in Life Cycle	Learning from Feedback	New/Old Target Groups	Demo-graphics	Atti-tudes	Behavior	Hetero-geneity	Key Metric	Acquisition Cost	Value Dynamics
Anderson and Simester (2004)	Acquisition channel: Price Promotions	–	–	–	–	–	Yes	–	Monetary Value	–	Yes
Lewis (2006a)	Acquisition channel: Price promotions	Yes	(Yes)	–	–	–	Yes	–	Tenure & Lifetime Revenue	–	Yes
Lewis (2006b)	Acquisition channel: Price Promotions (shipping fees)	–	–	–	–	–	Yes	–	Sales	Yes	–
Martínez-Jerez, Narayanan, and Brem (2003)	Acquisition channel: Internet	–	Yes	–	–	–	Yes	Yes	Profitability	Partly	Aggre-gate
Libai, Biyalogorski, and Gerstner (2003)	Acquisition Channel: Affiliate Marketing	–	–	–	–	–	–	–	Profitability	Yes	–

Chan, Wu, and Xie (2011)	Acquisition channel: Web Search	–	Yes	–	–	–	Yes	Yes	CLV	Yes	Yes
Kumar and Rajan (2012)	Acquisition channel: Social coupons	–	Yes	–	–	–	–	–	Net profit	–	Yes
Verhoef and Donkers (2005)	Acquisition channel: 6 online and offline media	Yes	Yes	–	Yes	–	Yes	–	Retention, Cross-buy	–	Partly
Choi, Bell, and Lodish (2012)	Acquisition channels: 4 online and offline media	Yes	Yes	Old	Yes	–	–	–	Number of new buyers	Yes	–
Breuer and Brettel (2012)	Acquisition channel: 4 online advertising	Yes	–	–	–	–	–	–	Sales	–	Yes

Table 10.1 (continued)

Selected Studies	Communication Channels			Acquisition/Sales Channels		Data		Insights	
	Offline	Online	Synergy	Online	Offline	B2B/B2C	Industry	Research	Managerial
Dwyer (1997)	–	–	–	–	–	B2C	NPO, Media	Types of CLV models	Importance of CLV and dynamics
Gupta, Lehmann, and Stewart (2004)	–	–	–	–	–	B2C	Retail, Banking	Customer value is a good proxy for firm value	Acquisition invest lowest, retention invest highest value
Coviello, Winklhofer, and Hamilton (2006)	–	–	–	–	–	B2C	Tourism	Both, transactional and relationship orientation prevail in acquiring customers in this industry	
Pfeiffer and Ovchinnikov (2011)	–	–	–	–	–	Analytical		CLV might not be the key metric when number of customers to serve is limited	
Thomas (2001)	–	–	–	–	–	B2B	Airline	Acquisition and retention are not independent processes; Tobit Model with selection prediction	Importance of accounting for dependence between acquisition and retention

Reinartz, Thomas, and Kumar (2005)	Yes	Yes	Yes	–	–	B2B	IT	Flat maximum of balance between acquisition and retention spending on customer value; asymmetric implications for budgeting	How to allocate spendings across communi-cation channels
Schweidel, Fader, and Bradlow (2008)	–	–	–	–	–	B2C	TV subscription	Acquisition and retention spells of customers are not independent	Acquisition timing indicates customer value
Voss and Voss (2008)	–	–	–	–	–	B2C	Theater	In competitive environments acquisition more profitable than retention efforts	In competitive environments profits are even more sensitive to right balance of spending allocation
Musalem and Joshi (2009)	–	–	–	–	–	n.a.	n.a.	Under competition, focusing on moderately responsive and valuable customers is preferable	
Grewal, Chandrashekaran, and Citrin (2010)	–	–	–	–	–	B2C B2B	Airline	Customer satisfaction levels and their heterogeneity influence	Return on satisfaction to shareholder value

Table 10.1 (continued)

Selected Studies	Communication Channels			Acquisition/Sales Channels		Data		Insights	
	Offline	Online	Synergy	Online	Offline	B2B/B2C	Industry	Research	Managerial
Grewal et al. (2010) cont'd –								shareholder value through their indirect impact on customer acquisition, retention, and service costs	decreases by almost 70% going from low to high satisfaction heterogeneity
Dong, Yao, and Cui (2011)	–	–	–	–	–	Analytical		Proper incentives when acquisition and retention effort are intertwined	Direct selling preferable when both processes are independent
Stahl et al. (2012)	–	–	–	–	–	B2C	Car	Brand Equity impacts average CLV directly and indirectly via marketing efforts	Components of brand equity exert differential effects on acquisition, retention, and profit
Martín-Herrera, McQuitty, and Sigué (2012)	–	–	Yes	–	–	Analytical		Synergy between offensive and defensive marketing strategies	Relative strength and market position determine

								gives firms full range of strategic options	strategy and allocation mix
Hansotia and Wang (1997)	–	–	–	–	–	Analytical		Use break-even analysis at individual prospect level based on acquisition packages	Simple rules for prospect selection under different scenarios
Bodapati and Gupta (2004)	Mailing	–	–	–	Yes	B2C	Retail	Binary selection approach better in frequent circumstances	Expand candidate models for prospect selection with binary
Cao and Gruca (2005)	Yes	–	–	Yes	–	B2C	Mail Order	Adverse selection of acquiring unprofitable customers can be reduced	Use additional information
Von Wangenheim and Bayōn (2007); Schmitt, Skiera and Van den Bulte (2011)	–	–	–	–	Yes	B2C B2B	Energy	Satisfaction–WOM link is non-linear and moderated by involvement dimensions	Better spending allocation when link is understood
Villanueva, Yoo, and Hanssens (2008)	–	–	–	Referral		B2C	Web Hosting	WOM customers almost double value compared to marketing-induced acquisitions	Resource allocation should accommodate WOM-marketing

Table 10.1 (continued)

Selected Studies	Communication Channels			Acquisition/Sales Channels		Data		Insights	
	Offline	Online	Synergy	Online	Offline	B2B/B2C	Industry	Research	Managerial
Kumar, Petersen, and Leone (2010)	–	–	–	Referral		B2C	Banking	Customer Referral Value adds to CLV	Targeting improvement with positive feedback on referrer
Schmitt, Skiera, and van den Bulte (2011)	–	–	–	Referral		B2C	Banking	Referred customers higher contribution, but difference is eroding over time; they have higher retention rates, are more valuable short and long term	Importance of referral programs for acquisition, but value differential varies by segment: use selective approach for referral programs
Kumar, Petersen, and Leone (2013)	Yes	Yes	–	Referral		B2B	TelCo, Financial	Business reference value (BRV) differently distributed from CLV; components	Reference influence increases with size and tenure of reference, as it does with

								and their measurement of BRV	richer media and higher congruency of reference
Libai, Muller, and Peres (2013)	–	–	–	Referral		Analytical		Expansion and acceleration dynamics in referral programs work differently depending on scenario	Design implications for referral programs
Anderson and Simester (2004)	Catalog	–	–	–	Yes	B2C	Mail Order	Use long-run elasticities to set prices	Price promotions are good for new, but bad for old customers
Lewis (2006a)	–	–	–	–	–	B2C	Newspaper, Online grocery	Customer uncertainty explains customer promotion response and subsequently lower CLV	Acquisition discount is negatively related to repeat buying and CLV
Lewis (2006b)	–	Yes	–	Yes	–	B2C	Online Retailer	Shipping fees as acquisition promotions are sensitive to order size, for retention base-line fees important	Shipping fees providing incentives for bigger order sizes may outperform free shipping promotions

Table 10.1 (continued)

Selected Studies	Communication Channels			Acquisition/Sales Channels		Data		Insights	
	Offline	Online	Synergy	Online	Offline	B2B/B2C	Industry	Research	Managerial
Martínez-Jerez, Narayanan, and Brem (2003)	–	–	–	Yes	Yes	B2C	Banking		Acquisition channel differentiates between customer behavior
Libai, Biyalogorski, and Gerstner (2003)	–	Yes	–	Yes	–	Analytical		Affiliate marketing: Pay-for-performance better for large number of affiliates, Pay-for-lead better in direct relationship	
Chan, Wu, and Xie (2011)	–	–	–	Yes	Yes	B2B	Chemicals	Google customers with higher CLV than offline acquisition customers	Evaluating long-term profit impact from search advertising
Kumar and Rajan (2012)	–	Yes	–	Yes	–	B2C	Restaurant	Social coupon success depends on context of firm	Need to check for appropriate environment before usage

Verhoef and Donkers (2005)	–	–	–	Yes	Yes	B2C	Insurance	Acquisition channels indicate different customer behavior & value	Analyzing acquisition channels matters
Choi, Bell, and Lodish (2012)	–	–	–	Yes	Yes	B2C	Online Retailer	WOM online & offline performance is driven by target customer density in ZIP, acquisition channels contribute differently to customer pool	Optimize location-based commun-ication and acquisition channel mix
Breuer and Brettel (2012)	–	Yes	–	Yes	–	B2C	Online Retailer	Online advertising channels work differently for new and existing customers in sales response	Here coupon/loyalty advertising better for existing, price comparison advertising for new customers

whether the study deals with customer acquisition for a new product or the firm, or how customer acquisition delineates itself from the other life cycle stages. This makes it sometimes difficult to put the generated insights into context or compare them to the results of other studies. However, many studies take some sort of feedback or long-term view along the customer life cycle into account. Employing the CLV metric, for instance, by definition ensures including the further evolution of a prospect (e.g., Dwyer, 1997; Gupta et al., 2004).

Target group or prospect characterization

Here we need to distinguish between new and old target groups, as existing customer profiles may be employed for penetrating well-known target groups, but new target groups may require entirely different approaches. Next, we need to assess how the target groups are characterized in the respective studies.

Only a few studies explicitly describe whether the firm is trying to penetrate current customer target segments (e.g., Gupta et al., 2004; Reinartz et al., 2005; Choi et al., 2012) or trying to lure new customer segments. Hence, for most studies we have to assume that the firm actually tries to penetrate existing target groups of customers.

With respect to profiling prospects or new customers, most studies included here refer to behavioral variables from the database. A few studies also include demographics (e.g., Bodapati and Gupta, 2004; Cao and Gruca, 2005). Attitudinal variables are rarely included. Here only four studies take them into account (e.g., Von Wangenheim and Bayón, 2007; Grewal et al., 2010). There is no study that incorporates all three types of characteristics, thus taking a holistic approach to target group or new customer description.

With respect to accounting for heterogeneity, about half of the reviewed studies account for some form of heterogeneity (e.g., Thomas, 2001; Kumar et al., 2010, 2013). This is surprising given that several of these studies are explicitly set in the context of customer relationship management.

Value orientation

Within this topic of customer acquisition, we need to know what key metric is ensuring the value orientation of the respective customer acquisition process. Specifically, we also assess whether acquisition costs informing the key metric are included, and whether the key metric accounts for value dynamics.

CLV is by now the most common measure of value to the firm across all studies reviewed. When other focal measures like net present value (e.g., Thomas, 2001), profitability (e.g., Coviello et al., 2006; Grewal et al., 2010), net profit (Kumar and Rajan, 2012), tenure (e.g., Schweidel et al., 2008), or referral value (e.g., Kumar et al., 2010, 2013) are used, then these measures are either similar to CLV or represent an alternative focus of the analysis (tenure, referral value). Accordingly, some form of acquisition cost and value dynamics are incorporated in most studies. However, acquisition costs are usually only included for the defined acquisition channel of the new customer before signing up (e.g., Pfeiffer, Haskins, and Conroy, 2005), not the total sequence of previous touches.

Another relevant issue when allocating acquisition channel costs to individual customers is that in many cases acquisition channels are also used as retention or win-back channels (i.e., direct mails; Rust and Verhoef, 2005; Feld et al., 2013). When simultaneous use occurs, firms should be aware that they should not allocate all channel or campaign costs to acquisition but assign them proportionally. Hence, the customer life cycle stages inform how channel costs are to be allocated across individual customers.

Communication, sales, and acquisition channels

Different channels have received considerable attention within existing research. Specific questions we deem relevant are: (1) what type of channel definition is studied; (2) does the study investigate 'classic' online communication channels (e.g., banner or search) or social media; and (3) does the study account for media synergies across communication channels?

Only a few studies distinguish between communication and sales channels. Often the final communication touch (e.g., catalog, referral), marketing approach (e.g., affiliate marketing, price promotion) or first sales channel (e.g., Internet) used by the new customer is labeled as an acquisition channel. It appears that there is no common understanding in the literature on what actually constitutes an acquisition channel. In fact, many studies do not account for the communication history of new customers, although this is in contrast with recent developments in practice with respect to online media analysis, for example, customer journey evaluations. Additionally, with the exception of Kumar and Rajan (2012) research on the role of social media in customer acquisition is scarce. We assume that this is mainly due to the fact that many touches are unobservable to the firm. When studies incorporate communication history, with the exception of Reinartz et al. (2005), media synergies (see Naik and Raman, 2003; Naik, 2007; Naik and Peters, 2009) are rarely accounted for.

Context of studies

Within the context of the study, the type of market that the acquisition channel study covers is highly relevant. From the empirical studies most studies employ data from B2C settings. Only five studies include B2B datasets (e.g., Von Wangenheim and Bayón, 2007; Chan et al., 2011; Kumar et al., 2013). Given the importance of B2B marketing and its affinity for CRM, this finding seems surprising.

MAIN FINDINGS OF THE STUDIES

Acquisition and CLV

Dwyer (1997) shows that different CLV models might be appropriate, depending on whether 'always-a-share' or 'lost-for-good' relationships are eminent in the context studied (e.g., Berger and Nasr, 1998). His analysis also underlines the importance of accounting for the dynamics of customer evolution. This distinction and the importance of dynamics should also hold when using this metric in the context of customer acquisition. We include the studies by Coviello et al. (2006) and Pfeiffer and Ovchinnikov (2011) because they discuss possible exceptions to this recommendation. In their analytical approach, Pfeiffer and Ovchinnikov (2011) suggest that CLV may not be the key metric when the number of potential customers for the firm is limited because of capacity constraints. In their study of small firm behavior in the tourism industry, Coviello et al. (2006) show that a mix of transactional and interaction-based relational customer acquisition approaches is more successful than either alone. However, as transaction marketing in this study also involves satisfying (new) customers, this approach might be not so distinct from CRM after all. Gupta et al. (2004) show that the sum of all CLVs across customers, customer equity, is a reasonable proxy for firm value. They also show that acquisition efforts may usually have a lower leverage on customer equity than retention efforts. Although this may imply that customer retention creates more value to the firm, natural attrition in the long run requires continuous efforts on customer acquisition.

Acquisition vs Retention Efforts

Marketing is ultimately the art of attracting *and* keeping profitable customers (Kotler and Armstrong, 1996). The results of all studies with a focus on the balance between acquisition and retention efforts indicate that these processes are intertwined (e.g., Thomas, 2001; Reinartz et al., 2005; Schweidel et al., 2008). Not accounting for the interdependence of these two processes leads to biased results in customer retention and acquisition analysis, usually favoring retention compared to acquisition efforts

(Thomas, 2001). Hence, acquisition is linked to retention and the optimal trade-off between these marketing investments along the customer life cycle cannot be easily determined. Although Reinartz et al. (2005) suggest that a flat maximum exists around the optimal split of efforts, sub-optimal retention spending hurts profits more than sub-optimal acquisition investment. This finding supports earlier findings of Gupta et al. (2004). The balance seems to be favorable to acquisition in more competitive environments (Voss and Voss, 2008), where firms should then better focus on moderately responsive and moderately profitable customers (Musalem and Joshi, 2009). Martín-Herrán et al. (2012) suggest that in competitive scenarios the balance between both investments may also be moderated by the relative position of the firm in the market, their comparative advantages in acquisition versus retention marketing, and the relative cost effectiveness of both instruments.

Dong et al. (2011) analyze the importance of proper incentive mechanisms when both processes are interdependent, and the firm has the option to acquire the new customers itself versus delegating customer acquisition. In their scenario, direct selling is only preferable when both processes are independent. Reinartz et al. (2005) derive a more detailed allocation approach for the communication channels of a firm, also accounting for interaction effects across the channels. For their B2B firm they can show that usually interpersonal and interactive communication channels are more effective when the firm initiates the contact.

Linking brand equity to CLV, Stahl et al. (2012) find that brand equity generated through communication efforts impacts average CLV directly as well as indirectly. However, it is important to manage the components of brand equity right as they exert differential effects on acquisition, retention, and, ultimately, firm profits. These findings again underline the importance of including communication efforts when evaluating acquisition and retention efforts (e.g., Blattberg and Deighton, 1996). Grewal et al. (2010) add another aspect of customer relationship management to the picture: customer satisfaction levels, which often represent a core metric in many firms, and their heterogeneity influence shareholder value through their indirect impact on customer acquisition, retention, and service costs. They find, however, that return on satisfaction to shareholder value decreases by almost 70 percent when going from low to high satisfaction heterogeneity. Today, most firms do not pay a lot of attention to the variance of customer satisfaction levels, but these findings suggest they should. Finally, recent research suggests that dynamics in relationship constructs may actually be more important than their states (e.g., Palmatier et al., 2013), but this has not yet been investigated in the context of customer acquisition and retention.

Prospect Selection

When targeting prospects many firms experience a dilemma, as quotes cited in Cao and Gruca (2005, abstract) underline: 'In many businesses, the customers most likely to sign on are precisely the worst customers you could possible find (Reichheld, 1996, p. 76)'. This adverse selection problem forces firms to qualify the recipients of their offer better in order to avoid unprofitable customers. Cao and Gruca (2005) show how internal data on prospects augmented by information from third party vendors may allow firms to reduce this adverse selection problem via better targeting procedures. Bodapati and Gupta (2004) suggest using a direct binary selection approach where the discretized response variable indicates the surpassing of a certain threshold. Comparing it to previous indirect approaches, which first estimate the metric response variable and then apply a binary selection, they can show theoretically that their new approach is superior when the response model is mis-specified and the sample size is large. Both conditions are frequently given in direct marketing. Hansotia and Wang (1997) take an analytical approach to prospect selection. In their most advanced scenario, each customer has different response probabilities depending on the acquisition package and a differential CLV. They then show how employing a combination of models for response likelihood and CLV allows firms to calculate the break-even point for each prospect. This allows the firm to improve its targeting of prospects as only prospects above a certain contribution level may be contacted with a certain offer package.

Acquisition Channels

We acknowledge above that the term 'acquisition channel' is not clearly defined in the literature. Accordingly, marketing instruments as well as communication or sales channels are labeled as such. Disregarding this definition problem of acquisition channels, we sort the selected contributions by 'type of acquisition channel'. This enables us to extract the collective insights per channel, namely 'referrals', 'price promotions', 'online (Internet, affiliate, web search, social coupons)', and studies comparing 'several acquisition channels'.

Referrals

Recently, referral programs became ubiquitous in many industries, offering incentives for existing customers to bring in new customers through referrals. Villanueva et al. (2008) find in their application that customers acquired through these 'word-of-mouth' activities are about two times as valuable compared to customers acquired through common marketing activities. Their findings also support the notion of dependence between acquisition and retention: they suggest that if quick customer growth is wanted, higher marketing budgets are needed – but also higher retention expenditures later on. Firms that grow their customer base organically may experience slower but more profitable growth. Accordingly, adapting the marketing resource allocation towards referral marketing yields higher returns for the firm in the long run. Schmitt et al. (2011) find that referred customers indeed have at least a 16 percent higher contribution than customers acquired through different means, but that this differential is eroding over time even though the retention rates for referred customers are consistently higher. Assigning the referred customers to predefined value segments their analysis further suggests that not all referred customers are equal, and that firms should offer more program alternatives than just a one-fits-all referral program.

Kumar et al. (2010, 2013) distinguish between CLV, the customer's own value contribution to the firm, and customer referral value (CRV for B2C) and business reference value (BRV for B2B) as additional components. Separating these contributions at the individual customer level enables them to show that these value components have distinct patterns. Customers with medium CLV have the highest CRV or BRV scores compared to high and low CLV customers. Relating CRV and BRV to a selection of drivers they demonstrate how to leverage these insights to generate higher profits for the respective firms. Von Wangenheim and Bayón (2007) analyze how the success of referrals is non-linearly linked to customer satisfaction and how this relationship is moderated by several dimensions of involvement.

Libai et al. (2013) decompose the value contributions of referral programs into market expansion, that is, customers who would not have bought, and acceleration, that is, customers who would have bought only later without referral. Under general conditions, the contribution ratio between expansion and acceleration is around 70:30. But both of these drivers are affected by competition, program targeting, retention, and profit decline. Knowing and measuring these drivers allows firms to optimize their referral program structure for higher profits.

Finally, Kumar and Rajan (2012) elaborate, simulate, and test social coupons as a hybrid instrument between classic instruments and social networks. They find that their application yields benefits only under certain circumstances. However, their research points to a larger issue: how can we account for third party or user-generated content as well as other users interacting with prospects? Currently, firms are hardly able to systematically capture this increasing source of indirect brand exposure, let alone analyze and link it to customer acquisition efforts. Extrapolating this trend, it may become ever more difficult to disentangle the isolated effect of a separate touch or referral. The literature is silent on these issues.

Price promotions

Price promotions are a tool frequently employed for customer acquisition. However, managers may fear that this increased acquisition success may come back to haunt them in the long term in several ways. For example, in a large Dutch energy firm they experience that providing attractive deals to prospects not only fosters churn, but also creates dissatisfaction among existing customers who did not receive the attractive offer (Snoeck, 2012). Lewis (2006a, 2006b) investigates the impact of price discounts and shipping fees on the long-term composition of a firm's customer portfolio. In a conceptual model he shows how customer uncertainty may explain why promotionally acquired customers have lower repurchase rates and smaller CLVs. In empirical field tests he shows on two occasions that discount depth is negatively related to repeat purchase rates and CLV, for example, acquiring a customer with a 35 percent discount leads to a 50 percent lower CLV. In his second contribution, Lewis (2006b) investigates the effect of shipping fees on order incidence, order volume, and customer acquisition. He shows that higher shipping fees reduce online store traffic and that order size incentives result in larger order sizes. Customer acquisition seems especially sensitive to higher shipping fees regarding order incidence as well as order size, whereas existing customers react more sensitively to base shipping fee levels. This has important implications for balancing customer acquisition versus retention efforts. It again underlines the connection between acquisition efforts and long-term profitability of the customer portfolio. Anderson and Simester (2004) also analyze the connection between price promotions and long-run effects on new and existing customers. They generate evidence that several long-term effects are at play: forward buying, selection, customer learning, and increased deal sensitivity. As short-run metrics ignore changes in deal sensitivity, long-run demand estimations for existing customers are biased upwards, leading to prices that are too low for this customer group. Among new customers, however, short-run metrics underestimate the total demand increase and prices will be too high for them. This finding is different from Lewis (2006a) in that deeper promotions increase repeat purchase rates for new customers. They suggest that as prospective customers have limited information, lower initial prices favorably shape future expectations. The difference between the studies may be explained by the difference in channels investigated: Lewis (2006a) studies an online channel, compared to classic mail order in the Anderson and Simester (2004) study.

Internet

In an early contribution, Martínez-Jerez et al. (2003) show in a case study on a Spanish bank that customers acquired over the Internet show higher activity rates and are more profitable to the bank compared to customers of the branches. Chan et al. (2011) investigate the performance of web search as a separate acquisition channel. They can show that, after accounting for offline spillover purchases from search advertising, customers acquired through this channel are more valuable to the firm, short and long term. With respect to online affiliate marketing, Libai et al. (2003) show analytically that pay-per-performance schemes should be preferable when a large number of affiliates are used, whereas pay-per-lead is advantageous when direct relationships with affiliates exist. Kumar and Rajan (2012) investigate social coupons and find that social coupons, when compared to traditional ones, suffer from various deficiencies. These result in social coupons representing a customer acquisition instrument that is difficult to balance in terms of short- and long-term profitability to the firm.

Comparing acquisition channels

For an insurance firm, Verhoef and Donkers (2005) compare the long-term performance of eight offline and online acquisition channels across several services on customer retention and cross-buying. Their analysis reveals a complex picture across acquisition channels and services. Overall, direct mailings and mass media like TV as final contacts before signing up as a customer indicate poorer performance on retention. In contrast, websites perform better, supporting the findings of Martínez-Jerez et al. (2003) in a similar industry. However, subsequent cross-buy rates are lower. Combining analysis on retention and cross-buying, outbound calls and print magazines appear more attractive than judging each effect alone. The opposite effect results for TV, radio, and print advertising. Overall, the differences between acquisition channels dilute with time. Choi et al. (2012) compare offline (referral, magazine) and online (referral, search) acquisition channels for an online retailer. When optimizing the acquisition media mix at the ZIP level, they find that the respective media mix per ZIP is mainly driven by potential customer density. Offline and online word-of-mouth works better in areas with high density, compared to magazine advertising, which works better in lower-density areas. Breuer and Brettel (2012) compare the performance of several online acquisition channels, namely search engine, banner, price comparison, and coupon/loyalty advertising for an online book retailer. Again, search has the longest duration effect, followed by banners, coupons, and price comparison advertising. Coupons work better for existing customers and price comparison advertising attracts new customers better. However, they evaluate performance on sales rather than profit.

Next, we extract managerial insights from this review and prioritize identified research gaps.

MANAGERIAL IMPLICATIONS AND FUTURE DIRECTIONS FOR RESEARCH

Managerial Implications and Challenges

Taking all these aspects of customer acquisition into account, implementing a sound customer acquisition process at the individual customer level is far from easy. From the review, we extract several valuable recommendations for practitioners:

- A firm needs to delineate the customer life cycle stages for its operations properly, that is, what is a new customer and when do failed win-back efforts move a lapsed customer back to the acquisition pool? Additionally, some firms may prefer to look at 'new' customers by divisions or products. Although this does not fit a customer-centric approach, in practice firms often keep old habits even though this may compromise efficient customer relationship management.
- Describe your target groups and prospects on demographics, attitudes, and behavior to enable most effective acquisition efforts.
- Measure brand equity, its components, and related intangible metrics. When linking them to performance measures like CLV and customer equity, make sure to include heterogeneity and metric dynamics.
- Use CLV or PLV as metrics guiding your acquisition efforts. Be careful to check whether CLV is the appropriate metric in the specific environment and which type of CLV model to employ, always-a-share or lost-for good.
- The acquisition rate or probability is not independent of acquisition efforts. The connection between effort and likelihood may vary by prospect and over time. A firm should investigate and monitor those links closely to inform its resource allocation for customer acquisition.
- When selecting prospects for acquisition campaigns, use augmented data and include long-term performance indications to avoid adverse customer selection phenomena.
- A firm should ensure measurement of acquisition efforts at a

reasonable and appropriate level: acquisition efforts are often difficult to attribute to individual prospects, especially for classic mass media, unaddressed bulk mailings, pseudo-anonymous targeted online banners, search, and the like. Valuable media synergies may further complicate the picture.

- Distinguish communication and sales channels adequately to enable their adequate measurement.
- Employ differentiated customer referral programs and measure CRV or BRV as an additive component to CLV.
- Use price promotions with care and differentiate between new and existing customers.
- Evaluate acquisition channels in the light of previous efforts and subsequent performance of customers in later stages of their life cycle. Labeling the final communication or sales channel as acquisition channel may systematically overrate the impact of that particular channel when evaluating a particular channel in isolation from other acquisition efforts, leading to biased perceptions and subsequently misguided resource allocation within the firm.
- Use dynamic approaches to evaluate acquisition performance along the life cycle and within the acquisition process over time to avoid myopic decision-making.
- Install feedback processes and continuously learn from these dynamics to improve acquisition efforts. Adopting a customer-centric approach often leads to reorganization in the marketing and sales department along the customer life cycle stages (e.g., departments for acquisition, retention, and win-back). This in turn often inhibits proper learning from unfolding value dynamics after acquiring the customer, because the firm does not systematically connect the long-term outcomes to acquisition efforts.
- Rely on analytics and specifically account for dynamics (e.g., Dickson, Farris, and Verbeke, 2001). Overall, we posit that customer and market analytics are of ever more strategic importance (Reinartz et al., 2004; Ramani and Kumar, 2008; Verhoef and Lemon, 2013). These analytics may drive both strategic marketing and brand decisions, but also may directly guide tactics, such as media channel use. Analytics can provide firms with a unique advantage over competitors as already shown by firms such as Amazon, Walmart, Zappos, and Tesco, and will become even more important in an era of big data (Davenport, 2006; Davenport and Patil, 2012).

Adding these recommendations to our integrated picture for the literature review, we arrive at the customer acquisition framework shown in

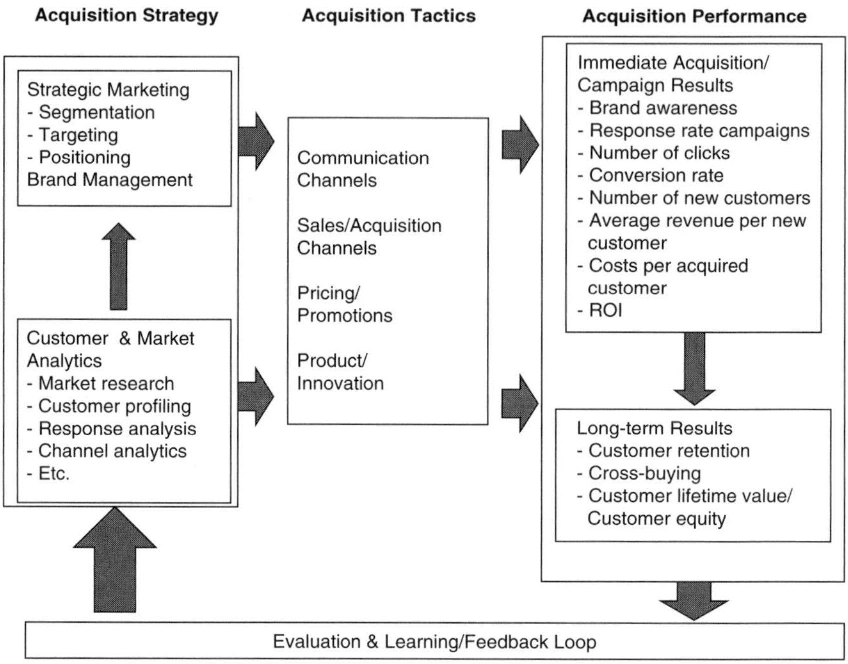

Figure 10.2 Strategic framework for customer acquisition

Figure 10.2. This framework may guide firms in managing their acquisition efforts better. In practice, first services are successfully offered that track a (new) customer's journey from contact initiation to actual sales conversion across several online media as shown in Figure 10.3. This tracking of (anonymous) customers allows better allocation of resources across media at the individual (profile-based) as well as aggregate level, e.g., for planning purposes. Subsequently, the transparency of such profiles initiates the discussion about attribution models, i.e., how to best reward media performance across a sequence of touches (see lower part of Figure 10.3). Whereas currently 'last-click-wins' schemes dominate in industry, the new insights from customer journey tracking now encourage the consideration of several alternatives that accommodate the branding funnel better. Currently, the 'bath-tub' model seems to emerge as the front runner, where the initial and last medium used by customers actually get a slightly larger share, while all intermediate touches get a lower share of advertising budget. Based on the tracking data, the actual distribution of the detailed scheme across media may in future depend on the brand, the CRM process (i.e., acquisition, retention, win-back), and customer value (segment) considered.

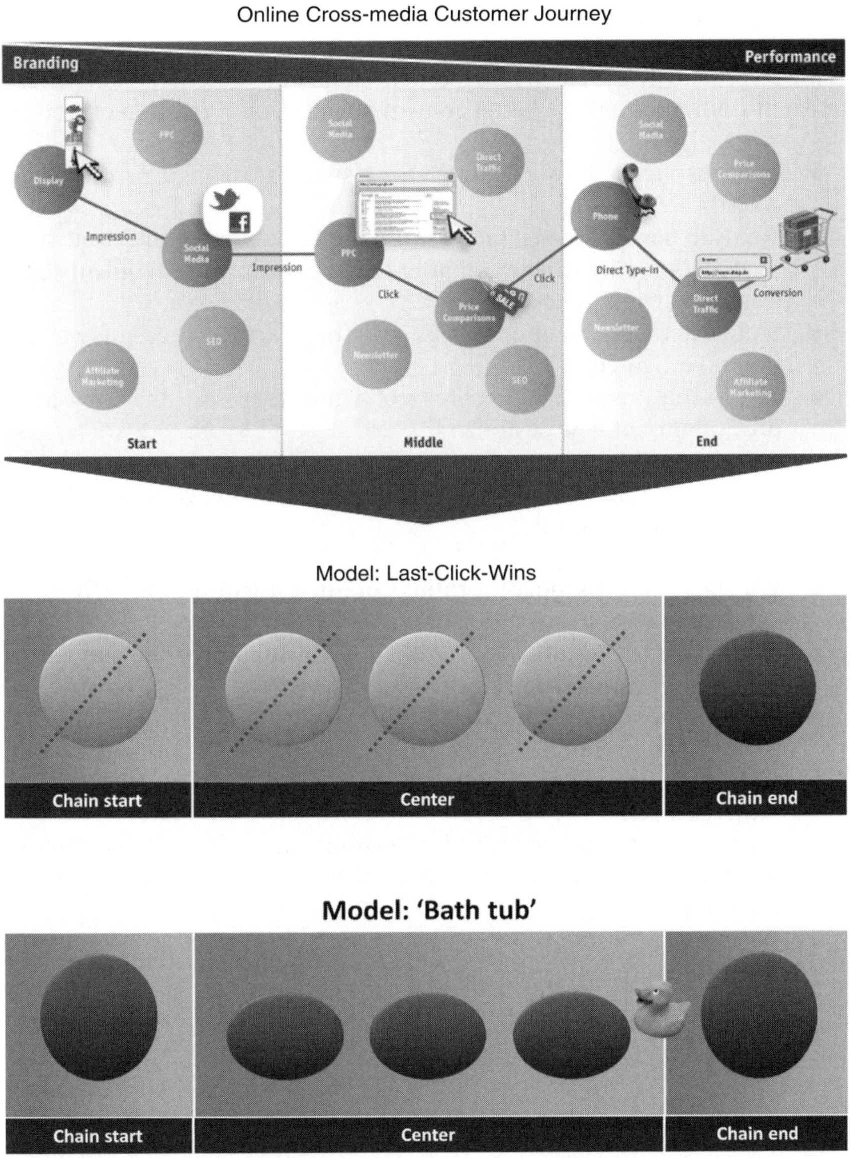

Source: Adapted from IntelliAd (2013).

Figure 10.3 Attribution models for customer acquisition journeys

Directions for Research

With respect to research, we suggest the following guidelines for future research studies to enable a better comparison and interpretation of results:

- State explicitly how customer acquisition is defined in the study context.
- Analyze new customer (acquisition) performance in the context of related dynamics before and after their acquisition to avoid any bias stemming from 'myopia'.
- State whether current or new target groups are the focus of acquisition investigated.
- If possible, use CLV or prospect value to assess the long-term profitability of new customers.

From the literature review, we also infer the following gaps that should be addressed in future studies:

- We lack studies incorporating or investigating the effect of communication history on new customer acquisition.
- Also, media synergies may require more attention in the context of new customer acquisition.
- We encourage more analytical studies on customer acquisition. Our insights on potential leverage of customer acquisition along the customer life cycle involving pre- and post-acquisition dynamics are rather scarce.
- We need more studies analyzing customer acquisition in the context of B2B environments, especially when buying centers may be involved.
- We need to develop a common understanding of what actually constitutes an acquisition channel.
- We encourage more studies involving attitudinal and personality-related information on prospects and new customers, linking them dynamically to outcomes along the customer life cycle such as CLV, CRV, and the like.

In this chapter, we aim to provide a deeper and enriched understanding of customer acquisition strategies. We specifically pledge for the understanding of and accounting for prospect heterogeneity and cross-media as well as dynamic effects when developing acquisition strategies. In sum, we believe that customer acquisition remains a highly relevant research topic in the area of customer equity management. At the same time, we hope that this chapter will be useful for marketing practice when developing acquisition strategies and for academic researchers to direct their future research.

BIBLIOGRAPHY

Anderson, E.T. and D.I. Simester (2004), 'Long-run effects of promotion depth on new versus established customers: three field studies', *Marketing Science*, **23**(1), 4–20.

Ansari, A. and C.F. Mela (2003), 'E-customization', *Journal of Marketing Research*, **40**(1), 131–45.

Berger, P.D. and N.I. Nasr (1998), 'Customer lifetime value: marketing models and applications', *Journal of Interactive Marketing*, **12**(1), 17–30.

Blattberg, R.C. and J. Deighton (1996), 'Manage marketing by the customer equity test', *Harvard Business Review*, **74**(4), 136–44.

Bodapati, A. and S. Gupta (2004), 'A direct approach to predicting discretized response in target marketing', *Journal of Marketing Research*, **41**(3), 73–85.

Bolton, R.N., K.N. Lemon and P.C. Verhoef (2004), 'The theoretical underpinnings of customer asset management: a framework and propositions for future research', *Journal of the Academy of Marketing Science*, **32**(3), 271–92.

Breuer, R. and M. Brettel (2012), 'Short- and long-term effects of online advertising: differences between new and existing customers', *Journal of Interactive Marketing*, **26**(3), 155–66.

Bruce, N.I., K. Peters and P.A. Naik (2012), 'Discovering how advertising grows sales and builds brands', *Journal of Marketing Research*, **49**(4), 793–806.

Bügel, M.S., P.C. Verhoef and A.P. Buunk (2011), 'Customer intimacy to relationships with five firms in five different sectors: preliminary evidence', *Journal of Retailing and Consumer Services*, **18**(4), 247–58.

Cao, Y. and T.S. Gruca (2005), 'Reducing adverse selection through customer relationship management', *Journal of Marketing*, **69**(4), 219–29.

Chan, T.Y., C. Wu and Y. Xie (2011), 'Measuring the lifetime value of customers acquired from Google Search advertising', *Marketing Science*, **30**(5), 837–50.

Choi, J., D.R. Bell and L.M. Lodish (2012), 'Traditional and IS-enabled customer acquisition on the internet', *Management Science*, **58**(4), 754–69.

Coviello, N., H. Winklhofer and K. Hamilton (2006), 'Marketing practices and performance of small service firms', *Journal of Service Research*, **9**(1), 38–58.

Davenport, T.H. (2006), 'Competing on analytics', *Harvard Business Review*, **84**(1), 99–107.

Davenport, T.H. and D.J. Patil (2012), 'Data scientist: the sexiest job of the 21st century', *Harvard Business Review*, **90**(10), 70–76.

Dhar, R. and R. Glazer (2003), 'Hedging customers', *Harvard Business Review*, **81**(5), 86–92.

Dickson, P.R., P.W. Farris and W.J.M.I. Verbeke (2001), 'Dynamic strategic thinking', *Journal of the Academy of Marketing Science*, **29**(3), 216–37.

Dong, Y., Y. Yao and T.H. Cui (2011), 'When acquisition spoils retention: direct selling vs. delegation under CRM', *Management Science*, **57**(7), 1288–99.

Dorotic, M., T.H.A. Bijmolt and P.C. Verhoef (2012), 'Loyalty programmes: current knowledge and research directions', *International Journal of Management Reviews*, **14**(3), 217–37.

Dorotic, M., D. Fok, P.C. Verhoef and T.H.A. Bijmolt (2011), 'Do vendors benefit from promotions in a multi-vendor loyalty program?', *Marketing Letters*, **22**(4), 341–56.

Dowling, G.R. and M. Uncles (1997), 'Do customer loyalty programs really work?', *Sloan Management Review*, **38**(4), 71–82.

Dwyer, F.R. (1997), 'Customer lifetime evaluation to support marketing decision making', *Journal of Direct Marketing*, **11**(4), 6–13.

Feld, S., H. Frenzen, M. Krafft, K. Peters and P.C. Verhoef (2013), 'The effects of mailing design characteristics on direct mail campaign performance', *International Journal of Research in Marketing*, **30**(2), 143–59.

Grewal, R., M. Chandrashekaran and A.V. Citrin (2010), 'Customer satisfaction heterogeneity and shareholder value', *Journal of Marketing Research*, **47**(4), 612–26.

Gupta, S. and D.R. Lehmann (2003), 'Customers as assets', *Journal of Interactive Marketing*, **17**(1), 9–24.

Gupta, S. and V. Zeithaml (2006), 'Customer metrics and their impact on financial performance', *Marketing Science*, **25**(6), 718–39.

Gupta, S., D.R. Lehmann and J.A. Stewart (2004), 'Valuing customers', *Journal of Marketing Research*, **41**(1), 7–18.

Hansotia, B.J. and P. Wang (1997), 'Analytical challenges in customer acquisition', *Journal of Direct Marketing*, **11**(2), 7–19.

Hanssens, D.M., D. Thorpe and C. Finkbeiner (2008), 'Marketing when customer equity matters', *Harvard Business Review*, **86**(5), 117–23.

Hoffman, D.L. and T.P. Novak (2000), 'How to acquire customers on the web', *Harvard Business Review*, **78**(3), 179–88.

IntelliAd (2013), 'An insight into multichannel-optimization along the customer-journey', Presentation at the NAMO Distance Selling Conference 2013, Moscow, 29 May.

Keller, K.L. (1993), 'Conceptualising, measuring, and managing customer-based brand equity', *Journal of Marketing*, **57**(1), 1–22.

Kotler, P. and G. Armstrong (1996), *Principles of Marketing*, 7th edition, Englewood Cliffs, NJ: Prentice-Hall.

Kumar, V. and B. Rajan (2012), 'Social coupons as marketing strategy: a multifaceted perspective', *Journal of the Academy of Marketing Science*, **40**(1), 120–36.

Kumar, V. and D. Shah (2009), 'Expanding the role of marketing: from customer equity to market capitalization', *Journal of Marketing*, **73**(6), 119–36.

Kumar, V., M. George and J. Pancras (2008), 'Cross-buying in retailing: drivers and consequences', *Journal of Retailing*, **84**(1), 15–27.

Kumar, V., J.A. Petersen and R.P. Leone (2010), 'Driving profitability by encouraging customer referrals: who, when, and how?', *Journal of Marketing*, **74**(5), 1–17.

Kumar, V., J.A. Petersen and R.P. Leone (2013), 'Defining, measuring, and managing business reference value', *Journal of Marketing*, **77**(1), 68–86.

Kumar, V., R. Venkatesan and W.J. Reinartz (2006), 'Knowing what to sell, when, and to whom', *Harvard Business Review*, **84**(3), 131–45.

Kumar, V., I. Dalla Pozza, J.A. Petersen and D. Shah (2009), 'Reversing the logic: the path to profitability through relationship marketing', *Journal of Interactive Marketing*, **23**(2), 147–56.

Kumar, V., R. Venkatesan, T. Bohling and D. Beckmann (2008), 'The power of CLV: managing customer lifetime value at IBM', *Marketing Science*, **27**(4), 585–99.

Lavidge, R.J. and G.A. Steiner (1961), 'A model for predictive measurements of advertising effectiveness', *Journal of Marketing*, **25**(3), 59–62.

Leeflang, P.S.H., T.H.A. Bijmolt, J. van Doorn, D.M. Hanssens, H.J. van Heerde, P.C. Verhoef and J.E. Wieringa (2009), 'Creating lift versus building the base: current trends in marketing dynamics', *International Journal of Research in Marketing*, **26**(1), 13–20.

Leenheer, J., H.J. van Heerde, T.H.A. Bijmolt and A. Smidts (2007), 'Do loyalty programs really enhance behavioral loyalty? An empirical analysis accounting for self-selecting members', *International Journal of Research in Marketing*, **24**(1), 31–47.

Leone, R.P., V.R. Rao, K.L. Keller, A.M. Luo, L. McAlister and R. Srivastava (2006), 'Linking brand equity to customer equity', *Journal of Service Research*, **9**(2), 125–38.

Lewis, M. (2006a), 'Customer acquisition promotions and customer asset value', *Journal of Marketing Research*, **43**(2), 195–203.

Lewis, M. (2006b), 'The effect of shipping fees on customer acquisition, customer retention, and purchase quantities', *Journal of Retailing*, **82**(1), 13–23.

Libai, B., E. Biyalogorski and E. Gerstner (2003), 'Setting referral fees in affiliate marketing', *Journal of Service Research*, **5**(4), 303–15.
Libai, B., E. Muller and R. Peres (2013), 'Decomposing the value of word-of-mouth seeding programs: acceleration vs. expansion', *Journal of Marketing Research*, **50**(2), 161–76.
Martínez-Jerez, A.A., V.G. Narayanan and L. Brem (2003), 'Internet customer acquisition strategy at Bankinter', *Journal of Interactive Marketing*, **17**(3), 62–74.
Martín-Herrán, G., S. McQuitty and S.P. Sigué (2012), 'Offensive versus defensive marketing: what is the optimal spending allocation?', *International Journal of Research in Marketing*, **29**(2), 210–19.
Moe, W.W. and P.S. Fader (2004), 'Dynamic conversion behavior at e-commerce sites', *Marketing Science*, **50**(3), 326–35.
Murray, K.B. (1991), 'A test of services marketing theory: consumer information acquisition activities', *Journal of Marketing*, **55**(1), 10–25.
Musalem, A. and Y.V. Joshi (2009), 'How much should you invest in each customer relationship? A competitive strategic approach', *Marketing Science*, **28**(3), 555–65.
Naik, P. (2007), 'Integrated marketing communications: provenance, practice, and principles', in G.J. Tellis and T. Ambler (eds), *Handbook of Advertising*, London: Sage Publications, pp. 35–53.
Naik, P.A. and K. Peters (2009), 'A hierarchical marketing communications model of online and offline media synergies', *Journal of Interactive Marketing*, **23**(4), 288–99.
Naik, P.A. and K. Raman (2003), 'Understanding the impact of synergy in multimedia communications', *Journal of Marketing Research*, **40**(4), 375–88.
Palmatier, R.W., M.B. Houston, T.P. Dant and D. Grewal (2013), 'Relationship velocity: toward a theory of relationship dynamics', *Journal of Marketing*, **77**(1), 13–30.
Pauwels, K., D.M. Hanssens and S. Siddarth (2002), 'The long-term effects of price promotions on category incidence, brand choice, and purchase quantity', *Journal of Marketing Research*, **39**(4), 421–39.
Pfeiffer, P.E., M.E. Haskins and R.M. Conroy (2005), 'Customer lifetime value, customer profitability, and the treatment of acquisition spending', *Journal of Managerial Issues*, **17**(1), 11–25.
Pfeifer, P.E. and A. Ovchinnikov (2011), 'A note on willingness to spend and customer lifetime value for firms with limited capacity', *Journal of Interactive Marketing*, **25**(3), 178–89.
Ramani, G. and V. Kumar (2008), 'Interaction orientation and firm performance', *Journal of Marketing*, **72**(1), 27–45.
Reichheld, F.F. (1996), *The Loyalty Effect: The Hidden Force Behind Growth, Profits, and Lasting Value*, Boston, MA: Harvard Business School Press.
Reinartz, W., M. Krafft and W.D. Hoyer (2004), 'The customer relationship management process: its measurement and impact on performance', *Journal of Marketing Research*, **41**(3), 293–305.
Reinartz, W., J.S. Thomas and V. Kumar (2005), 'Balancing acquisition and retention resources to maximize customer profitability', *Journal of Marketing*, **69**(1), 63–79.
Rust, R.T. and P. Verhoef (2005), 'Optimizing the marketing interventions mix in intermediate-term CRM', *Marketing Science*, **24**(3), 477–89.
Rust, R.T., K.N. Lemon and V.A. Zeithaml (2004), 'Return on marketing: using customer equity to focus marketing strategy', *Journal of Marketing*, **68**(1), 109–27.
Schmitt, P., B. Skiera and C. van den Bulte (2011), 'Referral programs and customer value', *Journal of Marketing*, **75**(1), 46–59.
Schweidel, D.A., P.S. Fader and E.T. Bradlow (2008), 'A bivariate timing model of customer acquisition and retention', *Marketing Science*, **27**(5), 829–43.
Shah, D. and V. Kumar (2012), 'The dark side of cross-selling', *Harvard Business Review*, **90**(12), 21–3.
Shah, D., V. Kumar, Y. Qu and S. Chen (2012), 'Unprofitable cross-buying: evidence from consumer and business markets', *Journal of Marketing*, **76**(3), 78–95.
Shin, J., K. Sudhir and D.-H. Yoon (2012), 'When to "fire" customers: customer cost-based pricing', *Management Science*, **58**(5), 932–47.

Smith, T.M., S. Gopalakrishna and R. Chatterjee (2006), 'A three-stage response model of integrated marketing communications at the marketing–sales interface', *Journal of Marketing Research*, **43**(4), 564–79.
Snoeck, S.J.M. (2012), 'Customer value models in the energy market: understanding the role of acquisition and retention effects', Dissertation, University of Groningen, accessed 15 September 2014 at http://irs.ub.rug.nl/ppn/353218669.
Stahl, F., M. Heitmann, D.R. Lehmann and S.A. Neslin (2012), 'The impact of brand equity on customer acquisition, retention, and profit margin', *Journal of Marketing*, **76**(4), 44–63.
Thomas, J.S. (2001), 'A methodology for linking acquisition to customer retention', *Journal of Marketing Research*, **38**(2), 262–8.
Thomas, J.S., R.C. Blattberg and E.J. Fox (2004), 'Recapturing lost customers', *Journal of Marketing Research*, **41**(1), 31–45.
Thomas, J.S., W. Reinartz and V. Kumar (2004), 'Getting the most out of all your customers', *Harvard Business Review*, **82**(7/8), 116–23.
Tokman, M., L.M. Davis and K.N. Lemon (2007), 'The WOW factor: creating value through win-back offers to reacquire lost customers', *Journal of Retailing*, **83**(1), 47–64.
Vakratsas, D. and T. Ambler (1999), 'Discovering how advertising works', *Journal of Marketing*, **63**(1), 26–43.
Van Diepen, M., B. Donkers and P.H. Franses (2009), 'Dynamic and competitive effects of direct mailings: a charitable giving application', *Journal of Marketing Research*, **46**(1), 120–33.
Vaughn, R. (1986), 'How advertising works: a planning model revisited', *Journal of Advertising Research*, **26**(1), 57–66.
Venkatesan, R. and V. Kumar (2004), 'A customer lifetime value framework for customer selection and resource allocation strategy', *Journal of Marketing*, **68**(4), 106–25.
Verhoef, P.C. and B. Donkers (2005), 'The effect of acquisition channels on customer loyalty and cross-buying', *Journal of Interactive Marketing*, **19**(2), 31–43.
Verhoef, P.C. and F. Langerak (2002), 'Eleven misconceptions about customer relationship management', *Business Strategy Review*, **13**(4), 70–76.
Verhoef, P.C. and K.N. Lemon (2013), 'Successful customer value management: key lessons and emerging trends', *European Management Journal*, **31**(1), 1–15.
Villanueva, J., S. Yoo and D.M. Hanssens (2008), 'The impact of marketing-induced versus word-of-mouth customer acquisition on customer equity growth', *Journal of Marketing Research*, **45**(1), 48–59.
Von Wangenheim, F. and T. Bayón (2007), 'The chain from customer satisfaction via word-of-mouth referrals to new customer acquisition', *Journal of the Academy of Marketing Science*, **35**(2), 233–49.
Voss, G.B. and Z.G. Voss (2008), 'Competitive density and the customer acquisition–retention tradeoff', *Journal of Marketing*, **72**(6), 3–18.
Wiesel, T., K. Pauwels and J. Arts (2011), 'Marketing's profit impact: quantifying online and offline funnel progression', *Marketing Science*, **30**(4), 604–11.

11. The chain of effects from customer satisfaction to customer profitability

Timothy L. Keiningham, Lerzan Aksoy, Yuliya A. Komarova and Mohammad Nejad

INTRODUCTION

The goal of all public companies is to generate reliable profits year in and year out. Consistently delivering profits requires having customers who not only initially buy, but consistently re-buy. As a result, to ensure that customers want to repurchase, the measurement and management of customer satisfaction has become critical to managers and researchers. The obvious reason is that they believe that customer satisfaction represents the starting point of a positive chain of effects, which ultimately leads to desirable business outcomes (e.g., profitability, market share, shareholder returns).

The core chain of effects can be thought of as follows: satisfaction → purchase intention → repurchase → firm performance. In fact, the seminal models in the service marketing literature that attempt to link customer satisfaction to firm performance either explicitly or implicitly presume this chain of effects, that is, SERVQUAL (Parasuraman, Zeithaml, and Berry, 1988), return on quality (ROQ) (Rust, Zahorik, and Keiningham, 1995), the service profit chain (Heskett, Sasser, and Schlesinger, 1997), and the satisfaction–profit chain (Anderson and Mittal, 2000). Without question, there is both theoretical and empirical support for the above chain of effects.

The relationship between satisfaction and behavioral intentions is grounded in the theory of planned behavior (Ajzen and Madden, 1986) an offshoot of the theory of reasoned action (Ajzen and Fishbein, 1980; Ajzen, 2001). The theory argues that intentions (and subsequent behaviors) are influenced by three factors: attitudes, subjective norms, and perceived behavioral control. Specifically, a consumer's favorable or unfavorable evaluation of the behavior, in combination with the influence of other people with respect to the behavior (i.e., the 'norm'), are the primary determinants of a consumer's intention to perform a behavior as long as the consumer perceives himself or herself to have complete control regarding performing the behavior.

Implicit in the theory of reasoned actions is the following chain of effects: attitudes → intentions → behaviors. In other words, the effect of attitudes on behaviors is mediated by intentions (Edwards and Lambert, 2007). Although satisfaction is generally viewed as perception (as opposed to an attitude), this chain of effects reflects the generally accepted view of how satisfaction ultimately influences consumer purchase decisions (Mittal and Frennea, 2010). The notion that consumer purchase decisions (i.e., consumption) impact firm performance is core to the foundation of economic theory (Friedman, 1957). Furthermore, the importance of consumer purchase process to economic theory has given rise to the field of behavioral economics (Thaler, 1980; Tversky and Kahneman, 1981).

There is also a wide body of research that supports empirically (at least in part) each of the links in the chain of effects from satisfaction to business performance. With regard to satisfaction, researchers have examined *direct* links between satisfaction and purchase intention (Mittal, Ross, and Baldasare, 1998), satisfaction and actual purchase/repurchase (Mittal and Kamakura, 2001) and satisfaction and financial outcomes (Aksoy et al., 2008). With regard to the repurchase intention → repurchase link, this relationship has been examined by a number of researchers (Jamieson and Bass, 1989; Bemmaor, 1995; Mittal and Kamakura, 2001; Morwitz, Steckel, and Gupta, 2007). And with regard to the repurchase → firm performance link, Reichheld and Sasser (1990) first demonstrated the financial benefits of retention. Currently, the link between retention (i.e., repurchase) and positive financial outcomes is widely accepted among managers and researchers.

In addition to examination of discrete links in the chain, researchers have also examined the mediating and moderating role of repurchase intention on the relationship between satisfaction and firm financial performance. In particular, Aksoy et al. (2012) find that for the overwhelming majority of firms, repurchase intention fully mediates the relationship between satisfaction and firm financial performance.

Yet, despite empirical and theoretical support for each of the links in the chain of effects, the unfortunate reality for managers has been an inability to obtain consistent positive financial outcomes from efforts to improve customer satisfaction and retention. The goal of this chapter is to identify how and why the links in the proposed chain of effects from satisfaction to firm performance break down, and to propose methods that managers and researchers can use to help overcome these problems.

LEAKY CONNECTIONS

The theory of planned behavior clearly argues against a perfect correlation among satisfaction, intentions, and behaviors. Specifically, the theory argues that norms and perceived control impact the relationship between attitudes and intentions. Without question, empirical evidence demonstrates that the correlation between each link in the chain is not perfect. First, while satisfaction and purchase intention are typically highly correlated (Keiningham, Vavra et al., 2005, p. 100) it is not one-to-one, as intentions can be influenced by other factors.

Second, the relationship between intentions and actual purchase behavior is not perfect (Jamieson and Bass, 1989; Bemmaor, 1995). For example, Morwitz et al. (2007) find that the predictive accuracy of purchase intentions is higher (1) when asked regarding existing products, (2) for durable goods, (3) when using short time horizons, (4) when related to specific brands (as opposed to category level), (5) when measured in terms of trial rather than total sales, and (6) when collected in a comparative mode. Furthermore, with regard to the entire chain of effects from satisfaction to purchase behavior, Mittal and Kamakura (2001) found that customer characteristics directly impact this relationship, and that the shape of the relationship between satisfaction ratings and repurchase intentions differed from that between satisfaction ratings and actual repurchase.

Clearly, the loss (i.e., lack of a one-to-one connection) associated with each link in the chain negatively impacts managers' (and researchers') ability to drive financial outcomes through satisfaction. In the authors' opinion, however, this is not the primary issue. Rather, we argue that three primary issues override all others in connecting satisfaction to customer behavior and to financial performance: (1) share of category spending – not retention – is the most relevant behavioral metric, (2) satisfaction's impact can only be understood relative to competitive alternatives, and (3) not all customers can be profitably satisfied.

FROM RETENTION TO SHARE-OF-WALLET

Because satisfaction has been tied to retention, and retention to profits, the majority of researchers seeking to both conceptualize and operationalize the chain of effects from satisfaction to business outcomes have proposed models linking satisfaction to retention and retention to profits (for example, Heskett et al., 1994; Rust et al., 1995; Zeithaml, Berry, and Parasuraman, 1996; Anderson and Mittal, 2000). Zeithaml (2000), for instance, conceptualizes several ways that customer retention leads to profits. In this model, customer retention is viewed as resulting in lower costs, increased volume of purchases (i.e., share-of-wallet), the ability to charge premium prices, and/or increased word-of-mouth. Research by Reinartz and Kumar (2000, 2002), however, finds that customer retention does not result in loyal customers costing less to serve, paying higher prices for the same bundle of services, or their marketing of the company (through word-of-mouth). This implies that the primary path from retention to profitability using the Zeithaml model is through increased share-of-wallet (SOW).

This fits with a growing disenchantment by both researchers and practitioners with using retention as the ultimate measure of loyalty, or as the best means of projecting profitability from service improvement efforts (Carroll, 1991–92; Carroll and Rose, 1993; Coyles and Gokey, 2002; Reinartz and Kumar, 2002). Managers have discovered that for most firms, more customers change their spending patterns with a company than completely stop doing business. Therefore, efforts designed to manage customers' spending patterns with a firm tend to represent far greater opportunities to a firm than simply trying to maximize customer retention rates. McKinsey & Company argues that focusing on customers' spending patterns to improve customers' SOW in addition to customer retention can have as much as ten times greater value to a company than focusing on retention alone (Coyles and Gokey, 2002). Our own research similarly supports the greater impact of SOW relative to retention for industries where customers tend to use multiple brands within the same category (Figure 11.1).

On the basis of the theoretical frameworks proposed by Heskett et al. (1994), Rust et al. (1995), Zeithaml (2000), and Anderson and Mittal (2000), in conjunction with the above-mentioned research that addresses key linkages in their models, we propose that share-of-wallet be substituted in place of retention in the chain of effects (Figure 11.2). By its very nature, share-of-wallet encompasses retention (i.e., SOW = 0 percent for defecting customers, and SOW > 0 for retained customers). But SOW better captures the level of spending given to a brand relative to competitive offerings.

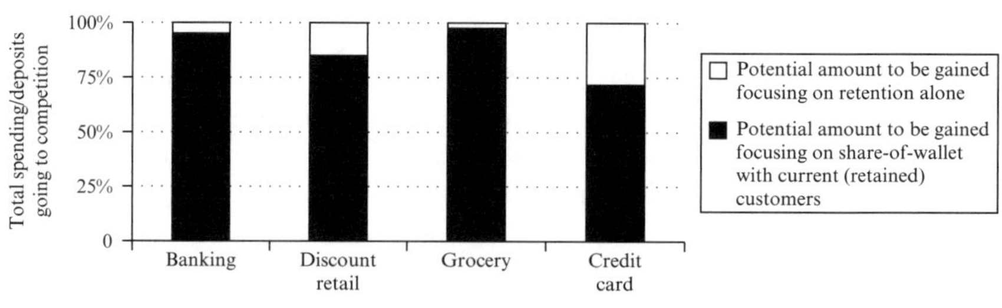

Figure 11.1 Total spending going to competition: a comparison of defecting customers and current customers

Figure 11.2 Proposed satisfaction–profit chain

SOW and level of product usage are also important to new product growth in the market and revenue as they are key predictors of one's influence on others' adoptions. In the case of high-involvement products and services, heavy users strongly influence others because of the knowledge they gain by using the product and because their recommendations are in line with their own behavior of using the product (Coulter, Feick, and Price, 2002; Iyengar, Van den Bulte, and Valente, 2011). Thus, recipients of messages from heavy users perceive them as experts and trustworthy, two key dimensions of source credibility (Pornpitakpan, 2004). Moreover, recommendations by these consumers are more likely to influence the recipients and lead to actual purchase because they serve as a proxy for adoption outcomes. Interestingly, volume usage of a product relative to that of other products in the category is a better predictor of one's influence on others than is self-reported opinion leadership (Iyengar et al., 2011).

SATISFACTION IS A WEAK PREDICTOR OF SHARE-OF-WALLET

Given management's interest in share-of-wallet, a number of researchers have examined the relationship between satisfaction and share-of-wallet. Using the share-of-wallet metric, studies find that customer satisfaction can lead to increased customer spending. For instance, a number of researchers proposed that satisfaction will have a positive influence on cross-buying, that is, customer satisfaction will be related to customers' allocation of a greater share of their spending to the firm (Parasuraman et al., 1988; Cronin and Taylor, 1992; Anderson and Sullivan, 1993; Zeithaml et al., 1996; Verhoef, Franses, and Hoekstra, 2001; Bolton, Lemon, and Verhoef, 2004). Additionally, Bowman and Narayandas (2004) and Keiningham, Perkins-Munn et al. (2005) conceptualize, operationalize, and find support for the satisfaction–profit chain proposed by Anderson and Mittal (2000), which initially linked satisfaction to retention as: (1) satisfaction, (2) share-of-wallet, (3) revenue, (4) profit. Despite the complex nature of the relationship, researchers have confirmed a link between satisfaction and share-of-wallet in a variety of industries (Silvestro and Cross, 2000; Mägi, 2003; Bowman and Narayandas, 2004; Baumann, Burton, and Elliott, 2005; Keiningham, Perkins-Munn et al., 2005; Perkins-Munn et al., 2005). Furthermore, in a longitudinal examination of the Canadian banking industry, Cooil et al. (2007) found that changes in satisfaction were positively related to changes in share-of-wallet.

The relationship between satisfaction and share-of-wallet, however, has been shown to be asymmetric and non-linear (Keiningham and Vavra, 2001). For example, Keiningham, Perkins-Munn, and Evans (2003) found that the relationship between satisfaction and share-of-wallet exhibited an S-shaped relationship, where extremes of satisfaction (high and low) corresponded to steep increases/decreases, respectively, in share-of-wallet. Between these two extremes, however, satisfaction showed little impact on SOW.

Furthermore, the problem for managers and researchers is that while research tends to support a positive relationship between satisfaction and share-of-wallet, the strength of the relationship is very weak. As Mägi (2003, p. 104) observes, 'Somewhat contrary to expectation, given the large emphasis on customer satisfaction in today's business world, customer satisfaction explained a fairly modest share of this variation [in share-of-wallet]'. For an excellent review of the research regarding the strength of the satisfaction–SOW relationship, see Hofmeyr et al., (2008).

The seemingly counter-intuitive weak satisfaction–SOW correlation results from a basic misunderstanding of how satisfaction and spending are related. Managers tend to believe that the better customers rate their level of satisfaction with the brand, the higher their corresponding share of category spending. This relationship, however, is not believed to be linear. Rather, there are believed to be three basic thresholds of satisfaction that have differing impacts on share of category spending (Keiningham and Vavra, 2001; Crotts, Pan, and Raschid, 2008): (1) zone of dissatisfaction (or zone of pain), (2) zone of mere satisfaction, (3) zone of delight (or zone of complete satisfaction) (see Figure 11.3).

Without question, there are different states of satisfaction. Researchers have clearly shown that the underlying drivers of low levels of satisfaction (i.e., dissatisfaction) differ from those that generate very high levels of satisfaction (e.g., delight) (Mittal, Ross, and Baldasare, 1998). Furthermore, there is an obvious logic regarding these different states of satisfaction (e.g., varying emotions and motives) and their resulting impact on share-of-wallet. However, increases in customers' share of category spending are not caused by increases in satisfaction ratings per se. Rather, it is because at some level, a higher state of satisfaction means a brand is more likely to be preferred to competitive brands that the customer also uses (Hofmeyr and Parton, 2010). So, it is not the absolute 'score' that a customer assigns to a brand that matters in terms of customers' share-of-wallet allocations.

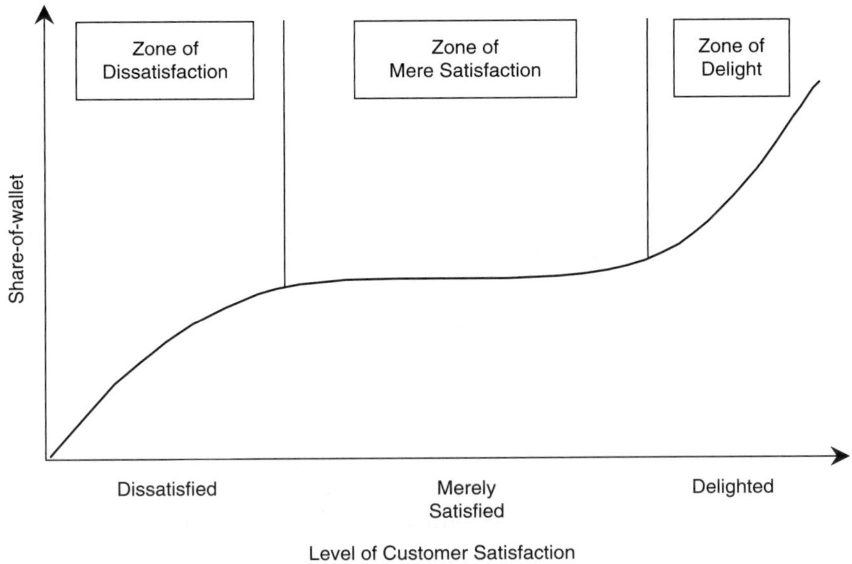

Figure 11.3 *The observed relationship between satisfaction levels and share-of-wallet*

Instead, it is a customer's perception of a brand's rank relative to competitive alternatives that impacts share of spending. Studies find a similar relationship between satisfaction and spreading word-of-mouth – a measure frequently used for behavioral intentions. Delighted customers and highly dissatisfied consumers spread more word-of-mouth about a product or a service compared to merely satisfied consumers (Anderson, 1998; Wangenheim and Bayón, 2007).

The problem with looking solely at a firm's satisfaction score is that it is a poor indicator of the relative ranking that customers assign to the brands they use. For example, imagine that a customer rated his or her satisfaction level a 9 on a 10-point scale. The most immediately relevant question to a manager is, 'Is this a good score or a bad score?' Most managers would argue that a '9' is a good satisfaction rating. But suppose the customer uses another brand in the category and for this brand he or she rates his or her satisfaction level a '10'. In this case, your competitor is the preferred brand. Similarly, imagine that another customer rated your brand an '8'. Most managers would not consider this to be a great score. But imagine the next highest score the customer assigns to a competitor brand is a '6'. In this case, your brand is the customer's preferred brand.

Researchers have shown that managers can strongly link these relative satisfaction rankings to share-of-wallet via power laws such as the Zipf distribution (Hofmeyr et al., 2008) and the wallet allocation rule (Keiningham et al., 2011). Therefore, if the goal is to link satisfaction to share-of-wallet, instead of focusing on absolute rating levels, researchers and managers need to understand what that score represents in terms of a customer's relative ranking of the various brands he or she uses.

REVENUE ≠ PROFIT

Let's assume that we have a good understanding of our customers' satisfaction, their share of category spending with our brand, and their total spending on our brand. Does this mean that obtaining highly satisfied, high share-of-wallet, high-revenue customers will be profitable? Unfortunately, the answer is a resounding NO! Even if managers can make the link from satisfaction to SOW to revenue, that does not ensure that the company will obtain the most important metric for sustaining operations – profits. The reality is that for most firms a significant percentage of high-satisfaction, high-SOW, high-revenue customers will be unprofitable.

Typically, the majority of customers do not generate acceptable financial returns. Simply put, there are three basic groups of customers in terms of profitability to the firm: those that make the company money, those that just break even, and those that lose the company money. While this sounds self-evident, the percentage of customers that fall into these three categories is not nearly so obvious. Research consistently finds that profitable customers tend to only constitute around 20 percent of all customers, break-even customers, around 60 percent, and unprofitable customers, around 20 percent in most businesses (Kaplan and Narayanan, 2001) (Figure 11.4). The problem with this distribution is that highly satisfied customers exist in each of these segments (Keiningham et al., 2005).

Part of the problem is that while improved satisfaction can increase sales revenue, the additional costs frequently outweigh the benefits (Souki and Filho, 2008; Keiningham et al., 2014). Managers typically have many competing alternatives for improving satisfaction. Not all of them will be profitable. Furthermore, not all customers can be profitably satisfied. Some are not willing to pay the necessary price for the level of service being offered. Other customers demand a level of service that more than offsets any revenue they provide. So why is it that consumers today are frequently difficult to please, and even when satisfied, often fail to turn profitable for companies?

Recalling our earlier brief discussion of the theory of planned behavior (Ajzen and Madden, 1986) and the theory of reasoned action (Ajzen and

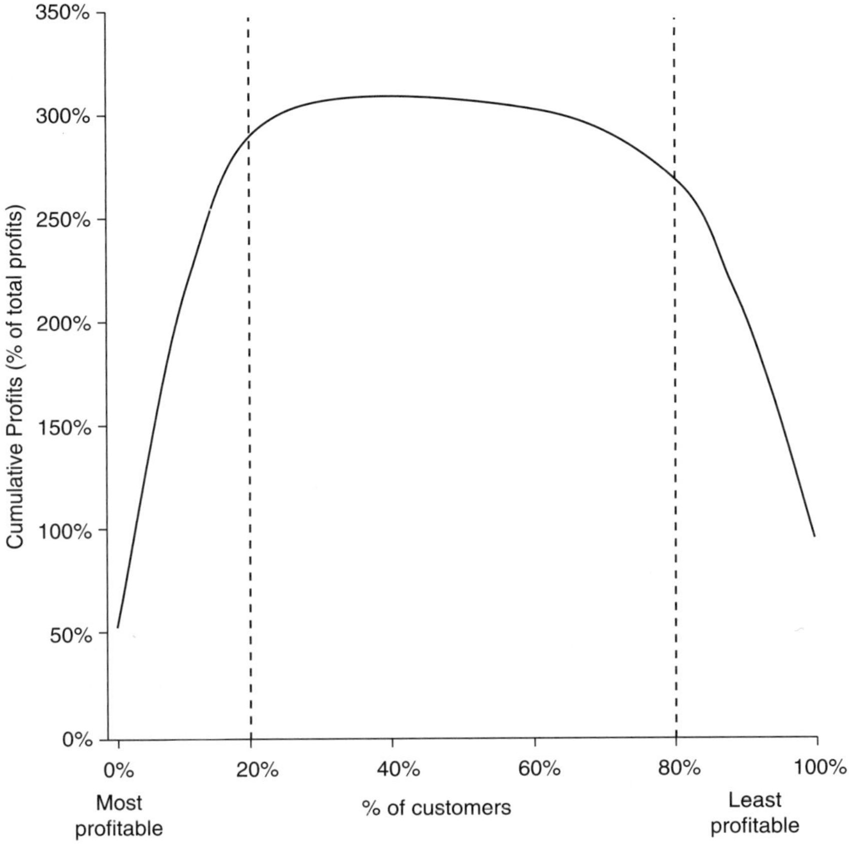

Figure 11.4 Cumulative customer profitability

Fishbein, 1980; Ajzen, 2001), behavioral intentions, and ultimately behavior, are not a simple function of consumer attitudes. They are also influenced by subjective norms and perceived behavioral control. Nevertheless, customer satisfaction, which is a perception, making it vulnerable to a multitude of factors (as opposed to a more stable attitude), has been the sole focus for many practitioners. Yet relatively recently, the consumption environment has evolved in a number of important ways, which have had a dramatic impact on what consumers consider normal and acceptable behavior (i.e., subjective norms) and their sense of power and control (i.e., perceived behavioral control).

In a March 2008 Special Report, *Bloomberg BusinessWeek* paints the reality of today's marketplace, where consumers, often dissatisfied with their consumption experiences for a valid reason or because of their unrealistic expectations, are particularly empowered by information and technology to stand up for themselves. Unfortunately, in many cases, distraught by consumption mishaps, consumers fight back in ways that are hardly appropriate. Consider, for instance, a '76-year-old retired nurse named Mona Shaw [who] smashed up a keyboard and a telephone in a Manassas (Va.) Comcast office after she says the cable operator failed to install her service properly' (McGregor, 2008). In fact, it appears that the modern day consumer vigilantes are the mainstream individuals who are driven to get results (i.e., fight back) by whatever means necessary (for an excellent review, please see Tian and Keep, 2002). As a Nielsen Online Strategic Services executive notes, 'companies can brush these off as being atypical, mutant consumers, or they can say there's a very important insight in [these consumers'] emotions' (McGregor, 2008).

While there isn't much companies can do to limit consumers' sense of empowerment and entitlement, they can, to some degree, leverage the level of expectations consumers hold for a product or service. Perhaps more importantly, they can carefully segment consumers not only based on their unique needs, but also on their potential for being profitable for the company. For instance, the research on customer fraud offers numerous illustrations of consumers repeatedly engaging in acts that promise little or no financial benefit to their perpetrators (e.g., repeatedly clicking on an online ad for the purpose of generating a charge per click without having actual interest in the target of the ad's link), require more effort than they may be worth (e.g., fraudulent product reviews), and involve high risk of getting caught (e.g., insurance fraud). Such consumer deviance results in multibillion-dollar annual losses with tremendous negative impact on individual businesses and the economy as a whole (e.g., Tian and Keep, 2002; McGregor, 2008). Moreover, the costs of consumer deviance are transferred to other consumers through increased product prices and service charges.

These and many other consumer acts that are detrimental to the bottom line are often driven and motivated by negative emotions, such as anger and frustration (Patterson et al., 2009). In line with the earlier discussion of the three zones of satisfaction, it seems reasonable to predict that dissatisfied customers may cause financial damage and this alone would support practitioners' sole focus on and efforts to improve satisfaction. Less intuitive, however, is a recent finding that negative emotions, as a consequence of a dissatisfactory experience with one firm, are actually very likely to spill over and hurt other companies, particularly if the latter share some degree of similarity with the source of dissatisfaction (e.g., offer similar products or services; Komarova, Haws, and Bearden, 2012). Furthermore, the merely satisfied and even delighted customers may posit a significant financial threat when something is even slightly wrong with a product or service – then, 'customer love turns into lasting hate' (Grégoire, Tripp, and Legoux, 2009). Hence, focusing on improving customer satisfaction alone not only fails to secure the bottom line, it may also actually hurt companies long term.

These findings point to an important truth about the relationship between customer satisfaction and customer profitability. While satisfaction and profitability are not mutually exclusive, they don't have to be aligned either. The bottom line is that there is no substitute for understanding the profit impact of your efforts to improve customer satisfaction. Armed with this information, managers can make the right – sometimes difficult – decisions for their businesses.

FIXING THE LINKS IN THE CHAIN

Creating and nurturing customer satisfaction is at the core of economic theory. As Harvard Professor Theodore Levitt noted three separate times in his book *The Marketing Imagination* (1986), 'the purpose of business is to create and keep a customer; all else is derivative'. Yet, while this truth is undeniable, its application is not straightforward. Satisfying customer needs and wants must be done at a sustainable profit.

Given the numerous obstacles in operationalizing the satisfaction–profit chain, what is a manager to do? The most important thing is to resist the calls of some to abandon the measurement and management of customer satisfaction (e.g., Gitomer, 1998; Reichheld, 2003).

The reality is that for many (if not most) firms, 'satisfaction is the only feasible goal for which they should strive' in the pursuit of customer loyalty (Oliver, 1999, p. 33). Managers must not, however, confuse the *necessity* of satisfaction to build sustainable profits with *sufficiency* – satisfaction in and of itself is not sufficient. Blindly pursuing satisfaction for its own sake is a recipe for disaster. For instance, the easiest way to improve satisfaction is to give away your product or service; customer satisfaction will definitely rise – at least until the firm is forced to close its doors. Fortunately, it is possible to repair the leaky connections in the satisfaction–profit chain, so that efforts to improve satisfaction result in positive financial returns.

Step 1: Satisfaction → share-of-wallet linkage

a. Survey customers and establish the number of brands (or stores or firms) customers use in the product category that you want to analyze.

b. Survey customers and obtain satisfaction ratings for each brand used by the customer.
c. Convert customers' satisfaction ratings into ranks. In the case of a tie, use the average for the ranks that they would have occupied had they not been tied. For example, if two firms are tied for second, use 2.5 (the average of ranks 2 and 3). In this case, the next highest rank will be 4, assuming no other ties exist.
d. Use a power law such as the wallet allocation rule (Keiningham et al., 2011) or the Zipf distribution (Hofmeyr et al., 2008) to transform these ranks into customer-level share-of-wallet projections.

Step 2: Align share-of-wallet, revenue, and profit objectives

While the goal of improving customer satisfaction is to positively impact share of category spending, it is not the only thing that matters. Researchers have found that as goods within a category are purchased with higher frequencies, the share that any one particular brand gets declines. For example, consumers who purchase the product only once clearly have given 100 percent of their share of spending to a single brand. Therefore, as the number of purchases in the category increases, there is a greater probability that customers will try alternatives, thereby lowering the average share for any one particular brand.

The first thing to note about the decline in share of spending is that it is not linear. As purchase frequency increases from 5 to 15, there is a sharp decline in share, a more moderate decline as purchases increase from around 16 to 60, and almost no decline when purchases exceed 60 (Stern and Hammond, 2004) (Figure 11.5). Researchers Stern and Hammond (2004, p. 5) observe, 'The pattern of apparent declining loyalty is largely a statistical artifact, dependent on the number of purchases used to calculate the measure'.

The fact is that greater purchase frequency in consumer goods typically means variety seeking, which offers the customer the pleasure of novelty, even if he or she is highly loyal (e.g., a loyal Coca-Cola customer will occasionally drink a Sprite, and may even sometimes consume a Pepsi). That is simply a function of the number of purchase opportunities. And even in business-to-business relationships, high purchase frequency is likely to mean critical need for a product, which often makes single sourcing too risky or economically unsound. Service providers, however, can fulfill consumers' need for variety seeking by increasing the breadth of services they provide and encouraging consumers to use these services, as opposed to switching to competitive providers. For example, cell phone companies have offered a wide range of services to consumers including data services such as web browsing, messaging, and fax; media services such as games, entertainment, and music; and information services such as weather and

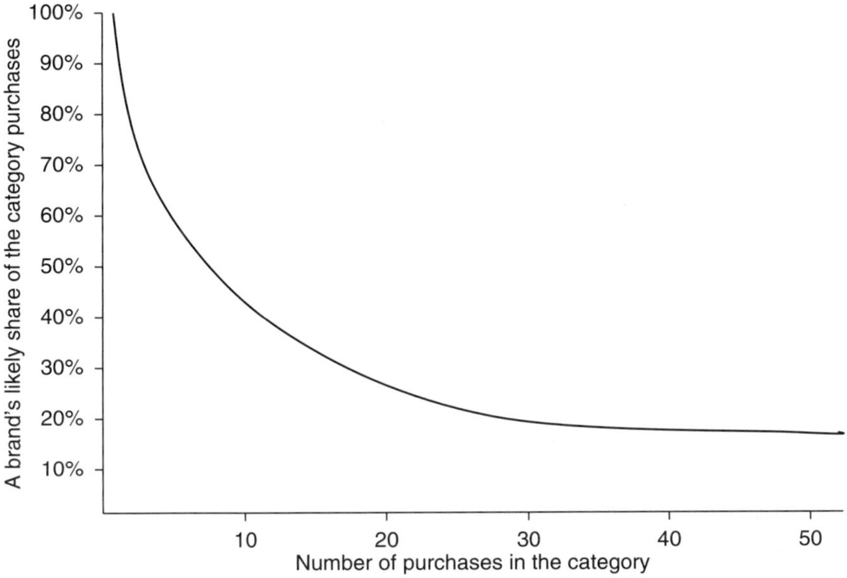

Figure 11.5 Purchase frequency and share of category spending

traffic (Chircu and Mahajan, 2009). While these services have fulfilled customers' novelty seeking, they have also provided cell phone service providers with significant additional revenue.

Therefore, we need to understand share-of-wallet in the context of the total volume (e.g., revenue) in the category purchased by customers. At a minimum, we must segment customers by the volume of business that they conduct in the category. Importantly, understanding the relationship between share-of-wallet and revenue is not enough. That is, because revenue is typically not a good predictor of profitability. Some of the largest customers are the most unprofitable. As Professors Kaplan and Narayanan (2001) of the Harvard Business School observe:

> A company cannot lose large amounts of money with small customers. It doesn't do enough business with a small customer to incur large (absolute) losses. Only a large customer, working in a particularly perverse way, can be a large loss customer. Large customers tend to be either the most profitable or the least profitable in the entire customer base. It's unusual for a large customer to be in the middle of the total profitability rankings.

The seemingly logical answer of ignoring revenue and focusing exclusively on return on investment (ROI), however, is typically not the best solution

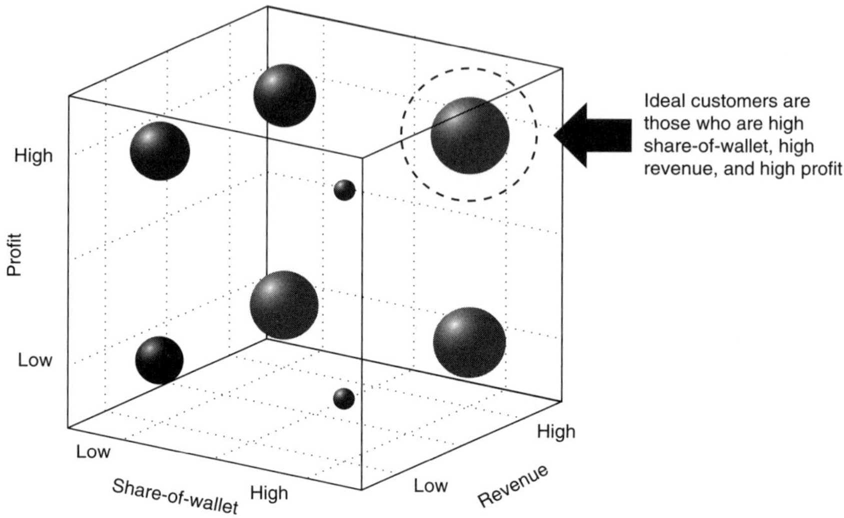

Figure 11.6 The share-of-wallet–revenue–profit cube

for a business either. That is because maximizing ROI often comes from too narrowly defining the customer pool to gain significant market share growth. Therefore, we recommend attempting to maximize share-of-wallet, revenue, and profitability simultaneously (Figure 11.6).

Using this framework, it becomes clear which customers are 'ideal' (i.e., they are high on share-of-wallet, revenue, and profit). It also becomes clear which of the firm's profitable customers are not providing the firm with the majority of their business in the category. Because revenue and profit are included in the framework, focusing on measures to improve the number of ideal customers keeps the goals of improving share-of-wallet and financial performance synchronized.

Knowing your percentage of ideal customers is a vital first step, but it is not the one and only number you need to know to manage your business. A focus on ideal customers also demands that managers recognize two fundamental truths about customers. First, not every customer is capable of being an ideal customer, nor does every customer want to be one. Some customers are only driven by price. Other customers make purchases so infrequently that unless this is the vast majority of the customer base, servicing and marketing to such customers yields too little in return. Another group of customers provide high revenue, but demand so much service and attention that their costs significantly outpace their contribution. Furthermore, ideal customers are almost always in the minority in terms of number of total customers. Strategies that neglect customer differentiation will inevitably tilt the scale toward break-even and unprofitable customers, as they represent the majority.

Second, ideal customers are not a monolith. Rather, they are a mix of many different types of customers who are drawn to the company for different reasons. For example, we can easily imagine that young families and seniors could be drawn to the same financial institution, be profitable, yet have very different needs from the relationship. Recognizing these distinctions in what drives ideal customer behavior is vital. It points to the danger in making decisions based upon the 'average' customer. Given the diversity within the population of ideal customers, prioritizing efforts designed to attract and retain these customers based upon averages often results in no customers getting what they really want from the relationship.

Once we know who has the greatest potential to become an ideal customer, we can begin to identify what it will actually take to make that happen. For most firms, improving the percentage of ideal customers is largely driven by increasing the share of spending that low-share customers provide to the firm. Here, the key is to understand why these low-share customers are allocating the bulk of their spending elsewhere. Inevitably, common themes will emerge among potential ideal customers regarding what is restraining their spending. Armed with this information, it becomes relatively simple to prioritize efforts to improve loyalty based upon whether the issue is easy or difficult to address versus the expected impact.

CONCLUSION

Customer satisfaction is a cornerstone of profitable business growth. Nevertheless, managers and researchers need to recognize that it isn't that simple, as happy customers may or may not lead to firm growth and profits. Simply investing in improvement efforts without an understanding of their likely impact is a recipe for failure. Therefore, it is imperative to understand which elements of the customer experience have measureable, positive downstream consequences.

For managers, this means that they must ensure that their efforts to improve profitability through customer satisfaction are done with sufficient force, so as to shift the rank of the firm vis-à-vis competitors to achieve a higher share of category spending by customers. This cannot be an occasional level of achievement. The research is also clear that achieving these thresholds cannot be hit-or-miss – consistency of the customer experience is no less vital (Keiningham et al., 2006). Therefore, it is essential to have a realistic model of what ultimately enhances future financial performance if managers are to focus their scarce resources appropriately. The rewards, however, are worth the effort. In fact, our own research found that firms with high customer satisfaction had annual abnormal returns of 10.56 percent (Aksoy et al., 2008)!

REFERENCES

Ajzen, I. (2001), 'Nature and operation of attitudes', *Annual Review of Psychology*, **52**(1), 27–58.
Ajzen, I. and M. Fishbein (1980), *Understanding Attitudes and Predicting Social Behavior*, Englewood Cliffs, NJ: Prentice Hall.
Ajzen, I. and T.J. Madden (1986), 'Prediction of goal-directed behavior: attitudes, intentions, and perceived behavioral control', *Journal of Experimental Social Psychology*, **22**(5), 453–74.
Aksoy, L., B. Cooil, C. Groening, T.L. Keiningham and A. Yalçin (2008), 'Long term stock market valuation of customer satisfaction', *Journal of Marketing*, **72**(4), 105–22.
Aksoy, L., T.L. Keiningham, B. Larivière, S. Mithas, F.V. Morgeson and A. Yalçin (2012), 'The satisfaction, repurchase intention and shareholder value linkage: a longitudinal examination of fixed and firm-specific effects', Fordham University Working Paper.
Anderson, E.W. (1998), 'Customer satisfaction and word of mouth', *Journal of Service Research*, **1**(1), 5–17.
Anderson, E. and V. Mittal (2000), 'Strengthening the satisfaction–profit chain', *Journal of Service Research*, **3**(2), 107–20.
Anderson, E. and M.W. Sullivan (1993), 'The antecedents and consequences of customer satisfaction for firms', *Marketing Science*, **12**(2), 125–43.
Baumann, C., S. Burton and G. Elliott (2005), 'Determinants of customer loyalty and share-of-wallet in retail banking', *Journal of Financial Services Marketing*, **9**(3), 231–48.
Bemmaor, A.C. (1995), 'Predicting behavior from intention-to-buy measures: the parametric case', *Journal of Marketing Research*, **32**(2), 176–91.
Bolton, R.N., K.N. Lemon and P.C. Verhoef (2004), 'The theoretical underpinnings of customer asset management: a framework and propositions for future research', *Journal of the Academy of Marketing Science*, **32**(3), 271–92.
Bowman, D. and D. Narayandas (2004), 'Linking customer management effort to customer profitability in business markets', *Journal of Marketing Research*, **41**(4), 433–47.
Carroll, P. (1991–92), 'The fallacy of customer retention', *Journal of Retail Banking*, **13**(4), 15–20.
Carroll, P. and S. Rose (1993), 'Revisiting customer retention', *Journal of Retail Banking*, **15**(1), 5–13.
Chircu, A.M. and V. Mahajan (2009), 'Revisiting the digital divide: an analysis of mobile technology depth and service breadth in the BRIC countries', *Journal of Product Innovation Management*, **26**(4), 455–66.
Cooil, B., T.L. Keiningham, L. Aksoy and M. Hsu (2007), 'A longitudinal analysis of customer satisfaction and share-of-wallet: investigating the moderating effect of customer characteristics', *Journal of Marketing*, **71**(1), 67–83.
Coulter, R., L. Feick and L. Price (2002), 'Changing faces: cosmetics opinion leadership among women in the new Hungary', *European Journal of Marketing*, **36**(11/12), 1287–308.
Coyles, S. and T.C. Gokey (2002), 'Customer retention is not enough', *The McKinsey Quarterly*, No. 2, 81–9.
Cronin, J.J. and S.A. Taylor (1992), 'Measuring service quality: a re-examination and extension', *Journal of Marketing*, **56**(3), 55–68.

Crotts, J.C., B. Pan and A.E. Raschid (2008), 'A survey method for identifying key drivers of guest delight', *International Journal of Contemporary Hospitality Management*, **20**(4), 462–70.

Edwards, J.R. and L.S. Lambert (2007), 'Methods for integrating moderation and mediation: a general analytical framework using moderated path analysis', *Psychological Methods*, **12**(1), 1–22.

Friedman, M. (1957), 'Introduction', in M. Friedman (ed.), *A Theory of the Consumption Function*, Princeton, NJ: Princeton University Press, pp. 1–6.

Gitomer, J. (1998), *Customer Satisfaction is Worthless, Customer Loyalty is Priceless*, Austin, TX: Bard Press.

Grégoire, Y., T.M. Tripp and R. Legoux (2009), 'When customer love turns into lasting hate: the effects of relationship strength and time on customer revenge and avoidance', *Journal of Marketing*, **73**(6), 18–32.

Heskett, J.L., W.E. Sasser and L.A. Schlesinger (1997), *The Service Profit Chain*, New York: The Free Press.

Heskett, J.L., T.O. Jones, G.W. Loveman, W.E. Sasser, Jr. and L.A. Schlesinger (1994), 'Putting the service–profit chain to work', *Harvard Business Review*, **72**(2), 164–74.

Hofmeyr, J. and G. Parton (2010) 'Rank matters', *Marketing Research*, **22**(3), 7–12.

Hofmeyr, J., V. Goodall, M. Bongers and P. Holtzman (2008), 'A new measure of brand attitudinal equity based on the Zipf distribution', *International Journal of Market Research*, **50**(2), 181–202.

Iyengar, R., C. van den Bulte and T.W. Valente (2011), 'Opinion leadership and social contagion in new product diffusion', *Marketing Science*, **30**(2), 195–212.

Jamieson, L.F. and F.M. Bass (1989), 'Adjusting stated intention measures to predict trial purchase of new products: a comparison of models and methods', *Journal of Marketing Research*, **26**(3), 336–45.

Kaplan, R.S. and V.G. Narayanan (2001), 'Customer profitability measurement and management', *Journal of Cost Management* (Sept/Oct), 5–15.

Keiningham, T.L. and T.G. Vavra (2001), *The Customer Delight Principle: Exceeding Customers' Expectations for Bottom-line Success*, New York: McGraw-Hill/American Marketing Association.

Keiningham, T.L., T. Perkins-Munn and H. Evans (2003), 'The impact of customer satisfaction on share-of-wallet in a business-to-business environment', *Journal of Service Research*, **6**(1), 37–50.

Keiningham, T.L., L. Aksoy, A. Buoye and B. Cooil (2011), 'Customer loyalty isn't enough. Grow your share-of-wallet', *Harvard Business Review*, **89**(10), 29–31.

Keiningham, T.L., S. Gupta, L. Aksoy and A. Buoye (2014), 'The high price of customer satisfaction', *MIT Sloan Management Review*, **55**(3), 37–46.

Keiningham, T.L., T. Perkins-Munn, L. Aksoy and D. Estrin (2005), 'Does customer satisfaction lead to profitability? The mediating role of share-of-wallet', *Managing Service Quality*, **15**(2), 172–81.

Keiningham, T.L., T.G. Vavra, L. Aksoy and H. Wallard (2005), *Loyalty Myths: Hyped Business Strategies That Will Put You Out of Business*, Hoboken, NJ: John Wiley and Sons.

Keiningham, T.L., L. Aksoy, B. Cooil, K. Peterson and T.G. Vavra (2006), 'Longitudinal examination of the asymmetric impact of employee and customer satisfaction on retail sales', *Managing Service Quality*, **16**(5), 442–59.

Komarova, Y., K. Haws and W. Bearden (2012), 'Punishing the guilty and the guiltless: an investigation of consumer deviance inspired by company wrongdoing', Fordham University Working Paper.

Levitt, T.M. (1986), *The Marketing Imagination*, new expanded edition, New York: The Free Press, pp. 5, 7, 19.

Mägi, A.W. (2003), 'Share-of-wallet in retailing: the effects of customer satisfaction, loyalty cards and shopper characteristics', *Journal of Retailing*, **79**(2), 97–106.

McGregor, J. (2008), 'Consumer vigilantes. Memo to corporate America: hell now hath no fury like a customer scorned', *Bloomberg BusinessWeek*, 3 March, Issue No. 4073, 38–42.

Mittal, V. and C.M. Frennea (2010), *Customer Satisfaction: A Strategic Review and Guidelines for Managers*, Fast Forward Series, Cambridge, MA: Marketing Science Institute.
Mittal, V. and W. Kamakura (2001), 'Satisfaction, repurchase intent, and repurchase behavior: investigating the moderating effect of customer characteristics', *Journal of Marketing Research*, **38**(1), 131–42.
Mittal, V., W.T. Ross Jr. and P.M. Baldasare (1998), 'The asymmetric impact of negative and positive attribute-level performance on overall satisfaction and repurchase intentions', *Journal of Marketing*, **62**(1), 33–47.
Morwitz, V.G., J.H. Steckel and A. Gupta (2007), 'When do purchase intentions predict sales?', *International Journal of Forecasting*, **23**(3), 347–64.
Oliver, R.L. (1999), 'Whence consumer loyalty', *Journal of Marketing*, **63** (Special Issue), 33–44.
Parasuraman, A., V.A. Zeithaml and L.L. Berry (1988), 'SERVQUAL: a multiple-item scale for measuring consumer perceptions of service quality', *Journal of Retailing*, **64**(1), 12–40.
Patterson, P.G., J.R. McColl-Kennedy, A.K. Smith and Z. Lu (2009), 'Customer rage: triggers, tipping points, and take-outs', *California Management Review*, **52**(1), 6–28.
Perkins-Munn, T., L. Aksoy, T.L. Keiningham and D. Estrin (2005), 'Actual purchase as a proxy for share-of-wallet', *Journal of Service Research*, **7**(3), 245–56.
Pornpitakpan, C. (2004), 'The persuasiveness of source credibility: a critical review of five decades' evidence', *Journal of Applied Social Psychology*, **34**(2), 243–81.
Reichheld, F.F. (2003), 'The one number you need to grow', *Harvard Business Review*, **81**(12), 46–54.
Reichheld, F.F. and W.E. Sasser, Jr. (1990), 'Zero defections: quality comes to services', *Harvard Business Review*, **68**(5), 105–11.
Reinartz, W. and V. Kumar (2000), 'On the profitability of long-life customers in a non-contractual setting: an empirical investigation and implications for marketing', *Journal of Marketing*, **64**(4), 17–35.
Reinartz, W. and V. Kumar (2002), 'The mismanagement of customer loyalty', *Harvard Business Review*, **80**(7), 86–94.
Rust, R.T., A.J. Zahorik and T.L. Keiningham (1995), 'Return on quality (ROQ), making service quality financially accountable', *Journal of Marketing*, **59**(2), 58–70.
Silvestro, R. and S. Cross (2000), 'Applying the service profit chain in a retail environment', *International Journal of Service Industry Management*, **11**(3), 244–68.
Souki, G. and C.G. Filho (2008), 'Perceived quality, satisfaction and customer loyalty: an empirical study in the mobile phones sector in Brazil', *International Journal of Internet and Enterprise Management*, **5**(4), 298–314.
Stern, P. and K. Hammond (2004), 'The relationship between customer loyalty and purchase incidence', *Marketing Letters*, **15**(1), 5–9.
Thaler, R. (1980), 'Toward a positive theory of consumer choice', *Journal of Economic Behavior and Organization*, **1**(1), 39–60.
Tian, K. and B. Keep (2002), *Customer Fraud and Business Responses: Let the Marketer Beware*, Westport, CT: Quorum Books.
Tversky, A. and D. Kahneman (1981), 'The framing of decisions and the psychology of choice', *Science*, **211**(4481), 453–8.
Verhoef, P.C., P.H. Franses and J.C. Hoekstra (2001), 'The impact of satisfaction and payment equity on cross-buying: a dynamic model for a multi-service provider', *Journal of Retailing*, **77**(3), 359–78.
Wangenheim, F.V. and T. Bayón (2007), 'The chain from customer satisfaction via word-of-mouth referrals to new customer acquisition', *Journal of the Academy of Marketing Science*, **35**(2), 233–49.
Zeithaml, V.A. (2000), 'Service quality, profitability, and the economic worth of customers: what we know and what we need to learn', *Journal of the Academy of Marketing Science*, **28**(1), 67–85.
Zeithaml, V.A., L.L. Berry and A. Parasuraman (1996), 'The behavioral consequences of service quality', *Journal of Marketing*, **60**(2), 31–46.

12. Customer-lifetime-value–based resource allocation

Rajkumar Venkatesan

INTRODUCTION

The field of customer relationship management (CRM) gained prominence among academics and practitioners with the advent of large customer databases. Technological improvements allowed firms to store and retrieve information on their interactions with individual customers and theoretically provided them with the ability to develop an ongoing and customized relationship with each customer. Driven by the belief that they could improve profits by developing these relationships, organizations made significant investments in CRM systems during the mid- to late 1990s. However, reality differed from expectations, and firms were not able to realize or measure the return on their investments.

Development of a resource allocation strategy that fully utilized the individual customer-level information was identified as key factor that would determine the success of CRM investments. Today, CRM activities in several firms are driven by the systematic and differential allocation of resources to individual customers to grow their profits.

Most theorists, including experts in the field of CRM, initially believed that loyal customers would provide firms with more profits and that customer profits increased over the span of a relationship (Reichheld, 1996). This idea was based on data pertaining to average profits for all customers in a cohort or contractual situations for financial services. However, the correlation between customer loyalty (i.e., customer tenure) and customer profitability weakens when customer heterogeneity is taken into consideration and in non-contractual settings, such as catalogs, business-to-business firms, and retailing (Reinartz and Kumar, 2002). Therefore, the objective of customer resource allocation should be to maximize lifetime customer profits. To this end, customer resource allocation models adopt long-term customer profitability, particularly customer lifetime value (CLV), rather than one-time transactional sales or profits as their criterion for resource allocation (Blattberg and Deighton, 1996; Berger and Bechwati, 2001).

A five-step process can be used to implement customer-focused resource

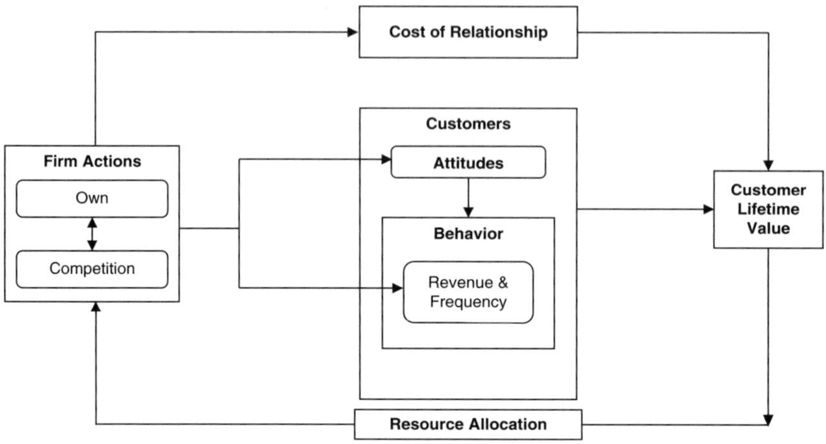

Figure 12.1 Framework for customer-level resource allocation

allocation based on CLV. A general framework for the process is proposed in Figure 12.1, but nuances within individual industries also exist.

THE FIVE-STEP RESOURCE ALLOCATION PROCESS

Two methods of determining resource allocations to customers have been widely adopted: decision calculus and econometric methods. The decision calculus approach is suitable for determining the optimal level of marketing resources to devote to an entire population or segments of customers, as well as the optimal balance of resources between activities, such as acquisition and retention (Blattberg and Deighton, 1996). The econometric approach is flexible and has been used to determine optimal resource allocation levels for both entire customer populations and individual customers.

In the five-step process of implementing customer-focused resource allocation, a firm must first determine the CLV formulation it would like to maximize. The second step is to break CLV down into its components. The third step in the process of using it to allocate resources is to determine the components of the metric that are unknown for future time periods and must be predicted. In general, the unknowns in a CLV formulation would include monthly margin (M), monthly cost of retention (R), and monthly retention rate (r).

The fourth step of the process is to develop a predictive model for each of the unknown components (e.g., M, R, and r). In order to assist in resource allocation and marketing strategy, the predictive model must connect the unknown CLV components to marketing mix elements, product, price, place, and promotion. This relationship can be established through econometric analysis of past data, experimentation, or decision calculus (i.e., informed intuition). Typically, econometric analysis is used to generate hypotheses about the relationships, field experiments are developed to test the hypotheses, and decision calculus finds the still unknown aspects of the relationships. For example, a log linear regression could be used to connect past sales and contribution margins to product characteristics, price, and advertising investments. The coefficients in the regression for price would provide an estimate of consumer price elasticity. A field experiment could then be designed to fine-tune the causal link (or elasticity) between price and sales/contribution margins. Finally, because the econometric analysis is based on a past ecosystem and experiments are generally conducted holding the ecosystem constant, decision calculus could be used to adjust the price elasticity based on contingencies (e.g., competitive reactions).

Once the relationships between the marketing mix and unknown components are established, the final step of the process is to find the optimal level of marketing resources to maximize CLV while considering factors such as budget constraints. A specific illustration of the five-step process is as follows:

Step 1 Specify the objective function for resource allocation. For example, Venkatesan and Kumar (2004) specify CLV as:

$$CLV_i = \sum_{y=1}^{T_i} \frac{CM_{i,y}}{(1+r)^{\frac{y}{frequency_i}}} - \sum_{l=1}^{n} \frac{\sum_m c_{i,m,l} * x_{i,m,l}}{(1+r)^{l-1}}, \tag{12.1}$$

where:
CLV_i = lifetime value of customer i;
$CM_{i,y}$ = predicted contribution margin from customer i (computed from a contribution margin model) in purchase occasion y, measured in dollars;
r = discount rate for money (15 percent annual rate in this case);
$c_{i,m,l}$ = unit marketing cost for customer i in channel m and year l (the formulation of CLV does not change if l is used to represent time periods other than one year);
$x_{i,m,l}$ = number of contacts to customer i in channel m and year l,
$frequency_i$ = predicted purchase frequency for customer i;
n = number of years to forecast; and
T_i = predicted number of purchases made by customer i before the end of the planning period.

Step 2 Specify a model that predicts the aspects of customer behavior required to determine the objective function. For example, equation (12.1) predicts purchase frequency and contribution margin and specifies models for predicting customer behavior as a function of marketing decision variables ($x_{i,d}$), covariates ($x_{i,cov}$), and unknown response parameters (b*s*). The marketing decision variables in the model include the number of contacts made through salesperson, direct, and telesales channels.

Step 3 Use calibration data to estimate the unknown response parameters, given observed customer behavior, marketing decision variables, and covariates.

Step 4 Given the response parameters estimated in Step 3 and the model specified in Step 2, predict customer behavior for the planning period. The predicted customer behaviors measure the objective function. The marketing decision variables ($X_{i,d}$) and covariates ($X_{i,cov}$) also can be predicted or assumed at their most recent values.

Step 5 Vary the marketing decision variables to determine the maximization level for the objective function. Venkatesan and Kumar (2004) vary the number of contacts made through salesperson, direct, and telesales channels to determine the best way to maximize CLV via contacts.

This framework can be adapted to customize resource allocation decisions for an entire population, a segment of customers, or each customer, based on the level at which the response parameters (b*s*) are specified and estimated in Steps 2 and 3. Firms will obtain the maximum benefits of this approach over time if they develop a systematic method of monitoring their resource allocation decisions, observed consumer and competitive reactions, and observed metric improvements. Continually capturing feedback is essential for constant process improvement and gathering information for future econometric analyses, experiments, and decision calculus.

MEASURING CUSTOMER LIFETIME VALUE

CLV captures the potential value or profits a customer provides a firm. Interest in CLV is driven by several factors. First, it eliminates the need to use a proxy for customer profitability, such as loyalty, which might introduce a weak correlation (Reinartz and Kumar, 2002). Second, it can help an organization transition from product- to customer-centricity (Gupta and Lehmann, 2008). Third, it is a forward-looking metric that enables firms to target customers they expect to be profitable in the future rather than focusing solely on customers who provided profits in the past (Venkatesan and Kumar, 2004). Finally, it provides managers with a venue for connecting marketing actions to financial returns, both cash and stock market valuation (Kumar and Shah, 2009; Rust, Lemon, and Zeithaml, 2004).

CLV refers to the net present value of future profits a customer will provide over the life of his or her business with a firm. In its basic form, CLV is defined by (Gupta, Lehmann, and Stuart 2004):

$$CLV = m * \frac{(1+d)}{1+d-r}, \qquad (12.2)$$

where:
m = margin provided by a customer;
r = retention rate; and
d = risk-free discount rate for the firm.

The formulation assumes margins and retention rates are constant over a customer's lifetime and use an infinite time horizon. Variants are obtained by relaxing these assumptions. For example, companies in some industries calculate four- to five-year customer values. Over such a short time, changes in the technological or competitive landscape do not affect CLV. In mature industries with more stable consumer preferences, less competitive intensity, and slower technological advances, it is recommended that managers use the infinite horizon. In other words, the infinite time horizon is more appropriate for companies in industries with a smaller discount rate; a shorter time horizon is more appropriate for calculating CLV when the discount rate is higher (Farris et al., 2010).

Some researchers have recommended calculating CLV as the value of future cash flows, assuming no changes in the current macro-environment and including a prescient value (*CLV-p*) to accommodate major changes in the external environment (Berger et al., 2006). Equation (12.2) does not include a customer's acquisition cost. In this form, CLV provides the upper limit of that cost. Other formulations have proposed subtracting acquisition cost from equation (12.2) to calculate CLV (e.g., Gupta and Lehmann, 2008).

A comprehensive CLV formulation may be represented as (Reinartz and Venkatesan, 2008):

$$CLV_i = \sum_{t=1}^{T} \frac{\sum_{p=1}^{P} M_{tp} * Q_{itp} - \sum_{j=1}^{J} CS_{itj}}{(1+r)^t} - \sum_{i=1}^{n} AC_i, \qquad (12.3)$$

where:
M_{tp} = gross margin (i.e., net of cost of goods sold) for product p in period t;
Q_{itp} = quantity purchased of product p by customer i in period t;
CS_{itj} = cost to serve customer i in period t through channel j;
AC_i = acquisition cost for customer i;
r = discount rate; and
n = number of customers of the firm in time $t = 0$.

Accommodating Uncertainty

Venkatesan et al. (2007) and Gönül and Ter Hofstede (2006) have proposed using Bayesian decision theoretic frameworks to optimize customer-focused resource allocation. Bayesian theory is ideally suited to problems in which a decision must be made with substantial parameter or modeling uncertainty.

Bayesian decision theory postulates that a firm can take certain actions (e.g., set marketing decision variables, select customers) in various uncertain states (e.g., quantity and timing of purchases by customers), which lead to consequences (e.g., profits). Bayes' theorem combines prior distributions for uncertain states with data to obtain posterior distributions that reduce uncertainty. The optimal action maximizes expected profit. Expected profits are computed with respect to the predictive distribution of future states (e.g., purchase quantity and timing). In the context of customer-focused resource allocation, the manager must determine the level of resources to allocate in each communication channel to maximize future profits using predictive distribution about future customer behavior, such as purchase timing and quantity.

Venkatesan et al. (2007) use Bayesian decision theory to address the uncertainty in customer response to marketing actions. Gönül and Ter Hofstede (2006) address the uncertainty of managers' risk preference. Further research is needed to develop Bayesian decision theoretic frameworks to accommodate uncertainty in customer-level resource allocation decisions.

PREDICTING STOCHASTIC COMPONENTS OF CLV

Because it is a forward-looking metric, CLV is influenced by multiple unknown components. These components must be predicted in order to use CLV to optimize customer resource allocation.

Acquisition Cost

Strategic acquisition emphasizes finding the right customers rather than as many customers as possible (Reinartz, Thomas, and Kumar, 2005). To that end, CRM acquisition campaigns attempt to acquire prospects that resemble current best customers. The two major objectives of such models are (1) to review historical acquisition campaigns to understand effective strategies and predict the acquisition probability and cost for a customer, and (2) identify the right customers from a prospect pool.

Logit or probit models are typically used to predict the acquisition probability of individual customers (Thomas, 2001). Vector autoregressive models are appropriate for identifying interactions among different marketing instruments and estimating the long-term profitability of customers acquired through each channel. However, they are focused on acquisition rates of a group of customers at each period rather than individual differences in acquisition probability and costs. Because a firm does not have access to a prospect's transaction history with the competition and a prospect does not have any interaction with the firm, current customers and prospects are typically matched according to profile variables provided by secondary data services, such as Dun & Bradstreet. Prospect scoring (Bodapati and Gupta, 2004) and clone marketing (Steenburgh, Ainslie, and Engebretson, 2003) have been proposed in this context.

Retention Rate

Two questions are critical when determining the appropriate model for predicting retention rate. First, is the customer–firm relationship contractual or non-contractual? Second, when customers end the relationship, are they lost for good, or is there always a share that can be obtained from them? Contracts dominate business-to-business transactions, but non-contractual situations also exist, such as the relationship between a firm and an office supplies vendor. In contractual settings in which customer defection is observed, a binary logit or probit model can be used as a function of covariates to predict retention.

Neslin and colleagues (2006) review several models submitted by academics and practitioners as part of a 'churn tournament'. The researchers find that logit and tree analyses outperform discriminant, explanatory, and novice approaches. Moreover, the more than 40 entries in the tournament yielded significant differences in terms of incremental profits, indicating the importance of model quality and predictive accuracy.

Three of the key models for non-contractual situations are negative binomial distribution (NBD)/Pareto (Schmittlein, Morrison, and

Colombo, 1987), hazard/purchase timing (Allenby, Leone, and Jen, 1999), and Markov switching (Pfeifer and Carraway, 2000). The NBD/Pareto model assumes customers buy at a steady, stochastic rate and then become inactive. Specifically, it uses the Pareto (exponential-gamma mixture) timing model and repeat buying behavior to represent time before dropout and the NBD (Poisson-gamma mixture) counting model to represent activity. Using simple and widely available data points (e.g., time since first purchase, time from first purchase until most recent purchase, total number of transactions), the NBD/Pareto model predicts the likelihood a customer is still active or retained at some period t. Applications of the model include Reinartz and Kumar (2002, 2003) and Fader et al. (2005).

Hazard models assume that changes (increases) in customers' inter-purchase times signal a threat of ending the relationship and therefore require intervention. Inter-purchase times are also assumed to follow a generalized gamma distribution, and the models must allow for both cross-sectional and temporal heterogeneity. In the context of investment decisions, Allenby et al. (1999) find three segments that vary according to activity status: super-active accounts, active accounts, and inactive accounts. In practice, firms can use these models to identify customers likely to become less active by recalculating their probability estimates and monitoring them periodically. Recent applications include Lewis (2006), Venkatesan and Kumar (2004), and Venkatesan et al. (2007).

The final alternative for modeling retention rates in non-contractual situations uses Markov switching matrices (Pfeifer and Carraway, 2000; Rust et al., 2004). Markov models assume the customer–firm relationship progresses through different states and determine the probabilities of a customer in a certain state moving to another state. While early Markov models estimated transition using observed states more applicable in contractual situations, recent frameworks examine customer transitions through unobserved states, making them applicable to non-contractual settings (Netzer, Lattin, and Srinivasan, 2008).

When weighing the effects of a customer ending a relationship, Rust et al. (2004) argue the lost-for-good approach understates CLV because it does not allow a defected customer to return. Other researchers argue that this is not a significant concern because customers can be treated as a renewable resource (Drèze and Bonfrer, 2009). The NBD/Pareto and hazard model frameworks are applicable to either the lost-for-good or always-a-share assumption. Markov models are more appropriate for the always-a-share assumption. In some cases, the choice between the assumptions is context specific. Gupta and Lehmann (2008) propose that business-to-business relationships call for the always-a-share approach because business customers tend to have relationships with multiple service providers.

Margins

Researchers have proposed using average past margins, regression models, and probability models to predict future customer-provided margins. With some exceptions (Niraj, Gupta, and Narasimhan, 2001; Venkatesan et al., 2007), most models focus on predicting future revenue and apply a constant gross margin and retention cost. However, retention costs can vary substantially among customers, and it can be challenging to assemble information on customer interactions across multiple divisions of a firm. Research on calculating and predicting customer margins is scarce partly because of the challenges in accurately estimating the costs of serving a customer.

MARKETING ACTIONS

Venkatesan et al. (2007) show that customer selection and resource allocation improves if lifetime value models also build a model to predict marketing costs rather than use average marketing costs in the calibration sample as an estimate of future marketing costs. Manchanda, Rossi, and Chintagunta (2004) indicate that firms do not assign marketing calls to customers in a random manner. Typically, firms would at least assign more marketing calls to customers who they expect to (1) spend more and (2) are more responsive to marketing. They show that not acknowledging the non-random nature of marketing calls can result in biased estimates of the customer-specific intercept and coefficient of marketing calls. The bias arises because in a typical customer-level sales model customer-specific random effects (the intercept) and the responsiveness to marketing mix (the slope coefficient) are correlated with the error term. This intercept and slope endogeneity can be accommodated by jointly estimating the sales model and the marketing calls model. The random effects intercept and slope coefficients are also used as predictors in the marketing calls model because these parameters capture managers' expectations of the customer that are unobserved by the researcher.

MAXIMIZING CUSTOMER LIFETIME VALUE

In addition to formulating CLV and breaking it down into components, managers must take practical steps to maximize the value each customer provides. Managers using CLV-based resource allocation face two critical questions for maximizing their returns: (1) Which customers should be selected for marketing intervention? (2) How should the firm interact with the selected customers?

Several different approaches for managing profits from individual customers have been explored (for a representation of customer profits, see equation (12.2)). Research thus far has focused on growing or maintaining customer profits by (1) managing the marketing costs allocated to the customers, (2) increasing the number of products purchased by customers (i.e., cross-selling), and (3) optimizing product returns.

Managing Marketing Costs

If we assume the revenue provided by a customer does not increase over time, the customer contribution margin (revenue minus cost of goods sold and cost to serve) can be improved by managing the costs allocated to the customer. Various marketing costs are customizable, including price promotions, direct mail (e.g., catalogs), email, and salesperson and telesales contacts. All models rely on estimating future customer revenue and maintaining the level of marketing resources allocated to the customer below the estimated gross margin.

The cost allocation management strategy originated in the catalog industry, where marketers must determine the number of catalogs a customer must receive before making a purchase. Catalog mailing decisions are based primarily on the customer's propensity to make a purchase and the expected revenue from the customer given a purchase. The proliferation of CRM systems has led to the adoption of this approach in industries where customer interaction is not restricted to direct mail.

Khan, Lewis, and Singh (2009) investigate the financial benefits of one-to-one marketing and computational challenges involved using data from an online grocery and drug retailer. Like many retailers, the firm uses multiple promotional instruments, including discount coupons, free shipping offers, and a loyalty program. Through a series of policy simulations, the researchers show that customizing promotions leads to a significant increase in profits relative to utilizing uniform promotions. The effectiveness of promotions varies because of cross-sectional differences in consumers and within-consumer heterogeneity due to purchase cycle factors. For instance, the researchers find that free shipping tends to be the preferred instrument for reacquiring lapsed customers, whereas an across-the-board price cut via a discount coupon is the most effective tool for managing the most active customers. The study concludes that the use of customized promotions maximizes increases in revenue and profitability when both individual preferences and purchase cycle information are used for guiding marketing tactics.

Venkatesan and Kumar (2004) apply an econometric approach to link a high-technology business-to-business firm's marketing expenditures to CLV (equation (12.1)) and evaluate the level of marketing resources for each customer that would maximize CLV resource allocation. The researchers specify the marketing decision variables (contacts through various channels) and covariates to influence customer purchase frequency and contribution margin. The response parameters in the model are specified for each individual customer, which enables customized marketing resource allocation. After estimating the models and calculating CLV for each customer, the researchers use a genetic algorithm to determine the optimal level of marketing contacts to maximize CLV.

Venkatesan and Kumar (2004) suggest that improvements in CLV can be made by considering each customer's responsiveness to marketing communication, along with the costs of each communication channel, when making resource allocation decisions. Moreover, the researchers find that the influence of marketing communication on customer interpurchase time is non-linear and that too much communication can have dysfunctional consequences, in some cases leading to a reduction in CLV. Customer profits increase when marketers incorporate the differences in individual responsiveness to various communication channels and determine the optimal level of marketing communications to maximize CLV. The resource allocation strategy thus can act as a basis for evaluating the potential benefits of CRM implementation in organizations and provide accountability for strategies geared toward managing customer assets.

The econometric approach for resource allocation has also been adopted to determine the balance between acquisition and retention investments that maximizes customer profitability (Reinartz et al., 2005). The level of marketing contacts undertaken before a customer is acquired can predict the acquisition model; the level of contacts made during the relationship serves the retention and profitability models.

The level of investment that maximizes the acquisition rate for a firm differs from the investment level required to maximize retention rates and customer profitability (ibid.). Underspending is more detrimental and results in smaller returns on investment than overspending, and a sub-optimal allocation of retention expenditures has a greater impact on long-term customer profitability than sub-optimal acquisition expenditures.

Both Venkatesan and Kumar (2004) and Reinartz et al. (2005) find highly interpersonal salesperson contacts are more effective than less interpersonal or standardized modes of communication.

Elsner, Krafft, and Huchzermeier (2004) apply the econometric approach to resource allocation in the direct-mail catalog industry and a dynamic multilevel modeling (DMLM) approach to determine the optimal (1) segmentation of customers on the basis of customer–firm interaction covariates, (2) number of customer segments to receive each direct mail campaign (i.e., size of a campaign), and (3) number of direct marketing campaigns to maximize customer response and order size. The DMLM approach consists of three levels. First, the order size and response elasticity for an additional promotion is calculated through a regression analysis. Second, the size of each segment (i.e., the number of catalogs sent for each campaign) and minimum order size necessary for a profitable campaign is calculated through analytical derivations of the objective function. Third, the covariates related to the customer–firm interactions available in the firm's database are used to segment the customers and determine which segments should receive each campaign. The DMLM approach is repeated over a rolling window of one year after responses to the current campaign have been integrated into the customer database.

Elsner et al. (2004) also provide empirical evidence of the benefits of this approach to resource allocation. Prior to intervention, the study company Rhenania only sent catalogs to customers whose expected revenue exceeded the total cost of merchandising, order fulfillment, and mailing. Implementing the DMLM approach resulted in a 43 percent improvement in sales, despite 15 percent market growth during the same period. As a consequence, Rhenania was able to acquire two major competitors. The company then extended its DMLM approach to a multivariate setting, such that a single firm handled three separate catalogs selling different brands. The updated approach explicitly accommodated global customer response to the number of catalogs obtained from Rhenania and the number of catalogs from each brand.

Rust and Verhoef (2005) show that response to marketing interventions is highly heterogeneous, response heterogeneity varies among marketing interventions, and response heterogeneity to marketing interventions may be partially explained by customer-specific variables related to customer characteristics and past interactions with a company. Because of these moderating effects, relationship-oriented interventions are more effective with loyal customers, while action-oriented interventions are more effective with disloyal customers. Their results indicate that accounting for heterogeneity in individual customer response to marketing mix would yield a significant increase in intermediate-term profitability compared to assuming heterogeneity only across customer segments.

Venkatesan et al. (2007) show that using a Bayesian-decision-theory–based customer selection framework leads to better identification of profitable customers than single-period expected profits in high-tech business-to-business industries. The researchers also find managers should focus on the expected costs of serving a customer during the customer selection process in addition to quantity. Further, the optimal costs derived from customers' historic responsiveness to marketing communication provide a good estimate of the expected costs of serving a customer and aid in identifying profitable customers.

Using a Bayesian decision theoretic framework accommodating for managers' risk preferences, Gönül and Ter Hofstede (2006) show that expected profits and utility are higher when a longer time horizon is employed than a single-period horizon. Within a catalog mailing context, the study's findings imply that current mailing decisions affect future decisions and that mailing strategies should update periodically and optimize over a longer time horizon. Catalog companies focused on household items typically send seasonal catalogs and an additional edition during the holidays. However, Gönül and Ter Hofstede (2006) show the traditional recency-frequency-monetary value (RFM) procedure will lead catalogers to send more catalogs than optimal.

Venkatesan, Reinartz, and Ravishanker (2012) evaluate the value of including customer mindset metrics in customer-level resource allocation. Using data collected from the pharmaceutical industry, the researchers show that companies can explain and predict customer responses (e.g., sales, retention, CLV) better if they include information about attitudes in their behavioral CLV models. Better attitudes positively affect CLV predominantly through retention. This result has important implications for the metrics and performance indicators that firms use to diagnose their customer base. For example, firms should consider retention rates rather than sales for any campaign focused on improving a customer's emotional attachment.

The resource allocation exercise examining mindset also suggests that firms would be better able to identify candidates for sales increase or decrease when they include attitudes in their CLV models. CLV models that do not include attitudes recommend increasing sales calls primarily to the top quintile and decreasing sales calls to all others. On the other hand, models including attitudes recommend sales call increases and decreases to individual customers across all quintiles.

By including attitudes, firms will have the potential to improve profits by investing in mid- and lower-tier customers in addition to the top tier.

In existing CRM literature, the mid- and lower-tier segments are the least discussed customer group. Attitudes act as a diagnostic for these customers in CLV models and can yield benefits because their mindset provides a forward-looking measure that identifies mid- and lower-tier customers with the potential to grow.

Venkatesan et al.'s (2012) results also imply that firms may be overspending on some top-tier customers. When customer attitude information is available, firms can improve the return on investment (ROI) of their CRM campaigns by balancing resources across customers in all profit tiers. This finding has not previously emerged in CRM literature and differs from an existing perspective that states that resources should be devoted primarily to top-tier customers (Rust et al., 2004).

Finally, the positive returns of attitudinal information in CLV modeling must be considered in light of the cost of collecting additional data. If an optimal sales call strategy that considers customer attitudes provides an incremental lift of 11 percent compared to an optimal strategy not considering attitudes, it translates to about \$520 in incremental annual profits per customer given an average CLV of \$14207. The incremental return estimate from the holdout sample can then be projected for the entire customer base (more than 150000) to obtain the financial ROI for the data collection. If a \$1 million investment is required to collect customer attitudes, the annual return is 51 times the investment. In this case, the benefits of using customer attitudes justify the investment.

Cross-selling

One method of growing customer profits is increasing the number of products a customer purchases, or cross-selling. Where cross-selling is a firm-side action employed to broaden customer relationships, its counterpart on the demand side, cross-buying, refers to a customer's propensity to make cross-category purchases. Cross-selling yields both immediate profit and the potential to deepen an existing relationship, which increases the switching costs associated with purchasing from another vendor.

Kumar, George, and Pancras (2008) and Verhoef, Franses, and Hoekstra (2001) show that marketing efforts can have a significant effect on the cross-buying behavior of customers. Research shows cross-buying increases initially in the catalog sales industry with an increase in direct mailings. However, direct mailings have a negative effect on cross-buying beyond a certain threshold. This inverted U-shaped relationship between the number of catalogs sent per quarter and cross-buying indicates that firms must optimize their mailing efforts to maximize cross-buying.

Two observations motivate managers to predict product acquisition patterns and provide the opportunity to cross-sell additional products or services. First, to implement a cross-selling strategy efficiently firms must understand the purchasing patterns of each individual customer across various product categories. Second, customers have predictable life cycles and purchase certain items before others (Li, Sun, and Wilcox, 2005). Markets in which consumers' wants or needs evolve after some preliminary consumption, in which consumers face some uncertainty about the quality of the product or service offering, or in which consumer learning is required to receive the full benefit of the product, are especially prone to this behavioral regularity.

Li et al. (2005) find that bank customers usually invest more aggressively in financial instruments that promise stable returns (e.g., CDs, money market) after they obtain basic financial services (e.g., checking, savings, debit, credit, loan) and pursue high-risk, high-return brokerage accounts last. However, household movement on the demand maturity continuum varies according to customer demographics, and the impact of customer satisfaction is greatest for advanced financial products (e.g., brokerage).

Firms naturally focus on determining the product or category with the highest purchase likelihood for each customer. This recently attracted attention in the context of CRM strategies (Verhoef et al., 2001; Knott, Hayes, and Neslin, 2002). Knott et al. (2002) attempt to uncover the benefits derived from a targeted cross-selling strategy (measured in profits) with a field experiment and propose a next-product-to-buy (NPTB) model. The researchers use a series of independent logistic regressions (one per product category *j*) to model a customer's propensity to purchase a product that he or she has not previously purchased, given current product ownership and demographic characteristics. They illustrate the effectiveness of the NPTB approach with a field test for a loan offer in a retail bank. In the experiment, customers were assigned to one of three groups: heuristic, NPTB, or prospect. Customers in the heuristic test group owned homes valued at greater than $100 000 or were unknown (i.e., a simple customer-demographic–based rule of thumb). Those in the NPTB group were expected to purchase the loan product. The prospect group consisted of potential customers obtained from a list source. Customers in each group were randomly assigned to a test or control sample. Those in the test sample received a promotional mailing for the loan offer. The experiment results showed that customers in the NPTB test group provided more profits than customers in the heuristic test group, and profits were incrementally better than sales that would have occurred through other channels (i.e., the test group customers provided more revenue than the control group). In addition, the heuristic and NPTB test group customers provided more revenues than the prospect group, which indicates that targeting retained customers is more profitable than targeting prospects.

Cross-selling can have detrimental impacts on a customer–firm relationship. First, frequent and incorrectly targeted sales efforts are likely to increase customer resentment, which results in relationship termination by the customer in the worst case. Second, unsuccessful attempts to increase a customer's product range indicate resource misallocation. Shah et al. (2012) caution that on average 10–35 percent of a firm's customers who cross-buy are unprofitable and account for a significant proportion (39–88 percent) of the firms' total loss from customers. The authors caution against using the cross-buy ratio (i.e., the number of different products or product categories sold to a customer) to evaluate the success of cross-selling campaigns, as it can contribute to indiscriminate cross-selling. Given that not all cross-buying leads to profitable outcomes, a better measure of cross-buy success would be a metric that measures the growth in customer profitability per additional product sold.

It is critical to know not only what customers are most likely to buy next but also when they will buy the product of highest affinity (Kumar, Venkatesan, and Reinartz, 2008). Knott et al. (2002) suggest a strong potential for improvement in models integrating a customer's purchase time and category choice decisions.

Product Returns

Petersen and Kumar (2009) show that product returns from a customer are inevitable but not always unprofitable. The researchers show that a customer's product return behavior positively affects his or her future buying behavior, up to a threshold. Their analysis of product returns and retailer profits showed that the optimal percentage of product returns is 13 percent. In their test case, the firm's existing level of product returns was only 3 percent from optimal, and the optimal amount of product returns was not close to 0 percent. Decreases in product returns beyond 13 percent decreased profits slowly. At 1 percent, or 15 percent below the test firm's existing level of returns, profit was \$63 937. Increases in product returns significantly decreased profits. At 31 percent, or 15 percent above the test firm's existing level of returns, profit was –\$20 608.

VALIDATING RESOURCE ALLOCATION RECOMMENDATIONS

The future projections and marketing mix recommendations of resource allocation models may turn out to be inaccurate. Firms must evaluate the accuracy of resource allocation recommendations because they suggest changes to managerial actions without considering customer response information. Experiments and holdout sample comparisons provide mechanisms for evaluating their accuracy.

In experiments, managers can implement resource allocation recommendations in a selected sample within a test group and compare its results with a control group that continues with the status quo strategy. Results from the experiment can provide impetus for broader implementation of the recommendations.

Kumar, Venkatesan, and Reinartz (2008) evaluate the benefits of targeted cross-selling offers to customers through a field experiment in a business-to-business firm. Customers assigned to a test group were contacted using the strategy recommended by a timing and category choice model. Those in the control group were contacted using the supplier's existing strategy, in which dedicated salespeople proactively pitched only products in their own product categories. For each product category, the company typically used a binomial logit model to predict a customer's propensity for action in the next year. The sales force for each product category used these propensity scores to prioritize customers for sales calls in the next year. The timing of calls within the year typically occurred at the discretion of the sales force, and, in general, customers with higher propensity scores were contacted before customers with lower scores.

Across all product categories in the year the experiment was conducted, a significant increase in customer profitability and ROI was observed for the test group compared with the pre-experimental period. No significant difference in profits was found among the control group. Customers in the test group provided significantly higher profits and ROI than did customers in the control group.

Zhang and Krishnamurthi (2004), Khan et al. (2009), and Venkatesan et al. (2012) use holdout sample comparisons to evaluate the accuracy of resource allocation recommendations. To implement the approach properly, researchers must identify customers in a sample for whom the marketing actions taken by management match the optimal marketing mix recommendations of the resource allocation exercise. The difference in realized profits between the customers whose marketing mix matches these recommendations and all others provides an estimate of the accuracy of the recommendations. To eliminate the possibility of an a priori guarantee the matched cases would render higher profits, the holdout data cannot be used for model estimation comparison.

In the context of customizing an online retailer's promotions to individual customers, Zhang and Krishnamurthi (2004) use a holdout sample comparison to show that the average actual profit was positive in the matched cases and negative in the non-matched cases. The proposed customization method was found to improve the effectiveness of the retailer's existing practice by focusing promotion spending on purchasing opportunities that would yield profit. The improvement was especially pronounced for consumers who purchased the target brand on their previous purchase occasion.

Khan et al. (2009) compare the recommendations of a customer-level resource allocation model with a holdout sample wherein an online retailer provided a discount coupon to all customers. The researchers show that holdout customers, for whom the discount coupon was the recommended policy, yielded the highest per-customer profit of $3.90. Per-customer profits were lowest ($1.68) when the recommended policy was 'no promotion' but the customers were offered the discount coupon. Offering a discount coupon when free shipping was the optimal policy also yielded a reduced average profit ($2.25). The policy recommendations were validated because per-customer profits were highest for the group whose policy recommendation matched actual practice.

Venkatesan et al. (2012) classify customers in a holdout sample into matched, missed, and wasted resources categories, according to the difference between the modeled optimal monthly sales calls (including or excluding attitudes) and the observed average monthly sales calls allocated during the first 12 months (16–28) of the prediction window. Customers who fell within the 40th to 60th percentiles were categorized as matched, as their optimal and observed sales calls were similar (a difference close to 0). The remaining customers were classified as missed if more sales calls would have been optimal (a difference above the 60th percentile) or wasted if the difference fell below the 40th percentile.

In the model including attitudes, the average profit for the 451 customers classified as matching was higher ($7866) than for the missed or wasted opportunities groups ($4870). A t-test indicated that the profit of the matched group was significantly higher than the ROI of the missed groups for the model, including attitudes. The average profit for the matched group ($6378) in the model excluding attitudes was significantly higher than for the missed or wasted opportunities group ($5545). The profit for the matched group in the model including attitudes ($7866) was significantly higher than the profit for the matched group in the model excluding attitudes ($6378). The profit from the matched group of the model including attitudes was significantly higher than from the missed or wasted opportunities groups for customers with observed attitudes and imputed attitudes. However, the difference in profit between the matched and missed or wasted opportunities groups was higher for customers with observed attitudes. For the sub-group of customers with observed attitudes, the profit from the matched group of the model excluding attitudes was lower (–$444) than from the customers in the missed or wasted opportunities group.

In this case, the holdout sample data validated the resource allocation recommendations because the highest profits among all groups came from customers for whom the actual firm practice matched the recommendations of the model including attitudes. Thirty-five percent of customers in the matched group of the model including attitudes also belonged to the matched group of the model excluding attitudes. This implied that the two models provided substantially different targeting recommendations. Further, the validation exercise showed that using imputed values whenever attitude information is not available is better than using only behavioral data.

IMPLEMENTING CRM AND USING CLV

Building and managing strong customer relationships requires a cross-functional integration of processes, people, operations, and marketing capabilities not strictly within the marketing department, but in all areas of a business (Reinartz, Krafft, and Hoyer, 2004; Becker, Greve, and Albers, 2009). Although systematically interfacing with a large number of customers requires a sound technological basis (e.g., software), technology alone does not suffice. Best-in-class organizations exhibit senior management commitment, synergy between strategic and analytic capabilities, and alignment of incentives with customer profits.

Using CLV to optimize customer-focused resource allocation requires a strategic shift from product-centric to customer-centric marketing. A product-centric approach is designed to sell products to whoever is willing to buy them; a customer-centric approach is designed to serve a portfolio of customers and build relationships. Customer-centricity is achieved by viewing customers as both a source of business and a potential business resource. This is achieved by (1) making more customer-level decisions than aggregate decisions, (2) responding to customer needs effectively and efficiently, (3) providing customers with a rich and varied experience, (4) encouraging customer-to-customer interactions, and (5) encouraging customer feedback (Ramani and Kumar, 2008). Administering changes to workforce elements is necessary for complete implementation. Company-wide human resource changes include (1) communicating the change through employee-targeted messages, (2) increasing employee interest and participation to support the change, (3) establishing transparency in sharing information and insights about the change process, (4) enabling stakeholders to implement the change on a daily basis, and (5) incorporating monitoring and evaluation procedures to keep the change in place.

A four-step process can be used to help managers communicate customer-centric concepts and strategies and garner the support of their organization's decision-makers. The first step involves presenting the concepts and strategies to decision-makers. Concepts such as using CLV to predict the profit brought in by a customer must be presented in a way that helps management obtain a deeper understanding of customer loyalty and adopt the proposed concepts.

The second step is to present best practices of companies that have successfully implemented the proposed concepts. The third step involves analyzing a sample dataset and comparing it with existing practices in order to empirically demonstrate the benefits of adopting the proposed concepts. For example, when the performance of CLV was compared with that of recency, frequency, and monetary value and past customer value for a high-tech business-to-business manufacturer, the net value generated by the customers was 45 percent greater (Venkatesan and Kumar, 2004).

The fourth step involves conducting a pilot study on a segment of the firm's customers to demonstrate the concepts and strategies. Alternatively, the pilot study can be done on one business unit of the firm. For example, when IBM wanted to (1) know which customers to select for targeting, (2) determine the level of resources to be allocated to the customers, and (3) select the right customers to be nurtured, it implemented a pilot study for 35000 of its mid-market customers. The study led to the reallocation of resources for approximately 14 percent of the customers and a tenfold increase in revenue of $20 million without any changes in marketing investment (Kumar, Vekatesan, Bohling et al., 2008).

THE POWER OF FEEDBACK LOOPS

Allocating resources based on CLV is one way that businesses can increase profits from the right types of customers. By (1) determining how best to formulate CLV for their business, (2) breaking the formulation down into its component parts, (3) finding the unknowns that accompany the forward-looking metric, (4) linking the formulation to relevant marketing strategies, and (5) varying marketing actions until the optimal mix is found, firms can directly tie the way they deploy their resources to the value a customer will bring them. However, managers must continuously capture feedback from their resource allocation decisions to maximize the results of the process.

In a 2011 *Wired* magazine article, Thomas Goetz explores the idea of 'feedback loops' in the context of dynamic speed displays (signs posted on roadsides that give drivers real-time information about how fast they are driving). According to Goetz, the signs have been effective at reducing speeding in a number of settings because they give people an opportunity to change their actions and push them toward better behaviors – 'action, information, reaction'.

CLV-based resource allocation can be used by business managers in much the same way. First, the results of resource allocation decisions must be measured and recorded. Second, managers must view the dataset in an understandable context, such as how it affects CLV. Third, they must understand the consequences of their decisions and recognize they can proceed in a manner that improves CLV. Finally, they can reallocate marketing resources to improve the value they derive from their customer mix. Once these actions have been taken, they must again be measured so the feedback loop can continue until the firm reaches its goals.

CLV gives companies a powerful analytical tool. But success will only be obtained if employees across all business functions adopt the process of allocating resources based on CLV and examine its results to continually improve.

REFERENCES

Allenby, G.M., R.P. Leone and L. Jen (1999), 'A dynamic model of purchase timing with application to direct marketing', *Journal of American Statistical Association*, **94**(446), 365–74.

Becker, J.U., G. Greve and S. Albers (2009), 'The impact of technological and organizational implementation of CRM on customer acquisition, maintenance, and retention', *International Journal of Research in Marketing*, **26**(3), 207–15.

Berger, P.D and N. Nasr-Bechwati (2001), 'The allocation of promotion budget to maximize customer equity', *The International Journal of Management Science*, **29**(1), 49–61.

Berger, P.D., N. Eechambadi, M. George, D.R. Lehmann, R. Rizley and R. Venkatesan (2006), 'From customer lifetime value to shareholder value: theory, empirical evidence, and issues for future research', *Journal of Service Research*, **9**(2), 156–67.

Blattberg, R. and J. Deighton (1996), 'Manage marketing by the customer equity test', *Harvard Business Review*, **74**(4), 136–44.

Bodapati, A. and S. Gupta (2004), 'A direct approach to predicting discretized response in target marketing', *Journal of Marketing Research*, **41**(1), 73–85.

Drèze, X. and A. Bonfrer (2009), 'Moving from customer lifetime value to customer equity', *Quantitative Marketing and Economics*, **7**(3), 289–320.

Elsner, R., M. Krafft and A. Huchzermeier (2004), 'Optimizing Rhenania's direct marketing business through dynamic multilevel modeling (DMLM) in a multicatalog-brand environment', *Marketing Science*, **23**(2) 192–206.
Fader, P.S., B.G.S. Hardie and K.L. Lee (2005), 'Counting your customers the easy way: an alternative to the Pareto/NBD model', *Marketing Science*, **24**(2), 275–84.
Farris, P.W., N.T. Bendle, P.E. Pfeifer and D.J. Reibstein (2010), *Marketing Metrics: The Definitive Guide to Measuring Marketing Performance*, 2nd edition, Philadelphia, PA: Wharton School Publishing.
Goetz, T. (2011), 'Harnessing the power of feedback loops', *Wired Magazine*, 19 June.
Gönül, F. and F. Ter Hofstede (2006), 'How to compute optimal catalog mailing decisions', *Marketing Science*, **25**(1), 65–74.
Gupta, S. and D. Lehmann (2008), 'Models of customer value', in B. Wierenga (ed), *Handbook of Marketing Decision Models*, New York: Springer, pp. 255–90.
Gupta, S., D. Lehmann and J.A. Stuart (2004), 'Valuing customers', *Journal of Marketing Research*, **41**(1), 7–18.
Khan, R., M. Lewis and V. Singh (2009), 'Dynamic customer management and the value of one-to-one marketing', *Marketing Science*, **28**(6), 1063–79.
Knott, A., A.F. Hayes and S. Neslin (2002), 'Next-product-to-buy models for cross-selling applications', *Journal of Interactive Marketing*, **16**(3), 59–75.
Kumar, V. and D. Shah (2009), 'Expanding the role of marketing: from customer equity to market capitalization', *Journal of Marketing*, **73**(6), 119–36.
Kumar, V., M. George and J. Pancras (2008), 'Cross-buying in retailing: drivers and consequences', *Journal of Retailing*, **84**(1), 15–27.
Kumar, V., R. Venkatesan and W. Reinartz (2008), 'Performance implications of adopting a customer focused sales campaign', *Journal of Marketing*, **72**(5), 50–68.
Kumar, V., R. Venkatesan, T. Bohling and D. Beckmann (2008), 'The power of CLV: managing cusomter lifetime value at IBM', *Marketing Science*, **27**(4), 585–99.
Lewis, M. (2006), 'Customer acquisition programs and customer asset value', *Journal of Marketing Research*, **43**(2), 195–203.
Li, S., B. Sun and R.T. Wilcox (2005), 'Cross-selling sequentially ordered products: an application to consumer banking', *Journal of Marketing Research*, **42**(2), 233–9.
Manchanda, P., P.E. Rossi and P. Chintagunta (2004), 'Response modeling with nonrandom marketing-mix variables', *Journal of Marketing Research*, **41**(4), 467–78.
Neslin, S.A., S. Gupta, W. Kamakura, J. Lu and C.H. Mason (2006), 'Defection detection: measuring and understanding the predictive accuracy of customer churn models', *Journal of Marketing Research*, **43**(2), 204–11.
Netzer, O., J. Lattin and S. Srinivasan (2008), 'A hidden Markov model of relationship dynamics', *Marketing Science*, **27**(2), 185–204.
Niraj, R., M. Gupta and C. Narasimhan (2001), 'Customer profitability in a supply chain', *Journal of Marketing*, **65**(3), 1–16.
Petersen, J.A. and V. Kumar (2009), 'Are product returns a necessary evil? Antecedents and consequences', *Journal of Marketing*, **73**(3), 35–51.
Pfeifer, P.E. and R.L. Carraway (2000), 'Modelling customer relationships as Markov chains', *Journal of Interactive Marketing*, **14**(2), 43–55.
Ramani, G. and V. Kumar (2008), 'Interaction orientation and firm performance', *Journal of Marketing*, **72**(1), 27–45.
Reichheld, F. (1996), *The Loyalty Effect*, Boston, MA: Harvard Business Press.
Reinartz, W. and V. Kumar (2002), 'The mismanagement of customer loyalty', *Harvard Business Review*, **80**(7), 86–94.
Reinartz, W. and V. Kumar (2003), 'The impact of customer relationship characteristics on profitable lifetime duration', *Journal of Marketing*, **67**(1), 77–99.
Reinartz, W. and R. Venkatesan (2008), 'Decision models for customer relationship management (CRM)', in B. Wierenga (ed.), *Handbook of Marketing Decision Models*, New York: Springer, pp. 291–326.
Reinartz, W., M. Krafft and W.D. Hoyer (2004), 'The customer relationship management

process: its measurement and impact on performance', *Journal of Marketing Research*, **41**(3), 293–305.
Reinartz, W., J.S. Thomas and V. Kumar (2005), 'Balancing acquisition and retention resources to maximize customer profitability', *Journal of Marketing*, **69**(1), 63–79.
Rust, R.T. and P.C. Verhoef (2005), 'Optimizing the marketing interventions mix in intermediate-term CRM', *Marketing Science*, **24**(3), 477–89.
Rust, R.T., K.T. Lemon and V. Zeithaml (2004), 'Return on marketing: using customer equity to focus marketing strategy', *Journal of Marketing*, **68**(1), 109–27.
Schmittlein, D., D.G. Morrison and R. Colombo (1987), 'Counting your customers: who are they and what will they do next?', *Management Science*, **33**(1), 1–24.
Shah, D., V. Kumar, Y. Qu and S. Chen (2012), 'Unprofitable cross-buying: evidence from consumer and business markets', *Journal of Marketing*, **76**(3), 78–95.
Steenburgh, T., A. Ainslie and P.H. Engebretson (2003), 'Revealing the information in zip-codes: Bayesian massively categorical variables and aggregated data in direct marketing', *Marketing Science*, **22**(1), 40–57.
Thomas, J.S. (2001), 'A methodology for linking customer acquisition to customer retention', *Journal of Marketing Research*, **38**(2), 262–8.
Venkatesan, R. and V. Kumar (2004), 'A customer lifetime value framework for customer selection and resource allocation strategy', *Journal of Marketing*, **68**(4), 106–25.
Venkatesan, R., V. Kumar and T. Bohling (2007), 'Optimal customer relationship management using Bayesian decision theory: an application to customer selection', *Journal of Marketing Research*, **44**(4), 579–94.
Venkatesan, R., W. Reinartz and N. Ravishanker (2012), 'Role of attitudinal information in CLV-based customer management', Marketing Science Institute report, accessed 15 September 2014 at http://www.msi.org/reports/the-role-of-attitudinal-information-in-clv-based-customer-management/.
Verhoef, P.C., P.H. Franses and J.C. Hoekstra (2001), 'The impact of satisfaction and payment equity on cross-buying: a dynamic model for multi-service provider', *Journal of Retailing*, **77**(3), 359–78.
Zhang, J. and L. Krishnamurthi (2004), 'Customizing promotions in online stores', *Marketing Science*, **23**(4), 561–78.

13. Customer mindset metrics and firm performance

Shuba Srinivasan

INTRODUCTION

Increasing demands for marketing accountability have created a new sense of urgency for marketers to obtain and analyze the right customer metrics to drive performance growth and demonstrate marketing's value in a consistent manner (Ambler, 2003). Increasingly, top management requires that marketing view its ultimate purpose as contributing to building customer assets and brand assets that lead to the enhancement of shareholder returns (Srinivasan and Hanssens, 2009).

The majority of marketing researchers and practitioners remain divided by the battle line between hard numbers and softer measures such as consumer perceptions, attitudes, and intentions, collectively referred to as customer mindset metrics. For marketers striving to justify their expenditures to top management, quantitative marketing mix models provide solid measures of accountability, with measures such as 'advertising elasticity' and 'return on investment'. Not surprisingly, such models linking marketing to quarterly sales results, long-term profitability, and stock prices generally go further in convincing CFOs and CEOs of marketing's value than research into metrics of customers' minds and hearts. More than five years ago, Gupta and Zeithaml (2006, p. 2) stated that, 'while senior executives realize the importance of customers, they still rely heavily on finance measures because customer metrics are not clearly defined'. But this emphasis on financial metrics leaves the drivers of marketing success or failure unexplored. To the extent that the drivers change, past performance impact of marketing may not be the best predictor of its future impact. In recent years, there has been growing academic interest in quantifying the linkage between a firm's marketing actions, customer mindset metrics, and firm performance (Srinivasan, Vanhuele, and Pauwels, 2010; Hanssens et al., 2012; Stahl et al., 2012). In this chapter, I build on these approaches to provide a conceptual framework that investigates the effects of customer mindset metrics on firm performance and integrates emerging insights from the literature into a framework that helps enumerate and explain the customer metrics/firm financial performance link.

A number of authors have developed a marketing value chain that links marketing to firm performance (e.g., Lehmann, 2004; Lehmann and Reibstein, 2006; Srinivasan and Hanssens, 2009; Srinivasan et al., 2010; Hanssens et al., 2012). Building on these approaches, the conceptual framework in Figure 13.1 in this chapter highlights that the impact of marketing actions on firm performance can occur through changes in customer mindset metrics and customer equity as well as directly without such changes.

Figure 13.1 also provides a roadmap for this chapter. I begin by discussing firm performance measures including shareholder value (i.e., stock returns and risks) as a metric to assess the performance impact of customer assets. Next, I propose that the routes by which these effects can occur are the 'mindset route' where marketing actions influence customer mindset metrics, which in turn influence brand and firm performance, or the 'transactions route' where marketing actions influence brand and firm performance directly without influencing the intermediate mindset metrics. Turning to the contribution of customer mindset metrics to firm value creation, I also posit two routes that make explicit this link. First is a direct route wherein the 'customers-as-assets' enhance firm value directly through the effects on cash flows. Second is an indirect route that considers the 'customers-as-information'. In this route, the customer mindset metrics act as a visible signal of the financial well-being of the firm. Focusing on the linkage between customer mindset metrics and firm performance, I provide a metrics overview and offer extant findings on the customer-mindset-metrics/financial-performance interface. The chapter concludes with an agenda for future studies that addresses gaps and challenges in this important and growing area of research.

CUSTOMER MINDSET METRICS AND FIRM PERFORMANCE

In this section I turn to the question of how customer mindset metrics affect firm performance. Figure 13.1 summarizes how firms' marketing actions influence firm performance through a series of intermediate steps. Firm performance includes the product market performance metrics of sales, top-line revenues, and bottom-line profits. The ultimate performance metric is shareholder value, which is determined by levels of stock returns and the volatility or risk associated with those returns (Srinivasan and Hanssens, 2009). Shareholder value not only balances the non-financial metrics but also matches the main objective of companies of maximizing shareholders' value on their equity (Rappaport, 1997). Commonly

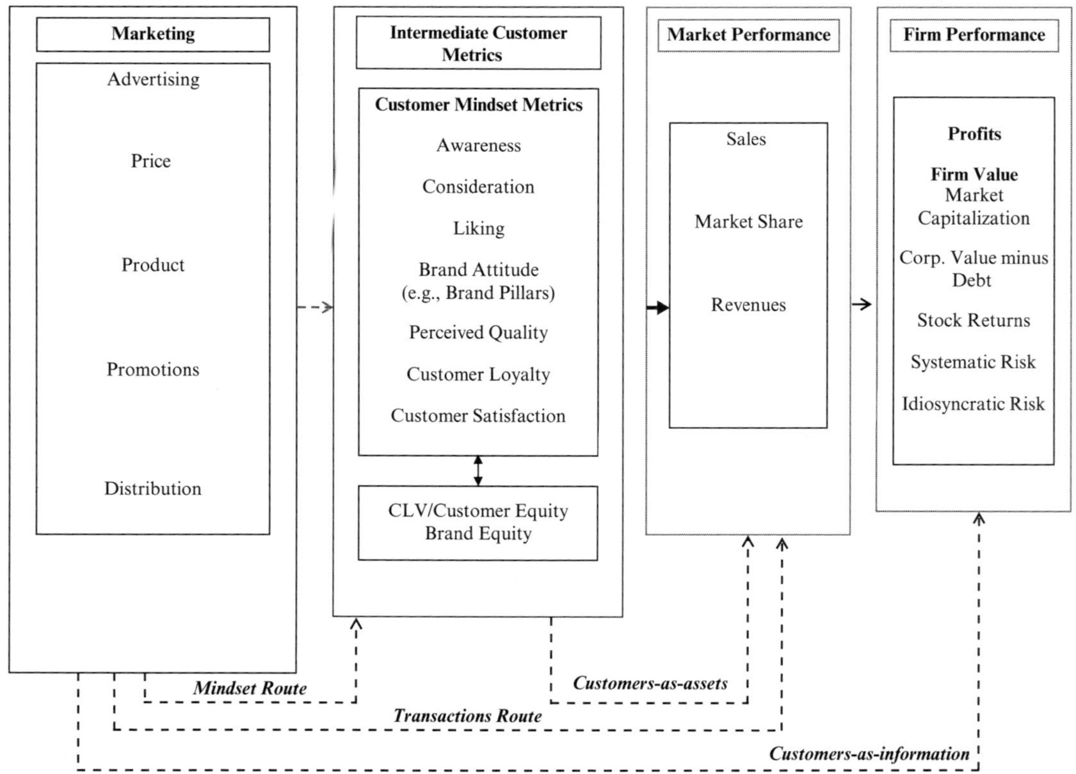

Figure 13.1 Linking marketing, customer mindset metrics, and performance

used metrics of shareholder value include stock returns, Tobin's q, equity value minus debt and market capitalization. The latter two shareholder value metrics are commonly used in finance and marketing respectively. Table 13.1 provides an overview of these common financial metrics, including their definitions and operationalization.

Shareholder Value and Firm Risk

The Carhart (1997) four-factor pricing model (FFPM) augments the Fama and French (1992) three-factor model with the fourth factor to correct for momentum, to estimate the components of firm value (see Srinivasan and Hanssens, 2009 or Srinivasan, Pauwels and Wiesel, 2011). The extended FFPM equation is estimated as follows:

$$R_{it} - R_{rf,t} = \alpha_i + \beta_i(R_{mt} - R_{rf,t}) + s_i SMB_t + h_i HML_t + u_i UMD_t + \varepsilon_{it} \tag{13.1}$$

where $\varepsilon_{it} \sim N(0, \sigma\varepsilon_{it})$, R_{it} is the stock return for firm i at time t, $R_{rf,t}$ is the risk-free interest rate of return in period t (e.g., three-month US Treasury bill), R_{mt} is the average market rate of return in period t, SMB_t is the difference between the return on a portfolio of small stocks and the return of big stocks, HML_t is the difference between the return on a portfolio of high book-to-market stocks and the return on a portfolio of low book-to-market stocks, UMD_t is the difference between the average return on two high prior-return portfolios and the average return on two low prior-return portfolios. These are referred to as the market factor, size factor, value factor, and momentum factor, respectively.

Abnormal return Abnormal return in FFPM is obtained from the intercept term (α_i). If the stock's performance is 'normal', the four factors in FFPM capture the variation and α_i would be zero.

Stock risk Stock risk is a fundamental metric in finance (Hamilton, 1994; Goyal and Santa-Clara, 2003). Greater risk, as a result of higher firm stock price volatility, may imply vulnerable and uncertain future cash flows, resulting in potential damage to firm valuation in the future. Total risk of a firm has two dimensions, namely systematic risk and idiosyncratic risk.

Systematic risk Systematic *market* risk is the part of the total risk that is explained by changes in overall market portfolio returns due to inflation, interest rate changes, and so on, that are common to all stocks (see, for example, McAlister, Srinivasan, and Kim, 2007) and is measured by the

Table 13.1 Relevant financial metrics for assessing customer mindset metrics

Financial Metrics	Definition	Measure	Feature	Illustrative Papers
Firm revenues	The amount of money a company receives from its normal business activities, usually from the sale of goods and services to customers	Obtained by multiplying the price at which goods or services are sold by the number of units or amount sold	Top-line performance from which costs are subtracted to determine net income	Pauwels et el. (2004)
Firm income	A company's total earnings (or profit)	It is computed as the top-line revenues minus expenses	Bottom-line performance that results from a company's operations	Pauwels et al. (2004)
Stock returns	Change in the total value of a stock over some period of initial investment	$\frac{Price_t + Dividend_t - Price_{t-1}}{Price_{t-1}}$	A stationary time series of stock returns is obtained as a dependent variable	Mizik and Jacobson (2003) Srinivasan et al. (2009)

Tobin's q (*q*-values)	Ratio of market value of the firm to the replacement cost of the firm's assets	$\frac{\textit{Market Price}}{\textit{Replacement cost of asset}}$	Tobin's q estimates have smaller average errors and greater correlation with true measures as compared with accounting rates of return	Simon and Sullivan (1993) Rao et al. (2004)
Market-to-book ratio (M/B)	Ratio of current share price to the book value per share	$\frac{\textit{Market Value}}{\textit{Book equity}}$	M/B > 1 signals firms creating value for its shareholders	Kerin and Sethuraman (1998) Pauwels et al. (2004)
Corporate value minus debt	The shareholder value is the corporate value minus the debt	*Corporate Value − Debt*	Estimates the economic value of an investment by discounting future cash flows by the cost of a capital	Rappaport (1997)
Market capitalization	Share price multiplied by the number of outstanding shares	*Stock Price × Number of shares outstanding*	Forward-looking measure, providing investor expectations of the firm's future profit	Fornell et al. (2006)

Table 13.1 (continued)

Financial Metrics	Definition	Measure	Feature	Illustrative Papers
Abnormal returns (α_i)	Difference between the expected return of a stock and the actual return	*Carhart four-factor Model:*	Positive abnormal returns indicates outperformance	Joshi and Hanssens (2009) Hsu et al. (2012)
Systematic risk (β_i)	The part of stock volatility that is explained by changes in average market portfolio returns	$R_{it} - R_{rf,t} = \alpha_i + \beta_i (R_{mt} - R_{rf,t}) + s_i SMB_t + h_i HML_t + u_i UMD_t + \varepsilon_{it}$ where $\varepsilon_{it} \sim N(0, \sigma_{\varepsilon_{it}})$	Cannot be mitigated through diversification	Osinga et al. (2011) Madden et al. (2006) McAlister et al. (2007)
Idiosyncratic risk ($\sigma_{\varepsilon_{it}}$)	The variability that is not explained by changes in average market portfolio return but instead by firm-specific events		Accounts for approximately 80% of total risk on average	Osinga et al. (2011)

Source: Adapted from Srinivasan, Hsu, and Fournier (2012) and Srinivasan and Hanssens (2009).

β_i in equation (13.1). By construction, the stock market as a whole has a β of 1.0. A stock whose return falls more than a drop in market return has a $\beta > 1.0$, and vice versa. Thus, β, a measure of the stock's sensitivity to market changes, is an important metric for publicly listed firms.

Idiosyncratic risk Idiosyncratic risk is that part of the risk that cannot be explained by changes in average market portfolio returns and is measured by the variance of the residuals in equation (13.1). Within the context of market-based assets such as customers, idiosyncratic risk is the risk associated with firm-specific circumstance or a characteristic such as customer switching after the market variation is accounted for. Insofar as marketing contributes to the stability or volatility of earnings, this has become an important area for marketing researchers to address (e.g., Rao and Bharadwaj, 2008; Hsu, Fournier, and Srinivasan, 2010).

Illustrative Customer Mindset Metrics

To assess the impact of customer mindset metrics on firm performance, marketers need a clear understanding of the associated metrics whereby it can be gauged. An excellent overview of customer metrics is provided in Gupta and Zeithaml (2006). This section of the chapter outlines key illustrative customer mindset metrics, summarized in columns 1 and 2 of Table 13.2.

Customer mindset metrics have a long history in marketing, especially in the advertising world and are also the building blocks of the hierarchy-of-effects model of advertising (Palda, 1966; Vakratsas and Ambler, 1999). The central idea of this model is that each advertisement exposure moves the consumer forward through a hierarchical sequence of events, including cognition (e.g., awareness, knowledge), affect (e.g., liking, desire) and, ultimately, behavior (purchase, sometimes measured as purchase intention). More recently, mindset metrics and the idea of this hierarchical sequence have also been used in the evaluation of brand performance from a customer's perspective (Keller and Lehmann, 2006). The commonly used mindset metrics include awareness (aided and unaided), consideration, liking, brand associations, perceived quality, and post-purchase metrics, notably customer loyalty and customer satisfaction. In what follows, I summarize these customer mindset metrics and set the stage for a review of research on customer-mindset-metrics/firm-value effects.

'Awareness' has been defined as the network of cognitive linkages that consumers hold in their memory regarding the brand (Keller, 1993). Awareness is typically the first stage in the customer purchase funnel (e.g., Srinivasan et al., 2010). Its measurement has typically depended on surveys

Table 13.2 Illustrative findings on the impact of customer mindset metrics on firm performance

Customer Mindset Metrics	Definition	Data Sources	Illustrative Papers	Illustrative Empirical Findings
Awareness	The network of cognitive linkages that consumers hold in their memory regarding the brand (Keller, 1993)	Kantar Worldpanel Y&R BAV model Survey data CRSP COMPUSTAT	Srinivasan et al. (2010) Hanssens et al. (2012) Mizik and Jacobson (2008)	There is a significant effect of awareness on brand sales; it serves as a leading indicator of sales Advertising moves the needle on brand awareness There is no significant direct effect of brand knowledge on stock return Brand knowledge indirectly affects the firm value through its influence on sales growth
Consideration	Consideration – the brands a customer considers when making a purchase, is a precursor to greater sales	Kantar Worldpanel Survey data	Srinivasan et al. (2010) Hanssens et al. (2012) Mizik and Jacobson (2008)	There is a significant effect of consideration on brand sales; it serves as a leading indicator of sales Promotions move the needle on the consideration mindset metric No known finding on its effect on firm value

Liking	The extent to which consumers like a brand and hold it in high regard	Kantar Worldpanel Y&R BAV model Survey data CRSP COMPUSTAT	Baldauf, Cravens, and Binder (2003) Hanssens et al. (2012) Frieder and Subrahmanyam (2005) Mizik and Jacobson (2008) Pahud de Mortanges and Van Riel (2003)	Sales moves with square root of liking Movements in liking are leading indicators of sales There is no direct effect of liking on stock return The investors perceive incremental effect of brand esteem in the future only when the increase in brand esteem enhances the profits
Brand attitude	Consumers' overall evalutions of a brand	Techtel Corporation Landor Image Power Survey Y&R BAV model	Aaker and Jacobson (2001) Lane and Jacobson (1995) Mizik and Jacobson (2009)	Consumer attitude is associatedwith stock return and provides incremental information contained in accounting measures Improvement in the predictive power of firm value by 16% in models that take into account brand metrics
Perceived brand differentiation	The ability of the brand to stand apart from its competitors	Y&R BAV model DataStream CRSP COMPUSTAT	Aaker (2003) Pahud de Mortanges and Van Riel (2003) Stahl et al. (2012)	Brand distinctiveness has positive influence on overall financial performance Brand differentiation is associated with higher customer profitability, but also with lower acquisition

Table 13.2 (continued)

Customer Mindset Metrics	Definition	Data Sources	Illustrative Papers	Illustrative Empirical Findings
				and retention rates, suggesting that highly distinct brands address fairly narrowly targeted segments whose members exhibit changing preferences over time Brand differentiation has lagged effect The positive relationship between brand equity and shareholder value is most pronounced when using the market-to-book value as an indicator of shareholder value
Perceived brand relevance	Personal relevance and appropriateness and perceived importance of the brand			Brand relevance provides incremental information and has the potential to generate higher future profit

Perceived brand energy	The ability of a brand to meet customers' needs and adapt to changing tastes in the future			Brand energy provides incremental information in explaining stock performance
Customer loyalty	A deep-held commitment to rebuy or repatronize a preferred product/service consistently in the future, thereby causing repetitive same-brand or same-brand-set purchasing, *despite* situational influences and marketing efforts having the potential to cause switching behavior	Annual reports 10-K statement 10-Q statement	Gupta, Lehmann, and Stuart (2004)	1% improvement in customer retention increases firm value by 5%
Perceived quality	Consumer's judgment about a brand's overall excellence or superiority	EquiTrend	Aaker and Jacobson (1994) Srinivasan et al. (2009)	Improved perceived quality has a positive impact on stock prices Perceived quality significantly increases brand profitability, market share, and customer perceived value
Customer satisfaction	Defined as the consumer's judgment that a product or service meets or falls short of expectations	Publicly available American Customer Satisfaction Index (ACSI) data	Ittner and Larcker (1998) Anderson, Fornell, and Mazvancheryl (2004)	A five-unit increase on a 0–100 scale (roughly one standard deviation from its mean) in the ACSI was associated with a 1% increase in cumulative abnormal returns

Table 13.2 (continued)

Customer Mindset Metrics	Definition	Data Sources	Illustrative Papers	Illustrative Empirical Findings
		Disaggregate firm/ product data available for certain industries (e.g., auto from JDPA)	Gruca and Rego (2005) Fornell et al. (2006); Mittal et al. (2005) Gupta and Zeithaml (2006) Luo and Bhattacharya (2006)	A 1% change in ACSI is associated with a 1.016% change in Tobin's q A one-point increase in the ACSI generates an additional growth in cash flows as well as a decrease in cash flow variability Highly satisfied customers generate positive returns There is a strong link between customer satisfaction, firm profitability and market value Customer satisfaction partially mediates the relationship between CSR and firm market value

Source: Adapted from Srinivasan, Hsu, and Fournier (2012) and Srinivasan and Hanssens (2009).

of consumer-aided and unaided recall. 'Consideration' – the brands a customer considers when making a purchase – is a precursor to greater sales. Generally, the more consideration sets a product is in, the greater the revenues. 'Liking' reflects the affective component of attitude, and measures the extent to which consumers like a brand and hold it in high regard. The differential consumer mindset metrics have been operationalized by a range of market research suppliers, each supporting different metrics (see column 3 in Table 13.2). Kantar Worldpanel brand performance tracker measures consideration by asking respondents to indicate in a list with all brands on the market 'the brands that you would consider buying'. The percentage of respondents who consider buying is the customer consideration metric. Liking is measured on a five-point scale ('like enormously', 'a lot', 'a little', 'not really', 'not at all') and an illustrative metric for liking is the average rating across consumers (ibid.).

'Brand attitude' is defined as consumers' integrated evalutions of a brand (Wilkie, 1986; Keller, 1993; Kumar, 2008). The differential consumer response to the brand is also measured using various operationalizations, models, and data providers (e.g., EquiTrend, Interbrand, Millward Brown, Young & Rubicam, Kantar Worldpanel). For example, Young & Rubicam's Brand Asset Valuator (Y&R BAV) model proposes dimensions of customer mindset metrics that include perceived brand differentiation, the relevance of the brand to the consumer, brand's esteem and consumer's knowledge (awareness and understanding) of the brand (Agres and Dubitsky, 1996). A recent addition is brand *energy*, which captures the degree to which the customer perceives the brand to be innovative and dynamic (Mizik and Jacobson, 2008).

'Perceived quality' is defined as the discrepancy between the consumer's perceptions and expectations about a product (Zeithaml, 1988; Gupta and Zeithaml, 2006). A measurement approach for service quality is SERVQUAL, a multiple-item measure operationalized as the difference between two scores – customers' expectations and perceptions of actual service performance for the attributes that respondents indicated are vital (Zeithaml and Parasuraman, 2004). Oliver (1999, p. 34) defines 'customer loyalty' as 'a deep-held commitment to rebuy or repatronize a preferred product/service consistently in the future, thereby causing repetitive same brand or same brand-set purchasing, despite situational influences and marketing efforts having the potential to cause switching behavior'. Loyalty has been traditionally measured through repeat purchase frequency rates as well as through repurchase intentions.

'Customer satisfaction' has been a central concept in marketing for decades (e.g., Howard and Sheth, 1969). It captures the reaction to a product or service after use (Parasuraman, Zeithaml, and Barry, 1998;

Zeithaml and Parasuraman, 2004) and is defined as the consumer's judgment that a product or service meets or falls short of expectations (Gupta and Zeithaml, 2006). Customer satisfaction as measured by the American Customer Satisfaction Index (ACSI) serves as a popular consumer attitudinal metric (Anderson, Fornell, and Lehmann, 1994). Researchers sometimes argue that customer satisfaction is a forward-looking measure of customer mindset and that it is a necessary but not sufficient condition for customer loyalty (Hallowell, 1996).

How Do Customer Mindset Metrics Affect Firm Performance?

In this section I turn to the question of how exactly customer mindset metrics affect firm value. In Figure 13.1, the first impact of marketing activities in column 1 is on the customer mindset measures of awareness, consideration, liking, attitude, and loyalty in column 2 (Lehmann and Reibstein, 2006). The next level of impact involves customer behavior in the product marketplace (in column 3) and includes metrics such as sales, share, average price paid, and so on. Finally, the ultimate performance outcomes are shown in column 4 and include measures such as profits, shareholder value, and firm risk.

It is important for the framework to allow for multiple hierarchies and for the idea that the impact of a marketing action on the customer mindset is neither immediate nor simultaneous but occurs in situationally varying and complex patterns of temporal precedence (Batra and Vanhonacker, 1988). Indeed, both prior brand familiarity and marketing actions can be expected to build connections in consumers' hearts and minds, resulting in subsequent purchase behavior and firm performance.

The conceptual framework displayed in Figure 13.1 highlights that the impact of marketing actions shown in column 1 on product market performance shown in column 3 and firm performance shown in column 4 can occur through changes in customer mindset and marketing asset metrics such as brand equity and customer equity shown in column 2 as well as directly without such changes. It is also important to note the interdependence between customer mindset metrics and marketing assets metrics. As an example, customer equity is positively influenced by customer loyalty through decreasing defection rates or an increasing customer contribution (more cross-buying, higher willingness to pay, etc.). Similarly, brand equity is positively influenced by attitudinal metrics of awareness, consideration, and liking. As for the link between marketing actions and firm performance, I argue that the routes by which these effects can occur are the 'mindset route' where marketing actions influence customer mindset metrics, which in turn influence brand and firm performance, or the 'transactions route' where marketing actions influence brand and firm performance directly without influencing the intermediate mindset metrics, following Hanssens et al. (2012). The framework therefore accounts for both generally accepted channels of marketing influence: through building the consumer attitudes that constitute the brand's health and/or through leveraging the brand's existing health.

Turning to the contribution of customer mindset metrics to firm value creation, I also posit two routes that make this link explicit. First is a direct route wherein the customer-as-asset enhances firm value directly through its effects on cash flows. Second is an indirect route that considers the customers-as-information. In this route, the customer assets act as a visible signal of the financial well-being of the firm. Given that the focus of the chapter is on the link between customer mindset metrics on firm value, I elaborate on these latter two routes.

Customers-as-assets

Investors perceive incremental information on customer assets as contributing to expectations of future cash flows (Srinivasan and Hanssens, 2009). Srivastava, Shervani, and Fahey (1998) propose that market-based assets such as customers can increase shareholder value through (1) an acceleration of cash flows, (2) an increase in the level of cash flows, (3) a decrease in the volatility and vulnerability of cash flows, and (4) an enhancement on the residual value of cash flows. Under the efficient markets model (Fama, 1970), stock prices at any point in time fully reflect available information and provide an expectation of discounted future cash flows.

Marketing actions that move the needle on customer mindset metrics can increase shareholder value by *enhancing the level* of cash flows (i.e., more cash), both by increasing revenues and lowering costs. The link between marketing actions to customer assets to firm performance and eventually to firm value can involve a hierarchy of effects. For instance, advertising by a brand can essentially move the consumer forward through a hierarchical sequence of events, including cognition (e.g., awareness, knowledge), affect (e.g., liking, desire) and, ultimately, behavior (purchase and loyalty) as argued by Vakratsas and Ambler (1999) and Srinivasan et al. (2010), leading to improved top-line, bottom-line performance and shareholder value.

Marketing can enhance shareholder value by *accelerating the receipt* of cash flows (i.e., faster cash). The acceleration of cash flows is affected by the faster speed of customer-value–generating marketing initiatives relative to competition. As an example, when customers with a preexisting favorable brand attitude adopt new products faster, this speeds up revenue generation, and the cash flow stream is accelerated. Srivastava et al. (1998) propose that acceleration in cash inflows influences shareholder value of the firm.

Finally, marketing investments can increase shareholder value by *lowering the vulnerability and volatility* of these cash flows (i.e., safer cash), which lowers the firm's systematic risk and results in a lower cost of capital or discount rate (e.g., Srivastava et al., 1998; McAlister et al., 2007). Improvements in a brand's customer mindset metrics can reduce volatility of cash flows because of higher customer perceptions of quality (Aaker and Jacobson, 1994), which can lower price sensitivity among consumers and in turn, protect cash flows (Sivakumar and Raj, 1997).

The residual of value of cash flows reflects the expected value of market-based brand assets beyond the market forecast (Rappaport, 1997). Cash flow residuals are positively influenced by customer loyalty and satisfaction, which accounts for a significant proportion of the net present value of a business.

Customers-as-information

The signaling framework (Spence, 1973; Myers and Majluf, 1984) contends that economic information that is uniquely known by management (e.g., improvements in customer mindset metrics) will be conveyed to shareholders through various signals, one of which is the customers-as-information. Consistent with this view of customers-as-information, when Dell's customer satisfaction rating dropped a steep 6.3 percent to 74 out of a possible 100, the biggest drop among major PC makers, its shares closed down from $41.79 to $36.58 (Di Carlo, 2005). Customer defections as a result of dissatisfaction can 'signal' possible weakness in the product portfolio leading to long-term sales losses.

Advocates of customer mindset metrics have hailed them as leading indicators or signals of performance successes and problems (Srinivasan et al., 2010). Their main argument is that, if marketing actions move customers closer to the buying decision in a series of mental steps, then measuring and interpreting the corresponding mindset metrics provides early signals on future performance (LaPointe, 2005), through the customers-as-information route. Specific actions that strengthen the competitive position of the brand in customers 'hearts and minds' may not translate into sales immediately but mindset metrics can verify that marketing moves customers in the right direction (Keller and Lehmann, 2006). In the case of performance problems, the consumer may not react immediately by switching to another brand, but mindset metrics may 'signal' a declined interest and offer a chance for remedial action before the bottom-line performance and firm value is affected.

Research has indeed suggested that movements in customer mindset metrics such as customer awareness through advertising can have a signaling or information value to investors (Joshi and Hanssens, 2009) and increase the firm's salience with individual investors who typically prefer holding stocks that are well known or familiar to them (Grullon, Kanatas, and Weston, 2004; Frieder and Subrahmanyam, 2005). These examples suggest that investors react quickly, rewarding firms with a higher stock price as information perceived as 'good news' becomes available, and vice versa.

ILLUSTRATIVE FINDINGS ON CUSTOMER MINDSET METRICS AND FIRM PERFORMANCE

The conceptual framework has been used in a number of studies on the impact of customer mindset metrics that allow us to formulate some empirical patterns, summarized in the last column of Table 13.2. I discuss, in turn, the findings on customer mindset metrics. The first three studies in this section focus on the impact of customer mindset metrics on product market performance, which has been previously demonstrated to significantly influence firm value (Pauwels et al., 2004; Srinivasan et al., 2009). These two studies find that when an unanticipated change in product market performance (e.g., revenues and earnings) occurs, investors view it as containing information not only about changes in current-term results but about future-term prospects as well. This information induces stock market participants to update their expectations about the firm's discounted future cash flows and revise stock price accordingly.

Srinivasan et al. (2010) analyzed the added explanatory value of including customer mindset metrics in a sales response model that already accounts for short- and long-term effects of the marketing mix using vector autoregressive modeling of the metrics for over 60 brands of four consumer goods. A key finding is the customer mindset metrics of 'awareness', 'consideration', and 'liking' translate into sales performance through the customers-as-assets route and help explain sales even in a model that accounts for long-term effects of own and competitive marketing mix actions. Interestingly, competitive and own mindset variables make a similar contribution to sales performance: the three consumer mindset metrics of awareness, consideration, and liking together account for 8.4 percent of the variation, while competitive mindset metrics account for an additional 7.9 percent of the variation in past sales. Liking has the highest cumulative sales elasticity (0.590), followed by consideration (0.374), and awareness (0.289). Analysis of wear-in times reveals that mindset metrics serve as leading indicators or signals through the customers-as-information route, allowing time for managerial action before market performance itself is affected. If the customer mindset metrics reveal a negative trend in consumer reactions, marketing can fine-tune their messaging or pull the plug on an advertising campaign before a significant decline in sales occurs. As a potential wellspring of strategic intelligence, mindset metrics could thus prove instrumental in expanding the role of marketing executives in corporate decision-making (Rust et al., 2004).

Hanssens et al. (2012) inform financially focused executives on how consumer attitudes influence sales, quantify the conditions under which the influence is strong or weak, the extent of marketing's role in it, and hence how this knowledge can be used to make sound marketing resource allocation decisions. They propose four key criteria – potential, responsiveness, stickiness, and sales conversion – that help determine and understand the connection between marketing actions, attitudinal metrics, and sales outcomes. In their study's context, cross-effects models (CREs) enable them to establish the extent to which the four criteria on translation of attitudes to behavior vary over time and across brands. In addition, longitudinal hierarchical linear models (HLM) allow an investigation of how marketing attitude and attitude–sales relations vary by brand, both in intercept and in the slope of the relationship. They find that marketing-customer and customer-mindset/sales-performance relationships are predominantly stable over time, but differ substantially across brands. For managers and researchers alike, they provide an empirically testable framework on the conditions under which consumer attitude movements result in sales movements, and hence improvements in firm performance.

In a related study, Stahl et al. (2012) examine a unique database from the US automobile market, comprising ten years of survey-based mindset measures as well as acquisition rates, retention rates, and customer profitability. They find that brand equity has a predictable and meaningful impact on all components of CLV, namely customer acquisition, retention, and profitability. Further, familiarity with the brand is positively related to all components of CLV. All three of these aforementioned studies have focused on the impact of customer mindset metrics on product market performance; future research needs to consider their impact on firm value.

Examining the stock return effects of esteem (or liking) and knowledge (or awareness), Mizik and Jacobson (2008) find that their effects are reflected in current-term accounting measures and in perceived brand relevance and energy. In addition, the findings suggest that firms increasing perceived differentiation might not receive abnormal returns contemporaneously but do so in subsequent periods. Recent research on brand identification and brand association trademarks as measures of brand awareness and associations, respectively, show positive effects on cash flows, Tobin's q, stock returns, and reduced cash flow variability (Krasnikov, Mishra, and Orozco, 2009). Brand attitude and brand name familiarity jointly influence the stock market in the context of brand extension announcements (Lane and Jacobson, 1995).

Lifts in customer mindset metrics such as awareness through marketing actions such as advertising can spill over onto investors (Joshi and Hanssens, 2009) and increase the firm's salience with individual investors who typically prefer holding stocks that are well known or familiar to them (Grullon, Kanatas, and Weston, 2004). Frieder and Subrahmanyam (2005), using brand perceptions data (i.e., brand familiarity and brand perceived quality) of 300 firms find that investors are more influenced by brand visibility than by perceptions of brand quality. The greater the familiarity with firms, the more evidence investors can have consistent with the view of customers-as-information; hence, investors exhibit a propensity toward companies with highly well-recognized brands.

Research supports that the customer metrics of perceived brand strength and brand stature, all measured by Young & Rubicam's Brand Asset Valuator model, are positively related to shareholder value (Pahud de Mortanges and Van Riel, 2003). Mizik and Jacobson (2008) assess the five 'pillars' that form the updated Y&R's Brand Asset Valuator model (i.e., differentiation, relevance, esteem, knowledge, and energy) and find that the customer asset metrics of perceived brand relevance and energy provide incremental information to accounting measures in explaining stock returns. Brand attitude in high-technology markets (e.g., Apple, Compaq, and IBM) has a significant influence on stock return comparable to that of unanticipated returns on equity (Aaker and Jacobson, 2001). In summary, the financial markets perceive that brand attitude affects expectations of future-term performance beyond those reflected in current-term earnings.

Changes in perceived quality are associated with changes in brand profitability, market share, customer perceived value, and stock returns. Investors view the perceived quality signal as providing useful information via the customers-as-information route about the future-term prospects of the firm (Aaker and Jacobson, 1994; Mizik and Jacobson, 2003). Moreover, research suggests that it takes innovation *and* quality assessment to improve stock performance. Srinivasan et al. (2009) assess the impact of unanticipated consumer-perceived quality scores in product innovations, while Tellis and Johnson (2007) focus on the impact of expert ratings of quality. Overall, movements in the customer mindset metric of perceived quality are positively related to firm value.

A stream of literature has worked to build better models of customer acquisition and retention (e.g., Gupta and Zeithaml, 2006) to estimate customer equity and finally market capitalization. The basic idea of this stream of literature is closer to the one of discounted cash flow models in finance and it has led to improved marketing decisions and greater marketing accountability (Wiesel, Skiera, and Villanueva, 2008). With respect to the metric of customer loyalty, Gupta, Lehmann, and Stuart (2004) find that that a 1 percent improvement in customer retention increases firm value by 5 percent whereas a 1 percent improvement in margin or acquisition cost generates improvements of only 1 percent and 0.1 percent in firm value, respectively. This result is consistent with suggestions of customer retention as the central metric to measure and monitor in customer relationship management (Thomas, Blattberg, and Fox, 2004). Recently, Schulze, Skiera, and Wiesel (2012) compared the influence of customer and financial metrics on shareholder value. They introduce a new leverage effect that can translate percentage changes in customer equity into shareholder value. The average leverage effect in more than 2000 companies across ten years is 1.55, which indicates that a 10 percent increase in customer equity is amplified to a 15.5 percent increase in shareholder value.

A vast stream of literature has shown a strong link between customer satisfaction and firm performance (see Gupta and Zeithaml, 2006; Srinivasan and Hanssens, 2009; Kumar and Umashankar, 2012 for a review). The linkage between customer satisfaction and firm financial performance such as profits has been well established (e.g., Rust, Zahorik, and Keiningham, 1995; Anderson, Fornell, and Lehmann, 1994; Kamakura et al., 2002; Bowman and Narayandas, 2004; Larivière, 2008). Changes in customer satisfaction are associated with increases in abnormal returns (Ittner and Larcker, 1998), increases in Tobin's q (Anderson et al., 2004), increases in cash flows, and decreases in cash flow variability (Gruca and Rego, 2005). Customer satisfaction partially mediates the relationship between corporate social responsibility and firm market value (Luo and Bhattacharya, 2006). Since cash flow volatility impacts the firm's cost of capital, this effect provides yet another source for stock price appreciation.

In cross-sectional analyses, Fornell et al. (2006) and Mittal et al. (2005) report that firms with highly satisfied customers usually generate positive returns. In addition, Fornell et al. (2006) report that changes in the ACSI are not immediately or fully incorporated in stock returns, creating an arbitrage opportunity for alert investors. Anderson et al. (2004), however, report that satisfaction growth is positively related to Tobin's q growth. Possible explanations for these different findings may include: different time periods used (1994 to 1997 versus 1994 to 2002), omission of an appropriate measure of unanticipated customer satisfaction in the stock return model, and an omitted variables problem (see Jacobson and Mizik, 2009 and Srinivasan and Hanssens, 2009 for a discussion).

Collectively, the extant literature shows that improvements in brand attitude, perceived quality, customer loyalty, and customer satisfaction have demonstrated positive effects on sales, revenues, profits, and the components of shareholder value. Evidence suggests that movements in upper funnel metrics such as brand awareness and brand consideration result in sales movements and hence improvements in firm performance. With respect to the potential impact of the upper funnel metrics on the components of firm value, there is a need for large-scale and systematic empirical study across a variety of contexts and settings. Improvements in customer mindset metrics such as brand awareness and brand liking can also potentially insulate a company from market downturns, grant protection from sales declines in the case of product failures and in the face of price increases; all of these could potentially lead to more stable cash flows and hence lower risk. As of now, little research exists on the linkage between firm risk and customer mindset metrics and this topic is ripe for future investigation.

CONCLUSIONS AND DIRECTIONS FOR FUTURE RESEARCH

This chapter has integrated research that highlights and demonstrates the importance of customer attitudinal metrics in building shareholder value. A key contribution of this chapter is the provision of an integrative conceptual framework that links marketing actions to the creation of customer mindset metrics and the identification of the mechanisms by which customer mindset metrics drive the components of shareholder value. Specifically, the cash flow perspective views customers-as-assets, with investments that move the needle on customer mindset metrics, enhancing firm value directly through their effects on cash flows, while the signaling perspective views customers-as-information with such movements serving as visible signals of the financial well-being of a firm. A second contribution is the development of a framework for organizing and relating customer mindset metrics to those of shareholder value creation, an activity that advances the marketing metrics imperative overall (Lehmann and Reibstein, 2006). A third contribution is the formulation of a rich, forward-looking research agenda on the customer-mindset-metrics/firm-value link. While much has been learned about the value-creating effects of customer metrics, significant gaps in our understanding remain:

- Is there a hierarchy governing the effects of marketing actions, consumer asset metrics, brand asset metrics, product market performance, and shareholder value generation?
- To what extent do customer mindset metrics mediate the relationship between marketing actions and firm performance?
- Do customer mindset metrics differentially impact the components of shareholder value, namely, returns, systematic and idiosyncratic risk? If so, which metrics are differentially relevant in influencing each of these components?
- What are the effects of the upper funnel customer mindset metrics, if any, on the components of shareholder value?
- More importantly, what are the empirical generalizations on the effect sizes of customer mindset metrics on firm performance?
- How are employee perceptual metrics and customer mindset metrics related to behavioral outcomes, customer lifetime value, and firm performance?
- Do the extant findings generalize to the impact of customer mindset metrics on firm performance in other regions and industries?
- Are some of the customer mindset metrics leading indicators of shareholder value losses and gains? Which of the two routes – customers-as-assets versus customers-as-information – dominate in influencing shareholder value?
- Are the Internet-derived attitudinal metrics measured in online consumer engagement (e.g., searches, social network sites, blogs, product reviews and the like) relevant from a shareholder value perspective and to what extent are they different from survey-based customer mindset metrics?
- Within the context of the ongoing tracking of customer mindset metrics, how do the profits gained from better decisions weigh against the cost of collecting attitudinal metrics? In other words, what is the ROI of tracking customer mindset metrics data?

In conclusion, given the growing availability of sophisticated consumer mindset metrics data combined with the need to understand which metrics best define marketers' contributions to corporate objectives, I hope that this chapter contributes to ongoing efforts by underscoring their impact on the firm's ultimate goal of enhancing firm performance. Future research should help bridge the gap between top management's increasingly intense focus on hard numbers and marketing's long-held view of customer knowledge as the fundamental driver of business success. I hope this chapter stimulates further research in integrating behavioral data with attitudinal data in models of firm financial performance to help managers demonstrate the importance of marketing actions in improving shareholder value. In the end, such efforts should help further strengthen marketers' case for building shares in customers' hearts and minds.

REFERENCES

Aaker, D.A. (2003), 'The power of the branded differentiator', *MIT Sloan Management Review*, **45**(1), 83–7.

Aaker, D.A. and R. Jacobson (1994), 'The financial information content of perceived quality', *Journal of Marketing Research*, **31**(2), 191–201.

Aaker, D.A. and R. Jacobson (2001), 'The value relevance of brand attitude in high-technology markets', *Journal of Marketing Research*, **38**(4), 485–93.

Agres, S.J. and T.M. Dubitsky (1996), 'Changing needs for brands', *Journal of Advertising Research*, **36**(1), 21–30.

Ambler, T. (2003), *Marketing and the Bottom Line*, 2nd edition, London: FT Press.

Anderson, E.W., C. Fornell and D.R. Lehmann (1994), 'Customer satisfaction, market share, and profitability: findings from Sweden', *Journal of Marketing*, **58**(3), 53–66.

Anderson, E.W., C. Fornell and S.K. Mazvancheryl (2004), 'Customer satisfaction and shareholder value', *Journal of Marketing*, **68**(4), 172–85.

Baldauf, A., K.S. Cravens and G. Binder (2003), 'Performance consequences of brand equity management: evidence from organizations in the value chain', *Journal of Product and Brand Management*, **12**(4/5), 220–34.

Batra, R. and W. Vanhonacker (1988), 'Falsifying laboratory results through field tests: a time-series methodology and some results', *Journal of Business Research*, **16**(4), 281–300.

Bowman, D. and D. Narayandas (2004), 'Linking customer management effort to customer profitability in business markets', *Journal of Marketing Research*, **41**(4), 433–47.

Carhart, M. (1997), 'On persistence in mutual fund performance', *Journal of Finance*, **52**(1), 57–82.

DiCarlo, L. (2005), 'Dell satisfaction rating takes deep dive', accessed 19 December 2007 at http://www.forbes.com/2005/08/16/dell-customersatisfaction-falls-cx_ld_0816dell.html.

Fama, E.F. (1970), 'Efficient capital markets: a review of theory and empirical work', *Journal of Finance*, **25**(2), 383–417.

Fama, E.F. and K. French (1992), 'The cross-section of expected stock returns', *Journal of Finance*, **47**(2), 427–65.

Fornell, C., S. Mithas, F. Morgeson and M.S. Krishnan (2006), 'Customer satisfaction and stock prices: high returns, low risk', *Journal of Marketing*, **70**(1), 3–14.

Frieder, L. and A. Subrahmanyam (2005), 'Brand perceptions and the market for common stock', *Journal of Financial and Quantitative Analysis*, **40**(1), 57–85.

Goyal, A. and P. Santa-Clara (2003), 'Idiosyncratic risk matters!', *Journal of Finance*, **58**(3), 975–1007.

Gruca, T.S. and L.L. Rego (2005), 'Customer satisfaction, cash flow, and shareholder value', *Journal of Marketing*, **69**(3), 115–30.

Grullon, G., G. Kanatas and J.P. Weston (2004), 'Advertising, breadth of ownership and liquidity', *The Review of Financial Studies*, **17**(2), 439–61.

Gupta, S. and V. Zeithaml (2006), 'Customer metrics and their impact on financial performance', *Marketing Science*, **25**(6), 718–39.

Gupta, S., D.R. Lehmann and J.A. Stuart (2004), 'Valuing customers', *Journal of Marketing Research*, **41**(1), 7–18.

Hallowell, R. (1996), 'The relationships of customer satisfaction, customer loyalty, and profitability: an empirical study', *International Journal of Service Industry Management*, **7**(4), 27–42.
Hamilton, J. (1994), *Time Series Analysis*, Princeton, NJ: Princeton University Press.
Hanssens, D., K. Pauwels, S. Srinivasan, M. Vanhuele and G. Yildirim (2012), 'Consumer mindset metrics for guiding marketing mix decisions', Working Paper, Anderson School of Management, UCLA.
Howard, J.A. and J.N. Sheth (1969), *The Theory of Buyer Behavior*, New York: John Wiley & Sons.
Hsu, L., S. Fournier and S. Srinivasan (2010), 'Brand portfolio strategy effects on firm value and risks', Working Paper, Boston University.
Jacobson, R. and N. Mizik (2009), 'The financial markets and customer satisfaction: re-examining the value implications of customer satisfaction from the efficient markets perspective', *Marketing Science*, **28**(5), 810–19.
Joshi, A.M. and D.M. Hanssens (2009), 'Movie advertising and the stock market valuation of studios: a case of "great expectations"?', *Marketing Science*, **28**(2), 239–50.
Ittner, C. and D. Larcker (1998), 'Are non-financial measures leading indicators of financial performance? An analysis of customer satisfaction', *Journal of Accounting Research*, **36**(3), Supplement, 1–35.
Kamakura, W.A., V. Mittal, F. de Rosa and J.A. Mazzon (2002), 'Assessing the service–profit chain', *Marketing Science*, **21**(3), 294–317.
Keller, K.L. (1993), 'Conceptualizing, measuring, and managing customer-based brand equity', *Journal of Marketing*, **57**(1), 1–22.
Keller, K.L. and D.R. Lehmann (2006), 'Brands and branding: research findings and future priorities', *Marketing Science*, **25**(6), 740–59.
Kerin, R. and R. Sethuraman (1998), 'Exploring the brand value–shareholder value nexus for consumer goods companies', *Journal of the Academy of Marketing Science*, **26**(4), 260–73.
Krasnikov, A., S. Mishra and D. Orozco (2009), 'Evaluating the financial impact of branding using trademarks: a framework and empirical evidence', *Journal of Marketing*, **73**(6), 154–66.
Kumar, V. (2008), *Managing Customers for Profits: Strategies to Increase Profits and Build Loyalty*, Upper Saddle River, NJ: Wharton School Publishing.
Kumar, V. and N. Umashankar (2012), 'Enhancing financial performance: the power of customer metrics', in S. Bharadwaj and S. Ganesan (eds), *Handbook of Marketing and Finance*, Cheltenham, UK and Northampton, MA, USA: Edward Elgar Publishing, pp. 155–203.
Lane, V. and R. Jacobson (1995), 'Stock market reactions to brand extension announcements: the effects of brand attitude and familiarity', *Journal of Marketing*, **59**(1), 63–77.
LaPointe, P. (2005), *Marketing by the Dashboard Light*, New York: ANA Press.
Larivière, B. (2008), 'Linking perceptual and behavioral customer metrics to multiperiod customer profitability: a comprehensive service–profit chain application', *Journal of Service Research*, **11**(1), 3–21.
Lehmann, D.R. (2004), 'Metrics for making marketing matter', *Journal of Marketing*, **68**(4), 73–5.
Lehmann, D.R. and D.J. Reibstein (2006), *Marketing Metrics and Financial Performance*, Cambridge, MA: Marketing Science Institute.
Luo, X. and C.B. Bhattacharya (2006), 'Corporate social responsibility, customer satisfaction, and market value', *Journal of Marketing*, **70**(4), 1–18.
Madden, J.T., F. Fehle and S. Fournier (2006), 'Brands matter: an empirical demonstration of the creation of shareholder value through branding', *Journal of the Academy of Marketing Science*, **34**(2), 224–35.
McAlister, L., R. Srinivasan and M. Kim (2007), 'Advertising, research and development, and systematic risk of the firm', *Journal of Marketing*, **71**(1), 35–48.

Mittal, V., E.W. Anderson, A. Sayrak and P. Tadikamalla (2005) 'Dual emphasis and the long-term financial impact of customer satisfaction', *Marketing Science*, **24**(4), 544–55.
Mizik, N. and R. Jacobson (2003), 'Trading off between value creation and value appropriation: the financial implications of shifts in strategic emphasis', *Journal of Marketing*, **67**(1), 63–76.
Mizik, N. and R. Jacobson (2008), 'The financial value impact of perceptual brand attributes', *Journal of Marketing Research*, **45**(1), 15–32.
Mizik, N. and R. Jacobson (2009), 'Valuing branded businesses', *Journal of Marketing*, **73**(6), 137–53.
Myers, S.C. and N.S. Majluf (1984), 'Corporate financing and investment decisions when firms have information that investors do not have', *Journal of Financial Economics*, **13**(2), 187–221.
Oliver, R.L. (1999), 'Whence consumer loyalty?', *Journal of Marketing*, **63** (Special Issue), 33–44.
Osinga, E.C., P.S.H. Leeflang, S. Srinivasan and J.E. Wieringa (2011), 'The effects of pharmaceutical direct-to-consumer advertising: a shareholder's perspective', *Journal of Marketing*, **75**(1), 109–24.
Pahud de Mortanges, C. and A. van Riel (2003), 'Brand equity and shareholder value', *European Management Journal*, **21**(4), 521–7.
Palda, K.S. (1966), 'The hypothesis of a hierarchy of effects: a partial evaluation', *Journal of Marketing Research*, **3**(1), 13–24.
Parasuraman, A., V. Zeithaml and L. Barry (1998), 'SERVQUAL: a multiple-item scale for measuring consumer perceptions of service quality', *Journal of Retailing*, **64**(1), 12–40.
Pauwels, K., J. Silva-Risso, S. Srinivasan and D.M. Hanssens (2004), 'New products, sales promotions, and firm value: the case of the automobile industry', *Journal of Marketing*, **68**(4), 142–56.
Rao, R.K.S. and N. Bharadwaj (2008), 'Marketing initiatives, expected cash flows, and shareholders' wealth', *Journal of Marketing*, **72**(1), 16–26.
Rao, V.R., M.K. Agarwal and D. Dahlhoff (2004), 'How is manifest branding strategy related to the intangible value of a corporation?', *Journal of Marketing*, **68**(4), 126–41.
Rappaport, A. (1997), *Creating Shareholder Value: A Guide for Managers and Investors*, New York: Free Press.
Rust, R.T., A.J. Zahorik and T.L. Keiningham (1995), 'Return of quality (ROQ): making service quality financially accountable', *Journal of Marketing*, **59**(2), 58–70.
Rust, R.T., T. Ambler, G.S. Carpenter, V. Kumar and R.K. Srivastava (2004), 'Measuring marketing productivity: current knowledge and future directions', *Journal of Marketing*, **68**(4), 76–89.
Schulze, C., B. Skiera and T. Wiesel (2012), 'Linking customer and financial metrics to shareholder value: the leverage effect in customer-based valuation', *Journal of Marketing*, **76**(2), 17–32.
Simon, C.J. and M.W. Sullivan (1993), 'The measurement and determinants of brand equity: a financial approach', *Marketing Science*, **12**(1), 28–52.
Sivakumar, K. and S.P. Raj (1997), 'Quality tier competition: how price change influences brand choice and category choice', *Journal of Marketing*, **61**(3), 71–84.
Spence, M. (1973), 'Job market signaling', *Quarterly Journal of Economics*, **87**(3), 355–74.
Srinivasan, S. and D.M. Hanssens (2009), 'Marketing and firm value: metrics, methods, findings, and future directions', *Journal of Marketing Research*, **46**(3), 293–312.
Srinivasan, S., L. Hsu and S. Fournier (2012), 'Branding and firm value', in S. Bharadwaj and S. Ganesan (eds), *Handbook of Marketing and Finance*, Cheltenham, UK and Northampton, MA, USA: Edward Elgar Publishing, pp. 155–203.
Srinivasan, S., K. Pauwels and T. Wiesel (2011), 'The marketing–finance interface', in J.E. Wieringa, P.C. Verhoef, and J.C. Hoesktra (eds), *Liber Amicorum of P. Leeflang*, University of Groningen, pp. 87–97.
Srinivasan, S., M. Vanhuele and K. Pauwels (2010), 'Mind-set metrics in market response models: an integrative approach', *Journal of Marketing Research*, **47**(4), 672–84.

Srinivasan, S., K. Pauwels, J. Silva-Risso and D.M. Hanssens (2009), 'Product innovations, advertising, and stock returns', *Journal of Marketing*, **73**(1), 24–43.
Srivastava, R.K., T.A. Shervani and L. Fahey (1998), 'Market-based assets and shareholder value: a framework for analysis', *Journal of Marketing*, **62**(1), 2–18.
Stahl, F., M. Heitmann, D.R. Lehmann and S.A. Neslin (2012), 'The impact of brand equity on customer acquisition, retention, and profit margin', *Journal of Marketing*, **76**(4), 44–63.
Tellis, G.J. and J. Johnson (2007), 'The value of quality', *Marketing Science*, **26**(6), 758–73.
Thomas, J.S., R.C. Blattberg and E.J. Fox (2004), 'Recapturing lost customers', *Journal of Marketing Research*, **41**(1), 31–45.
Vakratsas, D. and T. Ambler (1999), 'How advertising works: what do we really know?', *Journal of Marketing*, **36**(1), 26–43.
Wiesel, T., B. Skiera and J. Villanueva (2008), 'Customer equity – an integral part of financial reporting', *Journal of Marketing*, **72**(2), 1–4.
Wilkie, W. (1986), *Consumer Behavior*, New York: John Wiley & Sons, Inc.
Zeithaml, V.A. (1988), 'Consumer perceptions of price, quality, and value: a means-end model and synthesis of evidence', *Journal of Marketing*, **52**(3), 2–22.
Zeithaml, V.A. and A. Parasuraman (2004), *Service Quality*, Cambridge, MA: Marketing Science Institute.

PART 4

STRATEGIC MANAGEMENT OF CUSTOMER EQUITY

14. Risk considerations in the management of customer equity

Ruth N. Bolton and Crina O. Tarasi

INTRODUCTION

Customers are intangible assets of the firm. Customer equity represents the total value of the current and potential customers to the firm. When firms measure or manage customer equity, they typically focus on expected cash flow streams but not the risk associated with them. Hence, customer risk complements the concept of customer equity – it allows managers to assess whether the firm's returns from customers are commensurate with their risk. The notion of balancing customer equity and customer risk is rather new, so many scholars and managers have an incomplete understanding of how to manage the customer portfolio and optimize customer equity. The marketing literature has typically approached customer risk by considering the probability of customer defection and all other risks are assumed to be accounted for in the cost of capital. However, in the finance literature, risk is considered to arise from unanticipated variability in future cash flows. This chapter discusses the implications of this conceptualization of risk for managing customer equity. We argue that it is important to consider the variability or unpredictability of cash flows – arising from underlying customer interactions, consumption patterns, and purchase behavior, plus from organizational processes that support them. An understanding of customer variability is critical for firms to effectively forecast, allocate expenditures, manage customer equity, and prepare for the future via strategic investments.

Risk can be managed both at the individual customer level and the level of the overall customer portfolio. Although a customer portfolio and a financial portfolio seem similar, the underlying assumptions concerning risk are very different. Nevertheless, we can build upon the financial approach to measuring and managing risk to generate important insights for managing customer portfolios. A firm should identify customer risk that can (and should) be divested away because it does not reap higher returns for assuming this risk and (instead) suffers losses when conditions change (Gupta et al., 2006). Trade-offs between customer risk and return have become increasingly important due to the critical role of the marketing–finance interface in firms' strategic decisions and pressures to make marketing accountable.

A richer conceptualization of customer risk complements managers' and researchers' understanding of the drivers of individual customer value and customer equity. It adds many nuances to our understanding of customer cash flow levels and their variability and allows managers (and investors) to assess the true value of a customer in context. This chapter offers some practical guidance on how to incorporate risk into the measurement and management of customer equity and the customer portfolio.[1] In this way, firms can improve how they allocate expenditures and make investment decisions.

CUSTOMER EQUITY, FUTURE CASH FLOWS AND SOURCES OF RISK

The Customer Asset and Customer Equity

The concept of market-based assets, introduced in Srivastava, Shervani, and Fahey's (1998) landmark article, drew attention to customers as intangible assets of the firm. Their article sparked an important set of questions regarding the correct way to measure the value of customers as assets – that is, their value to the firm. These questions are important because the firm must select which customers or market segments to allocate resources to. Traditionally, the value of the customer asset, termed customer equity, was calculated by summing customers' net contribution margins over time, discounted to reflect the time value of money (Berger and Nasr, 1998). This calculation can be performed for an individual customer, a market segment or the entire customer base. For example, summing the discounted net contribution margins of a customer over his or her expected lifetime yields lifetime value.

This chapter uses the term 'customer equity' (CE) to refer to the long-term value of the firm's potential and existing customers (in aggregate), and the term 'customer lifetime value' (CLV) to refer to the long-term value of individual customers or customer segments. The availability of customer data has usually determined whether customer equity was calculated at the aggregate level (e.g., Gupta et al., 2004) – or whether CLV was calculated at the individual or segment level and then aggregated across customers to yield CE (e.g., Blattberg, Getz, and Thomas, 2001). Different methods of aggregation are possible. The simplest method of aggregation – which is not necessarily correct – simply sums across existing customers. However, distributional assumptions about purchasing patterns within and across (potential and existing) customers over time are required to develop a robust model that maximizes the value of the customer base (Ryals and Knox, 2005; Drèze and Bonfrer, 2009).[2] For a detailed classification of existing studies on customer portfolio management, see Table 14.1.

Early Emphasis on the Risk of Customer Defection

Blattberg and Deighton (1996) investigated the following question: what is the optimal level of spending on customer acquisition versus retention that maximizes CE? They explicitly recognized the importance of the risk associated with customer defection (or its inverse, customer retention). Hence, the treatment of defection risk became important to subsequent approaches to the calculation of CE – either at the individual, market segment or aggregate level. Approaches to CE calculations have made different assumptions about the probability of customer retention, margins, and the time horizon (Gupta et al., 2004; Kumar and George, 2007; Wiesel, Skiera, and Villanueva, 2008). For example, are customer retention rates and margins constant or changing? How are they distributed across customers? Short-run measures of profitability or CLV can be quite different than long-run measures. This observation raises another question: is the time horizon finite or infinite? Gupta and Lehmann (2003, 2005) showed that using an expected customer lifetime generally overestimates CLV.

Most approaches use the firm's weighted average cost of capital to calculate discounted cash flows. They usually do not recognize that defection risk – typically reflected in the firm's cost of capital – has been isolated and given explicit treatment in CLV or CE calculations. Hogan, Lehmann et al. (2002) note that cash flows from a customer segment should be discounted using a risk-adjusted rate that reflects customer-specific capital costs. Marketers (unlike their counterparts in finance) have been somewhat inattentive to how the firm's cost of capital is obtained because they focus on spending across customers or marketing activities. For example, Tirenni et al. (2007) describe a decision support system to optimize marketing planning for Finnair's frequent flyer program, accounting for customers' probability of responding to marketers' actions and generating profits.

Sources of Customer Risk

From a market-based asset perspective, each customer or market segment is associated with a stream of (unpredictable) future cash flows for the firm. Hence, the customer is a risky asset, with risk originating from

Table 14.1 Classification of existing studies of customer portfolio management

General Approach	Managerial Problem	Study Objective and Treatment of Risk	Exemplar Studies
Reach, frequency, monetary value models	Allocation of direct marketing effort	Maximize predicted purchases, profitability or CLV over the next T period(s) as a function of direct marketing activities. Emphasis on size and frequency of purchases	Dwyer (1997); Fader, Hardie, and Lee (2005); Borle, Singh, and Jain (2008)
Balancing spending allocated to acquisition versus retention	Allocating marketing spending between acquisition and retention and allocating across communication channels	Deterministic model (e.g., decision calculus model) that maximizes cumulative profits where customer profitability is measured as total revenue less direct costs (including acquisition and retention costs) over a fixed time period	Blattberg and Deighton (1996); Reinartz, Thomas, and Kumar (2005)
Relationship marketing or CRM models	Maximize CE, with an emphasis on customer retention and relationship bonds	Maximize duration of individual customer–firm relationships and/or revenues from existing customers using service strategies, loyalty programs etc.	Bolton (1998); Thomas (2001); Blattberg et al. (2001); Lemon, White, and Winer (2002); Bolton, Lemon, and Verhoef (2004); Reinartz et al. (2005); Bolton, Lemon, and Bramlett (2006); Tirenni et al. (2007)

Maximize CE by assessing return on marketing	Maximize CE through allocation of investments to all activities, recognizing controllable and non-controllable effects	Maximize CE with explicit consideration of defection to competitors through a brand-switching matrix	Rust, Lemon, and Zeithaml (2004)
Adding or subtracting segments[a]	Improve CE by reducing risk for a given level of return or increasing return for a given level of risk	Consider return and risk profiles of customers or segments, then add or subtract customers so that cash flow patterns complement each other	Dhar and Glazer (2003); Buhl and Heinrich (2008); Ryals (2002, 2003)
Re-weight portfolio of potential and/or existing segments[a]	Optimize CE by moving towards the efficient frontier of potential customer portfolios	Optimally re-weight the customer portfolio or incrementally adjust the customer mix to improve risk-return characteristics	Tarasi et al. (2011)
Strategies that modify consumption and cash flow patterns[a]	Decrease customer risk without adversely affecting CE	Adjust strategies for managing customers, service and marketing activities to reduce cash flow variability over time for a customer or market segment	Tarasi et al. (2013)

Note: a. These research streams account for how customer cash flows covary with each other.

different sources of variability inherent in cash inflows and outflows. Sources of variability include individual customer behavior dynamics: defection, product usage, and cross-buying. They are influenced by customer characteristics (Nitzan and Libai, 2011), competitive effects (Du, Kamakura, and Mela, 2007) and firm actions (Hogan, Lehmann et al., 2002, p. 27). Studies attempt to produce accurate CLV predictions by recognizing some sources of risk – but they do not quantify risk or assess risk–return trade-offs. In aggregate models of CE, customer behavior dynamics are often captured by an (expected) growth factor that influences margins – thereby recognizing uncertainty without quantifying it (Gupta et al., 2006). Approaches for capturing different sources of risk are displayed in the top half of Table 14.2 and approaches that recognize factors that influence these sources are displayed in the bottom half.

Sources of risk

Customer purchase behavior (trial, loyalty, usage, and cross-buying) generates cash flows and there is uncertainty about whether and when each of these potential behaviors will occur. Researchers have accounted for customer behavior dynamics in their CLV calculations in different ways. Measures of the recency, frequency, and monetary value (RFM) of purchases have been used to score customers for (short-run) targeting, calculate customer profitability and forecast CLV (Fader et al., 2005; Borle et al., 2008). Behavioral models can be used to predict the duration of customer–firm relationships, usage levels, and cross-buying based on customer characteristics and firm actions (Bolton, 1998; Bolton and Lemon, 1999; Lemon et al., 2002; Verhoef, 2003; Bolton et al., 2004), yielding CLV estimates. They can take into account customer expectations, experiences, and learning (e.g., Lewis, 2005). In general, CLV models out-perform RFM models. Researchers' goal has been to improve predictions of cash flows from individual customers or market segments, thereby yielding a measure of CE. They recognize sources of risk (via probabilities of outcomes), but they do not explicitly measure or manage customer risk.

Factors influencing sources of risk

Customer characteristics influence all sources of customer risk. To date, the most interesting work has studied how social effects – that is, organic word-of-mouth – influence customer behavior dynamics. Nitzan and Libai (2011) model how customers' social networks influence their defection from a service provider. Models of the factors driving win-back offer effectiveness – that is, re-acquiring lost customers – can take into account marketing and social effects (Thomas et al., 2004; Tokman et al., 2007).

Table 14.2 Sources of risk and their antecedents

	Approach	Metrics	Exemplar Studies
Source of risk			
Risk arising from customer behavior dynamics: acquisition, retention or gross margin (via cross-buying etc.)	Maximize CLV by maximizing the duration of individual customer–firm relationships or retention	Models of customer retention or length of customer–firm relationship as a function of marketing, service or channel strategies, with predictions (usually) linked to CLV calculations	Bolton (1998); Thomas (2001); Lemon et al. (2002); Reinartz et al. (2005); Bolton et al. (2006)
	Maximize revenues from existing customers by increasing sales or contributions per customer	Models of the amount purchased over time, including current product usage levels, upgrading and cross-buying.	Kamakura, Ramaswami, and Srivastava (1991); Bolton and Lemon (1999); Blattberg et al. (2001); Verhoef (2003); Bolton, Lemon, and Verhoef (2008)
Factors influencing sources of risk			
Customer characteristics (that are not controllable by the firm)[a]	Models that focus on maximizing retention or CLV	Customer behavior component emphasizes accounting for social effects, specifically positive and negative word-of-mouth or network effects	Ryals (2003); Tokman, Davis, and Lemon (2007); Nitzan and Libai (2011)

Table 14.2 (continued)

	Approach	Metrics	Exemplar Studies
Factors influencing sources of risk			
	Balance cash flows from groups of customers with different risk return profiles by adding or subtracting customers	The firm's goal is to forge multiple types of relationship marketing ties with customers. This approach entails assigning a risk metric to customers that reflects how their cash flows covary with cash flows of other customers. Proxy variables for customer risk in B2B markets include firm size and other firmographics; they include lifecycle, relationship, and other consumer characteristics in consumer markets	Dhar and Glazer (2003); Johnson and Selnes (2004); Buhl and Heinrich (2008); Tuli, Bharadwaj, and Kohli (2010)
Competitive factors	Models maximize CE or CLV, with specific focus on competitive effects	Customer behavior component includes brand switching to/from competitors or 'second lifetime value' obtained through win-back strategies that re-acquire and retain lapsed customers	Thomas, Blattberg, and Fox (2004); Rust et al. (2004); Nitzan and Libai (2011)

Firm factors especially product management decisions	Maximize CE through allocation of marketing investments that enhance the value of brand assets	Models emphasize the role of product R&D, channel support, promotion strategies, service strategies and loyalty programs	Bolton, Kannan, and Bramlett (2000); Bolton et al. (2004); Dong, Swain, and Berger (2007); Tarasi et al. (2013)
All forms of unanticipated cash flow variability or customer risk[a]	Optimal customer portfolio using financial portfolio risk metrics	Risk measured by deviations of actual cash flows from anticipated cash flows	Tarasi et al. (2011)
Environmental and market changes[a]	No current approach or metric in the customer portfolio literature because correlations between customers' cash flow patterns may not remain stable during periods of substantial environmental change		

Note: a. Non-routine risk.

Competitive factors Competitive factors influence customer acquisition, retention or switching, win-back and brand-switching probabilities – and each increase variability in cash flows or risk. An award-winning article by Rust et al. (2004) develops a comprehensive model of how marketing actions influence CE that incorporates competitive effects through a switching matrix. Choice models are especially useful for modeling competitive effects on CE when modelers have access to transactional data in direct marketing contexts (Kamakura et al., 2005). These studies do not consider customer risk (yet).

Firm factors A firm's decisions influence the product life cycle, new product development, and extension into other categories, as well as influencing customer behavior. For example, researchers have noted that channel quality has an important influence on customer acquisition and retention, so that resource allocation across communication channels influences individual customer profitability (Reinartz et al., 2005). This research typically focuses on maximizing individual customer profitability over a fixed time period, rather than CE. In a notable exception, Dong et al. (2007) present a model of CE that uses decision calculus methods to allocate resources across channels; they explicitly acknowledge that their model is deterministic, abstracting from risk.

Environmental and market changes Customers' cash flow patterns will not remain stable during periods of substantial environmental or market or economic changes. All three sources of customer risk (acquisition, retention, and margin) are likely to be altered in unanticipated ways. Moreover, the covariation among customers' cash flows is likely to change in ways that can't be predicted. Since our understanding of customer risk is based on historical cash flow patterns, there is no way to anticipate or manage this situation.

Risk and the Cost of Capital

Gupta et al. (2004) note that the cost of capital or discount rate is a critical variable in the evaluation of the net present value of a cash flow stream (p. 17); they suggest that it is useful to consider the sensitivity of any CE calculations to different rates of discount (p. 12). Sensitivity analyses are especially important if the cost of capital may change in the future. The value of the firm measures the true returns of shareholders after all costs – including all costs of capital (both cost of debt and equity) have been deducted. Since it is created only when returns exceed the cost of capital, the cost of capital should include risk inherent in uncertain future cash flow streams. Thus, cash flows in future periods should be adjusted for any customer-specific capital costs, such as marketing and customized services, as well as a discount rate.

Ryals and Knox (2005, p. 459) argue that managers should quantify the risk of targeting customer segments for investments intended to realize forecasted growth. Discount rates can vary substantially across firms utilizing different strategies. For example, Amazon's cost of capital is typically considered closer to the cost of capital of other Internet-based firms, rather than other book retailers. In business-to-business (B2B) markets, customer-associated risks can include risk of fraud, default risk, and risks due to competitive intensity or industry situation – as well as defection risk. There are standard financial methods for estimating the risk-adjusted cost of capital for a particular firm (Modigliani and Miller, 1958).

Synthesis

Many studies recognize sources of risk in predictions of customer profitability, CLV and CE. However, they typically recognize routine risk (Berger et al., 2006, p. 161) – that is, outcomes for which researchers or managers can assign a probability. Most studies cannot quantify unanticipated variability in cash flows from individual customers or groups of customers that arise from outcomes such as unanticipated firm actions or market changes that are not routine. Until recently, most research has not considered risk–return trade-offs in models of customer acquisition, retention, marketing spending, customer equity or customer portfolio optimization.

CUSTOMER EQUITY AND THE EVOLVING UNDERSTANDING OF RISK

Marketers have long known that the conceptualization of risk in financial portfolio theory is applicable to marketing decisions (Cardozo and Smith, 1983). However, they have been equally aware of the differences between financial and marketing investments (Devinney and Stewart, 1988). Dhar and Glazer (2003) highlighted the importance of measuring the riskiness of customers, termed customer beta. They demonstrated how a firm can maximize returns by acquiring or retaining particular customers or segments on the basis of how their spending patterns contribute to the diversification of the customer portfolio. Ryals (2002, 2003) examined the risk and return characteristics of a customer portfolio and described how a customer relationship scorecard could be used to assess customer risk.

Researchers have worked to integrate customer portfolio ideas within the well-developed stream of research focused on measuring CLV and CE (e.g., Von Wangenheim and Lentz, 2005). Buhl and Heinrich (2008) proposed a quantitative model based on financial portfolio theory that (1) considers CLV as well as the associated risks of customer segments and (2) provides a method for adding or subtracting market segments. They tested the model in a financial services context by using the average annual incomes of key customer segments (e.g., lawyers, physicians) as an indicator of future cash flows and demonstrated how they could create a customer portfolio with higher utility and better risk diversification than the existing portfolio.

From a conceptual standpoint, there are many ways for a firm to build more valuable customer portfolios: customer relationships in different stages can balance each other (Johnson and Selnes, 2004) and the number and types of ties a firm builds with its best customers can reduce their purchase variability (Tuli et al., 2010). By extension, the basic principle that underlies the creation of an optimal customer portfolio is fairly straightforward. By combining customers or market segments with different (i.e., offsetting) cash flow patterns, firms can reduce aggregate cash flow variability – so that risk is minimized for a given level of return.

Tarasi et al. (2011) extended these ideas by describing how the stream of future cash flows associated with a customer or group of customers is variable or risky because it is influenced by customer, competitive, and market conditions that are beyond the firm's control. The firm can (mathematically) optimize the portfolio or mix of customers by effectively allocating its resources across potential and existing customers – thus anticipating and managing future cash flow levels and their variability across the entire customer base. An optimal customer portfolio balances customer risk and return so that risk is minimized for the targeted level of return (or cash flow levels). Managing resources and customers to optimize (rather than maximize) customer equity enables managers to make decisions that are aligned with the long-term goals of the firm. It empowers managers to make the best decisions in a specific context – and evaluate the consequences for the customer portfolio and for (long-run) shareholder value.

This approach is a significant departure from early CLV and CE studies. Resource allocation decisions (such as acquiring versus retaining customers) were made by predicting the future cash flow streams of individual customers or groups of customers and making independent decisions about how to maximize the cash flow stream from a customer or group of customers. Managers and researchers didn't consider how individual customers' cash flows covary and potentially offset each other at the aggregate level. Hence, maximizing the CLV of individual customers or market segments can be sub-optimal from the firm or shareholder perspective (Drèze and Bonfrer, 2009). The magnitude and variability of aggregate future cash flows over the long run matter for the firm and they are influenced by the composition of the customer base. Shareholders require returns that are commensurate with risk.

Tarasi et al. (2011) pointed out that, in choosing what customers to add to a portfolio, a customer with purchasing behavior that is less similar to that of current customers will make a strong contribution to the stability of the customer portfolio (i.e., lowering customer risk); the more similar his or her behavior to existing customers, the weaker its contribution to lowering risk. Therefore, the attractiveness of a customer hinges not only on the size and frequency of purchases but also on the degree to which the customer's pattern of purchases covaries with those of other customers in the portfolio. A declining cash flow from one customer may be offset by increased returns from another. In other words, by re-weighting the customer portfolio it is possible to reduce the variability (risk) of the firm's future cash flows without decreasing its returns. They describe an approach for applying financial principles of diversification to a customer portfolio and show, in an empirical application, that a firm can substantially lower customer risk without decreasing CE. Through forward and backward testing, they demonstrated that an efficient customer portfolio built at one point in time consistently outperforms portfolios designed to (only) maximize profits, as well as outperforming current portfolios.

CONCEPTUALIZATION AND MEASUREMENT OF CUSTOMER RISK

We have introduced our key theoretical ideas without any mathematical exposition. However, at this point, it is useful to deepen our conceptualization of customer risk by describing how it can be measured and used. Each of the following measures goes beyond classical criteria such as average customer profitability, revenue or CLV to incorporate some measure of variability.

Variability or Standard Deviation

In the preceding discussion, we have been equating risk with variability in future cash flows over time.[3] However, it is important to note that we are specifically interested in unexpected variability in cash flows – that is, deviations from expected future cash flows. From this perspective, the standard deviation in cash flows over time – calculated relative to expected or predicted future cash flow levels – is a suitable measure of variability. Predictions might be generated from time series data using the average cash flow level, a trend line, or a more complex model. The standard deviation is also useful because it can be calculated at the level of the individual customer, the market segment or in aggregate. (Of course, we must assume that the variability in historical data will continue into the future.) Unfortunately, the standard deviation can be a misleading metric because higher average values are typically accompanied by higher standard deviations (Tarasi et al., 2013). This can be a significant drawback because managers are interested in determining whether returns (average cash flow levels) are commensurate with risk (the standard deviation of cash flows). Hence, we prefer that returns and variability are considered simultaneously – within a single metric – when making decisions to optimize CE. Hence, it is useful to consider alternative measures of variability.

Coefficient of Variation

The coefficient of variation (CoV) is calculated by dividing the standard deviation of cash flows (relative to an expected value) by the average or mean. It corrects for the relationship between average cash flow levels and corresponding variability, and provides a way to compare customers with different levels of risk and return. CoV has been used as dependent variable by Tarasi et al. (2013). They showed that CoV could be predicted reasonably well using simple descriptor variables such as satisfaction and relationship characteristics embedded in a model based on a semi-log function. A drawback is that CoV doesn't distinguish between large and small customers, whereas, in most real-world applications, managers prefer large customers – even if they are not particularly profitable – because they contribute to covering fixed costs. When using the CoV to evaluate the relative attractiveness of customers, managers can separately consider large and small customers, as shown by Tarasi et al. (2013).

Customer Beta as a Measure of Risk

In the finance literature, a common measure of risk is beta, or the correlation between an asset (e.g., a stock) and the market portfolio. The market portfolio consists of all assets, with the weight of each held in proportion to the total market value. In finance theory, it is assumed that the market portfolio is efficient because it accounts for all information in the market.

However, there is no (easily observed) equivalent to the efficient market portfolio in a customer portfolio context. Instead, we observe that a firm has a customer portfolio with specific risk and return characteristics; it is not efficient. Hence, for the purpose of calculating a customer beta (β), researchers replace the market portfolio with the firm's current customer portfolio (Ryals, 2002; Dhar and Glazer, 2003; Buhl and Heinrich, 2008; Tarasi et al., 2011).

We can calculate a customer β as the covariance between an individual customer's cash flow and the cash flows of all customers, divided by the variance of the cash flow for the overall customer cash flow. It represents the correlation between the cash flows of an individual customer and the cash flows of the focal firm's entire (existing) customer base.[4] A customer's β reflects how the customer responds to firm's actions compared to the existing customer base. A customer with a negative β is likely to reduce the overall variability of the portfolio, whereas a positive β means that the customer added to the portfolio may increase variability. Unlike beta in financial portfolio theory, customer β has no information regarding optimal risk.

Another assumption from financial portfolio theory that does not apply to customer portfolios is that the risk and return are independent from investors' actions. However, returns from customers are not independent of firm's actions. Expected variability due to firms' anticipated actions should not be considered risk. Risk is only the deviation from expected return (Buhl and Heinrich, 2008). We discuss this notion further in the next section.

Customer Reward Ratio

The customer reward ratio (RR) measures the reward for assuming customer risk, where a higher value can be interpreted as a greater reward. It was inspired by the Sharpe ratio (Sharpe, 1994), which measures the investor's reward for assuming risk above the risk-free rate. The RR is calculated as the expected return for a given customer less the return for a risk-free customer, divided by the variability in the customer's returns over time. In a customer portfolio context, a proxy for a risk-free customer asset may not exist. In this situation, the customer RR is calculated as the expected return for a given customer divided by the variability in the customer's returns over time, where variability is measured by the standard deviation (σ) of the returns (Tarasi et al., 2011, p. 6). The intuition about how to interpret the RR is fairly straightforward: for the same level of variability, the customer with the highest return will be preferred and, for the same level of return, the customer with the lowest variability will be preferred, all else being equal. However, when both risk and return are different, the customer with the highest RR is the most attractive customer. The customer RR has an advantage over the customer β in that managers can compare customers with various levels of returns and variability. For both customer β and RR, if measures of customer profitability are not available or reliable, using customer revenue is often acceptable (Wiesel et al., 2008).

Standard Deviation, CoV, Customer β or Customer RR: Which Measure is Best?

A comparison among the various measures for variability is presented in Hutt, Tarasi, and Walker (2012). Each of the four measures captures different aspects of variability and the optimal measure depends on the goals of the analysis. If the composition of the customer portfolio is a given, then adding customers with negative β and high RR is optimal. However, if the firm is in the process of building an optimal customer portfolio, identifying the customer segments first based on customer needs, behavior, and characteristics represents the logical first step. Then, an optimal customer portfolio can be identified using the efficient frontier. The efficient frontier identifies the portfolios for which variability is minimized for the desired level of return, or return is maximized for the desired level of risk. Next, steps are taken either to reduce risk or to increase return (or both) depending on the firm's strategic goals. The last step is to assess current and potential customers, identify their risk/return characteristics, and allocate resources accordingly to reach the desired levels of risk and return (Tarasi et al., 2011).

FOUR APPROACHES TO MANAGING CUSTOMER EQUITY

There are four approaches to managing CE. They differ in their treatment of customer risk, so they are suitable for different purposes. They are: (1) approaches to maximize CE by balancing resource allocations between acquisition and retention, (2) strategies to maximize CE that emphasize the management of customer relationships by allocating spending to different marketing activities, (3) optimization of CE by managing the composition of the customer portfolio (weighing customer groups), or (4) improving CE by managing underlying consumption, purchase, and organizational processes that yield cash flows. Only the last two approaches consider the trade-off between CE and customer risk.

Approach 1: Maximize CE by Balancing Spending on Acquisition versus Retention

Marketers have been keenly interested in how to allocate spending between acquisition and retention efforts. Blattberg and Deighton (1996) used a (deterministic) decision calculus model to allocate expenditures to customer acquisition versus retention. They assume that acquisition and retention are independent. Thomas (2001) developed a methodology for linking customer acquisition to retention, arguing that they are not independent processes. For example, customers acquired through a retail channel might have higher retention rates than those acquired through the Internet. (Ideally, we would like to model the mechanism that links acquisition and retention.) Blattberg et al. (2001) argue that firms should manage customer acquisition, retention, and cross-selling as interdependent activities. They provide an in-depth discussion of each of these activities, including statistical tools, models, and key performance indicators – as well as a CE balance sheet and cash flow statement. However, the quantification and management of risk in building CE is not explicitly addressed.

Managers frequently use short-term customer profitability measures as proxies when it is difficult to forecast CLV for an individual or segment. Reinartz et al. (2005) develop a 'bottom-up' model that is intended to be highly actionable for direct marketing firms. They investigate how to allocate spending across communication channels to acquisition and retention using the criterion of maximizing customer profitability over the projected duration of the customer–firm relationship. In their predictive model, individual customer profitability depends on customer, competitor, firm, and 'control variables' (such as relationship duration and cross-buying). Their approach to CE does not rely on financial analyses or behavioral models that quantify risk or consider risk–return trade-offs. They use predictive models and simulations of cumulative profitability to identify the best allocation of spending to modes of contact (including number of contacts). Interestingly, the profit-maximizing contact strategy doesn't maximize acquisition likelihood or relationship duration. Their results indicate that overspending is more profitable than underspending and, if there is a choice between spending a dollar on acquisition versus retention, firms are better off choosing retention. However, from a risk–reward standpoint, we must view these results with caution. In two consumer contexts, Tarasi et al. (2013) find that contact-intensive retention strategies (such as loyalty programs that rely on economic rewards) increase variability in cash flows without a commensurate increase in returns.

Approach 2: Customer Relationship Management Models that Maximize CE

Optimizing CE requires an understanding of how to create, build, and sustain relationships – thereby influencing customer behaviors that generate cash flows. Individual customers' CLV depends on the duration of the customer–firm relationship (i.e., retention), its depth as reflected in the amount/frequency of product usage over time, and its breadth as reflected in cross-buying or 'add-on' buying. These forecasts, combined with information about prices and costs, enable calculation of CLV and CE. CLV is also influenced by non-purchase behaviors, especially customer engagement dimensions, which are less well understood by marketers (e.g., Ittner and Larcker, 1998). For example, Hogan, Lemon, and Libai (2003) argue that when a firm loses a customer it also loses revenues attributable to the word-of-mouth influence of that customer.

Bolton et al. (2004) offer a comprehensive conceptual framework for understanding how these behavioral components – and their antecedents (price, service quality, communications etc.) – jointly influence CLV and CE. However, their primary focus is a theoretical specification of the behavioral components so that individual customers' future purchase behavior and cash flow streams can be predicted. They suggest that calculations of CLV and CE should be embedded in financial spreadsheets so that managers can conduct sensitivity analyses to evaluate alternative actions (p. 286). Their model does not explicitly quantify risk, but they emphasize that an improved understanding of dynamic and competitive effects (which influence the variability of cash flows) will be important to understanding CE.

In their article 'Return on marketing', Rust et al. (2004) develop a model for allocating marketing expenditures to maximize projected financial return and (consequently) increase in CE. Individual CLV measures are calculated from the frequency of category purchases, average quantity of purchase, and brand-switching patterns (measured from survey data), combined with the firm's contribution margin and an appropriate discount rate. These measures are summed across all customers in the industry, explicitly accounting for switching to a competitor and/or defection, to produce an estimate of return on investment for a particular firm action. This model addresses a major challenge in customer relationship management and CE: assessing the impact of the firm and competitors' actions (using cross-sectional data). By doing so, they recognize many aspects of 'routine' risk. The authors recognize that there will be variability in purchases over time – due to market expansion, cross-buying, changes in usage rates, and other non-routine events. However, a discussion of how to measure or manage the customer risk associated with CE is beyond the scope of their article.

Approach 3: Optimizing CE by Re-weighting the Portfolio of Potential and Existing Customers

The optimization of the customer portfolio is a relatively recent innovation. It is quite different from the profit-, CLV- or CE-maximizing objectives discussed above. The intuition behind customer portfolio methods is that combining customers or market segments with different (i.e., offsetting) cash flow patterns reduces aggregate cash flow variability or customer risk without eroding CE (Ryals, 2002, 2003; Dhar and Glazer, 2003; Ryals, Dias, and Berger, 2007; Buhl and Heinrich, 2008). Note that the concept of measuring the CLV of market segments is not new (e.g., Libai, Narayandas, and Humby, 2002); it is the way they are combined that is novel.

Tarasi et al. (2011) demonstrated their approach to optimizing the customer portfolio as follows. First, they segmented B2B customers (using cluster analysis) based upon monthly cash flow patterns over six years. Then, they observed that customers with similar cash flow patterns shared common characteristics, including purchase cycles, industry, and size of company. Their analyses confirmed that measures of cash flow variability complement traditional segmentation variables. Next, they identified the efficient frontier of customer portfolios by varying the weights of the market segments to reflect how much they are represented in the customer portfolio – where risk was minimized for each level of return. The firm's current customer portfolio was far below the efficient frontier. Last, they showed how managers could identify in which customers to invest and how much to invest, with the goal of moving closer to a portfolio with less risk, a higher return, or both. In this way, firms can move toward a better risk–return profile without 'firing' or 'stealing' customers or other actions with negative consequences.

Their findings are encouraging because simulation results (based on forward and backward testing on an extensive customer data base) showed that their optimal portfolio outperformed the current portfolio – that is, lower risk without lower revenue. However, re-weighting the customer portfolio implies that the firm acquires or divests certain customers and makes strategic changes in how it allocates resources across customer groups. Selnes (2011) argues that the customer portfolio perspective should take into account strategic considerations such as changes in the marketplace, industries, and relationship building.

Approach 4: Strategies that Modify Cash Flows by Changing Underlying Firm Processes

Most approaches to optimizing CE assume future cash flows are fixed characteristics of customers or segments. This assumption was carried over from financial portfolio analysis, where bonds and stocks have fixed cash flow characteristics. However, most firms use classic market segmentation, target marketing, and customer management methods that shape cash flow patterns and ultimately influence their efficient frontier of customer portfolios. Hence, it is useful for managers to ask whether cash flow patterns, the building blocks of the customer portfolio, can be changed through firms' strategies, tactics, and processes.

Since firm actions influence customer behavior and ultimately CE, it should not be surprising that they also influence the variability of cash flows. However, Tarasi et al. (2013) took this notion one step further: is there a way for firms to reduce aggregate variability in future cash flows of an existing customer portfolio without adversely affecting cash flow levels? They identified several possible ways to reduce the cash flow variability of customers or groups of customers while maintaining their cash flow levels and aggregate CE. They focused on simple approaches to managing the service consumption process through firm actions that influence customers. First, customer satisfaction and cross-buying affect the level and variability of future cash flows (e.g., Fornell, Rust, and Dekimpe, 2009). Hence, managers can design programs to improve customer satisfaction and cross-sell products so that they obtain more desirable cash flow patterns (i.e., lowering risk for a targeted level of return). Second, their study also suggests that loyalty programs should be changed from economic incentives to social incentives (e.g., Drèze and Nunes, 2009) because economic rewards are more likely to be associated with higher variability, but not necessarily with higher cash flow levels (ibid.)

Although the relationship characteristics identified by Tarasi et al. (2013) might not apply to all firms or industries, their general approach can be widely applied. For example, studies have investigated how customers engage in consecutive buying of financial products according to their 'financial maturity' (Li, Sun, and Wilcox, 2005). With this information, financial services firms are able to better manage the uncertainty related to customers' future purchases and (consequently) manage customer risk. This approach requires a significant change in managerial viewpoint. We must view risk as variability that we cannot foresee and for which we cannot prepare. Whereas, when we can foresee and prepare for variability, it is an operations management problem, as opposed to a risk management problem.

DIRECTIONS FOR FUTURE RESEARCH ON CUSTOMER RISK

Over the past decade, advances in marketing science and practice have substantially improved our ability to calculate CE and to consider how alternative resource allocation strategies might affect it. However, much more work is needed on the quantification and management of customer risk – to ensure it is commensurate with returns/CE.

Marketing Communications and Loyalty Programs

The direct marketing literature has made important contributions to our understanding of CE. However, it has typically focused on the effects of tactical (marketing communications) decisions on short-run customer profitability or CLV – leading to firm decisions that are sub-optimal in the long run (Godfrey, Seiders, and Voss, 2011). Hence, despite the adoption of the CLV concept, these approaches do not optimize CE. For example, marketing decisions related to loyalty programs – and especially the reward point system that is prevalent in the marketplace – may not be effective or efficient in the long term. Kivetz, Urminsky, and Zheng (2006) found that customers accelerate purchases in the short term to reach a target, but that behavior results in simple acceleration, without increased benefits for the firm (Tarasi et al., 2013). Loyalty programs that focus on social or status rewards and other similar customer targeting tactics are more likely to decrease variability in customers' cash flows. Marketers do not know, as yet, what specific features of loyalty programs are most likely to result in higher cash flow levels without unduly increasing risk in the long term.

Social and Network Effects

In a socially connected world, a customer's opinions and behavior influence other customers, thereby influencing their cash flows. In this way, the CLV of a customer partially depends on his or her connections in a social system (Goldenberg et al., 2009) – and these effects have become more powerful with the advent of social and digital media. We must recognize that social and network effects are likely to influence cash flow variability, as well as cash flow levels. Moreover, social influences can interact with customer satisfaction, such that they amplify variability in cash flows. Satisfied (dissatisfied) customers can attract (repel) other customers, where these effects are magnified by review sites, blogs, brand or user communities, and so on. Research suggests that investments in satisfaction reduce customer risk, but we have limited understanding of how social networks ultimately influence CE and customer risk (Gupta et al., 2006). Since social networks can be broad (e.g., extending across market segments), it is critical that we understand how they influence risk.

Firms' Strategic Decisions

Predictions of customers' cash flow streams are conditional on assumptions about a fixed set of firm decisions. (This feature often becomes apparent when marketers must reconcile predictions from two different models built to describe the same customer base.) However, this approach raises three important challenges. First, firms may change the timing of decisions (e.g., a new product launch) or there may be unanticipated changes in strategy or market conditions – which will alter cash flow levels (returns) and their variability (risk). We have been calling the former routine risk or uncertainty because it can be represented by a probability distribution. In contrast, uncertainty due to unanticipated changes in firm or market activities is not usually considered in current risk metrics. Second, researchers have a limited understanding of how managerial actions may reduce customer risk without adversely affecting CE (or increase CE without increasing customer risk). For example, how should managers think about (and quantify) the effects of a brand extension into a new market on customer risk? Several authors have argued that marketers should use real option theory to address these phenomena (Hogan, Lehmann et al., 2002; Berger et al., 2006). Third, we can reverse the way we think about this issue. How would considering customer risk change managers' decisions to introduce or eliminate channels or products from the product line? How does considering customer risk influence pricing and branding decisions?

Growth

The treatment of growth poses three challenges to customer portfolio methods. First, measures of variability in cash flows do not distinguish between variability due to growing versus declining sales or cash flows. Tarasi et al. (2013) provide some analyses of this issue in two industry contexts and identify some strategies that will reduce risk while maintaining CE. However, more work is needed. Second, growth (or lack of it) may be due to product management strategies. Hogan, Lehmann et al. (2002, p. 28) describe how firm expansion into a related product category or market has different consequences for anticipated growth (returns or CE) and uncertainty (customer risk). Expansion into an unrelated category or market may offer an opportunity for higher returns, but it will usually be associated with higher risk. Moreover, the firm's expansion beyond current categories and markets (and its concomitant uncertainty) is likely to influence the firm's cost of capital – the extent of the effect will depend on the firm's capabilities (Hogan, Lemon, and Rust, 2002, p. 7). Third, risk metrics (customer β or RR) are calculated on the basis of historical data – assuming that past volatility is a good predictor of future volatility. This assumption will be unsuitable when firms expand into new categories or markets.

Risk Adjustment and Risk Optimization

Recall that current approaches to calculating CE typically apply a single discount rate to all cash flows, usually the weighted average cost of capital. However, as we have discussed, the anticipated cash flows from a customer or market segment are at risk from customer behavior dynamics, social, competitive, and product management effects, and so on. When predicted cash flows are adjusted to reflect any of these sources of routine risk (e.g., by incorporating the probability of defection, competitors' actions or the outcomes of product management decisions), the weighted average cost of capital is no longer appropriate. Ideally, the discount rate should be adjusted to reflect the true risk of the particular customer segment that is being valued – by measuring the unanticipated variability in cash flows at a given point in time. Hogan, Lehmann et al. (2002) have suggested that the finance literature on stochastic valuation models could be helpful in this respect.

Another facet of this issue, termed 'risk optimization', arises when managers within firms make resource allocation decisions. What level of risk is recommended for which types or categories of customers? If segments have different responses to the firm's activities intended to alleviate risk, how can we take them into account before we build the customer portfolio? What is the optimal way to arrive at optimal customer portfolios, given that the response of the customer asset is not independent of the actions of the firm? Hutt et al. (2012) provide the steps for building a customer portfolio, but further research is needed to identify the ideal customer-targeted actions to match each of the steps.

Implementation Challenges

The effectiveness of strategies for managing customers depends on how well these strategies are implemented and integrated with firms' existing processes and capabilities, as well as coordinated across channels, technologies, customers, employees, and other stakeholders. Clearly, there is a need to reward managers from different functional areas and levels for behaving in ways that are optimal for the firm. However, the cross-functional and future-oriented nature of these tasks is emphasized insufficiently. To optimize CE, the firm must target and manage segments of the customer portfolio using an adaptive foresight process (Berger et al., 2006; Zeithaml et al., 2006). This process entails understanding how each segment will respond to firm actions (e.g., offering characteristics, value proposition, relationship and service management) and thereby influence future cash flows. Thus, the firm can manage the risk associated with the customer portfolio with greater effectiveness and efficiency, attracting and retaining customers that offer the highest promised return and profitability in the context of potential and existing customers. Selnes (2011, p. 20) proposes that 'customer portfolio theory would be advanced to a greater degree by incorporating customer purchase behaviors and marketing managers' multifaceted objectives'.

Big Data and Computer Science Models

In general, marketers have favored a decompositional approach to customer portfolio management. That is, they utilize cross-sectional or longitudinal data to predict the buying behavior of individual customers or market segments as a function of marketing activities. Then, they calculate CLV, customer risk, and CE – as well as identify the optimal customer portfolio – using financial identities (based on revenue and cost allocations to activities) and simulations of future outcomes (e.g., Villanueva, Yoo, and Hanssens, 2008; Rust, Kumar, and Venkatesan, 2011). With these methods, different modeling approaches (e.g., probability models versus econometric models) are appropriate depending on the buying context – for example, consumer package goods, services, catalog sales, and so forth. However, Gupta et al. (2006, p. 148) point out that there are also many computer science approaches to predictive modeling that could work well, especially for 'big data'. In the future, marketers should pay more attention to these models. They may be especially useful for forecasting future volatility in cash flows when market conditions are evolving. Firms that are successful in exploiting big data (by adopting technology, as well as investing in their employees and organizational learning) to better manage CE and risk will have a sustainable competitive advantage in the marketplace. In this way, non-relational assets leverage relational assets (Hogan, Lemon, and Rust, 2002).

Managing risk in the customer portfolio is a challenging task – both strategically and tactically. Some tools exist and there is the potential – through research – to identify additional tools to quantify and manage variability in cash flows from individual customers and in aggregate. Strategic goals prevail, but within the context of the greater purpose, balancing customer risk and return ensures long-term shareholder value. The CE framework enhances our understanding of the role of marketing in maximizing shareholder value.

NOTES

1. This chapter does not discuss models that link customer lifetime value or customer equity (CE) to shareholder value or the firm's stock price. However, there is substantial research on this topic. Gupta, Lehmann, and Stuart (2004) show that customer lifetime value calculations can be used to approximate shareholder value. Firms' strategic decisions (brand equity, satisfaction levels, etc.) are linked to shareholder value through its future cash flow patterns. Kumar and Shah (2009) show that the relationship between CE and market capitalization is moderated by risk factors in the form of volatility and vulnerability of cash flows from customers.
2. The Pareto/NBD (negative binomial distribution) model, first developed by Schmittlein, Morrison, and Columbo (1987) has been a very popular way of specifying purchase patterns, but there are many other alternatives.
3. We refer to cash flows throughout this chapter. Firms may use revenues or profits depending on their ability to correctly allocate costs.
4. We estimate the customer β for short-lived or potential customers by extrapolating from existing customers who are similar in terms of demographics, psychographics or responsiveness to marketing variables.

REFERENCES

Berger, P.D. and N.I. Nasr (1998), 'Customer lifetime value: marketing models and applications', *Journal of Interactive Marketing*, **12**(1), 17–30.

Berger, P.D., N. Eechambadi, G. Morris, D.R. Lehmann, R. Rizley and R. Venkatesan (2006), 'From customer lifetime value to shareholder value: theory, empirical evidence and issues for future research', *Journal of Service Research*, **9**(2), 156–67.

Blattberg, R.C and J. Deighton (1996), 'Manage marketing by the customer equity test', *Harvard Business Review*, **74**(4), 136–44.

Blattberg, R.C., G. Getz and J.S. Thomas (2001), *Customer Equity: Building and Managing Relationships as Valuable Assets*, Cambridge, MA: Harvard Business School Press.

Bolton, R.N. (1998), 'A dynamic model of the duration of the customer's relationship with a continuous service provider: the role of satisfaction', *Marketing Science*, **17**(1), 45–65.

Bolton, R.N. and K.N. Lemon (1999), 'A dynamic model of customers' usage of services: usage as an antecedent and consequence of satisfaction', *Journal of Marketing Research*, **36**(2), 171–86.

Bolton, R.N., P.K. Kannan and M.D. Bramlett (2000), 'Implication of loyalty programs and service experiences for customer retention and value', *Journal of the Academy of Marketing Science*, **28**(1), 95–108.

Bolton, R.N., K.N. Lemon and M.D. Bramlett (2006), 'The effect of service experiences over time on a supplier's retention of business customers', *Management Science*, **52**(1), 1811–23.

Bolton, R.N., K.N. Lemon and P.C. Verhoef (2004), 'The theoretical underpinnings of customer asset management: a framework and propositions for future research', *Journal of the Academy of Marketing Science*, **32**(3), 271–92.

Bolton, R.N., K.N. Lemon and P.C. Verhoef (2008), 'Expanding business-to-business customer relationships: modeling the customer's upgrade decision', *Journal of Marketing*, **72**(1), 46–64.

Borle, S., S.S. Singh and D.C. Jain (2008), 'Customer lifetime value measurement', *Management Science*, **54**(1), 100–112.

Buhl, H.U. and B. Heinrich (2008), 'Valuing customer portfolios under risk–return aspects: a model-based approach and its application in the financial services industry', *Academy of Marketing Science Review*, **12**(1), 1–32.

Cardozo, R.N. and D.K. Smith, Jr. (1983), 'Applying financial portfolio theory to product portfolio decisions: an empirical study', *Journal of Marketing*, **47**(2), 110–19.

Devinney, T.M. and D.W. Stewart (1988), 'Rethinking the product portfolio: a generalized investment model', *Management Science*, **34**(3/4), 1080–95.

Dhar, R. and R. Glazer (2003), 'Hedging customers', *Harvard Business Review*, **81**(3), 86–92.

Dong, W., S.D. Swain and P.D. Berger (2007), 'The role of channel quality in customer equity management', *Journal of Business Research*, **60**, 1243–52.

Drèze, X. and A. Bonfrer (2009), 'Moving from customer lifetime value to customer equity', *Quantitative Marketing and Economics*, **7**(3), 289–320.

Drèze, X. and J.C. Nunes (2009), 'Feeling superior: the impact of loyalty program structures on consumers' perceptions of status', *Journal of Consumer Research*, **35**(6), 890–905.

Du, R.Y., W.A. Kamakura and C.F. Mela (2007), 'Size and share of customer wallet', *Journal of Marketing*, **71**(2), 94–113.

Dwyer, F.R. (1997), 'Customer lifetime valuation to support marketing decision making', *Journal of Interactive Marketing*, **11**(4), 6–13.

Fader, P.S., B.G.S. Hardie and K.L. Lee (2005), 'RFM and CLV: using iso-CLV curves for customer base analysis', *Journal of Marketing Research*, **42**(4), 415–30.

Fornell, C., R.T. Rust and M.G. Dekimpe (2009), 'The effect of customer satisfaction on consumer spending growth', *Journal of Marketing Research*, **47**(2), 28–35.

Godfrey, A., K. Seiders and G.B. Voss (2011), 'Enough is enough! The fine line in executing multichannel relational communication', *Journal of Marketing*, **75**(4), 94–109.

Goldenberg, J., S. Han, D.R. Lehmann and J.W. Hong (2009), 'The role of hubs in the adoption process', *Journal of Marketing*, **73**(2), 1–13.

Gupta, S. and D.R. Lehmann (2003), 'Customers as assets', *Journal of Interactive Marketing*, **17**(1), 9–14.

Gupta, S. and D.R. Lehmann (2005), *Managing Customers as Investments. The Strategic Value of Customers in the Long Run*, Upper Saddle River, NJ: Wharton School Publishing.

Gupta, S., D.R. Lehmann and J.A. Stuart (2004), 'Valuing customers', *Journal of Marketing Research*, **41**(1), 7–18.

Gupta, S., D. Hanssens, B. Hardie, W. Kahn, V. Kumar, N. Lin, N. Ravishanker and S. Sriram (2006), 'Modeling customer lifetime value', *Journal of Service Research*, **9**(2), 139–55.

Hogan, J.E., K.N. Lemon and B. Libai (2003), 'What is the true value of a lost customer?', *Journal of Service Research*, **5**(3), 196–208.

Hogan, J.E., K.N. Lemon and R.T. Rust (2002), 'Customer equity management: charting new directions for the future of marketing', *Journal of Service Research*, **5**(1), 4–12.

Hogan, J., D. Lehmann, M. Merino, R. Srivastava, J. Thomas and P. Verhoef (2002), 'Linking customer assets to financial performance', *Journal of Service Research*, **5**(1), 26–38.

Hutt, M.D., C.O. Tarasi and B.A. Walker (2012), 'Financial portfolio theory and customer management: insights and research directions', in S. Ganesan (ed.), *Handbook of Marketing and Finance*, Cheltenham, UK and Northampton, MA, USA: Edward Elgar Publishing, pp. 86–107.

Ittner, C.D. and D.F. Larcker (1998), 'Are nonfinancial measures leading indicators of financial performance?', *Journal of Accounting Research*, **36** (Supplement), 1–35.
Johnson, M.D. and F. Selnes (2004), 'Customer portfolio management: toward a dynamic theory of exchange relationships', *Journal of Marketing*, **68**(2), 1–17.
Kamakura, W.A., S.N. Ramaswami and R.K. Srivastava (1991), 'Applying latent trait analysis in the evaluation of prospects for cross-selling of financial services', *International Journal of Research in Marketing*, **8**(4), 329–49.
Kamakura, W., C.F. Mela, A. Ansari, A. Bodapati, P. Fader, R. Iyengar, P. Naik, S. Neslin, B. Sun, P.C. Verhoef, M. Wedel and R. Wilcox (2005), 'Choice models and customer relationship management', *Marketing Letters*, **16**(3/4), 279–91.
Kivetz, R., O. Urminsky and Y. Zheng (2006), 'The goal-gradient hypothesis resurrected: purchase acceleration, illusionary goal progress and customer retention', *Journal of Marketing Research*, **43**(1), 39–58.
Kumar, V. and M. George (2007), 'Measuring and maximizing customer equity: a critical analysis', *Journal of the Academy of Marketing Science*, **35**(2), 157–71.
Kumar, V. and D. Shah (2009), 'Expanding the role of marketing: from customer equity to market capitalization', *Journal of Marketing*, **73**(6), 119–36.
Lemon, K.N., T.B. White and R.S. Winer (2002), 'Dynamic customer relationship management: incorporating future considerations into the service retention decision', *Journal of Marketing*, **66**(1), 1–14.
Lewis, M. (2005), 'Incorporating strategic consumer behavior into customer valuation', *Journal of Marketing*, **69**(4), 230–38.
Li, S., B. Sun and R.T. Wilcox (2005), 'Cross-selling sequentially ordered products: an application to consumer banking services', *Journal of Marketing Research*, **42**(2), 233–9.
Libai, B., D. Narayandas and C. Humby (2002), 'Toward an individual customer profitability model: a segment-based approach', *Journal of Service Research*, **5**(1), 69–76.
Modigliani, F. and M.H. Miller (1958), 'The cost of capital, corporation finance and the theory of investment', *American Economic Review*, **48**(3), 261–97.
Nitzan, I. and B. Libai (2011), 'Social effects on customer retention', *Journal of Marketing*, **75**(6), 24–38.
Reinartz, W., J.S. Thomas and V. Kumar (2005), 'Balancing acquisition and retention resources to maximize customer profitability', *Journal of Marketing*, **69**(1), 63–79.
Rust, R.T., V. Kumar and R. Venkatesan (2011), 'Will the frog change into a prince? Predicting future customer profitability', *International Journal of Research in Marketing*, **28**(4), 281–94.
Rust, R.T., K.N. Lemon and V.A. Zeithaml (2004), 'Return on marketing: using customer equity to focus marketing strategy', *Journal of Marketing*, **68**(1), 109–27.
Ryals, L.J. (2002), 'Measuring risk and returns in the customer portfolio', *Journal of Database Marketing and Customer Strategy Management*, **9**(3), 219–27.
Ryals, L.J. (2003), 'Making customers pay: measuring and managing customer risk and returns', *Journal of Strategic Marketing*, **11**(3), 165–75.
Ryals, L.J. and S. Knox (2005), 'Measuring risk-adjusted customer lifetime value and its impact on relationship marketing strategies and shareholder value', *European Journal of Marketing*, **39**(5), 456–72.
Ryals, L., S. Dias and M. Berger (2007), 'Optimizing marketing spend: return maximization and risk minimization in the marketing portfolio', *Journal of Marketing Management*, **23**(9), 991–1011.
Schmittlein, D.C., D.G. Morrison and R. Columbo (1987), 'Counting your customers: who are they and what will they do next?', *Management Science*, **33**(1), 1–24.
Selnes, F. (2011), 'A comment on balancing risk and return in a customer portfolio', *Journal of Marketing*, **75**(3), 18–21.
Sharpe, W.F. (1994), 'The Sharpe ratio', *Journal of Portfolio Management*, **21**(1), 49–58.
Srivastava, R.K., T.A. Shervani and L. Fahey (1998), 'Market-based assets and shareholder value: a framework for analysis', *Journal of Marketing*, **62**(1), 2–18.
Tarasi, C.O., R.N. Bolton, A. Gustafsson and B.A. Walker (2013), 'Relationship characteris-

tics and cash flow variability: implications for satisfaction, loyalty and customer portfolio management', *Journal of Service Research*, **16**(2), 121–37.
Tarasi, C.O., R.N. Bolton, M.D. Hutt and B.A. Walker (2011), 'Balancing risk and return in a customer portfolio', *Journal of Marketing*, **75**(3), 1–17.
Thomas, J.S. (2001), 'A methodology for linking customer acquisition to customer retention', *Journal of Marketing*, **38**(2), 262–8.
Thomas, J.S., R.C. Blattberg and E.J. Fox (2004), 'Recapturing lost customers', *Journal of Marketing Research*, **41**(1), 31–45.
Tirenni, G., A. Labbi, C. Berrospi, A. Elisseeff, T. Bhose, K. Pauro and S. Pöyhönen (2007), 'Customer equity and lifetime value management (CELM): Finnair case study', *Marketing Science*, **26**(4), 553–65.
Tokman, M., L. Davis and K.N. Lemon (2007), 'The WOW factor: a model of customer reacquisition', *Journal of Retailing*, **83**(1), 47–64.
Tuli, K.R., S.G. Bharadwaj and A.K. Kohli (2010), 'Ties that bind: the impact of multiple types of ties with a customer on sales growth and sales volatility', *Journal of Marketing Research*, **47**(1), 36–50.
Verhoef, P.C. (2003), 'Understanding the effect of customer relationship management efforts on customer retention and customer share development', *Journal of Marketing*, **67**(4), 30–45.
Villanueva, J., S. Yoo and D.M. Hanssens (2008), 'The impact of marketing-induced versus word-of-mouth customer acquisition on customer equity growth', *Journal of Marketing Research*, **45**(1), 48–59.
Von Wangenheim, F. and P. Lentz (2005), 'Customer portfolio analysis: applying financial risk and volatility measures to customer segmentation and risk-adjusted lifetime value determination', Working Paper, University of Dortmund.
Wiesel, T., B. Skiera and J. Villanueva (2008), 'Customer equity: an integral part of financial reporting', *Journal of Marketing*, **72**(2), 1–14.
Zeithaml, V.A., R.N. Bolton, J. Deighton, T.L. Keiningham, K.N. Lemon and J.A. Petersen (2006), 'Forward-looking focus: can firms have adaptive foresight?', *Journal of Service Research*, **9**(6), 168–83.

15. Co-managing brand equity and customer equity

Anita Luo, Donald R. Lehmann and Scott A. Neslin

INTRODUCTION

Despite the success that La-Z-Boy has enjoyed throughout its 80-year history, it has had trouble connecting with female shoppers. The image that La-Z-Boy had among female shoppers was the over-stuffed and old-fashioned recliner that women don't want in their living room (Dan, 2012). In order to change its brand image to appeal to female customers, La-Z-Boy hired Brooke Shields as its spokesperson in place of the traditional male celebrities. Brooke Shields's down-to-earth style attracted female shoppers. As a result, La-Z-Boy enjoyed success beyond stationary furniture. However, La-Z-Boy is not planning to change its name. As its president states, 'La-Z-Boy has an amazing equity of comfort, value and friendliness and almost 100 percent awareness. I felt that if we can just pivot it a little bit, we could build on these tremendous strengths instead of building a new brand from scratch' (Dan, 2012).

Rackspace, a young company in the business of cloud computing and hosting, found the key to growing business was to turn its customers into brand fanatics (Bain Insights, 2012) who recommend the company to their friends. As the CEO explains, 'Customers who are promoters are also more profitable, staying with us longer and buying more of our services . . . The creation of loyal promoters not only reduces customer acquisition costs, it improves retention rates and inspires our Rackers employees' (ibid.). This suggests that high customer equity can create brand equity through positive word-of-mouth.

The previous examples show that some companies manage their brands along with their customers. However, these two marketing assets do not necessarily move in the same direction or respond equally to marketing activities. Companies sometimes face the dilemma of whether they should invest in building brand equity or customer equity. This chapter attempts to help answer that question by examining the relationship between customer equity and brand equity.

We suggest that the mutual relationship between customer equity and brand equity calls for their co-management. Mismanaging either the brand or the customer feeds back negatively on the other. This can cause a damaging spiral, because if the brand is mismanaged that damages brand equity immediately, customer equity a little later, and then, through feedback, damages brand equity again in the long term.

In the following sections, we discuss what brand equity is, what customer equity is, the indirect effects of brand equity (via employees, channels, and partners) on customer equity, and current research on examining the effect of brand equity on customer equity. We then discuss the feedback effect of customer equity on brand equity, setting the stage for the 'co-management' of the two. We conclude with implications for research and management.

BRAND EQUITY

Brand equity has emerged as a key concept in marketing in the past 20 years. However, the exact definition of brand equity has generated considerable debate. Srinivasan (1979, p. 12) defines brand equity 'as the component of overall preference not explained by objectively measured attributes'. A somewhat different view, proposed by Keller (1998, p. 8), describes brand equity as 'the differential effect that brand knowledge has on consumer or customer response to the marketing of that brand'. While the exact definitions differ, the key theme is that brand equity encompasses benefits that accrue to a product due to customers identifying its brand name.

Keller and Lehmann (2003) suggest that there are three approaches for assessing brand equity: customer mindset (e.g., Aaker, 1996; Keller, 2008), product market (e.g., Kamakura and Russell, 1993; Swait et al., 1993; Park and Srinivasan, 1994), and financial market (e.g., Simon and Sullivan, 1993; Mahajan, Rao, and Srivastava, 1994). Customer mindset measures include consumers' brand knowledge and the structure of brand associations. The product market approach estimates brand equity by teasing out the effect of other marketing factors and attributing the residual to brand equity. Financial market measures assess the value of a brand as a financial asset, for example to gauge the price at which a brand should be sold.

Ailawadi, Lehmann, and Neslin (2003) review these three approaches and their strengths and weaknesses. Stahl et al. (2012) suggest that the financial market perspective often relies on subjective judgments or volatile measures when estimating future brand value, whereas product market measures capture brand value related to current marketing activity but generally don't capture future potential. Customer mindset metrics do not provide sufficient information about brand performance regarding market share or profitability, despite providing insights about brand strengths and weaknesses (ibid.). Despite these concerns, all three approaches capture value created by the brand and have a role to play in practice depending on the particular application.

CUSTOMER EQUITY

Definition of Customer Equity

Blattberg and Deighton (1996) first identified customer equity as a firm asset and defined a firm's customer equity as the sum of the customer lifetime values (CLVs) of the firm's current and future customers. Customer equity thus combines the firm's acquisition and lifetime value performance. Acquisition entails the attraction of new customers who generate long-run profits to the firm. High acquisition rates are not worthwhile if the customer contributes little value once he or she is acquired. Thus, customer equity requires success in both customer acquisition and CLV.

Research on CLV

The customer equity component pertaining to the CLV of current customers has received more attention than the acquisition/future-customers component. Many analyses assess the lifetime value of current customers at the individual level (Rust, Lemon, and Zeithaml, 2004). This includes the models developed by Berger and Nasr (1998), direct-marketing–motivated models (e.g., Blattberg and Deighton, 1996; Blattberg, Getz, and Thomas, 2001), and models that use longitudinal databases (e.g., Reinartz and Kumar, 2000; Bolton, Lemon, and Verhoef, 2004).

There are aggregate as well as disaggregate approaches to measuring CLV. The disaggregate-level approach calculates the CLV for each customer and then sums them. The disaggregate approach, represented by Venkatesan and Kumar (2004), predicts a customer's purchase frequency and contribution margin given his or her purchase history, and then calculates the individual's CLV given the predicted variables. The aggregate-level approach uses segment- or firm-level data to compute the average lifetime value of a customer (Gupta, Lehmann, and Stuart, 2004; Kumar and George, 2007). The aggregate approach has the advantage of simplicity and is therefore quite practical. However, it does not capture customer heterogeneity nor identify profitable vs unprofitable customers.

Research on Acquisition

An important stream of research shows that the channel through which the customer is acquired relates to their lifetime value. Villanueva, Yoo, and Hanssens (2008) suggest that customers acquired through marketing actions such as broadcast media and direct mail have more short-term value, while customers acquired through word-of-mouth have nearly twice as much long-term value to the firm. Verhoef and Donkers (2005) studied the effect of acquisition channels on customer loyalty and cross-buying. They found out that the direct mail acquisition channel performs poorly on retention and cross-selling, while radio and TV perform poorly in terms of retention. Customers acquired through telephone calls or magazines tend to cross-buy more, while websites have a negative effect on cross-buying behavior of customers (ibid.). This work provides a bridge between acquisition and CLV.

Research that Combines Customer Acquisition and CLV/CE

Studies that combine customer acquisition and CLV into a measure of CE are not very common. However, important work in this area includes: (1) Gupta, Lehmann, and Stuart (2004; GLS), (2) Blattberg, Getz and Thomas (BGT; 2001), and (3) Rust, Lemon, and Zeithaml/Rust, Zeithaml, and Lemon (RLZ; 2000, 2004). Total customer equity is calculated by summing the segment level CLVs in the BGT approach, while the other approaches use firm-level CLV to come up with customer equity (Kumar and George, 2007).

All these approaches except the GLS approach predict CLV for a finite time period. The BGT, RLZ, and VK (Venkatesan and Kumar) approaches do not assume a constant retention rate (Kumar and George, 2007) whereas the special case of GL does. Therefore, the different approaches have different assumptions that should be taken into account when employing them.

RESEARCH LINKING BRAND EQUITY TO CUSTOMER EQUITY

Brand equity enhances customer equity in several ways. It allows the firm to acquire customers more efficiently, for example, less expensively. It enhances CLV by allowing the firm to increase price and have higher profit margins, or decrease price and increase purchase volume. Brand equity also increases retention rates and allows the firm to develop the customer through enhanced cross-selling and up-selling. The underlying cause of these benefits can be visualized best by thinking of brand equity as a mindset that enables higher CLV and more efficient acquisition.

Leone et al. (2006), among others, suggest that the customer equity of an individual can be calculated based on the profits derived from estimates of acquisition and retention probabilities minus the marketing cost of acquisition and retention. Acquisition and retention probabilities are functions of brand equity and price. Leone et al. (2006) indicate that price can be reduced for strong brands through lower distribution costs, and that stronger brands may also have lower discount and higher retention rates. Therefore, brand equity increases customer equity by enhancing the productivity and effectiveness of marketing mix spending in general and customer relationship management efforts in particular.

For existing customers, the retention probability depends on the customer's purchase history and attitude toward the brand. For a potential customer, the acquisition probability is determined by factors such as market potential, brand knowledge, and customer characteristics (ibid.). While the objective of their model is to measure the impact of brand equity on customer equity at the individual and aggregate level, Leone et al. (2006) did not empirically test the model.

Stahl et al. (2012, p. 45) measured the impact of brand equity on customer acquisition, retention, and profit margin, the key components of customer lifetime value. They used Y&R's BAV measures, which assess brand equity in terms of four 'pillars': knowledge, relevance, esteem, and differentiation. Knowledge is defined as 'the extent to which customers are familiar with the brand'. Relevance is defined as 'the extent to which customers find the brand to be relevant to their needs'. Esteem is defined as 'the regard customers have for the brand's quality, leadership, and reliability'. Differentiation is defined as 'the extent to which the brand is seen as different, unique, or distinct'.

More specifically, Stahl et al. (2012) formulated a regression model to quantify (1) brand equity as a function of marketing activity, and (2) retention, acquisition, and profit margin as a function of brand equity and marketing activities. As expected, they found that knowledge is positively related to acquisition and retention. Esteem is positively related to customer retention, but not to acquisition, while relevance has a positive effect on acquisition, but no significant effect on retention. They also show that differentiation, relevance, and knowledge are both significantly and positively related to profit margin, while esteem is not. Somewhat counter-intuitively, they found that differentiation is negatively related to acquisition and retention. The explanation they offer for this result is that differentiation results in a non-central/niche positioning, which limits

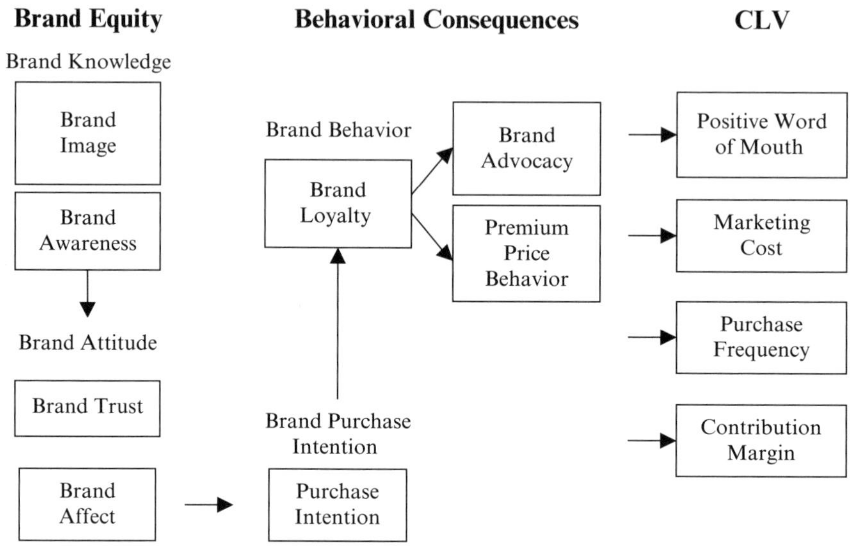

Source: Kumar et al. (2012).

Figure 15.1 Linking brand equity and its behavioral consequences to customer lifetime value

acquisition as well as retention as customers' tastes change over their lifetime.

In summary, Stahl et al. (2012) find that brand equity has an impact on customer acquisition, retention, and profitability, even after controlling for marketing activities. Importantly, the components of brand equity have different influences on acquisition, retention, and profit margins.

Kumar et al. (2012) proposed the framework shown in Figure 15.1 that links a customer's mindset brand equity to this customer's lifetime value through a structural equation model. Brand equity at the customer level is measured in terms of brand knowledge, brand attitude, and brand purchase intention. The behavioral consequences of this brand equity are manifested in brand loyalty and brand behavior. These behaviors then determine CLV.

Kumar et al. (2012) estimated the framework using data from a retailer firm selling home furnishing products. Their results suggested that brand trust and brand affect are driven by brand awareness and brand image. Brand image seems to be more positively related to brand attitude (brand trust and brand affect) than brand awareness is. Purchase intention is also positively related to brand trust and even more strongly to brand affect. Unsurprisingly, brand loyalty is positively related to brand purchase intention (ibid.). Brand advocacy and premium price are also positively related to brand loyalty. Finally, customer lifetime value is positively related to brand advocacy, but, surprisingly, not significantly related to premium price (ibid.). A possible explanation for this is that the brand studied was a value brand and therefore customers were not willing to pay a premium price for it. This result is consistent with the finding of Ailawadi et al. (2003) that brands take their brand equity mainly in the form of increased sales rather than as a price premium.

THE INTER-RELATIONSHIPS AMONG BRAND EQUITY, CUSTOMER EQUITY AND THE BRAND'S EMPLOYEES, CHANNELS AND PARTNERS

The firm's partners play an important role in creating both brand equity and customer equity, and brand equity can increase the effectiveness of the firm's partners. Brand equity can motivate or enable the firm's partners – employees, channel partners, and co-marketers – to create customer equity. For example, brands with high brand equity receive favorable in-store placement by retailers, which in turn improves customer acquisition and retention. Employees take pride in working for a brand with high equity – Burberry's, Saks Fifth Avenue, Mercedes, Apple, and so on – which improves morale and performance. Employees want to sell a brand they can believe in. Co-marketers recognize the enhanced marketing response for a brand with high brand equity, and hence are willing to expend more effort to co-market the brand.

Sawhney and Zabin (2002) suggested that a company's ability to build and leverage relationships between the firm and its partners, suppliers, and employees leads to 'relational equity' – the potential to create wealth through relationships with the brand's constituencies. They suggest that one type of relational equity can influence other types of relational equity. For example, motivated employees can build customer loyalty and therefore increase brand equity (Heskett, Saskett, and Schleslinger, 1997; Reichheld, 2001). Consequently, Sawhney and Zabin (2002, p. 315) argue that companies should manage customer, channel partners, supplier, and employee relationships as a system. According to them:

> [A] company's relationships with its stakeholders do not exist independent of each other. Rather, they mutually reinforce each other as employees interact with customers, customers interact with channel partners, suppliers interact with channel partners, and so on. The quality of employee relationships, supplier relationships, and channel partner relationships affects the quality of customer relationship and vice versa.

Lemon, Rust, and Zeithaml (2001) suggest that attitudes toward the brand are also determined by how well the firm can create emotional ties with its customers. Consumers' perceptions of a brand include evaluations of its corporate ethics, sustainable behaviors, community sponsorships, employee relations, and so on (ibid.). In this way, the non-customer constituencies of the firm can create brand equity. In fact, research suggests that employees constitute the original sources of brand equity. Therefore, the creation of brand equity inside the company is very important (Aaker and Joachimsthaler, 2002; Jones, 2005).

According to Delgado-Ballester and Munuera-Alemán (2005, p. 192), 'Stakeholder theory says that the firm is reliant on a network of relations in which it becomes obliged to the members of this network'. Therefore, as a relational market-based asset, brand equity should be analyzed in the context of relationships with other members of the value chain, including employees, suppliers, and so on. Successful brand management leads to successful relationships with the brand's stakeholders who are often key players in serving the customers.

THE FEEDBACK EFFECT – HOW CUSTOMER EQUITY IMPACTS BRAND EQUITY

While the previous discussion suggests that brand equity is a prime determinant of customer equity, there is some evidence, plus strong reason to believe, that customer equity feeds back to brand equity, as Leone et al. (2006) acknowledge. High customer equity can increase brand equity through several routes, including: (1) social influence, (2) market presence, (3) customer satisfaction, (4) advocacy, and (5) network externalities.

In terms of social influence, previous research indicates the consumption of certain products may depend more on their social meaning than their functional utility (Levy, 1959; Zaltman and Wallendorf, 1979). Consumers seeking social approval may value a brand more because many people are buying or using it. The consumer thinks, 'Everybody's using this brand – maybe I should too'. Keller (1993) suggested that social influence plays an important role in terms of user and usage attributes that consumers associate with it. For example, a consumer may favor a brand because his or her friends have a favorable image of it or by assuming there must be a good reason for a brand's popularity (Parker and Lehmann, 2011).

High customer equity implies the brand has a prominent, visible presence in the marketplace. This in turn can lead to favorable inferences about the brand, that is, enhanced brand equity. For example, when non-users see a lot of people using an iPad, they feel that the product is more relevant to them and have a more positive image of the product. This suggests that the design of a product should be distinctive so that it is easily recognized by non-users when they see it. In addition, choosing a popular brand of food or drink to serve at a party is a safe choice, one likely to appeal to most of the guests and the least likely to offend. Another good example of how market presence can enhance brand equity is Starbucks. A crowded Starbucks store, where customers are enjoying the free Wi-Fi and engrossed with their computers, helps the brand position itself as a third place other than home and work – a place to socialize.

High customer equity suggests the brand has more customers (through higher acquisition and retention rates). This provides the potential for more current customers to become more satisfied with the brand, which in turn increases its average brand equity among all customers in the market. Howard (1974) recognized that satisfaction in product usage influences post-purchase attitude. More usage of a product (and hence the higher the customer equity) improves brand attitude given the experience is positive.

A by-product of higher brand equity among current customers is that these customers are more likely to advocate the brand through word-of-mouth, and since there are more current customers, there is more word-of-mouth. According to Keller (1993, p. 4), 'User and usage imagery attributes can be formed directly from a consumer's own experiences and contact with brand users or indirectly through the depiction of the target market as communicated in brand advertising or by some other source of information (e.g., word of mouth)'. For example, if an environmentally conscious person finds out that the products at Whole Foods are environmentally friendly while shopping at Whole Foods or by hearing from other people shopping there, the brand image (equity) of Whole Foods will increase.

Customer equity can also influence brand equity through network externality, defined as 'products for which the utility that a user derives from consumption for the goods increases with the number of other agents consuming the good' (Katz and Shapiro, 1985, p. 424). There are many possible sources for positive network externality. For example, the utility a consumer derives from joining Facebook depends on the number of his or her friends joining Facebook. Similarly, a person planning to purchase a video game console may try to find out how many other consumers also purchase it because more video games will be provided if there are more users of that particular video game console. Also, a consumer who purchases a car may be concerned about the number of people who also purchase this car, because post-purchase service availability will depend on the size of the user network. Therefore, the larger the customer base is (higher customer equity), the more additional utility that users assign to it (higher brand equity).

CUSTOMER EQUITY VS BRAND EQUITY

Both brand equity and customer equity enhance short-term profits and long-term value in firms. However, sometimes these concepts either conflict or are not in synch with each other. This can cause a 'customer equity *versus* brand equity' mentality, rather than a 'customer equity *and* brand equity' mentality.

One example of this is the concept of brand-centric vs customer-centric management. Blattberg, Malthouse, and Neslin (2009) argue that the customer-centric approach is rooted in direct marketing while the brand-centric approach is typically tied to advertising. Similarly, Hogan, Lemon, and Rust (2002) suggest that customer equity management can be traced back to research on direct marketing, customer satisfaction and service quality, and relationship marketing.

As Rust, Zeithaml, and Lemon (2004) suggest, enhancing brand equity does not necessarily have a positive impact on customer equity. For example, Whole Foods has positive brand equity but relatively small customer equity due to its niche strategy. Some luxury brands like Gucci limit customer acquisition to retain an exclusive brand image. Blattberg et al. (2009) suggest that the customer-centric approach identifies profitable customers and manages marketing resources based on the needs of these customers, while the brand-centric approach emphasizes positioning a brand, often based on marketing research. These two approaches can lead to conflicting decisions. For example, the two approaches lead to different segmentation bases: attitudes, beliefs, and psychographics in the brand-centric approach, vs the components of customer value in the customer-centric approach (ibid.).

Rust, Zeithaml, and Lemon (2004) suggest that customer equity is a more important marketing concept than brand equity and that marketing practice should reflect the shift from brand-centric management to customer-centric management. They also suggest that brand management should be individualized because different customers may feel differently about a brand. According to them, brands can be managed in a better way if the goal is to grow customer equity not brand equity, and brands are managed at a highly individual level.

Gupta et al. (2004) demonstrated the link between customer equity and stock price and suggested how firms can increase shareholder value by understanding and managing customer assets. Hogan et al. (2002)

proposed that the shift towards customer equity management is driven by several factors: (1) marketing expenses are under increasing pressure to be financially accountable; (2) companies have access to rich information on individual customers in terms of their attitude, preferences, shopping behavior, and so on; (3) CRM and data warehouse technology enable companies to make strategic decisions based on customer data; (4) new technology allows companies to personalize products, services, communications, and pricing.

However, others argue that an emphasis on customer equity ignores important benefits resulting from brand equity management such as higher-quality employees and stronger distribution channels and supply chains (Leone et al., 2006). They further suggest that the customer equity perspective often leads to too much emphasis on the bottom-line financial value extracted from current customers and ignores long-term benefits generated from a strong brand. Leone et al. (2006) also note that the customer equity perspective does not give recommendations beyond marketing activities directed towards acquisition, retention, and cross-selling.

Hogan et al. (2002) point out that the brand strengthens the value of customer assets by serving as a tie between the customer and the firm for two principal reasons: (1) brands enable companies to have a higher share-of-wallet from existing customers by encouraging them to purchase the brand and its extensions; (2) brands attract new customers through their overall reputation in the marketplace. Therefore, they call for research to understand how brand equity and customer equity are related.

CO-MANAGING BRAND EQUITY AND CUSTOMER EQUITY

The findings that brand equity and customer equity mutually determine each other, coupled with the potential conflicts between the two concepts, suggest that brand and customer equity must be managed holistically, that is, strategically taking into account the effects of brand equity on customer equity and vice versa, and to avoid one concept from undermining the other.

Leone et al. (2006, p. 131) suggest that optimal marketing solutions emerge if brand equity and customer equity are managed simultaneously. As they point out, 'Brands serve as the "bait" that retailers and other channel intermediaries use to attract customers from whom they extract value. Customers serve as the tangible profit engine for brands to monetize their brand value'. By linking brand equity to customer equity, marketers can simultaneously grow brand equity as well as customer equity (Kumar et al., 2012), quantify the consequences of brand management, and uncover diagnostic information about drivers of customer lifetime value (Stahl et al., 2012).

Once the links between BE and CE are recognized, managers need to ask two key questions. First, how can I translate my current brand equity into higher customer equity? For example, can the company acquire and maintain customers by reinforcing a positive brand image through advertising? Second, how can I translate current customer equity into higher brand equity? The answer may be a social media campaign using the brand's current customers as seeds. Alternatively, the answer may be an ad campaign that emphasizes the network externalities that result when many customers use the brand. For a relatively unknown brand, the answer may be a logo stamped on each product that increases awareness and market presence.

Gupta et al. (2004) and Leone et al. (2006) propose that one way to help sort through these issues is to think of a matrix composed of brands and customers as rows and columns. Consider such a matrix for different products, featuring brand equity and attributes as columns and customers as rows. Figure 15.2 divides each product into two components. The first is 'brand equity', as defined earlier (Srinivasan, 1979; Park and Srinivasan, 1994; Keller, 1998). The second, labeled 'attribute equity', is the equity (revenue) generated due to the objectively defined functional attributes of the product. Brand and attribute equity combine to produce retention, contribution margin, and usage rates among current customers, and in addition, acquisition rates among potential customers, which in turn produce net revenues. We next have costs, which include product cost and acquisition cost or retention cost depending on whether the customer is a future customer or current customer. The far right column represents the discounted long-term profit stream of that consumer, that is, his or her CLV across all the firms' brands and products. Summing the columns represents the CLV of current customers and the potential CLV of potential customers if acquired. Summing the rows in the right-most column yields customer equity. Summing the left-most columns yields brand equity and attribute equity.

Figure 15.2 can serve as a diagnostic tool to enable managers to determine which brands are most crucial to determining total customer equity, which merit investment and which merit divestment, and which brands have brand equity problems and which have fundamental product deficiencies (low attribute equity). Because the matrix keeps track of retention, purchase, and acquisition rates, it can help guide resource allocation between acquisition and retention. It can also be used to evaluate the profitability of each customer segment.

		Current Products					(Potential) Brand Extensions					Overall Customer Value			
		Brand equity	Attribute equity	Brand equity	Attribute equity		Brand equity	Attribute equity	Brand equity	Attribute equity		Performance metrics	Net revenues	Costs	Long-term net revenues – costs
Current customers	1											Retention rates, contribution margin, and purchase frequency		Product + Retention	CLV of this customer
	2														
	…														
	N														
Potential customers (prospects)	1											Acquisition rates and retention, contribution margin, and purchase rates if acquired		Product + Acquisition +Retention	CLV of this prospect
	2														
	…														
	N														
		Brand equity of product 1	Attribute equity of product 1	Brand equity of product 2	Attribute equity of product 2		Brand equity of product 3	Attribute equity of product 3	Brand equity of product 4	Attribute equity of product 4			Total revenues	Total cost	Customer equity

Figure 15.2 Co-managing brand equity and customer equity

For example, the matrix may indicate that a particular brand has ample brand equity but low retention and acquisition rates. Perhaps the customer side of the business is not doing its job in translating a strong brand into high customer equity. This suggests an examination of the specific customer management efforts being used and the customer analytics that guide those efforts. Alternatively, it may mean that the brand equity is being created through a niche positioning (e.g., Harley Davidson), and inherently retention and acquisition rates will be low. This means that the brand needs to establish and emphasize different aspects of brand equity that take it out of its niche position, or perhaps the firm should offer brand extensions.

As another example, the matrix may reveal very high acquisition rates for all the firm's brands, coupled with low brand equity and low retention rates. This may suggest that the firm is overly investing in steep discounts to attract customers who do not stay with the brand once they are acquired, and in the process, the brand is eroding its brand equity. As we hope is apparent, populating the matrix with the necessary metrics would help practitioners 'co-manage' brand equity and customer equity.

Another application of the matrix entails customer segmentation. To calculate how profitable a customer segment is, a company needs to sum the customer lifetime value generated from this segment across all the brands a company offers. Companies should analyze why a customer segment is or is not profitable, which may be due to its brand image. If a customer segment has a low share-of-wallet, the company may try to grow this segment, by increasing brand equity, improving acquisition and retention efforts, or by launching a better-fitting brand. On the other hand, if the customer segment has limited profit potential, de-emphasizing it may be the best strategy. In general, companies should communicate to potential and existing customers a brand image that serves as 'bait' to attract and retain the right (profitable) customers.

Segmenting customers on the basis of customer lifetime value helps companies decide where to focus their marketing resources. For example, a company may divide customers into high-, median-, and low-CLV customers. Intuitively, companies should focus on high-CLV customers and grow the customer equity of this segment. However, in allocating resources optimally, what matters more than *current* customer CLV is *potential* CLV and how responsive each customer type is to marketing efforts. Reinartz and Kumar (2002, p. 93) segmented customers into Butterflies, True Friends, Strangers, and Barnacles. Then they suggest for True Friends, the need to 'delight these customers to nurture, defend, and retain them'. As to Butterflies, they suggest the company should 'milk the accounts as long as they are active'. For Strangers, they suggest a company 'make no investment in these relationships'. For Barnacles, they suggest it is important to 'measure both the size and share of wallet' and take actions, including 'firing' them, based on that information.

To increase a customer's purchase frequency and contribution margin, companies often encourage add-on and cross-selling. It helps when companies integrate their marketing communication (e.g., across their distribution channels) so that customers are not confused about the brand's image. Importantly, Shah et al. (2012) suggest that not all cross-buying behavior is profitable. Therefore, companies should be careful about who they cross-sell to and what they cross-sell. The contribution margin for a customer segment can be evaluated based on the brand equity, attribute equity and cost, while the purchase frequency is determined on the retention rate over time.

Because of the role of the brand's partners discussed earlier, companies also need to manage relationships with other stakeholders including employees, channel partners, and suppliers in an integrated system. As discussed earlier, companies with higher brand equity can leverage it to recruit better employees and have more negotiating power with channel partners and suppliers.

THE IMPACT OF THE MARKETING MIX ON BRAND EQUITY AND CUSTOMER EQUITY

This chapter has discussed the linkage between customer equity and brand equity and the impact of the marketing mix on them. We suggested how brand equity and customer equity can be co-managed by developing a customer management matrix, which encourages joint management of brands and customer segments. We then provided suggestions for how to manage customer equity and brand equity strategically to increase customer acquisition and retention, as well as drive up purchase frequency and contribution margins. The key to implementing these suggestions is recognizing that marketing efforts have a direct effect on brand equity, CLV, and acquisition, and an indirect effect on CLV and acquisition by influencing brand equity, which in turn influences CLV and acquisition.

Previous researchers found a positive relationship between marketing effort and a customer's lifetime value (Venkatesan and Kumar, 2004; Reinartz, Thomas, and Kumar, 2005). Rust, Lemon, and Zeithaml (2001, 2004) explicitly measured the impact of marketing variables on customer equity. Companies acquire new customers through a variety of marketing activities, including mass media and personalized contacts (Villanueva et al., 2008). Hanssens, Thorpe, and Finkbeiner (2008) suggest that advertising has a stronger impact on acquisition than on retention, suggesting that branding efforts induce customer trial while customer management drives repeat buying and margin expansion.

There is a consensus that advertising increases brand equity (Aaker and Biel, 1993; Kirmani and Zeithaml, 1993; Mela, Gupta, and Lehmann, 1997; Ailawadi et al., 2003). For example, Sriram, Balachander, and Kalwani (2007) found that advertising has a significant and positive effect on brand equity. However, the impact of sales promotions was not significant in both the toothpaste and the dish detergent markets they studied. Stahl et al. (2012) suggest that advertising and market presence significantly influence the components of brand equity. Product innovations are especially important for performance-based brands, since their brand equity primarily comes from product-related associations (Keller, 1998).

Sriram et al. (2007) suggest, based on learning theory, that promotions can lead to attitudinal loyalty, thereby enhancing brand equity. However, some studies (e.g., Jedidi, Mela, and Gupta, 1999) find that brand equity is negatively related to promotions in the long run. Blattberg, Briesch, and Fox (1995) propose that too much price promotion lowers brand equity by reducing the price premium. Similarly, Aaker (1996) suggests that while price promotions may increase sales, they may also damage brand equity by focusing too much attention on price.

CONCLUSIONS

The theme of this chapter is that brand and customer equity are interrelated and must be co-managed. Brand and customer equity are marketing assets that are essential to a company's success. Both concepts focus on the importance of customer loyalty and are in agreement on the premise that value is created through having customers who are willing to pay a premium price. However, the two concepts tend to focus on different approaches: the brand equity perspective emphasizes 'brand-building' marketing activities such as advertising, customer attitudes, and growth opportunities via brand extensions, while the customer equity perspective emphasizes bottom-line financial value generated through customer relationship management activities such as direct marketing, cross-selling, and churn management. Therefore, it is important to analyze both brand equity and customer lifetime value as well as how they are related.

Stahl et al. (2012) measured the impact of brand equity on customer acquisition, retention, and profit margin using Y&R's BAV measures, knowledge, relevance, esteem, and differentiation. Kumar et al. (2012) link customer brand equity to customer lifetime value through a structural equation model. Brand equity also affects customer equity through its impact on stakeholders such as employees, suppliers, and channel partners.

Unfortunately, there is limited research that examines the feedback effect of customer equity on brand equity. However, we believe it is quite plausible that customer equity affects brand equity through social influence, market presence, customer satisfaction, customer advocacy, and network externalities. We invite further research on this topic.

The mutual relationships between customer equity and brand equity, coupled with potential conflicts that can occur if the customer and brand sides of the business are out of synch, suggest the co-management of brand and customer equity. These two crucial components must be managed holistically, with a long-term view of how actions on one component may impact the other. The result can be a coherent, coordinated brand/customer strategy, and a more profitable business. Co-managing brand and customer equity is indisputably a challenge. Therefore, we recommend the development of a brand/customer performance matrix to gain a holistic perspective that can facilitate the co-management of brand and customer equity. By addressing this issue, we hope to stimulate future research in the field and enhance awareness about the concept among managers.

REFERENCES

Aaker, D. (1996), *Building Strong Brands*, New York: The Free Press.

Aaker, D. and A. Biel (1993), *Brand Equity and Advertising*, Hillsdale, NJ: Lawrence Erlbaum Associates.

Aaker, D. and E. Joachimsthaler (2002), *Brand Leadership*, Glasgow: Free Press Business.

Ailawadi, K.L., D.R. Lehmann and S. Neslin (2003), 'Revenue premium as an outcome measure of brand equity', *Journal of Marketing*, **67**(4), 1–17.

Bain Insights (2012), 'How Rackspace turns customers into brand fanatics', *Forbes*, accessed 5 November 2012 at http://www.forbes.com/sites/baininsights/2012/09/20/how-rackspace-turns-customers-into-brand-fanatics/.

Berger, P.D. and N.I. Nasr (1998), 'Customer lifetime value: marketing models and applications', *Journal of Interactive Marketing*, **12**(1), 17–29.

Blattberg, R. and J. Deighton (1996), 'Manage marketing by customer equity test', *Harvard Business Review*, **74**(4), 136–44.

Blattberg, R., R. Briesch and E.J. Fox (1995), 'How promotions work', *Marketing Science*, **14**(3), 122–32.

Blattberg, R., G. Getz and J.S. Thomas (2001), *Customer Equity: Building and Managing Relationships as Valuable Assets*, Boston, MA: Harvard Business School Press.

Blattberg, R., E. Malthouse and S. Neslin (2009), 'Customer lifetime value: empirical generalizations and some conceptual questions', *Journal of Interactive Marketing*, **23**(2), 157–68.

Bolton, R.N., K.N. Lemon and P.C. Verhoef (2004), 'The theoretical underpinnings of customer asset management: a framework and propositions for future research', *Journal of the Academy of Marketing Science*, **32**(3), 271–92.

Dan, A. (2012), 'Embracing brand equity while changing perception to reach new customers', *Forbes*, accessed 5 November 2012 at http://www.forbes.com/sites/avidan/2012/08/28/embracing-brand-equity-while-changing-perception-to-reach-new-customers/.
Delgado-Ballester, E. and J.L. Munuera-Alemán (2005), 'Does brand trust matter to brand equity?', *Journal of Product & Brand Management*, **14**(3), 187–96.
Gupta, S., D.R. Lehmann and J.A. Stuart (2004), 'Valuing customers', *Journal of Marketing Research*, **41**(1), 7–18.
Hanssens, D.M., D. Thorpe and C. Finkbeiner (2008), 'Marketing when customer equity matters', *Harvard Business Review*, **86**(5), 117–23.
Heskett, J., E. Sasser and L.A. Schleslinger (1997), *The Service Profit Chain*, New York: Free Press.
Hogan, J.E., K.N. Lemon and R.T. Rust (2002), 'Customer equity management: charting new directions for the future of marketing', *Journal of Service Research*, **5**(1), 4–12.
Howard, J.A. (1974), 'The structure of buyer behavior', in J.U. Farley, J.A. Howard and L.W. Ring (eds), *Consumer Behavior: Theory and Application*, Boston, MA: Allyn & Bacon.
Jedidi, K., C. Mela and S. Gupta (1999), 'Managing advertising and promotions long-run profitability', *Marketing Science*, **18**(1), 1–22.
Jones, R. (2005), 'Finding sources of brand value – developing a stakeholder model of brand equity', *Journal of Brand Management*, **13**(1), 10–32.
Kamakura, W.A. and G.J. Russell (1993), 'Measuring brand value with scanner data', *International Journal of Research Marketing*, **10**(1), 9–22.
Katz, M.L. and C. Shapiro (1985), 'Network externalities, competition, and compatibility', *The American Economic Review*, **75**(3), 424–40.
Keller, K.L. (1993), 'Conceptualizing, measuring, and managing customer-based brand equity', *Journal of Marketing*, **57**(1), 1–22.
Keller, K.L. (1998), *Equity*, Englewood Cliffs, NJ: Prentice Hall.
Keller, K.L. (2008), *Strategic Brand Management*, 3rd edition, Upper Saddle River, NJ: Pearson Prentice-Hall.
Keller, K.L. and D. Lehmann (2003), 'How do brands create value', *Marketing Management*, **12**(3), 26–31.
Kirmani, A. and V. Zeithaml (1993), 'Advertising, perceived quality and brand image', in D.A. Aaker and A. Biel (eds), *Brand Equity and Advertising*, Mahwah, NJ: Erlbaum, pp. 143–62.
Kumar, V. and M. George (2007), 'Measuring and maximizing customer equity: a critical analysis', *Journal of the Academy of Marketing Science*, **35**(2), 157–71.
Kumar, V., A. Luo and V. Rao (2012), 'Connecting brands with customers', Working Paper, Georgia State University.
Lemon, K.N., R.T. Rust and V.A. Zeithaml (2001), 'What drives customer equity?', *Marketing Management*, **10**(1), 20–25.
Leone, R.P., V.R. Rao, K.L. Keller, A.M. Luo, L. McAlister and R. Srivastava (2006), 'Linking brand equity to customer equity', *Journal of Service Research*, **9**(2), 125–38.
Levy, S.J. (1959), 'Symbols for sale', *Harvard Business Review*, **37**, 117–24.
Mahajan, V., V.R. Rao and R. Srivastava (1994), 'An approach to assess the importance of brand equity in acquisition decisions', *Journal of Product Innovation Management*, **11**(3), 221–35.
Mela, C., S. Gupta and D.R. Lehmann (1997), 'The long-term impact of promotion and advertising on consumer brand choice', *Journal of Marketing Research*, **34**(2), 248–61.
Park, C.S. and V. Srinivasan (1994), 'A survey-based method for measuring and understanding brand equity and its extendibility', *Journal of Marketing Research*, **31**(2), 271–88.
Parker, J. and D.R. Lehmann (2011), 'When shelf-based scarcity impacts consumer preferences', *Journal of Retailing*, **87**(2), 142–55.
Reichheld, F. (2001), *Loyalty Rules!*, Cambridge, MA: Harvard Business School Press.
Reinartz, W. and V. Kumar (2000), 'On the profitability of long-life customers in a

non-contractual setting: an empirical phase and implications for marketing', *Journal of Marketing*, **64**(4), 17–36.
Reinartz, W. and V. Kumar (2002), 'The mismanagement of customer loyalty', *Harvard Business Review*, **80**, 86–97.
Reinartz, W., J. Thomas and V. Kumar (2005), 'Balancing acquisition and retention resources to maximize customer profitability', *Journal of Marketing*, **69**(1), 63–79.
Rust, R.T., K.N. Lemon and V.A. Zeithaml (2001), 'Where should the next marketing dollar go?', *Marketing Management*, **10**(3), 24–8.
Rust, R.T., K.N. Lemon and V.A. Zeithaml (2004), 'Return on marketing: using customer equity to focus marketing strategy', *Journal of Marketing*, **68**(2), 109–27.
Rust, R.T., V.A. Zeithaml and K.N. Lemon (2000), *Driving Customer Equity: How Customer Lifetime Value is Reshaping Corporate Strategy*, New York: The Free Press.
Rust, R.T., V.A. Zeithaml and K.N. Lemon (2004), 'Customer-centered brand management', *Harvard Business Review*, **82**(9), 110–18.
Sawhney, M. and J. Zabin (2002), 'Managing and measuring relational equity in the network economy', *Journal of the Academy of Marketing Science*, **30**(4), 313–32.
Shah, D., V. Kumar, Y. Qu and S. Chen (2012), 'Unprofitable cross-buying: evidence from consumer and business markets', *Journal of Marketing*, **76**(3), 79–95.
Simon, C.J. and M.W. Sullivan (1993), 'The measurement and determinants of brand equity: a financial approach', *Marketing Science*, **12**(1), 28–52.
Srinivasan, V. (1979), 'Network models for estimating brand-specific effects in multi-attribute marketing models', *Management Science*, **25**(1), 11–21.
Sriram, S., S. Balachander and M. Kalwani (2007), 'Monitoring the dynamics of brand equity using store-level data', *Journal of Marketing*, **71**(2), 61–78.
Stahl, F., M. Heitmann, D.R. Lehmann and S.A. Neslin (2012), 'The impact of brand equity on customer acquisition, retention, and profit margin', *Journal of Marketing*, **76**(4), 44–63.
Swait, J., T. Erdem, J. Louvière and C. Dubelaar (1993), 'The equalization price: a consumer-perceived measure of brand equity', *International Journal of Research in Marketing*, **10**(1), 23–45.
Venkatesan, R. and V. Kumar (2004), 'A customer lifetime value framework for customer selection and resource allocation strategy', *Journal of Marketing*, **68**(4), 106–25.
Verhoef, P.C. and B. Donkers (2005), 'The effect of acquisition channels on customer loyalty and cross-buying', *Journal of Interactive Marketing*, **19**(2), 31–43.
Villanueva, J., S. Yoo and D.M. Hanssens (2008), 'The impact of marketing-induced versus word-of-mouth customer acquisition on customer equity growth', *Journal of Marketing Research*, **45**(1), 48–59.
Zaltman, G. and M. Wallendorf (1979), *Consumer Behavior: Basic Findings and Management Implication*, New York: Wiley.

16. Customer equity management – opportunities and threats from new technologies

Raji Srinivasan

INTRODUCTION

Diverse technological developments across information technology, biotechnology, nanotechnology, and satellite communications technology have the potential not only to affect products and services, but also consumers' behaviors and firms' strategic advantages, with important economic and policy implications. Emerging technologies dramatically affect firms, markets, and marketing as we know them. Indeed, while some firms have benefited from the effects of emerging technologies, other firms, and sometimes even industries, have been overwhelmed by them. Today, a wave of new technologies, enabled by the Internet, mobile telephony, and related technologies are forcing companies to completely re-engineer the way they conduct business generally, but also specifically in the way in which they manage customer relationships.

We propose that these emerging technologies, with their different distinctive characteristics, simultaneously pose opportunities and threats to markets, customers, customer relationships, and, in turn, for customer equity management. We further propose that these opportunities and threats to customer equity management generate strategic implications for customer equity management practice and generate opportunities for further research.

We structure the chapter as follows. First, we provide a definition and a conceptual framework on the factors influencing customer equity to provide the setting for describing the effects of emerging technologies on customer equity management. Following that, we identify dual effects (i.e., opportunities and threats) of technologies on various aspects of customer equity management and discuss the resultant implications for business practice. In the following section, we identify opportunities for academic research arising from the effects of technologies on customer equity management. We conclude with a discussion of the summary of the key highlights of the paper.

CUSTOMER EQUITY AND CUSTOMER EQUITY MANAGEMENT: AN OVERVIEW

Following Blattberg and Deighton (1996) and Rust, Zeithaml, and Lemon (2000, p. 4) we define customer equity as the net present value of the customer lifetime value, that is, profits (revenue less acquisition and retention costs) from the firm's current customers over their lifetimes. Over the past two decades, there has been a large body of conceptual, analytical, and empirical work on customer equity management, including its drivers and effects, which has identified and quantified the effects of various marketing program elements on customer equity. Please see Villanueva and Hanssens (2007) for a comprehensive review of the developments and findings in the marketing literature on the measurement and management of customer equity.

Following Rust et al. (2000), we view customer equity as being driven by value equity, brand equity, and relationship (retention) equity. Value equity is defined as the customer's objective assessment of the utility of a brand, based on perceptions of what is given up for what is received. Three key levers influence value equity: quality, price, and convenience. Brand equity is built through image and meaning and is defined as the customer's subjective and intangible assessment of the brand, above and beyond its perceived value. The key actionable levers of brand equity are: brand awareness, brand attitude, and corporate ethics. Relationship equity is the tendency of the customer to stay with the brand, above and beyond the objective and subjective assessments of the brand. The key levers under the firm's control that enhance relationship equity include loyalty programs, special recognition and treatment, affinity programs, and knowledge-building programs.

We use the framework for customer equity management conceptualized by Rust et al. (2000), which proposes that the firm's investments in marketing result in improvements in the three drivers of customer equity (value equity, brand equity, and relationship equity), which, in turn, improve customer perceptions. Customer perceptions affect customer attraction to the firm's products and retention, which should increase customer revenues, decrease costs, and increase customer lifetime values, and in turn, customer equity.

Further, we incorporate the idea proposed by Blattberg, Malthouse, and Neslin (2009) that exogenous customer characteristics also affect customer equity. We propose that the opportunities and threats arising from new technologies will affect the various elements of the framework including the levels (and cost) of marketing investments, the three drivers of customer equity (value equity, brand equity, and relationship equity), customer perceptions and associated customer attraction and retention, and customer lifetime values, in turn affecting customer equity.

DUAL EFFECTS OF NEW TECHNOLOGIES ON CUSTOMER EQUITY MANAGEMENT

As noted above, our review of the technology developments and business practices suggests dual effects of new technologies, creating both opportunities and threats for various aspects of customers, markets, and customer relationships, and, in turn, for customer equity management (Figure 16.1). We next identify the specific challenges and threats posed by new technologies along five broad dimensions: markets, customer relationships, the basis of competition, marketing program effectiveness, and customer value propositions – key aspects that feed into the customer equity management framework. Within each of these five dimensions, we further identify specific examples of how these technologies affect these five dimensions. Following the identification of the effects of technologies in each dimension, we discuss the specific opportunities and threats for customer equity management practice. New technologies are in a continuous state of change and their effects are therefore in a state of flux. Thus, in the interest of brevity, our goal in this section is to provide instances of the nature of the effects of new technologies for customer equity management, and not to be comprehensive about all the effects of new technologies.

Creating and Destroying Markets and Customers

Borderless markets

The hectic pace of development of Internet technologies including high-speed web and cloud computing capabilities has changed the face of markets. Specifically, the effects of the Internet with its abundant growth of information-based and e-commerce websites and search engines has given rise to borderless markets (e.g., a customer in North America can very easily order products online from a supplier in China); widespread disintermediation of distribution channels (e.g., companies that used to previously sell through channel partners are now selling their products through their website); and 24–7 worldwide connectivity (i.e., the store never shuts) not just for customers in developed markets, but also in emerging markets.

For customer equity management practice, this connectedness of firms to customers all over the world, suggests that the potential market for a

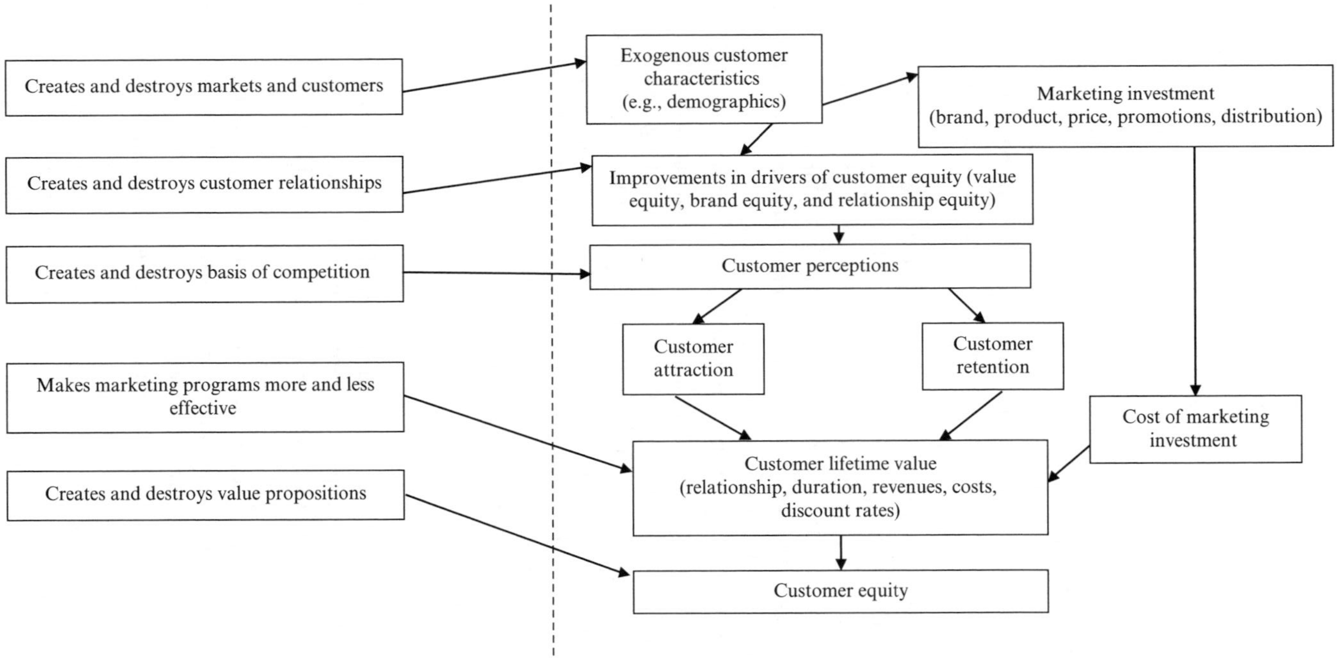

Figure 16.1 Multifaceted effects of new technologies on various aspects of customer equity management

firm's products increases dramatically, representing a tangible opportunity to increase the base of customers. In addition, the enhanced 24–7 connectivity of customers has resulted in the customer perception that companies and their customer relationship management systems are always open for business, increasing the pressures on companies to increase their customer service and retention efforts, with implications for their costs, and therefore, customer lifetime values.

Increasing price heterogeneity of customers

While the substantially increasing size of markets offers opportunities for market expansion, the ability and willingness to pay of customers in different parts of the world may vary with such a large number of customers. There may be individuals who never buy a company's product (i.e., because they cannot afford it) even as they are 'aspirant' customers. In other words, where previously most (excluding a few window shoppers) of a company's customers bought products from it, we now have a situation where we have customers who 'buy' and customers who 'do not buy' (lurker customers) but have the power to influence other customers through their presence on social media websites.

From a customer equity management perspective, the opening up of new markets and customers (with limited purchasing power) represents opportunities to develop new low-cost product offerings with a view to acquiring new customers, build a stream of revenues, and increase customer lifetime values. However, simultaneously companies must also be concerned about and manage the large increase in number of lurker customers, with the potential to have a substantial negative downside (e.g., flaming on online websites) despite not contributing any revenue or profits to the companies.

Converging product markets

With the advent of digital convergence, technological developments are combining previously separate technologies into new products, resulting in converging markets (Yoffie, 1997). As a result, there is a blurring of market boundaries. For example, *The New York Times* newspaper and its website now compete with pure-online websites such as the *Huffington Post*. Technological developments have also resulted in disruption to existing business models (e.g., if news content is available sooner online and at no cost, this threatens the existence of advertisement-based revenue models for traditional media such as newspapers and magazines). In addition to changing consumption practices of various converged digital products (e.g., music, news, and film) and changing the boundaries of previously watertight industries (e.g., news, media, film), technological convergence has destroyed revenue models and industries in the offline world (e.g., magazines, newspapers, and radio).

For customer equity management practice in industries facing digital convergence, managers must address issues related to product positioning and pricing. Traditional product offerings and pricing models based on pricing content are being overturned in favor of pricing models based on advertising (typically for a third party advertiser) where content is provided 'free' to customers. Managers have to examine issues related to how such a pricing model affects value equity (will customers value content if they do not pay for the content?), brand equity (how will advertising for a third party advertiser affect company's brand equity?), and retention equity (will customers who do not pay for a product have any reason to stay with the company?).

Creating and Destroying Customer Relationships

Always connected customers

New technologies have dramatic effects on the way companies manage their relationships with their customers. Because of the high connectivity and interactivity of Internet-based technologies and the 24–7 nature of online commerce operations, customers now expect to receive, and companies are forced to provide, 'real-time' instantaneous service. Easy access to social media websites where consumers can provide feedback on their experiences, without an opportunity for the firm to respond, increases the stakes for companies to manage and respond to customers in a timely manner, indeed, in real time. Moreover, given the computer/smartphone-based interface in online environments that increasingly constitute the storefront, companies can automate their front-office and customer service and farm out the work to their customers, which should result in substantial cost reductions even as it increases service levels and decreases response times.

For customer equity management practice, the higher expectations of customers combined with the accelerated pace of commerce (immediate billing, overnight and same-day delivery) have a clear benefit in terms of hastening customer purchase cycles and improving inventory management, both of which should decrease customer acquisition, management, and retention costs. However, at the same time, customers may also be habituated into expecting 'faster and faster' responses, even as they expect lower prices from companies, both of which will increase customer acquisition and retention costs, while simultaneously decreasing revenues and profits, thereby decreasing customer lifetime values.

Channel disintermediation

In addition, as more and more customers go online to purchase goods and services, many traditional retailers and suppliers will continue to struggle with issues relating to how the Internet will impact product pricing, promotion, and distribution. Given that buyers now have access to an endless source of information about any number of goods and services, retailers and suppliers are beginning to concede that a shift in power is underway. Indeed, the ubiquitous presence of online stores (many of them retailers) offers a real threat to existing customer relationships, as offline stores and manufacturers are disintermediated by the 'showrooming' effect where customers search for information about products in the offline stores and buy the products online (Bughin et al., 2011).

These changes have many implications for customer equity management. To start with, the low switching costs (essentially at the click of a mouse) enabled by the online e-commerce stores, increase choices for customers. In the face of such increased choices for customers, companies will have to invest in new product development, additional acquisition, and customer retention costs (Chen and Hitt, 2002). In addition, to ensure that customers remain loyal to them (in the face of price competition), companies may have to implement a real-time dynamic pricing management system (so that prices can be matched to those of suppliers) and ensure timely (even same-day) delivery to ensure that customers do not abandon them, increasing customer retention costs, decreasing customer profits, and decreasing customer lifetime values.

Changing ownership of customers

In addition, with the explosive growth of online e-commerce channels (e.g., eBay, Amazon, and Alibris), which maintain relationships with their customers who buy various products (e.g., consumer electronics, books, and groceries), an important question arises with respect to the 'ownership of customers'. For example, an Amazon customer who buys Gillette shaving products from Amazon has a dual relationship with both Amazon and Gillette. While one may argue that this type of divided loyalty also exists in offline environments (e.g., buying Gillette from Safeway Stores), we contend that because of the ease with which customers can move across online retailers, it is a more serious issue in online environments.

From a customer equity management perspective, for companies selling branded products through online retailers, their relationships with their customers are under threat, as information on their customers' purchase behaviors is available to Amazon, which can now use this information to develop competitive offerings for the next purchase cycle and/or offer cross-sell and up-sell offerings, having established a direct online channel with the customers.

Emergence of the social customer

Another key technology development are social media, whose growth depends on mobile and web-based technologies to create highly interactive platforms via which individuals and communities share, co-create, discuss, and modify user-generated content. They introduce substantial and pervasive changes to communication between organizations, communities, and individuals, and, in particular, to the relationships between companies and their customers (Kaplan and Haenlein, 2011).

In terms of opportunities, because of the openness of the social media platform, there is a transition from a one-to-one relationship between the company and its customers (whether online or offline) to a 'social network' of customers where one customer not only sees what other customers are saying about the company and its products but also how the company is responding to the customers' feedback.

From a customer equity management perspective, social media create a situation where even low-value customers become empowered (in the social media space), which may strengthen their bonds with the company, decreasing customer acquisition and retention costs. This provides companies with an opportunity to potentially leverage these low-value customers into higher-value customers, increase customer revenues, lifetimes and customer lifetime values.

Moreover, user-generated feedback from customers emerges as a strong and credible input for customers' decision-making so that, in essence, satisfied customers become 'champions' of the company, with implications for both value equity and brand equity drivers of customer equity. For customer equity management, this implies reallocation of resources from advertising in traditional channels to online reputation management systems, with attendant implications for customer acquisition, customer retention, and customer lifetime values.

For customer equity management, the unbridled provision of open customer feedback also creates some threats for companies. A low-value customer may have as much and sometimes even more influence (because of loud and strident negative feedback) than a high-value customer (Kaplan and Haenlein, 2010). This suggests that social media can create negative externalities for the firm's high-value customers, increasing its customer acquisition costs, not only for its low-value customers, but also for its high-value customers.

Creating and Destroying Basis of Competition

Geolocation technologies

A key feature of Internet technologies is their ability to offer information on products and prices to customers, and to allow customers to search for information across products (including their features and prices). Finally, the pertinent question for the next wave of social networking is 'Where are you?', and services like Foursquare, Gowalla, Brightkite, and Loopt want customers to use their smartphone to answer this question. The technology at the heart of this trend is geolocation, and with a GPS-enabled smartphone such as the Apple iPhone, HTC Google Nexus One, or BlackBerry, people can use it to let their friends (and the rest of the world) know where they are, or to find places recommended by people they know, or to check in remotely at clubs, bars, and restaurants (Vaughan-Nichols, 2009). Geolocation technologies now enable marketers to significantly enhance their ability to reach customers using both business type and consumer location data.

The emergence of mobile technologies and geolocation technologies (which use a geographic positioning system to pinpoint a customer's geographical coordinates) has further reduced search costs for customers, even as they have made browsing, searching, and online shopping even more convenient and cost-effective, enabling superior customer decision-making. In addition, because of the nature of search capabilities that allow sorting by features (including price), customers (or at least a segment of them) will become more price-sensitive as the ubiquity of Internet-enabled online shopping grows over time.

For customer equity management, by knowing precisely where someone is going online, firms can target audience segments based on business-type level, reaching customers in ways that were previously possible only through direct mail and billboards. In addition, this method of reaching customers is dynamic and can accommodate the development of dynamic product and price offerings, with the potential to increase both producers' and consumers' surpluses. In addition, based on customer purchase behaviors combined with their geolocation, companies can identify superior cross-sell and up-sell programs, with the potential to increase customer loyalty, even as they increase customer profits, and customer lifetime values.

Destroying entry barriers

Because of Internet technologies, small, hitherto unknown companies can enter new product markets that may not have been accessible to them in the pre-Internet time period.

For customer equity management, for these small companies, the

Internet, by virtue of the fact that an online product catalog constitutes a store, has allowed them to compete, effectively reducing market entry and customer acquisition costs. On the negative side, large companies are now being challenged by these small companies, and have to compete on price, which should increase their customer acquisition and retention costs, even as customer lifetime value decreases.

Making Marketing Programs More (and Less) Cost-effective

More effective and targeted marketing

Developments in new technologies are having profound impacts on both the content and the effectiveness of marketing programs. Developments in keyword search (both paid and organic search) offer a pinpointed, targeted method to target customers interested in a product. This approach to reach 'interested' customers (one at a time) represents a radical departure from the one-to-many efforts embodied in traditional media campaigns such as television, newspapers, and even direct marketing.

For customer equity management, these developments suggest a new way of developing marketing programs to acquire, manage, and retain customers, with implications for reorganizing firms' marketing function and customer management systems. For example, digital marketing (search engine optimization, online reputation management) requires very different skills (e.g., business analytics) from those required for offline marketing management (e.g., copy, creative, media buys, etc.). In addition to their implications for marketing organization, these changes can have substantial implications for customer acquisition costs and attendant effects on customer profits and customer lifetime value.

Increased customer control

E-commerce websites, by definition, are driven by customers' actions to visit the website, browse, search, and eventually place online orders for products. An important feature of online e-commerce websites is that the customer is the agent, whose actions determine whether or not there will be a product sale. Contrast this with the pre-online, offline commerce world, where customers would walk into a store (and meet with a salesperson), call the company (and speak with the salesperson), so that closure of the sale was driven, in part, by the company's salesperson and his or her efforts. In the online environment, customers are now adapted to performing all the work related to the sales transaction, and in a sense, this has wrested the action from the control of companies.

From a customer equity management perspective, there are clear opportunities for companies to outsource the work necessary for the sales transaction, reducing customer acquisition costs. But in doing so, companies also run the risk of not being able to control and manage their customers' interactions with them, which will result in increased customer defections, and decreasing customer profits and customer lifetime value.

Increased user-generated content

Social media impact customer purchase decision-making by making available user-generated consumer reviews and ratings. An important outcome from this is that companies' online reputations are now a novel, yet integral, component of the firm's brand equity.

From a customer equity management perspective, online reputation management becomes an important part of the firm's marketing program, to build and maintain the company's online reputation, with implications for organizing the firm's marketing function (see our discussion above), marketing budgets, and, in turn, value equity, brand equity, and retention equity drivers of customer equity.

Enhanced role for the crowd

Developments in social media technologies have resulted in the growth of crowdsourcing platforms (Howe, 2008) where the crowd consisting of individuals, not just the company's customers, can be used to develop new products and promotions.

The opportunity to crowdsource innovations offers many benefits to companies from a customer equity management perspective. To start with, companies are able not only to cost-effectively harness the 'creative' talents of the crowd, but also create products that, presumably, have better fit with their customers' needs, which should increase their chances of commercial success, so decreasing customer acquisition and retention costs, and increasing customer profits and lifetime value. Moreover, companies can also use the crowdsourcing exercise to build higher involvement and engagement with these individuals in the crowd, who may eventually become their customers. In addition, by becoming co-creators of goods and services with the companies, these crowds may be used as very credible brand/corporate ambassadors, again decreasing the company's marketing investments, and increasing its customer profits and lifetime value. From a customer equity management perspective, these new social media channels offer a timely supplement to traditional mechanisms of channel relationships, decrease customer acquisition and retention costs, in addition to adding superior value for its customers (increasing value equity), and increasing customer profits and customer lifetime value. The order of magnitude of gains in reach and efficiency of reaching are unprecedented when compared to the offline mechanisms to reach customers (e.g., television and newspaper advertising).

Growing role of big data

A clear consequence of the emergence of new technology-enabled e-commerce, including mobile marketing, is the growing amount of information that is being collected. Developments in data collection, computing, and storage (primarily Internet-based technologies), have resulted in the availability of big data (extremely large datasets) and associated developments in reductions in the cost and speed of computing (including cloud computing). These advances have created many opportunities for generating insights that can be useful to increase the efficiency and effectiveness of marketing programs in general, but customer equity management practice in particular.

For customer equity management practice, this growing information in big datasets, combined with lower costs of computing, provide opportunities to generate insights that will allow companies to develop and customize product offerings to not only increase customer value (i.e., value equity), but also increase customer profits and customer retention, increasing customer lifetime value. Further, an important threat from the growing availability of information in big data are issues related to secure handling of data, including privacy concerns, with implications for value equity, brand equity, and retention equity drivers of customer equity.

Creating and Destroying Value Propositions

Moving from boxes to the cloud

An important effect of new technologies for customers is the creation of new value propositions. For example, the emergence of direct communications between customers and manufacturers, combined with the advent of cloud computing has resulted in a direct engagement between manufacturers of information goods such as software (e.g., Adobe Inc., Microsoft Inc.) and their customers, which has resulted in a shift in business models from selling 'boxes' (i.e., packaged software) to 'annuity-based' relationship models where software is sold as a business service on a monthly subscription or pay-per-use model without the intervening marketing channels (i.e., value-added resellers).

From a customer equity management perspective, besides offering the key strategic advantage of building customer relationships, the foundational building block for customer equity management, this shift in business model allows companies to establish a direct link to customers, which allows them to understand their customers better, develop superior product offerings, and cross-sell and up-sell to the customers, so increasing customer lifetimes and customer lifetime values.

Free is the new price

Developments in new technologies have also resulted in the development of new business models based on advertising, and free pricing models with information on customers as sources of revenues and profits. In terms of threat, models based on advertising and free pricing, and customers as sources of revenues and profits, up-end traditional customer relationship management systems.

For customer equity management, this may mean a reassessment of how to build value equity, brand equity, and retention equity, along with implications for customer acquisition and retention costs, and customer profits and customer lifetime values.

RESEARCH OPPORTUNITIES ON THE EFFECTS OF TECHNOLOGIES ON CUSTOMER EQUITY MANAGEMENT

While there is a rich and insightful body of work on customer equity management, the changes in technologies and their effects on customer equity management have created opportunities for additional work using multiple approaches, including theoretical models, field studies, secondary data from various sources, and surveys to generate additional insights in this area, to improve both managerial practice and marketing theory. In this section, we identify a representative set (rather than an exhaustive list) of research opportunities that arise in the five domains where new technologies affect customer equity management discussed in the previous section.

Customers and Markets

Given the substantive effects of new technologies in generating new customers and new markets, and the interconnectedness of customers and companies across markets, it would be interesting to examine externalities across customers in different markets. For example, companies (e.g., Amazon Inc., Dell Inc.) have e-commerce websites catering to customers in different countries with different products, promotion, and pricing strategies. What, if any, are the externalities (both positive and negative) across the different markets?

Digital convergence has resulted in the emergence of converging markets (e.g., news) where the online media characterized by a 'free' pricing model and the offline media characterized by a pay model have to coexist within the same company (e.g., *New York Times*, *Wall Street Journal*). Interesting research questions in this area include the effects of different product and pricing bundles (e.g., subscription-based models, pay-per-use model) on customer retention and loyalty to extend the customer equity literature into the complex domain of mixed media markets.

Customer Relationships

Internet-technology–enabled e-commerce websites enable 24–7 connectivity with customers, which should strengthen a company's bonds with its customers. Yet, at the same time, the very nature of the technology allows customers to surf away from the website, increasing customer attrition. An important research question would be to examine the benefits and costs of such connectivity in terms of customers' expectations, satisfaction, and customer acquisition and customer retention costs. What website design features increase the stickiness of websites in increasing customer attraction and decreasing customer acquisition and retention costs?

Online e-commerce websites (e.g., Amazon, eBay) offer a real threat to companies higher in the value chain (e.g., manufacturers) because of possible disintermediation and to offline retailers (e.g., Best Buy) because of the showrooming effect. What are the 'net' effects of online e-commerce websites to manufacturers higher up in the value chain and to offline retailers? Are there any specific strategies that these affected companies can adopt that can stem customer switching and reduce their customer acquisition and retention costs? Further, in instances where the customer has purchase a branded product (say Gillette) through a branded online retailer (say Amazon.com), which is tantamount to co-branding, it would be useful to examine the effects of such co-branding on customers' loyalty to each of the brands and willingness to be cross-sold products in related product categories from the online retailer. This has direct implications for increasing customer acquisition and retention costs and customer lifetimes as well. While this phenomenon no doubt occurs in the offline world, the low costs of searching and placing online orders make the threats of switching to the online retailer more salient here.

With respect to the effects of social media on customer relationships, an interesting question that arises is, what are the effects of different types of social media activity (e.g., Facebook, Twitter) on customer satisfaction with the company's goods and services? Do customers who 'speak up' on social media websites (either company owned or third party) have different levels of customer satisfaction (or dissatisfaction) with the company and therefore have different levels of customer loyalty than customers who are 'lurkers' or do not go to social media websites? This research would be interesting as it would generate implications for whether companies should direct customers (which types?) to social media websites or company-provided forums (e.g., emails) to elicit feedback with a view to maximizing customer loyalty.

Effects on the Competitive Landscape

A key effect of digital convergence is the emergence of companies from different industries into the new market. For example, companies such as Dell Inc. (a hardware company), Google Inc. (an online search company), Amazon Inc. (an online retailer) and IBM Inc. (software services) have entered the market for cloud-based computing services. An interesting research question would be: are the costs of getting customers to switch from hardware to service (cross-product categories) or across services (within product categories) and then retain them, different?

One development from the emergence of social media is the simultaneous emergence of online competition based on customers' online feedback (e.g., when customers choose hotels, they will be looking at the ratings of all hotels in a given geographical region). A question that offers the opportunity to generate rich insight examines the effects of online competition (i.e., performance ratings) on a given firm's market performance, relative to the effects of the focal firm's online ratings. Are these effects asymmetric? Are the effects of different competitors dependent on the characteristics of the focal firm? How can companies take these effects into account as they devise marketing programs to attract, manage, and retain customers?

Marketing Program Effectiveness

A key consequence of online e-commerce technologies is that customers are increasingly taking on more responsibilities for completing the sales transaction. While this clearly represents a cost saving for the firm, there may be some collateral costs in terms of how customers feel about using self-service technologies, without any interactions with humans from the company's side. Further, instead of salespersons providing customer and sales support, the customer has to self-serve from the company's website, which would typically have user-generated content (i.e., postings of customers' online ratings and reviews). An interesting research opportunity would be to examine whether there are any interaction effects between user-generated content on the website and customers' self-service options.

While mobile technologies are growing dramatically in usage, it is not clear whether the insights on customer equity management that developed in more traditional contexts of catalog marketing and direct marketing (i.e., call centers) apply to mobile marketing contexts. Customers' attention spans are short, the information provision is sparse, and customers' approaches to processing information and purchasing the product are substantially different from what is prevalent in traditional contexts. Moreover, given the seemingly 'low' costs of searching and browsing (on smartphones), it would be interesting to examine whether customer loyalty in mobile marketing mediums is different from those in other online mediums (e.g., online e-commerce stores) and offline contexts (e.g., catalog marketing and call center marketing). In addition, it would be useful to examine whether and how mobile marketing channels work in tandem with other channels in a multi-channel marketing program to affect customer acquisition and retention costs and customer loyalty. And, search-based marketing, e-commerce marketing and, especially, mobile marketing are likely to result in the generation of a large volume of customer data on searching, browsing, and buyer behaviors. The availability of such large datasets would offer interesting possibilities to examine how customers develop preferences for brands and products, how these preferences evolve over time, and how they affect customer loyalty.

Customer Value Propositions

An important effect of these new technologies is the emergence of price-based competition from low-cost suppliers, which seriously disrupts the value propositions offered to customers. Do customers who seek low prices also have scaled-down expectations in terms of service and product quality? Or are their expectations still anchored on the higher prices of the products in the offline world?

Another effect of emerging technologies combined with cloud-based services is the movement to annuity-based customer relationships from a relationship based on selling boxes. Interesting questions that arise here include what types of customers to target for transitioning from boxes to services, what services offerings are mostly likely to be cross-sold and what pricing structures (e.g., subscription-based models, with add-on pricing) would be most effective from a customer loyalty-building perspective.

CONCLUSION

Over the past two decades, developments in a broad range of technologies have had substantial impacts on customer equity management. In this chapter, we argue that the effects of new technologies on customer equity management have been double-edged, creating tremendous opportunities and threats for firms. This chapter represents a starting point for generating further research on this topic, highlights important issues that academic researchers may consider exploring in future research, and attempts to generate some insights for managers involved in grappling with the issues at the intersection of technologies and customer equity management.

REFERENCES

Blattberg, R.C. and J. Deighton (1996), 'Manage marketing by the customer equity test', *Harvard Business Review*, **74**(4), 136–44.

Blattberg, R.C., E.C. Malthouse and S. Neslin (2009), 'Lifetime value: empirical generalizations and some conceptual questions', *Journal of Interactive Marketing*, **23**(2), 157–68.

Bughin, J., A.H. Byers and M. Chui (2011), 'How social technologies are extending the organization', *McKinsey&Company*, accessed 14 February 2013 at http://www.mckinseyquarterly.com/How_social_technologies_are_extending_the_organization_2888.

Chen, S. and L.M. Hitt (2002), 'Measuring switching costs and the determinants of customer retention in Internet-enabled businesses: a study of the online brokerage industry', *Information Systems Research*, **13**(3), 255–74.

Howe, J. (2008), *Crowdsourcing: How the Power of the Crowd is Driving the Future of Business*, New York: Random House.

Kaplan, A.M. and M. Haenlein (2010), 'Users of the world, unite! The challenges and opportunities of social media', *Business Horizons*, **53**(1), 59–68.

Kaplan, A.M. and M. Haenlein (2011), 'Two hearts in 3/4 time: how to waltz the social media–viral marketing dance', *Business Horizons*, **54**(3), 253–63.

Rust, R.T., V.A. Zeithaml and K.N. Lemon (2000), *Driving Customer Equity: How Customer Lifetime Value Is Reshaping Corporate Strategy*, New York: The Free Press.

Vaughan-Nichols, S.J. (2009), 'Will mobile computing's future be location, location, location?', *Computer*, **42**(2), 14–17.

Villanueva, J. and D.M. Hanssens (2007), *Customer Equity: Measurement, Management and Research Opportunities (Foundations and Trends in Marketing)*, Hanover, MA: Now Publishers Inc., pp. 1–65.

Yoffie, D.B. (1997), *Competing in the Age of Digital Convergence*, Boston, MA: Harvard Business School Press.

17. Stop grouping and start regulating: a new approach to social media marketing

V. Kumar, Nandini Krishnamoorthy and Gayatri Shukla

INTRODUCTION

Social media are bringing about a transformation in the world of business and marketing. While this transformation is indeed all-pervasive and does require businesses to realign their consumer outreach strategies and budgets, it has unfortunately also triggered knee-jerk reactions from some businesses that are yet to understand the potent power of social media as a marketing medium, which as a result, have ended up taking a rather myopic approach to social media marketing (SMM). In this chapter, we specifically look at how certain myopic but widely rampant strategies such as 'group purchase discounts' have backfired on some businesses and how this can be corrected by using a 'regulated' approach in SMM. The growth in SMM has been primarily due to technological innovation and the rapid customer adoption of new technologies. Web 2.0, innovative mobile technologies, and the proliferation of cell phones and tablet computers have been key contributors to this technology rush. It was natural for businesses to try to leverage this fertile playground for business benefit and thus social media marketing was born.

Web 2.0 is the term given to describe the second generation of the World Wide Web that is focused on the ability of people to collaborate and share information online.[1] Web 1.0 gave people web-search, emails and e-commerce opportunities. Web 2.0 took it all a step further to provide mashed content, rich content sharing, streaming, social networks, and device-based content adaptation.

Businesses have leveraged the location-based services (LBS) functionality of mobile devices to reach audiences with the right content at the right time in the right context. Such highly targeted marketing combined with the power of social networks can be used to get customers to influence other customers through word-of-mouth (WOM) marketing. Regulated social media marketing, as we will see later in this chapter, is a way of combining SMM, WOM, and mobile technologies to help businesses attract new customers and also influence repeat purchases from the same customer. Regulated SMM essentially enables businesses to activate its influential customers to act as WOM agents and this can be measured using a metric called 'mobile influence value' (MIV), which is a derivative of 'customer influence value' (CIV).

In this chapter, we will also see how 'group purchase discounts' fail to leverage a customer's influence on his or her own network and instead, how regulated SMM in the mobile environment benefits businesses in the long term much more than such one-time deals.

THE ORIGINS OF SOCIAL MEDIA

Merriam-Webster defines social media as forms of electronic communication (such as websites for social networking and microblogging) through which users create online communities to share information, ideas, personal messages, and other content such as videos.[2] Before the term 'social media' existed pioneers of the Internet were only able to communicate with each other over email, a messaging service that now almost everyone has used. Fast-forwarding to 2003, major innovations occurred – the social networks MySpace, and Del.icio.us (now Delicious), the business networking website LinkedIn, and the photo-sharing website Photobucket were all introduced. These websites dominated the Internet, registering hundreds of thousands of users worldwide, creating and popularizing bookmarking and social news sharing, among other things. Soon after, in 2004, the 'blue giant' Facebook was introduced as a closed network for Harvard students to communicate with each other. Other social sharing platforms such as Digg and Flickr were also introduced to help disseminate news, facilitate bookmarking, and photo sharing. Year on year thereafter, the number of social networking and media companies went on increasing, and names such as YouTube, Twitter, Facebook, and Groupon became household names. In 2008, Facebook overtook MySpace as the number one social network and in 2011 reached over 600 million[3] users through the introduction of social networking apps on smartphones. As an astounding analogy based on this milestone, if Facebook were a country, it would be the third largest in the world.[4] The birth and evolution of social media is depicted pictorially in Figure 17.1

SOCIAL MEDIA – LATEST TRENDS

Although a relatively recent phenomenon, social networks have changed the world to such an extent that some may question how people

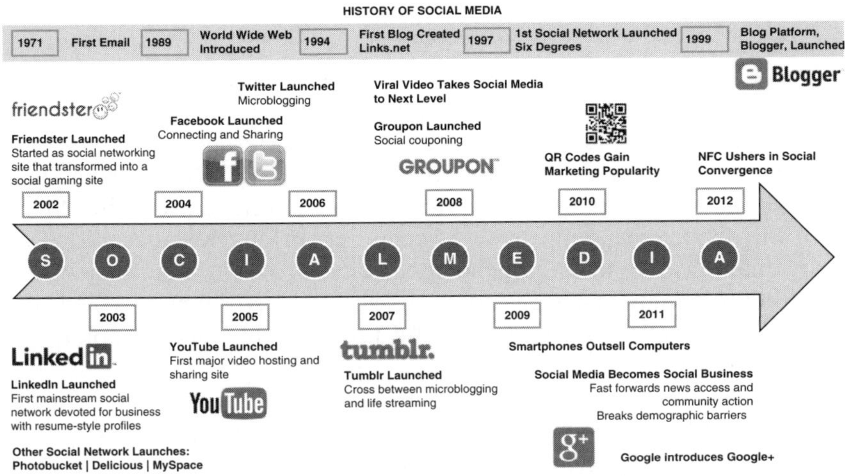

Source: Adapted from www.findandconvert.com/images/social-media-history.png.

Figure 17.1 Social media timeline

communicated effectively when these networks didn't exist. For a lot of people, it is hard to imagine a world without Facebook and Twitter. The convergence of social networks and mobile devices has made it extremely easy and compelling for people to stay connected. It is tough to ignore the silent calls on a vibrating phone in your pocket, to not check your Facebook newsfeed regularly, to not retweet a friend's funny tweet, not 'like' the latest picture he or she has posted, or not pin a meal choice on Pinterest. According to a recent HubSpot data survey, Internet users in the USA spent 27 percent of their Internet time on social networking sites; and social media generate two times the marketing leads compared to tradeshows, telemarketing, direct mail and so on.[5] From a survey by Market Force, 81 percent of US respondents indicated that posts from their friends directly influenced their purchase decisions.[6] Pew research stats on various social networks show that an astounding 67 percent of the population use Facebook.[7] Also, according to the *Huffington Post*, one million websites have integrated with Facebook and there are around 750 tweets on Twitter every second.[8] These statistics clearly indicate the enormous growth and power of social media. It is evident from the survey that social media have evolved from being just a way to stay connected, into a way to express oneself, and one's personal likes, dislikes, and lifestyle. This has created a golden opportunity for marketers to personalize offers based on the customer's location, who they know, and what they like, thereby maximizing the effectiveness of their marketing efforts.

There are many explanations as to why social networking is so addictive and has a majority of the population in its grasp. Its ability to identify, communicate, and connect with others literally at the push of a button and so quickly is a major contributor, along with its immediacy, the user interface, and how well it fits in with an average user's life. Along with the above, spending time on social networking websites allows an individual to unwind and spend their time browsing and keeping up with their digital persona. Spending time on these social networking sites is almost second nature and a second life to many users (Kaplan and Haenlein, 2010).

To stimulate and actively engage customers in their new social marketing initiatives, businesses are experimenting with novel ways of marketing and are willing to explore ideas that would help them acquire new customers, retain existing customers, and, in turn, increase their sales. One such recent trend has been the 'apparent' success of social coupons as offered by firms like Groupon, Scoutmob, and LivingSocial to name a few. An October 2012 survey by Manta, an online hub for entrepreneurs, showed that 82 percent of 1087 respondents do not intend to run daily deal promotions this year. Only 3 percent said such campaigns have garnered them repeat business, while 11 percent said they either lost money or made no money on the coupons[9] (Figure 17.2).

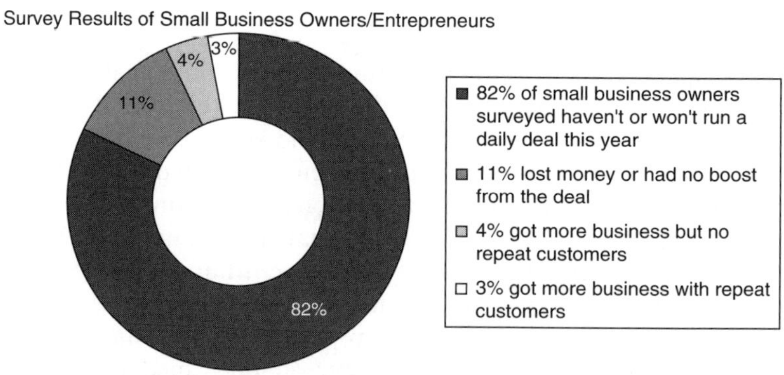

Source: Figure adapted from http://www.businessweek.com/printer/articles/84310-small-business-and-daily-deals-the-affair-is-over.

Figure 17.2 Daily deals: indifferent to unpopular

MIS-STEPS IN EARLY SOCIAL MEDIA MARKETING

One of the most commonly seen forms of social commerce these days is the concept of 'social coupons'. Social coupons have become very popular in new age marketing and many businesses include them in their marketing efforts. People need services from hair salons, nail salons, car washes, restaurants, and so on all the time. Companies like Groupon, LivingSocial and Scoutmob pitch coupon discounts on all these services through their websites on a daily basis. Groupon pioneered the evolution of the older concept of 'group discounts' into the social network realm by providing deep discounts for customers in return for providing businesses with guaranteed sales (provided certain minimums were met). This chapter explores the basics of Groupon-like networks, the pros and cons of such networks, what we think should be the next logical step to address the shortcomings of those networks, and the strategy of social couponing.

Groupon launched in late 2008, offering a daily deal on the best stuff to do, see, eat, and buy in 48 countries.[10] Groupon is the largest and most popular when it comes to social coupons, with 50+ million subscribers as of January 2011.[11] LivingSocial is the second largest daily deal site. Groupon's pitch to businesses was based on the premise that with deep discounts offered to consumers, it could expect to acquire new customers, thereby increasing sales instantly. The concept caught on with businesses and soon companies like LivingSocial were launched, with the same modus operandi of giving customers discounts on daily use services.

The attractiveness of Groupon lies in the huge advertising/publicity cost saves and increased reach it offers to businesses along with providing a targeted database (email) to send the coupons to. When it comes to social coupon campaigns, it is important to consider customer acquisition and retention from a customer management viewpoint. When making decisions about allocating resources for marketing efforts, it is common for businesses to target customers who are inexpensive (easier) to acquire and inexpensive (easier) to retain. Such decision-making has been shown to overemphasize the short-term gain of acquiring and retaining low-value customers over targeting the customers who are going to be most profitable in the long run (Thomas, Reinartz, and Kumar, 2004). In addition, by focusing on either short-term acquisition or retention, businesses fail to look at the bigger picture of balancing acquisition and retention together.

Businesses find social coupons appealing not just because they promote them, but also because they offer more control and measurability of the exact number of coupons sold, and a timely reimbursement for the number of coupons sold. In the case of Groupon, for example, if a hair salon offers $20 for a shampoo/cut service, but gives 50 percent off as a Groupon special, the customer buys the coupon for $10. But on an average, Groupon charges up to 50 percent of the revenue as the cost of advertising and acquiring customers. So, eventually, Groupon sends the business only a $5/coupon. In effect, the hair salon has to offer services worth $20 for a meager $5 for the entire promotional period, which is indeed a huge cut in its profits (although the immediate lift in sales appears misleadingly high).

Social coupon sites like Groupon do have the ability to introduce and expose a business to potential customers and entice them to come back if the redemption is handled properly. But a poorly designed coupon campaign can do serious harm to a business's profit margin. Social coupons can generate value for customers and the social coupon service providers themselves (who earn almost half of their revenue), but they can also cause problems for businesses. The issue of whether social coupons benefit businesses is covered in popular press articles while the academic literature on social coupons is rather scarce. A study by Edelman, Jaffe, and Kominers (2011) on the profitability and implications of online discount vouchers found that offering vouchers is more profitable for merchants who are relatively unknown and for merchants with low marginal costs. Research on businesses' experience with providing a deal via Groupon is studied in detail by Dholakia (2010). A more recent but similar research by the same author (Dholakia, 2011) studied 324 businesses that conducted a daily deal promotion between August 2009 and March 2011. Of those businesses, 55.5 percent reported making money, 26.6 percent lost money, and 17.9 percent broke even on their promotions. Although close to 80 percent of deal users were new customers, significantly fewer users spent beyond the deal's value or returned to purchase at full price. The study also concludes that it is highly possible that daily deal sites will not get as much revenue from businesses as they are getting now and it will be harder and more expensive for them to find viable businesses that are ready to give out discount coupons.

A recent detailed study (Kumar and Rajan, 2012) on social coupons answered several questions such as: (1) Do social coupons bolster or hurt small and medium businesses? (2) Can they help businesses acquire and retain more customers and nurture profits, or are the coupons more likely to cannibalize profitability? (3) How can businesses use social coupons effectively?

The research considered the impact of social coupon campaigns on the following three locally oriented businesses: (1) an ethnic restaurant that normally earned $2500 per month; (2) a car wash service that normally earned $6000 per month; and (3) a beauty salon that normally earned $6000 per month. All three businesses were located in a southeastern city in the United States. After their coupon launches, the businesses were tracked for one year. The research provided information on several business variables, including the revenue trend and acquisition and retention rates. This data helped the researchers develop a business model projecting the effect of coupon campaigns on both short-term and long-term performance. The study also focused on identifying the potential up-selling and/or cross-selling initiatives that could help the business attain more profits.

The findings revealed that social coupons do not guarantee customer acquisition and business profits if used in the way they are currently. The study did observe that all the three businesses acquired a significant number of new customers with their coupon offers and each of those businesses had considerable shortfalls in profits from their normal levels during the month the coupon campaigns were launched.

All three businesses had a retention rate of 70 percent. However, there were minor variations in the way the businesses retained the new customers. In effect, the following was found:

- The ethnic restaurant will not have any new customers to retain in Month 19 (assuming the coupon was launched in Month 1), and therefore will have to, if at all possible, recover the shortfall in profits by Month 18.
- The car wash business will have 140 new customers (that used the coupon) visit them every month.
- The beauty salon will have 31 new customers (that used the coupon) visit them every month. This in turn created significant financial losses.

In the case of the restaurant, the shortfall in the first month was more than \$7000; for the car wash service, it was \$6300, and for the spa it was \$18 360. If the businesses were able to achieve higher revenues and increased profits in future months, the financial burden could have been avoided. Even with the best marketing efforts, the three businesses had difficulty retaining most of the new customers who initially became customers just because of the deeply discounted coupons. The study projected that it would take the car wash business 15 months and the ethnic restaurant almost 18 months to recover from the financial burden. The worst of the lot was the case of the beauty salon. The projected recovery period (to recoup the money spent on discount coupons) for the beauty salon was more than 98 months, or eight years (keeping the coupon campaign at current business levels). Such long recovery periods for businesses will discourage them from attempting to relaunch similar social coupons in future. So to get the most of out of social couponing, businesses should prevent those coupons from cannibalizing their profits.

DEFINING A PROFITABLE SOCIAL MEDIA STRATEGY

After analyzing the results of the study, we present some general guidelines that can help businesses identify and avoid the pitfalls of social coupon campaigns. We strongly recommend that businesses limit the extent and frequency of their coupon promotions and design them painstakingly to ensure they are not giving too much away:

1 *Building deeper relationships with new customers*. Businesses usually initiate coupon promotions so they can acquire a large number of new customers in a single sweep. But as we saw in our study, the three businesses incurred losses as the 'new' customers did not continue their patronage and remained just one-time customers. Businesses should view the customer visits as an opportunity to up-sell/cross-sell rather than just letting the customer go with the one product or service they came for. Although most customers buy coupons for a specific purpose and redeem them for particular products, there may be some opportunities to broaden their spending horizon. For example, a hair salon can offer a shampoo and blow dry or hair color to the customer who has the discount coupon for a haircut and make money on the new services offered. Thus, the decision to launch a social coupon should be viewed as part of a broader business strategy rather than as a specific marketing tactic.

2 *Bringing strategy into the discount offers*. Customers purchase social coupons for their deep discounts and such offers make the customers deal prone. Deal-prone customers are the ones who usually buy items only when they are on sale or discounted, which causes losses to the firm. Businesses can fine-tune their strategy of when/how to offer coupons, for example, by limiting the offers to specific days of week (lean days versus busy days). Another strategy could also be to offer lesser/fewer discounts on peak service days such as weekends.

3 *Guarding against cannibalization of existing revenues*. As previously stated, most businesses introduce social coupons in the hope of acquiring new customers, but these coupons are typically also open to existing customers who would have otherwise paid the full price for the service had the discount not been offered. This encourages deal-seeking behavior among customers, and is harmful to the firm's bottom line. To eliminate the risk of existing customers using the discount coupons, businesses can restrict them to 'new customers only' or 'first-time buyers only'. This can bring in more new customers and avoid cannibalization of the revenues from current customers. Such a

trend is already seen in satellite TV providers as they only offer deals to new customers, for example, 'Offer valid for the first 15 months after signing up'.

4 *Leveraging coupon sites to introduce new products.* While it may not necessarily make sense for a new brand/product to be launched through online coupons/deals, firms may benefit from considering the promising synergies that coupon sites can offer a new brand/product, as a marketing channel. In a study[12] conducted by Millward Brown's Dynamic Logic and Coupons.com, it was found that engaging with visitors on a digital coupon site through strategically placed display advertising not only increases advertising recall and purchase intent significantly higher than the industry benchmark, but also helps the brand gain awareness among a group of consumers that has a proven track record of shopping more frequently and spending more than the average shopper.

Thus, by changing the nature of the coupon deals offered and the way coupon sites are used, businesses have the ability to control the number of new customers visiting them, the discount rate, and the number of existing customers who receive lower prices. The way that campaigns are calibrated can have tremendous implications for revenues and profits both in the short term and over extended periods of time.

Typically, the network effects of group buying are considered highly desirable. But unless managers are careful about how coupon programs are designed, network effects can become a negative force. This is not to deter businesses from using social media, but instead to encourage them to be careful and smart in planning their campaigns so that they can maximize the value they can reap from social media channels. Timing the content of the offer to duly meet consumer expectations, as well as choosing the right marketing channels and targeting the right audiences are keys to success in social media campaigns.

SOCIAL MEDIA: RELATED LITERATURE

The social networking movement that began on the web is moving faster than ever to mobile devices. The total time customers spend on social media through their mobile devices has increased substantially in recent years. Moreover, a majority of customers are using social networks via their mobile devices to communicate with each other. Around 55 percent of social networking content consumption occurs on a mobile device, providing marketers with a new platform through which to communicate with their customers.[13] Companies may risk losing customers if they do not cater to the customers who are ahead in the world of social networking on mobile devices.

Recent research has shown that by identifying the net influence wielded by a user in a social network and predicting the user's ability to generate the spread of viral information, businesses can determine the 'right' individuals to engage in social media conversations to promote WOM (Kumar et al., 2013). Further, by linking the generated WOM to the actual sales, research shows that social media can be used to generate sales, increase return on investment (ROI), induce positive WOM, and spread brand knowledge. Many existing research articles explore the effect of WOM on the recipient. However, most of these existing studies fail to capture the network effect in WOM propagation (Christiansen and Tax, 2000). The marriage between social media and mobile marketing is inevitable. Shankar and Balasubramanian (2009, p. 118) define mobile marketing as 'the two-way communication and promotion of an offer between a firm and its customers'. This allows marketers to not only reach their customers anytime, anywhere but also observes the customer's response to the offer. A recent study defines mobile social media as 'a group of mobile marketing applications that allow the creation and exchange of user-generated content' (Kaplan, 2012, p. 131). (Smith, Fischer, and Yongjian, 2012) found that brand-related, user-generated content differed across various social media sites for businesses that managed their brands proactively. In addition to this, businesses also need to focus on the relevance of the information provided to customers to invoke a response. Depending on the customer's physical space, their needs tend to vary (Rao and Minakakis, 2003). A study by Dickinger and Kleijnen (2008), reveals that firms should not overwhelm consumers with mobile coupons and recommends considering the usability of those coupons in deciding the right offer, and educating consumers on using the mobile coupons. Many mobile devices feature geolocation software and a lot of apps that transform them 'into location-sensitive broadcasters and receivers, endowed with knowledge of the individual's patterns, preferences, and social network' (Deighton and Kornfeld, 2009, p. 7). A study on mobile marketing indicates that message content is of greatest relevance for the perceived advertising value, while a high frequency of message exposure has a negative impact on it (Haghirian, Madlberger, and Tanuskova, 2005). Another study on location-based mobile and Internet marketing reveals that the benefit of browsing for geographically close matches is higher on mobile phones (Ghose, Goldfarb, and Han, 2012). Hence, there is a pressing need for marketers to reach out to their customers with a refined message based on the customer's location and preference. Mobile devices can help marketers provide location-based services along with facilitating real-time feedback/responses from consumers, thereby making a relevant two-way communication possible, which is not possible via desktops or PCs. In the past, there have been studies on developing a framework for mobile marketing and advertising applications that deliver promotional information to consumers based on their preferences and location (Kurkovsky and Harihar, 2006), but it is not integrated with social media.

Location-based services have enhanced a customer's experience with relevant content delivered on mobile devices but the security and privacy of users have been identified as the most prominent of the challenges in mobile social media (Kaplan, 2012). The continuing concerns on privacy leakages may restrict users from exploring the maximum potential of mobile social networking (Krishnamurthy and Wills, 2010). The perceived risk of losing information by using mobile technologies determines a user's usage pattern on mobile devices (Tao, 2008). When using social media, consumers usually share/receive deals only to/from trusted networks that they have voluntarily joined, and this reduces privacy concerns. Recent research on mobile local search behavior (Teevan et al., 2011) indicates that people are usually interested in immediately using the information they find during mobile local searches, and thus the availability of local search results is very important. Additionally, the study states that 63 percent of the participants who searched for local vendors/events were using mobile gadgets and they conducted their search on social media (in a trusted network). Kumar and Sundaram (2012) discuss the concept of regulated social media marketing (regulated SMM) where customers control the environment of the delivery and spread of the message, possibly as a foreseeable solution to these problems. However, the topic of regulated SMM in a mobile setting is yet to be explored. In spite of the fact that mobile social networking is gaining tremendous popularity, there has been limited research conducted in this field. The proposed study in this chapter suggests a framework for integrating the rich combination of social, mobile, and location-based information in a regulated SMM setting to successfully engage customers.

SOCIAL MEDIA CAMPAIGNS BASED ON WORD-OF-MOUTH

Social networks are aptly suited for word-of-mouth (WOM) marketing. Social networks and mobile platforms make WOM spread almost instantaneously, as compared to yesteryear when WOM was primarily in-person or over the phone, as a result of which messages took a good bit of time to spread to a larger group. The power of social networks lies in the extent

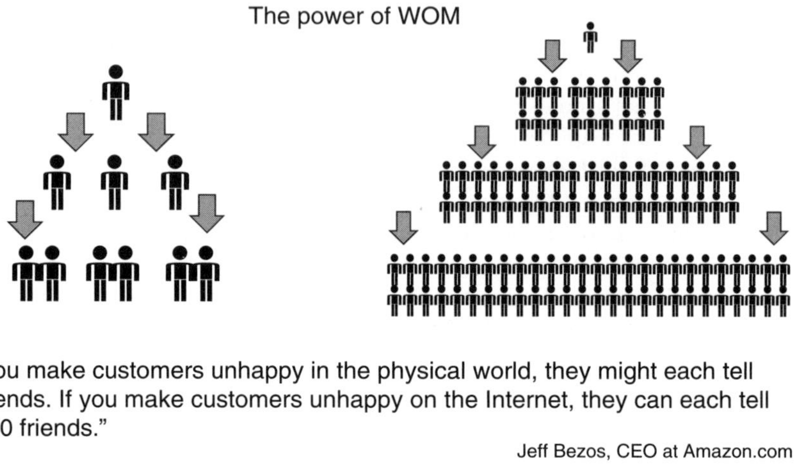

Source: www.ebusinessjournals.com/2011/01/25/dave-carroll-versus-united-airlines-when-customer-service-fails.

Figure 17.3 Power of WOM

of reach and the immediacy of interaction. Mobile devices further enhance this reach by allowing users to stay connected and interact. Traditional WOM wasn't always interactive. Social networks on the other hand, allow rich interaction and mobile devices allow that interaction to happen anywhere and anytime.

As people like to talk to each other and share ideas, social media channels serve as a medium for them to interact and interconnect. This interaction also includes consumer-to-consumer communication pertaining to products. This has led to the creation of a new form of WOM marketing that is consumer managed (see Figure 17.3).

The interaction happens via exchanging online messages, pictures, and videos across a network of consumers (mostly like-minded groups). Consumers are now communicating with each other about every facet of information regarding products/services and firms, thereby creating a brand image that is more lasting and impactful than that created by any traditional marketing efforts. Social media have become the information source of choice for consumers to compare competitive offerings, read other consumers' experiences, and also share their own experiences for the benefit of the general populace.

To know the extent of influence a person has on another and to measure this influence (if it is quantifiable), we define two metrics, which are given below.

CUSTOMER INFLUENCE EFFECT (CIE) AND CUSTOMER INFLUENCE VALUE (CIV)

This study introduces two new metrics called customer influence effect (CIE) and customer influence value (CIV) to help measure and quantify customer influence on others. (1) CIE measures the net spread and influence of a message from a particular individual (Kumar et al., 2013). (2) CIV calculates the monetary gain or loss realized by a firm that is attributable to a customer (influence), through his or her spread of positive or negative WOM. Using this metric, it is possible to craft strategies that can weed out or appreciate propagators of negative or positive WOM based on the probable damage or gain that they can cause to an organization in terms of lost or won sales (ibid.).

By identifying the net influence of a user in a social network and predicting the user's ability to spread viral information, businesses can be selective about who to engage in social media conversations for effective WOM. Once businesses associate the generated WOM to the actual sales, they could create strategies to effectively use social media to enhance brand awareness, influence positive WOM among customers, and spawn sales, thereby resulting in increased ROI.

The next section provides a framework that was implemented for an India-based ice cream retailer Hokey Pokey. This framework can be implemented by any business to create successful social media campaigns based on the principles detailed above.

HOKEY POKEY – A DETAILED STUDY

Hokey Pokey, a popular 'super premium' ice cream retailer in India, has 12 locations in Mumbai and seven outside Mumbai spread over multiple cities and rapidly expanding all over India. It offers 'customized mix-in' flavors, which are very popular among their strong customer base of young people in the 16 to 25 age group. The concept of customized food services was fairly new in the region, and the retailer felt the need to create a strong brand identity through its passionate customers. Most of Hokey Pokey's existing brand advocates fell within the 'Millennial' demographic cohort (today's 18- to 29-year-olds – members of the so-called Millennial Generation or Gen Y). The fact that over 50 million consumers in this segment in India actively spent over three hours on average on various online social networks underscored the pertinent relevance of social media marketing in the context of Hokey Pokey. The key business question then was 'How do we measure the return on investment of social media marketing?'

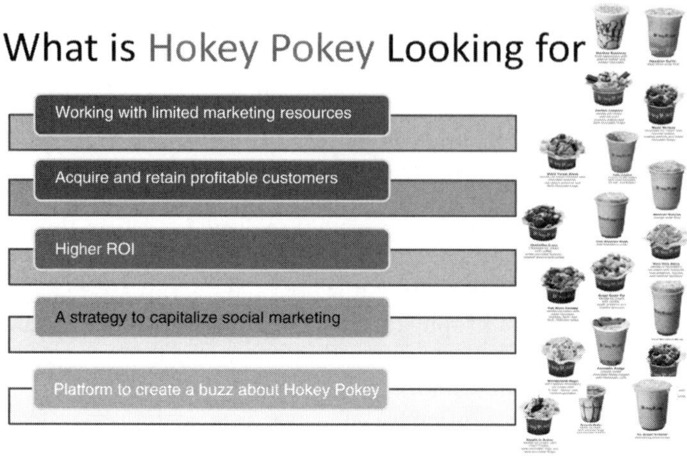

Figure 17.4 Goals for social media campaign of Hokey Pokey

Hokey Pokey decided to implement a seven-step framework (Kumar et al., 2013) to formulate its social media strategy (Figure 17.4) to optimize its shoestring marketing budget and realize a lift in sales and ROI.

Based on the framework, Hokey Pokey was able to develop an innovative strategy to market to customers in a targeted way. 'Creations on the wall' and 'Share your brownies' campaigns were the two pillars of Hokey Pokey's success in its outreach to customers. 'Creations on the wall' allowed customers to create their own ice cream flavors and display them on the 'Creations Wall' inside the store. Researchers encouraged customers to market their creations with their online peers on social platforms and 'share their brownies'. This portion of the campaign rewarded customers for marketing their creations and influencing others to try them, or to create their own. These customer creations stayed up on the wall for an extended period of time, allowing other customers to view and choose from these creations. Throughout this period, they monitored the frequency at which certain ice cream creations were selected by other customers. Hokey Pokey identified the most popular creations from the 'Creations on the wall' and added them to the permanent menu at all HokeyPokey locations.

The seven-step implementation was implemented at Hokey Pokey over a 30-month period. The project timeline is shown in Figure 17.5 and implementation details are given in Table 17.1.

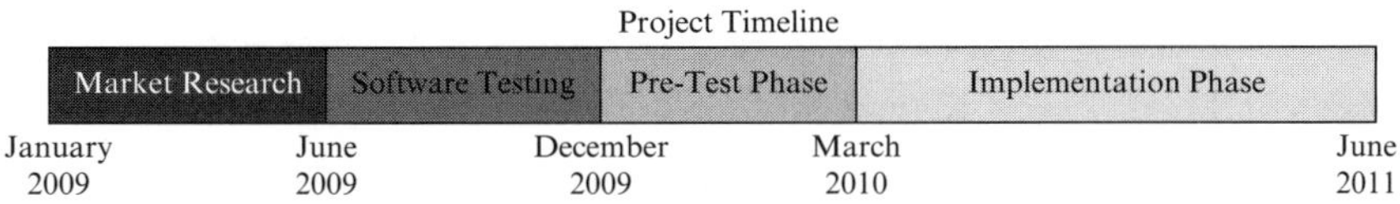

Figure 17.5 Hokey Pokey social media marketing project timeline

Campaign Results and Implications

The Hokey Pokey campaign was successful at both the company and the customer level. At the company level, the main impact was in the area of social media accountability. While most companies are still grappling with social media accountability, the use of the CIE and CIV metrics gave Hokey Pokey a distinct competitive edge. At the customer level, being able to calculate the value of an individual's influence in a network and measure the monetary value of customer influences made it possible for Hokey Pokey marketers to greatly enhance the efficiency of their social media campaign. In evaluating the performance of the framework, we benchmarked Hokey Pokey's revenue (based on ice cream sales) generated through Facebook and Twitter against specific performance metrics of the previous 16 months, including sales growth rate, ROI, number of positive and negative conversations, and number of repeat visits. Out of the total revenue generated from the 'Share your brownies' campaign, about 23 percent was attributable to the conversations on Twitter and about 80 percent was attributable to Facebook, with a 3–8 percent overlap between the two social networks. Overall, the campaign was a huge success: Hokey Pokey realized increases of 49 percent in brand awareness, 83 percent in ROI, and 40 percent in the sales revenue growth rate (Figure 17.6). The lack of robust methodologies to measure the impact of social media efforts is also addressed in this study. It provides tangible metrics and a robust methodology to measure the effectiveness of social media marketing spend and to maximize the ROI of social media campaigns.

This study provides a robust, first-of-its-kind methodology to measure the effectiveness of social media marketing spends and a guideline for maximizing the ROI of social media campaigns. The key takeaways from this study are:

- *Measurement and accountability of the tangible and intangible value of social media.* The new metrics CIV and CIE will help marketing managers project future monetary gains more accurately than metrics such as hits and page views.

Table 17.1 Implementation of the 'seven step social media marketing framework' at Hokey Pokey

Seven Step Framework for Successful Social Media Campaign	Brief Description	Framework Implementation
1. Monitor the conversations	Businesses must monitor social media conversations while conforming to legal and social norms to understand the pulse of their customers	The trial implementation was done for Hokey Pokey – an ice cream retailer in Mumbai, India. 825 091 conversations involving 1736 individuals were monitored
2. Identify the influential individuals	Next step is to identify those 'right' individuals who have the potential to influence others	Influencers were picked from networks that had at least 1000 users. The individuals chosen were connected to at least 20% of the network in the same location and wielded good influence on the others. Social networks were also chosen that provided for easy data collection
3. Identify factors shared by influential individuals	Common factors shared by the identified influencers need to be identified and listed to form profiles of influencers. For example, message spread, # of times messages are passed around by friends of influencers and the impact created (number of comments, likes, tweets etc.) are a few of the common factors	Activeness – number of times that an individual interacts with a message Clout – number of people an individual is connected to and being actively followed by Talkativeness (of the receiver) – number of times an individual's message gets repeated Likemindedness – a factor revealed by common interests and similarities between individuals

4. Locate all potential influencers with relevant interests	Businesses must locate all individuals who satisfy the influencer profile and who share the common factors that have been identified	Stickiness Index (SI) was used to locate influencers by finding the ratio of conversations related to ice creams to the total conversations they had over a certain period and individuals with high SI were selected
5. Recruit influencers	Enlist identified influencers to spread positive WOM	The previous four steps resulted in a list of influencers who were then sought after and recruited
6. Incentivize influencers for positive WOM	Engage enlisted influencers to actively spread positive WOM by providing them with incentives	Two intangible incentives and one tangible incentive were given to the influencers. The intangibles included 'Creations on the wall' and 'Share your brownies'. 'Creations on the wall' allowed influencers to come up with their own combo recipes that were displayed in a special location in ice cream parlors and 'Share your brownies' allowed them to 'tweet' about their creations actively in exchange for T-shirts
7. Reap the rewards	Associate conversations and positive WOM to brand recognition and sales to measure results	The final step involved converting social media measures like 'comments' and 'conversations' into quantified financial metrics by computing customer influence value (CIV) modeled upon the abstract social media measures

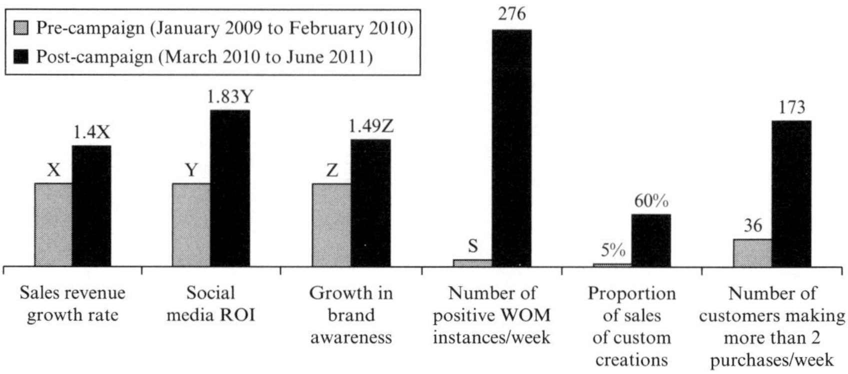

Note: X, Y, and Z represent base values for the sales revenue growth rate, social media ROI, and growth in brand awareness respectively.

Source: Figure partially adapted from Kumar and Mirchandani (2012).

Figure 17.6 ROI of social media

- *Customer cloning and effective targeting.* Social media can help businesses identify and activate 'clones' of their most influential (and therefore valuable) customers. This is one of the key ways that social media channels can leverage the rich databases of customer demographic, psychographic, and attitudinal information that are at their disposal. Initiatives such as, for instance, Facebook's 'graph search'[14] functionality in conjunction with its advertising solutions platform is designed to address this very need.
- *Generalizable and scalable evaluation framework.* While the Hokey Pokey project was focused on Facebook and Twitter, the framework it is based on is universally applicable and it is a scalable model that can be directly applied in the context of any retailer, any WOM campaign and for the promotion of any product/service and in any market that is actively influenced by social media.

Based on the results of the Hokey Pokey campaign, it is very easy to see how WOM marketing and social media are interrelated. We have also seen how they deliver mutual benefits to businesses and customers (Kumar and Sundaram, 2012). While the marriage of WOM and social media is successful, it needs to be able to offer the following for long-term sustainability:

- ability to adapt to different kinds of businesses;
- ability to cater to different sizes of business organizations;

- capability of operating in a mode where the identification of the 'right influencers' happens automatically in the chosen social media channels once the 'right message' is provided;
- ability to provide customers with the freedom to share or not share, but still be part of the network without any negative impact on transactions and on the social network;
- ability for the network to track WOM messages at the message level as well as influencer level.

With the above observations, we now proceed to the next section, which builds upon the strength of relationship between WOM and social media and also addresses the above 'wants' to make the marriage a long-term success. There should be a mechanism through which consumers feel encouraged to refer businesses to their social networks. These referrals, when acted upon by the friends of consumers, should be traceable to the respective sourcing consumers. The business can then reward those consumers for their referrals. The businesses can also choose to reward the friends that transacted/acted upon the referral and encourage them to, in turn, provide referrals.

NEWER APPROACH – REGULATED SOCIAL MEDIA MARKETING (REGULATED SMM)

To achieve all of the above, we introduce the concept of regulated social media marketing (regulated SMM). Regulated SMM deals with the aspect of sharing a customer's transaction experience in a social network, which will be perceived in the right spirit without making customers become clueless spam agents, but who instead through their own goodwill feel they are contributing to society.

These are the salient points of the framework:

- Provide the customer with the choice to share or not share messages from businesses.
- Allow the customer to have a choice of network where the message can be shared.
- Messages must be traceable throughout the activity cycle.
- Timing of the messages is key to making sure that the customer and customer's friends get actively engaged in social conversation.
- As with any marketing campaign, the message content is of paramount importance.

Merits of Regulated Social Marketing – a Comparative Look

Table 17.2 depicts the benefits of regulated SMM compared to other online/social media marketing strategies.

Regulated SMM provides a way by which the network automatically identifies and selects influencers and also regulates message sharing without having to expend time and money on these steps. This helps lower the cost of social media campaigns while simultaneously increasing efficiencies and lift in sales promised by social media marketing. Overall, regulated SMM achieves the increase in ROI as expected by businesses, which will translate into an increase in profits and shareholder value.

As in the case of the Hokey Pokey campaign as well as other regulated SMM campaigns, users are required to connect via social networks. One of the noticeable aspects of user behavior in social networks is the 'immediacy' or 'urgency' shown by people in their interactions. There seems to be a sense of thrill or instant gratification that people get in responding and sharing information the instant they think of it. This has been made possible by a timely successful marriage of social networks and mobile networks. Smartphones and tablets (like Apple's iPad and Android tablets) have made it easy for users to capture, share, and interact anytime and anywhere. It is only logical for marketers to reach out to customers via these new channels. This leads us to the topic of mobile marketing in the next section.

MOBILE MARKETING

A unique aspect about mobile marketing is the willingness of the mobile phone owner to be presented with marketing messages, promotions, and advertisements on his or her mobile device. Unlike other channels including in-browser advertising for online channels (which can also be presented on mobile phones) where advertisements can be targeted based on browsing history, context and cookies, mobile devices are essentially very personal and private devices. If somebody were to allow messages to reach them on such a personal device that in itself is a very big win.

There are three back-end technological enablers that help mobile device owners to stay connected with each other – cellular access, cellular data access, and/or through Wi-Fi. Cellular access provides Voice, SMS (TXT) and USSD (Unstructured Supplementary Service Data) channels while cellular data/Wi-Fi provides email, apps and browser interactions.

Mobile tariffs vary from country to country based on how they have evolved in the mobile communication world and the nature of the market in

Table 17.2 Comparative study

Evaluation Criteria	Broadcast Social Media Marketing	Selective Social Media Marketing	Regulated Social Media Marketing
Ability to track message chain	Low (almost none)	High (can track at customer level)	High (can track at message and customer level)
Ability to track customer acquisition	Low (only on an aggregate basis)	High (ability to track message effectiveness and influencers)	High (ability to track message effectiveness and influencers)
Ability to track customer retention	Low (only on an aggregate basis)	Medium (since repeat purchases are not directly traceable to customers but will have to be traced through messages)	High (ability to track repeat purchases which can be tracked to customers as well as messages)
Cost	Medium (while per message/ impression cost might be low, since broadcast is in volume, it might drive up the overall cost)	High (needs selection of media, message and customers)	Medium (no selection of customers but would require some initiation cost to get customer to opt in/register)
Flexibility in choice of social media	High (Facebook, Twitter, Google+, Groupon, LivingSocial, etc.)	Medium – customer privacy controls from networks like Facebook may constrain ways of selecting customers or groups of customers to share	Medium – requires customer intervention to opt in, which can help overcome customer privacy concerns and controls; allows for cross-network sharing

Table 17.2 (continued)

Evaluation Criteria	Broadcast Social Media Marketing	Selective Social Media Marketing	Regulated Social Media Marketing
Media-rich messages	Medium (depends on the channel of broadcast)	Medium (since there is a selection process involved, the richness of message would depend on the channel)	High (same message can be adopted to different formats since cross-channel or cross-network sharing is allowed)
Ability to regulate based on customer activity	Low (almost zero customer level usage tracking)	Medium (ability to track at message level with possibility of extending to customer level with some difficulty)	High (flexible enough to track at both customer level and message level)
Ability for customer to control and maintain his or her privacy, freedom and profile	Low (mostly pushed by business)	Medium (can choose to share or not and also whom to share with)	High (can choose to share or not, whom to share with, which network to share on and also share on a 'trusted but anonymous basis')

that country. Generally in the USA, voice calls are rated by the minutes and SMS/TXT messages are charged per message. Though unlimited calling and TXT plans abound, there is no way for marketers to know whether a user is on an unlimited or limited plan unless the user explicitly shares the information and agrees to receive voice calls and TXT messages for promotions on his or her cell phone. So the traditional and basic cellular access does not provide much of a leeway in the USA for marketing. In countries like India, incoming calls and TXT messages are not charged to the mobile user, which is a great boon for marketers to use the mobile channel for multicasting purposes like emails. This option is not available in the USA and hence protects the general public but doesn't help marketers.

Email marketing has existed for a long time and since email is one of the most used applications by any mobile user, the surge in mobile technology has enhanced the ability of email marketing to connect with users quickly.

The more effective channels for mobile marketing have been mobile applications (apps), which are downloaded and installed by users from app providers via formalized channels like Apple's App store and Google's Play store. The app installation process requires a mandatory user action for accepting the Terms & Conditions listed by the app provider and store for a successful install. Most of these applications operate on a model where ads are periodically displayed if the app is installed for free, whereas if the app is paid for, then the app experience is either ad-free or ads are less frequent and less intrusive in ad displays. Since there are a lot of freeware applications, marketers have a huge playground to play in. Also, since the app installs have a lot of user demographic details in them, the ads can be well targeted.

Social networking and gaming apps are the most commonly used applications on mobile devices. According to a recent article, 79 percent of mobile usage time is spent on social media and games.[15] Since both these genres of apps involve sharing and word-of-mouth marketing, the mobile device has become a fertile channel for mobile marketing. SMM on mobile is a very effective and fast-growing space within the marketing industry.

Mobile marketing has a few advantages over marketing in other channels:

- Immediacy – of connection and probable transaction.
- Relevance – ads can be context- and content-sensitive when driven through apps and mobile browsers.
- Cross-sell – mobile devices have the capability to provide location information (via GPS or cell tower location), time-of-day information and app context information. All of these are extremely useful inputs to create targeted ads for cross- and up-sell deals.

- Direct – since the user is addressed directly through their mobile device based on the context, time-of-day and location information, it is more likely that the user will find the ads more useful than on other channels.

Let us see how Facebook, which is one of the forerunners of the industry, uses mobile marketing to its advantage. Recently Facebook started concentrating more on mobile marketing and is already reaping the benefits of this new initiative. As of the fourth quarter in 2012, mobile ad revenues had rocketed up to $305 million, or 23 percent of overall ad sales.[16] The way it works is if a person is logged into Facebook, ads appear in the newsfeed and not in the side bar as we normally see on the desktop. This experience is designed around the idea of integrating ads with the natural user activity of checking their newsfeed and thereby maximizing the chances of the user noticing and clicking on the ads. According to Simon Mansell, CEO of Facebook marketing firm TBG Digital, with 'any mobile apps where you're logged into Facebook, you can run Sponsored Stories related to those apps. The ads also will be especially appealing to local businesses such as restaurants that appeal to people on the go'.[17]

Facebook's mobile marketing success story has unraveled the prevalent misconceptions in the mobile marketing industry about users not noticing the ads that are smaller than the flashy ads on their desktop browsers. Another myth that has exploded is of people not liking the idea of seeing ads that appear in the middle of their friends' posts. But tests that Facebook ran found that the insertion of ads reduced comments, likes, and other interaction with newsfeed updates by 2 percent, a small decline that the company deems acceptable.[18] It is also to be noted that Facebook still trails far behind mobile ad revenue leader Google – which earned $2.2 billion from mobile searches and ads in 2012.

Mobile marketing is the trend in emerging markets as well. Recently in Hong Kong, an interactive campaign developed by the McCann World Group for Coca-Cola became a hit, and sent viewers at home, in cinemas and in front of giant outdoor screens into a frenzy.[19] As part of the campaign, catchy ads were aired on television, prompting thousands of viewers to grab their phones and start shaking them frantically to virtually 'catch' the falling bottle caps on the screen and win instant prizes. Nine million people saw the ad, with 380000 downloading that specific app in the first month. Success stories of this kind indicate that marketers are right in directing ads at consumers via mobile phones.

All of the above developments in social networks and mobile technologies along with the explosion of context- and location-based data (Social-Location-Mobile – SoLoMo) provide a fertile landscape for targeted and effective social media campaigns. Coupons/discounts sent via social networks can be even more appealing, attractive, and effective, with marketers cashing in on the unique advantage of mobile phones by sending a targeted ad to the person specific to the location he or she is in. Human interaction on social networks makes it very evident that people are ready to share and exchange information if (1) they trust the information source and (2) are comfortable sharing that information with the group they want to share with. Businesses like Groupon extend steep discounts on purchases based on committed group purchases, but lack the concept of social networks in bringing friends, family, and trusted relationships together in a social environment.

Regulated SMM, on the other hand, brings group purchases, social networks, and mobile devices together into play, thus creating a 'trusted network environment' that is mutually beneficial for both customers and businesses. Since regulated SMM enables businesses to stay in touch with their customers beyond the shopping and transaction window, it helps the customer recall the shopping experience, the store or website, brands and so on. This is targeted as well as customized marketing at its best. Such marketing efforts have the potential to increase the percentage of repeat purchases, thereby extending and increasing the customer lifetime value (CLV) of customers (Kumar and Shah, 2011), where CLV is defined as the discounted net present value of expected future cash flows from a customer.

A great example of regulated SMM is the recent 'Tweet-a-Coffee' Twitter program[20] by Starbucks that lets consumers gift a $5 Starbucks eGift card to their friends via Twitter. To join the program, consumers have to add their credit card information to their Starbucks account and link it up with their Twitter account. To send the eGift to a friend, consumers simply have to put '@tweetacoffee' and the friend's handle in a tweet. This program was launched in late October 2013 and by early December 2013 it not only provided Starbucks with a completely trackable social media campaign, but also a rich database of 54,000 registered users with appended Twitter IDs and mobile phone numbers. On the revenue front, the program has resulted in $180,000 in purchases as of, 6 December 2013. Moreover, thanks to the rich social data reaped through this campaign, Starbucks will be able to also identify the key social media influencers from among the registered program users, as well as their interests, preferences, and consumption trends, and potentially target them in a focused manner. This campaign will also help Starbucks bring in new customers through the doors, and potentially get them to make additional purchases while in the store – all in return for the miniscule financial investment they have made on Twitter.

Thus, regulated SMM not only helps increase a customer's repeat purchase potential, but also builds on the customer's influence to bring in additional customers (new and returning). This increases the future revenue potential of the business, which in turn lowers the perceived risk involved with social media campaigns. Willingness to repurchase in the future is one of the factors that contribute to the measurement of a customer's propensity to defect in the future, thereby making future cash flow vulnerable to this defection. Response data from a regulated SMM campaign can provide the business with the ability to more accurately project future purchase behavior and cash flow, thereby reducing the perceived risk involved with social media campaigns. This directly contributes to the CLV computation, and marketing initiatives that are based on CLV computations have been proven to lift the stock prices of a company, since the perceived risk is lower (Kumar and Shah, 2011). By implementing regulated SMM in the mobile setting, the weak link between social media marketing and the target consumers is made stronger than ever before. We can target the right customers with the right message, at the right time. This is what we call 'focused marketing', which reduces volatility and vulnerability. With focused marketing, customer purchases are maximized, resulting in higher CLV, which in turn paves the way for higher firm-level revenue and profits, thereby predicting a higher shareholder value. The schematic diagram in Figure 17.7 represents how regulated social media marketing in a mobile setting impacts the shareholder value (by better predicting cash flow).

In conclusion, we state that regulating a business's marketing initiatives particularly in social media campaigns using regulated SMM techniques will provide it with a better outlook of future revenue and hence contribute to the bottom line of the company that is seen by the market. Regulated SMM brings in more efficiency in maximizing ROI for marketing campaigns, resulting in increased profits and a longer-lasting relationship with the customer, thereby increasing CLV. By calculating the CLV, firms can target customers with higher profit potential, manage and reward existing customers based on their profitability, and invest in high-profit customers (Kumar, 2008).

While group purchase discounts like Groupon are meant to deliver on the promise of committed purchases in the future, they don't offer an insight into repeat purchases and CLV of customers. Group purchase discount networks offer immediate revenue lifts to businesses, but fail to address customer buying behavior when group discounts aren't available, thereby negatively affecting the CLV of a customer. In contrast, regulated SMM takes a strategic and long-term view of helping businesses stay connected with customers, and increasing their repeat purchases and CLV, thereby empowering businesses to take control of their bottom lines.

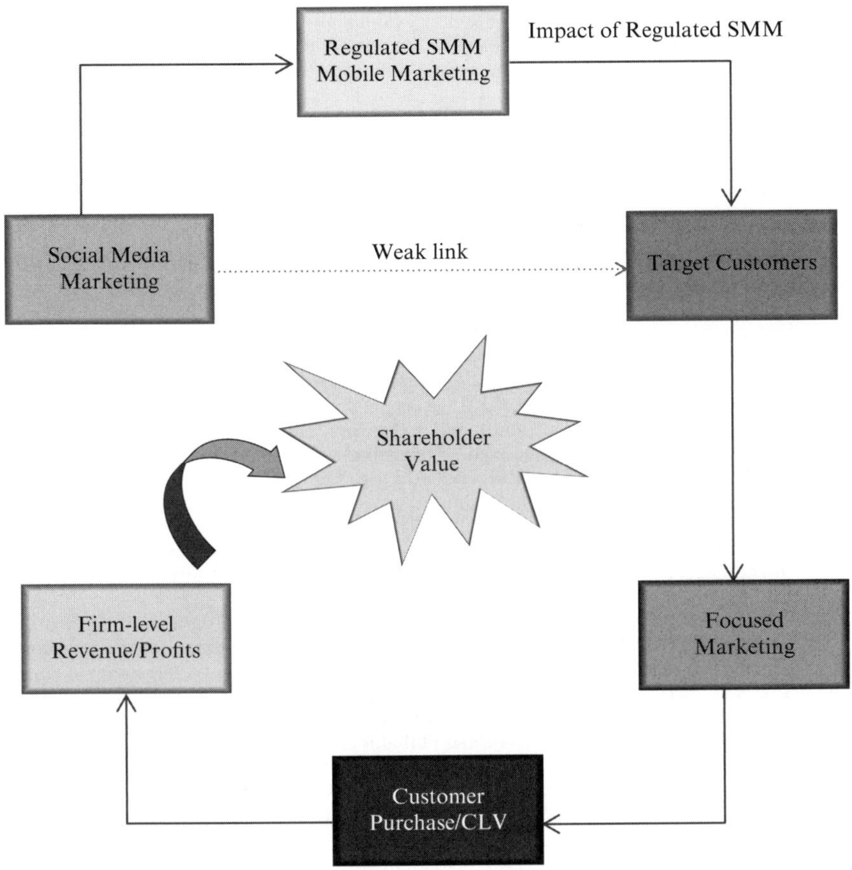

Figure 17.7 Impact of regulated SMM

NOTES

1. 'Web 2.0', *Webopedia*, accessed 7 May 2013 at http://www.webopedia.com/TERM/W/Web_2_point_0.html.
2. 'Social media', *Merriam-Webster*, accessed 7 May 2013 at http://www.merriam-webster.com/dictionary/social percent20media.
3. 'Social media: a history', *Boston University College of Communication*, accessed 8 May 2013 at http://www.bu.edu/com/comment/library/downloads/2012_comment.pdf.
4. Williams, R. (2012), 'Revealed: the third largest "country" in the world – Facebook hits one billion users', *Independent*, 4 October 2012, accessed 14 September 2014 at http://www.independent.co.uk/life-style/gadgets-and-tech/news/revealed-the-third-largest-country-in-the-world--facebook-hits-one-billion-users-8197597.html.
5. Soskey, G. (2013), '18 fresh stats about the current state of social media marketing',

HubSpot, 30 April 2013, accessed 15 September 2014 at http://blog.hubspot.com/18-fresh-stats-about-social-media-marketing.
6. Olenski, S. (2012), 'Are brands wielding more influence in social media than we thought?', *Forbes MarketShare*, 7 May 2012, accessed 15 September 2014 at http://www.forbes.com/sites/marketshare/2012/05/07/are-brands-wielding-more-influence-in-social-media-than-we-thought/.
7. Ferenstein, G. (2013), 'Fresh stats on social networks', *Techcrunch.com*, 17 February 2013, accessed 18 September 2014 at http://techcrunch.com/2013/02/17/social-media-statistics-2012/.
8. Honigman, B. (2012), '100 fascinating social media statistics and figures from 2012', *Huffington Post*, 29 November 2012, accessed 18 September 2014 at http://www.huffingtonpost.com/brian-honigman/100-fascinating-social-me_b_2185281.html.
9. Klein, K. (2012), 'Small businesses and daily deals: the affair is over', *BusinessWeek*, 29 November 2012, accessed 18 September 2014 at http://www.businessweek.com/printer/articles/84310-small-business-and-daily-deals-the-affair-is-over.
10. 'About us', *Groupon*, accessed 10 May 2013 at http://www.groupon.com/about.
11. Groupon Press, accessed 10 May 2013 at http://www.groupon.com/pages/press-kit.
12. 'New research explores attitudes and behaviors of digital coupon users' (2013), *Coupons.com*, 9 May 2013, accessed 9 December 2013 at http://www.couponsinc.com/corporate/OurCompany/PressReleases/2013.aspx?udt_678_param_detail=587.
13. '55 percent of social networking consumption occurs on a mobile device' (2013), *Marketing Charts*, 27 February 2013, accessed 29 April 2013 at http://www.marketingcharts.com/online/55-of-social-networking-consumption-occurs-on-a-mobile-device-27327/.
14. 'Introducing Graph Search', *Facebook*, accessed 7 May 2013 at http://www.facebook.com/about/graphsearch.
15. Alexander, A. (2012), 'Mobile app usage statistics 2012 [Infographic]', *Ansonalex.com*, 9 August 2012, accessed 7 May 2013 at http://ansonalex.com/infographics/mobile-app-usage-statistics-2012-infographic/.
16. Hof, R. (2013), 'How Facebook slew the mobile monster', *MIT Technology Review*, 6 March 2013, accessed 18 September 2014 at http://www.technologyreview.com/news/511781/how-facebook-slew-the-mobile-monster/.
17. Hof, R. (2012), 'Facebook launches first mobile only ads – right in your news feed', *Forbes.com*, 5 June 2012, accessed 18 September 2014 at http://www.forbes.com/sites/roberthof/2012/06/05/facebook-launches-first-mobile-ads-right-in-your-news-feed/.
18. See note 16.
19. Holton, K. and L. Abboud (2013), 'Chok! Chok! Chok! Ad shakes up mobile marketing', *Reuters.com*, 28 February 2013, accessed 18 September 2014 at http://www.reuters.com/article/2013/02/28/us-mobile-world-advertising-idUSBRE91R13I20130228.
20. Wasserman, T. (2013), 'Starbucks "Tweet-a-Coffee" campaign prompted $180,000 in purchases', *Mashable.com*, 6 December 2013, accessed 18 September 2014 at http://mashable.com/2013/12/05/starbuckss-tweet-a-coffee-180000/?utm_cid=mash-com-fb-bus-link.

REFERENCES

Christiansen, T. and S.S. Tax (2000), 'Measuring word of mouth: the questions of who and when?', *Journal of Marketing Communications*, **6**(3), 185–99.

Deighton, J. and L. Kornfeld (2009), 'Interactivity's unanticipated consequences for marketers and marketing', *Journal of Interactive Marketing*, **23**(1), 4–10.

Dholakia, U.M. (2010), 'How effective are groupon promotions for businesses?', *Social Science Research Network*, accessed 18 September 2014 at http://papers.ssrn.com/sol3/papers.cfm?abstract_id=1696327.

Dholakia, U. (2011), 'How businesses fare with daily deals: a multi-site analysis of Groupon, Livingsocial, Opentable, Travelzoo, and BuyWithMe promotions', *Social*

Science Research Network, accessed 18 September 2014 at http://papers.ssrn.com/sol3/papers.cfm?abstract_id=1863466.
Dickinger, A. and M. Kleijnen (2008), 'Coupons going wireless: determinants of consumer intentions to redeem mobile coupons', *Journal of Interactive Marketing*, **22**(3), 23–39.
Edelman, B., S. Jaffe and S. Kominers (2011), 'To groupon or not to groupon: the profitability of deep discounts', Harvard Business School NOM Unit Working Paper No. 11–063.
Ghose, A., A. Goldfarb and S.P. Han (2012), 'How is the mobile Internet different? Search costs and local activities', *Information Systems Research, Articles in Advance*, 1–19, accessed 18 September 2014 at http://pages.stern.nyu.edu/~aghose/mobilePC.pdf.
Haghirian, P., M. Madlberger and A. Tanuskova (2005), 'Increasing advertising value of mobile marketing – an empirical study of antecedents', in *System Sciences, 2005. HICSS'05. Proceedings of the 38th Annual Hawaii International Conference*, IEEE.
Kaplan, A.M. (2012), 'If you love something, let it go mobile: mobile marketing and mobile social media 4x4', *Business Horizons*, **55**(2), 129–39.
Kaplan, A.M and M. Haenlein (2010), 'Users of the world, unite! The challenges and opportunities of social media', *Business Horizons*, **53**(1), 59–68.
Krishnamurthy, B. and C.E. Wills (2010), 'Privacy leakage in mobile online social networks', in *Proceedings of the 3rd Conference on Online Social Networks: USENIX Association.*
Kumar, V. (2008), *Managing Customers For Profit*, Philadelphia, PA: Wharton School Publishing.
Kumar, V. and R. Mirchandani (2012), 'Increasing the ROI of social media marketing', *MIT Sloan Management Review*, **4**(1), 55–61.
Kumar, V. and B. Rajan (2012), 'The perils of social coupon campaigns', *MIT Sloan Management Review*, **53**(4), 15–16.
Kumar, V. and B. Sundaram (2012), 'Winning with social media marketing: an evolutionary roadmap', *Marketing Research*, **24**(2), 4–7.
Kumar, V. and D. Shah (2011), 'Can marketing lift stock prices?', *MIT Sloan Management Review*, **52**(4), 24–6.
Kumar, V., V. Bhaskaran, R. Mirchandani and M. Shah (2013), 'Creating a measurable social media marketing strategy for hokey pokey: increasing the value and ROI of intangibles and tangibles', *Marketing Science*, **32**(2), 194–212.
Kurkovsky, S. and K. Harihar (2006), 'Using ubiquitous computing in interactive mobile marketing', *Personal and Ubiquitous Computing*, **10**(4), 227–40.
Rao, B. and L. Minakakis (2003), 'Evolution of mobile location-based services', *Communications of the ACM*, **46**(12), 61–5.
Shankar, V. and S. Balasubramanian (2009), 'Mobile marketing: a synthesis and prognosis', *Journal of Interactive Marketing*, **23**(2), 118–29.
Smith, A.N., E. Fischer and C. Yongjian (2012), 'How does brand-related user-generated content differ across YouTube, Facebook, and Twitter?', *Journal of Interactive Marketing*, **26**(2), 102–13.
Tao, Z. (2008), 'The impact of privacy concern on m-commerce user acceptance', Grid and Pervasive Computing Workshops, 2008, IEEE.
Teevan, J., A. Karlson, S. Amini, A.J. Brush and J. Krumm (2011), 'Understanding the importance of location, time, and people in mobile local search behavior', in *Proceedings of the 13th International Conference on Human Computer Interaction with Mobile Devices and Services*, ACM.
Thomas, J.S., W. Reinartz and V. Kumar (2004), 'Getting the most out of all your customers', *Harvard Business Review*, **82**(7–8), 116–23.

PART 5

IMPLEMENTING CUSTOMER EQUITY IN FIRMS

18. Implementing marketing return metrics in organizations

Donald E. Sexton

INTRODUCTION

Marketing managers face pressure from outside their organizations – from customers with unparalleled abilities to find information on products and services and competitors with technologies that can provide superior products and services, all in an environment where the rules and expectations change every moment. Meanwhile, within their organizations, marketing managers face pressure from colleagues – managers in finance and in other functions who expect marketing actions that will deliver more positive results for revenue and for profits. One widely circulated study found that 24 percent of finance managers described marketing as 'wasteful and unfocused' and that 35 percent of senior managers considered marketing an expense rather than an investment (ANA, 2007c). No recent studies suggest that those attitudes toward marketing have changed.

In such unfriendly circumstances, marketing managers need as much guidance as possible from metrics. Yet many marketing managers – apparently and surprisingly – admit that they employ few metrics in their decision-making. In a recent study of metrics in marketing decision-making, Rogers and Sexton found that many marketing managers claimed to use only a single metric – brand awareness – as a basis for making marketing decisions (Rogers and Sexton, 2012).

Those marketing managers not employing some type of marketing return metric to make decisions – what do they do? More than half of marketing managers surveyed have stated that they set their budgets in line with what they spent last year (ANA, 2005; Prophet, 2007). Budget persistence is a simple procedure but not one that builds confidence in marketing decision-making. The impact on market forecasts? In several studies conducted over several years, the findings were constant – only about 20 percent of marketing managers believed that they could predict the impact of a 10 percent cut in their budget (ANA, 2005, 2006, 2007b, 2008, 2009). It is not surprising that many non-marketing managers also do not have confidence that marketing managers can predict the impact of their activities on sales (ANA, 2007c; Marshall, 2008).

In contrast, in a recent study, those marketing managers satisfied with their marketing return metrics were found to be much *less* likely to rely simply on what they spent last year as the basis for their budgeting and, presumably, invested more thought in the likely consequences of their budget decisions (Rogers and Sexton, 2012).

What appears astonishing is the slow rate of change in the use of marketing return metrics, including customer equity metrics, by marketing managers. While there are marketing managers who employ marketing return metrics to their advantage, many managers are still not even using return metrics as well known and straightforward as customer lifetime value (Gupta and Lehmann, 2005). A variety of studies by researchers over the past decade have consistently shown these same results: little change in the use of marketing metrics (APQC, 1998; ANA, 2004, 2007a; Beaman, Guy, and Sexton, 2008; Conference Board, 2008; Sexton and ANA, 2012).

What are the reasons for this disappointing state of affairs and what can be done to improve the use of marketing metrics within organizations? This chapter draws on more than ten years of surveys – including the most current surveys publically available – to consider the following questions:

- What marketing metrics are in use and how effective are they considered to be?
- Why are marketing return metrics, including customer lifetime value, not more commonly used?
- What criteria might be applied to evaluate marketing metrics and are there examples of metrics that satisfy those criteria?

MARKETING METRICS AND THEIR USE

Many metrics are currently used by marketers (Box 18.1). Generally, they fall into three major categories:

- *marketing inputs*, including advertising expenditures, promotions, customer acquisition costs;
- *marketing outputs*, including awareness, attitude, conviction to purchase, retention, share;
- *business outputs*, including revenue, contribution, profit, cash flow, return on investment (ROI).

BOX 18.1 EXAMPLES OF MARKETING METRICS

Marketing Inputs	**Marketing Outputs**		**Business Outputs**
Advertising spend	Awareness	Frequency	Sales
Promotion spend	Knowledge	Price premium	Margin
Digital spend	Liking	Promotion lift	Channel margins
Customer service spend	Preference	Coupon redemption rates	Product profits/volume
Selling expense	Conviction	Average deal depth	Contribution per unit
	Trial	Pass-through price	Target volume
	Purchase	Website visits	Net profit
	Usage	Costs per click	Return on sales
	Loyalty/retention	Sales force effectiveness	Return on investment
	Satisfaction	Customer profit	Return on marketing investment
	Recommend	Customer acquisition costs	Economic value-added
	Perceived value	Customer lifetime value	Payback
	Consideration set	Market share	Net present value
	Complaints	Reach	Internal rate of return

Marketing Metrics Employed

Over time, studies indicate that the metrics most often used by marketing managers appear to be – with the major exceptions of sales and share – metrics focused specifically on marketing inputs or outputs and *not* on usual business objectives such as contribution or ROI (ANA, 2004, 2007a, 2007b). Marketing return metrics that link marketing inputs or outputs to business outputs are employed relatively rarely.

Instead of business measures of return such as profit or cash flow, marketers often use marketing output metrics as a measure of return. In a recent study, when managers were asked to define marketing return on investment (MROI), 37 percent of those surveyed made *no* mention of financial measures (Rogers and Sexton, 2012). Unaided MROI definitions provided by the respondents included 'reach and frequency', 'brand awareness', and 'market share' as well as 'anything that makes my boss happy'.

Inertia in Choice of Metrics

There seems to be considerable inertia in the choice of marketing metrics. The measures used by organizations to monitor marketing activities are generally measures that have been employed for many years, despite reservations regarding their effectiveness. For example, one study found that the most commonly tracked measures were sales, market share, brand recognition, and customer satisfaction, each tracked by more than 75 percent of the respondents, but not one of those measures was considered effective by the majority of respondents (ANA, 2007b). Overall, there was a lack of association between the effectiveness of measures and the frequency with which they were tracked – a correlation of only 40 percent. Only two of the top five measures considered most effective were among the top five measures that were tracked – suggesting that managers persist in collecting many metrics that are not considered particularly informative.

One of the most comprehensive studies of the use of marketing metrics was conducted by the Marketing Investment Research Working Group of The Conference Board (Beaman et al., 2008). Among the 36 marketing metrics examined, not one single metric was considered by the managers surveyed to be particularly effective in monitoring marketing return – no measure averaged above 4 on the 5-point effectiveness scale (and the much-hyped net promoter score did not even score above 3 on the effectiveness scale).

Among the customer measures investigated in the Conference Board study, loyalty/satisfaction and retention – similar measures of an important component of customer lifetime value – were reported as the *most* used measures. However, customer lifetime value was reported to be one of the *least* used measures. That suggests that while many managers may be aware of customer lifetime value – or its most important component, relatively few have actually calculated the lifetime value of their customers.

One recurring example of inertia in the use of marketing metrics is 'brand awareness'. There is nearly a decade of evidence suggesting that many organizations commonly utilize brand awareness as one of their top two or three measures to evaluate their marketing efforts (ANA, 2004, 2007a; Beaman et al., 2008; Rogers and Sexton, 2012; Sexton and ANA, 2012). Using brand awareness in concert with other measures can be useful. However, in a recent survey of over 200 marketing managers, 22 percent of those responding claimed that brand awareness was the *only* measure that they used to evaluate their marketing efforts (Rogers and Sexton, 2012). If brand awareness is a trailing indicator of marketing performance as some researchers believe, then reliance on brand awareness as a sole measure may lead to decision mistakes (Agres, 1996).

It is understandable that marketers would be concerned with marketing inputs and marketing outputs since that is what is in their field of vision. However, predominant use of such marketing-oriented metrics may keep marketers separate from the decision-making in the rest of the organization where business objectives such as contribution or cash flow are of more common concern.

Linkage of Marketing Output Metrics to Business Objectives

Many of the marketing-oriented metrics are flawed due to their reliance on obsolete models such as the purchase funnel, also referred to as the hierarchy of effects model or DAGMAR (Defining Advertising Goals for Measured Advertising Results) (Colley, 1961; Lavidge and Steiner, 1961). The purchase funnel model was proposed more than 50 years ago in a world where consumer behavior perhaps seemed simpler, yet it is still used by some marketing managers.

The purchase funnel suggests that purchasers go through stages such as awareness, knowledge, liking, preference, and conviction, before taking action such as making a purchase. While the purchase funnel seems to have face validity, researchers have found that the order of these stages may vary with the purchase situation. In a noteworthy article examining more than 200 studies of purchase behavior, Vakratsas and Ambler found that in some cases, cognitive behavior (or learning) was at the beginning of the purchase process; in some cases, affective behavior (or feeling) was at the beginning of the process; and in some cases, action (or doing) was at the beginning of the process (Vakratsas and Ambler, 1999).

While the purchase funnel may be an appropriate model for some situations, it is not appropriate for all situations (Vaughn, 1980). Employing measures such as brand awareness in solitary fashion decouples marketing decisions from financial outcomes since there is no argument consistently linking brand awareness *alone* to financial results.

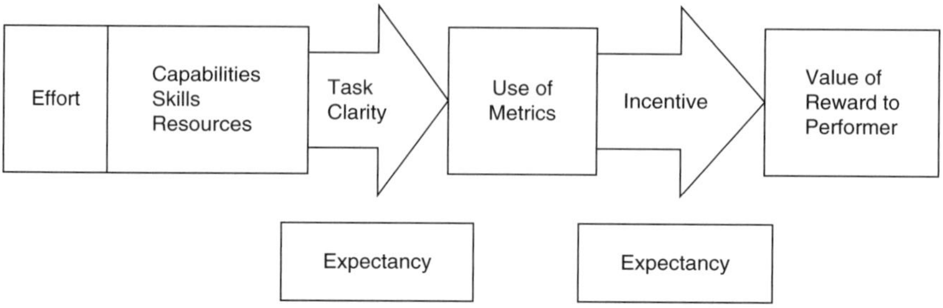

Figure 18.1 Expectancy model and use of metrics

CHANGE MANAGEMENT AND THE USE OF MARKETING METRICS

The expectancy theory models (Figure 18.1) for managing change provide insights into why marketing metrics linked to financial return are not more widely used. In its simplest form, expectancy theory posits that people behave according to what they believe are the likelihoods of the various possible outcomes of their actions and the extent to which they value those outcomes (Vroom, 1964; Porter and Lawler, 1968; Green, 1992; Green and Hayes, 1993; Isaac, 2001; Sexton, 2007; Hayes, 2010; *Harvard Business Review*, 2011; Kotter, 2012). Expectancy theory can serve as a structure to evaluate the findings from many surveys over the past several years and spotlight the drivers encouraging the more widespread use of marketing return metrics and the barriers preventing or retarding their utilization.

In expectancy theory, 'task clarity' links the effort of a performer to results such as the use of marketing return metrics. Task clarity answers the question, 'Do I know what I am supposed to do?' Once a manager understands the metrics required, they must expend 'effort' to employ them. However, effort without analytical skills, data, and other resources will not be productive – 'capabilities' to develop and implement metrics must be available. Leaders must create an 'environment' – organizational culture, prioritization of goals – that supports the use of market return metrics. If the environment does not support the use of metrics, then metrics will not be used effectively even if the necessary efforts are made and the necessary capabilities exist.

Finally, one must consider whether or not a performer is given an 'incentive' for succeeding in employing marketing return measures and, if so, whether or not the reward has 'value' to them.

In sum, if the use of marketing metrics is to be encouraged, managers must be able to answer 'Yes' to *each* of the following questions:

1 Do I know what metrics are needed?
2 If I put in the effort, can I develop those metrics?
3 Do I have the capabilities to develop those metrics?
4 Does leadership create an environment that supports the development and continual use of metrics?
5 If I develop effective metrics, will I be rewarded and do I value that reward?

These questions describe a system with two expectancies: effort to results and results to reward.

Task Clarity

Do I know what metrics are needed? Many managers do not seem to have a clear idea of what metrics should be used to connect marketing actions to business objectives. Many managers appear to believe that it is important to agree on a definition of marketing return – 57 percent of managers surveyed considered it 'important or very important' to agree on a marketing ROI definition (ANA, 2007b). Yet, only 12 percent of those respondents were 'satisfied or very satisfied' with agreement on a marketing ROI definition in their organization – only 22 percent thought that their senior management had communicated a clear definition of marketing accountability.

In a variety of studies, managers report numerous measures used to evaluate the effectiveness of marketing efforts in their organizations (ANA, 2004, 2007a, 2007b; Beaman et al., 2008, Sexton and ANA, 2012). Among the marketing metrics cited, generally, brand-related measures appear to be used most frequently while return-based measures such as MROI and customer lifetime value are used much less frequently. For example, in a survey of global managers conducted during Fall, 2012, of those responding, nearly 50 percent regularly collected brand awareness measures from their country markets, while fewer than 25 percent collected marketing return measures such as brand value and customer lifetime value (Table 18.1).

Effort

If I put in the effort, can I develop those metrics? Success in developing marketing metrics comes by having people work on marketing accountability issues. In one Conference Board study of the determinants of success in employing marketing metrics, half the organizations reporting no progress on marketing ROI had *no* people working on marketing

Table 18.1 Global collection of measures

Measure	Collected in All or Most Countries[a]
Profitability	73%
Market share	59%
Customer satisfaction	49%
Brand awareness	49%
Brand knowledge	40%
Brand attitudes	35%
Brand value	23%
Recommendation	22%
Perceived value	20%
Change in brand value	18%
Lifetime customer value	17%
MROI	17%

Note: a. Top two rows, 1–7 scale.

Source: Sexton and ANA (2012).

ROI (Beaman et al., 2008). 'Time spent working' does not mean meetings but actual hands-on work. In the Conference Board study, number of meetings and number of communications about marketing ROI were *not* significantly associated with progress in measuring marketing ROI, but personal days per month were significantly (0.05 level) associated (ibid.).

Of all the possible drivers of success in developing marketing return metrics that have been studied, time spent working on marketing return metrics seems to be the most important single indicator of success in measuring MROI. The relationship between time spent and metric success was the most statistically significant finding in the Conference Board study (ibid.). More recently, Rogers and Sexton found a very clear relationship between time spent working on marketing ROI and reported success in evaluating MROI (Rogers and Sexton, 2012).

Capabilities

Do I have the capabilities to develop those metrics? Effort without capabilities does not result in progress. Capabilities include both analytical skills and data.

Skills

Many managers seem to feel that their organizations lack the analytical skills to develop useful marketing return metrics. A survey by the

Lenskold Group found that only 17 percent of those responding considered that their marketing ROI measurement abilities were as 'good as they needed to be or a real source of leadership' while 43 percent thought their marketing ROI measurement capabilities were 'a long way' from what they needed (Lenskold, 2008).

The Conference Board study found that all the skills that respondents considered important for marketing ROI were consistently rated lower with respect to their availability. Overall, the majority of Conference Board study respondents agreed that, 'lack of technology and appropriate infrastructure have affected significantly, or a lot, the ability to measure marketing ROI' (Beaman et al., 2008).

Data

Having appropriate data to utilize forward-looking marketing measures is an ongoing issue for many organizations. In 2008, a study of marketing executives by the CMO Council found that less than 7 percent of the respondents believed that they had excellent knowledge of their customers with regard to demographic, behavioral, psychographic, and transactional information (Clemente, 2008). Some gurus seem to believe that 'big data' will fill in the informational gaps. However, a recent study found that companies are struggling to extract insights from 'big data' (Rogers and Sexton, 2012).

Even if data appropriate for marketing metrics are available, data may not be shared due to silos within the organization. In one ANA survey, only 24 percent of managers surveyed were 'satisfied or very satisfied' with their access to internal data (ANA, 2007b), a finding corroborated in later studies by the CMO Council (2007) and Rogers and Sexton (2012) (Table 18.2).

Table 18.2 Biggest challenges to use of 'big data'[a]

Challenge	
We have too little or no customer/consumer data	29%
Our data are collected too infrequently or are not real-time enough	39%
The lack of sharing data across our organization is an obstacle to measuring the ROI of our marketing	51%
We are not able to link our data together at the level of individual customers	42%

Note: a. Top two rows: strongly agree or somewhat agree, 1–5 scale.

Source: Rogers and Sexton (2012).

Table 18.3 Importance of drivers and barriers to measuring MROI

Measuring Marketing	Driver[a]	Barrier[a]
Culture	72.4%	52.2%
Leadership	88.9%	51.0%
Data	87.3%	52.6%
Tools	81.8%	56.1%
Skills	77.1%	47.1%
Process	81.5%	49.0%

Note: a. Top two rows: very important and somewhat important, 1–5 scale.

Source: Rogers and Sexton (2012).

Not only customer data but also availability of relevant cost data may be an issue when developing marketing metrics. Many accounting systems still appear to have difficulty sorting out variable or incremental costs from fixed costs. Instead, some form of average costs are often used to develop financial performance metrics, which means that those measures will necessarily be misleading when used to make decisions. In addition, many organizations assign overhead costs in ways that cause further distortion of decision-making.

Environment

Does leadership create an environment that supports the development and continual use of metrics? Along with lack of analytical tools and data, cultural resistance and lack of leadership can be major barriers to progress on developing marketing metrics (Table 18.3) (Rogers and Sexton, 2012). An earlier study by the Conference Board reached similar conclusions – leadership, data, and technology were found to be the most important drivers of improved marketing metrics (Beaman et al., 2008).

Rewards

If I develop effective metrics, will I be rewarded and do I value the reward? The American Productivity and Quality Center found that almost 60 percent of firms they surveyed made *no* use of the utilization of marketing ROI metrics in determining compensation (APQC, 2003). Including marketing ROI in performance reviews and recognition programs has been found to be highly associated with progress in measuring marketing ROI (Beaman et al., 2008). Most recently, Rogers and Sexton found that

Table 18.4 Incentives and satisfaction with ability to measure MROI

	Satisfied[a]	Dissatisfied[a]
Incentives employees		
Performance objectives	46.3%	24.2%
Compensation plans	63.1%	15.5%
Recognition programs	54.8%	25.8%
Incentives marketing vendors		
Performance objectives	67.8%	13.6%
Compensation plans	60.0%	11.4%

Note: a. Top two rows: very satisfied and somewhat satisfied. Bottom two rows: dissatisfied and very dissatisfied, 1–5 scale.

Source: Rogers and Sexton (2012).

managers who were very satisfied with their marketing return metrics were much more likely to provide recognition or compensation incentives for improved metrics to *both* their employees and their vendors (Table 18.4) (Rogers and Sexton, 2012).

Summary of Drivers and Barriers to Use of Marketing Metrics

The use of marketing metrics, especially those concerning marketing return, at this point still seems very much a work in progress. There seems to be dissatisfaction with many current measures but, with some exceptions among companies, also a lack of interest and energy for exploring new metrics or improving current metrics. Many managers state that they would like improved marketing measures but seem unwilling to move very far from the metrics they are familiar with – even though they judge many of those metrics as unsatisfactory.

Surveys of managers over the past ten years do not indicate that the five questions posed earlier are commonly being answered positively in many organizations. Improving the use of marketing return metrics requires examining the answers to these questions within a specific organization and, if the answers are found to be negative, considering what can be done to make them positive.

PROPOSED CUSTOMER EQUITY METRICS

The most common difficulty with many marketing metrics currently in use by marketers is that they focus much too early in the buying process to be effective return measures. For example, brand awareness may be at the very start of the buying process for certain products and services, but it may be distant from the actual purchase. Not employing marketing metrics that are clearly linked to business objectives makes it difficult to make decisions based on marketing return.

Suggested Criteria for Marketing Return Metrics

To be useful to marketing managers, marketing metrics should meet several criteria (Sexton, 2005). They should be:

- *understandable:* make sense and not be unduly complicated;
- *measurable:* based on data and objective statistical analysis;
- *linked to financial returns:* show financial impact of marketing investments;
- *comprehensive and flexible:* applicable to all marketing decisions;
- *supported conceptually:* have a foundation in disciplines such as economics and psychology;
- *actionable:* be related to and guide managerial decisions;
- *capable of providing steering control:* allow prediction of financial impact of marketing actions;
- *proven in application:* have been shown to work.

Suggested Metrics

Two universal metrics that meet all of the above criteria are 'perceived value' and 'customer value-added' (CVA®). Both measures are explained in detail in Sexton (2009b). Perceived value, in particular, is a well-known but under-utilized measure of customer equity. The perceived value metric has been endorsed by managers on the Council on Corporate Brand Management (Sexton, 2005). Perceived value or a similar construct has been shown in studies by Aaker, Jacobson, Agres, Buzzell, Gale, and Gregory – different researchers and different datasets – to be a *leading* indicator of business outputs such as revenue, profit, cash flow, and stock price (Buzzell and Gale, 1987; Aaker and Jacobson, 1994, 2001; Agres, 1996; Gregory, 2005). Nonetheless, most surveys cited in this chapter do not even mention perceived value as a marketing metric.

Perceived value is the maximum the customer is willing to pay for a unit of the product or service (Figure 18.2). Perceived value is *not* price but the ceiling on price and it varies by customer and by purchase situation (Sexton, 2009a, 2009b, 2010a, 2010b). Marketing can be defined as 'managing perceived value'. Customer value-added is the difference between

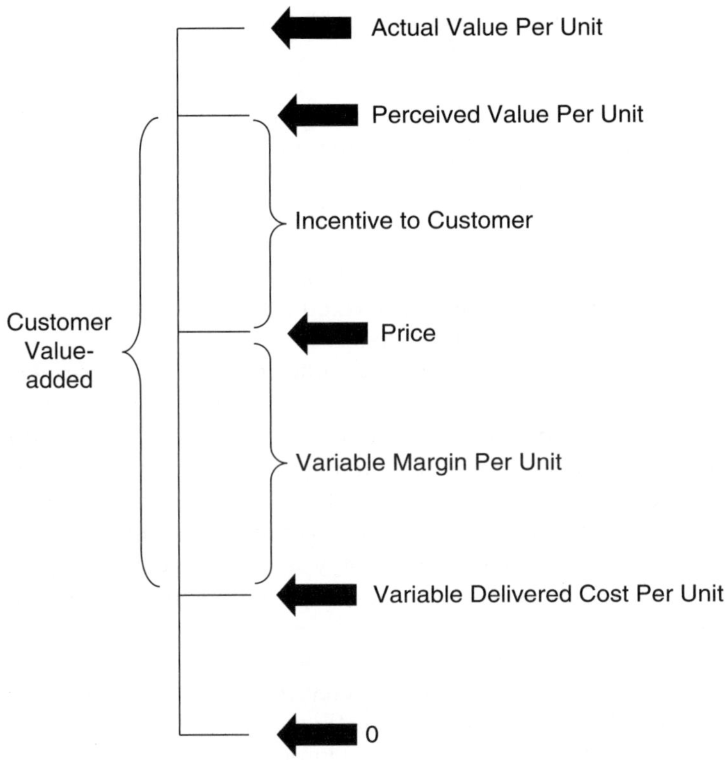

Source: Sexton (2009b).

Figure 18.2 Perceived value and customer value-added

perceived value and variable cost per unit (Figure 18.2) and so represents the *net* perceived value per unit produced that an organization provides society.

These two metrics – perceived value and customer value-added – satisfy all the criteria for effective metrics of customer equity.

Understandable Perceived value is the value that the potential customer assigns a unit of the product or service. Customer value-added then is the perceived value an organization contributes to society, net of incremental costs.

Measurable A variety of methods – value-in-use, constrained choice models, regression analysis – are available to estimate perceived value.

Accounting information systems should be capable of providing variable cost estimates.

Linked to financial returns Perceived value is linked to unit sales through the demand curve. Demand curves represent the perceived value for each of the customers in the market. When marketing activities affect perceived values – either positively or negatively – the demand curve shifts and units sold at any price shift. Customer value-added is linked to contribution through the demand curve and through the variable cost per unit. Again, if the demand curve shifts due to marketing efforts that affect perceived values, then overall contribution changes.

Any price divides the customer value-added range into two parts (Sexton, 2006). The distance from the price to the variable delivered cost per unit is the 'variable margin per unit' or 'incremental profit per unit'. The distance from the perceived value per unit to the price is the 'incentive per unit' for the customer to purchase. The higher that incentive, the more the 'unit demand'. A low price strategy yields high volume with a low margin – what is known as a 'penetration pricing strategy'. A high price strategy yields low volume with a high margin – what is known as a 'skimming pricing strategy'.

Comprehensive and flexible Any action in marketing involves perceived value in some way. Therefore, both perceived value and customer value-added can be employed to measure the financial return related to any marketing effort, such as product or service improvements or communications via traditional or non-traditional media.

Supported conceptually Perceived value is based on consumer psychology. The relationship of perceived value to revenue and that of customer value-added to contribution is based on microeconomics.

Actionable Since marketing is 'managing perceived value', then perceived value and customer value-added can be affected by any marketing action (Sexton, 2009b).

Capable of providing steering control Predicted changes in perceived value and customer value-added indicate the likely financial returns from any marketing actions.

Proven in application In company situations, perceived value has been shown to predict revenue with high accuracy (R-squared values above 0.90). Customer value-added has been shown to predict contribution with high accuracy (R-squared values above 0.90). Predictions with both metrics were found to be more accurate than predictions made with past revenue and past contribution, respectively (Gaber et al., 2009; Sexton, Sen, and Gorti, 2010).

Perceived value and customer value-added (CVA®) meet the suggested criteria for marketing return metrics. Their use might solve many of the issues involved with employing marketing return metrics.

IMPROVING THE UTILIZATION OF MARKETING METRICS

Some specific suggestions:

1 Improve task clarity – make sure everyone knows what marketing metrics are needed.
2 Manage expectancies – reward efforts to develop useful marketing metrics.
3 Develop capabilities – acquire the data and the analytical skills needed to conduct the appropriate analyses of marketing and branding activities.
4 Consider the use of universal metrics such as perceived value and customer value-added to manage and monitor marketing actions.
5 Get engaged – the data and skills will improve.

CONCLUSIONS

Over many decades, considerable efforts have been made to improve measures of marketing return. While some organizations have met with success, managers in many organizations are still seeking measures that will deliver greater insights for developing their strategies and will provide them with steering control for their decisions. Inertia appears to have slowed down some of these efforts but there are opportunities to improve marketing return metrics if the organizations will create an environment where such efforts can succeed and if they will consider employing metrics beyond those that have been in use for a long while.

REFERENCES

Aaker, D.A. and R. Jacobson (1994), 'The financial information content of perceived quality', *Journal of Marketing Research*, **31**(2), 191–201.

Aaker, D.A. and R. Jacobson (2001), 'The value relevance of brand attitudes in high-technology markets', *Journal of Marketing Research*, **38**(4), 485–93.
Agres, S. (1996), 'Brand Asset Valuator® presentation', New York: Columbia Business School.
American Productivity and Quality Center (APQC) (1998), *The State of Brand Building and Communications*, Houston, TX: APQC.
American Productivity and Quality Center (APQC) (2003), *Managing Marketing Assets*, Houston, TX: APQC.
Association of National Advertisers (ANA) (2004), *The State of Marketing Accountability Study*, New York: ANA.
Association of National Advertisers (ANA) (2005), *The Path to Marketing Accountability Study*, New York: ANA.
Association of National Advertisers (2006), *The Path to Marketing Accountability Study*, New York: ANA.
Association of National Advertisers (2007a), *Brand Deterioration Study*, New York: ANA.
Association of National Advertisers (2007b), *Marketing Accountability Study*, New York: ANA.
Association of National Advertisers (ANA) (2007c), *Survey of Finance Managers*, New York: ANA.
Association of National Advertisers (2008), *Marketing Accountability Study*, New York: ANA.
Association of National Advertisers (2009), *Marketing Accountability Study*, New York: ANA.
Beaman, K., G.R. Guy and D.E. Sexton (2008), *Managing and Measuring Return on Marketing Investment, Research Report 1435–08-RR*, New York: The Conference Board.
Buzzell, R.D. and B.T. Gale (1987), *The PIMS* Principles*, New York: Free Press.
Clemente, M. (2008), *Business Gain from How You Retain: A CMO Council Study on Making the Most of Your Existing Customers*, Chicago, IL: American Marketing Association.
CMO Council (2007), *Marketing Outlook*, Palo Alto, CA: CMO Council.
Colley, R. (1961), *Defined Advertising Goals for Measured Advertising Results*, New York: Association of National Advertisers.
Conference Board, The (2008), *Measurement and ROI*, New York: TCB.
Gaber, S., D.E. Sexton, K. Sen and A. Lin (2009), 'Customer input: incorporate perceived value into marketing strategy', *Business Digest*, October, 3–9.
Green, T. (1992), *Performance and Motivation Strategies for Today's Workforce: A Guide to Expectancy Theory*, New York: Quorum Books.
Green, T. and M. Hayes (1993), *The Belief System*, Winston-Salem, NC: Beechwood Press.
Gregory, J.R. (2005), *Driving Brand Equity*, New York: ANA.
Gupta, S. and D. Lehmann (2005), *Managing Customers as Investments*, Upper Saddle River, NJ: FT Press.
Hayes, J. (2010), *The Theory and Practice of Change Management*, New York: Palgrave Macmillan.
Harvard Business Review (2011), *HBR's 10 Must Reads on Change Management*, Cambridge, MA: Harvard Business Review Press.
Isaac, R.G. (2001), 'Leadership and motivation: the effective application of expectancy theory', *Journal of Managerial Issues*, **23**(2), 212–22.
Kotter, J. (2012), *Leading Change*, Cambridge, MA: Harvard Business Review Press.
Lavidge, R. and G. Steiner (1961), 'A model for predictive measurements of advertising effectiveness', *Journal of Marketing*, **25**, 137–69.
Lenskold, J. (2008), *Lenskold Group/Kneebone 2008 Marketing ROI and Measurement Study*, Manasquan, NJ: Lenskold.
Marshall, J. (2008), 'Finance: friend or foe?', paper presented at the 2008 ANA Marketing Conference, July, 14.
Porter, I. and E.E. Lawler (1968), *Managerial Attitudes and Performance*, New York: Irwin.
Prophet (2007), *State of Marketing Survey*, Chicago, IL: Prophet.

Rogers, D. and D.E. Sexton (2012), *Marketing ROI in the Era of Big Data*, New York: BRITE and NY AMA.
Sexton, D.E. (2005), 'Building the brand scorecard', *The Advertiser*, February 2005.
Sexton, D.E. (2006), 'Pricing, perceived value, and communications', *The Advertiser*, April, 56–8.
Sexton, D.E. (2007), 'Managing the performance of individuals', *Effective Executive*, August, 34–9.
Sexton, D.E. (2009a), 'Achieving marketing nirvana', *The Advertiser*, 15 July 2009.
Sexton, D.E. (2009b), *Value Above Cost: How CVA® Drives Financial Performance, the Most Important Metric You've Never Used*, Upper Saddle River, NJ: FT Press (Wharton School Publishing).
Sexton, D.E. (2010a), 'Competing with customer value-added', *Effective Executive*, February, 44–7.
Sexton, D.E. (2010b), 'How to determine marketing ROI', *ANA Insight Brief*, New York: ANA.
Sexton, D.E. and ANA (2012), *Global Brand Equity Survey Results*, New York: ANA.
Sexton, D.E., K. Sen and V. Gorti (2010), 'Determining marketing accountability', *Journal of Marketing Trends*, **1**(4), 39–45.
Vakratsas, D. and T. Ambler (1999), 'Advertising effects', *Journal of Marketing*, **63**(1), 26–43.
Vaughn, R. (1980), 'How advertising works', *Journal of Advertising Research*, **1**, 57–66.
Vroom, V.H. (1964), *Work and Motivation*, London: Wiley.

19. Interaction orientation and complaint handling

Girish Ramani and George Knox

INTRODUCTION

Marketing managers have had to reinvent their marketing practices during the last decade to cope with the disruption of traditional marketing channels and the reintegration of customers and marketers in an Internet-driven eco-system. The marketing environment today allows unprecedented and often uncontrolled levels of interactivity. A new breed of companies such as Google, Amazon, Apple, and Facebook can be credited with possessing the necessary firm-wide interaction orientation that has helped them set new rules of business and emerge as the exemplars of modern marketing. Several other firms, like Zappos (subsequently acquired by Amazon; Weisul, 2005), Express Scripts, and Whole Foods owe at least some of their success to their adopting an interaction orientation.

The objectives of this chapter are, (1) to review the concept of interaction orientation, (2) to trace the adoption of interaction orientation by firms, (3) to examine the linkages between interaction orientation and complaint handling, and (4) discuss the implications for customer equity.

An interaction orientation reflects a firm's ability to interact with its individual customers and take advantage of information obtained from them through successive dyadic interactions in order to achieve profitable customer relationships (Ramani and Kumar, 2008). The interaction orientation construct consists of an organization's fundamental belief (the customer concept) and the relevant processes (interaction response capacity) and practices (customer empowerment and customer value management) that supplement this fundamental belief.

The customer concept The customer concept is the organizational belief that the unit of analysis of every marketing action and reaction should be the individual customer. Considering the individual customer central (Fader, 2012) and as an exogenous driver of the marketing mix (Leeflang, 2011) enables marketers accurately to measure the return on marketing activities.

Interaction response capacity The interaction response capacity represents the degree to which the firm offers successive products, services and relationship experiences to each customer by dynamically incorporating feedback from previous behavioral responses of that specific customer, and of other customers collectively. A superior interaction response capacity is known to enhance existing and new customer performance (Ramani and Kumar, 2008; Köhler et al., 2011)

Customer empowerment Customer empowerment reflects the extent to which a firm provides its customers avenues to (1) connect with the firm and actively shape the nature of transactions, and (2) connect and collaborate with each other by sharing information, praise, criticism, suggestions, and ideas about its products, services, and policies. Customer empowerment strategies have been shown to increase willingness to purchase new products as long as the customers selected perceive themselves to be competent decision-makers (Fuchs, Prandelli, and Schreier, 2010).

Customer value management Customer value management represents the extent to which the firm is able to define and dynamically measure individual customer value and use it as its guiding metric for marketing resource allocation. Measures such as customer lifetime value and customer equity are key indicators of marketing efficiency (Rust, Lemon, and Zeithaml, 2004; Venkatesan and Kumar, 2004).

THE ADOPTION OF THE ELEMENTS OF INTERACTION ORIENTATION

When the notion of interaction orientation was proposed and empirically validated with data from practicing managers (Ramani and Kumar, 2008), there were far fewer companies in 2005 than today that could have been naturally classified as possessing such an orientation. We now trace how some well-known firms have adopted the elements of interaction orientation into their marketing approaches over the last few years. We pick as examples top-ranked firms from the Fortune World's Most Admired Companies 2012 (Colvin, 2013) as well as firms classified as 'customer-centric' (Fader, 2012) to illustrate the adoption of the elements of an interaction orientation.

Most firms traditionally approached marketing on the basis that customers need to be grouped into segments and that marketing efforts should be tailored to serve the needs of the average customer in each of these segments. However, some firms began to recognize that there were efficiencies to be gained in viewing a market as consisting of unique individual customers. Over time, the marketing function in an organization has become responsible for managing all the relational processes that take place with each and every individual customer. Hence, marketing efforts need to focus on providing customer solutions by managing each individual customer's relational experiences with the firm. Firms that recognize the importance of this role of marketing are implicitly accepting that the 'customer concept' should be at the core of their marketing efforts. The idea of the customer concept has shaped Amazon's marketing strategies for over a decade. Amazon realized early on that its customer base was a much larger asset than its inventory of books and CDs. In contrast to selling the same product to several customers, Amazon began exploring ways to sell several products to the same customer. In other words, the unit of analysis for marketing effectiveness shifted from the product to the individual customer. For example, Amazon sells its physical product, the Kindle, at cost because it believes that real long-term profits lie with future interactions with its customers when people continue to use the Kindle device (Farber, 2012).

Firms are taking cognizance of the fact that standardized relational processes are likely to be experienced differently and uniquely by each customer of the firm. It is the customer who determines the degree of importance and relevance of an interaction with the firm and not the other way around (see Heffernan and LaValle, 2006). Therefore, a firm needs to be able to develop kinship with its key customer base by being different things to different customers and at different points in time. Firms have been investing in building in-depth customer knowledge and developing dynamic capabilities for interacting with their customers. This component of an interaction orientation, termed 'interaction response capacity', is an ability that differentiates the more nimble companies from the slower ones. A comparison of Google and Microsoft illustrates the importance of this ability. Google has invested in personnel and skills to capture and analyze customer interactions in order to develop products and services catering to a variety of customer needs. Despite the challenges that firms face while customizing products and services where customers are either unable or unwilling to explicitly indicate their preferences (Franke, Keinz, and Steger, 2009), Google has pioneered the use of customer interaction knowledge to target products and services from their sponsors to their users. Google Glass, an evolving project, promises a hands-free user experience for several smartphone and computer functions (Ante, 2012). On the other hand, Microsoft and its hardware partners, at least in the last decade, have stayed largely self-oriented. Microsoft, in recent times, appears to have decided to react to the changing customer landscape by launching physical products, such as the Surface tablet that take advantage of newer generation touch compatible software. More recently, Microsoft also seems to be considering developing its own version of augmented reality glasses (Arthur, 2012). However, the extent of success that Microsoft achieves in the future would depend on how comprehensively its products and services can envelop the consumer in both the physical and the virtual sense. In other words, Microsoft would need to continually invest in improving its interaction response capacity.

The third element of an interaction orientation, 'customer empowerment', refers to the extent to which companies treat customers as their partners and promote customer-to-customer interactions. To develop customized products and services by soliciting customer or channel participation, companies need to relinquish some control over the process of co-creation. Marketing is becoming less a matter of domination and control, and more a matter of fitting in (Deighton and Kornfeld, 2009). One of the reasons for Google's late but successful entry into the smartphone market is that the operating system Android was first made available freely to software developers; this ensured that consumers were able to access a range of applications even the Google researchers had not originally imagined. But the temptation to sustain short-term competitiveness by keeping technology proprietary is compelling. If and when Google Glass is available to the general population, the approach of opening out the development of applications to their licensees and software developers would be consistent with the idea of customer empowerment.

The fourth component of an interaction orientation, termed 'customer value management', reflects the firms' ability to link customer-directed marketing activities with current and future profits from individual customers. In other words, this refers to the companies' ability to predict the lifetime value of individual customers and, in the aggregate, the overall customer equity for a firm or brand, and employ these metrics for making dynamic marketing and business-level decisions. Amazon charges its Prime members an annual fee and in return offers free expedited shipping and access to online books and movies. While the pricing of the Prime membership is currently uniform across its customers, Amazon has the ability to track individual customer-level profitability and use this information for investing in future services and products, both at the individual customer level and aggregate market segment levels. Firms that report customer-level metrics either as a service to the investor community or as required by the Securities and Exchange Commission are more likely to understand the usefulness and importance of these metrics in guiding their decisions (Wiesel, Skiera, and Villanueva, 2008). For example, Netflix has demonstrated that its ability to measure and manage its customer profitability on a continuous basis has resulted in its continuous growth in clout with its content providers. At the same time Netflix (and its investors) can easily track the impact of marketing decisions on customer profitability and therefore take augmentative or corrective action whenever necessary.

INTERACTION ORIENTATION AND COMPLAINT MANAGEMENT

Companies with a greater interaction orientation have better customer relationships and higher customer profits (Ramani and Kumar, 2008). One critical way in which this link varies across firms is how central the customer–firm relationship is to the customer. Since genuine customer relationships require that the customer acknowledge them as such (Peppers and Rogers, 2004), one measure that captures the centrality of the relationship as perceived by the customer is the level of customer-initiated contacts. Contacts such as these contain rich information on customer needs and how firms can tailor their responses to their needs (Bowman and Narayandas, 2001).

Ramani and Kumar (2008) show that the higher the level of customer-initiated contacts, the stronger the positive link between interaction orientation and customer-based profit performance. While this moderated effect finding helps establish the boundary conditions of adopting an interaction orientation, customer-initiated contacts include several disparate categories, particularly negative customer complaints, and more neutral contacts, such as inquiries about product use or availability.

We extend the original interaction orientation framework to discuss the role of customer complaints and the company's ability to recover from product and service failures (Grönroos, 1988). We first consider how companies that adopt an interaction orientation are likely to have superior recovery capabilities by examining the links between these concepts. Next, we discuss the consequences for customer-based performance metrics.

Customer complaints are an especially rich source of information about the customer base, especially in situations where the firm does not observe the time at which a customer becomes inactive – what Schmittlein, Morrison, and Colombo (1987) call a 'non-contractual' setting (Reinartz and Kumar, 2003). In these cases 'there is no way to know when a customer stops being a customer . . . When they stop being a customer . . . they just silently attrite' (Mason, 2003, p. 55). Hence, part of the informational value of complaints is that they break this silence.

One perspective on complaints is that they are observed indicators of latent customer dissatisfaction and potential defection (Blattberg, Kim, and Neslin, 2008; Van Oest and Knox, 2011). While complaints and complaint management have long been studied in economics (Hirschman, 1970) and marketing (Fornell and Wernerfelt, 1987), the focus has been less on the implications of customer–firm interactions. In part, technology has changed the landscape by making it possible for firms to collect massive amounts of data on their customers to track their movements; at the same time, there are many channels through which customers may now complain, view other customer's complaints, and even join 'customer insurgencies'. We discuss several implications of complaint management and interaction orientation below.

Firms with a strong interaction orientation start with the individual customer, develop a capacity to respond to customers, empower customers to connect with the firm and other customers, and can differentially tailor their response based on predicted future customer profitability. We now consider how each of the four components of an interaction orientation relates to complaints and complaint management. Figure 19.1 shows a graphical conceptual overview.

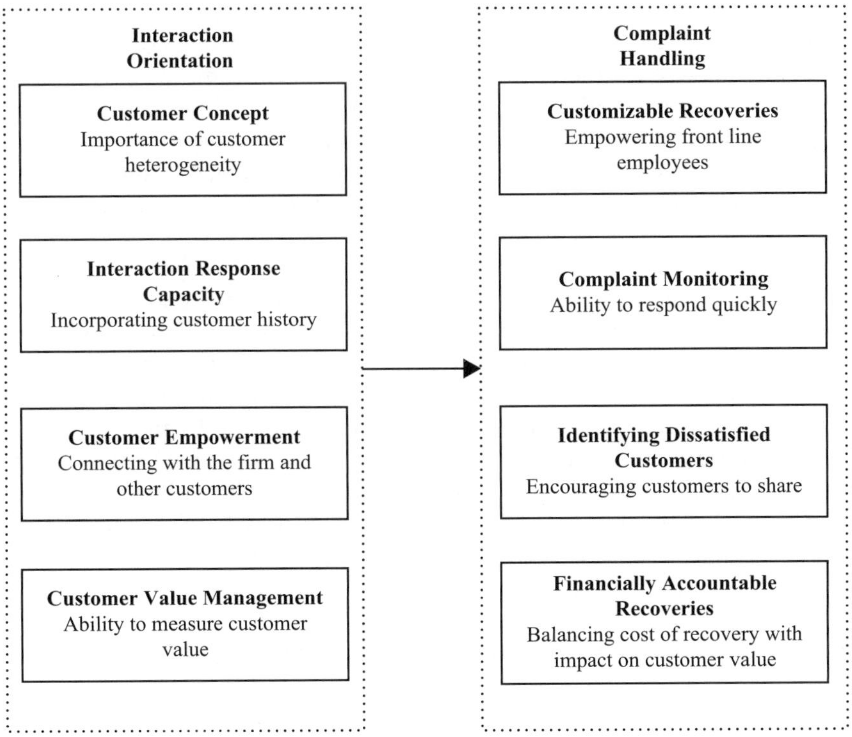

Figure 19.1 Conceptual framework

The Customer Concept and Complaint Management

The customer concept is in part the belief in the importance of customer heterogeneity – that each customer is unlikely to be satisfied with the same products and services. Customer complaints arise as a direct result of heterogeneity in individual customer needs as well as possible heterogeneity in the service experienced itself (Rust and Chung, 2006). Although most dissatisfied customers do not complain, complaints are managerially useful because they are observed indicators of latent customer dissatisfaction and potential defection (Blattberg, Kim, and Neslin, 2008; Van Oest and Knox, 2011); they are opportunities to strengthen the relationship with customers (Berry and Parasuraman, 1991) because a good recovery can lead to increased satisfaction and retention (Smith and Bolton, 1998). Some authors argue that a company's worst complainers are easier to convert to company champions than non-complainers (e.g., Peppers, 2012).

Similarly, customer heterogeneity carries over to customer reaction to the firm's recovery from the product or service failure as well. Companies that believe in the customer concept will be more likely to understand that not every customer will be satisfied with the same recovery. Such companies are more likely to empower front-line employees to appropriately handle customer complaints rather than adopting a 'one-size-fits-all' approach to recovery. Training employees and empowering the front line allows for recoveries to be better tailored to specific customer needs (e.g., Hart, Heskett, and Sasser 1990).

The e-tailer Zappos provides a good example of this approach. Its stated mission is 'to provide the best customer service possible' and its culture is to treat each customer as an individual (Thompson, 2010). Customer-service representatives undergo several weeks of training (along with every new employee). They are not pressured to keep calls short and they do not use scripts. There are several stories of the great lengths their customer service representatives go to in order to win over their customers. For example, in 2011, Zappos sent flowers to a woman who ordered six different pairs of shoes because her feet were damaged by harsh medical treatments (Edwards, 2012). Such individually tailored responses are likely to lead to greater customer loyalty.

Interaction Response Capacity and Complaint Management

Interaction response capacity, which is the ability to adapt products, services, and relationships to previous responses of the customer as well as other customers, is also a key ingredient for developing successful recovery capabilities. This involves pooling information from different sources, monitoring customer complaints and certain areas of the organization, and even developing a service quality information system (e.g., Berry and Parasuraman, 1997). Processes such as these are necessary to anticipate failures before they escalate (Hart et al., 1990; Gilly, Stevenson, and Yale, 1991).

Gatorade, the sports drink manufactured by PepsiCo, perhaps best epitomizes the trend of developing sophisticated capabilities to monitor its customers. Its mission is to become the 'largest participatory brand in the world' and it has created a social media 'war room' to monitor its brand across social media (Divol, Edelman, and Sarrazin, 2012). It has developed capabilities to analyze and track brand mentions over time, created dashboards related to its brand, sponsored athletes and its competitors. This information helps Gatorade tailor its marketing campaigns.

At the same time, several platforms have emerged that let anyone monitor customer chatter for free. Websites like SocialMention.com allow anyone to monitor social media websites like Twitter, Facebook, and blogs for any mention of a company or its products. For a given search, it produces several analytics related to how negative the mentions are (sentiment), the relative proportion of new authors mentioning the brand (reach), the proportion of people who repeatedly mention it (passion), and the number of phrase mentions in the past 24 hours divided by the total number of possible mentions (strength).

McDonald's provides a good example of how a company can use monitoring tools to prevent a customer uprising from escalating. In 2011, at the start of the weekend, someone posted a photo on Twitter claiming that McDonald's was charging African American customers an extra $1.50 as an 'insurance measure' against robberies (Forbes, 2011). More people re-tweeted the photo using the hashtag '#SeriouslyMcDonalds' to express their disgust; it became a leading entry on Twitter's trend list. In less than a day, however, McDonald's, which actively monitors social media, was able to issue a response that the photo was a hoax. The company reinforced its message and responded to concerned individuals on Twitter over the weekend, thereby preventing a public relations disaster (Divol et al., 2012). In the words of CBS *Early Show* technology expert, Katie Linendoll:

> If you're a big corporation and something viral . . . happens against you, you have to formulate a plan and respond quickly. In this case, McDonald's handled it correctly; they used the medium Twitter they were accused on. Time is of the essence. We're not going to the press in the morning. We're going to the press in real time, when it comes to social networking. You have to respond and respond quickly. Well-handled. (CBS News, 2011)

Customer Empowerment and Complaint Management

Product and service failures are often thought of as potential turning points in the customer–company relationship (Van Doorn and Verhoef, 2008), because a good recovery can secure a customer's loyalty. At the same time, very few customers complain. Most customers silently defect. By empowering customers, and thereby making it easier for them to complain, companies have more of an opportunity to break the silence and save the customer by choosing an effective recovery (Fornell and Wernerfelt, 1987). Hence, companies that excel at empowering their customers, the third component of interaction orientation, are likely to identify more dissatisfied customers.

Increasingly customers are turning to Twitter to voice their complaints. The chief reason is speed, though most companies are unwilling to admit that they respond faster on Twitter (Bachelor, 2012). Complaining customers are advised to 'get clout behind' their complaint, by putting their Twitter handles in the tweet and finding journalists who will pick up the story. Hence, it appears that one of the most beneficial attributes of Twitter as a channel for customer complaints is that others can see the complaint.

Empowering customers not only makes it easier for customers and potential customers to share opinions with or complain to a company, but it also encourages customers to share information with each other. The dramatic rise of YouTube, Twitter, Yelp, Facebook and other rating and social media sites has made it possible for customers to amplify their complaints to an extent unparalleled in history. *New York Times* columnist Thomas Friedman says that customers are now 'hyper-connected and super-empowered'. One famous example of a complaint propagated by social media is the 'United breaks guitars' incident, in which a disgruntled musician created the eponymous music video about how his $3500 guitar was broken during a flight on United Airlines in 2008. The video, which was posted on YouTube and within six months garnered more than 10 million hits, was picked up by CNN, BCC, and *The Wall Street Journal*, and proved to be a public relations disaster for United Airlines. The airline eventually responded months later by offering to pay the cost of repairing his guitar and flight vouchers worth $1200; the musician told the airline to donate the sum to charity (Tran, 2009).

Another consequence of the increasing customer empowerment has been the online organization and collective action of customers to influence companies to respond to complaints. A recent example is Bank Transfer Day, which called on bank customers to switch from commercial banks to non-profit credit unions on 5 November 2011. Gallery owner

Kristen Christian, a customer dissatisfied with Bank of America's customer service and recently introduced debit card fees, created a Facebook event that borrowed from the rhetoric of the Occupy Wall Street movement. The event received over 8000 likes and over 35 000 people indicated they would join it. An online petition started by Molly Katchpole, a 22-year-old woman from Washington, opposing Bank of America's fee attracted more than 300 000 signatures on the website Change.org, which allows consumers to start grassroots campaigns (Carrns, 2011). Despite an outpouring of complaints online and offline, the bank at first refused to rescind its newly introduced fee. On the day of the event, the Credit Union National Association estimated that approximately 40 000 customers with $80 million in funds opened new accounts with credit unions. In the month leading up to the event, credit unions saw an increase of 50 percent in new customers. Bank of America cancelled its debit card fee approximately a month after it had first announced it.

Not surprisingly some experts have advised businesses to deal with 'customer insurgencies' by drawing parallels to the military's experience fighting terrorism (Bonchek and Fussell, 2012). They advocate monitoring social media to listen in for customer complaints, developing a network of brand advocates to help neutralize any customer insurgency, and restructuring departments to be able to respond quickly to any customer uprising. Hence, empowering customers to complain is a risky strategy unless the company can respond quickly and effectively (as in the case of McDonald's).

Customer Value Management and Complaint Management

Companies that practice good customer value management – companies that can predict how much their customers will contribute to profits in response to their recovery – are likely to have good recovery capabilities. Financially accountable recovery strategies should balance customers' future potential (e.g., customer lifetime value) and costs to the company for providing different types and levels of recovery resources (Parasuraman, 2006). In general, companies tend to think about immediate costs of the recovery instead of the long-term value of a customer saved from defecting (Day and Moorman, 2010).

Though both academics (Rust and Chung, 2006) and managers (Goodman, 2006) have called for making recovery strategies more financially accountable, there is scant work that investigates the long-run financial impact of failures and recoveries. Luo (2007) matches complaints lodged with the US Department of Transportation (DOT) against the airline industry and stock market prices and finds that higher levels of complaints reduce firm-idiosyncratic stock returns. For example, the study finds that a 1 percent increase in DOT complaints lodged against Southwest Airlines would lead to a $262 million loss in their market capitalization. Knox and Van Oest (2014) examine a panel of 20 000 customers over two years and find that approximately one-third to two-thirds of customer lifetime value (CLV) is destroyed following a complaint; there is considerable variation in the negative effects across different types of complaints and positive effects of recoveries on CLV; and that providing recovery works up to a certain point. They also show that the 'memory' for past complaints fades quickly, so a successful recovery not only retains the customer but also resets the relationship. More research is needed to quantify the dynamic effects on customer behavior.

COMPLAINT MANAGEMENT AND FIRM PERFORMANCE

In order to establish the significance of an interaction orientation for firms Ramani and Kumar (2008) identified two composite customer-centric performance measures, (1) customer-based profit performance consisting of the successful identification of profitable customers, efficiency of the acquisition and retention process, and the conversion of unprofitable customers to profitable ones, and (2) customer-based relational performance consisting of customer satisfaction, customer ownership, and positive word-of-mouth. Having discussed how companies that adopt an interaction orientation are likely to have superior recovery capabilities, we now consider the consequences for customer-based profit and relational performance.

While Ramani and Kumar (2008) recorded the positive impact of an interaction orientation on customer-based profit performance measures in the presence of customer-initiated contacts, they did not consider the implications of customer-initiated contacts that arise out of product or service failures. The first component of customer-based profit performance has to do with the company's ability to predict customer profitability. Companies have invested heavily in CRM systems and now obtain much wider access to data that track customer–firm interactions over time (Bolton, Lemon, and Bramlett, 2006). These non-transactional elements increasingly collected by companies are sometimes thought of as behavioral manifestations of customer engagement, and often include, in addition to customer complaints, word-of-mouth referrals, participation in the firm's activities, suggestions for service improvements, among other things (Bijmolt et al., 2010). Though there is scant research in this area, there is a growing recognition that incorporating these individual-level, non-transactional elements in general leads to significantly better predictions of future customer purchasing and lifetime value (Blattberg et al., 2008; Van Oest and Knox, 2011). One reason for this is that complaints add more information about the state of the customer than purchases. While complaints are indicators of dissatisfaction, they also signal that the customer still feels involved with the company (Maute and Forrester, 1993) and provide the firm with an opportunity to recover. On the other hand, observing a purchase (rather than a complaint) may result from one of two stories with very different implications for the state of the customer relationship: the customer does not complain because he or she is either satisfied, or is dissatisfied and prefers abandoning the firm rather than voicing this dissatisfaction. Insights such as these are possible only by incorporating complaints into a customer base model (Van Oest and Knox, 2011). Hence, one would expect that companies that keep careful track of customer complaints and other non-transactional elements would be better at identifying profitable customers in the long run.

The last two components of profit performance have to do with customer profitability in the long run and converting unprofitable customers. Converting unprofitable customers to profitable ones in the context of complaint management has a special name, the 'service recovery paradox'. Though there are several variants of the paradox, the basic idea is that customers who experience a service failure respond so well to a company's recovery, that their satisfaction, loyalty, or repurchase levels are higher as a result than they would have been had no failure occurred (Smith and Bolton, 1998). Customers have expectations about the quality of service, but also about the quality of the company's recovery following a service failure (McCollough, Berry, and Yadav, 2000). If the recovery is worse than expected, then there is a 'double deviation' from expectations, resulting from both the service failure and the company's unsuccessful recovery (Bitner, Booms, and Tetreault, 1990). If the recovery is better than expected, then the paradox may emerge.

Hart, Heskett, and Sasser (1990) present several examples as empirical support of the paradox. They write: 'A good recovery can turn angry, frustrated customers into loyal ones. It can, in fact, create more goodwill than if things had gone smoothly in the first place' (p. 148). Hence, they argue that service recovery should be treated as a strategic company resource. Smith and Bolton (1998) use hypothetical failure and recovery scenarios, and find that excellent recoveries lead to increased satisfaction and repurchase intentions, but only when the quality of the recovery itself is judged by customers as being very high. A meta-analysis conducted on 24 studies examining the service recovery paradox shows that the paradox effect exists for satisfaction, but not on repurchase intentions, word-of-mouth, and corporate image (De Matos, Henrique, and Rossi, 2007). The importance of accounting for unobserved differences across customers, other moderators of the process, such as previous service experiences, and low effect sizes represents challenges to identifying the service paradox effect in practice. Hence, failures are still to be avoided as much as possible.

The implications of this discussion suggest that companies with superior recovery skills are likely to see a stronger link between interaction orientation and customer satisfaction. An interaction orientation entails that firms are able to assess customer heterogeneity, and have documented past customer service experiences. This means that their complaint management efforts would result in more consistent service recovery performance. The above discussion also suggests that an interaction orientation can improve a firm's performance with its customers by softening the negative impact of customer complaints. While further empirical validation is necessary, it is reasonable to propose that an interaction orientation can prevent the erosion of long-run profitability, and restore customer ownership and positive word-of-mouth in situations arising out of customer complaints.

INTERACTION ORIENTATION, COMPLAINT MANAGEMENT AND CUSTOMER EQUITY

Customer equity and customer lifetime value have emerged as popular performance metrics. Customer equity and customer lifetime value are conceptually appealing to top managers as they bridge the gap between marketing and finance. However, managers must note that customer lifetime value and customer equity are not one-size-fits-all standardized metrics. The computation of individual customer lifetime value is usually a mathematical model specific to a firm's context and considerable modeling skills are required to arrive at the best prediction of future profitability. An interaction orientation, by definition, subsumes customer value management as one of its components. The development of customer lifetime value and customer equity models is therefore necessary for the practice of customer value management. Firms must invest in computing resources and skills to develop accurate customer lifetime models. However, access to computing skills alone does not guarantee success unless customer value management is integrated with the other elements of interaction orientation, notably customer empowerment and interaction response capacity.

To illustrate the importance of complaint management in driving customer equity, we provide an example from the hotel industry. Hampton Inn instituted a 100 percent satisfaction guarantee that provided a full refund to any complaining customer; furthermore, the guarantee could be enacted by any employee from any level, from janitor to desk clerk (Rust, Subramanian, and Wells, 1992). The policy made it more likely that a complaining customer would see his or her problem resolved, but it also made it more likely that customers would voice their complaints (Heskett, Sasser, and Scheslinger, 1997). Hence, this led to an increase in the number of customer complaints.

The increase in the number of complaints could have been an attempt by Hampton Inn's customers to 'game' the system and collect their refund, causing a large loss for Hampton Inn. On the other hand, the increase could have resulted from genuinely dissatisfied customers who would have otherwise silently defected. If this were true, the policy would be profitable for Hampton Inn, because the gains in customer retention and patronage would outweigh the loss of providing refunds. After a year of instituting the policy, Hampton Inn calculated that the policy generated $8.2 million in incremental revenue and $800 000 in incremental cost, from issuing refunds and advertising the guarantee. Survey results indicated that 45 percent of the complaining customers who invoked the guarantee returned, and 61 percent of these customers indicated they returned because of the guarantee; this represented an additional $1 million in revenue for Hampton Inn (ibid.). These results indicate the role of strong complaint handling policies driving customer equity, and, in line with the findings of Knox and Van Oest (2014), that the impact of past complaints fades quickly: an effective recovery not only retains the customer but also resets the relationship.

INTERACTION ORIENTATION AND THE FUTURE

Front-line employees have been instrumental in acquiring and deploying customer need knowledge with the intention of improving customer satisfaction and willingness to pay (Homburg, Wieseke, and Bornemann, 2009). However, in many businesses today interactions between front-line employees and customers have become infrequent. The onus of obtaining customer need knowledge has shifted to automated processes that piece together digital information. Companies need to make trade-offs in terms of resource allocation between employee training and developing scoring and/or predictive models. The process of aligning front-line employees with customers is giving way to aligning online experiences with customers' dynamic needs. Hence, firms cannot limit themselves to depending on metrics that measure front-line employees' performance to drive profits. Instead, firms need to continually invest in customer value management metrics such as customer lifetime value and customer equity in order to align their marketing activities with their individual customer needs and capabilities.

Firms have long used patents and trademarks to protect their long-term interests. However, an over-dependence on a patents-based strategy has been shown to have a negative impact on a firm's level of interaction orientation (Ramani and Kumar, 2008). Product life cycles are increasingly determined by the emergence of replacement technologies and not by the duration of protection that patents can provide. When companies enter into protracted litigation against each other on the grounds of patent infringement they end up diverting resources away from understanding customers and from continuous product innovation to legal expenses (Guglielmo, 2012). Future studies could further explore the long-term impact of the dependence on patents on a company's interaction orientation and performance.

Monitoring of interactions in the age of the Internet is now commonplace. Monitoring increases the amount of information available to marketers but does not alter the capacity to construct meaning (Deighton and Kornfeld, 2009). The way people took to social media suggests that individuals benefit by using their social groups (1) as sources of information on products, services, and places, (2) to satisfy human curiosity about their peers, and (3) to reinforce self-identities. The power of customer empowerment is obvious today when we see how brands such as Red Bull and Zynga facilitate customer interactions in social networks and concurrently develop insights, and test and promote products and services at the level of granularity of the individual customer. Social media have grown primarily from the inside out as people expanded their groups and became part of other groups. The challenge for the marketer is to be able to approach social connections from the outside in while recognizing the individuality of each customer. For instance Coca-Cola was able to build a social reward system into its marketing efforts by adding thousands of recycle bin locations in Israel into the Facebook Places system and announcing the Recycling King award to the person who checks into the most recycling sites. Facebook users checked into their accounts every time they recycled a bottle and posted their pictures putting the plastic in the bins. Understanding the role of social media and developing an outside-in application helped elevate Coca-Cola customers from mere users to empowered social citizens.

Marketers have recognized the importance of their role as facilitators of interactions that result in social exchange. The next step for marketers is to establish themselves as enablers of cultural interactions (ibid.). Hence, fostering an interaction orientation would be a continuing priority for managers.

REFERENCES

Ante, S.E. (2012), 'Hype and hope: test driving Google's new glasses', *Wall Street Journal*, 11 September 2012, accessed 19 September 2014 at http://online.wsj.com/article/SB10000872396390443779404577643981045121516.html.

Arthur, C. (2012), 'Microsoft gets patent on augmented reality glasses as "AR Wars" start', *The Guardian*, 27 November 2012, accessed 19 September 2014 at http://www.theguardian.com/technology/2012/nov/27/microsoft-augmented-reality-glass-google-apple.

Bachelor, L. (2012), 'Complain on Twitter for an instant response', *The Guardian*, 12 May 2012, accessed 20 September 2014 at http://www.guardian.co.uk/money/2012/may/12/complaint-air-on-twitter.

Berry, L.L. and A. Parasuraman (1991), *Marketing Services: Competing Through Quality*, New York: Free Press.

Berry, L.L. and A. Parasuraman (1997), 'Listening to the customer – the concept of a service-quality information system', *Sloan Management Review*, **38**(3), 65–76.

Bijmolt, T.H.A., P.S.H. Leeflang, F. Block, M. Eisenbeiss, B.G.S. Hardie, A. Lemmens and P. Saffert (2010), 'Analytics for customer engagement', *Journal of Service Research*, **13**(3), 341–56.

Bitner, M.J., B.H. Booms and M.S. Tetreault (1990), 'The service encounter: diagnosing favorable and unfavorable incidents', *Journal of Marketing*, **54**(1), 71–84.

Blattberg, R.C., B.-D. Kim and S.A. Neslin (2008), *Database Marketing: Analyzing and Managing Customers*, New York: Springer.

Bolton, R.N., K.N. Lemon and M.D. Bramlett (2006), 'The effect of service experiences over time on a supplier's retention of business customers', *Management Science*, **52**(12), 1811–23.

Bonchek, M. and C. Fussell (2012), 'Countering a customer insurgency', *Harvard Business Review Blog Network*, 15 October 2012, accessed 20 September 2014 at http://blogs.hbr.org/cs/2012/10/countering_a_customer_insurgen.html.

Bowman, D. and D. Narayandas (2001), 'Managing customer-initiated contacts with manufacturers: the impact on share of category requirements and word-of-mouth behavior', *Journal of Marketing Research*, **38**(3), 281–97.

Carrns, A. (2011), 'Petition on debit card fee attracts 200,000 supporters', *The New York Times*, 13 October 2011, accessed 20 September 2014 at http://bucks.blogs.nytimes.com/2011/10/13/petition-on-debit-card-fee-attracts-200000-supporters/.

CBS News (2011), 'McDonald's racist Twitter message was hoax', *CBS News*, 15 June 2011, accessed 10 June 2013 at http://www.cbsnews.com/stories/2011/06/15/earlyshow/leisure/gamesgadgetsgizmos/main20071227.shtml.

Colvin, G. (2013), 'The world's most admired companies: built for brilliance', *Fortune Magazine*, 28 February 2013, accessed 20 September 2014 at http://fortune.com/2013/02/28/the-worlds-most-admired-companies-built-for-brilliance/.

Day, G.S. and C. Moorman (2010), *Strategy from the Outside In*, New York: McGraw Hill.

Deighton, J. and L. Kornfeld (2009), 'Interactivity's unanticipated consequences for marketers and marketing', *Journal of Interactive Marketing*, **23**(1), 4–10.

De Matos, C.A., J.L. Henrique and C.A.V. Rossi (2007), 'Service recovery paradox: a meta-analysis', *Journal of Service Research*, **10**(1), 60–77.

Divol, R., D. Edelman and H. Sarrazin (2012), 'Demystifying social media', *McKinsey Quarterly*, April 2012, accessed 20 September 2014 at http://www.mckinseyquarterly.com/Demystifying_social_media_2958.

Edwards, J. (2012), 'Check out the insane lengths Zappos customer service reps will go to', *Business Insider*, 9 January 2012, accessed 19 September 2014 at http://articles.businessinsider.com/2012–01–09/news/30606433_1_customer-service-zappos-center-services.
Fader, P. (2012), 'Customer centricity and why it matters', *Knowledge@Wharton*, 18 November 2012, accessed 19 September 2014 at http://knowledge.wharton.upenn.edu/article.cfm?articleid=2875.
Farber, D. (2012), 'Amazon's Jeff Bezos admits Kindles are sold at cost', *CNET*, 11 October 2012, accessed 20 September 2014 at http://news.cnet.com/8301–1023_3–57530924–93/amazons-jeff-bezos-admits-kindles-are-sold-at-cost/.
Forbes, P. (2011), 'McDonald's isn't charging African Americans $1.50 extra', *Eater.com*, 13 April 2011, accessed 20 September 2014 at http://eater.com/archives/2011/06/13/mcdonalds-isnt-actually-charging-african-americans-150-extra.php.
Fornell, C. and B. Wernerfelt (1987), 'Defensive marketing strategy by customer complaint management: a theoretical analysis', *Journal of Marketing Research*, **24**(4), 337–46.
Franke, N., P. Keinz and C. Steger (2009), 'Testing the value of customization: when do customers really prefer products tailored to their preferences?', *Journal Of Marketing*, **73**(5), 103–21.
Fuchs, C., E. Prandelli and M. Schreier (2010), 'The psychological effects of empowerment strategies on consumers' product demand', *Journal of Marketing*, **74**(1), 65–79.
Gilly, M.C., W.B. Stevenson and L.J. Yale (1991), 'Dynamics of complaint management in the service organization', *Journal of Consumer Affairs*, **25**(2), 295–322.
Goodman, J. (2006), 'Manage complaints to enhance loyalty', *Quality Progress*, February, 28–34.
Grönroos, C. (1988), 'Service quality: the six criteria of good perceived service quality', *Review of Business*, **9**(3), 10–13.
Guglielmo, C. (2012), 'Apple, Samsung patent war puts future of innovation at risk', *Forbes*, 23 August 2012, accessed 20 September 2014 at http://www.forbes.com/sites/connieguglielmo/2012/08/23/apple-samsung-patent-war-puts-future-of-innovation-at-risk/.
Hart, C.W.L., J.L. Heskett and W.E. Sasser (1990), 'The profitable art of service recovery', *Harvard Business Review*, **68**(4), 148–56.
Heffernan, R. and S. LaValle (2006), 'Advocacy in the customer focused enterprise: the next generation of CRM done right', *IBM Global Business Services*, accessed 19 September 2014 at http://www-935.ibm.com/services/us/gbs/bus/pdf/ibv-ge510–6284–00.pdf.
Heskett, J.L., E.W. Sasser and L.A. Schlesinger (1997), *The Service Profit Chain*, New York: Free Press.
Hirschman, A.O. (1970), *Exit, Voice, and Loyalty: Responses to Decline in Firms, Organizations and States*, Cambridge, MA: Harvard University Press.
Homburg, C., J. Wieseke and T. Bornemann (2009), 'Implementing the marketing concept at the employee–customer interface: the role of customer need knowledge', *Journal of Marketing*, **73**(4), 64–81.
Köhler, C., A. Rohm, K. de Ruyter and M. Wetzels (2011), 'Return on interactivity: the impact of online agents on newcomer adjustment', *Journal of Marketing*, **75**(2), 93–108.
Knox, G. and R. van Oest (2014), 'Customer complaints and recovery effectiveness: a customer base approach', *Journal of Marketing*, **78**(5), 42–57.
Leeflang, P. (2011), 'Paving the way for distinguished marketing', *International Journal of Research in Marketing*, **28**(2), 76–88.
Luo, X. (2007), 'Consumer negative voice and firm-idiosyncratic stock returns', *Journal of Marketing*, **71**(3), 75–88.
Mason, C.H. (2003), 'Tuscan Lifestyles: assessing customer lifetime value', *Journal of Interactive Marketing*, **17**(4), 54–60.
Maute, M. and W.R. Forrester (1993), 'An investigation into the structure and determinants of consumer complain intentions and behavior', *Journal of Economic Psychology*, **14**(2), 219–47.
McCollough, M.A., L.L. Berry and M.S. Yadav (2000), 'An empirical investigation of

customer satisfaction after service failure and recovery', *Journal of Service Research*, **3**(2), 121–37.
Parasuraman, A. (2006), 'Modeling opportunities in service recovery and customer-managed interactions', *Marketing Science*, **25**(6), 590–93.
Peppers, D. (2012), 'The untapped value of complainers', *Fast Company*, 2 August 2012, accessed 19 September 2014 at http://www.fastcompany.com/1844420/untapped-value-complainers.
Peppers, D. and M. Rogers (2004), *Managing Customer Relationships: A Strategic Framework*, Hoboken, NJ: John Wiley and Sons.
Ramani, G. and V. Kumar (2008), 'Interaction orientation and firm performance', *Journal of Marketing*, **72**(1), 27–45.
Reinartz, W.J. and V. Kumar (2003), 'On the profitability of long-life customers in a non-contractual setting: an empirical investigation and implications for marketing', *Journal of Marketing*, **64**(4), 17–35.
Rust, R.T. and T.S. Chung (2006), 'Marketing models of service and relationships', *Marketing Science*, **25**(6), 560–80.
Rust, R.T., K.N. Lemon and V.A. Zeithaml (2004), 'Return on marketing: using customer equity to focus marketing strategy', *Journal of Marketing*, **68**(1), 109–27.
Rust, R.T., B. Subramanian and M. Wells (1992), 'Making complaints a management tool', *Marketing Management*, **1**(3), 40–45.
Schmittlein, D.C., D.G. Morrison and R. Colombo (1987), 'Counting your customers: who they are and what will they do next?', *Management Science*, **33**(1), 1–24.
Smith, A.K. and R.N. Bolton (1998), 'An experimental investigation of customer reactions to service failure and recovery encounters: paradox or peril?', *Journal of Service Research*, **1**(1), 65–81.
Thompson, B. (2010), 'Case study: thinking outside the (shoe) box', *Frost & Sullivan GIL eBulletin*, **3**(3), accessed 19 September 2014 at http://www.growthconsulting.frost.com/web/images.nsf/0/18A17AA365A9331F652576EE004BA993/$File/GILMAR_Case%20Study.pdf.
Tran, M. (2009), 'Singer gets his revenge on United Airlines and soars to fame', *The Guardian News Blog*, 23 July 2009, accessed 20 September 2014 at http://www.guardian.co.uk/news/blog/2009/jul/23/youtube-united-breaks-guitars-video.
Van Doorn, J. and P. Verhoef (2008), 'Critical incidents and the impact of satisfaction on customer share', *Journal of Marketing*, **72**(4), 123–42.
Venkatesan, R. and V. Kumar (2004), 'A customer lifetime value framework for customer selection and resource allocation strategy', *Journal of Marketing*, **68**(4), 106–25.
Van Oest, R. and G. Knox (2011), 'Extending the BG/NBD: a simple model of purchases and complaints', *International Journal of Research in Marketing*, **28**(1), 30–37.
Weisul, K. (2005), 'A shine on their shoes', *Bloomberg BusinessWeek*, 4 December 2005, accessed 20 September 2014 at http://www.businessweek.com/stories/2005–12–04/a-shine-on-their-shoes.
Wiesel, T., B. Skiera and J. Villanueva (2008), 'Customer equity: an integral part of financial reporting', *Journal of Marketing*, **72**(2), 1–14.

20. Customer equity reporting*

Thorsten Wiesel and Bernd Skiera

INTRODUCTION

Nowadays, managers and investors are confronted with an overload of information. This mass of information has to support managers running their company and investors making good investment decisions. Although gathering company information is very time consuming, structuring the available information in such a way that it provides value for the company may prove to be even more difficult. Numerous metrics evaluating managers' performance tend to reflect past performance rather than future performance. As such, they provide limited guidance for long-term–oriented management and short-term–oriented management was certainly also responsible for the breakout of the financial crisis (Skiera, Bermes, and Horn, 2011).

Consider, for example, the profitability analysis in Figure 20.1 that was done for two consecutive periods, evaluating management's performance in a company with contractual relationships, such as a bank, an insurance company or a telecommunications provider (Wiesel, Skiera, and Villanueva, 2008). The results clearly indicate that management has done an excellent job: all metrics increased substantially and profit rose by more than 30 percent. So why bother?

The problem is that these profitability metrics are short-term oriented. They mirror 2014's results, but do not outline what is likely to happen in the coming years. What is worse, they might even provide incentives for short-term–oriented management like reducing advertising spending in order to improve profitability at the expense of diminishing consumers' awareness and their intention to buy in the future. How can such behavior be avoided?

We recommend reporting customer equity. Customer equity measures the long-term value of a company's current customer base, which is the discounted profit that a company will make with its current customers – now and in the future. This idea is illustrated by using additional information to that shown in Figure 20.1. Table 20.1 includes the additional numbers that drive the results displayed in Figure 20.1. Considering the number of acquired and lost customers enables us to calculate the churn rates by dividing the number of lost customers by the average number of customers

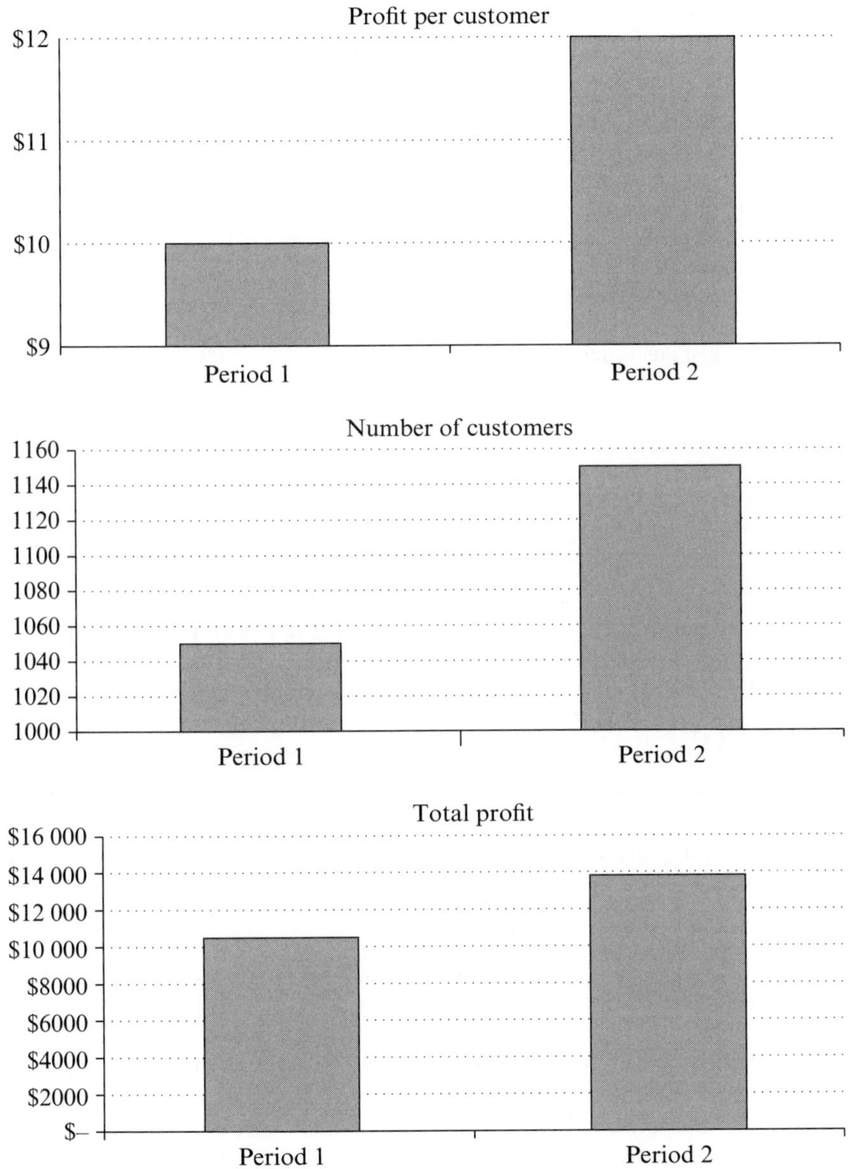

Figure 20.1 Profitability analysis in numerical example

Table 20.1 Data of numerical example

	Period 1	Period 2	Change (%)
Profit per customer (in $)	10.00	12.00	20.00
Total profit (in $)	10 500	13 800	31.43
Total number of customers, in 1000s (beginning of period)	1000	1050	5.00
Total number of customers, in 1000s (end of period)	1050	1150	9.52
Number of acquired customers, in 1000s (during the period)	150	300	100.00
Number of lost customers, in 1000s (during the period)	100	200	100.00
Churn rate (%)	9.76	18.19	86.37
Retention rate (%)	90.24	81.81	−9.34
Customer lifetime value (in $)	55.67	46.83	−15.89
Customer equity (in K$)	58 451	53 848	−7.87
Change in customer equity (in K$)		−4602	

in the given period. The latter is simply the average number of customers at the beginning and end of the respective period. Unfortunately, this churn rate increased dramatically (+86.37 percent), which sheds some doubts on the management effectiveness. Still, evaluating whether management has done a good job is quite difficult by just considering the first eight rows of Table 20.1. Some metric changes are positive, whereas others are negative, yet the overall effect remains unclear.

Using the available information to estimate an easily applicable model of customer lifetime value (CLV) such as:

$$CLV_t = m_t \frac{1+i}{1+i-r_t} \tag{20.1}$$

where m_t is the profit per customer in period t, i is the discount rate (here 10 percent) and r_t is the retention rate in period t, indicates that CLV diminished by 15.89 percent. Please note that the CLV formula shown here is a simple and an easily applicable version. More complex versions (e.g., with varying retention rates and/or discount rates over time or segments) are available, might improve accuracy, and are briefly discussed in Schulze, Skiera, and Wiesel (2012) and in a later section of this chapter (see 'Outlook on Future Developments with Respect to Customer Equity Reporting').

Customer equity, here defined as CLV multiplied by the number of customers, also decreased by 7.87 percent (−\$4602.54). Given that customer equity denotes the present value of all current and future profits of current customers, this decrease represent bad news. Hence, it would appear this manager has increased the profit margin at the expense of the retention rate – in terms of short-term profit a wise decision, but not in terms of the long-term success of the company. Instead of congratulating the manager for increasing the current period's profitability by 31.43 percent, we should ask why he or she has destroyed so much long-term value.

Enlarging on this example, we would like to stress the importance of tracking future-oriented customer metrics and reporting customer equity internally as well as externally. Customer equity allows for better company management and value creation, but it also tackles the increasing demand for additional information that facilitates investors' decision-making. Thereby we focus particularly on companies with contractual relationships (e.g., Internet service providers, financial service providers, telecommunication companies, energy suppliers, pay-TV broadcasters, online movie rental services), which can easily determine the number of existing and lost customers at a particular point in time.

CUSTOMER EQUITY REPORTING

The marketing discipline is concerned with managing customer relationships as key intangible assets that a company possesses (Schulze, Skiera, and Wiesel, 2012). Customer equity is the value of those customer assets and equals the sum of the customer lifetime values. Therefore, marketing aims to manage the customer assets in order to maximize customer equity and, subsequently, shareholder value. Customer equity reporting is the process of preparing information about the value and management of companies' customer assets in accordance with generally accepted principles. Wiesel et al. (2008) propose that the information prepared comprises two main elements with respect to the value of companies' customer assets: the customer equity statement and the customer equity flow statement. The customer equity statement reports customer equity (i.e., the value of the customer base) and its components in a single, clear display, thus revealing the value of the existing customer base. The customer equity flow statement describes changes in customer equity and its components between two points in time and reports the influence of any changes in customer metrics on customer equity.

For the specific purpose of reporting, Wiesel et al. (2008) define customer equity as the sum of the CLVs (after marketing expenditure) of all of the companies' current customers in period 1. CLVs before marketing expenditure result from several customer metrics, such as profit per customer and the duration of a customer's relationship with the company, known as customer lifetime. To retain or acquire customers, a company must invest money; the measures of retention and acquisition costs per customer reflect those investments. Combining customer metrics with an appropriate discount rate provides a calculation of the present value of all profits of a customer (CLV before marketing expenditure) and the present value of all costs necessary for retaining a customer (lifetime retention expenditure). These metrics are labeled as customer value metrics because they determine the value of a particular customer. Altogether, they determine each customer's CLV after marketing expenditure. For simplicity, we do not distinguish here between different segments of customers, but the requirements for doing so are fairly straightforward.

The number of customers at the end of a period equals the number of customers at the beginning of a period plus the number of customers acquired minus the number of customers lost. To understand these customer movements, we use the number of existing customers (at the beginning of a period) and the number of new and lost customers (during a period) as customer quantity metrics. Multiplying the CLV of an average customer before marketing expenditure by the number of existing, new, or lost customers provides the corresponding value of existing, new, or lost customers before marketing expenditure. A similar calculation for acquisition and retention expenditures is equally valid.

These various combinations of customer value and quantity metrics provide several different components of customer equity. As illustrated in Figure 20.2, customer equity can be broken down according to the various kinds of customers (existing, new, or lost) or the value components, which is the net present value of customer cash flows, retention expenditure, and acquisition expenditure.

APPLICATION: THE NETFLIX CASE

Wiesel et al. (2008) apply this reporting technique to Netflix.com. Netflix.com's principal activity at this point in time was to provide online movie rental services through access to more than 55000 movies, television, and other entertainment titles. The standard subscription plan gives customers up to three titles at the same time with no due dates, late fees, or shipping charges. Shipping and receiving centers throughout the United States deliver the DVDs through the US Postal Service free of charge to customers. Wiesel et al. (2008) use publicly available, quarterly data from annual reports, 10-K and 10-Q statements, and other company reports

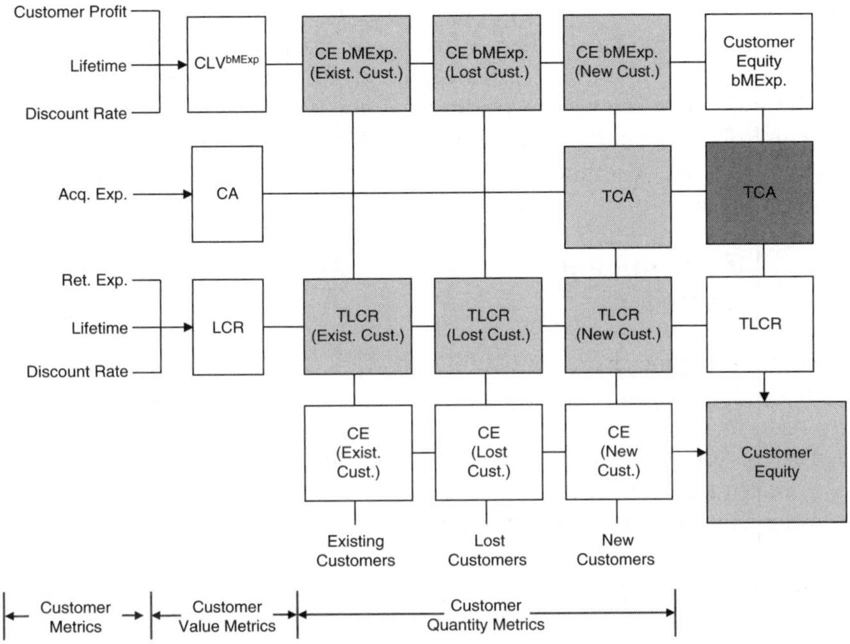

Note: Acq. Exp.: acquisition expenditure; Ret. Exp: retention expenditure; CE: customer equity; $CLV^{bMExp.}$: customer lifetime value before marketing expenditure; CA: acquisition expenditure; LCR: lifetime retention expenditure per customer; MExp.: marketing expenditure; TCA: total cost of acquisition; TCLR: total cost of lifetime retention.

Source: Wiesel et al. (2008).

Figure 20.2 Customer equity breakdown

from September 2001–September 2006. The data for each quarter include the number of customers, average monthly churn rate, gross subscriber additions, subscription revenue, subscription costs of revenue, operating expenses (without marketing expenditure), acquisition cost per customer and marketing expenditure.

This information enables them to calculate the necessary customer metrics (see Table 20.2). The company provides no information about its discount rate, so Wiesel et al. (2008) chose an annual discount rate of 10 percent (the quarterly discount rate amounts to 2.41 percent).

Figure 20.3 illustrates the value and changes of customer metrics over a certain period of time. On the positive side, Netflix.com increased its number of customers and its retention rate, as well as the profit per customer in 2006 after suffering a drop in 2005. However, its acquisition

Table 20.2 Calculation of customer metrics

	Calculation Method or Data Source	Q4 2005	Q1 2006	Q2 2006	Q3 2006
Number of customers (in thousands)	Reported number of customers per quarter (source: financial statements)	4179	4866	5169	5662
Number of new customers (in thousands)	Reported number of gross subscribers additions (source: financial statements)	1156	1377	1070	1310
Number of lost customers (in thousands)	Difference in number of customers between the current and the previous quarter + number of gross additions in the current quarter	569	690	767	817
Quarterly profit per customer (in $)	(Subscription revenue – subscription cost of revenue – operating expenses without marketing)/number of customers	9.97	10.84	11.87	12.60
Retention rate	1 – (number of lost customers during quarter/[number of customers at the beginning of quarter + number of customers at the end of quarter]/2)	0.83	0.84	0.85	0.85
Retention expenditure per customer (in $)	(Reported marketing expenditure – reported acquisition cost per customer × number of new customers)/(number of customers – number of new customers)	0.00	0.00	0.00	0.00
Acquisition expenditure per customer (in $)	Reported acquisition cost per customer (source: financial statements)	38.08	38.47	43.95	45.32

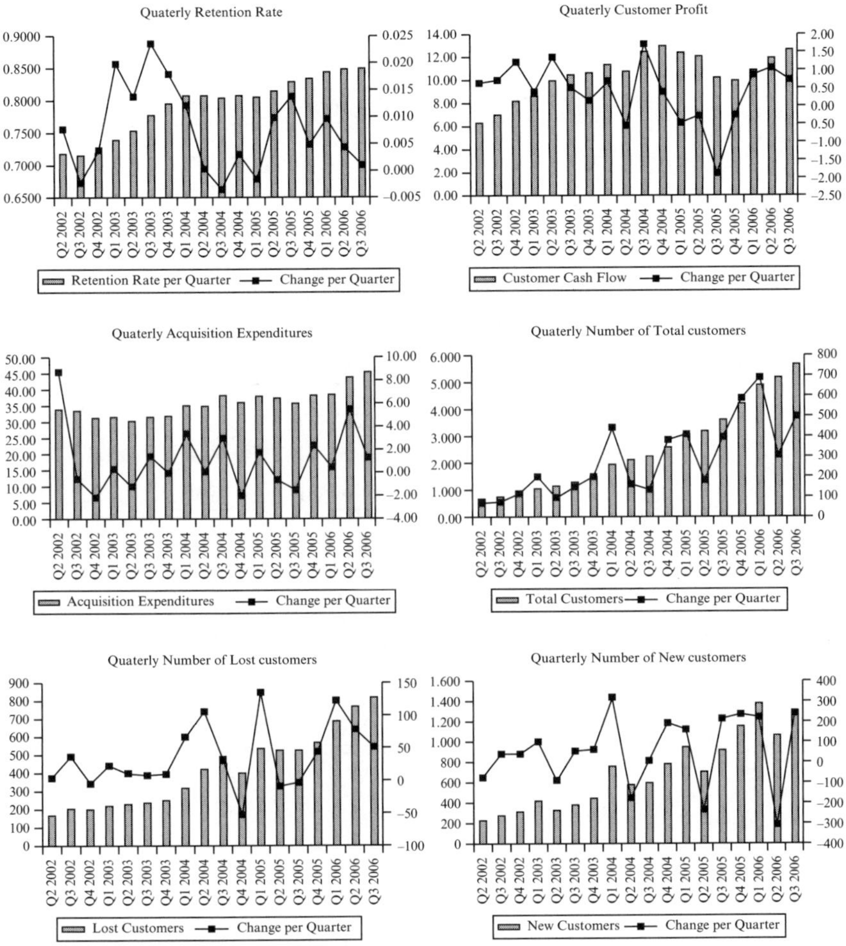

Source: Wiesel et al. (2008).

Figure 20.3 Development of customer metrics over time

expenditures increased. Therefore, these measures do not provide a clear picture of the overall value of the customer base.

Netflix.com's Customer Equity Statement

Wiesel et al. (2008) select the easily applicable CLV specification shown in equation (20.1). Based on this formulation, Figure 20.4 depicts

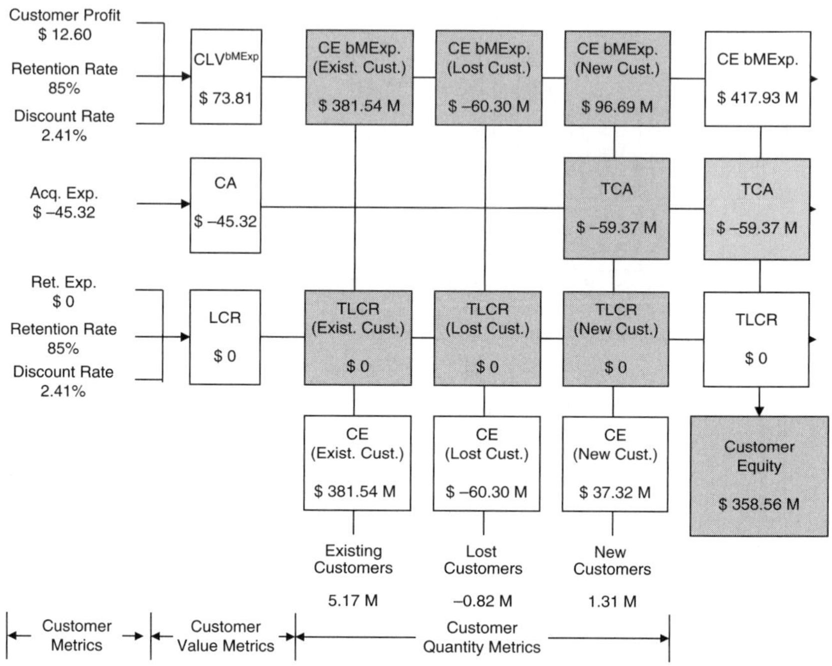

Note: Acq. Exp.: acquisition expenditure; Ret. Exp.: retention expenditure; CE: customer equity; $CLV^{bMExp.}$: customer lifetime value before marketing expenditure; CA: acquisition expenditure; LCR: lifetime retention expenditure per customer; MExp.: marketing expenditure; TCA: total cost of acquisition; TCLR: total cost of lifetime retention.

Source: Wiesel et al. (2008).

Figure 20.4 Netflix.com's customer equity statement (Q3 2006)

Netflix.com's customer equity statement for Q3 2006. Customer equity yields $358.56 million in Q3 2006, according to the customer equity without marketing expenditure for existing customers ($381.54 million), lost customers (−$60.30 million), new customers ($96.69 million), and total lifetime acquisition expenditure ($59.37 million). Because all Netflix.com's marketing expenditure is for acquiring new customers, the total lifetime retention expenditure is always zero. It also shows the breakdown according to groups of new, lost and existing customers in Figure 20.4.

The customer equity statement monitors customer equity over a given period of time. Therefore, it provides information about the value of the customer base and its components as well as an illustrative overview of customer metrics. However, it does not indicate the sources of change in customer equity over a certain period of time. It would enhance any analysis by giving insights into how much the value of the customer base has changed due to whichever metric. More detailed statements about the companies' customer management activities appear in the customer equity flow statement.

Netflix.com's Customer Equity Flow Statement

Following on from Figure 20.4, Figure 20.5 depicts Netflix.com's total change in customer equity, its components and its customer metrics in Q2–Q3 2006. Customer equity changed by $48.10 million, which reflects

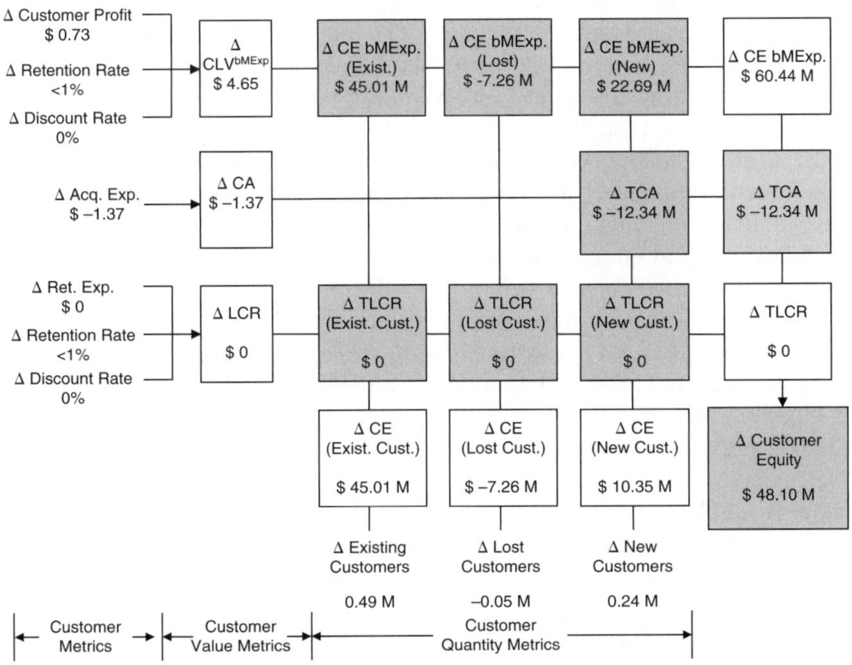

Note: ΔAcq. Exp.: change in acquisition expenditure per customer; ΔRet. Exp.: change in retention expenditure; ΔCE: change in customer equity; $\Delta CLV^{bMExp.}$: change in customer lifetime value before marketing expenditure; ΔCustomer Profit: change in customer profit; ΔCA: change in cost of acquisition per customer; ΔLCR: change in lifetime cost of retention per customer; MExp.: marketing expenditure; ΔTCA: change in total cost of acquisition; ΔTCLR: total cost of lifetime retention.

Source: Wiesel et al. (2008).

Figure 20.5 Netflix.com's customer equity flow statement (Q2–Q3 2006): components view

a shift in customer equity before marketing expenditure of $60.44 million and a change in total lifetime acquisition expenditure of −$12.34 million (i.e., total lifetime acquisition expenditure increase). The change in customer equity before marketing expenditure comprises three components: change in customer equity before marketing expenditure of existing customers ($45.01 million), lost customers (−$7.26 million) and new customers ($22.69 million).

Furthermore, Figure 20.5 indicates the changes in customer metrics, customer value metrics, and customer quantity metrics, thus summarizing what has happened during this period and the future-oriented effects of those changes, for example, in customer equity. Another good example is that Netflix.com increased its existing customers (0.49 million) in Q3 2006, but lost more customers than in Q2 2006 (−0.05 million), yet gained more customers than in Q2 2006 (0.24 million). Therefore, it increased the value of the whole customer base, primarily because its average profit per customer ($0.73) rose during that period.

In addition to breaking down changes in customer equity for several components, managers and investors might want to know which metrics caused those changes. These results are provided in Table 20.3, which includes the total effect (total change), value effects (changes due to shifts in customer value metrics), quantity effects (changes due to the number of existing, lost, and new customers) and interaction effects (changes due to simultaneous changes in customer value and quantity metrics).

According to Table 20.3, the major sources of Netflix.com's increased customer equity in Q2–Q3 2006 ($48.10 million) are positive value effects ($22.35 million) and quantity effects ($23.22 million), which indicate that Netflix.com boosted the value of its customer base by having more and more valuable customers. Furthermore, the change in profit per customer raised customer equity by $21.98 million, supported by the increase in customer lifetime ($1.84 million) but this was partly compensated for by higher acquisition expenditure (−$1.47 million). The positive interaction effects ($2.53 million) also suggest that the profit per customer and retention rate increases prompt positive customer lifetime value effects for existing and new customers, but also lead to a more negative effect for the lost customers.

In order to understand Table 20.3 better, it also includes trends compared with the previous quarter. These trends show that management has done a good job in reducing the increase of acquisition costs. However, management was not able to sustain the robust increase in profit per customer and customer lifetime value any longer. The value of new customers is still significantly higher than the value of lost customers. Yet, management could not continue supporting the growth in new customers,

Table 20.3 Netflix.com's customer equity flow statement (Q2–Q3 2006): effects view

	Q1–Q2 2006	Q2–Q3 2006	Trend
Total effect	63 543.91	48 098.98	↓
Value effects	27 342.54	22 351.47	↓
Customer profit	28 515.11	21 977.91	↓
Customer lifetime	6373.39	1839.46	↓
Acquisition expenditure	−7545.96	−1465.90	↑
Quantity effects	32 166.34	23 218.85	↓
Lost customers	−4769.71	−56 503.34	↓
New customers	79 436.05	79 722.18	↑
Interaction effects	4035.03	2528.67	↓
Lost customers	−5664.43	−3782.62	↑
Customer profit	−4494.67	−3473.78	↑
Customer lifetime	−1169.75	−308.84	↑
New customers	8949.88	6179.70	↓
Customer profit	6270.28	5569.95	↓
Customer lifetime	2679.60	609.75	↓
Lifetime customer profit	1753.39	234.67	↓
Other	−1003.80	−103.08	↑

Note: In $1000s.

but unfortunately now faces an increase in the value of lost customers. If these trends continue, then management will quickly face a situation in which the value of new and lost customers will be comparable, indicating that management is no longer able to improve growth by the number of customers.

OUTLOOK ON FUTURE DEVELOPMENTS WITH RESPECT TO CUSTOMER EQUITY REPORTING

In this final part of our chapter, we aim to provide an outlook on future developments with respect to customer equity reporting. In doing so, we tackle how to best use company-reported summary data to estimate the CLV and customer equity of a companies' current customers, briefly review new developments with respect to customer equity reporting, and discuss marketing disclosure in general.

Company-reported Summary Data

The measurement of CLV and customer equity has been discussed in numerous articles and books (e.g., Dwyer, 1997; Berger and Nasr, 1998; Pfeifer and Carraway, 2000; Lehmann and Gupta, 2005; Blattberg, Kim, and Neslin, 2008; Schulze et al., 2012). Kumar and George (2007) discuss features with respect to the measurement of CLV and customer equity using both aggregate- and disaggregate-level approaches as well as a hybrid approach. These approaches allow companies to choose an appropriate methodology depending on their objective and data availability. Stakeholders who do not have access to internal company databases (e.g., analysts, investors) need to rely on company-reported summary data to estimate the CLV and customer equity of companies' current customers. Alternatively, they can receive this information through customer equity reporting.

The means to calculate and report CLV and customer equity for external stakeholders as presented in this chapter might be further fine-tuned and complicated (see also Schulze et al., 2012). In a recent series of articles in the *Journal of Interactive Marketing*, Fader and Hardie (2011), Neslin (2011), Pfeifer (2011), Venkatesan (2011), and Wiesel, Skiera, and Villanueva (2011) examine how to best use company-reported summary data for doing so. Several aspects for revising current calculations are proposed. For example, the receipt of the first cash flow, timing of the cash flows, and estimation of retention rate (Neslin, 2011; Pfeifer, 2011), the assumption of a constant retention rate (Fader and Hardie, 2011), inclusion of competition, and calculation of the discount rate (Wiesel et al., 2011). Additionally, Venkatesan (2011) explores company and market conditions that are especially relevant for the modifications to customer equity measurement.

Overall, these papers contribute to a more accurate evaluation of a company's customer base based on externally available data and hence assist in improving customer equity reporting. However, all authors state that more research is needed in building more complete approaches that are still applicable for reporting purposes.

Nevertheless, some parts of the financial community may actually prefer a diverse set of customer metrics (quantitative and qualitative) in management commentaries on financial reports opposed to CLV or customer equity and its components. Doing so will enable users of financial reports to do their own calculations rather than relying on the aggregate numbers disclosed by the companies. In any case, disclosing more marketing-related information should help users of financial reports in their decision-making and is also implicitly supported by the finance community (IASB, 2005).

New Developments in Customer Equity Reporting

Persson and Ryals (2010) state that providing information about customer equity and its components are useful indicators of the success of a company's marketing and customer management, but criticize that it provides an incomplete view. In their view, additional qualitative information with respect to the actual drivers of customer equity would be more fruitful in order to predict companies' future earnings potential. Based on this, they propose that a customer equity scorecard be included in companies' financial reporting.

Besides customer equity, the customer equity scorecard includes measures such as customer satisfaction, attitudinal loyalty, average number of different products/services bought/held, and average customer risk score. Persson and Ryals (2010) believe that the identification of the drivers and components of customer equity serves both the marketing managers in their efforts to manage customer relationships profitably and investors' decision-making by increasing transparency about the value of intangible assets.

We agree that companies should provide explanations with respect to the reasons for the changes in customer equity, its components, and its underlying customer metrics. For that reason, we argue that 'in addition to this decomposition, management can explain why those changes occurred and what it plans to do in the future' (Wiesel et al., 2008, p. 11).

In extension to the ideas presented in Wiesel et al. (2008), we propose to add a third element to our customer equity reporting approach: *the customer equity notes*. The customer equity notes should contain information with respect to the management of companies' customer assets and explanations for values included in the customer equity (flow) statement (e.g., drivers of customer equity, reasons for changes in customer equity and/or its drivers, future outlook with respect to management of companies' customer assets and expected effects on customer equity).

A new quantitative customer measure to be included in companies' financial reporting was recently proposed by Skiera et al. (2011). Their customer equity sustainability ratio (CESR) compares the future value of an existing customer base with current earnings, and complements customer equity reporting. It creates more transparency about the consequences of securitization for future earnings and the riskiness of the underlying business model. In an empirical study of 38 banks in ten countries, Skiera et al. (2011) demonstrate the limited transparency of long-term value creation in financial statements and outline the adequacy of CESR for creating more transparency in banking and non-banking industries.

Discussion about Disclosure of Marketing Information

Besides the particular reporting of the value of a company's customer assets, a broader discussion about marketing information disclosure has recently begun (for a summary of arguments see Srinivasan and Sihi, 2012). Specifically, Mizik and Nissim (2011) review accounting principles related to the reporting of marketing activities. They recommend expanding the disclosure of marketing information as a solution to further preventing the decline in marketing's standing within companies and the rise of myopic management. Contrarily, they do not consider balance sheet recognition of marketing assets such as customers and brands as a remedy. Instead, they argue that consistent disclosures about marketing expenditures better facilitates the performance evaluation, forecasting, and valuation by outsiders. They urge marketers to understand the importance of financial reporting and the effect of the distortions introduced by the current accounting system on marketing's standing. As such, they call for a dialog and specific research among marketers and across the marketing and accounting disciplines to improve the financial reporting model with respect to marketing-related activities.

Similarly, the Marketing Accountability Standards Board (www.themasb.org) aims to partner with the financial reporting/investment community to improve the accounting and/or reporting rules related to marketing. The MASB has started a project called 'Improving Financial

Reporting (IFR)', which aims to equip marketers with 'generally accepted brand investment and valuation standards' in order to be prepared once the reporting of brands is required for internally developed brands.

CONCLUSIONS

In this chapter, we emphasize that reporting future-oriented customer metrics assists managers in leading their company and taking strategic decisions as well as helping investors making investment decisions. It builds upon the current shortcoming that financial reporting and most management instruments are primarily backward oriented because they focus primarily on reporting business that was realized in the past. The major idea is to provide more information about the long-term value of the companies' customers. Therefore, customer equity reporting measures the long-term value of a company's customer base, which should support management decisions that are long-term rather than short-term–value oriented. As such, it avoids the risk that management increases short-term profits at the expense of long-term value creation. Therefore, customer equity reporting also faces the demand for additional information that facilitates investors' decision-making. Moreover, it contributes to the discussion about marketing accountability and may support marketing's re-entry into the boardroom, because it aligns customer management with corporate goals and the investor's perspective.

NOTE

* Some parts of this chapter are based on Wiesel et al. (2008). Parts of Wiesel et al. (2008) are already summarized in Wiesel et al. (2010).

REFERENCES

Berger, P.D. and N.I. Nasr (1998), 'Customer lifetime value: marketing models and applications', *Journal of Interactive Marketing*, **12**(1), 17–30.

Blattberg, R.C., B.-D. Kim and S.A. Neslin (2008), *Database Marketing: Analyzing and Managing Customers*, New York: Springer.

Dwyer, F.R. (1997), 'Customer lifetime valuation to support decision making', *Journal of Direct Marketing*, **11**(4), 6–13.

Fader, P. and B. Hardie (2011), 'A comment on "On estimating current-customer equity using company summary data"', *Journal of Interactive Marketing*, **25**(1), 18–19.

International Accountings Standards Board (IASB) (2005), 'Discussion paper: management commentary', London: IASB.

Kumar, V. and M. George (2007), 'Measuring and maximizing customer equity: a critical analysis', *Journal of the Academy of Marketing Science*, **35**(2), 157–71.
Lehmann, D. and S. Gupta (2005), 'Managing customers as investments: the strategic value of customers in the long run', Philadelphia, PA: Wharton School Publishing.
Mizik, N. and D. Nissim (2011), 'Accounting for marketing activities: implications for marketing research and practice', MSI Working Paper No. 11–103.
Neslin, S.A. (2011), 'On estimating current-customer equity using company summary data: comment', *Journal of Interactive Marketing*, **25**(1), 15–17.
Persson, A. and L. Ryals (2010), 'Customer assets and customer equity: management and measurement issues', *Marketing Theory*, **10**(4), 417–36.
Pfeifer, P.E. (2011), 'On estimating current-customer equity using company summary data', *Journal of Interactive Marketing*, **25**(1), 1–14.
Pfeifer, P.E. and R.L. Carraway (2000), 'Modeling customer relationships as Markov chains', *Journal of Interactive Marketing*, **14**(2), 43–55.
Schulze, C., B. Skiera and T. Wiesel (2012), 'Linking customer and financial metrics to shareholder value: the leverage effect in customer-based valuation', *Journal of Marketing*, **76**(2), 17–32.
Skiera, B., M. Bermes and L. Horn (2011), 'Customer equity sustainability ratio: a new metric for assessing a firm's future orientation', *Journal of Marketing*, **75**(3), 118–31.
Srinivasan, R. and D. Sihi (2012), 'Marketing information disclosure: a review and research agenda', in S. Ganesan and K. Eller (eds), *Handbook of Marketing and Finance*, Cheltenham, UK and Northampton, MA, USA: Edward Elgar Publishing.
Venkatesan, R. (2011), 'A comment on "On estimating current-customer equity using company summary data"', *Journal of Interactive Marketing*, **25**(1), 23–4.
Wiesel, T., B. Skiera and J. Villanueva (2008), 'Customer equity – an integral part of financial reporting', *Journal of Marketing*, **72**(2), 1–14.
Wiesel, T., B. Skiera and J. Villanueva (2010), 'My customers are better than yours! On customer equity reporting', *Marketing Intelligence Review*, **2**(1), 43–53.
Wiesel, T., B. Skiera and J. Villanueva (2011), 'Customer lifetime value and customer equity models using company-reported summary data', *Journal of Interactive Marketing*, **25**(1), 20–22.

21. The future of customer equity

V. Kumar and Gayatri Shukla

INTRODUCTION

As seen in the previous chapters, customer equity (CE) brings acquisition, retention, and customer lifetime value (CLV) together under the same umbrella. While the definition and concept of customer equity more or less have common consensus, there is scope for fine-tuning and standardizing the measurement, implementing optimization strategies, and evolving the concept of CE in the era of big data, globalization, and the technologically charged times of the future, which will involve not only multiple marketing channels, but also a plethora of devices and platforms of access.

This chapter sheds light upon the road ahead for the metric and management of CE, and the contextual relevance of the concept in the future to firms that are interested in maximizing the value their customers bring to the firm.

WHAT IS THE FUTURE OF CUSTOMER EQUITY?

In this chapter, we discuss at length how the concept and metrics of customer equity will evolve in future, and identify the three main areas that this evolution will encompass:

1 *Modeling/forecasting and implementation.* The frameworks for modeling, measuring, forecasting, and implementing customer management strategies will evolve in tandem with the changing business landscape and the ever-increasing focus on customer engagement as a direct path to profitability. The emphasis will be not only on accurate value assessment and forecasting for each individual customer, but also on ensuring firm-wide management and dissemination of this information, from the boardroom to the frontline.

2 *Big data and technology.* Technology has facilitated an exponential increase in the depth and granularity of information that can be recorded on the profile, behavior and preferences of customers. This data explosion is also known as 'big data', and is also the ammunition that has been propelling firms further towards the era of hyper-personalized

targeting and communications, and one that is gradually facilitating the convergence of marketing and technology. The business landscape of the future will be characterized by the ready availability of real-time, holistic, and actionable big data on not only the firm's own customer base, but also its competitors, partners, and the industry it operates in. The beauty of the big data of the future will also lie in the level of richness of the information that it will encompass, as well as the integration of location-based, conversation-based, device-based, and context-based information. Big data and technology will together help firms move towards the era of predictive and intuitive marketing.

3 *Adoption and applicability*. Globalization and the surge in technology and communications is leading the world towards a relatively more unified global business landscape. That said, there continue to be differences between the diffusion/adoption rates of the strategies and implementation of CRM, CE, and other customer management frameworks, across various countries. Subsequently, we recommend that multinational firms take special notice of these factors before entering new global markets, so as to ensure optimal allocation of their customer management budgets. In the business landscape of the future, we expect the measurement and management of CE to be more pervasive, and universally applicable across contractual and non-contractual settings and geographic diversities.

Modeling/Forecasting and Implementation

Measurement

The metrics used to measure CE will evolve over time to include acquisition, retention, and referral, as well as win-back perspectives. In addition to measuring these perspectives, CE will also incorporate efficient estimation techniques to enhance not only the accurate knowledge of the various cost components of CE, but also the actual extent, frequency, and timing of these components.

Currently, the CE metric enables us to quantify the value that an acquired customer on average can bring to the firm, but with more data on customers' behavior and profile – the CE metric will help predict the aggregate value that a customer will spend in a particular category over a period of time. Firms are beginning to realize the importance of having the capability to predict this value. Walmart, for instance, learnt the importance of the metric the hard way in 2008, when it saw that while its C-Sat metric was improving, its same-store sales and category share-of-wallet were steadily declining.[1]

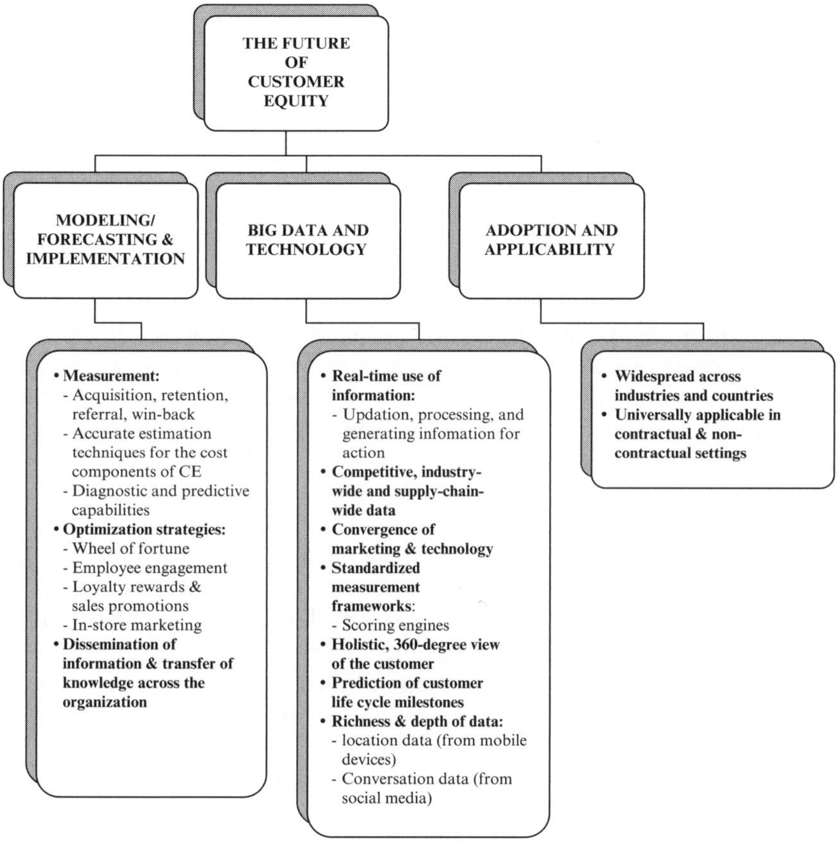

Figure C.1 The future of customer equity

Further, in order to invest adequate focus and resources towards value measurement and ensure that it has the necessary diagnostic and predictive customer management capabilities to harvest this intelligence, Walmart has acquired Inkiru, an analytics firm that provided Walmart with a predictive analytics platform and a customizable decision engine to measure, manage and forecast the CE and CLV of its present and future customers.[2]

The future will also bring with it enhanced accuracy and timeliness in the predictive analytics capabilities of firms, which will be especially useful to industries such as fast fashion, technology, and so on that are characterized by ever-changing consumer preferences and intense competitive pressure. Abercrombie & Fitch (A&F), one such fast fashion retailer that is currently plagued by declining sales and competitiveness, is attempting to salvage its fortunes by introducing a new predictive analytics-based approach.[3] Most retailers in the fast fashion industry usually follow the norm of testing new apparel concepts in stores, identifying the bestsellers, and then producing more of the bestselling pieces. Needless to say, there is a substantial lead time involved between the time the bestsellers are identified and the time additional stocks reach the stores. In order to minimize this lead time and optimize its operational investments, A&F has recently chosen to enlist predictive analytics software from First Insight Inc. to feed all its product/design ideas (including cut, color and pricing) to an algorithm using sentiment analysis, and identify the ideas that have the most success/popularity potential and are thus most suitable for production and launch. This approach is expected to not only generate substantial cost saves for A&F, but also shave three months off the typical nine-month production cycle.

Optimization strategies

We have seen how implementing strategies such as the 'wheel of fortune' framework developed by Kumar (2008)[4] can help firms maximize both CE and CLV. In the future, as CE evolves, the need of the hour will be to identify more such strategies and tactics that can help firms maximize their CE. We present below a few such optimization strategies that can be implemented by firms to augment their CE efforts:

Employee engagement A recent study on employee engagement (EE) defined EE as a multi-dimensional construct that comprises of all the different facets of the attitudes and behaviors of employees towards the organization.[5] This study that was aimed at determining the components of EE and identifying its impact on the firm's bottom line interviewed HR and marketing managers from across the globe, and from various industries ranging from hotels, telecom, airlines, to retail and banking, to derive practitioners' insights and perspective on the topic. The study identified the following components of EE: employee training and orientation, branding, employee satisfaction, employee identification, employee commitment, employee loyalty, and employee performance. EE was measured on a four-level scale, with the scores ranging from 20 to 100, partitioned as quartiles (in intervals of 20) in a frequency distribution. A quartile score of 20–39 (disengaged), indicated the lowest levels of EE and the least engaged employees, 40–59 (somewhat engaged) indicated that the employees are relatively less engaged with the firm, thereby implying a scope for immediate improvement. A quartile score of 60–79 (moderately engaged) indicated a moderate level of EE, whereas a quartile score of 80–100 (super-engaged) indicated the highest level of EE, a benchmark that every firm should strive to achieve. The results from this study indicated that while all the companies in the United States scored between 50 and 80, the companies from the emerging markets ranked much lower (between 30 and 60) in their overall score of EE.

Engaged employees can be passionate advocates of the firm and ensure a superior customer experience, and are in turn bound to impact not only customer acquisition, retention, referrals and win-backs, but also the cost components of CE. Thus, while analyzing, forecasting, and optimizing the relationship between CE and firm performance, it is in the interest of the businesses of the future to be mindful of the effect of EE. By optimizing EE, firms can maximize their returns from their customers and, as a result, their CE and overall firm profitability.

Loyalty rewards and sales promotions Firms can accelerate their sales by offering targeted promotions and incentives to reward their customers/prospects for taking the desired action – be it purchase, repurchase, referrals, cross-buying, or switchovers. While one may argue that these promotional initiatives would involve incremental costs, the fact is that the corresponding increase in the value that the customer would bring back to the firm would help the firm recoup those costs and, in fact, enhance overall customer satisfaction and equity.

In-store marketing In-store marketing (ISM), has been defined as the use of information and communication-related retail marketing instruments within the physical outlets of a retailer.[6] Strategic ISM can help any customer-facing firm enhance its in-store interactions with the customer, as well as better understand its shopping behavior and decision-making criteria and process. This will positively influence the in-store experience for customers and help firms ensure a more needs-based selling approach to the development and dissemination of the firm's targeted offers and communications, a factor that will gain even more relevance given the continued proliferation of smartphones and handheld/wearable connected devices. This, in turn, will enhance the responsiveness and also the value/equity of customers towards the firm.

Dissemination of information and transfer of knowledge across the organization

Accurate data and predictive insights on the behaviors, preferences, profiles, and lifetime values of customers can empower firms to design the right targeting, segmentation, and marketing strategies. However, if this information is not duly percolated across every level of the organization, down to their customer-facing frontline staff, they may not be able to offer differentiated service in order to enhance the value that the customer provides to the firm, resulting in sub-optimal interactions and experiences with the customer, in turn causing service failures.

In order for CE to be leveraged to its true potential, firms will need to ensure complete dissemination/transfer of information from the management to the frontline – this will also help the firm create fewer instances of service recovery (due to fewer service failures). For example, a cable company service technician visiting a high-value customer should be aware of how valuable the customer is to the firm and, accordingly, be able to deliver a customized, superior service experience to him or her. From the customer's viewpoint, this service technician is the face of the organization, and hence all the knowledge and data that the firm possesses on the equity of the customer is worthless unless it actually translates into action and service at the ground level.

Big Data and Technology

We are already in the big data age where firms can access intelligence on every aspect of the behavior, preferences, and interactions of customers in order to derive actionable insights. The future heralds even more accuracy in the predictive power of data, and also signals the imminent, *real-time* updation, processing, and generation of actionable insights. This real-time information will in turn empower firms to measure the lifetime value and equity of customers on a real-time basis, as opposed to at periodic discrete intervals of time, as is the case currently.

The biopharmaceutical industry is an excellent case in point of the emergence of real-time use and processing of actionable insights.[7] Biopharmaceutical companies are leveraging big-data–based insights on a real-time basis to improve the patient data quality in clinical trials, as well as the relationships between the pharmaceutical company and trial physicians. Additionally, such insights enable investigators to monitor clinical trials real time, as well as reduce costs by identifying any adverse events at an earlier stage in the trial process, thereby being able to bring innovative drugs to market sooner.

In spite of the tremendous capabilities of the existing big data platforms and tools, a big shortcoming of these is the lack of competitive (as well as industry-wide and supply-chain–wide) intelligence on CE and CLV related performance that can help firms fine-tune their customer targeting and management strategies. The future will belong to those who are able to integrate information and cull actionable insights such as disparate sources of data, and take calculated decisions to act on all this information. An encouraging sign is that firms are gradually beginning to realize the importance of the big-picture view and the role that big data and technology can play in it. For instance, Citigroup developed transaction data models for a big-picture view of the cash flows and income streams in different parts of its business, and was then able to identify some inefficiencies in its payment mechanism that made its client experience not so positive. Through these integrated models, Citigroup was able to provide its clients with solutions to address this issue and better support their overall transactions.[8]

Advancements in technology and the increased focus on developing CE-based strategies will help bring about a convergence in marketing and technology. Marketing has already risen up the ranks to become the fundamental driver of technology purchases by a firm, and, in fact, Gartner has predicted that by 2017, chief marketing officers (CMOs) will spend more on technology than chief information officers (CIOs).[9] This imminent convergence will proliferate standardized measurement frameworks such as scoring engines for ranking customers based on the true value they contribute to the firm. This will also help firms customize their service strategy, acquisition/retention tactics, and implement other customer management strategies.

The future will also bring in a more holistic, 360-degree view of not only every individual customer of the firm along with his or her life cycle, spending habits, and multi-channel behavior with the firm, and in the category – but also all such information about his or her household. This will help firms ensure more prudent use of their marketing communications, and a more accurate assessment of the customer's actual and potential lifetime value. For instance, big-data–savvy financial services firms like American Express follow the astute business practice of monitoring their high-value customers' spending patterns on a regular basis. If there is a change in this pattern that suggests that the customers are not using their Amex card as their primary card anymore, Amex proactively reaches out to these customers to understand how it can win them back. While this is a great CE-oriented practice, the future of CE lies in being able not only to identify such customers who may be on their way out, but also gain a more holistic perspective on their CE by integrating the CLV of the other members of their household who may be customers with the firm, and offering incentives on their spending categories and channels of preference. This will also help firms amplify the predictive power of the available data, and better predict various customer life cycle milestones such as churn, renewal, buying rate, usage, and so on (in contractual settings), as well as the average basket size, inter-purchase time, number of products purchased, and so on (in non-contractual settings).

Finally, as a direct result of the continuing surge in technology and data science, the recordable information on customer transactions, behaviors, and preferences will evolve exponentially – both in terms of its richness and level of detail, and also in terms of the complexity and multiple layers it will render to the CE metric by default. With the advent of technology and the ever-increasing rate of customer transactions through mobile devices, location-based data will play a larger role in the accurate computation of CE, as will conversation-based and contextual data from social media. For instance, in the not-so-distant future, a supermarket will be able to identify when you are in the physical vicinity of its store, and layer that piece of information with specific indicators such as your preferred brands, refill/repurchase pattern, previous transaction history and in-store experience, and so on, based on your customer profile on its database, and further link it up with contextual cues about your preferences, brand perceptions, and experiences by sifting through your conversations across social media and online review sites. This detailed, multi-layered information will help the retailer design a hyper-targeted offer/message that is most likely to appeal to you, and deliver it to your handheld device just when you are in the vicinity of the store, to induce you to walk in to that store and act upon the message immediately.

Adoption and Applicability

As technology progresses even further and data science becomes exponentially more detailed, the measurement and management of CE will become even more widespread across industries and countries. In fact, of the 524 CMOs who were interviewed from across 19 industries in the *IBM Global C-suite Study 2013*, 94 percent expected to use more advanced (predictive) analytics, and 89 percent expected to use CRM tools more extensively in the subsequent three-to-five-year timeframe.

A 2011 study[10] on the diffusion of truly global customer relationship management (GCRM) quantifies the growth of CRM on a global scale through a diffusion modeling framework. The results of this framework provide solid evidence of the untapped market potential in the GCRM market, with varying adoption patterns across regions. Also, while the study determines that the diffusion of CRM and its technological platforms is imminent, it cautions firms against using a generic GCRM strategy across global markets, and emphasizes the importance of customizing this strategy in view of country-specific regulations, privacy laws, consumption patterns, and consumer characteristics. Another recommendation of the authors is to establish the practice of designing a CRM strategy at the highest management level, while providing adequate room for customization at the international subsidiary level.

This study also emphasizes the importance of managers being cognizant of the differences in the rates of CRM adoption across different countries, so that they can allocate their resources and schedule their marketing plan appropriately. For instance, it identifies multinational corporations in the Asia-Pacific region as having lower levels of innovation but higher growth potential in GCRM, as compared to North America, and also finds a lead-lag learning effect between North America and the Asia-Pacific region, North America and Europe, and Europe and the Asia-Pacific region. The study also identifies that while CRM strategies are currently more prevalent and popular in North America, the coming years will witness a surge in the adoption and implementation of CRM strategies, tools and technology in the Asia-Pacific region, to keep pace with the growing relevance of the region in the world economy.

Thus, we recommend that in order for multinational firms to be truly successful at managing their customers in the future, they should understand the applicable lags in the adoption/diffusion rates of customer management strategies like CE, as well as all other differences in customer characteristics, behavior, consumption patterns, and country-specific laws and regulations before entering a country, in order to ensure optimal investment of their CRM resources.

Lastly, several estimates indicate that close to 95 percent of consumer purchases are non-contractual in nature, and only about 5 percent are contract-bound. In order to be truly comprehensive and *universally applicable* as a standard metric for future businesses, the measurement of CE will need to be transferable in the straightforward context of contractual settings where the customer's lifetime is fixed in duration, as well as in non-contractual settings where the duration will have to be estimated on the basis of historical buying and consumption patterns, use cases, industry trends, and so on.

In conclusion, as technology advances and competition intensifies, efficient customer management will increasingly become crucial. As a result, the importance of CE will only continue to grow further, and its measurement and predictive power will become all the more accurate and reliable. Moreover, the generation, updation and use of the metric will be on a real-time basis, in order to duly support the lightning-fast speed of the businesses of the future. Firms will increasingly focus on developing and implementing strategies to maximize CE. As the world continues to shrink with the convergence of communications and technologies, the principles and practice of CE, CLV, and other customer management strategies will spread to the emerging markets as well, making the concept of CE globally relevant. Finally, this universal applicability and relevance will ensure that the CE metric is ingrained in the DNA of every business, worldwide.

NOTES

1. Keiningham, T.L., L. Aksoy, A. Buoye and B. Cooil (2011), 'Customer loyalty isn't enough. Grow your share of wallet', *Harvard Business Review*, October, article posted on http://hbr.org/2011/10/customer-loyalty-isnt-enough-grow-your-share-of-wallet/ar/1; last accessed 21 September 2014.
2. Demery, P. (2013), 'Wal-Mart buys a "predictive analytics" firm', *Internet Retailer*, 10 June 2013, article posted on http://www.internetretailer.com/2013/06/10/wal-mart-buys-predictive-analytics-firm; last accessed 21 September 2014.
3. Boulton, C. (2014), 'Abercrombie & Fitch turns to predictive analytics to spark sales cool factor', *CIO Journal*, 10 March, article posted on http://blogs.wsj.com/cio/2014/03/10/abercrombie-fitch-turns-to-predictive-analytics-to-spark-sales-cool-factor/; last accessed 20 September 2014.
4. Kumar, V. (2008), *Managing Customers for Profit: Strategies to Increase Profits and Build Loyalty*, Upper Saddle River, NJ: Prentice Hall Professional, pp. 75–280.
5. Kumar, V. and A. Pansari (2014), 'The construct, measurement, and impact of employee engagement: a marketing perspective', *Customer Needs and Solutions*, **1**, 52–67.
6. Zentes, J., D. Morschett and H. Schramm-Klein (2011), *Strategic Retail Management: Text and International Cases*, 2nd edition, Wiesbaden: Gabler Verlag.
7. Dutton, G. (2014), 'How big data is transforming drug development', *Forbes*, 17 March, article posted on http://www.forbes.com/sites/emc/2014/03/17/how-big-data-is-transforming-drug-development/; last accessed 21 September 2014.
8. Rashik, P., I. Mackenzie, D. Cohn and D. Gann (2014), 'The new patterns of innovation', *Harvard Business Review*, **92**(1/2), 86–95.
9. Arthur, L. (2012), 'Five years from now, CMOs will spend more on IT than CIOs do', *Forbes*, 8 February 2012, article posted on http://www.forbes.com/sites/lisaarthur/2012/02/08/five-years-from-now-cmos-will-spend-more-on-it-than-cios-do; last accessed 21 September 2014.
10. Kumar V., S. Sunder and B. Ramaseshan (2011), 'Analyzing the diffusion of global customer relationship management: a cross-regional modeling framework', *Journal of International Marketing*, **19**(1), 22–39.

Index